COMPUTABILITY

George J. Tourlakis
York University
Atkinson College
Downsview, Ontario

Reston Publishing Company, Inc.
A Prentice-Hall Company
Reston, Virginia 22090

για την δεσποινα,

τον γιαννη,

και την μαρινα

Library of Congress Cataloging in Publication Data

Tourlakis, George J.
  Computability.

  Includes bibliographical references.
  1. Computable functions.   2. Recursive functions.
I. Title.
QA9.59.T68 1983   511.3   83-9751
ISBN 0-8359-0876-3

© 1984 by **Reston Publishing Company, Inc.**
        A Prentice-Hall Company
        Reston, Virginia 22090

10   9   8   7   6   5   4   3   2   1

Interior design and production: **Jack Zibulsky**

Printed in the United States of America

Contents

Preface

Computability, or Recursion Theory, came into being as a result of the research of several logicians (e.g., Gödel, Church, Post, Kleene) in the 1930s.

The motivation for this research was largely due to Hilbert's belief that the *Entscheidungsproblem* (or *decision problem*) of formal mathematical theories was solvable by "algorithmic" methods. Clearly, there was a need for a mathematically precise counterpart to the intuitive notion of "algorithmic" method (or function) in order to investigate Hilbert's conjecture.

Computability "attempts" to provide such a mathematical characterization of "algorithm" and "algorithmic" function. It is noteworthy that all the different formalisms proposed to date for this purpose have been proved to be equivalent (i.e., they all describe the same set of algorithmic functions).

Within a formal theory of algorithmic functions, the question "is such and such problem algorithmically solvable?" becomes mathematically meaningful and amenable to mathematical investigation. The Entscheidungsproblem was among the first problems so investigated. It is unsolvable by algorithmic methods expressible in any of the presently known formalisms,(*) thus the validity of Hilbert's conjecture is quite unlikely.(*) (The unsolvability of the Entscheidungsproblem was first explicitly proved by Church, but it was also "buried between the lines" of Gödel's earlier result on the incompleteness of consistent formal arithmetic. See Section 9.6.)

Unfortunately, most of the important results in Recursion Theory are of such "negative" nature, that is, that such and such task cannot be carried out algorithmically.

The advent and subsequent proliferation of computers has made Computability an area of study also relevant to computer scientists. After all, a computer "program" is just an algorithm expressed in a "language" understandable to a computer installation. Thus, the need to investigate, with mathematical tools, the theoretical limitations of computers, translates to the need for studying what *can,* and what *cannot,* be done, once again, with algorithmic methods. For example, we can prove (Chapter 7) that there is no "method" by which we can tell whether a given program belongs to a class of programs which share some fixed "specification" or, in more catchy jargon, "checking a program for correctness is a task beyond a computer's capability".

This book provides a uniformly mathematically rigorous introduction to (ordinary) Recursion Theory. Intuitively appealing "handwaving" arguments,

---

(*)The qualifications are due to the fact that we cannot possibly have a "proof" that the intuitive notion of "algorithmic method" is completely captured by the known formalisms. (See Section 7.1 for further discussion on this point.)

however, supplement formal proofs. Very often pseudo-PL/1 programs provide the vehicle for these informal arguments. The "pseudo" aspect should make these programs easily readable; no programming background is assumed or needed on the part of the reader.

The only prerequisite expected from the reader is a mix of patience and "mathematical maturity". The less one has of the latter the more one should have of the former.

As the reader who studies Chapter 10 will discover, Computability is founded, essentially, on two technical results—the *universal function* and the *S-m-n* theorems. This being the case, any single formalism for algorithmic functions is as good as any other for the foundation and development of the theory.(†) Yet, on the one hand, different instructors who will use this book will have different opinions as to which is (pedagogically) "the best" formalism, while on the other hand, the exposure of the reader to the foundational work of the pioneers of the subject will certainly enhance his or her understanding of the theory. For these reasons I fully discuss in this book practically all the known alternative formalisms for algorithmic functions.

The following is an outline of the material in the various chapters.

Chapter 1 is for reference only to the relevant results from Set Theory, in particular, and mathematics in general. It will normally be skipped, or read quickly, by the reader who has a reasonable mathematical background.

Chapter 2 studies the primitive recursive functions ($\mathcal{PR}$), which provided, historically, the first approximation to a formalism for the notion of "algorithmic" (or "computable") functions. Both a number theoretic and a programming formalism for $\mathcal{PR}$ are given.

Chapter 3 introduces the partial recursive, and recursive, functions via Kleene's number theoretic $\mathcal{P}$-formalism (μ-recursiveness). Along with "direct" proofs of the universal function and *S-m-n* theorems, carried out in a unique way in this book, avoiding the digression of "arithmetizations" or "machines" at this stage, the reader gets a first flavor of provably "unsolvable" problems.

The road leading to the deeper results of the theory is now open, and the reader may skip to Chapter 7 with no disruption of continuity.

Chapter 4 presents (deterministic) formalisms, as alternatives to that of $\mathcal{P}$, including the popular Turing Machine. Without recourse to arithmetizations, all the formalisms in this chapter are shown to be equivalent to the $\mathcal{P}$-formalism.

Most of this chapter could be covered before Chapter 2, if so required.

In Chapter 5, most of Bennett's results on *rudimentary* predicates are presented, apparently for the first time in textbook form, and a rudimentary

---

(†)This statement should not blur the fact that certain specialized results, for example of combinatorial nature such as the unsolvability of the *word problem* for semigroups, are more easily handled within a formalism such as those of Turing or Markov, which are founded combinatorially and explicitly rather than axiomatically.

arithmetization of Turing Machines is carried out, providing alternative proofs to those in Chapter 3 for the universal function and *S-m-n* theorems. A particularly "simple" arithmetical predicate, which describes the behavior of Turing Machines, is obtained in the process.

Chapter 6 includes nondeterministic formalisms such as Turing Machines, Thue-like systems, Smullyan's elementary formal systems, and Kleene's systems of equations. All these are proved to be equivalent. A brief introduction to *Formal Languages* naturally belongs here, and is pursued up to the proof of the two "pumping lemmas", thus enabling us to prove that the Chomsky hierarchy is proper on the one hand, and on the other to prove that one or two meaningful problems in language theory are "solvable" (this is good for the reader's morale(!), since "most" meaningful problems are "unsolvable").

The unsolvability of the word problem for semigroups, and of the *ambiguity problem* for context free grammars (the latter via the unsolvability of *Post's correspondence problem*) are also proved.

With Chapters 7, 8, and 9, which are formalism-independent and directly accessible from Chapter 3, the development of the general theory resumes. Here our stock of unsolvable problems is enlarged, and recursive and recursively enumerable sets are discussed in depth, including the Rice-Myhill-McNaughton-Shapiro theorem, productiveness and recursive isomorphisms. Chapter 9 concludes with a generous dosage of *first-order logic,* which, along with material on formal mathematical systems à la Smullyan, leads to self-contained rigorous proofs of Gödel's and Rosser's incompleteness theorems (a few informal "proofs" of Gödel's results are also included).

Relativized Computability is introduced in Chapter 10 as a derived notion from that of *computability of functionals.* The latter is presented via *Kleene-schemata,* departing from the traditional approach of books at this level.

Apart from its mathematical elegance and its ability to lead to substantial results rigorously *and* quickly, the Kleene-schemata approach presents the reader with a bridge from Ordinary Recursion Theory to recursion in "higher types". Two cornerstone results of recursion in higher types, the *Ordinal Comparison* and *Selection* theorems are proved, and are applied in order to circumvent the technical difficulties arising from the inclusion of *nontotal* functions as arguments of functionals.

The Friedberg-Muchnik theorem, and an introduction to the arithmetical hierarchy are included.

Chapter 11 includes some ad hoc results of the theory proved with the aid of Kleene's *second* recursion theorem, recursive operators and an indexing-independent approach to the computability of functionals, and the *first* recursion theorems of Kleene and Moschovakis.

Chapter 12 covers the five basic results of Blum-style Complexity.

Chapter 13: Complexity of "down to earth" total functions. Here, under one cover for the first time, the reader will find Cobham's number theoretic characterization of "feasibly" computable functions, Grzegorczyk, loop-program, and Axt hierarchies and their comparison à la Schwichtenberg, Tsichritzis' characterizations of K_1 and $\mathcal{L}_1$.

It has been the author's experience that most of the material in Chapters 2, 3, 4, 6, 7, and 13 can be covered in a one-semester course. It should be feasible to include all chapters in a two-semester course.

Each instructor will, of course, adjust the amount and sequence of coverage according to the particular teaching situation he or she is in.

Most of the end-of-chapter problems extend the theory developed in the text. Since the most difficult among them are provided with enough "hints", I made no attempt to rate the problems according to difficulty.

Named topics (e.g., Theorem, Example, Problem, Definition) are referenced by an "address" x.y.z, where x is the chapter number, y is the section number, and z is the number of the topic itself. If the topic is in the same chapter (resp. section) where the reference to it originated, then "x." (resp. "x.y.") is omitted. Exceptions to this rule are references to problems and bibliography, both appearing at the end of chapters. Thus, bibliographical references use one or more integers separated by commas in [ ]-brackets, whereas problem references use a single integer (if the target and the origin of the reference are in the same chapter), or an integer prefixed by the chapter number, followed by ".".

The symbol // marks ends of examples, definitions, and proofs.

*   *   *   *   *

I would like to express my gratitude to all those who have taught me over a period of several years and thus have indirectly contributed to the writing of this book. In this group of people I would especially like to mention my parents, Yannis Ioannidis, and in more recent times, Derek Corneil, John Lipson and John Mylopoulos. Special thanks are due to Dennis Tsichritzis, Steve Cook and Alan Borodin who introduced me to the subjects of Computability and Complexity.

I would also like to acknowledge the influence I have received from the books of Péter, Davis, Rogers and Hinman and from the paper of Grzegorczyk, all of which furthered my understanding of Recursion Theory.

Miss Kieh Wong prepared the Index, Mrs Bonnie McKee and Miss Marian Rigon have typed the first five chapers; I thank them all. I am also particularly indebted to Mrs. Teresa Miao, who with remarkable speed and accuracy typed Chapters 6 through 13, and to Ben Wentzell and Jack Zibulsky of Reston Publishing Company whose cooperation and help at all times made the publication process as painless as practicable.

George J. Tourlakis

CHAPTER 1

Introduction

In this first chapter, we lay out the notation we shall use throughout this book. Moreover, the notions from general Set Theory which are relevant to our development are introduced in a more or less self-contained manner.

The main topics in this chapter are the diagonal method of Cantor, the Schröder-Bernstein theorem, and the principle of inductive definition of functions on partially ordered sets. They all have counterparts in *Recursion Theory*. Of the three, the Cantor diagonalization technique is perhaps the most important; indeed we may say that Recursion Theory is the art of diagonalizing.

§1.1 SETS

It is beyond our scope to attempt an axiomatic approach to Set Theory. Our approach will be informal, and this section is to serve as a brushing-up tool for the reader who has already had an exposure to "naive set theory" (good introductions to the subject are Halmos [3] and Kamke [4]).

The term *set* will not be defined; if it is helpful to intuition, it should be taken as synonymous with "class", "aggregate", and "collection". Intuitively, a set is a "container along with its contents" (we shall see below that it is convenient to have a set-theoretic counterpart to an empty container). As for any mathematical object, it is helpful to use letters (no restriction to upper- or lowercase; both are allowed) as *names* of sets and their contents. The most basic relation in set theory is that of *belonging* denoted usually by the symbol "∈". Thus "$x \in y$", intuitively, captures the meaning of the phrase "the object (named) x is in the set (named) y". Note that x itself may very well be (a name for) a set.

We shall freely use notation borrowed from Mathematical Logic to avoid overburdening our arguments with prose. Thus, "&" stands for "and", and "∨" stands for "(non-exclusive) or", and "¬" stands for "not".

The symbol "⇒" stands for "implies". Both "⇔" and "≡" are symbols for two-way implication. Often we shall use "iff" ("if and only if") instead of ⇔ or ≡.

Finally, let $S(x)$ be "shorthand" for a statement involving x such that for some "values" of x the statement may be "true", whereas for others it may be "false".

Then $(\forall x)S(x)$ stands for "for all values of x, $S(x)$ is true"; thus, for it to be a true statement, $S(x)$ had better be true for *all* (relevant) values of x!

Also, $(\exists x)S(x)$ stands for "for some value of x, $S(x)$ is true"; again, for $(\exists x)S(x)$ to be true, *some* value of x *must* exist that makes $S(x)$ true, otherwise $(\exists x)S(x)$ is false.

It is intuitively clear that $(\exists x)S(x)$ has the same "truth-value" or meaning as $\neg(\forall x)\neg S(x)$.

The ∀ symbol is called *universal quantifier* (whereas ∃ is called an *existential quantifier*). Some variations of these symbols are useful: $(\exists! x)S(x)$ means "there exists exactly one x such that $S(x)$ is true". Of course "∃!" can be simulated by ∀ and ∃ (or, for that matter, by ∃ alone, given that $\exists \equiv \neg\forall\neg$) since $(\exists! x)S(x) \equiv (\exists x)[S(x) \, \& \, (\forall y)(y \neq x \Rightarrow \neg S(y))]$, however the ∃! is more compact.

Further, $(\forall x)_{x \in A} S(x)$ [resp. $(\exists x)_{x \in A} S(x)$] stands for $(\forall x)(x \in A \Rightarrow S(x))$ [resp. $(\exists x)(x \in A \, \& \, S(x)$]. Again, on many occasions we will find the compacter notation more convenient than its equivalent expanded form.

It should be noted that "⇒" is the "material implication" that we use in classical (non-intuitionist) mathematics; i.e., the statement (implication) $A \Rightarrow B$ is true *unless* A is true but B is false.

After this short digression regarding the use of logical symbols in our arguments, let us resume our discussion of sets.

In all discussions involving sets, it is logically desirable to have around a *universal* or *reference set*, say U, such that all the sets we consider are "parts" of U. Thus, in such a context, whenever $(\forall x)$ [or $(\exists x)$] notation is used, x "runs" (or varies) over the elements of the universal set. In symbols, $(\forall x)$ [resp. $(\exists x)$] stands for $(\forall x)_{x \in U}$ [resp. $(\exists x)_{x \in U}$]; i.e., the subscript "$x \in U$" of the quantifier is omitted since it is understood from the context.

Let then A and B be two sets; we write $A \subset B$ as shorthand for $(\forall n)(n \in A \Rightarrow n \in B)$ [in words: "for all n in the universal set, if n is in A, then it is in B"].

We say that A is a *subset* of B. If $A \subset B$ *and* $B \subset A$, then we say that A is equal to B and write $A = B$. This adoption of the notion of equality is obviously natural, as it says that the two sets A and B contain exactly the same elements.

Suppose that $A \subset B$ and also $\neg(B \subset A)$ (also written as $B \not\subset A$). Then, by the definition of equality, we have $A \subset B \ \& \ A \neq B$ (since $A = B$ would imply $B \subset A$). We write in this case $A \subsetneq B$ and say that A is a *proper subset* of B.

Incidentally, to show that $B \not\subset A$ one usually(*) has to exhibit a *particular* x such that $x \in B \ \& \ x \notin A$ [**Note:** $x \notin A$ is the usual notation for the statement $\neg(x \in A)$], as it follows from the definition of "$\subset$".

The *empty set* is a strange but very convenient set. Its symbol, traditionally, is $\varnothing$, and by definition it has *no* elements. That is, the statement "$x \in \varnothing$" is false for *every choice* of x (equivalently, $x \notin \varnothing$ is true for *every choice* of x). Note that neither uniqueness nor existence of $\varnothing$ is guaranteed by the above definition.

Existence of $\varnothing$ within the context of some given reference set U amounts to the existence of a statement $S(x)$ which is false for every $x \in U$. (See Problem 1.) Uniqueness of $\varnothing$ will be proved shortly. (See Example 3 below.)

We have seen that letters are used to denote sets; they represent names of sets in the same way that letters stand for numbers in "high school algebra".

How about denoting *particular* sets? If a set is (intuitively) *finite* and moreover it has not too many elements we denote it by *listing* its elements (the order is immaterial—see definition of equality!) inside braces, "$\{,\}$". For example, $\{2\}$ is the set that contains the symbol 2, $\{\#,7\}$ is the set that contains the symbols $\#$ and 7. The set $\{a,a\}$ is the same as the set $\{a\}$ (see the definition of equality); thus, when denoting sets by listing their elements, elements are not duplicated.

When a set is finite but too big to be listed conveniently, or when it is not finite(†), an alternative method (if possible) is used to denote it: If $P(x)$

(*)Sometimes $B \not\subset A$ is proved by assuming $B \subset A$ and eventually deriving a contradiction of some sort; this is known as "proof by contradiction".

(†)The notions of finite and infinite will be formalized shortly. For the time being, the reader is asked to attach to them the ordinary intuitive meaning.

denotes a "property" that is shared by all its elements but by no other elements, then the set can be concisely denoted by $\{x \mid P(x)\}$, read "the set of all x satisfying $P(x)$". For example, $\{x \mid x < 2 \ \& \ x \ is \ real\}$ is the set of all real numbers strictly less than 2. $\{x \mid x \ is \ an \ integer \ \& \ x = 0 \ mod \ 2\}$ is the set of all *even* integers. $\{x \mid 0 \le x \le 1000000 \ \& \ x \ is \ an \ integer\}$ is the set of all integers between 0 and 1000000, inclusive.

As a final example, let S be any set. Then $S = \{x \mid x \in S\}$; here the property $P(x)$ is "$x \in S$". We see then that to each set corresponds a *defining property* (for S, the property is "$x \in S$") and conversely, to each property $P(x)$ a set $P = \{x \mid P(x)\}$. Of course, if we do not know much about a set S, then we do not know any more about the corresponding property $x \in S$.

Definition 1   The union (resp. intersection) of two sets A and B is denoted by $A \cup B$ (resp. $A \cap B$) and is defined by $\{x \mid x \in A \vee x \in B\}$ (resp. $\{x \mid x \in A \ \& \ x \in B\}$).//

Example 1   For any A and B, $A \cap B \subset A$ and $A \subset A \cup B$. Indeed, let x be *any* element in U, the reference set. We consider two cases:

(1) $x \in A \cap B$. Then (definition of $A \cap B$) $x \in A \ \& \ x \in B$. That is, $x \in A \cap B \Rightarrow x \in A$.(*) (We discard the extra conclusion that $x \in B$ as we are not interested in it.)

(2) $x \notin A \cap B$. Then $x \in A \cap B \Rightarrow x \in A$, as "$x \in A \cap B$" is false. Hence, under both cases (1) and (2) we have $x \in A \cap B \Rightarrow x \in A$. Since x was arbitrary, we have shown $(\forall x)(x \in A \cap B \Rightarrow x \in A)$—i.e., $A \cap B \subset A$.

It is left to the reader to prove similarly that $A \subset A \cup B$.//

Note: It is clear that we could have confined our arbitrary x in $A \cap B$.

Example 2   For any set A, $\varnothing \subset A$. Indeed, let x be *any* arbitrary element. We know that $x \in \varnothing$ is false. Thus, $x \in \varnothing \Rightarrow x \in A$ is true. Since x was arbitrary, it follows that $(\forall x)(x \in \varnothing \Rightarrow x \in A)$; i.e., $\varnothing \subset A$.//

Example 3   There is only one empty set. Let $\varnothing_1$ and $\varnothing_2$ be empty sets. Then $\varnothing_1 \subset \varnothing_2$. But also $\varnothing_2 \subset \varnothing_1$; that is, $\varnothing_1 = \varnothing_2$.//

Definition 2   Let A and B be sets. Then $A - B$, the *difference* of A and B in the denoted order, is defined to be $\{x \mid x \in A \ \& \ x \notin B\}$. If $A = U$, the reference set, then $U - B$ is denoted by $-B$ or $\overline{B}$ and is called the *complement* of B.//

(*)When in the course of an argument, we say " . . . i.e., $S(x)$ . . ." This is abuse of language for ". . . i.e., $S(x)$ *is true* . . ."

Example 4   (De Morgan laws). $-(A \cup B) = (-A) \cap (-B)$ and $-(A \cap B) = (-A) \cup (-B)$.
Indeed, let $x \in -(A \cup B)$. Then the following statements are implied in the given sequence:

$x \notin A \cup B$ [Definition of "$-$"]

$\neg(x \in A \vee x \in B)$ [def. of "$\cup$"]

$x \notin A$ & $x \notin B$ [$P \vee Q$ is true iff at *least* one of P,Q is true.

Thus, $P \vee Q$ is false iff *both* P and Q are false.]

$x \in -A$ & $x \in -B$ [definition of "$-$"]

$x \in (-A) \cap (-B)$ [definition of "$\cap$"]

Since x was arbitrary, the above establishes $-(A \cup B) \subset (-A) \cap (-B)$. The opposite inclusion is also true; the reader may easily establish that the above sequence of implications may be reversed.
As for the second De Morgan law, it suffices to prove that $A \cap B = -[(-A) \cup (-B)]$, since, trivially, $-(-X) = X$ for any X. This amounts to $(-(-A)) \cap (-(-B)) = -[(-A) \cup (-B)]$ which is true by the first law.
As a by-product, we saw that the first De Morgan law implies the second. A minute's reflection should satisfy the reader that the converse is true. That is, the two De Morgan laws are (logically) equivalent.//

Definition 3   *The power set.* For any set A, $\mathcal{P}(A)$ or 2^A, stands for $\{x \mid x \subset A\}$; that is, the set of all (nonproper) subsets of A.//

Example 5 $2^\varnothing = \{\varnothing\}$, $2^{\{1\}} = \{\varnothing, \{1\}\}$, $2^{\{0,1\}} = \{\varnothing, \{0\}, \{1\}, \{0,1\}\}$.//

We conclude this section with the important notions of ordered pairs and Cartesian products of sets.

Definition 4   Let x and y be any objects. Then the symbol (x,y) is the *ordered pair* of x and y, x is the first and y the second component of the pair. Equality of pairs is defined by $(x,y) = (u,v)$ iff $x = u$ and $y = v$.//

Note: An ordered pair is an *ordered* set—that is, we distinguish elements not only by their "value" but also by their *position*. Thus, $\{a\}$ is *not* the same as (a,a) because in the latter the two a's have different position attributes and thus we are not allowed to collapse them to a single a.

Definition 5   *The ordered n-tuple (n-vector).* The symbol $(x_1, \ldots, x_n)$ is defined inductively on n by:

$$(x_1) \stackrel{\text{def}}{=} x_1$$

$$(x_1, \ldots, x_{n+1}) \stackrel{\text{def}}{=} ((x_1, \ldots, x_n), x_{n+1})$$

For $1 \le i \le n$, x_i is called the ith component of $(x_1 \ldots, x_n)$.//

Exercise 1   Show that $(x_1, \ldots, x_n) = (y_1, \ldots, y_n)$ iff $(\forall i)_{1 \le i \le n} \, x_i = y_i$.

Definition 6   Let A_i, $i = 1 \ldots, n$, be sets. Then the Cartesian product of $A_1, A_2, \ldots, A_n$, in that order, in symbols $X_{1 \le i \le n} A_i$, is the set $\{(x_1, \ldots, x_n) \mid (\forall i)_{1 \le i \le n} x_i \in A_i\}$.//

Note: $X_{1 \le i \le n} A_i$ is sometimes written $A_1 \times \ldots \times A_n$. If $n = 2$, we usually write $A_1 \times A_2$ instead of $X_{1 \le i \le 2} A_i$. If $A = A_1 = \ldots = A_n$, we may write A^n.

Example 6 $\{0,1\}^2 = \{(0,0), (0,1), (1,0), (1,1)\}$.//

Example 7 $\{0\} \times \{1\} = \{(0,1)\}$. $\{1\} \times \{0\} = \{(1,0)\}$. Thus, "$\times$" is not commutative (unlike $\cup, \cap$)—i.e., in general $A \times B \ne B \times A$.//

Note: In the sequel, we reserve the symbols $\mathcal{N}, \mathcal{Z}$ for the set of natural numbers $\{0,1, \ldots\}$ and set of integers $\{\ldots, -1,0,1, \ldots\}$, respectively. We also write $\vec{x}_n$ for $(x_1, \ldots, x_n)$ or simply $\vec{x}$ if n is understood or unimportant.

§1.2 RELATIONS AND FUNCTIONS

Definition 1   An *n-ary Relation* (or *Predicate*) on $X_{1 \le i \le n} A_i$, where $n \ge 1$, is a subset of $X_{1 \le i \le n} A_i$. If $A_1 = \ldots = A_n = A$, we say that the *n*-ary relation is on A rather than A^n.//

Notes: (1) In Mathematical Logic, a *predicate* is a more abstract object than what we defined above. A *relation* is then a concrete realization of a predicate. We follow the usual practice, adopted in most of the literature on Recursion Theory, to use the terms *predicate* and *relation* as synonymous.

(2) Intuitively, a relation is a list of *n*-tuples, each *n*-tuple containing "related" elements.

(3) There is no distinction between sets and relations as Definition 1 makes clear! The new term introduces some convenience in our notation. Instead of writing $(x_1, \ldots, x_n) \in S$ (set-theoretic notation), we may (and often

do) write $S(x_1, \ldots, x_n)$. Both assert that $(x_1, \ldots, x_n)$ is an element of the relation (set) S.

(4) If S is an 1-ary relation, it is called *unary*. If it is a 2-ary relation, it is called *binary*. There are a lot of important classes of binary relations. For binary relations R, one often writes xRy instead of $(x,y) \in R$ or $R(x,y)$.

Example 1 $\varnothing$ is the empty relation. $\{1,2,3\}$ is a unary relation on the integers. $\{1,2,(1,2)\}$ is a unary relation on the set $\{1,2,(1,2)\}.//$

(5) We shall often say " ... let $S(\vec{x}_n)$ be a relation ... " instead of " ... let S be an n-ary relation ... " This is consistent with the fact that $S = \{\vec{x}_n \,|\, \vec{x}_n \in S\} = \{\vec{x}_n \,|\, S(\vec{x}_n)\}$. In other words, in our notation, we identify the set S with its "defining property" $S(\vec{x}_n)$.

E.g. let $R(x)$ and $Q(y,z)$ be two relations on $\mathcal{N}$. Then $R(x)\ \&\ Q(y,z)$ stands for the set $(R \times \mathcal{N}^2) \cap (\mathcal{N} \times Q)$. $R(x) \vee Q(x,y)$ stands for $(R \times \mathcal{N}) \cup Q$.

(6) It will also occur frequently that we wish to obtain a new relation from $R(\vec{y}_n)$ by identifying, transposing, renaming, "freezing" (i.e., substituting constants into) some of the y_i's or finally, by adding some new "dummy" (or "don't care") variables.

For example, let $R = \{(x,y) \,|\, x < y\}$ be on $\mathcal{N}$. We may wish to refer to $S = \{(x,y,z) \,|\, x < z\}$ or $T_y = \{x \,|\, x < y\}$. S [or $S(x,y,z)$] can be denoted by $\lambda xyz.R(x,z)$. T_y can be denoted by $\lambda x.R(x,y)$. The following definition summarizes:

Definition 2   Let $R(\vec{y}_m)$ be an m-ary relation. Then, $S = \lambda \vec{x}_n.R(\vec{w}_m)$, where for all i if w_i is not an x_j then it is a constant, satisfies:

$\vec{a}_n \in S$ iff $\vec{b}_m \in R$, where each of the b_i's is an a_j if $w_i = x_j$, a constant w_i otherwise. S is called an *explicit transform* of R (Smullyan [6], Bennett [1]).//

Example 2   Let $x + y = 3$ be a binary relation on $\mathcal{Z}$. It denotes the set $\{(x,y) \,|\, x + y = 3\}$. Now, $\lambda x.x + y = 3$ is $\{x \,|\, x + y = 3\}$ for some unspecified, but fixed, y. Similarly, $\lambda x.x + x = 3$ is $\{x \,|\, 2x = 3\} = \varnothing.//$

Example 3   Let $R(x,y)$ be on $\mathcal{N}$. Then $(\forall x)_{\leq z} R(x,y)$ in "set notation" is $\bigcap_{i=0}^{z} R_i$, where $R_i = \{y \,|\, R(i,y)\} = \lambda y.R(i,y)$. The compactness of the "relational notation" is evident.//

Definition 3   Let R be a binary relation on $A \times B$. Then $\mathrm{dom}(R) = \{x \in A \,|\, (\exists y)(x,y) \in R\}$; this is the *domain* of R. Also, the *range* of R, range(R), is the set $\{y \in B \,|\, (\exists x)(x,y) \in R\}$.

If $a \in A$, then $R(a) = \{y \,|\, (a,y) \in R\}$. If $X \subset A$, then $R(X) = \{y \,|\, (\exists a)a \in X\ \&\ y \in R(a)\}$. $R(X)$ is the *image* of X under R.

If $Y \subset B$, then $R^{-1}(Y) = \{x \mid R(x) \cap Y \neq \varnothing\}$. $R^{-1}(Y)$ is the *inverse image* of Y under R. If $y \in B$, then instead of $R^{-1}(\{y\})$ we write $R^{-1}(y)$.//

Note: If R is a binary relation, R^{-1}, "the converse" of R, is $\{(x,y) \mid R(y,x)\}$. The context will not allow any confusion due to the use of R^{-1} as both converse relation and inverse-image operator.

Example 4   Let $R = \{(1,2), (1,3), (2,1)\}$ be a binary relation on $\mathcal{N}$. Then, $R(1) = \{2,3\}$, $R(2) = \{1\}$, $R(3) = \varnothing$. $R^{-1}(1) = \{2\}$, $R^{-1}(2) = \{1\}$, $R^{-1}(\{2,3\}) = \{1\}$, $R^{-1}(\{0,1,2,3,4\}) = \{1,2\}$, $R^{-1}(4) = \varnothing$.//

Definition 4   Let f be a binary relation on $A \times B$, such that, for all x in A, $f(x)$ contains at most one element.

A relation such as f is called *single-valued* and if $y \in f(x)$, for some $x \in A$, we usually write $y = f(x)$ rather than $\{y\} = f(x)$. We say that f is a *partial function* or *partial map* from A to B, in symbols $f : A \to B$.

If $\mathrm{dom}(f) = A$, then f is a *total* function (map). If $\mathrm{dom}(f) \subsetneq A$, then f is *nontotal*. We write $f(a)\downarrow$ if $a \in \mathrm{dom}(f)$, $f(a)\uparrow$ otherwise. We say that f is *defined* (or *converges*) at a if $f(a)\downarrow$; otherwise f is *undefined* (or *diverges*) at a.//

Note: Clearly, the notions *partial* and *total* are not mutually exclusive: Every function (even a total one) is partial; however, some partial functions are *nontotal*. Thus, from now on the qualification "partial" will be omitted.

Given a map $f : A \to B$, whenever A and B are understood from the context or are unimportant for the argument at hand, the symbols $x \mapsto f(x)$ or $\lambda x.f(x)$ or f, all denote $f : A \to B$.

If $f : A \to B$ and $g : C \to B$ are two functions, then $f(a) = g(b)$ is equivalent to the statement "$(f(a)\uparrow \& g(b)\uparrow) \vee (f(a)\downarrow \& g(b)\downarrow \& f(a) = g(b))$". If $f : A \to B$ and $g : A \to B$ are such that $g \subset f$, then g is a *restriction* of f and f is an *extension* of g. The empty relation, $\varnothing$, on $A \times B$ is the *totally undefined* (or *empty*) map $\varnothing : A \to B$. Every function $f : A \to B$ is an extension of $\varnothing : A \to B$. If $f : A \to B$ and $C \subset A$, then $f \mid C$ denotes $f \cap (C \times C)$, the *restriction* of f on C.

Example 5   If $A = B = \mathcal{N}$ is understood, then $\lambda x.1$, or $x \mapsto 1$, stand for the map $\{(x,1) \mid x \in \mathcal{N}\}$.//

Example 6   Let $R(x,y)$ be single-valued on $\mathcal{N}$. Then $f = \lambda uv.R(u)$ is the function $\{((u,v),y) \mid v \in \mathcal{N} \ \& \ (u,y) \in R\}$.

$$\mathrm{dom}(f) = \mathrm{dom}(R) \times \mathcal{N}, \ \mathrm{range}(f) = \mathrm{range}(R).$$

Note that $((u,v),y)$ is the same as (u,v,y) [Definition 1.1.5].//

Note: λ-notation is very useful as, with minimal prose, we may draw attention to the number and names of the variables of a function and also give the rule by which the value is obtained, at the same time avoiding the use of the symbol $f(x,y)$ to stand for the function f. In some contexts, this symbol might be ambiguous (is $f(x,y)$ the *value* of f at (x,y), or is it the whole thing—i.e., $\{(x,y,z) \mid (x,y,z) \in f\}$?).

One usually wants to combine relations (functions) to build more "complicated" ones. Since a relation is a set, operations on them such as $\cup$, $\cap$, $-$, $\times$ are readily understood.

Example 7 Let $\lambda x.f(x)$, $\lambda x.g(x)$ be two functions from A to B. Then $\lambda xy.(f(x),g(y))$ is a function from A^2 to B^2. Clearly, $\lambda xy.(f(x),g(y)) = \{(x,y,u,v) \mid f(x) = u \And g(y) = v\}$ Let the relation $R(x,y,u,v)$ denote the righthand side above. Consider the explicit transform $S = \lambda xuyv.R(x,y,u,v)$ of R. That is, $(x,u,y,v) \in S$ iff $(x,y,u,v) \in R$. S contains exactly the same "information" as R and we may view the two as interchangeable. This suggests the notation $f \times g$ for $\lambda xy.(f(x), g(y))$.//

Some special types of functions are important for our purposes:

(1) A function $f{:}A \to B$ is *onto* (resp. *1-1*) iff range$(f) = B$ (resp. $f^{-1}(y)$ contains at most one element for each $y \in B$).

(2) If $f{:} A \to B$ is *total*, *onto*, and *1-1*, then it is called a *1-1 correspondence* between A and B.

Exercise 1 If $f{:} A \to B$ is a 1-1 correspondence between A and B, then f^{-1} is a 1-1 correspondence between B and A.

(3) Let I, X be two sets. A *total* map $f{:} I \to \mathcal{P}(X)$ is called an *indexed family* of sets (subsets of X) and is usually denoted as $(S_a)_{a \in I}$, where $f = \lambda a.S_a$.

Exercise 2 In what sense does an arbitrary relation R on $I \times X$ define a family of subsets of X?

We often need to operate on families of sets. Thus,

$$\bigcup\nolimits_{a \in I} S_a \overset{\text{def}}{=} \{x \in X \mid (\exists a)\, a \in I \And x \in S_a\},$$

$$\bigcap\nolimits_{a \in I} S_a \overset{\text{def}}{=} \{x \in X \mid (\forall a)\, a \in I \Rightarrow x \in S_a\}$$

$$X_{a \in I}\, S_a \overset{\text{def}}{=} \{\text{total } g \mid g{:}I \to \bigcup\nolimits_{a \in I} S_a \And (\forall a)_{a \in I}\, g(a) \in S_a\}$$

Note that the last definition is the natural extension of the operator $X_{1 \le i \le n}$, as g may be thought of, intuitively, as the "sequence" $(g(a))_{a \in I}$; e.g., if $I = \{1, \ldots, n\}$, then $(g(a))_{a \in I}$, i.e., g, is $(g(1), g(2), \ldots, g(n))$.

The *Axiom of Choice* is stated here for completeness only, as we shall not rely on it in the text. It says that if $(\forall a)_{a \in I} \, S_a \neq \varnothing$, then $X_{a \in I} \, S_a \neq \varnothing$.

Exercise 3   Show that $\bigcup_{a \in I} S_a = \varnothing$ and $\bigcap_{a \in I} S_a = X$ if $I = \varnothing$, where X is the reference set.

We shall next briefly investigate under what conditions a function $f \colon A \to B$ is a 1-1 correspondence.

Definition 4   Let $f \colon A \to B$ and $g \colon B \to C$ be functions. Then $g \circ f$ is defined to be the relation $\{(x,y) \mid (\exists z)(x,z) \in f \,\&\, (z,y) \in g\}$ on $A \times C$.//

It is clear that $g \circ f$ is a function from A to C as it is clearly single-valued. (See Problem 6.) We say it is obtained from f and g through *composition*. Note that $g \circ f = \lambda x.g(f(x))$. Clearly, if $f(a)\uparrow$, then $g(f(a))\uparrow$.

Definition 5   Let $f \colon A \to B$ and $g \colon B \to A$ be such that $f \circ g = \lambda x.x$ (on B). Then f is *a left inverse* of g, and g is *a right inverse* of f.//

Note: The *total* function $\lambda x.x \colon A \to A$ we denote by $\mathbf{1}_A$ and call it the *identity function* on A.

Exercise 4   Show associativity of composition; that is, $f \circ (g \circ h) = (f \circ g) \circ h$ for all composable functions f,g,h.

Exercise 5   Show that composition of functions from A to A is *not* commutative; that is, in general, $f \circ g \neq g \circ f$.

Example 8   Let $A = \{1,2,3,4\}$, $B = \{a,b\}$. Let $f_1 = \{(1,a), (2,a), (3,b)\}$, $f_2 = \{(1,b), (2,a), (3,b)\}$, $g_1 = \{(a,2), (b,3)\}$, and $g_2 = \{(a,1), (b,3)\}$. Clearly, f_1, f_2 are functions $A \to B$ and g_1, g_2 are functions $B \to A$. Moreover, $f_1 \circ g_1 = f_1 \circ g_2 = f_2 \circ g_1 = \mathbf{1}_B$. We conclude that neither the left nor the right inverses are unique in general. Also, note that the f's are not 1-1 and the g's are not onto. Finally, the f_1, and f_2 need not be total, but the g_1, and g_2 must (why?).//

Proposition 1   If f and g are as in Definition 5, then f is *onto* and g is *1-1*.

Proof (1) Let $b \in B$. Then $g(b) \in A$ (why is $g(b)\downarrow$?). But $b = f \circ g(b) = f(g(b))$, hence f is onto.
(2) Let $g(a) = g(b)$. Then $f(g(a)) = f(g(b))$ i.e., $a = b$. Thus, $g^{-1}(y)$ has at most one element for each $y \in A$; i.e., g is 1-1.//

Note: In general, g is not onto, and f is not 1-1 as seen in Example 8.

Corollary 1   Not every function $f: A \to B$ has a left (resp. right) inverse.

Corollary 2   Functions $f: A \to B$ exist which have neither left nor right inverses.

Proposition 2 $f: A \to B$ is a 1-1 correspondence iff it has both a left and a right inverse.

>   ***Proof*** Left as an exercise.//

Proposition 3   If $f: A \to B$ has both left and right inverses, then they are equal and unique.

>   ***Proof.*** Let $g \circ f = \mathbf{1}_A$ and $f \circ h = \mathbf{1}_B$. Then $g \circ (f \circ h) = g \circ \mathbf{1}_B = g$.
>   By associativity, lefthand side $= (g \circ f) \circ h = \mathbf{1}_A \circ h = h$. Hence, $g = h$.
>   As for uniqueness of g, $(x,y) \in g \Rightarrow y = g(x) \Rightarrow f(y) = f(g(x)) = \mathbf{1}_B(x) = x$
> $\Rightarrow (y,x) \in f \Rightarrow (x,y) \in f^{-1}$—i.e., $g \subset f^{-1}$.
>   Conversely, the symmetry of the g-f relationship implies $f \subset g^{-1}$—i.e., f^{-1}
> $\subset g$. Hence, $g = f^{-1}$.//

In many circumstances, we may have to deal with many functions $f_i: A_i \to B_i$ at the same time. As with Geometry, diagrams can help us to sort out the various cases and also to have a "picture" of the situation at all times. By a *function diagram*, we understand a pair (S,F), where S is a finite set of sets, and F is a finite set of functions, with the property $f: A \to B$ is in F only if $A \in S$ and $B \in S$. For example, here is a diagram:

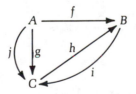

where $S = \{A,B,C\}$, $F = \{f,h,g,i,j\}$. Note that we write $A \overset{f}{\longrightarrow} B$ on the diagram instead of $f: A \to B$.

A *chain* from A to B in a diagram, where A and B are in S, is a sequence of functions $f_i: A_i \to B_i$ for $i = 1, \ldots, k$ such that $A_1 = A$, $B_k = B$ and, for $i = 1, \ldots, k - 1$, $A_{i+1} = B_i$.

A diagram is *commutative* iff for every pair of its sets A and B, and for every chain $f_1, \ldots, f_n$ from A to B $f_n \circ f_{n-1} \circ \ldots \circ f_1$ depends only on A and B.

Example 9

Commutative diagrams:

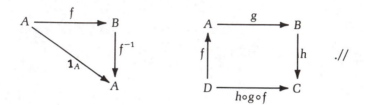

§1.3 SOME SPECIAL RELATIONS

This paragraph discusses binary relations which have some interesting properties.

Definition 1   Let R be a binary relation on A. It is *reflexive* iff $(\forall a)_{a \in A}\, aRa$. It is *symmetric* iff $(\forall a,b)_{(a,b) \in A^2}(aRb \Rightarrow bRa)$. It is *antisymmetric* iff $(\forall a,b)_{(a,b) \in A^2}(aRb\ \&\ bRa \Rightarrow a = b)$. It is *transitive* iff $(\forall a,b,c)_{(a,b,c) \in A^3}\ (aRb\ \&\ bRc \Rightarrow aRc).//$

Exercise 1   Show that the above properties of binary relations are independent—i.e., no combination of them implies any of the others.

Exercise 2   Show that the symmetry rule is equivalent to the rule $(\forall a,b)_{(a,b) \in A^2}\ (aRb \Longleftrightarrow bRa).$

Example 1   The total function $\lambda x.x{:}A \to A$ is a reflexive relation on $A.//$

Example 2   The relation "=" on $\mathcal{N}$ has all the properties of Definition 1.//

Example 3   The relation "<" on $\mathcal{N}$ has the last property only (transitivity).//

Example 4   The relation "≤" on $\mathcal{N}$ has all the properties in Definition 1 but symmetry.//

Example 5   The relation "$\equiv$ (mod m)" has all the properties in Definition 1 but antisymmetry.//

The relations in the last two examples are representatives of two important classes of relations; partial order relations and equivalence relations.

Definition 2   A binary relation on A is called an *equivalence relation* iff it is reflexive, symmetric, and transitive.//

Example 6   "=" and "$\equiv$ (mod m)" on $\mathcal{N}$ are both equivalence relations on $\mathcal{N}$.//

If we have an equivalence relation R on a set A we naturally "identify" the equivalent elements. This idea is captured by the notion of equivalence classes $[x]_R$: $[x]_R$ stands for the set of all elements equivalent to x—that is, $\{y \in A \mid xRy\}$.

It is easy to see that nonequivalent elements are in different equivalence classes. More specifically, $[x]_R = [y]_R$ iff xRy; also $[x]_R \cap [y]_R \neq \varnothing \Rightarrow [x]_R = [y]_R$. It is also clear that, for all x, $[x]_R \neq \varnothing$ and finally that $\bigcup_{x \in A} [x]_R = A$. In other words, the set of equivalence classes is a *partition* on the set A, as captured by the following definition:

Definition 3   Let $(F_a)_{a \in I}$ be a family of sets from A such that all the following hold:

(i) $(\forall a)_{a \in I} F_a \neq \varnothing$

(ii) $(\forall a,b)_{(a,b) \in I^2} F_a \cap F_b \neq \varnothing \Rightarrow F_a = F_b$

(iii) $\bigcup_{a \in I} F_a = A$

Then $(F_a)_{a \in I}$ is a *partition* of A.//

Partitions and equivalence relations are related so that one creates the other. (See Problem 16.)

Another important binary relation is now introduced; we shall study it further in the next section.

Definition 4   A relation R on A is a *partial order* iff it is reflexive, antisymmetric, and transitive. If for all x and y in A at least one of xRy or yRx holds, then R is also a *total* (or *linear*) order.//

Example 7   "=", "$\leq$" and "$\geq$" on $\mathcal{N}$ are partial orders.//

Example 8   "$\subset$" on $\mathcal{P}(X)$ is a partial order on $\mathcal{P}(X)$ (recall that "$\subset$" is not a proper inclusion). Note that not all elements of $\mathcal{P}(X)$ are "comparable". This justifies the use of the term "partial" in "*partial order*".

Partial orders are so important that a set equipped with one has a special name: *partially ordered set* or *poset*. We usually write a poset as (A, R), where A is the set and R is the partial order. Also, traditionally, the arbitrary partial order is denoted by "$\leq$", whenever confusion will not arise in contexts where the particular "$\leq$" on $\mathcal{N}$ (or on numbers in general) is also employed.//

§1.4 INDUCTION

Definition 1   Let $(A, \leq)$ be any poset. An element $a \in A$ is *minimal* iff $(\forall x)_{x \in A} x \leq a \Rightarrow x = a$. a is *maximal* iff $(\forall x)_{x \in A} a \leq x \Rightarrow x = a$.//

Note: Instead of "$a \leq b$ & $a \neq b$" we shall often write $a < b$. Thus, a is minimal if there is no x such that $x < a$.

Example 1   Let $A = \mathcal{P}(X) - \{\varnothing\}$ and $\leq = \subset$. Then each $\{x\}$, where $x \in A$, is minimal in $(A, \leq)$.//

Example 2   Let $A = \{x \text{ real} \mid 0 < x < 1\}$ and "$\leq$" be the usual order on reals. Then $(A, \leq)$ has neither a minimal nor a maximal element.//

Definition 2   Let $(A, \leq)$ be a poset. Then, (i) $(A, \leq)$ satisfies the *minimal condition* iff each nonempty subset B of A has a minimal element with respect to $\leq$. (ii) $(A, \leq)$ satisfies the *inductiveness condition* iff for any unary relation $R(x)$ on A, $R(x)$ holds for all $x \in A$ whenever $R(a)$ holds for all the minimal elements a of A and $R(b)$ holds whenever $R(x)$ holds for all $x < b$.//

Example 3 $(\mathcal{N}, \leq)$, where $\leq$ is the ordinary order on the natural numbers, satisfies both the minimal and inductiveness conditions.//

Theorem 1   The inductiveness and the minimal conditions are equivalent.

Proof   (1) (i) $\Rightarrow$ (ii). Let $(A, \leq)$ satisfy the minimal condition. Let $R(x)$ on A be such that $R(a)$ holds for all the minimal elements of A and $R(b)$ holds whenever $R(x)$ holds for all $x < b$.

Claim. $A \subset R$; that is, $(\forall x) \ x \in A \Rightarrow R(x)$. Let instead $A - R \neq \varnothing$. Then, let $a \in A - R$ be minimal. Now a is not minimal in A (all such elements are in R by hypothesis), hence elements $x \in A$ such that $x < a$ exist; for all such x, $R(x)$ holds (if $\neg R(x)$, then $x \in A - R$, violating the minimality of a). Hence, $R(a)$ holds by assumptions on R. We have a contradiction to the choice of a. The claim is proved.

(2) (ii) $\Rightarrow$ (i). Let $(A, \leq)$ satisfy the inductiveness condition. Moreover, instead of what we hope to establish, assume that there is a B, $\varnothing \neq B \subset A$, which has no minimal elements. Let $C = A - B$. Observe:

The minimal elements of A are in C (why?).

For any c in A, if all x such that $x < c$ are in C, then so is c (if instead $c \in B$, then c is minimal in B). By the inductiveness condition $A \subset C$, hence (since also $C \subset A$) $A = C$, yielding the contradiction $B = \emptyset$.//

Example 4   The poset $(\mathcal{N}^{n+1}, \leq)$, where $(x, \vec{x}_n) \leq (y, \vec{y}_n)$ iff $x \leq y$ and $\vec{x}_n = \vec{y}_n$ satisfies the minimal (and hence the inductiveness) condition. For each nonempty subset B of $\mathcal{N}^{n+1}$ the minimal elements are those with the smallest first component.//

The minimal (inductiveness) condition, besides being useful for proving statements about posets "by induction" (the reader is certainly familiar with the technique of "proof by induction" on $\mathcal{N}$), can be also used to define functions "inductively", or "recursively". The next theorem asserts that if $(A, \leq)$ satisfies the minimal condition, then a unique function $f:A \to X$ exists (X being any set), which is inductively defined by the "recurrence" relations:

(i) $f(a) = h(a)$ on all minimal elements of A, where $h:A \to X$ is some specific function.

(ii) $f(b)$, for any non-minimal b, is obtained from $\{f(x) \mid x < b\}$ according to some specific "rule".

Example 5   The well-known Fibonacci sequence is an inductively defined function $F: \mathcal{N} \to \mathcal{N}$. The recurrence relations are:

(i) $F(0) = 0$.

(ii) $F(x) = $ **if** $x = 1$ **then** 1 **else if** $x > 1$ **then** $F(x - 1) + F(x - 2)$.

Here, $h: \mathcal{N} \to \mathcal{N}$ is $\lambda x.x$ and the "rule" for (ii) is "if $n = 1$, then ignore $\{F(x) \mid x < n\}$ and answer 1; else if $n > 1$, answer the sum of the last two terms of $\{F(x) \mid x < n\}$; ignore all other terms".//

In preparation for the proof of the next theorem, we must introduce some notation in order to formalize the notion of "rule" in this context. We expect that rule will be formalized as a certain function.

Let X^Y denote the set of all *total* functions $f:Y \to X$. If $Y = \mathcal{N}$ we may view $X^{\mathcal{N}}$ as the set of all *sequences* with elements in X, since we may identify $f:\mathcal{N} \to X$ with $(f(n))_{n \in \mathcal{N}}$.

The symbol $\mathcal{P}(Y;X)$ denotes the set of *all* functions $f:Y \to X$ (i.e., including nontotal functions). Clearly, $A \subset Y$ implies $f \mid A \in \mathcal{P}(Y;X)$. Also, $X^Y \subset \mathcal{P}(Y;X)$.

For each $a \in Y$, we consider the ath projection function $\lambda f.f(a): \mathcal{P}(Y;X) \to X$. We reserve the symbol p_a for $\lambda f.f(a)$ assuming X and Y are understood from the context. Note that the term "ath projection" is in harmony with intuition, as when $Y = \{0,1,2, \ldots, n\}$ $p_i(x) = x(i)$ [or x_i], where $x:Y \to X$ is in $\mathcal{P}(Y;X)$. Also, clearly, $p_a(f)\!\uparrow$ iff $f(a)\!\uparrow$; moreover $p_a \mid X^Y$ is total for all a (why?).

Let now $(A, \leq)$ be a poset; if $a \in A$, then S_a denotes $\{x \in A \mid x \leq a\}$ and $\mathring{S}_a$ denotes $\{x \in A \mid x < a\}$.

We may now formalize the rule (ii) in the recurrence relations defining a function f by requiring the rule to consult with $f \mid \mathring{S}_b$ in order to provide $f(b)$.

Theorem 2   Let $(A, \leq)$ be a poset which satisfies the minimal condition, and let X be any set. Let $h: A \to X$ and $g: A \times \mathcal{P}(A;X) \to X$ be two functions.

Then a unique function $f{:}A \to X$ exists which satisfies the following inductive definition:

$$f(a) = h(a) \text{ for all the minimal elements } a \in A$$

$$f(b) = g(b, f \mid \mathring{S}_b), \text{ for all nonminimal elements } b \in A$$

Proof It is convenient to prove uniqueness first. So let f and j be two functions from A to X satisfying the inductive definition.

Observe that $f(a) = j(a)(*)$ for all the minimal elements $a \in A$.

For any b, assume $f(x) = j(x)$ for all $x < b$ (this is the induction hypothesis). This means $f \mid \mathring{S}_b = j \mid \mathring{S}_b$. Hence, $f(b) = j(b)$. By inductiveness condition on $(A, \leq)$, $A \subset \{x \mid f(x) = j(x)\}$.

Next, we prove existence of f: by induction on $(A, \leq)$ we prove that for any $a \in A$ a function $f_a: S_a \to X$ exists satisfying the given recurrence relations, except that we substitute S_a for A.

First, (basis of induction) if $a \in A$ is minimal, then $S_a = \{a\}$ and we are done $[f_a = \{(a, h(a))\}]$.

Next (induction hypothesis), let $f_x{:}S_x \to X$ exist for all $x < b$, which satisfies the recurrence relations. We observe that:

(1) If $x \leq y$, then $S_x \subset S_y$, hence $f_x \subset f_y$ as $f_y \mid S_x$ satisfies the same recurrence relations as f_x on S_x [uniqueness invoked!].

(2) If $w \in S_x \cap S_y$, then $f_x(w) = f_y(w)$, since $S_w \subset S_x \cap S_y$. Now, set $f_b: S_b \to X = \{(b, \tilde{b})\} \cup (\bigcup_{x<b} f_x)$, where $\tilde{b} = g(b, \bigcup_{x<b} f_x)$. Since $f_b \mid \mathring{S}_b = \bigcup_{x<b} f_x$ and, for all $x < b$, f_x satisfies the recurrence relations (induction hypothesis), it follows that $f_b: S_b \to X$ satisfies the recurrence relations.

This proves the contention that functions $f_a{:}S_a \to X$ exist satisfying the recurrence relations, for all $a \in A$.

Finally, define $f{:}A \to X$ by $f = \bigcup_{a \in A} f_a$. It is easy to see that f satisfies the recurrence relations. (See Problem 22.)//

Example 6   (Example 5 revisited.) For the Fibonacci sequence, take $\lambda bf.g(b,f) \overset{\text{def}}{=} \textbf{if } b = 1$ **then** 1 **else if** $b > 1$ **then** $p_{b-1}(f) + p_{b-2}(f).//$

(*)Recall that if $f(a){\uparrow}$, then $f(a) = j(a)$ iff $j(a){\uparrow}$.

We conclude this section by considering a fixed-point theorem which will reappear under various guises in the sequel.

Definition 3   Let $(A,\leq)$ be a poset and $\varnothing \neq B \subset A$. Then $a \in A$ is an *upper bound* of B iff $(\forall x)x \in B \Rightarrow x \leq a$. a is a *least upper bound* of B, in symbols, $a = \text{lub}(B)$ iff a is an upper bound of B and for any upper bound c of B $a \leq c.//$

Example 7   Let $(Q,\leq)$ be the poset of rational numbers with the usual ordering $\leq$. Let $B = \{x \in Q \mid x < \sqrt{2}\}$. Then $\text{lub}(B)$ does not exist in this poset.$//$

Example 8   Let X be any set; consider the poset $(\mathcal{P}(X),\subset)$. Then for any $B \subset \mathcal{P}(X)$, $\text{lub}(B)$ exists; indeed, $\text{lub}(B) = \bigcup_{x \in B} x.//$

Example 9   Let $B \subset \mathcal{P}(X;Y)$. If $\bigcup_{x \in B} x \in \mathcal{P}(X;Y)$, then $\text{lub}(B) = \bigcup_{x \in B} x$ in $(\mathcal{P}(X;Y),\subset)$. Note that $\bigcup_{x \in B} x \in \mathcal{P}(X;Y)$ iff it is single-valued on $X \times Y.//$

Note: A nonempty subset B of $(A,\leq)$ is called a *chain* iff $(\forall x,y)_{(x,y)\in B^2} x \leq y \vee y \leq x$ (i.e., all the elements of B are comparable) and $B = \text{range}(f)$, $\text{dom}(f) = \mathcal{N}$. An element m of A is *minimum* iff, for all $x \in A$, $m \leq x$. The terms *minimum* and *minimal* should not be confused.

Example 10 $(A,\leq)$ can have at most one minimum element. Indeed, let both m,n be minima. Then $m \leq n$ (since m is minimum) and $n \leq m$ (since n is minimum), thus $m = n$ by antisymmetry.$//$

Definition 4   Let $f:A \rightarrow B$ be a function, where $(A,\leq_1)$ and $(B,\leq_2)$ are posets. It is *monotone* (or *order-preserving*) iff $f(x)\leq_2 f(y)(*)$ whenever $x \leq_1 y$. It is *continuous* iff for *any* chain $X \subset A$, whenever $\text{lub}(X)$ exists then so does $\text{lub}(f(X))$, and $\text{lub}(f(X)) = f(\text{lub}(X)).//$

Theorem 3   (Knaster-Tarski) Let $(A,\leq)$ be a poset with a minimum element m, in which every chain has a lub. Let $f: A \rightarrow A$ be monotone and continuous.(†) Then there is an $a \in A$ such $f(a) = a$. Such an a is called a *fixed point of f*.

Proof Define by induction a sequence $(a_n)_{n \in \mathcal{N}}$:

$$a_0 = m$$

$$\text{For } n > 1: a_n = f(a_{n-1}).$$

(*)If $f(x)\uparrow$, then we always have $f(x) \leq_2 f(y)$ by convention.

(†)Continuity implies monotonicity, for if $x \leq y$, then $y = \text{lub}(\{x,y\})$, and $\{x,y\}$ is a chain. Moreover, $f(y) = \text{lub}(\{f(x),f(y)\})$; that is, $f(x) \leq f(y)$.

Since $(\mathcal{N},\leq)$ satisfies the minimal condition, Theorem 2 ensures that the sequence $(a_n)_{n\in\mathcal{N}}$ exists (and is unique). Observe that for all n, $a_n \leq a_{n+1}$. [By induction on n: $n = 0$, $a_0 \leq a_1$; this is true as $a_0 = m$, the minimal element in A. For $n < k$, let $a_n \leq a_{n+1}$ (induction hypothesis). *Claim:* $a_k \leq a_{k+1}$. Now, $a_{k-1} \leq a_k$ by induction hypothesis. As f is monotone, $f(a_{k-1}) \leq f(a_k)$—i.e., $a_k \leq a_{k+1}$.] By assumption, $\text{lub}(\bigcup_{n\in\mathcal{N}}\{a_n\})$ exists, let it be $a \in A$. *Claim:* $f(a) = a$. Indeed by continuity of f, $f(a) = \text{lub}(\bigcup_{n\in\mathcal{N}}\{f(a_n)\})$. Since $a_{n+1} = f(a_n)$ for all n and $a_{n+1} \leq a$ by definition of a, it follows that a is an upper bound of $\bigcup_{n\in\mathcal{N}}\{f(a_n)\}$. Let c be any upper bound of $\bigcup_{n\in\mathcal{N}}\{f(a_n)\}$. Then, for all n, $f(a_n) \leq c$—that is, for all n, $a_{n+1} \leq c$. Since also $a_0 \leq c$ (why?), we have $a_n \leq c$ for all $n \in \mathcal{N}$; thus, c is an upper bound of $\bigcup_{n\in\mathcal{N}}\{a_n\}$ as well, hence $a \leq c$. It follows that $a = \text{lub}(\bigcup_{n\in\mathcal{N}}\{f(a_n)\}) = f(a).//$

Corollary   The fixed point a defined in the proof above is *the least* fixed point of f.

Proof   Let c be any fixed point of f—that is, $f(c) = c$. For all n, $a_n \leq c$. Indeed, $a_0 \leq c$ trivially. Let $a_n \leq c$ for $n \leq k$.

Consider a_{k+1}: $a_{k+1} = f(a_k) \leq f(c)$ [monotonicity of f], hence $a_{k+1} \leq c$ [$f(c) = c$]. Hence, $a = \text{lub}(\bigcup_{n\in\mathcal{N}} a_n) \leq c.//$

§1.5 CARDINALITY

In the sequel, we shall only be interested in sets which can be written as finite or infinite sequences.

Definition 1   A set A is *finite* if it is either $\varnothing$ or there exists a 1-1 correspondence $f:\{0, \ldots, n\} \to A$ for some n. Otherwise, it is *infinite.//*

Note: If $f:\{0, \ldots, n\} \to A$ is a 1-1 correspondence, then we say that A has $n + 1$ *elements* and write $|A| = n + 1$.

The following propositions show that the above definition for a finite set is intuitively appealing.

Proposition 1   If $0 \leq m < n$, then there is no *onto* map $f:\{0, \ldots, m\} \to \{0, \ldots, n\}$.

Proof   Induction on n.
Basis: $n = 1$: The only maps $f:\{0\} \to \{0,1\}$ are $\varnothing$, $\{(0,0)\}$ and $\{(0,1)\}$. Clearly, none of them is onto.
Assume that if $n \leq k$, then there is no onto map $f:\{0, \ldots, m\} \to \{0, \ldots, n\}$ for any $m < n$.

Claim: Situation is the same when $n = k + 1$. We proceed by contradiction; so let $f:\{0, \ldots, m\} \to \{0, \ldots, k + 1\}$ be onto, where $1 \le m < k + 1$. (*) Then, for some $h \in \{0, \ldots, m\}$, $f(h) = k + 1$. Clearly, if $g = f \mid (\{0, \ldots, m\} - \{h\})$, then $g: \{0, \ldots, m\} - \{h\} \to \{0, \ldots, k\}$ is onto. Now, $\{0, \ldots, m\} - \{h\} = \{0, 1, \ldots, h - 1, h + 1, \ldots, m\}$. Define $j: \{0, \ldots, m - 1\} \to \{0, \ldots, m\} - \{h\}$ by $j(x) = $ **if** $0 \le x \le h - 1$ **then** x **else** $x + 1$. Clearly, $g \circ j: \{0, \ldots, m - 1\} \to \{0, \ldots, k\}$ is onto and we have just contradicted our induction hypothesis.//

Corollary 1   If $0 \le m < n$, then there is no 1-1 correspondence $f: \{0, \ldots, m\} \to \{0, \ldots, n\}$. Corollary 1 sometimes is called the "pigeon-hole principle".

Corollary 2   If A and B are finite sets, then a 1-1 correspondence $f: A \to B$ exists iff $|A| = |B|$.

Proof First, let $|A| = |B| = n$. Then, if $h: \{0, \ldots, n\} \to A$ and $g: \{0, \ldots, n\} \to B$ are 1-1 correspondences, then $f = g \circ h^{-1}$ will be a 1-1 correspondence between A and B.

Next, let $|A| = m < n = |B|$, and $f: A \to B$ be a 1-1 correspondence.

Let $h: \{0, \ldots, m\} \to A$ and $g: \{0, \ldots, n\} \to B$ be 1-1 correspondences. Then $g^{-1} \circ f \circ h: \{0, \ldots, m\} \to \{0, \ldots, n\}$ is a 1-1 correspondence; a contradiction. Hence, if $f: A \to B$ is a 1-1 correspondence, then $|A| = |B|$.//

Proposition 2   There is no onto map $f: \{0, \ldots, n\} \to \mathcal{N}$.

Proof Induction on n.

Basis: $n = 0$; the result is immediate.

Let the assertion be true for $n \le k$ (induction hypothesis). By contradiction, assume that $f: \{0, \ldots, k + 1\} \to \mathcal{N}$ is onto. Let $h \in \{0, \ldots, k + 1\}$ be such that $f(h) = 0$. Define $j: \{0, \ldots, k\} \to \{0, \ldots, h - 1, h + 1, \ldots, k + 1\}$ by $j(x) = $ **if** $0 \le x \le h - 1$ **then** x **else** $x + 1$; clearly, j is onto. Define $i: \mathcal{N} - \{0\} \to \mathcal{N}$ by $i = \lambda x. x - 1$; clearly, i is onto. Thus, $i \circ g \circ j: \{0, \ldots, k\} \to \mathcal{N}$ is onto, where $g = f \mid \{0, \ldots, h - 1, h + 1, \ldots, k + 1\}$. We have just contradicted the induction hypothesis.//

Corollary $\mathcal{N}$ is infinite.

Definition 2   A set S is *enumerable* (or *countable*) iff there is a 1-1 correspondence $f: \mathcal{N} \to S$. A set S is at *most enumerable* if it is either finite or enumerable.//

(*)$m = 0$ cannot lead to onto functions as we have just seen, so it is not considered.

Example 1   For any n, $\{0, \ldots, n\}$ and $\{1, \ldots, n\}$ are finite. $\mathcal{N}$ is enumerable. The set E $= \{2n \mid n \in \mathcal{N}\}$ is enumerable, as $f: \mathcal{N} \rightarrow E$ defined by $f = \lambda x.2x$ is a 1-1 correspondence.//

Theorem 1   An infinite subset A of $\mathcal{N}$ is enumerable.

Proof   We shall inductively define a function $f:\mathcal{N} \rightarrow A$ which is 1-1, onto and total:

$$f(0) = \text{minimum element of } A \text{ (recall } A \subset \mathcal{N})$$

if $n > 0$, $f(n) = $ minimum element of $A - \{f(0), \ldots, f(n-1)\}$. Observe:
(1) f is total (otherwise, for some $n > 0$, $f(n)\uparrow$—i.e., $\min(A - \{f(0), \ldots, f(n-1)\})$ does not exist. This can only be if $A - \{f(0), \ldots, f(n-1)\} = \varnothing$. Let n be the smallest such that $f(n)\uparrow$, then $A = \{f(0), \ldots, f(n-1)\}$—i.e., A is finite; contradiction).
(2) f is strictly increasing (hence, 1-1). Indeed, $f(n) < f(n+1)$ for $n \in \mathcal{N}$ [by induction: if $n = 0$, then it is obvious. Let $f(n) < f(n+1)$, $n \leq k$. Then $f(0) < f(1) < \ldots < f(k+1)$.
Claim: $f(k+1) < f(k+2)$. Now $f(k+1) = \min(A - \{f(0), \ldots, f(k)\})$, hence, $f(k+1) \leq f(k+2)$ as $f(k+2) \notin \{f(0), \ldots, f(k)\}$. But $f(k+1) \neq f(k+2)$ as $f(k+2) \notin \{f(0), \ldots, f(k), f(k+1)\}$].
(3) Range$(f) = A$.
That range$(f) \subset A$ is obvious. Let then $A - $range$(f) \neq \varnothing$, and m be the minimum element in it. By (1) and (2) above, $f(n) \geq n$ for all n [$f(0) \geq 0$ and $f(n+1) > f(n) \geq n$ implies $f(n+1) \geq n+1$]. Thus, n exists such that $m \leq f(n)$. Let n_0 be smallest such n. Then $\{m, f(n_0)\} \subset A - \{f(0), \ldots, f(n_0-1)\}$ and $m < f(n_0)$ [why was $m \leq f(n_0)$ sharpened?]. On the other hand, $f(n_0) = \min(A - \{f(0), \ldots, f(n_0-1)\})$; we have a contradiction.
Thus, range$(f) = A$, hence, $f:\mathcal{N} \rightarrow A$ is 1-1 and onto.//

Corollary 1 A is at most enumerable iff there is an onto map $f:\mathcal{N} \rightarrow A$.(*)

Proof   *if* part. If A is finite, then we are done. So let it be infinite and show it is enumerable. The plan is to show that there is a 1-1 correspondence between A and an infinite subset of $\mathcal{N}$.
Define a function $g:A \rightarrow \mathcal{N}$ by $g(x) = $ minimum $(f^{-1}(x))$, for all $x \in A$. Observe:

(1) g is total, as $f^{-1}(x) \neq \varnothing$ for all $x \in A$ (why?).

(*)We state, for emphasis, that f need not be total.

(2) g is 1-1 as $g(x) = g(y) \Rightarrow \min(f^{-1}(x)) = \min(f^{-1}(y)) \Rightarrow f^{-1}(x) \cap f^{-1}(y) \neq \varnothing \Rightarrow f$ is not single-valued, unless $x = y$; hence, $x = y$.

(3) range(g) is infinite. [For if instead $h:\{0, \ldots, n\} \to$ range(g) is a 1-1 correspondence, then $w \circ h:\{0, \ldots, n\} \to A$ is a 1-1 correspondence, making A finite. w is g^{-1}: range(g) $\to A$].

By Theorem 1, there is a 1-1 correspondence $v:\mathcal{N} \to$ range(g), hence, $w \circ v:\mathcal{N} \to A$ is a 1-1 correspondence. The *only if* part is trivial.//

Corollary 2   If A is at most enumerable, then so is B, where $B \subset A$.

   Proof There is an obvious onto map $f:A \to B$.//

Corollary 3   If A and B are at most enumerable, then so are $A \cap B$ and $A - B$.

Proposition 3   Let A_i, $i = 1, \ldots, n$, be at most enumerable. Then so is $X_{i=1}^n A_i$.

   Proof Let $f_i:\mathcal{N} \to A_i$ be onto, for $i = 1, \ldots, n$ (Corollary 1, Theorem 1). Then $(f_1, \ldots, f_n):\mathcal{N}^n \to X_{i=1}^n A_i$ is onto. It therefore suffices to show that $\mathcal{N}^n$ is enumerable.
   To this end, define $f:\mathcal{N}^n \to \mathcal{N}$ by $f = \lambda \vec{x}_n.p_1^{x_1+1}p_2^{x_2+1} \ldots p_n^{x_n+1}$ where p_i is the ith prime number ($p_0 = 2$, $p_1 = 3$, etc.). By the unique factorization theorem, f is 1-1, hence, $g = \{(x,y) \mid (y,x) \in f\}$ is single-valued on $\mathcal{N} \times \mathcal{N}^n$. $g:\mathcal{N} \to \mathcal{N}^n$ is clearly onto. Thus, $(f_1, \ldots, f_n) \circ g:\mathcal{N} \to X_{i=1}^n A_i$ is onto. Now invoke Corollary 1 of Theorem 1.//

Corollary 1   If A_i is at most enumerable, $i = 1, \ldots, n$, then so is $\bigcup_{i=1}^n A_i$.

   Proof Let $f_i:\mathcal{N} \to A_i$ be onto, $i = 1, \ldots, n$.
   Define $g:\mathcal{N} \times \{1, \ldots, n\} \to \bigcup_{i=1}^n A_i$ by $g(x,i) = f_i(x)$ for all x,i. Clearly, g is onto. Further, there is an obvious onto map $h:\mathcal{N} \times \mathcal{N} \to \mathcal{N} \times \{1, \ldots, n\}$. Thus, $g \circ h:\mathcal{N}^2 \to \bigcup_{i=1}^n A_i$ is onto. Now invoke Proposition 3.//

Corollary 2   If A_i is at most enumerable, $i = 1, 2, \ldots$, then so is $\bigcup_{i=1}^{\infty} A_i$.

   Proof Let $f_i:\mathcal{N} \to A_i$ be onto, for $i = 1, 2, \ldots$. Define $g:\mathcal{N} \times \mathcal{N} \to \bigcup_{i=1}^{\infty} A_i$ by $g(x,i) = f_i(x)$ for all x,i. Clearly, g is onto. Result follows from Proposition 3.//

Corollary 3 $\bigcup_{n=1}^{\infty} \mathcal{N}^n$ is enumerable.

Proof This set is obviously infinite ($\mathcal{N} \subset \bigcup_{n-1}^{\infty} \mathcal{N}^n$). Next, let $d:\mathcal{N} \to \bigcup_{n-1}^{\infty} \mathcal{N}^n$ be defined by $d(x) = (x_0, \ldots, x_m)$ iff m is smallest such that $x = p_0^{x_0} p_1^{x_1} \ldots p_m^{x_m}$. Clearly, d is onto.//

Theorem 2   (Schröder-Bernstein.) Let the maps $f:A \to B$ and $g:B \to A$ be 1-1 and total. Then there exists an 1-1 correspondence $h:A \to B$.

Proof If either f or g is also onto, then the result is trivial. We shall therefore proceed without such an assumption.

Assume for a minute that we managed to show that X,Y,X',Y' exist so that $X \cap Y = \varnothing$ and $X \cup Y = A$, $X' \cap Y' = \varnothing$ and $X' \cup Y' = B$ and $X' = f(X)$ and $Y = g(Y')$. Then $h = \lambda x.$**if** $x \in X$ **then** $f(x)$ **else if** $x \in Y$ **then** $g^{-1}(x)$, where g^{-1} is the single-valued relation $\{(x,y) \mid (y,x) \in g\}$ on $A \times B$, is a 1-1 correspondence $A \to B$.

Let us establish existence of X,Y,X',Y': We use the figure below as a guide, first in an *intuitive* argument. Note that X cannot be $A - g(B)$ as there is no room left in B for $X' = f(X)$, since then $Y = g(B)$—that is, $B = Y'$ and $X' = \varnothing$. However, the sought-after X would contain $A - g(B)$ and $g{\circ}f(X)$. [That X should contain $A - g(B)$ is clear, so that $X' = \varnothing$ in B is "enlarged". That it must also be $X \supset g{\circ}f(X)$ follows from the requirement $X \cap Y = \varnothing$ and the observation that $g{\circ}f(X) = g(X')$ and $g(X') \cap g(Y') = \varnothing$; that is, $g(X') \cap Y = \varnothing$, hence $g(X') \subset X$.] End of informal discussion.

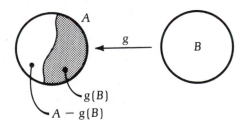

Now let $X = \min(Q)$, where $Q = \{S \mid S \supset (A - g(B)) \cup g{\circ}f(S)\}$; i.e., $X = \bigcap_{S \in Q} S$.

Note that $Q \neq \varnothing$ ($A \in Q$ trivially) and X—i.e., $\min(Q)$—is in Q, for $X \supset (A - g(B))$, since $(\forall S)S \in Q \Rightarrow S \supset A - g(B)$, and $X \supset g{\circ}f(X)$ since $\bigcap S \supset \bigcap g{\circ}f(S) = g{\circ}f(\bigcap S)$ (the last equality since $g{\circ}f$ is 1-1).

We claim that X is the sought-after set. So, let $Y = A - X$, $X' = f(X)$ and $Y' = B - X'$. We must prove that $Y = g(Y')$.

First, recall that we assumed f not onto, hence $Y' \neq \varnothing$. Next,
(1) $X \supset (A - g(B)) \cup g{\circ}f(X)$ implies (why?) $A - X \subset A - [(A - g(B)) \cup$

$g \circ f(X)] = g(B) \cap (A - g \circ f(X))$ (the equality by De Morgan law). Hence, $A - X \subset g(B) - g \circ f(X)$.

(2) *Claim*: $g(B) - g \circ f(X) \subset A - X$. Let instead $a \in [g(B) - g \circ f(X)] - (A - X)$. Hence, $a \in g(B)$ & $a \notin g \circ f(X)$ & $a \in X$. Set $\tilde{X} = X - \{a\}$.

We have $g \circ f(\tilde{X}) \subset g \circ f(X) \subset X$ (the latter inclusion from the definition of X). Since $a \notin g \circ f(X)$, it follows that $g \circ f(\tilde{X}) \subset \tilde{X}$ ... (3). Next, $A - g(B) \subset \tilde{X}$, since $A - g(B) \subset X(= \tilde{X} \cup \{a\})$ and $a \in g(B)$. This and (3) implies $\tilde{X} \supset (A - g(B)) \cup g \circ f(\tilde{X})$, contradicting the minimality of X.

We have shown that $Y(= A - X) = g(B) - g \circ f(X) = g(B - f(X))$ (the last equality since g is 1-1); that is, $Y = g(Y')$.//

Note: The above proof has used an idea found in Dieudonné [2], p. 10. We now turn our attention to *uncountable* sets.

Definition 3 A set is *nonenumerable* or *uncountable* if it is neither finite nor enumerable.//

Sets referred to in Definition 3 exist. Let us demonstrate this, as Cantor did, by showing that the set of real numbers x such that $0 < x < 1$ is uncountable.

By contradiction, let $n_0, n_1, n_2, \ldots$, be an enumeration of $\{x \text{ real} \mid 0 < x < 1\}$—i.e., $\lambda i.n_i$ is a total onto map.

We view each number $0 < x < 1$ as a *binary* fraction— that is, a string of the form $.a_0 a_1 a_2 \ldots$, where "." is the "binary" (in analogy to "decimal") point and each $a_i \in \{0,1\}$.

As some numbers x accept two expansions (for example, $.1 = .0111\ldots$), $\underset{\text{1's forever}}{}$ we decide to accept only the infinite expansion. Thus, essentially, $n_0, n_1, \ldots$, is an enumeration of all infinite-length strings composed of 1's and 0's. The assumed enumerability gives rise to the following infinite matrix:

$$n_0 = .\, n_0^0 n_0^1 n_0^2 \ldots$$

$$n_1 = .\, n_1^0 n_1^1 n_1^2 \ldots$$

$$\vdots \qquad \vdots$$

$$n_i = .\, n_i^0 n_i^1 n_i^2 \ldots$$

$$\vdots \qquad \vdots$$

where n_j^i is the ith binary digit of the jth number.

Cantor observed that if he considers first the number $.n_0^0 n_1^1 n_2^2 \ldots$—i.e., the one composed of the digits on the *major diagonal*—then he can derive from it a number *not* in the matrix, namely $.\overline{n}_0^0\overline{n}_1^1\overline{n}_2^2\ldots$, where $\overline{n}_i^i = 1 - n_i^i$, as the new number for all i differs from n_i at the ith digit!

This contradicts the assumption that *all* the numbers $x{:}0 < x < 1$ were in the matrix.

Note: (A small technicality.) It is easy to argue that it can be arranged that $.\overline{n}_0^0\overline{n}_1^1\overline{n}_2^2\ldots$ is neither such that $\overline{n}_i^i = 0$ (or $\overline{n}_i^i = 1$) for all i. This case would annoy us, as then $.\overline{n}_0^0\overline{n}_1^1\ldots$ is *not* in the interval $0 < x < 1$. However, without loss of generality (why?), we assume that $n_0^0 = 0$ and $n_1^1 = 1$.

For obvious reasons, the above technique is called *diagonalization*, or a *diagonal* argument, method, or technique. In the sequel, it will reoccur alarmingly often under various guises. Here are some further preliminary examples:

Proposition 4   The set $\mathcal{P}(\mathcal{N})$ is not enumerable.

Proof *Version 1.* There is a 1-1 correspondence between $\mathcal{P}(\mathcal{N})$ and the set of all infinite binary strings [incidentally, if the reader is uneasy about the notion of an infinite binary string s, it is nothing more than an element of $\{0,1\}^{\mathcal{N}}$]. Indeed, if $S \in \mathcal{P}(\mathcal{N})$ then let $s \in \{0,1\}^{\mathcal{N}}$ correspond to S iff $s(n) =$ **if** $n \in S$ **then** 0 **else** 1 [$\lambda n.s(n)$ is known as the *characteristic* function of S]. However, we have just seen that $\{0,1\}^{\mathcal{N}}$ is uncountable since it is, essentially, (a representation of) the set $\{x \mid 0 < x < 1\}.//$ of *Version 1*.

Version 2. Let $S_0, S_1, \ldots$ be an enumeration of $\mathcal{P}(\mathcal{N})$. Then, the set $S = \{x \mid x \notin S_x\}$ is in $\mathcal{P}(\mathcal{N})$ but *not* in the list (enumeration) as the assumption that $S = S_i$, for some i, leads to $i \in S_i \Longleftrightarrow i \notin S_i.//$ of Version 2.

Note: The second version is essentially identical to the first although, on the surface, this is not obvious. [Recall the 1-1 correspondence $s_i \leftrightarrow S_i$. The characteristic function of the "diagonal set" S is $s = \lambda n.1 - s_n(n)$.

Let us assume for a minute that the *diagonal* object (the Russel-paradox "set") $S = \{x \text{ a set} \mid x \notin x\}$ is a set. Then, it is allowable to substitute S in x to obtain $S \in S \Longleftrightarrow S \notin S$. Thus, S is *not* a set. This was a diagonal argument over all sets.

PROBLEMS

1. Let U be a given reference set. Show that $\varnothing$ exists in the context of U. (*Hint*: Consider the statement $S(x) \overset{\text{def}}{=} x \neq x$.)

   In what follows, U will be a fixed reference set, and all sets will be subsets of U.

2. Show that for any sets A, B, and C, each of the following holds:

   (a) $A = A \cup A$ $\qquad\qquad\qquad\qquad A = A \cap A$

   (b) $A = A \cup (A \cap B)$ $\qquad\qquad A = A \cap (A \cup B)$

   (c) $A \cap (B \cup C) = (A \cap B) \cup (A \cap C)$ $\qquad A \cup (B \cap C) = (A \cup B) \cap (A \cup C)$

   (d) $A \subset B$ iff $A = A \cap B$ $\qquad\qquad A \subset B$ iff $B = A \cup B$

   (e) $A \cup (B \cup C) = (A \cup B) \cup C$ $\qquad A \cap (B \cap C) = (A \cap B) \cap C$

   (f) $--A = A$

3. Prove that $\{a,\{a,b\}\} = \{c,\{c,d\}\}$ iff $a = c$ and $b = d$. (Thus, $\{a,\{a,b\}\}$ "implements" the pair (a,b).)

4. Prove that for any A, B, $\varnothing = A \times B$ iff $A = \varnothing$ or $B = \varnothing$.

5. Show that for any A, B, and C, $(A \cup B) \times C = (A \times C) \cup (B \times C)$.

6. Prove: For any functions $f{:}X \to Y$ and $g{:}Y \to Z$, $g{\circ}f$ is a function.

7. Prove that for any $f{:}X \to Y$ and $A \subset Y$, $B \subset Y$,

   (a) $f^{-1}(A \cup B) = f^{-1}(A) \cup f^{-1}(B)$

   (b) $f^{-1}(A \cap B) = f^{-1}(A) \cap f^{-1}(B)$

   (c) if $A \subset B$, $f^{-1}(B - A) = f^{-1}(B) - f^{-1}(A)$.

   Is this last equality true if $A \not\subset B$?

8. Show that for any $f{:}X \to Y$ and $A \subset X$, $B \subset X$,

   (a) $f(A \cap B) \subset f(A \cap B)$

   (b) $f(A \cup B) = f(A) \cup f(B)$

   (c) if $A \subset B$, $f(B - A) \supset f(B) - f(A)$.

   Can the above inclusions be sharpened into equalities?

9. Prove that for any $f: X \rightarrow Y$ and $A \subset X, B \subset Y$,

(a) $f(f^{-1}(B)) \subset B$

(b) $f^{-1}(f(A)) \supset A$

Show by examples that the above inclusions cannot in general be improved to equalities.

10. Show that $\bigcup_{a \in \varnothing} S_a = \varnothing$ and $\bigcap_{a \in \varnothing} S_a = U$.

11. Let $(S_a)_{a \in I}$ be a family of subsets of X. Let $f: X \rightarrow Y$ and $g: Z \rightarrow X$ be two functions. Prove that

(i) $g^{-1}(\bigcap_{a \in I} S_a) = \bigcap_{a \in I} g^{-1}(S_a)$

(ii) $g^{-1}(\bigcup_{a \in I} S_a) = \bigcup_{a \in I} g^{-1}(S_a)$

(iii) $f(\bigcup_{a \in I} S_a) = \bigcup_{a \in I} f(S_a)$

(iv) $f(\bigcap_{a \in I} S_a) \subset \bigcap_{a \in I} f(S_a)$ (Can $\subset$ be replaced by $=$?)

12. Show that for any set A and family of sets $(S_a)_{a \in I}$,

(a) $A - \bigcup_{a \in I} S_a = \bigcap_{a \in I} (A - S_a)$

(b) $A - \bigcap_{a \in I} S_a = \bigcup_{a \in I} (A - S_a)$

13. Let R be a reflexive binary relation on a set A such that for all a, b, c in A, R satisfies $aRb \& aRc \Rightarrow bRc$ Show that R is an equivalence relation.

14. *False* "theorem": Every symmetric-transitive relation is an equivalence relation. "*Proof*". Let R on A be as specified. Then aRb implies bRa by symmetry. But then $aRb \& bRa \Rightarrow aRa$ by transitivity. Thus aRa, i.e., R is reflexive. End of "proof".

(1) Give an example of a symmetric-transitive relation which is *not* reflexive.

(2) What is wrong with the above "proof"?

15. Let R be an equivalence relation on a set A. Prove that

(i) $[x]_R = [y]_R$ iff xRy

(ii) $x \in [y]_R$ iff $[x]_R = [y]_R$

(iii) $[x]_R \cap [y]_R \neq \varnothing \Rightarrow [x]_R = [y]_R$

(iv) $A = \bigcup_{x \in A} [x]_R$

(v) $[x]_R \neq \varnothing$, for all $x \in A$

(iii)–(v) say that $([x]_R)_{x \in A}$ is a partition of A.

16. Let $(F_a)_{a \in I}$ be a partition of a set A. Prove that

   (i) The relation F defined by

$$xFy \overset{\text{def}}{\equiv} (\exists a)(a \in I \,\&\, x \in F_a \,\&\, y \in F_a)$$

   is an equivalence relation.

   (ii) For any $x \in A$ and $a \in I$, $[x]_F \cap F_a \neq \varnothing \Rightarrow [x]_F = F_a$.

   (iii) For any $x \in A$ (resp. $a \in I$), there is an $a \in I$ (resp. $x \in A$) such that $[x]_F \cap F_a \neq \varnothing$. [In words, $(F_a)_{a \in I}$ induces an F such that $(F_a)_{a \in I} = ([x]_F)_{x \in A}$]

17. Show that a binary relation R is transitive iff $R^2 \subset R$, where $R^2 = R \circ R$.

   Note: If $R \subset A \times B$ and $S \subset B \times C$, then for all (x,y), $xR \circ Sy \Longleftrightarrow (\exists z)\, xRz$ $\&\, zSy$.

18. *Transitive closure.* A transitive closure of a binary relation R on A is any relation S on A satisfying

   (i) S is transitive and $R \subset S$.

   (ii) If T (on A) is transitive and $R \subset T$, then $S \subset T$. (In words, S is a *smallest* transitive relation including R.)

   Prove:

   (a) For any R, its transitive closure (if it exists) is *unique;* thus, we talk of *the* transitive closure of R, usually denoted by R^+.

   (b) For any R, R^+ *exists.* (*Hint:* Let R be on A. Now $A \times A$ is transitive and $R \subset A \times A$. Thus, $F = \{S \mid R \subset S \text{ and } S \text{ is transitive}\}$ is a nonempty family. Argue that $R^+ = \bigcap_{S \in F} S$. Alternatively, see the next problem.)

19. (a) Show that for any R on A,

$$R^+ = \bigcup_{i \geq 1} R^i,$$

   where $R^1 = R$ by definition, and $R^{i+1} = R \circ R^i$ for $i \geq 1$.

   (b) Show that if A has n elements,

$$R^+ = \bigcup_{1 \leq i \leq n} R^i.$$

   Also show, by an example, that all the powers R^i ($1 \leq i \leq n$) are, in general, necessary.

   (c) Show that if A has n elements and R is reflexive, $R^+ = R^{n-1}$.

20. Let $f:X \to Y$ be 1-1 and $g:X \to Y$ be onto. Let $A \subset X$ and $B \subset Y$. Prove that $f^{-1}(f(A)) = A$ and $g(g^{-1}(B)) = B$. Then prove appropriate converses of the above statements.

21. Let $f:X \to Y$ be 1-1, and let $A \subset B \subset X$.

   (a) Show that $f(B - A) = f(B) - f(A)$.

   (b) Prove a converse as well.

   (c) What can you say if $A \not\subset B$?

22. Complete the proof of Theorem 1.4.2.

23. Let A be infinite and B enumerable. Show that there is a 1-1 correspondence between $A \cup B$ and A. (*Hint:* Use the Axiom of Choice to prove that A has an enumerable subset.)

24. Prove that the set of finite subsets of $\mathcal{N}$ is enumerable. (*Hint:* With the help of characteristic functions, identify each finite subset of $\mathcal{N}$ with an infinite sequence of 0 and 1 elements such that all but a finite number of terms are 1's. Then identify such sequences with rational numbers. Alternatively, code each finite subset by some integer. [Think in base 2.])

25. Prove: If X has n elements, then $\mathcal{P}(X)$ has 2^n elements. (*Hint:* Use induction on n.)

26. Show that for any set X, there is no 1-1 correspondence between X and $\mathcal{P}(X)$.

27. (Dedekind's definition of "infinite set".) Prove that a set X is infinite iff it is in 1-1 correspondence with a *proper* subset. (*Hint:* for the *only if* part, use the Axiom of Choice to find an enumerable subset A of X. Then X and $X - A$ are in 1-1 correspondence by Problem 23.)

REFERENCES

[1] Bennett, J. *On Spectra*. Ph.D. dissertation, Princeton University, 1962.

[2] Dieudonné, J. *Foundations of Modern Analysis*. New York: Academic Press, 1960.

[3] Halmos, P. *Naive set Theory*. Princeton, N.J.: Van Nostrand, 1960.

[4] Kamke, E. *Theory of Sets*. New York: Dover Publications, 1950.

[5] Kurosh, A.G. *Lectures on General Algebra*. New York: Chelsea Publishing Co., 1963.

[6] Smullyan, R. *Theory of Formal Systems*. Annals of Mathematics Studies, No. 47. Princeton, N.J.: Princeton University Press, 1961.

CHAPTER 2

Primitive Recursive Functions

Beginning with the present chapter, the rest of the book will be devoted, broadly speaking, on one hand to the formalization of the intuitive notion of "algorithm" and, on the other hand, once such formalism is available, to the general question of how much of the theory of algorithms can be dealt with using algorithmic methods. (For example, we shall later ask and answer the question "Is there an algorithmic method by which we can decide whether or not any two given algorithms define the same function?")

Intuitively, an algorithm, as encountered in various branches of mathematics, satisfies the following conditions:

(a) It consists of a finite set of well-defined (unambiguous) instructions.

(b) It operates on a well-defined set of possible inputs.

(c) It generates an output for every input after a finite number of steps, a "step" being the execution of any one of the permissible instructions.

(d) The algorithm is "deterministic"—i.e., given an input, each step (during execution of the algorithm) is followed by a uniquely determined next step.

Some fairly well-known examples of algorithms are:

(i) the Euclidean algorithm, which, given any two integers a and b, calculates their greatest common divisor, $\gcd(a,b)$;

(ii) an algorithm for the calculation of the derivative function of a polynomial;

(iii) an algorithm for the calculation of the number of connected components of an undirected graph;

(iv) an algorithm which tests whether an integer is a prime or not.

Note that an algorithm defines a function from the set of all possible inputs to the set of all possible outputs; we naturally call such a function *algorithmic* or *computable*. One should be careful to distinguish between *algorithm* and *algorithmic function* for, clearly, an algorithmic function may have more than one algorithm which define it.(*) For example, the function $(a,b) \mapsto \gcd(a,b)$ may be calculated by algorithms which employ different instructions than those in the Euclidean algorithm. (See Problem 1.)

We also note that: example (i) defines a function from Z^2 to $\mathcal{N}$; example (ii) defines a function from the set of polynomials to the set of polynomials; example (iii) defines a function from the set of finite undirected graphs to $\mathcal{N}$, and finally, example (iv) defines a function from Z to a two-element set $\{a,b\}$ ($a \neq b$), where a is being output if the input is a prime, whereas b is being output otherwise.

We shall see in the sequel that there is no loss of generality if we restrict our study to "number theoretic functions"—that is, functions from $\mathcal{N}^k$ to $\mathcal{N}$ for every k;(†) we shall therefore consider only such functions in the rest of this book unless otherwise specified.

Historically, early attempts to formalize the notion of algorithmic function led to the definition of the class of "Recursive Functions" (Dedekind [3], Gödel [4]) which is known in the modern literature as the class of "Primitive Recursive Functions". In this chapter, we will present the elements of the theory of primitive recursive functions.

§2.1 DEFINITIONS AND SIMPLE EXAMPLES

The class of primitive recursive functions is defined as the *smallest* class which contains certain *initial* functions and is closed under certain *operations on functions*. The initial functions are chosen to be very simple and, intuitive-

(*)We shall later see that every algorithmic function has countably infinite different algorithms which define it.

(†)Under some reasonable restrictions, objects outside $\mathcal{N}$ can be "coded" by elements of $\mathcal{N}$; more on this later.

ly, algorithmic (e.g.,$\lambda x.x + 1, \lambda x.0$); so are the functional operators. Thus, intuitively, repeated application of the latter will always produce algorithmic functions.

One possible way of defining primitive recursive functions is by using an *infinite* set of initial functions and two operations, *composition* and *primitive recursion*.

Composition in this context is restricted so that the resulting function has its range in $\mathcal{N}$—i.e., is always number-theoretic.

Definition 1   The function $\lambda \vec{y}_m.h(\vec{y}_m)$ is obtained from the functions $\lambda \vec{x}_n.f(\vec{x}_n)$ and $\lambda \vec{y}_m.(g_1(\vec{y}_m), \ldots, g_n(\vec{y}_m))$ by *composition* iff

$$h = f \circ (g_1, g_2, \ldots, g_n).//$$

Example 1   Let $u = \lambda x.x$, $a = \lambda xy.x + y$. Then if $d = \lambda x.2x$, we have $d = a \circ (u,u).//$

Example 2   With a, d and u as in Example 1, it is clear that also $d = \lambda x.a(x,u(x))$ or even $d = \lambda x.a(x,x)$. Note, however, that the operation of substituting y of $a(x,y)$ by $u(x)$ or the operation of identifying x and y is *not* composition.$//$

Example 3   Let $f: \mathcal{N}^2 \to \mathcal{N}$ and $g: \mathcal{N} \to \mathcal{N}$ be two functions; let us also have available the functions $u_1^2 = \lambda xy.x$ and $u_2^2 = \lambda xy.y$.

Then the function $h = \lambda xy.f(x,g(y))$ can be obtained by a sequence of compositions:

Set $h_1 = \lambda xy.g(u_2^2(x,y))$. Then $h = f \circ (u_1^2, h_1).//$

We shall now introduce the second allowed operation, *primitive recursion*, so that finally we can define the class of primitive recursive functions.

Definition 2   The operation of *primitive recursion* assigns *a* number-theoretic function $\lambda x \vec{y}_m.f(x,\vec{y}_m)$ to the pair of number-theoretic functions $\lambda \vec{y}_m.h(\vec{y}_m)$ and $\lambda x \vec{y}_m z.g(x,\vec{y}_m,z)$ iff for all x and $\vec{y}_m$ the following equations hold (the *schema of primitive recursion*):

$$f(0,\vec{y}_m) = h(\vec{y}_m)$$
$$f(x + 1,\vec{y}_m) = g(x,\vec{y}_m,f(x,\vec{y}_m)).//$$

Proposition 1   The schema of primitive recursion defines a *unique* function f, thus the italicized "a" of Definition 2 can be replaced by "*the*".

Proof   Since f is inductively defined on the poset $(\mathcal{N}^{m+1}; \geq)$, where $(x,\vec{u}_m) \geq (y,\vec{v}_m)$ iff $x \geq y$ (in the usual sense) and $\vec{u}_m = \vec{v}_m$, the result follows from Theorem 1.4.2.$//$

Note: The function defined from the primitive recursive schema has one argument *more* than h (the *recursion basis*) and one argument *less* than g (the *iterated element*). Thus, we cannot define functions of one argument by just a primitive recursion (this is because we do not allow functions of 0 number of arguments).

Example 4   Consider the schema

$$A \begin{cases} 0 \div 1 = 0 \\ (x + 1) \div 1 = x \end{cases}$$

which defines the predecessor function $p = \lambda x.x \div 1$. ("$\div$" denotes *proper subtraction*: $\lambda xy.x \div y \overset{\text{def}}{=}$ **if** $x < y$ **then** 0 **else** $x - y$). This schema is *not* a (formal) primitive recursion for reasons outlined in the note above. However, let $\hat{p}$ be defined by:

$$B \begin{cases} \hat{p}(0,y) = z(y) \\ \hat{p}(x + 1,y) = u_1^3(x,y,\hat{p}(x,y)), \end{cases}$$

where $z = \lambda x.0$, $u_1^3 = \lambda xyz.x$. Schema B is obviously a primitive recursion. Thus, p is obtained from z, u_1^3 and u ($u = \lambda x.x$) by primitive recursion and composition, since $p = \lambda x.\hat{p}(u(x),u(x))$.//

Example 5 $\lambda x.x + 1$ *cannot* be defined by primitive recursion; such a definition would be circular due to the appearance of "$x + 1$" in the primitive recursive schema.//

Example 6   Let $d = \lambda xy.x \div y$.
Clearly,

$$A \begin{cases} x \div 0 = x \\ x \div (y + 1) = (x \div y) \div 1 \end{cases}$$

Schema A can be made into a proper primitive recursive schema B:

$$B \begin{cases} \hat{d}(0,y) = u(y) \\ \hat{d}(x + 1,y) = p(u_3^3(x,y,\hat{d}(x,y))) \end{cases}$$

where $u_3^3 = \lambda xyz.z$. Clearly, for all x,y, $\hat{d}(x,y) = y \div x = d(y,x)$.
     Thus, $d = \lambda xy.\hat{d}(u_2^2(x,y),u_1^2(x,y))$; in other words, d is defined from u,u_1^2,u_2^2,u_3^3,p by primitive recursion and composition. It is important to note that in passing from $\hat{d}$ to d we switched the position of two arguments (with the help of u_1^2,u_2^2).//

Example 7 The function $\overline{sg} = \lambda x.1 \doteq x$ is defined from $\hat{d}$ by $\overline{sg} = \lambda x.\hat{d}(u(x),s(z(x)))$—that is, it is defined through composition from $\hat{d},u,s = \lambda x.x + 1$ and $z.//$

Example 8 Consider the function $\lambda x.x \bmod 2$. We shall define it by primitive recursion and composition from functions already encountered in our previous examples.

First observe that

$$ A \begin{cases} 0 \bmod 2 = 0 \\ (x + 1) \bmod 2 = \overline{sg}(x \bmod 2) \end{cases} $$

Formally, let k be defined by the primitive recursive schema B:

$$ B \begin{cases} k(0,y) = z(y) \\ k(x + 1,y) = \overline{sg}(u_3^3(x,y,k(x,y))) \end{cases} $$

Clearly, $\lambda x.x \bmod 2 = \lambda x.k(u(x),u(x))$ and, hence, the mod 2 function is defined from $\overline{sg},u,u_3^3,z$, by primitive recursion and composition.$//$

Note: The reader, no doubt, will by now have conjectured that, although the primitive recursive schema as well as the operation of composition are quite rigidly defined, we can make them very flexible with the help of functions such as u,u_1^2,u_3^3, etc., and the use of composition. In particular, the technique of defining a function of *one* argument by primitive recursion was to introduce a second *dummy argument* (i.e., one that does not contribute "information" for the determination of the function-value (see Examples 4 and 8)—trivial-looking iterated elements were "dressed up" to the requirements of Definition 2 with the help of u-functions [Examples 4(A), 6(A), and 8(A).] Finally, the substitution of one argument by a function (see Example 3), of one argument by a constant [Example 7; $s(z(x)) = 1$ for all x], the identification of two arguments (see Example 1), the interchange of two arguments (see Example 6), are all *derived operations* of composition *if* the u-functions, s, and z are available. Thus the reader, unless it is otherwise required for the purpose of exercise, should free himself from the rigidity of Definitions 1 and 2 and freely use schemas like those in Example 4(A), Example 6(A), and Example 8(A), as well as schemas of substituting variables by functions or constants, schemas of identification of variables and schemas of switching two variables. (See also Problem 2.)

We shall now define the class of primitive recursive functions.

Definition 3 The class of *primitive recursive functions*, $\mathcal{PR}$, is the *closure* of $I = \{\lambda x.x + 1, \lambda x.0,((\lambda \vec{x}_n.x_i)_{1 \leq i \leq n})_{n \geq 1}\}$ under the operations of *composition* (Definition 1) and *primitive recursion*.$//$

Note that, generally, "closure" of A under operations $O_1, O_2, \ldots$ is the *smallest* class $\mathcal{C}$ containing A and such that $O_1, O_2, \ldots$ operating on elements of $\mathcal{C}$ yield elements of $\mathcal{C}$. The infinite class of functions I is the class of *initial* functions for $\mathcal{PR}$.

The following important theorem establishes the validity of the method of *proof by induction with respect to* $\mathcal{PR}$. In other words, in full analogy with the mathematical induction on $\mathcal{N}$, if we want to prove a property $\mathcal{P}$ for every $f \in \mathcal{PR}$ it suffices to prove that:

(a) Every $f \in I$ has the property $\mathcal{P}$.

(b) $\mathcal{P}$ propagates with composition [i.e., if the number-theoretic functions $f, g_1, \ldots, g_n$ have property $\mathcal{P}$, so does $f \circ (g_1, \ldots, g_n)$].

(c) $\mathcal{P}$ propagates with primitive recursion [that is, if the number-theoretic functions h and g have the property $\mathcal{P}$, so does the function f defined from h and g by primitive recursion].

Theorem 1　If each function of I has a certain property $\mathcal{P}$, and if the property $\mathcal{P}$ propagates with composition and primitive recursion, then every function in $\mathcal{PR}$ has property $\mathcal{P}$.

Proof Let $\hat{P} = \{f \mid f \text{ is a number-theoretic function with property } \mathcal{P}\}$. By assumption,

(a) $I \subset \hat{P}$.

(b) $\hat{P}$ is closed under composition (why?).

(c) $\hat{P}$ is closed under primitive recursion (why?).

From the definition of $\mathcal{PR}$ it follows that $\mathcal{PR} \subset \hat{P}$ (why?)—i.e., every f in $\mathcal{PR}$ has property $\mathcal{P}$.//

Proposition 1　$\mathcal{PR}$ contains only *total* functions.

Proof Problem 3.//

Theorem 1 is useful in proving properties of the whole class $\mathcal{PR}$, whereas the following Theorem 2 (quite often given as the definition of $\mathcal{PR}$) is mostly useful in proving that such and such a function is primitive recursive. It [Theorem 2] says, intuitively, that a function is primitive recursive iff it is definable from a finite subset of the initial functions through a *finite* number of applications of composition and/or primitive recursion.

Theorem 2 A number-theoretic function f is primitive recursive iff there is a finite sequence of functions $(g_1, \ldots, g_k)$ such that $g_k = f$ and for all i, $1 \le i \le k$, g_i is either in I or it is obtained from *previous* functions in the sequence through the operations of composition or primitive recursion.

Note: A sequence such as $(g_1, \ldots, g_k)$ is called a *derivation* of g_k.

Proof Let us call $\hat{\mathcal{PR}}$ the class of functions defined as described in the Theorem 2. We shall show that $\hat{\mathcal{PR}} = \mathcal{PR}$.

(1) $\mathcal{PR} \subset \hat{\mathcal{PR}}$. This will be shown by induction with respect to $\mathcal{PR}$.

Let the property $\mathcal{P}$ be: A function f has property $\mathcal{P}$ iff it is in $\hat{\mathcal{PR}}$.

(1.1) Each function in I has property $\mathcal{P}$, since it has a one-element derivation.

(1.2) If $\lambda \vec{x}_n.f(\vec{x}_n)$ and $\lambda \vec{y}_m.g_1(\vec{y}_m), \ldots, \lambda \vec{y}_m.g_n(\vec{y}_m)$ have property $\mathcal{P}$, then so does $f \circ (g_1, \ldots, g_n)$, for let $(h_1^f, \ldots, h_l^f)$ be a derivation for f, $(h_1^i, \ldots, h_{k_i}^i)$ a derivation for g_i, $i = 1, \ldots, n$; then $(h_1^f, \ldots, h_l^f, h_1^1, \ldots, h_{k_1}^1, \ldots, h_1^n, \ldots, h_{k_n}^n, f \circ (g_1, \ldots, g_n))$ is a derivation for $f \circ (g_1, \ldots, g_k)$ (why?). Thus $f \circ (g_1, \ldots g_k)$ has property $\mathcal{P}$.

(1.3) If $\lambda \vec{y}.h(\vec{y})$ and $\lambda x \vec{y} z.g(x, \vec{y}, z)$ have property $\mathcal{P}$, so does $\lambda x \vec{y}.f(x, \vec{y})$ defined from them by primitive recursion [this is shown as in (1.2)].

Thus, by Theorem 1, all of $\mathcal{PR}$ has property $\mathcal{P}$—i.e., (1) is proved.

(2) $\hat{\mathcal{PR}} \subset \mathcal{PR}$. This will be shown by induction with respect to the *length* of a derivation in $\hat{\mathcal{PR}}$. Although we are interested in the last function of each derivation, it will be easier to prove more: "Every function of a derivation in $\hat{\mathcal{PR}}$ is also in $\mathcal{PR}$." Let l be the length of a derivation $(g_1, \ldots, g_l)$. For $l = 1$: then (g_1) is the sequence, hence $g_1 \in I \subset \mathcal{PR}$.

Assume that for $l = k$, each g_i of the derivation $(g_1, \ldots, g_l)$ is in $\mathcal{PR}$.

Let $l = k + 1$. Clearly, $(g_1, \ldots, g_{l-1})$ is a derivation (why?) of length k. Thus, by induction hypothesis, $g_1, \ldots, g_{l-1}$ are all in $\mathcal{PR}$. Since g_l is either in I, or obtained from some of the $g_1, \ldots, g_{l-1}$ by either composition or primitive recursion, it follows that $g_l \in \mathcal{PR}.//$

It follows from Theorem 2 that all the functions defined in Examples 1 to 8 are in $\mathcal{PR}$.

Note: The reader familiar with the elements of Mathematical Logic will observe the similarity between a primitive recursive *derivation* and a (Logical) *proof*. Recall that a proof is a *finite* sequence of "well-formed formulas" (wff), $(F_1, F_2, \ldots, F_k)$ such that for each $1 \le i \le k$, either F_i is an *axiom* (analogy in $\mathcal{PR}$: initial function) or is obtained from previous wff's by use of accepted "rules of inference" (in $\mathcal{PR}$, the analogous "rules" are the primitive recursive and composition schemas).

Let us give some more examples.

Example 9 $\lambda xy.x + y \in \mathcal{PR}$.
Set

$$A = \lambda xy.x + y.$$

Then

$$A(0,y) = y$$

$$A(x + 1,y) = A(x,y) + 1.//$$

Example 10 $\lambda xy.xy \in \mathcal{PR}$.
Set

$$M = \lambda xy.xy.$$

Then

$$M(0,y) = 0$$

$$M(x + 1,y) = M(x,y) + y.//$$

Example 11 $\lambda x.2^x \in \mathcal{PR}$.
Set Twotothe $= \lambda x.2^x$. Then,

$$\text{Twotothe}(0) = 1$$

$$\text{Twotothe}(x + 1) = 2 \cdot \text{Twotothe}(x).//$$

Example 12 $\lambda x.2^{2^{\cdot^{\cdot^{\cdot 2}}}} \big\rangle x \in \mathcal{PR}$.
Set $g = \lambda x.2^{2^{\cdot^{\cdot^{\cdot 2}}}} \big\rangle x$. Then

$$g(0) = 1$$

$$g(x + 1) = \text{Twotothe}(g(x)).//$$

Example 13 $\lambda xy.x^y \in \mathcal{PR}$ (resolve 0^0 by setting $0^0 = 1$).
Then

$$x^0 = 1$$

$$x^{y+1} = x \cdot x^y.//$$

We see that $\mathcal{PR}$ is quite rich in functions; moreover, it contains some quite huge functions. (See Example 12.) The reader will appreciate why a theory of Computability (or theory of Algorithmic Functions) cannot be based on technology (present or foreseeable). Such a theory would certainly require $\lambda x.2^{2^{\cdot^{\cdot^{\cdot^{2}}}}}_{x}$ to be *nonalgorithmic* (no machine, no matter how fast it may be, can compute $2^{\cdot^{\cdot^{\cdot^{2}}}}_{x}$ for large x in a reasonable amount of time); yet, theoretically, there is a very simple procedure by which, given enough time (!) and pencil and paper, one can compute $2^{\cdot^{\cdot^{\cdot^{2}}}}_{x}$ for any x. In this connection we shall give in to the temptation (prematurely in our development in this book so far) to argue that, *intuitively*, every function in $\mathcal{PR}$ is *algorithmic*. By induction with respect to $\mathcal{PR}$: Each function of I is *intuitively* computable; further, the property of being *intuitively* computable propagates with composition (composition corresponds to *program superposition* or nonrecursive procedure calls) and primitive recursion (which corresponds to FORTRAN-like *do-loops*, more of which in §2.7).

Example 14 $\lambda xyz.\ \textbf{if } x = 0 \textbf{ then } y \textbf{ else } z \in \mathcal{PR}.//$

Exercise 1   For each of the Examples 9 through 14, provide the formal definitions (according to Definitions 1 and 2) showing the functions to be in $\mathcal{PR}$.

Before we proceed to more involved aspects of the theory, we shall present an alternative definition of $\mathcal{PR}$.

Definition 4   *The rules of substitution.* (Grzegorczyk [5].)
The following rules (a) through (e) are the rules of substitution:
(a) *Substitute* x_i in $g(\vec{x}_n)$ by $f(\vec{y}_m)$ to obtain

$$\lambda x_1 \ldots x_{i-1} x_{i+1} \ldots x_n \vec{y}_m . g(x_1, \ldots, x_{i-1}, f(\vec{y}_m), x_{i+1}, \ldots, x_n)$$

(b) *Substitute* x_i in $g(\vec{x}_n)$ by 0 to obtain from g the function

$$\lambda x_1 \ldots x_{i-1} x_{i+1} \ldots x_n . g(x_1, \ldots, x_{i-1}, 0, x_{i+1}, \ldots, x_n)$$

(c) *Interchange* variables x_i and x_j, $i < j$, in $g(\vec{x}_n)$ to obtain from g the function

$$\lambda \vec{x}_n . g(x_1, \ldots, x_{i-1}, x_j, x_{i+1}, \ldots, x_{j-1}, x_i, x_{j+1}, \ldots, x_n)$$

(d) *Identify* two variables x_i and x_j in $g(\vec{x}_n)$ to obtain the function

$$\lambda x_1 \ldots \hat{x}_j \ldots x_n . g(x_1, \ldots, x_{i-1}, x_i, x_{i+1}, \ldots, x_{j-1}, x_i, x_{j+1}, \ldots, x_n)$$

where $\hat{x}_j$ means that the variable x_j is missing

(e) Introduce *dummy variables* $\vec{y}$ to obtain

$$f = \lambda \vec{x}_n \vec{y}.g(\vec{x}_n)$$

from $\lambda \vec{x}_n.g(\vec{x}_n).//$

Note: If in (b) we allow the substitution of *any* constant [call this rule (b')], then the rules (b'), (c), (d), (e) constitute the rules of *explicit transformation*. (See, for example, Ritchie [11], Smullyan [15], Bennett [1].)

Theorem 3 $\mathcal{PR}$ is the closure of $\hat{I} = \{\lambda x.x + 1, \lambda x.x\}$ under *substitution* and *primitive recursion*.

Exercise 2   Prove Theorem 3. (*Hint*: Let $\hat{\mathcal{PR}}$ be the class defined in Theorem 3. Show that $I \subset \hat{\mathcal{PR}}$. Show that $\hat{\mathcal{PR}}$ is closed under composition. Conversely, show that $\mathcal{PR}$ is closed under substitution, using Problem 2).

In what follows, we shall employ the characterization of Theorem 3 in our study of $\mathcal{PR}$ instead of the original definition.(*)

Example 15 $u_1^2 = \lambda xy.x$ is obtained from $\lambda x.x$ by Definition 4, rule (e).//

Example 16 $z = \lambda x.0$ is obtained from $\lambda x.x$ by Definition 4, rule (b).//

Example 17 $\lambda x.3$ is obtained from $\lambda x.x + 1$ by Definition 4, rules (a) and (b). [First obtain $\lambda x.((x + 1) + 1) + 1$ (a); then substitute x by 0 (b).]//

§2.2 PRIMITIVE RECURSIVE PREDICATES

To facilitate our further study, we shall now introduce the notion of *primitive recursive relations* or *predicates*. Recall that a *relation (predicate)* is a synonym for a set. We shall be interested only in relations on $\mathcal{N}^k$ (for various $k \geq 1$), i.e., in relations on natural numbers.

Definition 1   The *characteristic function* c_R of a relation $R(\vec{x}_n) \subset \mathcal{N}^n$ is defined by: $c_R = \lambda \vec{x}_n.$ **if** $R(\vec{x}_n)$ **then** 0 **else** $1.//$

Note: In some of the literature, the roles of 0 and 1 are reversed.

(*)There is a strong tendency in the literature to define $\mathcal{PR}$ by Definition 3 rather than by the equivalent statement of Theorem 3.

Definition 2 A relation is *primitive recursive* iff its characteristic function is in $\mathcal{PR}$. The set of all primitive recursive relations is denoted by $\mathcal{PR}_*$.//

Lemma 1 The function sg defined by $sg = \lambda x.\text{if } x = 0 \text{ then } 0 \text{ else } 1$ is in $\mathcal{PR}$.

Proof Problem 6.//

Theorem 1 $R(\vec{x}) \in \mathcal{PR}_*$ iff there is a function f in $\mathcal{PR}$ such that

$$R(\vec{x}) \equiv f(\vec{x}) = 0.(*)$$

Proof *if* part. Let $f \in \mathcal{PR}$ and $R(\vec{x}) \equiv f(\vec{x}) = 0$. But $c_R = \lambda \vec{x}.\, sg(f(\vec{x})) \in \mathcal{PR}$. Hence, $R(\vec{x}) \in \mathcal{PR}_*$.

only if part. Let $R(\vec{x}) \in \mathcal{PR}_*$. Clearly, c_R plays the role of f, since $c_R \in \mathcal{PR}$ and $R(\vec{x}) \equiv c_R(\vec{x}) = 0$.//

Note: The statement of Theorem 1 is taken as the definition of a primitive recursive predicate in Péter [9] and Grzegorczyk [5].

Theorem 2 $\mathcal{PR}_*$ is closed under the Boolean operations.

Proof It suffices to show closure under $\neg$ and $\vee$.

(1) Let $R(\vec{x}) \in \mathcal{PR}_*$.
Then $\neg R(\vec{x}) \in \mathcal{PR}_*$ since $c_{\neg R} = \lambda \vec{x}.1 \div c_R(\vec{x})$.

(2) Let $R(\vec{x})$ and $Q(\vec{y})$ be in $\mathcal{PR}_*$.

Then, $c_{R \vee Q} = \lambda \vec{x}\vec{y}.c_R(\vec{x}) \cdot c_Q(\vec{y}) \in \mathcal{PR}$.//

Example 1 The relations $x \geq y, x > y, x = y, x \neq y$ are in $\mathcal{PR}_*$, for

(a) $x \geq y \equiv y \div x = 0$

(b) $x > y \equiv \neg(y \geq x)$

(c) $x = y \equiv x \leq y \,\&\, y \leq x$

(d) $x \neq y \equiv \neg(x = y)$.//

Example 2 $\lambda xy.|x - y| \in \mathcal{PR}$.
For,

$$|x - y| = x \div y + y \div x.//$$

(*)The symbol "$\equiv$" denotes the equality (equivalence) of predicates.

Example 3   Let $\lambda \vec{x}.f(\vec{x})$ and $\lambda \vec{y}.g(\vec{y})$ be in $\mathcal{PR}$.
Then

$$\lambda \vec{x} \vec{y}.f(\vec{x}) = g(\vec{y}) \in \mathcal{PR}_*.$$

For

$$f(\vec{x}) = g(\vec{y}) \equiv |f(\vec{x}) - g(\vec{y})| = 0.//$$

Theorem 3   Let $R(\vec{x}_n) \in \mathcal{PR}_*$ and $\lambda \vec{y}.f(\vec{y}) \in \mathcal{PR}$.
Then

$$\mathrm{S}(\vec{y},x_1, \ldots, \hat{x}_i, \ldots, x_n) \equiv R(x_1, \ldots, x_{i-1}, f(\vec{y}),x_{i+1}, \ldots, x_n) \in \mathcal{PR}_*.$$

Note: $\hat{x}_i$ means here that x_i is missing.

Proof Problem 8.//

Note: A trivial corollary of Theorem 3 is that if $\lambda \vec{x}.f(\vec{x}) \in \mathcal{PR}$, then $\lambda y \vec{x}.y = f(\vec{x})$ is in $\mathcal{PR}_*$. The reader should be *very* careful not to conjecture the converse, because it is false. (See § 3.3 for a counter-example to such a conjecture).

Definition 3   (Péter [9].) Given a number-theoretic function $\lambda x \vec{y}.f(x,\vec{y})$, then $\Sigma_{i \leq z}$ and $\Pi_{i \leq z}$ are defined by:

$$\Sigma_{i \leq z} f(i,\vec{y}) = f(0,\vec{y}) + \ldots + f(z,\vec{y})$$

and

$$\Pi_{i \leq z} f(i,\vec{y}) = f(0,\vec{y}) \cdot \ldots \cdot f(z,\vec{y}).//$$

Lemma 2   (Péter [9], Grzegorczyk [5].) $\mathcal{PR}$ is closed under $\Sigma_{i \leq z}$ and $\Pi_{i \leq z}$.

Proof Let $\lambda x \vec{y}.f(x,\vec{y}) \in \mathcal{PR}$.

(a)   For Σ: $\Sigma_{i \leq 0} f(i,\vec{y}) = f(0,\vec{y})$
      $\Sigma_{i \leq z+1} f(i,\vec{y}) = f(z + 1,\vec{y}) + \Sigma_{i \leq z} f(i,\vec{y})$

(b)   For Π: $\Pi_{i \leq 0} f(i,\vec{y}) = f(0,\vec{y})$
      $\Pi_{i \leq z+1} f(i,\vec{y}) = f(z + 1,\vec{y}) \cdot \Pi_{i \leq z} f(i,\vec{y}).//$

Theorem 4 $\mathcal{PR}_*$ is closed under $(\exists y)_{\leq z}$ and $(\forall y)_{\leq z}$.

Proof Let $R(y,\vec{x}) \in \mathcal{PR}_*$.
Then,

$$(\exists y)_{\leq z} R(y,\vec{x}) \equiv \Pi_{i \leq z}\, c_R(i,\vec{x}) = 0$$

$$(\forall y)_{\leq z} R(y,\vec{x}) \equiv \neg(\exists y)_{\leq z} \neg R(y,\vec{x}).//$$

Note: The case for $\forall$ can be also obtained directly:

$$(\forall y)_{\leq z} R(y,\vec{x}) \equiv \Sigma_{i \leq z} c_R(i,\vec{x}) = 0.$$

Exercise 1   Show Theorem 4 without the help of Σ- or Π-operators.

Theorem 5   *Definition by cases.* Let the function f be defined by:(*)

$$\text{For all } \vec{x} : f(\vec{x}) = \begin{cases} f_1(\vec{x}) \textbf{ if } R_1(\vec{x}) \\ f_2(\vec{x}) \textbf{ if } R_2(\vec{x}) \\ \quad \vdots \\ f_k(\vec{x}) \textbf{ if } R_k(\vec{x}) \end{cases}$$

If $f_i \in \mathcal{PR}$, $i = 1, \ldots, k$ and $R_i(\vec{x}) \in \mathcal{PR}_*$, $i = 1, \ldots, k$ and if moreover $R_1(\vec{x}) \vee \ldots \vee R_k(\vec{x})$ holds for all $\vec{x}$, then $f \in \mathcal{PR}$.

Proof For all $\vec{x}$:

$$f(\vec{x}) = f_1(\vec{x}) \cdot \overline{sg}(c_{R_1}(\vec{x})) + \ldots + f_k(\vec{x}) \cdot \overline{sg}(c_{R_k}(\vec{x})).//$$

Exercise 2   Show Theorem 5 without the use of "+".
We shall now define a very important operator, the *bounded-μ* or *bounded search* operator.

§2.3   BOUNDED SEARCH

Definition 1   *Bounded search.* Let f be number-theoretic. Then,

$$(\mu y)_{<z} f(y,\vec{x}) \overset{\text{def}}{=} \begin{cases} \min\{y|\, y < z \;\&\; f(y,\vec{x}) = 0\} \\ z \text{ if the } \textit{min} \text{ does not exist.}// \end{cases}$$

(*)The relations R_i, $i = 1, \ldots, k$ are mutually exclusive.

Note: There is an alternative definition which, instead of z, returns 0 (Péter [9], Grzegorczyk [5]) if the *min* does not exist. We find the present definition more convenient since, when $\neg(\exists y)_{<z} f(y,\vec{x}) = 0$, it returns a value which is *outside* the range $(0,1,2,\ldots,z-1)$ of the search for a y. Incidentally, the name *bounded search* is derived from the fact that (for an algorithmic f) $(\mu y)_{<z} f(y,\vec{x})$ can be "computed" by searching in the bounded range $\{0,1,\ldots,z-1\}$ for a possible y. Intuitively, $(\mu y)_{<z}$ is an algorithmic operator.

Theorem 1 $\mathcal{PR}$ is closed under $(\mu y)_{<z}$.

Proof (Grzegorczyk [5], Péter [9].) Let

$$\lambda y \vec{x}.f(y,\vec{x}) \in \mathcal{PR}.$$

Then

$$(\mu y)_{<z} f(y,\vec{x}) = \Sigma_{i<z}\, sg(\Pi_{j\leq i} f(j,\vec{x})),$$

where

$$\Sigma_{i<z}\, h(i,\vec{x}) = 0,$$

by definition, if $z = 0$.

Clearly, if $h \in \mathcal{PR}$, then $\lambda z\vec{x}.\Sigma_{i<z} h(i,\vec{x}) \in \mathcal{PR}$ since

$$\Sigma_{i<z} h(i,\vec{x}) = \begin{cases} 0 \textbf{ if } z = 0 \\ \Sigma_{i\leq z \doteq 1} h(i,\vec{x}) \textbf{ if } z \neq 0.// \end{cases}$$

Note: Theorem 1 can be also proved without the use of Σ and Π: Set

$$g = \lambda z\vec{x}.(\mu y)_{<z} f(y,\vec{x}).$$

Then,

$$g(0,\vec{x}) = 0$$

$$g(z+1,\vec{x}) = \textbf{if } g(z,\vec{x}) \neq z \textbf{ then } g(z,\vec{x})$$
$$\textbf{else if } f(z,\vec{x}) = 0 \textbf{ then } z \textbf{ else } z + 1.$$

Corollary 1 $\mathcal{PR}$ is closed under $(\mu y)_{\leq z}$.

Proof $y \leq z$ iff $y < z + 1$; moreover $\lambda z.z + 1 \in \mathcal{PR}.//$

Definition 2   Let $R(y,\vec{x})$ be a predicate. Then $(\mu y)_{<z}R(y,\vec{x})$ [resp. $(\mu y)_{\le z}R(y,\vec{x})$] is defined to be $(\mu y)_{<z}c_R(y,\vec{x})$ [resp. $(\mu y)_{\le z}c_R(y,\vec{x})$].//

Corollary 2   If $R(y,\vec{x}) \in \mathcal{PR}_*$, then $\lambda z\vec{x}.(\mu y)_{<z}R(y,\vec{x})$ [resp. $\lambda z\vec{x}.(\mu y)_{\le z}R(y,\vec{x})$] is in $\mathcal{PR}$.

Example 1   The following functions and predicates are primitive recursive.

(a) $\lambda xy.\lfloor x/y \rfloor$; $(\lfloor x/0 \rfloor \overset{\text{def}}{=} 0)$
[For, $\lfloor x/y \rfloor = (\mu z)_{\le x}(x < (z + 1) \cdot y \lor z = 0 \,\&\, y = 0)$.]

(b) $\lambda xy.x \bmod y$; [for $x \bmod y = x \mathbin{\dot-} [x/y] \cdot y$]

(c) $\lambda x.\lfloor \sqrt{x} \rfloor$; [for, $\lfloor \sqrt{x} \rfloor = (\mu y)_{\le x}(y + 1)^2 > x$]

(d) $\lambda xy.(x + y)(x + y + 1)/2 + y$

(e) $\lambda xy.x \,|\, y$ ($\equiv \lambda xy.x$ divides $y \equiv \lambda xy.(y \bmod x = 0)$)

(f) $\pi = \lambda x.$ number of prime numbers $\le x$.

(g) $\mathbf{Pr}(x) \equiv x$ is a prime

(h) $\lambda xy.y = \pi(x)$

(i) $\lambda ny.y = p_n \equiv \lambda ny.y$ is the nth prime [for, $y = p_n \equiv \mathbf{Pr}(y)\,\&\, \pi(y) = n + 1$]

(j) $\lambda xy.\max(x,y)$ [for, $\max(x,y) = x \mathbin{\dot-} y + y$]

(k) $\lambda xy.\min(x,y)$

(l) $\lambda xy.\exp(x,y) = \lambda xy.$ "exponent of the xth prime in the prime number decomposition of y".//

Exercise 1   Show that the functions (predicates) in (d), (f), (g), (h), (k), and (l) are primitive recursive.

§2.4 PAIRING FUNCTIONS

We are now ready to deal with the very important technique of "algorithmically" *coding finite sequences* of numbers by a *single* number and conversely *decoding* any code number back to a sequence, also algorithmically. As a start, we shall deal with the case of sequences of length 2.

Definition 1   A total 1-1 number-theoretic map $J:\mathcal{N}^2 \to \mathcal{N}$ is called a *pairing function*.//

Note: The definition does *not* require J to be *onto*.

Definition 2   A total map $\lambda z.(K(z),L(z))$ such that $(K,L) \circ J(x,y) = (x,y)$ for all x,y is *a pair of projection functions* of J.//

Note: Unless J is *onto*, K and L are *not* unique. However, the restriction of (K,L) on the set $J(\mathcal{N}^2)$ *is* unique. Traditionally, the letters K and L are used to denote first and second projections of J.

There are a lot of pairing functions in $\mathcal{PR}$ with projections in $\mathcal{PR}$:

Example 1 $J = \lambda xy.2^x 3^y$ is a pairing function. We *usually* take $K = \lambda z.\exp(0,z)$, $L = \lambda z.\exp(1,z)$, where 2 is the 0th and 3 the 1st prime.

Note that $K(z)$ for odd z is 0. Clearly, a "code" z (i.e., a z in $J(\mathcal{N}^2)$) is not divisible by any prime greater than 3. For a "noncode" z (i.e., $z \notin J(\mathcal{N}^2)$)—for example, $z = 20$ $(=2^2 \cdot 5)$—we could equally well define $K(z)$ to be 1 or 2 or anything else! We adopted $\exp(0,z)$ for all z for reasons of convenience; it would be rather extravagant (and purposeless) although correct to, say, define K by:

$$\text{For all } z, K(z) = \begin{cases} 81 \text{ if } \left\lfloor \dfrac{z}{2^{\exp(0,z)} 3^{\exp(1,z)}} \right\rfloor > 1 \\ \\ \exp(0,z) \text{ otherwise} \end{cases}$$

Still, $K \circ J(x,y) = x$ for all x,y. Similar comments for L.//

Example 2 $J = \lambda xy.2^{x+y+2} + 2^{y+1}$ is a pairing function (Minsky [8], Schwichtenberg [14]). It is *not* onto.//

Exercise 1   Prove the claims of Examples 1 and 2; for Example 2, find suitable K and L.

Example 3 $J = \lambda xy. (x + y) \cdot (x + y + 1)/2 + y$ is pairing; indeed this J is also *onto*. Let us show this:

Consider a listing of $\mathcal{N}^2$ which is defined as follows: Let $A_i = \{(x,y) \in \mathcal{N}^2 \mid x + y = i\}$. List now $\mathcal{N}^2$ by listing A_i's in ascending order of i and in each A_i list (x,y)'s in ascending order of y. Since A_i has $i + 1$ elements, for each i, the position of (x,y) in the listing is $1 + 2 + \ldots + (x + y) + y = (x + y)(x + y + 1)/2 + y$, thus $J:\mathcal{N}^2 \to \mathcal{N}$ is the function which calculates the position of (x,y) in the defined listing above. From the discussion [each (x,y) has a *unique* position in the listing; *each position* on the other hand will be occupied since $\mathcal{N}^2$ is infinite] it follows that J is *1-1* and *onto*. Thus it has *unique* K and L. J being in $\mathcal{PR}$ (trivial), so are K and L [for example, $K = \lambda z.(\mu x)_{\leq z}(\exists y)_{\leq z}((x + y)(x + y + 1) + 2y = 2z)]$. Davis [2] gives a direct proof that J is 1-1 and onto by finding closed forms for K and L.//

Example 4 $\lambda xy.2^x \cdot (2y + 1)$ is pairing (Grzegorczyk [5]).
For all z, take $K(z) = \exp(0,z)$, where $p_0 = 2$ (the 0th prime).
$L(z) = \lfloor (\lfloor z/2^{K(z)} \rfloor \dot{-} 1)/2 \rfloor$ .//

Examples 1 through 4 have provided us with instances of primitive recursive pairing functions which also have primitive recursive projection functions. A primitive recursive pairing function "codes", in an "algorithmic" way, any pair of numbers in $\mathcal{N}$. We now turn to the problem of coding (primitive recursively) an n-tuple $(n > 2)$ of numbers by a single number.
We require a 1-1 function $I_n : \mathcal{N}^n \to \mathcal{N}$ and its projections $\Pi_i^n : \mathcal{N} \to \mathcal{N}$, $i = 1, \ldots, n$ so that I_n and $\Pi_i^n, i = 1, \ldots, n$ are primitive recursive.

Note: Although the projections Π_i^n are *not* unique when I_n is not onto, now that the point has been made, we shall feel free to use expressions like "its projections . . ." as if they were unique. In all cases of nonuniqueness the least extravagant alternative is preferred. As such, usually, is taken the "natural" extension of I_n^{-1} (defined on $I_n(\mathcal{N}^n) \subset \mathcal{N}$) to $\mathcal{N}$. (See Example 1.) Finally the reader should *not* automatically assume the validity of $I_n \circ (\Pi_1^n, \ldots \Pi_n^n)(z) = z$ if I_n is *not* onto!
From now on we reserve the letters J, K, L for an unspecified pairing function and its projections.

Definition 3 Define $I_1 = \lambda x.x$ and $\Pi_1^1 = \lambda x.x$.
Assuming that I_n, Π_i^n, $i = 1, \ldots, n$ have been defined, $n \geq 1$, define
$I_{n+1} = \lambda \vec{x}_{n+1}.J(I_n(\vec{x}_n), x_{n+1})$

$$\Pi_i^{n+1} = \Pi_i^n \circ K, i = 1, \ldots, n; \Pi_{n+1}^{n+1} = L.//$$

Lemma 1 For each $n \geq 2$, for all $\vec{x}_n$ and $i = 1, \ldots, n$ $\Pi_i^n \circ I_n(\vec{x}_n) = x_i$. Moreover, I_n, Π_i^n, $i = 1, \ldots, n$ are in $\mathcal{PR}$.

Note: This establishes I_n as *1-1* and each Π_i^n as *onto*.

Proof Problem 17.//

Corollary If, moreover, J is also *onto*, then so is each $I_n, n \geq 2$.

Definition 4 (Rogers [13], Grzegorczyk [5].) For each $n \geq 1$, we define the symbol $\langle x_1, \ldots, x_n \rangle$ or $\langle \vec{x}_n \rangle$ to stand for $I_n(\vec{x}_n)$ where I_n is an (unspecified but fixed in any discussion) *coding* function (*not* necessarily *onto*) defined according to Definition 3.

Example 4 Define for any (fixed) n
$c_n(\vec{x}_n) \stackrel{\text{def}}{=} p_1^{x_1} \cdot p_2^{x_2} \cdot \ldots \cdot p_n^{x_n}$, where p_i is the ith prime.

The function $\lambda \vec{x}_n . c_n(\vec{x}_n)$ is *coding* (1-1 but *not* onto) with ith projection $\lambda z . \exp(i,z)$. Note that for a number z to be "legal" code it must not have any primes other than $p_1, \ldots, p_n$ in its prime factorization.//

We shall discuss later the method of coding *variable length sequences*, $\vec{x}_n$ (the coding function has n as argument). But first let us deal with some fancy primitive recursions.

§2.5 SOME SPECIAL RECURSION SCHEMAS

Definition 1   The functions $\lambda xy\vec{y}.f_i(x,\vec{y})$, $i = 1, \ldots, n$ are obtained from $\lambda \vec{y}.h_i(\vec{y})$ and $g_i(x,\vec{y},\vec{z}_n)$, $i = 1, \ldots, n$ by the *schema of simultaneous primitive recursion* iff for all i,x and $\vec{y}$:

$$A \begin{cases} f_i(0,\vec{y}) = h_i(\vec{y}) \\ f_i(x+1,\vec{y}) = g_i(x,\vec{y}, f_1(x,\vec{y}), \ldots, f_n(x,\vec{y})).// \end{cases}$$

The following theorem, due to Hilbert and Bernays [6], shows that if the h_i and g_i, $i = 1, \ldots, n$ are in $\mathcal{PR}$ so are the f_i, $i = 1, \ldots, n$.

Theorem 1   If h_i and g_i, $i = 1, \ldots, n$ are primitive recursive, then so are the f_i, $i = 1, \ldots, n$.

Proof   We shall first present the idea of the proof informally: Schema A defines the vector-valued function $\lambda xy\vec{y}.(f_1(x,\vec{y}), \ldots, f_n(x,\vec{y}))$ from the vector-valued functions $\lambda \vec{y}.(h_1(\vec{y}), \ldots, h_n(\vec{y}))$ and $\lambda xy\vec{y}\,\vec{z}_n.(g_1(x,\vec{y},\vec{z}_n), \ldots, g_n(x,\vec{y},\vec{z}_n))$ by what looks like ordinary primitive recursion on vector-valued functions.(*) If, therefore, we code the vectors into single numbers we might end up with an ordinary recursion as defined in Definition 2.1.2. Let us try this:

Set $F = \lambda xy\vec{y}.(f_1(x,\vec{y}), \ldots, f_n(x,\vec{y}))$. If I can show $F \in \mathcal{PR}$, then, since $f_i = \Pi_i^n \circ F$ and $\Pi_i^n \in \mathcal{PR}$, also $f_i \in \mathcal{PR}$, $i = 1, \ldots, n$.

Indeed, with F defined as above, we obtain

$$F(0,\vec{y}) = \langle h_1(\vec{y}), \ldots, h_n(\vec{y}) \rangle$$

$$F(x+1,\vec{y}) = \langle g_1(x,\vec{y},f_1(x,\vec{y}), \ldots, f_n(x,\vec{y})), \ldots,$$

$$g_n(x,\vec{y}, f_1(x,\vec{y}), \ldots, f_n(x,\vec{y})) \rangle =$$

(*)By Theorem 1.4.2, a unique such $(f_1, \ldots, f_n)$ exists.

$$\langle g_1(x,\vec{y},\Pi_1^n(F(x,\vec{y}))), \ldots, \Pi_n^n(F(x,\vec{y}))), \ldots,$$

$$g_n(x,\vec{y},\Pi_1^n(F(x,\vec{y}))), \ldots, \Pi_n^n(F(x,\vec{y}))))\rangle$$

To put this in familiar form, let

$$H = \lambda\vec{y}.\langle h_1(\vec{y}), \ldots, h_n(\vec{y})\rangle$$

$$G = \lambda x\vec{y}z.\langle g_1(x,\vec{y},\Pi_1^n(z)), \ldots, \Pi_n^n(z)), \ldots, g_n(x,\vec{y},\Pi_1^n(z)), \ldots, \Pi_n^n(z))\rangle$$

Thus, F is given by

$$F(0,\vec{y}) = H(\vec{y})$$

$$F(x + 1,\vec{y}) = G(x,\vec{y},F(x,\vec{y})).$$

Since H,G are in $\mathcal{PR}$ (why?) so is F, and hence each f_i, $i = 1, \ldots, n.//$

In preparation of showing that another special recursion does not lead outside $\mathcal{PR}$, we shall now show the technique of coding *variable-length* vectors by single numbers.

Example 1   We have seen (Definition 2.4.3) that the projections of I_{n+1}, $n \geq 1$, are given by $\Pi_i^{n+1} = \Pi_i^n\circ K$, $i = 1, \ldots, n$; $\Pi_{n+1}^{n+1} = L$.

A simple induction on n shows that $\Pi_1^{n+1} = K^n$ and for $i = 2, \ldots, n + 1$ $\Pi_i^{n+1} = L\circ K^{n+1-i}$, where an exponent in a function letter denotes composition $(K^n = \overbrace{K\circ K\circ\ldots\circ K}^{n})$.

To code a variable-length vector, the length information must be included in the code since it is not going to be always the same. Thus, we code $(x_1, \ldots, x_n)$ by $\langle x_1, \ldots, x_n, n\rangle$

To decode a number z,(*) apply first L to find $n = L(z)$; then $x_1 = K^{L(z)}(z)$ and, for $i = 2, \ldots, n$, $x_i = L\circ K^{L(z)+1-i}(z)$

To sum up: *code* $\vec{x}_n$ as $\langle\vec{x}_n,n\rangle$, *decode* z as $K^{L(z)}(z), L\circ K^{L(z)-1}(z), \ldots, L\circ K(z).//$

Note: We cannot talk anymore about a *coding function* since the number of arguments is variable; even the *number* of *decoding functions* (*projections*) depends on z.

Exercise 1   Show that the projections $\lambda z.K^{L(z)}(z)$ and $\lambda zi.L\circ K^{L(z)+1-i}(z)$ are in $\mathcal{PR}$.

Example 2 $\lambda x.p_x$ (the xth-prime function) is very popular as a stepping stone for coding variable-length vectors (Davis [2], Péter [9], Grzegorczyk [5]).

(*)Illegal codes will lead to nonsense.

Let us code $(x_0, x_1, \ldots, x_n)$ by $\Pi_{i \leq n} p_i^{x_i}$.

The ith projection function is $\lambda iz.\exp(i,z)$ ($\in \mathcal{PR}$). Under this code, the sequences (1), (1,0), (1,0,0) have the same code: 2.

It is clear, then, that in order to use this coding scheme effectively, the length of the coded sequence should be known from the context.

An alternative method is to code $(x_0, \ldots, x_n)$ by $\Pi_{i \leq n} p_i^{x_i+1}$ or by $2^{n+1} \cdot \Pi_{1 \leq i \leq n+1} p_i^{x_{i-1}}.//$

Example 3 Start with $J = \lambda xy.2^x 3^y$. Then every number divisible by primes other than (exclusively) 2 and 3 is illegal code when coding according to Example 1. Any power of 3 is also illegal (why?).//

So far, primitive recursive schemas have been defining the value of a function f at the point $x + 1$ if we know the value of f at the point x.

There is a scheme called *course-of-values* recursion (Péter [9]) which defines the value of a function f at $x + 1$ by using *more than one* previous point. Thus, in general, $f(x + 1, \vec{y})$ is defined in terms of (part or all of) the sequence $f(0,\vec{y}), f(1,\vec{y}), \ldots, f(x,\vec{y})$ rather than just in terms of $f(x,\vec{y})$. The $(x + 1)$-element set $\{f(0,\vec{y}), \ldots, f(x,\vec{y})\}$ is the *history* of f at $(x,\vec{y})$. Let $H = \lambda x\vec{y}.code(\{f(0,\vec{y}), \ldots, f(x,\vec{y})\})$, where "*code*" is any scheme with primitive recursive projections; for example, (to fix our ideas) let it be $\langle \vec{x}_n, n \rangle$.

Definition 2 The general *course-of-values recursion schema* is A below: for all $x, \vec{y}$:

$$\text{A} \begin{cases} f(0,\vec{y}) = h(\vec{y}) \\ f(x + 1, \vec{y}) = g(x, \vec{y}, H(x,\vec{y})) \end{cases}$$

where, as we agreed, $H = \lambda x\vec{y}.\langle f(0,\vec{y}), \ldots, f(x,\vec{y}), x + 1 \rangle.//$

Theorem 2 If f is defined from schema A above, and if h and g are in $\mathcal{PR}$, then f is in $\mathcal{PR}$.(*)

Proof Clearly, it suffices to show that H is in $\mathcal{PR}$ (why?) Observe that,

$$H(0,\vec{y}) = \langle f(0,\vec{y}),1 \rangle = \langle h(\vec{y}), 1 \rangle,$$

(*)A unique such f exists (Theorem 1.4.2).

and

$$H(x + 1,\vec{y}) = \langle\, f(0,\vec{y}), \ldots, f(x,\vec{y}), f(x + 1,\vec{y}), x + 2\rangle$$
$$= \langle\langle\langle\, f(0,\vec{y}), \ldots, f(x,\vec{y})\rangle, f(x + 1,\vec{y})\rangle, x + 2\rangle$$
$$= \langle\langle K{\circ}H(x,\vec{y}), f(x + 1,\vec{y})\rangle, x + 2\rangle$$
$$= \langle\langle K{\circ}H(x,\vec{y}),g(x,\vec{y},H(x,\vec{y}))\rangle\, x + 2\rangle$$

or

$$\begin{cases} H(0,\vec{y}) = b(\vec{y}) \\ H(x + 1,\vec{y}) = r(x,\vec{y},H(x,\vec{y})) \end{cases}$$

where

$$b = \lambda\vec{y}.\langle h(\vec{y}),1\rangle$$
$$r = \lambda x\vec{y}z.\langle\langle K(z),g(x,\vec{y},z)\rangle, x + 2\rangle.$$

Since b, r are in $\mathcal{PR}$, H (and hence f) is in $\mathcal{PR}.//$

   Note: In Péter [9] the reader can find a proof identical to the above in all respects except the use of H; Péter uses the coding scheme of Example 2. Also note that the length of the sequence $\{f(0,\vec{y}), \ldots, f(x,\vec{y})\}$ is known ($=x + 1$), since H is a function of x; thus, in this case, it was an overkill to include $x + 1$ as the last component in $\langle\ \rangle$.

Example 4   The *Fibonacci* function F is defined (informally) by

$$F(0) = 0, F(1) = 1, \text{ and } F(n + 1) = F(n) + F(n - 1), \text{ for } n \geq 1$$

This defines the *Fibonacci sequence* 0,1,1,2,3,5,8, ... We claim that $F \in \mathcal{PR}$. To see this, let us transform its definition into a course-of-values definition (Péter [9]).

$$F(0) = 0$$
$$F(n + 1) = \overline{sg}(n) + F(n) + F(n \doteq 1).$$

Let $H(n) = \Pi_{i \leq n}p_i^{F(i)}$ be the method of coding the history of F at n. Set $g = \lambda nz.\overline{sg}(n) + \exp(n,z) + \exp(n \doteq 1,z).$

Then,

$$F(0) = 0$$

$$F(n + 1) = g(n, H(n))$$

which is a schema of the form A of Definition 2.

Note that (as it is expected) the use of the coding function $\hat{H} = \lambda n \cdot \langle F(0),$ $\ldots, F(n), n + 1 \rangle$ works as well:

Set

$$\hat{g} = \lambda n z.\textbf{if } n = 0 \textbf{ then } 1 \textbf{ else}$$

$$\textbf{if } n = 1 \textbf{ then } L{\circ}K(z) + K^2(z)$$

$$\textbf{else } L{\circ}K(z) + L{\circ}K^2(z)$$

Then

$$F(0) = 0$$

$$F(n + 1) = \hat{g}(n, \hat{H}(n)). //$$

We have got a flavor of the power of the technique of coding finite sequences. There will be quite a few more occurrences of this technique in this book. In the meanwhile, we shall close this paragraph by giving a preview of the technique of "arithmetization". In what follows, our coding scheme (for number sequences) will be fixed but unspecified, $(\vec{x}_n, n) \mapsto \langle \vec{x}, n \rangle$, (*) subject to the constraints:

(a)   There is a function L in $\mathcal{PR}$ such that $L(\langle \vec{x}_n, n \rangle) = n$ for all $\vec{x}_n \in \mathcal{N}^n$.

(b)   The projections $\lambda z i n.\Pi_i^{n+1}(z)$—that is, the functions such that $\Pi_i^{n+1}(\langle \vec{x}_n, n \rangle) = x_i, i = 1, \ldots, n$—are all in $\mathcal{PR}$.

(c)   The predicate $\text{Code}(y)$, meaning "y is legal code", is in $\mathcal{PR}_*$.

(d) $\lambda x n.\langle \underbrace{x, x, \ldots, x}_{n \ x's}, n \rangle$ is in $\mathcal{PR}$ and $\lambda x_i.\langle \vec{x}_n, n \rangle$ is increasing for all i.

(e) $n \neq 0 \Rightarrow \langle \vec{x}_n, n \rangle \neq 0$.

The above properties (a) through (e) are all shared by each of the specific codings introduced earlier and are, intuitively, reasonable requirements for a coding to be called "primitive recursive".

(*)For the balance of this paragraph, $\langle \ \rangle$ need not be defined from some pairing function J according to Definition 2.4.3.

Definition 3   Let A be a finite set; we shall call it an *alphabet*. A *string* or *word* of length n *over* A is an n-vector $(a_1, \ldots, a_n) \in A^n$. We usually use the notation $a_1 a_2 a_3 \ldots a_n$ in this context (that is, the vector components are placed one after the other without separating commas and without enclosing them in brackets). By the symbol Λ we shall understand the unique element of A^0—i.e., the *empty* string or string of length 0.

A^* stands for $\bigcup_{i=0}^{\infty} A^n$ and A^+ for $\bigcup_{i=1}^{\infty} A^n$.//

Using the technique of coding sequences of numbers by a single number, we shall give numerical counterpart for any set A^*.

Definition 4   Let $A = \{\sigma_1, \ldots, \sigma_m\}$. Consider each σ_i, $i = 1, \ldots, m$, as a one-letter *sequence* and assign to it the number code i. Thus, $num(\sigma_i) \stackrel{\text{def}}{=} \langle i, 1 \rangle$ for $i = 1, \ldots, m$.

Let $x = a_1 a_2 \ldots a_n$, $n > 1$, be in A^*, where $a_i \in A$ for $i = 1, \ldots, n$. Define $num(a_1 a_2 \ldots a_n)$ to be $\langle num(a_1), \ldots, num(a_n), n \rangle$. Finally, define $num(\Lambda) = 0$.//

Note: By (e) above, Λ is the only element of A^* whose code is 0.

Proposition 1   The function $num: A^* \to \mathcal{N}$ of Definition 4 is 1-1.

Proof Let $num(a_1 a_2 \ldots a_n) = num(b_1 b_2 \ldots b_l)$. Hence, $\langle num(a_1), \ldots, num(a_n), n \rangle = \langle num(b_1), \ldots, num(b_l), l \rangle$. Applying the L function to both sides of the above equality we obtain $n = l$ and, hence, for $i = 1, \ldots, n \ (= l)$, we have $num(a_i) = num(b_i)$ (why?) Applying Π_1^2 to the above equalities $(i = 1, \ldots, n)$ we obtain $a_i = b_i$.//

Definition 4 and Proposition 1 have established an "arithmetization" of A^*. Thus, statements about A^* and its various subsets have exact counterparts in statements about $\mathcal{N}$ and its corresponding subsets. We can see now that our study of algorithmic functions, which we started in this chapter, does not lose in generality due to our restricting attention to number-theoretic functions, since to each word-function we can assign a number-theoretic function and vice versa.

Definition 5   Let $A = \{\sigma_1, \sigma_2, \ldots, \sigma_m\}$. Then $f:(A^*)^k \to A^*$ is primitive recursive, in symbols $f \in \mathcal{PR}(A)$, iff $g:\mathcal{N}^k \to \mathcal{N}$ is in $\mathcal{PR}$, where

$$g(\vec{x}_k) = \begin{cases} 0 \textbf{ if for some } i, 1 \le i \le k, x_i \notin num(A^*) \\ num \circ f(num^{-1}(x_1), \ldots, num^{-1}(x_k)) \textbf{ otherwise.}// \end{cases}$$

The above definition can be made clearer through the following diagram (*num* has been established to be 1-1; thus, num^{-1} is defined on the set $num(A^*)$, which contains range(g)):

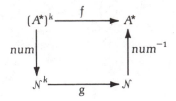

The above diagram is easily seen to be commutative, as it follows from Definition 5; that is, $f(\vec{w}_k) = num^{-1} \circ g(num(w_1), \ldots, num(w_k))$. (See Problem 21.) That is, to "compute" f for "input" $\vec{w}_k$, first convert the input to "numerical" format via a primitive recursive procedure (*num*), then compute the primitive recursive function g on the converted input and, finally, convert the result into a string using the (primitive recursive) algorithm num^{-1}.

Before we can feel completely comfortable with Definition 5, however, we must first ensure that it is independent of the choice of the coding "*num*"; indeed, it would be counterintuitive if a function $f: A^* \rightarrow A^*$ existed along with two different *primitive recursive* codings num_1 and num_2 such that $g_1 = \lambda x.\textbf{if}$ $x \notin num_1(A^*)$ **then** 0 **else** $num_1(f(num_1^{-1}(x)))$ is in $\mathcal{PR}$, whereas $g_2 = \lambda x.\textbf{if}$ $x \notin num_2(A^*)$ **then** 0 **else** $num_2(f(num_2^{-1}(x)))$ is not. We shall now see that this cannot happen.

Proposition 2   Let $f: A^* \rightarrow A^*$ be such that the function g_1 defined above is in $\mathcal{PR}$. Then so is g_2.

Proof Let $A = \{\sigma_1, \ldots, \sigma_m\}$. Without loss of generality, we assume that both num_1 and num_2 agree on A—that is, $^1\Pi_1^2 \circ num_1(\sigma_i) = {}^2\Pi_1^2 \circ num_2(\sigma_i)$, $i = 1,$ $\ldots, m$, where $^i\Pi$ denotes the projections associated with $^i\langle \ \rangle$, the $\langle \ \rangle$-coding used in num_i; the corresponding L-function will be denoted by ^iL.

As an aid to our reasoning, consider the following diagram:

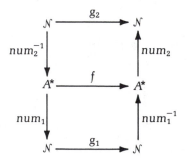

We observe that for any $x \in \mathcal{N}$,

$$g_2(x) = \textbf{if } x \notin num_2(A^*) \textbf{ then } 0 \textbf{ else } num_2(f(num_2^{-1}(x))).$$

Also (lower part of the diagram; see our remarks following Definition 5), for all y in A^*, $f(y) = num_1^{-1}(g_1(num_1(y)))$.

It follows that for all x in $\mathcal{N}$,

$$g_2(x) = \textbf{if } x \notin num_2(A^*) \textbf{ then } 0 \textbf{ else } num_2 \circ num_1^{-1} \circ g_1 \circ num_1 \circ num_2^{-1}(x)$$

To show $g_2 \in \mathcal{PR}$, we must show:

(a) $\lambda x.x \in num_2(A^*) \in \mathcal{PR}_*$

(b) $\lambda x. \textbf{ if } x \notin num_1(A^*) \textbf{ then } 0 \textbf{ else } num_2 \circ num_1^{-1}(x)$ is in $\mathcal{PR}$.

(c) As in (b), but with num_1 and num_2 interchanged.

As for (a),

$$x \in num_2(A^*) \equiv Code(x) \,\&\, \{x = 0 \vee (^2L(x) = 1 \,\&\, {}^2\Pi_1^2(x) \le m \,\&\, {}^2\Pi_1^2(x) \ge 1 \vee$$

$${}^2L(x) > 1 \,\&\, (\forall i)_{\le {}^2L(x)} \,[i > 0 \Rightarrow {}^2L \circ {}^2\Pi_i^{{}^2L(x)+1}(x) = 1 \,\&\,$$

$${}^2\Pi_1^2 \circ {}^2\Pi_i^{{}^2L(x)+1}(x) \le m \,\&\, {}^2\Pi_1^2 \circ {}^2\Pi_i^{{}^2L(x)+1}(x) \ge 1])\},$$

which is clearly in $\mathcal{PR}_*$.

As for (b), set $h(x) = \textbf{if } x \notin num_1(A^*) \textbf{ then } 0 \textbf{ else } num_2 \circ num_1^{-1}(x)$.

Now,

$$y = h(x) \equiv x \notin num_1(A^*) \,\&\, y = 0 \vee y = num_2(num_1^{-1}(x))$$

Also,

$$y = num_2(num_1^{-1}(x)) \equiv x \in num_1(A^*) \,\&\, y \in num_2(A^*) \,\&\, \{x = 0 \,\&\, y = 0 \vee$$

$$(^1L(x) = {}^2L(y) \,\&\, (\forall i)_{\le {}^1L(x)}[i > 0 \Rightarrow {}^1\Pi_1^2 \circ {}^1\Pi_i^{{}^1L(x)+1}(x) = {}^2\Pi_1^2 \circ {}^2\Pi_i^{{}^1L(x)+1}(y)])\}.$$

Thus, $y = h(x) \in \mathcal{PR}_*$, given the result in (a) above.

Define

$$l(x) = \textbf{if } x \notin num_1(A^*) \textbf{ then } 0 \textbf{ else } {}^2\langle \underbrace{k, \ldots, k}_{{}^1L(x) \text{ times}}, {}^1L(x) \rangle$$

where

$$k = \max_{1 \le i \le m}(num_2(\sigma_i))$$

Using the assumption (d) for admissible $\langle\ \rangle$-codings, it follows that $l \in \mathcal{PR}$. Further,

$$h = \lambda x.(\mu y)_{\leq l(x)}(y = h(x)); \text{ hence, } h \in \mathcal{PR}$$

This proves (b).

(c) has a proof identical to that of (b).//

Corollary   Definition 5 is independent of the choice of a primitive recursive code-scheme $\langle\ \rangle$.

The above ideas carry over to relations. First, let us define what we mean by a relation being primitive recursive over A:

Definition 6   A relation $R(\vec{x}_k)$ on $(A^*)^k$ is *primitive recursive*, in symbols $R(\vec{x}_k) \in \mathcal{PR}(A)_*$, iff, for some $\lambda\vec{x}_k.f(\vec{x}_k) \in \mathcal{PR}(A)$, $R(\vec{x}_k) \equiv f(\vec{x}_k) = \Lambda$.//

Note: This definition is the counterpart of the one for the number-theoretic case. Indeed, if we said that $R(\vec{x}_k)$ is primitive recursive iff its characteristic function $c_R = \lambda\vec{x}_k.$**if** $R(\vec{x}_k)$ **then** w_1 **else** w_2 (where $w_1 \neq w_2$ are arbitrary but fixed elements of A^*) is in $\mathcal{PR}(A)$, then the following proposition would still be true, (see Problem 23), making this alternative definition equivalent to Definition 6.

Proposition 3   A relation $R(\vec{x}_k)$ on $(A^*)^k$ is in $\mathcal{PR}(A)_*$ iff $R(num^{-1}(y_1) \ldots, num^{-1}(y_k))$ & $\vec{y}_k \in (num(A^*))^k$ on $\mathcal{N}^k$ is in $\mathcal{PR}_*$.

Proof Let $R(\vec{x}_k)$ be in $\mathcal{PR}(A)_*$. Then, for some $\lambda\vec{x}_k.f(\vec{x}_k) \in \mathcal{PR}(A)$,

$$R(\vec{x}_k) \equiv f(\vec{x}_k) = \Lambda.$$

Now a $g \in \mathcal{PR}$ exists such that for all $\vec{y}_k \in \mathcal{N}^k$

$$g(\vec{y}_k) = \begin{cases} 0 \text{ if for some } i \ \ y_i \notin num(A^*) \\ num(\,f(num^{-1}(y_1), \ldots, num^{-1}(y_k))) \text{ otherwise} \end{cases}$$

Hence,

$$R(num^{-1}(y_1), \ldots, num^{-1}(y_k)) \,\&\, \vec{y}_k \in (num(A^*))^k \equiv$$
$$f(num^{-1}(y_1), \ldots, num^{-1}(y_k)) = \Lambda \,\&\, \vec{y}_k \in (num(A^*))^k \equiv$$

$$num(f(num^{-1}(y_1), \ldots, num^{-1}(y_k))) = 0 \ \& \ \vec{y}_k \in (num(A^*))^k \equiv$$

$$g(\vec{y}_k) = 0 \ \& \ y_1 \in num(A^*) \ \& \ldots \& \ y_k \in num(A^*)$$

which is in $\mathcal{PR}_*$, since $g \in \mathcal{PR}$ and $\lambda z.z \in num(A^*)$ is in $\mathcal{PR}_*$.

Conversely, let $R(\vec{x}_k)$ on $(A^*)^k$ be such that $R(num^{-1}(y_1), \ldots, num^{-1}(y_k))$ $\& \ \vec{y}_k \in (num(A^*))^k$ is in $\mathcal{PR}_*$.

Then a $\lambda \vec{y}_k.h(\vec{y}_k) \in \mathcal{PR}$ exists such that range$(h) = \{0,m\}$, where $0 \neq m \in num(A^*)$ and $R(num^{-1}(y_1), \ldots, num^{-1}(y_k)) \ \& \ \vec{y}_k \in (num(A^*))^k \equiv h(\vec{y}_k) = 0$.

Define $f:(A^*)^k \to A^*$ by $f = \lambda \vec{x}_k.num^{-1}(h(num(x_1) \ldots, num(x_k)))$.(*)

We have the following equivalences:

$$R(\vec{x}_k) \equiv (\exists \vec{y}_k)[y_1 = num(x_1) \ \& \ldots \& \ y_k = num(x_k) \ \& \ R(num^{-1}(y_1),$$

$$\ldots, num^{-1}(y_k))] \equiv (\exists \vec{y}_k)[y_1 = num(x_1) \ \& \ldots \& \ y_k = num(x_k) \ \& \ h(\vec{y}_k) = 0] \equiv$$

$$h(num(x_1), \ldots, num(x_k)) = 0 \equiv num^{-1} \circ h(num(x_1), \ldots, num(x_k)) = \Lambda \equiv$$

$$f(\vec{x}_k) = \Lambda$$

Now, from the definition of f, we have, for all $\vec{y}_k$ in $(num(A^*))^k$, that $h(\vec{x}_k) = num \circ f(num^{-1}(y_1), \ldots, num^{-1}(y_k))$.

We also know that $h \in \mathcal{PR}$ and $\lambda z.z \in num(A^*) \in \mathcal{PR}_*$.

Hence,

$$g = \lambda \vec{y}_k.\textbf{if } y_1 \in num(A^*) \ \& \ldots \& \ y_k \in num(A^*) \textbf{ then}$$

$$h(\vec{y}_k) \textbf{ else } 0 \text{ is in } \mathcal{PR}. \text{ But also, } g = \lambda \vec{y}_k.\textbf{if } y_1 \in num(A^*) \ \&$$

$$\ldots \& \ y_k \in num(A^*) \textbf{ then } num \circ f(num^{-1}(y_1), \ldots, num^{-1}(y_k))$$

$$\textbf{else } 0,$$

which proves that $f \in \mathcal{PR}(A)$. This along with $R(\vec{x}_k) \equiv f(\vec{x}_k) = \Lambda$ proves $R(\vec{x}_k) \in \mathcal{PR}(A)_*.//$

Definition 5 and Propositions 2 and 3 are intuitively appealing. They say in effect that if we (tentatively!) adopt primitive recursiveness (on number-theoretic functions) as the formal counterpart to the notion "algorithmic", then a word-function (resp. word-predicate) is "algorithmic" iff it can be algorithmically transformed (through *num*) to a number-theoretic function (resp. predicate) which is algorithmic.

Example 5 Davis [2] uses the following coding in his arithmetizations of Turing Machines and Logical Theories: Let $A = \{\sigma_0, \sigma_1, \ldots, \sigma_m\}$. Assign the code $2i + 1$ to each σ_i

(*)The restriction on the range of h ensures that dom$(f) = (A^*)^k$.

$(i = 0, \ldots, m)$. Now define $num(a_1a_2 \ldots a_n) = \Pi_{i=1}^n p_i^{num(a_i)}$, where $n > 1$ and $a_i \in A, i = 1, \ldots, n$.

Also set $num(\Lambda) = 0$.

For this code, one can easily verify that:

(a) $num(a)$ is odd if $a \in A$; it is even if $a \in A^+$ and has length > 1.

(b) A primitive recursive function L exists such that for all z which are codes, $L(z) =$ "the length of the coded sequence".

(c) If we set $\langle x_1, \ldots, x_n, n \rangle \overset{\text{def}}{=} \Pi_{i=1}^n p_i^{x_i}$, then $\langle \ \rangle$ is onto $\mathcal{N}$ and $\lambda x n. \underbrace{\langle x, \ldots, x, n \rangle}_{n \ x\text{'s}}$ is in $\mathcal{PR}$. For all i, $\lambda x_i.\langle \vec{x}_n, n \rangle$ is increasing.

(d) The projections are $\lambda iz.\exp(i,z)$, $1 \le i \le L(z)$ and are clearly in $\mathcal{PR}$.

(e) $n \ne 0 \Rightarrow \langle \vec{x}_n, n \rangle \ne 0$.

Thus all the intuitively appealing properties of codings we required (prior to Definition 3) are satisfied for this particular coding.//

Example 6   The following arithmetization is not defined in terms of arithmetic coding functions. Let $A = \{\sigma_1, \ldots, \sigma_m\}$ be the alphabet. Assign the code i to σ_i, $i = 1, \ldots, m$. Assign the code 0 to Λ.

Let $x \in A^*, a \in A$; then assign the code $m \cdot code(x) + code(a)$ to $xa \in A^+$, where "." is the multiplication operation on numbers.

This is the m-adic coding (Smullyan [15], Bennett [1]) by which each string in A^+ is interpreted as an m-adic representation of an integer, where the m-adic digits are the elements of A. Note that 0 is *not* an m-adic digit.

It is clear that this arithmetization is *onto* $\mathcal{N}$. We shall return to this arithmetization in Chapter 5.//

Example 7   Consider a function $f = (f_1, \ldots, f_l):\mathcal{N}^k \to \mathcal{N}^l$, $l > 1$. We shall say that f is primitive recursive iff each of its projections $f_1, \ldots, f_l$, is in $\mathcal{PR}$. Note that this is equivalent to saying that $\lambda \vec{x}_k.\langle f_1(\vec{x}_k), \ldots, f_l(\vec{x}_k) \rangle$ is in $\mathcal{PR}$. How about functions of variable number of arguments and variable number of "outputs"—i.e., functions $f:\mathcal{N}^* \to \mathcal{N}^*$? (Here the alphabet is $\mathcal{N}$—that is, infinite.)

To each $(x_1, \ldots, x_n) \in \mathcal{N}^*$ we associate the number $code(\vec{x}_n)$ equal to $\langle x_1, \ldots, x_n, n \rangle$,(*) where $\langle \ \rangle$ is a coding of variable-length sequences in $\mathcal{N}$, satisfying the properties (a)–(e) mentioned prior to Definition 3.

Then $f:\mathcal{N}^* \to \mathcal{N}^*$ is primitive recursive, by definition, iff an $\hat{f} \in \mathcal{PR}$ exists such that for all $y \in \mathcal{N}$

$$\hat{f}(y) = \textbf{if } y \text{ is not of the form } \langle \vec{x}_n, n \rangle \textbf{ then } 0$$

$$\textbf{else } code(\ f(code^{-1}(y))).$$

(*)Assign 0 to Λ.

The reader can see that this definition is independent of the choice of $\langle\ \rangle$ (See Problem 24).

A *polynomial* in x with coefficients in $\mathcal{N}$ is, formally, a vector (of coefficients) $(a_0, \ldots, a_n)$—i.e., an element of $\mathcal{N}*$.

Thus, a function h that takes us from polynomials to polynomials (e.g., formal differentiation) is a function $h\colon \mathcal{N}* \to \mathcal{N}*$. It follows that the question of whether or not such an h is primitive recursive is reduced to the question of whether or not the associated (through *code*) number-theoretic function $\hat{h}\colon \mathcal{N} \to \mathcal{N}$ is in $\mathcal{PR}$.//

§2.6 ITERATION

Definition 1 The following schema A is the schema of *pure iteration* (Robinson [12].)

$$A \begin{cases} f(0,y) = y \\ f(x + 1,y) = g(\,f(x,y)).// \end{cases}$$

It is clear that if $g \in \mathcal{PR}$ then $f \in \mathcal{PR}$ since A is a particular case of primitive recursion. The name *iteration* comes from the fact that $f(x,y) = g^x(y)$, where, as usual, $g^0(y) = y$ for all y; that is, g is iterated x times.

Theorem 1 In the presence of pairing functions, primitive recursion and iteration are equivalent.

Proof Let J be a pairing function with projections K and L.

Thus, for all $x,y,K\circ J(x,y) = x$ and $L\circ J(x,y) = y$ (J is *not* assumed to be onto).

Let f be defined as in schema D below (primitive recursion).

$$D \begin{cases} f(0,\vec{y}_n) = h(\vec{y}_n) \\ f(x + 1,\vec{y}_n) = g(x,\vec{y}_n, f(x,\vec{y}_n)) \end{cases}$$

Set

$$F = \lambda xy.f(x,\Pi_1^n(y), \ldots, \Pi_n^n(y)), \text{ thus } f(x,\vec{y}_n) = F(x, \langle \vec{y}_n \rangle)$$

Hence,

$$F(0,y) = h(\Pi_1^n(y); \ldots, \Pi_n^n(y))$$

$$F(x + 1,y) = g(x, \Pi_1^n(y), \ldots, \Pi_n^n(y), F(x,y))$$

or, setting

$$H = \lambda y.h(\Pi_1^n(y), \ldots, \Pi_n^n(y)),$$

$$G = \lambda xyz.g(x, \Pi_1^n(y), \ldots, \Pi_n^n(y), z)$$

we obtain F from schema C below:

$$C \begin{cases} F(0,y) = H(y) \\ F(x + 1,y) = G(x,y,F(x,y)) \end{cases}$$

That is, in the presence of J,K,L, schema C is as powerful as D. We shall now eliminate x and y from G.

Set $\tilde{F} = \lambda xy.\langle x,y,F(x,y)\rangle$; thus, $F = \Pi_3^3 \circ \tilde{F}$:

$$\tilde{F}(0,y) = \langle 0,y,F(0,y)\rangle = \langle 0,y,H(y)\rangle$$

$$\tilde{F}(x + 1,y) = \langle x + 1,y,F(x + 1,y)\rangle$$

$$= \langle \Pi_1^3 \circ \tilde{F}(x,y) + 1, \Pi_2^3 \circ \tilde{F}(x,y), G(\Pi_1^3 \circ \tilde{F}(x,y), \Pi_2^3 \circ \tilde{F}(x,y), \Pi_3^3 \circ \tilde{F}(x,y))\rangle$$

Thus, setting $\tilde{H} = \lambda y.\langle 0,y,H(y)\rangle$,

$$\tilde{G} = \lambda z.\langle \Pi_1^3(z) + 1, \Pi_2^3(z), G(\Pi_1^3(z), \Pi_2^3(z), \Pi_3^3(z))\rangle,$$

$$B \begin{cases} \tilde{F}(0,y) = \tilde{H}(y) \\ \tilde{F}(x + 1,y) = \tilde{G}(\tilde{F}(x,y)) \end{cases}$$

Thus, B is as powerful as C [and hence D] in the presence of J,K,L.

Finally, the schema

$$A \begin{cases} \tilde{\tilde{F}}(0,y) = y \\ \tilde{\tilde{F}}(x + 1,y) = \tilde{\tilde{G}}(\tilde{\tilde{F}}(x,y)) \end{cases}$$

is as powerful as B [and hence D] since

$$\tilde{F}(x,y) = \tilde{G}^x(H(y)) \text{ and } \tilde{\tilde{F}}(x,y) = \tilde{\tilde{G}}^x(y).//$$

Example 1   Consider the schema

$$A \begin{cases} f(0,y) = h(y) \\ f(x + 1,y) = f(x,g(y)) \end{cases}$$

Thus,

$$f(1,y) = f(0,g(y)) = h(g(y))$$
$$f(2,y) = f(1,g(y)) = h(g^2(y))$$
$$\vdots$$

By induction on x we can show that $f(x,y) = h \circ g^x(y)$. Hence, schema A does not lead outside $\mathcal{PR}$.//

Example 2   Consider the schema

$$B \begin{cases} f(0,y) = h(y) \\ f(x + 1,y) = f(x, f(x,y)) \end{cases}$$

Observe that

$$f(1,y) = f(0, f(0,y)) = f(0,h(y)) = h^2(y)$$
$$f(2,y) = f(1, f(1,y)) = h^2 \circ h^2(y) = h^4(y).$$

By induction on x we can show that $f(x,y) = h^{2^x}(y)$. Hence, schema B does not lead outside $\mathcal{PR}$.//

We are in the position now to give yet another interesting characterization of the class $\mathcal{PR}$. This is stated below as Theorem 2.

Theorem 2 $\mathcal{PR}$ is the closure of $\{\lambda x.x + 1, \lambda x.x \dot- 1\}$ under *substitution, bounded search* $(\mu y)_{\leq z}$ and *pure iteration* (Definition 1).

Proof   It suffices to show that the class defined above contains pairing functions. This will result to closure of that class under primitive recursion (by Theorem 1), from which the result will follow trivially. Let $\widehat{\mathcal{PR}}$ be the class defined in Theorem 2.

The following functions are all in $\hat{\mathcal{PR}}$:

(a) $\lambda xy.x + y$ (this is $\lambda xy.s^y(x)$, where $s = \lambda x.x + 1$)

(b) $\lambda xy.x \doteq y$ (this is $\lambda xy.p^y(x)$, where $p = \lambda x.x \doteq 1$)

(c) $\lambda x.2x$ ($\lambda xy.x + y$ and substitution)

(d) $\lambda xy.2^y x$ (this is $\lambda xy.d^y(x)$, where $d = \lambda x.2x$)

(e) $\lambda xy.|x - y|$ ($= \lambda xy.x \doteq y + y \doteq x$)

(f) $\lambda x.2^x$ (substitution and $\lambda xy.2^y x$)

(g) $J = \lambda xy.2^{x+y+2} + 2^{y+1}$ (substitution and the $x + 1, 2^x, x + y$ functions)

We recognize J as a pairing function. We shall obtain its projections in $\hat{\mathcal{PR}}$.

Note that $2^{x+y+2} = \text{max power of } 2 \leq J(x,y)$.

(h) $\lambda z.\lfloor \log_2 z \rfloor$ ($\lfloor \log_2 z \rfloor = (\mu y)_{\leq z}(2^{y+1} > z)$), or,

setting $f = \lambda yz.z + 1 \doteq 2^{y+1}$,

which is in $\hat{\mathcal{PR}}$, $\lfloor \log_2 z \rfloor = (\mu y)_{\leq z} f(y,z)$. Note that $\lfloor \log_2 0 \rfloor = 0$ according to this scheme.)

Observe that $2^{x+y+2} = 2^{\lfloor \log_2 J(x,y) \rfloor}$ and $2^{y+1} = J(x,y) \doteq 2^{x+y+2}$. Thus,

$$L = \lambda z.\lfloor \log_2(z \doteq 2^{\lfloor \log_2 z \rfloor}) \rfloor \doteq 1$$

$$K = \lambda z.\lfloor \log_2 z \rfloor \doteq (L(z) + 2)$$

which are in $\hat{\mathcal{PR}}$.//

It is fairly straightforward for a reader with some programming background to see that any primitive recursive function can be programmed on a computer.

In particular, it follows from Theorem 2 that instructions of the following formats would be sufficient to compute any function in $\mathcal{PR}$:

(a) $x = x + 1$

(b) $x = x \doteq 1$

(c)   ⌈**repeat** x **times**
     ⌊**end**

(d)   ⌈**repeat** x **times while** $y \neq 0$
     ⌊**end**

Indeed, $\lambda x.x + 1$ and $\lambda x.x \doteq 1$ can be computed using instructions of types (a) and (b), respectively.

Moreover, assuming that we already know how to program a function $\lambda x.f(x)$ we can program $\lambda xy.f^y(x)$ using the *loop* of type (c):

repeat y **times**
    compute $f(x)$
    store result in x
end; x holds the final result

Similarly, assuming that we already know how to program $\lambda \vec{x}y.f(\vec{x},y)$, we can program $\lambda \vec{x}z.(\mu y)_{<z} f(\vec{x},y)$ by a loop of type (d):

$y = 0$
compute $f(\vec{x},y)$; store in w
repeat z **times while** $(w \neq 0)$
    $y = y + 1$
    compute $f(\vec{x},y)$
    store in w
end
  y holds the final result.

In the following paragraph, we shall make the above discussion precise. It will be seen, moreover, that the implied programming formalism computes no more than functions of $\mathcal{PR}$.

§2.7 A PROGRAMMING FORMALISM FOR $\mathcal{PR}$.

The main result(*) in this paragraph states in effect that the subset of PL/1 whose only *control instruction* is of the form **repeat** X **times** [**while** $Y \neq 0$](†) . . . **end** computes exactly the functions of $\mathcal{PR}$.

Definition 1   The set of *Loop Programs*, L:
    The *alphabet* of the language L, A, is the set $A = \{$**Loop, end, while,**

add, sub, X, 1, ;$\}$

(*)This paragraph contains results which originally appeared in Meyer and Ritchie [7] and Ritchie [10] under slightly different form.

(†)[A] means that A is optional.

A *variable* of L is defined by BNF(*) as: $\langle\text{var}\rangle::\,= X\,|\,\langle\text{var}\rangle 1$

Examples: X, $X1$, $X111$, $X1^n$, where $1^n = \overbrace{11\ldots 1}^{n}$, are variables of L.
An *instruction* of L is:

$\langle\text{instruction}\rangle::\,=$ **add** $\langle\text{var}\rangle\,|\,$**sub** $\langle\text{var}\rangle\,|\,$**end**
$|\,$**Loop**$\langle\text{var}\rangle\,|\,$**Loop** $\langle\text{var}\rangle$ **while** $\langle\text{var}\rangle$

Examples: **add** $X1$, **sub** $X1111$, **end, Loop** $X11$ **while** $X1$, **Loop** X are all valid instructions.
A *statement* of L is:

$\langle\text{statement}\rangle::\,=$ **add** $\langle\text{var}\rangle\,|\,$**sub** $\langle\text{var}\rangle\,|$
   Loop $\langle\text{var}\rangle;\,\langle\text{program}\rangle;\,$**end**$\,|$
   Loop $\langle\text{var}\rangle$ **while** $\langle\text{var}\rangle;\langle\text{program}\rangle;\,$**end**

L is the set of *programs*, where $\langle\text{program}\rangle$ is defined as:

$\langle\text{program}\rangle::\,= \langle\text{statement}\rangle\,|\,\langle\text{program}\rangle;\langle\text{statement}\rangle.//$

Intuitively, we understand **add** X as the assignment statement $X = X + 1$, **sub** X as $X = X \doteq 1$. We also understand

$\left[\begin{array}{l}\textbf{Loop } X1^n \textbf{ while } X1^m \ (n \text{ can very well be } = m) \\ \quad \langle\text{program}\rangle \\ \textbf{end}\end{array}\right.$

to mean "repeat $\langle\text{program}\rangle$ as many times as the 'contents' of $X1^n$ as long as the contents of $X1^m$ are nonzero". Further, our intention is that if the contents of $X1^n$ are equal to 0 initially, then the loop is skipped. Moreover, the value of $X1^n$ is allowed to change inside the loop but this is *not* to affect the maximum number of repetitions which must equal the *initial* value $X1^n$ had upon entry to the loop.

We shall now formalize our intentions for the semantics of the programs in L. Before we do that, let us agree to slightly abuse the notational requirements of Definition 1 whenever this will improve clarity. In particular,

---

(*)BNF stands for Backus-Naur- or Backus-Normal-Form. The name is angular brackets to the left of the "::=" symbol is defined to be what is found to the right of "::=". The symbol "$|$" means "or". E.g. $\langle\text{var}\rangle::=X\,|\,\langle\text{var}\rangle 1$ says that a "var" is either a single X, or (recursively!) a "var" followed by a "1".

we shall freely use any capital letters of the alphabet, with or without subscripts, as variables. Further, instead of **add** X, **sub** X we shall use the more explicit $X = X + 1$ and $X = X \mathbin{\dot{-}} 1$, respectively.

Note: It would unnecessarily complicate our grammar if we formally defined the instruction schemas $X = X + 1$ and $X = X \mathbin{\dot{-}} 1$ in L.

Definition 2   Let $P \in L$ and let **Loop** $X1^n$ [**while** $X1^m$]; S; **end** be a substring of P such that $S \in L$. We shall then say that the **Loop** and **end** keywords above *correspond*. Further, the pair of the **Loop**-instruction and its corresponding **end** is called a *do-loop* and S is the *inside* of the loop, while $X1^n$ is its *bound*.//

Note: The correspondence between **Loop** and its "matching" **end** is similar to the correspondence between "(" and its matching ")" in an arithmetic expression.

Before we embark on the following rather lengthy definition, observe that if $P \in L$ then P involves a finite number of different instructions; hence, also, it involves a finite number of different variables. We shall call the number of instructions of P the *length of P*.

Definition 3   Let $P \in L$ and let $X_1, \ldots, X_n$ be the involved variables. Let there be m *do-loops* in P and let $X_{i_1}, \ldots, X_{i_m}$ be the *bounds*

Note: The X_{i_j}'s need *not* be distinct! It may also be $m > n$.

An *ID-frame* is the $(n + m + 1)$-tuple of *symbols* $(X_1, \ldots, X_n; B_1, \ldots, B_m; I)$.

An *instantaneous description (ID)* of a computation(*) of P is an $(n + m + 1)$-tuple in $\mathcal{N}^{n+m+1}, (x_1, \ldots, x_n; b_1, \ldots, b_m; i)$, where $1 \le i \le \text{length}(P) + 1$.

We understand the x_j's as the *contents* of the *registers* X_j of a hypothetical *fixed-program machine* M_p (the program is P).

The registers $B_j, j = 1, \ldots, m$, hold the values of the bounds upon *entry* to the corresponding *do-loops*, thus the X_{i_j}'s can change their contents without affecting the contents of the B_j's.

Finally, I holds i, the *instruction number* of the *current instruction* which is to be performed. The x_j and b_j values are the ones held in X_j and B_j, respectively, immediately *before* the execution of instruction i. Note that the instructions of P are *implicity numbered sequentially*. If P has q instructions then $q + 1$ labels the *null* instruction after P.//

We shall now formally define what we mean by "a program P in L

(*)There is no circularity here! To define ID we need not know what a computation is. In fact we shall define the latter in terms of the former.

computes". Our definition will incorporate all our intentions stated after Definition 1.

Definition 4   Let P be as in Definition 3. Let $I = (x_1, \ldots, x_n; b_1, \ldots, b_m; i)$ and $\tilde{I} = (\tilde{x}_1, \ldots, \tilde{x}_n; \tilde{b}_1, \ldots, \tilde{b}_m; \tilde{i})$ be two ID's of P.

  We say that I *yields* $\tilde{I}$, in symbols $I \rightarrow_P \tilde{I}$, or simply $I \rightarrow \tilde{I}$, if P is understood, iff one of the following conditions hold:

  (a) i labels **Loop** X_{i_s} [**while** X_t]; $\tilde{i}$ labels the corresponding **end** instruction and $x_i = \tilde{x}_i$ $(i = 1, \ldots, n)$, $b_j = \tilde{b}_j$ $(j = 1, \ldots, m, j \neq s)$, $\tilde{b}_s = x_{i_s}$.

  (b) i labels the **end** instruction corresponding to **Loop** X_{i_l} [**while** X_k] (for some l and k), whose instruction number is, say, i_0, and exactly *one* of the following alternatives (b.1) or (b.2) holds:

  (b.1) $\tilde{i} = i_0 + 1$, $x_k \neq 0$, $b_l \neq 0$, $\tilde{b}_l = b_l - 1$, $x_i = \tilde{x}_i$ $(i = 1, \ldots, n)$, $b_j = \tilde{b}_j$ $(j = 1, \ldots, m, j \neq l)$.

  (b.2) $\tilde{i} = i + 1$, $x_k.b_l = 0$, $x_i = \tilde{x}_i$ $(i = 1, \ldots, n)$, $b_j = \tilde{b}_j (j = 1, \ldots, m)$.

  (c) i labels **add** X_l, $\tilde{i} = i + 1$, $\tilde{x}_l = x_l + 1$, $x_i = \tilde{x}_i$ $(i = 1, \ldots, n, i \neq l)$, $b_j = \tilde{b}_j$ $(j = 1, \ldots, m)$

  (d) i labels **sub** X_l, $\tilde{i} = i + 1$, $\tilde{x}_l = x_l \dot{-} 1$, $x_i = \tilde{x}_i$ $(i = 1, \ldots, n, i \neq l)$, $b_j = \tilde{b}_j$ $(j = 1, \ldots, m)$.

  (e) i labels the *null* instruction (i.e., $i = \text{length}(P) + 1$) and $I = \tilde{I}$ as $(n + m + 1)$-vectors.//

  Intuitively, (a) says that executing a **Loop** X_{i_s} [**while** X_t] instruction involves initializing the loop-counter b_s and transferring to the matching **end** instruction.

  It is clear from (b) that unless the **while**-clause causes an exit from a loop, the number of iterations is governed by the loop-counter—that is, the *initial* value of the loop bound; the loop-bound variable can change value inside the loop without affecting the number of iterations. Further, (a) and (b) satisfy our intention that if the initial value of a loop-bound variable is 0, then the loop is skipped.

  (e) is a mathematical artifice and shouldn't be interpreted that we force loop-computations to be *essentially* unending (after all, nothing changes no matter how many times the null instruction is executed). Our purpose is to ultimately view the contents of the variables as functions of "time" (see Chapter 13); thus we want them to be defined for *all* possible values of time. That would not be the case if a loop-program stopped computing after some finite value of time.

Definition 5   Let $P \in L$. A *P-computation* is a finite sequence of P-IDs I_1, $\ldots, I_t, t \geq 1$, such that for $i = 1, 2, \ldots, t - 1$ $I_i \rightarrow_P I_{i+1}$ and the last component of I_t labels the *null* instruction. For any $i = 1, 2, \ldots, t - 1$, the act $I_i \rightarrow_P I_{i+1}$ is *one step* of the computation. t is the *length*(*) or *complexity of the computation*. I_1 is the *initial* ID, I_t is the *final* ID. The initial ID, I_1, is *normal* iff all the *B*-components of I_1 are 0.//

Proposition 1   Let $P \in L$ and let I be a *P*-ID. Then there exists a *unique* *P*-ID J such that $I \rightarrow J$.

Proof It involves no more than a rephrasing of Definition 4. (See Problem 26.)//

We say that programs in L are deterministic (i.e., each ID has a *unique* successor).

Theorem 1   Let $P \in L$ and let I_1 be a *P*-ID. Then there exists a computation $I_1, \ldots, I_t$.

Proof Existence of a computation is shown by induction with respect to the formation of L. It is straightforward to see that the BNF definitions amount to the inductive definition that

(a)   **add** X is a program (X is any variable)

(b)   **sub** X is a program

(c)   *if* P and Q are programs so is $P;Q$

(d)   if P is a program so is **Loop** X [**while** Y]; P; **end**

We will prove that programs of type (a) and (b) have the *property* of defining computations for any initial ID, *normal* or not, and that the property propagates with operations on programs like (c) (*concatenation*) and (d) (*loop-closure*).
(1) *Induction basis*
Let $P = $ **add** X be a program.
   Let (x,i) be a *P*-ID. Then $i = 1$ or 2 (else it is not a *P*-ID).
   Case 1. $i = 1$. Then $(x,1) \rightarrow (x + 1, 2)$ and $(x,1)$, $(x + 1, 2)$ is a computation.
   Case 2. $i = 2$. Then $(x,2)$ is a computation. Similarly, if $P = $ **sub** X.
(2) The *property* propagates with *concatenation*. Let P and Q in L have the property.
   Consider the program $P;Q$ and let I_1 be a $(P;Q)$-ID. We assume without

---

(*)We must not confuse *length* of *program* with *length* of *computation*.

loss of generality that

$$I_1 = (x_{1,1}^{(P)}, \ldots, x_{n,1}^{(P)}, x_{1,1}^{(P,Q)}, \ldots, x_{m,1}^{(P,Q)}, x_{1,1}^{(Q)}, \ldots, x_{l,1}^{(Q)};$$
$$b_{1,1}^{(P)}, \ldots, b_{k,1}^{(P)}, b_{1,1}^{(P,Q)}, \ldots, b_{r,1}^{(P,Q)}, b_{1,1}^{(Q)}, \ldots, b_{s,1}^{(Q)}; i_1)$$

where the (P) superscript indicates values of X and B symbols relevant to P *only*, while the (P,Q) superscript indicates relevance to *both* P and Q and the (Q) superscript indicates relevance to Q only. i_1 is the instruction number; finally, the second subscript indicates the *I-subscript*.

Case 1. $i_1 \le$ number of instructions of P. Then

$$I_1^{(P)} = (x_{1,1}^{(P)}, \ldots, x_{n,1}^{(P)}, x_{1,1}^{(P,Q)}, \ldots, x_{m,1}^{(P,Q)}; b_{1,1}^{(P)}, \ldots, b_{k,1}^{(P)}, b_{1,1}^{(P,Q)}, \ldots, b_{r,1}^{(P,Q)}; i_1^{(P)})$$

where $i_1^{(P)} = i_1$, is a P-ID. By induction hypothesis there is a (unique by Proposition 1) computation $I_1^{(P)}, \ldots, I_t^{(P)}$ that involves the null instruction only *once*, where $I_t^{(P)} = (x_{1,t}^{(P)}, \ldots, x_{n,t}^{(P)}, x_{1,t}^{(P,Q)}, \ldots, x_{m,t}^{(P,Q)}; b_{1,t}^{(P)}, \ldots, b_{k,t}^{(P)}, b_{1,t}^{(P,Q)}, \ldots, b_{r,t}^{(P,Q)}; i_t^{(P)})$ and $i_t^{(P)} = \text{length}(P) + 1$. The Q-ID's are of the form

$$I_j^{(Q)} = (x_{1,j}^{(P,Q)}, \ldots, x_{m,j}^{(P,Q)}, x_{1,j}^{(Q)}, \ldots, x_{l,j}^{(Q)}; b_{1,j}^{(P,Q)}, \ldots, b_{r,j}^{(P,Q)}, b_{1,j}^{(Q)}, \ldots, b_{s,j}^{(Q)}; i_j^{(Q)}).$$

By induction hypothesis, any Q-ID defines a computation (unique, if the null instruction label appears only once). In particular then there is a Q-computation $I_1^{(Q)}, \ldots, I_v^{(Q)}$ whose initial ID is

$$I_1^{(Q)} = (x_{1,t}^{(P,Q)}, \ldots, x_{m,t}^{(P,Q)}, x_{1,1}^{(Q)}, \ldots, x_{l,1}^{(Q)}; b_{1,t}^{(P,Q)}, \ldots, b_{r,t}^{(P,Q)}, b_{1,1}^{(Q)}, \ldots, b_{s,1}^{(Q)}; i_1^{(Q)})$$

where $i_1^{(Q)} = 1$. It is clear then that if we add an appropriate P-only invariant part to Q-IDs (resp. Q-only invariant part to P-IDs), we obtain a (P,Q)-computation

$$\tilde{I}_1^{(P)}, \ldots, \tilde{I}_t^{(P)}[= \tilde{I}_1^{(Q)}], \ldots, \tilde{I}_v^{(Q)}$$

where

$$\tilde{I}_j^{(P)} = (x_{1,j}^{(P)}, \ldots, x_{n,j}^{(P)}, x_{1,j}^{(P,Q)}, \ldots, x_{m,j}^{(P,Q)}, x_{1,1}^{(Q)}, \ldots, x_{l,1}^{(Q)};$$
$$b_{1,j}^{(P)}, \ldots, b_{k,j}^{(P)}, b_{1,j}^{(P,Q)}, \ldots, b_{r,j}^{(P,Q)}, b_{1,1}^{(Q)}, \ldots, b_{s,1}^{(Q)}; i_j^{(P)}),$$

for $j = 1, \ldots, t$ [i.e., the Q-only part is invariant] and

$$\tilde{I}_j^{(Q)} = (x_{1,t}^{(P)}, \ldots, x_{n,t}^{(P)}, x_{1,t+j-1}^{(P,Q)}, \ldots, x_{m,t+j-1}^{(P,Q)}, x_{1,j}^{(Q)}, \ldots, x_{l,j}^{(Q)};$$
$$b_{1,t}^{(P)}, \ldots, b_{k,t}^{(P)}, b_{1,t+j-1}^{(P,Q)}, \ldots, b_{r,t+j-1}^{(P,Q)}, b_{1,j}^{(Q)}, \ldots, b_{s,j}^{(Q)};$$
$$i_j^{(Q)} + \text{length}(P)) \text{ for } j = 1, \ldots, v$$

where the $x^{(P)}$ and $b^{(P)}$ vectors are invariant and equal to the corresponding vectors of $\tilde{I}_t^{(P)}$, and the following vector equalities hold: $\vec{x}_{t+j-1}^{(P,Q)}$ (in $\tilde{I}_j^{(Q)}$) = $\vec{x}_j^{(P,Q)}$ (in $I_j^{(Q)}$) $\vec{b}_{t+j-1}^{(P,Q)}$ (in $\tilde{I}_j^{(Q)}$) = $\vec{b}_j^{(P,Q)}$ (in $I_j^{(Q)}$); further in $i_j^{(Q)}$ + length(P) $i_j^{(Q)}$ refers to the value of the i-component in $I_j^{(Q)}$.

Case 2. $i_1 >$ number of instructions of P. Then the $(P;Q)$-computation starting with I_1 is essentially the Q-computation starting with I_1 without the P-only part. The reader is asked to fill in the details.

(3) The *property* propagates with loop closure. So let P have the property and consider the program Q:

$$Q \begin{cases} \textbf{Loop } X_1 \\ \quad P \\ \textbf{end} \end{cases}$$

We shall show that there is a Q-computation starting with any Q-ID

$$I_1 = (x_{1,1}, \ldots, x_{n,1}; b_{1,1}, \ldots, b_{m,1}; i_1).$$

We shall assume without loss of generality that B_1 corresponds to X_1 in the ID-frame. Further, we assume that X_1 is in P (the other case is similar). The number of times P executes depends on $b_{1,1}$ if $i_1 > 1$, on $x_{1,1}$ if $i_1 = 1$ (i.e., on $b_{1,2}$ in this case where $i_2 = $ length(Q)). It is clear then that the case $i_1 > 1$ subsumes that of $i_1 = 1$; we shall consider only the former ($i_1 > 1$) in what follows.

We show by *induction on* $b_{1,1}$ that, if P has the property, then there is a computation starting with I_1; that is, a sequence of Q-ID's $I_1, \ldots, I_t$ such that

(1) I_t is final;

(2)   If $t > 1$, then $I_i \rightarrow_Q I_{i+1}$ for $i = 1, \ldots, t - 1$.

Now, if I_1 is already final, then there is nothing to prove. So assume in what follows that I_1 is *not* final.

Basis. Let $b_{1,1} = 0$. We assume that $i_1 <$ length(Q) as otherwise I_1, I_2 is a computation, where I_1 and I_2 are identical except for the i components which are length(Q), length(Q) + 1, respectively.

Let $I_1^{(P)}, \ldots, I_k^{(P)}$ be a P-computation (induction hypothesis applied!) where $I_j^{(P)} = (x_{1,j}^{(P)}, \ldots, x_{n,j}^{(P)}; b_{2,j}^{(P)}, \ldots, b_{m,j}^{(P)}; i_j^{(P)})$, $j = 1, \ldots, k$, and $I_1^{(P)}$ coincides with I_1 except that (a) $i_1 = i_1^{(P)} + 1$; (b) $I_1^{(P)}$ does not have a $b_{1,1}^{(P)}$ component.

Then $I_1, \ldots, I_{k+1}$ is a Q-computation where for $j = 1, \ldots, k$ I_j coincides with $I_j^{(P)}$ except that (a) $i_j = i_j^{(P)} + 1$ (b) $I_j^{(P)}$ does not have a $b_{1,j}^{(P)}$ component. (Note that $b_{1,j} = 0$ for each $I_j, j = 1, \ldots, k$.) Finally, I_{k+1} is the same as I_k except that its i-component is length(Q) + 1.

Assume now that the claim is true for $b_{1,1} = l$.

Consider the case $b_{1,1} = l + 1$:

Case 1. $i_1 = \text{length}(Q)$.

Then I_2 exists such that $I_1 \to_Q I_2$ and $b_{1,1} = l + 1$, $i_1 = \text{length}(Q)$, $b_{1,2} = l$, $i_2 = 2$, the remaining corresponding components of I_1, I_2 being identical. By induction hypothesis, there is a Q-computation $I_2, I_3, \ldots, I_t$, hence $I_1 I_2, I_3, \ldots$, I_t is a Q-computation.

Case 2. $i_1 < \text{length}(Q)$.

Consider once more the P-computation $I_1^{(P)}, \ldots, I_k^{(P)}$ we discussed when $b_{1,1}$ was 0. Consider also the Q-ID's $I_1, \ldots, I_k$ derived from the $I_j^{(P)}$'s as above (now $b_{1,j} = l + 1, j = 1, \ldots, k$). Clearly,

(a) $I_j \to_Q I_{j+1}, j = 1, \ldots, k - 1$

(b)   there exists a Q-ID I_{k+1} such that $I_k \to_Q I_{k+1}$.

Necessarily, $b_{1,k+1} = l, i_{k+1} = 2$. By induction hypothesis, there is a Q-computation $I_{k+1}, \ldots, I_t$. Thus, $I_1, \ldots, I_k, I_{k+1}, \ldots, I_t$ is a Q-computation. The case for loop-closure with *while-clause* is similar. The details are left to the reader.//

The proof of Theorem 1 is long but very simple. The point in giving most of it here was to show that our formalism agreed with our intuitive goals for the behavior of loop-programs.

Let us now define in what sense a program $P \in L$ computes a function.

Definition 6   Let $P \in L$ and let $X_1, \ldots, X_m$ be its variables. Let $Y_{\vec{x}_n}$ be one of the X_i's.

By Theorem 1, for all $\vec{x}_n$ in $\mathcal{N}^n$, $n \leq m$, there is a unique $y_{\vec{x}_n}$ in $\mathcal{N}$ such that $I_1, I_2, \ldots, I_t$ is a P-computation where

(1) I_1 is *normal*.

(2) $I_1 = (x_1, \ldots, x_n, 0, \ldots; \overbrace{0, \ldots, 0}^{b\text{-vector}}; 1)$ (i.e., all x-values except those for X_1 through X_n are 0).

(3)   The Y-value of I_t is $y_{\vec{x}_n}$. The function $\lambda \vec{x}_n . y_{\vec{x}_n}$ we denote by $P_Y^{\vec{X}_n}$. *It is the function computed by P with input variable(s) $\vec{X}_n$ and output variable Y. If $m = n$, we write P_Y.//*

Example 1   Consider the loop-program P below:

```
     ┌─ Loop X       /* variables X,Y */
P:   │      Y = Y + 1
     └─ end
```

Clearly,

$$P_Y^{X,Y} = \lambda XY.X + Y.//$$

Example 2 Consider the loop-program P below:

$$
P: \quad
\begin{array}{l}
\text{\textbf{Loop} } X \qquad \text{/* variables } X,Y,Z \text{ */} \\
\qquad \text{\textbf{Loop} } Y \\
\qquad\qquad Z = Z + 1 \\
\qquad \text{\textbf{end}} \\
\text{\textbf{end}}
\end{array}
$$

Clearly

$$P_Z^{X,Y} = \lambda XY.X \cdot Y.//$$

Exercise What is P_Z, where P is as in Example 2?

Example 3. The program P below computes $\lambda X.X + 1$

$$P: \quad X = X + 1 \qquad \text{/* variables } X \text{ */}$$

That is, $P_X^X = \lambda X.X + 1.//$

Example 4 The program P below computes $\lambda X.X \div 1$

$$P: \quad X = X \div 1 \qquad \text{/* variables } X \text{ */}$$

That is, $P_X^X = \lambda X.X \div 1.//$

Example 5 Consider the loop-program P below:

$$
P: \quad
\begin{array}{l}
\text{\textbf{Loop} } X \qquad \text{/* variables } X \text{ */} \\
\qquad X = X \div 1 \\
\text{\textbf{end}}
\end{array}
$$

Clearly, $P_X^X = \lambda X.0.//$

Example 6   Consider the program P below:

$$P: \quad \begin{bmatrix} \textbf{Loop } Y & \quad /* \text{ variables } X,Y */ \\ \quad X = X + 1 \\ \textbf{end} \end{bmatrix}$$

Clearly, $P_X^Y = \lambda Y.Y.//$

Example 7   Consider the program P below:

$$P: \begin{cases} X_1 = X_1 + 1;\ldots;X_{i-1} = X_{i-1} + 1; X_{i+1} = X_{i+1} + 1;\ldots,X_n = X_n + 1; \\ \begin{bmatrix} \textbf{Loop } X_i & \quad /* \text{ variables } X_1,\ldots,X_n,Y \\ \quad Y = Y + 1 & \quad \text{The irrelevant \textbf{add}-instructions} \\ \textbf{end} & \quad \text{above the loop are only used to} \\ & \quad \text{reference all variables at least} \\ & \quad \text{once. Without them, } P \text{ would have} \\ & \quad \text{only } Y \text{ and } X_i \text{ as variables.*/} \end{bmatrix} \end{cases}$$

Clearly, $P_Y^{\vec{X}_n} = \lambda \vec{X}_n.X_i.$

Example 8   The program

$$\begin{bmatrix} \textbf{Loop } X \\ \quad X = X \dot- 1 \\ \textbf{end} \end{bmatrix}$$

simulates the instruction of PL/1 $X = 0$
   The program

$$\begin{cases} \begin{bmatrix} \textbf{Loop } X \\ \quad X = X \dot- 1 \\ \textbf{end} \end{bmatrix} \\ \begin{bmatrix} \textbf{Loop } Y \\ \quad X = X + 1 \\ \textbf{end} \end{bmatrix} \end{cases}$$

simulates the PL/1 instruction $X = Y.//$

Example 9   The following program P simulates $X = k$

$$P: \left\{ \begin{array}{l} \left[\begin{array}{l} \textbf{Loop } X \\ \qquad X = X \div 1 \\ \textbf{end} \end{array}\right. \\[1em] k \text{ identical} \\ \quad \text{instructions} \end{array} \right. \qquad \left\{ \begin{array}{l} X = X + 1 \\ X = X + 1 \\ \quad \cdot \\ \quad \cdot \\ \quad \cdot \\ X = X + 1 \end{array} \right.$$

for k constant.

Another way of viewing it is that $P_X^X = \lambda X.k.//$

Proposition 2   The instruction sets $\{X = X + 1, \ X = X \div 1\}$ and $\{X = 0, \ X = Y, \ X = X + 1\}$ are equivalent in the presence of the **Loop** $X \ldots$ **end** instruction pair.

Proof From Examples 5 and 8 we already know that the first set is at least as powerful as the second.

We need only establish that with (possibly) the help of **Loop** $X \ldots$ **end** instruction pair, the second set can simulate $X = X \div 1$.

This is indeed the case as the following program shows:

$Y = 0$

Loop X

   $X = Y$

   $Y = Y + 1$

end

The final value of X equals the original $\div 1.//$

Note: Traditionally, the set of assignment statements in the loop-program formalism is taken to be $\{X = 0, X = Y, X = X + 1\}$[7,10]. We see from Proposition 2 that this is equivalent to the set $\{X = X + 1, X = X \div 1\}$. Some differences do arise if we look at the system more closely. For example, using the second set we can compute $\lambda xy.x \div y$ with one level of loop nesting. This is not possible for the set $\{X = 0, X = Y, X = X + 1\}$ as it follows from results in Tsichritzis [16].

Thus, although (by Proposition 2) we can compute exactly the same functions with either set of instructions, some things are "easier" to do in one set (e.g., $\lambda x.0$ in the first set), some others are easier to do in the other (e.g., $\lambda xy.x \dotdiv y$ in the second set).

Proposition 3 The instructions **Loop** X **while** Y and **Loop** X are equivalent in the loop-program formalism.

Proof Let $P \in L$ and let Q be the following program:

$$\left[\begin{array}{l} \textbf{Loop } X \quad \textbf{while } Y \\ \qquad P \\ \textbf{end} \end{array}\right.$$

Informally, this is the same as:

Loop X

      if $Y \neq 0$ **then** P /* if $Y \neq 0$ do P *once*, else skip P. */

end

"**if** $Y \neq 0$ **then** P" has the following counterpart in the loop-program formalism if we do not use the **while**-clause:

```
Z = 0
Loop Y            /* This is "if Y ≠ 0 then Z = 1 else Z = 0".
    Z = 0            "Z = 0" can be replaced by
    Z = Z + 1        the appropriate program, (see Example 5)
end                if "Z = 0" is not in the instruction
Loop Z             set */
    P              /* Without loss of generality, we assume
end                that Z does not occur in P */
```

We are done.//

Note: Traditionally, the **while**-clause is not included in the loop-program formalism. By Propositions 3 and 2 this does not affect the computing power of loop programs.

Our only reason of introducing the loop-program formalism the way we did was that these language constructs were immediately suggested by Theorem 2.6.2. We shall therefore continue using the same formalism for the balance of this paragraph. We shall now show that the loop programs compute all (and only) the $\mathcal{PR}$-functions.

Definition 7 $\mathcal{L} = \{f \mid \text{for some } P \in L \text{ and variables } X_1, \ldots, X_n, Y \text{ of } P,$
$f = P_Y^{\vec{X}_n}\}.$

Lemma 1 $\mathcal{PR} \subset \mathcal{L}.$

   Proof Induction with respect to $\mathcal{PR}$. For $\mathcal{PR}$ we use the characterization given in Theorem 2.6.2.
   (1) *Basis.* $\lambda x.x + 1$ and $\lambda x.x \div 1$ are in $\mathcal{L}$. This has already been established. (See Examples 3 and 4.)
   (2) Show that the operations of substitution leave "loop programmability" invariant (that is, show that $\mathcal{L}$ is closed under substitution). Let $\lambda \vec{X}_n Z.f(\vec{X}_n, Z)$ and $\lambda \vec{Y}_m.g(\vec{Y}_m)$ be in $\mathcal{L}$. We shall show that $\lambda \vec{X}_n \vec{Y}_m.f(X_n, g(\vec{Y}_m)) \in \mathcal{L}$. Let $f = P_W^{\vec{X}_n, Z}$ $g = Q_Z^{\vec{Y}_m}$, where without loss of generality, none of the variables of P other than Z occur in Q. Clearly, if R is the program $Q;P$ then $\lambda \vec{X}_n \vec{Y}_m.f(\vec{X}_n, g(\vec{Y}_m)) = R_W^{\vec{X}_n, \vec{Y}_m}$. In view of what we proved and Example 5, $\mathcal{L}$ is closed under substitution of a variable by 0. The other cases of substitution are trivial.
   (3) $\mathcal{L}$ is closed under pure iteration. Indeed, let $\lambda X.f(X) \in \mathcal{L}$ where $f = P_X^X$. Then $\lambda XY. f^Y(X) = Q_X^{X,Y}$ where Q is

$$
\begin{array}{l}
\ulcorner \ \textbf{Loop } Y \\
\ \ P \\
\ \ X_1 = 0 \\
\ \ X_2 = 0 \\
\ \ \quad \vdots \\
\ \ X_k = 0 \\
\llcorner \ \textbf{end}
\end{array}
$$

where $X_1, X_2, \ldots, X_k$ are all the variables of P other than X. The instructions $X_i = 0$ are shorthand for subprograms of our formalism (see Example 5).
   (4) $\mathcal{L}$ is closed under bounded search: Let $\lambda X \vec{Y}_n. f(X, \vec{Y}_n) \in \mathcal{L}$, where $f = P_Y^{X \vec{Y}_n}$. We further assume without loss of generality that $Y \neq Y_i, i = 1, \ldots, n$.
Then, $\lambda Z \vec{Y}_n.(\mu X)_{<Z} f(X, \vec{Y}_n) = Q_X^{Z, \vec{Y}_n}$

where Q is

$$
Q: \begin{cases}
\begin{array}{l}
X = 0 \\
P \\
\lceil \textbf{Loop } Z \quad \textbf{while } Y \\
\;\; X_1 = 0 \\
\;\; X_2 = 0 \\
\qquad \vdots \\
\\
\;\; X_m = 0 \\
\;\; X = X + 1 \\
\;\; P \\
\lfloor \textbf{end}
\end{array}
\end{cases}
$$

where $X_1, \ldots, X_m$ are all the variables of P other than $X, \vec{Y}_n$. Further, we have assumed, without loss of generality, that P does not change X and $\vec{Y}_n$ during the computation.//

We now prepare to prove that $\mathcal{L} \subset \mathcal{PR}$.

Lemma 2 $\qquad P \in L$ implies $P_{\vec{Y}^n}^{\vec{X}_n} \in \mathcal{PR}$, for any variables $\vec{X}_n, Y$ of P.

Proof It will be simpler to show that $P \in L \Rightarrow P_Y \in \mathcal{PR}$ for all variables Y in P. The lemma will follow through substitution of 0 to non-input variables.

We proceed by induction with respect to L.

(i) *Basis.* $P \in L \Rightarrow P$ is one of $X = X + 1$, $X = X \dotminus 1$. The defined functions are $\lambda X. X + 1$, $\lambda X. X \dotminus 1$ respectively, all in $\mathcal{PR}$.

(ii) Let P and Q in L have the *property* that for *all* variables Y (of P) and Z (of Q) P_Y and Q_Z are in $\mathcal{PR}$. Let $\vec{Y}_n$ (resp. $\vec{Z}_m$) be all the variables of P (resp. Q). Consider $(P;Q)_W$.

Case 1. P and Q do not share variables
Case 1.1. W is in P. Then $(P;Q)_W = \lambda \vec{Y}_n \vec{Z}_m . P_W$.
Case 1.2. W is in Q. Then $(P;Q)_W = \lambda \vec{Y}_n \vec{Z}_m . Q_W$.
Case 2. P and Q do share variables. Without loss of generality let Z_1 be the only such variable.
Set

$$
f = P_{Z_1}^{\vec{Y}_n}, \; g = Q_W^{\vec{Z}_m}. \text{ Then } (P;Q)_W = \lambda \vec{Y}_n \vec{Z}_m . g(f(\vec{Y}_n), Z_2, \ldots, Z_m),
$$

which is in $\mathcal{PR}$, since f and g are.

(iii)   Let P in L be such that $P_Y \in \mathcal{PR}$ for all Y in P. Let $\vec{X}_n$ be all the variables of P and consider **Loop** $X; P;$ **end.** Call this program Q.

Case 1. X is in P, say $X = X_1$.
Set

$$f_i = P_{X_i} \text{ and } g_i = Q_{X_i} \text{ for } i = 1, \ldots, n.$$

Then,

$$\begin{cases} g_1(0,X_2, \ldots, X_n) = 0 \\ \text{for } i = 2, \ldots, n \ \ g_i(0,X_2, \ldots, X_n) = X_i \end{cases}$$

and

$$\text{for } i = 1, \ldots, n \ \ g_i(X_1 + 1, X_2, \ldots, X_n) = f_i(g_1(\vec{X}_n), \ldots, g_n(\vec{X}_n))$$

Thus, since $f_i \in \mathcal{PR}$, $i = 1, \ldots, n$, also $g_i \in \mathcal{PR}$, $i = 1, \ldots, n$ because of the above simultaneous primitive recursion.

Case 2. X is *not* in P.
Set

$$g_0 = Q_X, \text{ and for } i = 1, \ldots, n, f_i = P_{X_i}, g_i = Q_{X_i}$$

Then,

$$\begin{cases} g_0(0,\vec{X}_n) = 0 \\ \text{for } i = 1, \ldots, n \ \ g_i(0,\vec{X}_n) = X_i \end{cases}$$

and

$$\begin{cases} g_0(X + 1, \vec{X}_n) = X + 1 \\ \text{for } i = 1, \ldots, n \ \ g_i(X + 1, \vec{X}_n) = f_i(g_1(X, \vec{X}_n), \ldots, g_n(X, \vec{X}_n)) \end{cases}$$

We draw the same conclusion for the g_i's as above. The case for the *loop-while closure* being reducible to the simple *loop closure* (See Proposition 3) shall not be considered.//

Theorem 2 $\mathcal{PR} = \mathcal{L}$.

Proof By Lemmas 1 and 2.//

We shall probe deeper into the properties of loop programs in Chapter 13. In the meantime, we shall use them to argue informally that the primitive recursive functions is an incomplete formalism of the notion "algorithmic function".

To begin with, observe that L, a finitely generated set, is necessarily enumerable, and the same must be true of $\mathcal{PR}$ $(= \mathcal{L})$ since there is an onto map $m:L \to \mathcal{PR}$. Thus, since $\{f \mid f:\mathcal{N} \to \{0,1\}\}$ is uncountable, clearly there exist number-theoretic functions not in $\mathcal{PR}$. With little additional effort, we can show that an (intuitively) algorithmic function exists which is not in $\mathcal{PR}$. The following argument, starting with Lemma 3, is necessarily informal.

Lemma 3 L can be "algorithmically enumerated".

Argument A program P in L is a string over $A = \{$**add,sub,** $X,1,$**Loop,end,while,;**$\}$ satisfying some very simple syntactical rules. We can parse algorithmically any given string of A^+ to decide whether it is in L or not.

Thus, to enumerate L, enumerate A^+ in *List 1* according to string-length and in each length group lexicographically (according to an arbitrarily chosen order of A). Every time a new string is placed in *List 1*, test to see if it belongs to L. If yes, place it also in *List 2.//*

Lemma 4 $\{f \mid f:\mathcal{N} \to \mathcal{N}$ & $f \in \mathcal{PR}\}$ can be algorithmically enumerated.

Argument Form algorithmically *List 3* as follows: For each P placed in *List 2*, place (P,X,Y) in *List 3* for all pairs of variables X,Y of P (including the case $X = Y$). Order the (P,X,Y) according to length(XY);(*) for equal lengths, lexicographically. Interpret (P,X,Y) as P_Y^X.//

Theorem 3   (*Informal*). There is an algorithmic function $f:\mathcal{N} \to \mathcal{N}$ such that $f \notin \mathcal{PR}$.

Argument Let us call g_i the ith object of *List 3*.

Set (Diagonalization!)

$$f = \lambda x.g_x(x) + 1.$$

Clearly, $f \notin \mathcal{PR}$, since $f \in \mathcal{PR}$ implies $f \in \mathcal{L}$ hence, for some i, $f = g_i$. Then,

(*)Recall that a variable is of the form $X1^n$ for some $n \geq 0$.

$f(i) = g_i(i)$ (since $f = g_i$). Also, $f(i) = g_i(i) + 1$ (by definition of f), a contradiction.

f on the other hand is intuitively algorithmic, for, given x, we algorithmically find g_x (xth item of *List 3*); $g_x = P_W^Z$ for some $P \in L$. Apply P for $(Z) = x$ and take as $f(x)$ the final contents of W plus 1.//

Problems 35 and 36 give alternative methods of arguing the validity of Theorem 3.

Corollary   There is an algorithmic relation $R(x)$ which is not in $\mathcal{PR}_*$.

Argument $R(x) \equiv 1 \dot- g_x(x) = 0$ will do.//

Note: The relation $g_x(x) = 0$ is not in $\mathcal{PR}_*$ either (why?).

PROBLEMS

1. Let q be the quotient and r the remainder of the division a/b, i.e., $a = b \cdot q + r$ and $0 \le r < b$.

   (a) Show that $\gcd(a,b) = \gcd(b,r)$. Derive from this observation an algorithm (Euclidean algorithm) for the computation of $\gcd(a,b)$, given any integers a and b.

   (b) Show that (i) $\gcd(a,a) = a$, (ii) if $a > b$, then $\gcd(a,b) = \gcd(a - b,b)$, (iii) $\gcd(a,b) = \gcd(b,a)$. Derive from (i)-(iii) an alternative method to compute $\gcd(a,b)$ which relies *only* on the arithmetic operation of subtraction.

2. Show that, in the presence of $\lambda x.x + 1$ and $\lambda \vec{x}_n.x_i$ (all $n \ge 1$ and $1 \le i \le n$), the following operations on functions can be effected by compositions:

   (a) Substitution of a variable by a function

   (b) Substitution of a variable by a constant

   (c) Identification of two variables

   (d) Permutation of two variables

   (e) Introduction of new (dummy) variables.

3. Prove Proposition 2.1.1.

4. Prove Theorem 2.1.3.

5. Prove that $\lambda n.n! \in \mathcal{PR}$.

6. Prove that $\lambda xyz.\text{if } x = 0 \text{ then } y \text{ else } z$ is in $\mathcal{PR}$.

7. *Without* using Problem 6, show that $\lambda xy.\min(x,y)$ is in $\mathcal{PR}$.

8. Prove Theorem 2.2.3.

9. Do Exercise 2.2.1.

10. Do Exercise 2.2.2.

11. Define

$$(\mathring{\mu}y)_{<z}f(y,\vec{x}) \overset{\text{def}}{=} \begin{cases} \min\{y \mid y < z \ \& \ f(y,\vec{x}) = 0\} \\ 0 \text{ if the } min \text{ does not exist} \end{cases}$$

Prove that $\mathcal{PR}$ is closed under $(\mathring{\mu}y)_{<z}$.

12. Do Exercise 2.3.1.

13. Prove that if a class of functions $\mathcal{C}$ is closed under $(\mu y)_{\leq z}$ and substitution, then $\mathcal{C}_*$ is closed under $(\exists y)_{\leq z}$.

14. Let us call the class of *polynomials* the *closure* of $\{\lambda xy.x + y, \lambda xy.xy, \lambda xy.x \dotdiv y\}$ under *substitution*. Find a *1-1* and *onto* polynomial pairing function in $\mathcal{PR}$ with projections in $\mathcal{PR}$. Explicitly give the projections (i.e., in closed form, without using $(\exists y)_{\leq z}$ and/or $(\mu y)_{\leq z}$). (*Hint:* Consider, for example, an enumeration of $\mathcal{N} \times \mathcal{N}$ by groups G_i, $i = 0,1 \ldots$, where the G_i group is enumerated as

$$(0,i),(1,i), \ldots , (i-1,i),(i,i),(i,i-1),(i,i-2), \ldots , (i,0).$$

A different enumeration is suggested in Warkentin[17].)

15. Do Exercise 2.4.1.

16. Find the closed forms of K and L of the pairing function in Example 2.4.3.

17. Prove Lemma 2.4.1.

18. Let $\lambda\vec{x}.h(\vec{x})$, $\lambda y\vec{x}z.g(y,\vec{x},z)$ and $\lambda y\vec{x}.j(y,\vec{x})$ be in $\mathcal{PR}$, and let $j(y,\vec{x}) \leq y$ for all $\vec{x},y$. Then the function f defined by the schema

$$f(0,\vec{x}) = h(\vec{x})$$
$$f(y+1,\vec{x}) = g(y,\vec{x},f(j(y,\vec{x}),\vec{x}))$$

is in $\mathcal{PR}$. (*Hint:* This is a special case of course-of-values recursion.)

19. Prove that $\lambda x.\lfloor 10^x \sqrt{2} \rfloor \in \mathcal{PR}$. (*Hint:* Use bounded search.)

20. Relying on Problem 19, show that $\lambda x.[\text{the } (x+1)\text{st digit in the decimal expansion of } \sqrt{2}]$ is in $\mathcal{PR}$.

21. Show that the diagram following Definition 2.5.5 is commutative.

22. Show that if $\lambda\vec{x}y.y = f(\vec{x}) \in \mathcal{PR}_*$, f is total, $f(\vec{x}) \leq g(\vec{x})$ for all $\vec{x}$, and $g \in \mathcal{PR}$, then $f \in \mathcal{PR}$.

23. Prove the claim made in the Note following Definition 2.5.6.

24. Fill in the details in Example 2.5.7.

25. (Robinson [12]) Prove that $\mathcal{PR}$ is the closure of $\{\lambda xy.x + y,\ \lambda xy.x \doteq y,\ \lambda x.\lfloor\sqrt{x}\rfloor\}$ under *substitution* and the operation $\lambda x.f(x) \longmapsto \lambda x.f^x(0)$.

    (*Hint:* [12] Call the above-mentioned closure $\mathcal{PR}'$. Show that the pairing function $\lambda xy.((x + y)^2 + y)^2 + x$ is in $\mathcal{PR}'$ and so are its projections, K and L. Prove that for all z, $K(z) = z \doteq \lfloor\sqrt{z}\rfloor^2$ and $L(z) = \lfloor\sqrt{z}\rfloor \doteq \lfloor\sqrt{\lfloor\sqrt{z}\rfloor}\rfloor^2$, concluding that $K(0) = L(0) = 0$ and that $K(z + 1) \neq 0$ implies $K(z + 1) = K(z) + 1$ and $L(z + 1) = L(z)$. Under those circumstances, show that $\lambda x.f(x) \longmapsto \lambda x.f^x(0)$ is as powerful as schema (A) of the proof of Theorem 2.6.1. Conclude that $\mathcal{PR}'$ is closed under primitive recursion.)

26. Prove Proposition 2.7.1.

27. Write a loop program which computes $\lambda x.\lfloor x/k\rfloor$.

28. Write a loop program which computes $\lambda x.\mathrm{rem}(x,k)$.

29. Write a loop program which computes $\lambda x.\lfloor\sqrt{x}\rfloor$.

30. Prove Lemma 2.7.2 without the help of Proposition 2.7.3. In particular, deal with loop closure of the type **Loop** X **while** Y explicitly.

31. Prove that if $\lambda x.h(x)$, $\lambda xyz.g(x,y,z)$, $\lambda x.i(x)$ and $\lambda y.j(y)$ are in $\mathcal{PR}$ and if $j(y) < y$ for all $y > 0$, then the function f defined for all x,y by the schema

$$f(x,0) = h(x)$$
$$f(x,y) = g(x,y,f(i(x),j(y)))$$

is also in $\mathcal{PR}$. (*Hint:* Find a loop program for f. A further suggestion to this end: Try first to calculate $f(x,y)$ "by hand".)

    Note: This is a course-of-values recursion (simplified), where "substitutions are made for parameters" (Péter). The "parameter" here is, of course, x.

32. As in Problem 31, but with the following exception: Instead of $j(y) < y$ for $y > 0$, we are given that

    (1) The equation $j^x(y) = 0$ has an x-solution for all y.

    (2) There is a $\lambda y.b(y) \in \mathcal{PR}$, such that for all y, $\min\{x \mid j^x(y) = 0\} \leq b(y)$.

33. Show that if g and h are in $\mathcal{PR}$, then so is f, defined by the schema

$$f(0,y) = g(y)$$
$$f(x + 1,y) = h(x,f(x,y + 1)).$$

34. In regard to the function $\lambda ix.g_i(x)$ of Theorem 2.7.3, show

(a) For all $\lambda x.h(x) \in \mathcal{PR}$, there is an i such that $h(x) < g_i(x)$ for all x.

(b) Base an argument that $\lambda ix.g_i(x) \notin \mathcal{PR}$ on property (a).

35. Consider the alphabet $A = \{1,;,U,Z,S,C,c,R,r\}$. Define a subset $\mathcal{F}$ of A^* inductively, as follows:

(a) Z has *rank* 1 and is in $\mathcal{F}$.

(b) S has rank 1 and is in $\mathcal{F}$.

(c) $U1^n;1^m$ has rank n and is in $\mathcal{F}$ if $1 \le m \le n$ (where 1^n is $\underbrace{11 \ldots 1}_{n}$, as usual).

(d) If F of rank n and $G_i, i = 1, \ldots, n$, all of rank m, are in $\mathcal{F}$, then so is the string $CFG_1G_2 \ldots G_nc$, which has rank m.

(e) If H of rank m and G of rank $m + 2$ are in $\mathcal{F}$, then so is $RGHr$, which has rank $m + 1$.

(f) No string is in $\mathcal{F}$, unless it can be shown to be constructed by a finite number of applications of (a)-(e).

Show

(i) $\mathcal{F}$ can be effectively ("algorithmically") listed.

(ii) There is a 1-1 algorithmic correspondence between $\mathcal{F}$ and primitive recursive derivations.

(iii) $\mathcal{PR}$ can be effectively listed.

(iv) The set of unary functions of $\mathcal{PR}$ can be effectively listed, say, by a function $\lambda ix.\psi(i,x)$.

(v) Conclude that ψ is algorithmic but not in $\mathcal{PR}$.

36. (This is a "numerical" version of Problem 35.) Let $\langle x_0, \ldots, x_n \rangle$ stand for $\Pi_{i \le n} p_i^{x_i+1}$

Define $\tilde{\mathcal{F}}$ to be the *smallest* set satisfying (a) through (e) below:

(a) $\langle 0,1,0 \rangle$ is in $\tilde{\mathcal{F}}$ (it "stands" for Z).

(b) $\langle 0,1,1 \rangle$ is in $\tilde{\mathcal{F}}$ (it stands for S).

(c) $\langle 0,n,m \rangle$ is in $\tilde{\mathcal{F}}$ iff $1 \le m \le n$ (it stands for $U1^n;1^m$).

(d) Let $a, b_1, \ldots, b_n$ be in $\tilde{\mathcal{F}}$ so that $\exp(1,a) = n + 1$ and $\exp(1,b_i) = m + 1$ for $i = 1, \ldots, n$ ($n \ge 1$, $m \ge 1$). Then $\langle 1,m,a,b_1, \ldots, b_n \rangle \in \tilde{\mathcal{F}}$. (Note that for any $a \in \tilde{\mathcal{F}}$, $\exp(1,a) - 1$ plays the role of "rank", in the sense of Problem 35.)

(e) Let a and b be in $\tilde{\mathcal{F}}$ so that $\exp(1,a) = m + 3$, $\exp(1,b) = m + 1$, and $m \geq 1$. Then $\langle 2,m + 1,a,b \rangle \in \tilde{\mathcal{F}}$.

Show that

(i) There is a "natural" effective 1-1 correspondence between $\tilde{\mathcal{F}}$ and $\mathcal{F}$. (Thus, the problem $x \in \tilde{\mathcal{F}}$ is "algorithmically" solvable.)

(ii) There is a *unique* map $M:\tilde{\mathcal{F}} \to \mathcal{PR}$ such that

$$M(\langle 0,1,0 \rangle) = \lambda x.0$$
$$M(\langle 0,1,1 \rangle) = \lambda x.x + 1$$
$$M(\langle 0,n,m \rangle) = \lambda \vec{x}_n.x_m \text{ for } 1 \leq m \leq n$$
$$M(\langle 1,m,a,b_1, \ldots, b_n \rangle) = M(a) \circ (M(b_1), \ldots, M(b_n))$$

where $\exp(1,a) = n + 1$, $\exp(1,b_i) = m + 1$, $i = 1, \ldots, n$, $m \geq 1$, $n \geq 1$. $M(\langle 2,m + 1,a,b \rangle)$ is $\lambda x \vec{y}_m.f(x,\vec{y}_m)$, defined by

$$f(0,\vec{y}_m) = h(\vec{y}_m)$$
$$f(x + 1,\vec{y}_m) = g(x,\vec{y}_m,f(x,\vec{y}_m))$$

where $M(a) = g$ and $M(b) = h$. (*Hint:* $\tilde{\mathcal{F}}$ is inductively defined, hence it can be naturally partially ordered so that, as a poset, it satisfies the minimal condition. The result follows from Theorem 1.4.2.)

(iii) M is onto. (*Hint:* Induction with respect to $\mathcal{PR}$.)

(iv) Define Univ^m, for $m \geq 1$, by

$$\mathrm{Univ}^m(x,\vec{y}_m) = \begin{cases} M(x)(\vec{y}_m) \text{ if } x \in \tilde{\mathcal{F}} \text{ and } \exp(1,x) = m + 1 \\ 0 \text{ otherwise} \end{cases}$$

Show that Univ^m is "algorithmic" for all $m \geq 1$; but for *no* $m \geq 1$ is $\mathrm{Univ}^m \in \mathcal{PR}$. (*Hint:* Univ^m enumerates, algorithmically, the m-ary functions of $\mathcal{PR}$.)

37. Show that $\tilde{\mathcal{F}} \in \mathcal{PR}_*$.

38. Show that for each $f \in \mathcal{PR}$, $M^{-1}(f)$ is infinite. (M is as in Problem 36.)

39. Find members a,b of $\tilde{\mathcal{F}}$ such that $M(a) = \lambda xy.x \doteq 1$ and $M(b) = \lambda xy.x \doteq y$. ($M$ is as in Problem 36.)

REFERENCES

[1] Bennett, J. *On Spectra*. Ph.D. dissertation, Princeton University, 1962.

[2] Davis, M. *Computability and Unsolvability*. New York: McGraw-Hill, 1958.

[3] Dedekind, R. *Was sind und was sollen die Zahlen?* Braunschweig (1888). In English translation by W.W. Beman, *Essays on the Theory of Numbers*. Chicago: Open Court, 1901.

[4] Gödel, K. "Über formal unentscheidbare Sätze der Principia Mathematica und verwandter Systeme I". *Monatshefte für Math. und Physik* 38 (1931): 173–198. In English translation by M. Davis, *The Undecidable*. New York: Raven Press, 1965: 5–38.

[5] Grzegorczyk, A. "Some Classes of Recursive Functions". *Rozprawy Matematyczne* 4. Warsaw, 1953: 1–45.

[6] Hilbert, D., and Bernays, P. *Grundlagen der Mathematik. Vol. 1* Heidelberg: Springer-Verlag, 1934 and 1968: 333–334.

[7] Meyer, A.R., and Ritchie, D.M. *Computational Complexity and Program Structure*. IBM Research Report RC-1817, 1967.

[8] Minsky, M.L. *Computation: Finite and Infinite Machines*. Englewood Cliffs, N.J.: Prentice-Hall, 1967.

[9] Péter, R. *Recursive Functions*. New York: Academic Press, 1967.

[10] Ritchie, D.M. *Program Structure and Computational Complexity*. Ph.D. dissertation, Harvard University, 1968.

[11] Ritchie, R.W. "Classes of Predictably Computable Functions". *Trans. Amer. Math. Soc.* 106 (1963):139–173.

[12] Robinson, R.M. "Primitive Recursive Functions". *Bull. Amer. Math. Soc.* 53 (1947):925–942.

[13] Rogers, H. *Theory of Recursive Functions and Effective Computability*. New York: McGraw-Hill, 1967.

[14] Schwichtenberg, H. "Rekursionszahlen und die Grzegorczyk-Hierarchie". *Arch. Math. Logik* 12 (1969):85–97.

[15] Smullyan, R. *Theory of Formal Systems*. Annals of Mathematics Studies 47. Princeton, N.J.:Princeton University Press, 1961.

[16] Tsichritzis, D. "The Equivalence Problem of Simple Programs". *Journal of the ACM* 17 (1970):729–738.

[17] Warkentin, J.C. *Small Classes of Recursive Functions and Relations*. Dept. of Appl. Analysis and Comp. Science Research Report CSRR 2052, University of Waterloo, 1971.

CHAPTER 3

Partial Recursive and Recursive Functions

It should be suspected by now, because of the informal arguments at the end of §2.7, that any formal (i.e., purely syntactic) scheme which defines a class of "computable" total functions will be open to an obvious (diagonal) incompleteness argument.

This is not the case for formal schemes that define partial functions—that is, functions which are not necessarily defined everywhere on $\mathcal{N}^k$ (for the relevant k). Indeed, for such functions a situation like $f_i(i) = f_i(i) + 1$ is not necessarily absurd since both sides of the equality sign may very well be undefined, in which case we cannot say they are not equal!

In our desire to formalize completely the informal notions of "algorithm" on one hand and "algorithmic" or "computable" function on the other, we relax from now on the requirement that an algorithm must give an answer for every "legal" input.(*) This should not be regarded as an artificial technical

---

(*)Legal inputs, for example, for an algorithm that differentiates (formally) polynomials are only polynomials. We can always test (algorithmically) for input legality.

convenience. In everyday programming, we may all produce programs (whether intentionally or not) which may get into an "infinite loop" for some (legal) inputs. Even though some such programs may not be doing what they should, they still are completely mechanical processes and we must, justifiably, consider that they express "algorithms".

Example 1   Consider the function f defined as follows:

$$f(x) = \begin{cases} \text{the position of the first "7" in the first maximal} \\ \text{block of } x \text{ 7's in the decimal expansion of } \pi. \end{cases}$$

Given the fact that there are algorithms which can keep producing digits of π, there certainly exists an "algorithm" which for every input x computes $f(x)$, *if $f(x)$ is defined, and runs forever otherwise.//*

Example 2   Let $F(X)$ be a PL/1 function of one parameter X. Assume that X can accept only values from $\mathcal{N}$.

Consider the function g defined as follows:

$$g(Y) = \begin{cases} 0 \text{ \textbf{if} } F(X) \text{ returns } Y \text{ for some value of } X \\ \text{undefined \textbf{otherwise}}. \end{cases}$$

Allowing for the possibility that F may get into an "infinite loop" (either by design or by accident) for some (legal) values of X, we can argue that g is "algorithmic". A mechanical process for the computation of g is:

```
read Y
I = 0
repeat
    I = I + 1
    Simulate I steps of the computations F(0), ... ,
    F(I − 1); if, during the simulation, F(i) (for some i:
    0 ≤ i ≤ I − 1) returns Y, then goto L
end repeat
L: write 0.//
```

Note: The technique used in computing F above is called "dovetailing".

Example 3   If $F(X)$ of Example 2 is known to return a value for every input from $\mathcal{N}$ then an alternative way to compute g is:

```
    read Y
    I = 0
┌─► repeat
│   if F(I) = Y then goto L
│   I = I + 1
└── end repeat
    L: write 0
```

What is wrong with this algorithm if the assumptions of Example 2 hold?//

Example 4 Under the assumptions of Example 2, define the function h as follows:

$$h(Y) = \begin{cases} 0 \text{ if } F(X) \text{ returns } Y \text{ for some value of } X \\ 1 \text{ otherwise} \end{cases}$$

We shall see that for some choices of $F(X)$ there is *no existing* formal counterpart of the notion "algorithmic function" in which h is definable.//

In order to enlarge the class of primitive recursive functions with the addition of partial functions, we shall introduce an operator which can lead to such functions. This will be obtained from the Bounded Search operator $(\mu y)_{<z}$ by removing the bound "$<z$". Intuitively, this amounts to adding the instruction **Loop while** $(X \neq 0)$ to our programming formalism of §2.7. In the next paragraph, we present the formal realization of this idea.

§3.1 DEFINITIONS AND SIMPLE EXAMPLES

Definition 1 *Unbounded Search or μ-Operator.* Let $\lambda y \vec{x}.g(y,\vec{x})$ be a *partial* function.

$$(\mu y)g(y,\vec{x}) \text{ stands for } \begin{cases} \min\{y \mid g(y,\vec{x}) = 0 \ \& \ (\forall z)_{<y} \, g(z,\vec{x})\downarrow \} \\ \uparrow \text{ if the } \textit{min} \text{ does not exist.}// \end{cases}$$

Note: In much of the literature (notably Davis [3] and Rogers [13]) μ is defined on *total* functions (see, however, Brainerd and Landweber [2], Manna [8], and Kleene [6]) according to the following definition.

Definition 2   *Unbounded search on **total** functions.* Let $\lambda y\vec{x}.g(y,\vec{x})$ be a *total* function.

$$(\tilde{\mu}y)g(y,\vec{x}) \text{ stands for } \begin{cases} \min\{y\,|\,g(y,\vec{x}) = 0\} \\ \uparrow \text{ if the } min \text{ does not exist.}// \end{cases}$$

Note: (1) μ and $\tilde{\mu}$ coincide on *total* functions.

(2) In order to apply $\tilde{\mu}$ on a function $\lambda y\vec{x}.g(y,\vec{x})$ "given" by a finite description (say, a "program") we must know that this program never gets into an infinite loop. Unfortunately, this cannot be tested by another "program", as we shall eventually see.

(3) If $G(Y,\vec{X})$ is a PL/1 procedure which computes $\lambda y\vec{x}.g(y,\vec{x})$ then the following program computes $\lambda\vec{x}.(\mu y)g(y,\vec{x})$

```
    Y = 0
┌─► do while (G(Y,X⃗) ≠ 0)
│     Y = Y + 1
└── end /* Y holds (μy)g(y,x⃗) */
```

Because of (1), the above program also computes $\lambda\vec{x}.(\tilde{\mu}y)g(y,\vec{x})$ *if* g is total. Note, however, that if, say, $g(0,\vec{x})\uparrow$ and $g(1,\vec{x}) = 0$, then the above program does *not* compute $\lambda\vec{x}.(\tilde{\mu}y)g(y,\vec{x})$, because it is in an infinite loop for $y = 0$. In fact, we shall show eventually that there is a PL/1 programmable nontotal function $\lambda xy.\psi(x,y)$ such that $\lambda x.(\tilde{\mu}y)\,\psi(x,y)$ is *not* PL/1 programmable.(*)

In summary, we can say that μ is a syntactic operator—that is, it can be applied to a function, given by a "finite description", as soon as the finite description is found to be syntactically correct, whereas in order to apply $\tilde{\mu}$ we must probe into the semantics of the finite description. Further, by (3) above, μ has an intuitively simple programming language counterpart whereas $\tilde{\mu}$ (in general) has none.

Because of the above, we shall present the definition of the partial recursive functions in terms of μ.

Definition 3   The class of *partial recursive functions, $\mathcal{P}$,* is the *closure* of $\{\lambda x.x + 1, \lambda x.x \doteq 1\}$ under *substitution, pure iteration,* and *unbounded search* (μ). The class of *recursive functions, $\mathcal{R}$,* is the class of all *total* functions of $\mathcal{P}.//$

(*)We resist the temptation to say that "there exists an 'algorithmic' nontotal function $\lambda xy.\psi(x,y)$ such that $\lambda x.(\tilde{\mu}y)\psi(x,y)$ is not algorithmic", hence, our reference to PL/1. If "algorithmic", though, is to be taken as the layman's term for "partial recursive", then the statement in quotes ". . ." can be rigorously proved.

Theorem 1 $\mathcal{R}$ and $\mathcal{P}$ are closed under primitive recursion.

    Proof Clearly $\mathcal{R}$ is closed under *pure iteration*, since $\lambda x.f(x)\in\mathcal{R}$ $\Rightarrow \lambda xy.f^y(x)\in\mathcal{P}$, but $\lambda xy.f^y(x)$ is *total*. It suffices to show (Theorem 2.6.1) that $\mathcal{R}$ (hence, $\mathcal{P}$) contains pairing functions with their projections. As in the proof for Theorem 2.6.2, we conclude that $\lambda x.2^x$, $\lambda xy.x \doteq y$ and $J = \lambda xy.2^{x+y+2} + 2^{y+1}$ are in $\mathcal{R}$. Similarly, $\lambda z.\lfloor \log_2 z \rfloor \in \mathcal{R}$ since $\lfloor \log_2 z \rfloor = (\mu y)(z + 1 \doteq 2^{y+1})$.

    (Note: $\lfloor \log_2 0 \rfloor$ is taken to be 0.) Hence, the projections K and L of J are also in $\mathcal{R}$.//

Corollary 1 $\mathcal{PR} \subset \mathcal{R} \subsetneq \mathcal{P}$.

    Proof $\mathcal{PR} \subset \mathcal{R}$ is immediate since $\lambda x.x \in \mathcal{R}(*)$ and because of Theorem 1 (see Theorem 2.1.3). Now let $\psi = \lambda xy.x + 1$. Then $\lambda x.(\mu y)\psi(x,y) \in \mathcal{P} - \mathcal{R}$.//

Corollary 2 $\mathcal{R}$ is closed under $(\mu y)_{<z}$.

    Proof Let $\lambda \vec{x}y.f(\vec{x},y) \in \mathcal{R}$. Then $(\mu y)_{<z} f(\vec{x},y) = (\mu y)|y - z|\cdot f(\vec{x},y)$. Since $\mathcal{PR} \subset \mathcal{R}$, $\lambda yz.|y - z|$ and $\lambda xy.xy \in \mathcal{R}$.//

    It will soon be seen that $\mathcal{PR} \subsetneq \mathcal{R}$. To this end, we shall need some more machinery.

§3.2 RECURSIVE PREDICATES

Definition 1   A predicate $P(\vec{x})$ is *recursive* (in symbols, $P(\vec{x})\in\mathcal{R}_*$) iff its characteristic function c_p defined by $c_p(\vec{x}) = $ **if** $P(\vec{x})$ **then** 0 **else** 1 is recursive.//

    Note: Intuitively, $P(\vec{x})$ is recursive iff there exists an "algorithm" (that for the "algorithmic" c_p) which decides membership in P.

Theorem 1 $P(\vec{x}) \in \mathcal{R}_*$ iff there is a $\lambda\vec{x}.f(\vec{x}) \in \mathcal{R}$ such that $P(\vec{x}) \equiv f(\vec{x}) = 0$.

    Proof Same as that of Theorem 2.2.4.//

Theorem 2 $\mathcal{R}_*$ is closed under the Boolean operations.

(*)$\lambda x.x = \lambda x.(x+1) \doteq 1.$

Proof See Theorem 2.2.5.//

Example 1   The relations of $\mathcal{PR}$ are in $\mathcal{R}_*$. In particular, $x \geq y, x > y, x = y, x \neq y$ are all in $\mathcal{R}_*$.//

Theorem 3   Let $P(\vec{x}_n) \in \mathcal{R}_*$ and $\lambda \vec{y}.f(\vec{y}) \in \mathcal{R}$.
Then $P(x_1, \ldots, x_{i-1}, f(\vec{y}), x_{i+1}, \ldots, x_n) \in \mathcal{R}_*$.

Proof Problem 2.//

Note: A trivial corollary of Theorem 3 is that $\lambda \vec{x}.f(\vec{x}) \in \mathcal{R}$ implies $\lambda xy.y = f(\vec{x}) \in \mathcal{R}_*$. Note that the converse is true, for $f(\vec{x}) = (\mu y)|y - f(\vec{x})| = (\mu y) \, c_f(y,\vec{x})$, where $c_f \in \mathcal{R}$ is the characteristic function of $\lambda xy.y = f(\vec{x})$. Contrast this with the case for functions in $\mathcal{PR}$ (see the Note after the proof of Theorem 2.2.6 and Corollary of Theorem 3.3.2).

Definition 2   Let $P(y,\vec{x})$ be a predicate. $(\mu y)P(y,\vec{x})$ denotes $(\mu y)c_p(y,\vec{x})$, where c_p is the characteristic function of $P(y,\vec{x})$.//

Theorem 4 $\mathcal{R}_*$ is closed under $(\exists y)_{\leq z}$ and $(\forall y)_{\leq z}$.

Proof Let $P(y,\vec{x}) \in \mathcal{R}_*$. Then $(\forall y)_{\leq z} P(y,\vec{x}) \equiv z+1 = (\mu y)(y = z+1 \vee \neg P(y,\vec{x}))$.//

Note: Another way of putting it is $z+1 = (\mu y)_{\leq z} \neg P(y, \vec{x})$, since $\mathcal{R}$ is closed under $(\mu y)_{\leq z}$ (Corollary 2 of Theorem 3.1.1). Also, since $\mathcal{R}$ is closed under primitive recursion, then it is also closed under $\Sigma_{\leq z}$ and $\Pi_{\leq z}$ (Lemma 2.2.2) and hence Theorem 4 can also be proved exactly as Theorem 2.2.7. However, the above proof will be useful later (Lemma 3.9.5).

Theorem 5 $\mathcal{P}$ is closed under definition by cases.

Proof See proof of Theorem 2.2.8.//

§3.3 ACKERMANN'S FUNCTION

We shall now establish the proper inclusion $\mathcal{PR} \subsetneq \mathcal{R}$.

Definition 1   *A version of Ackermann's function.*
In the sequel A_n shall denote the nth function of the sequence $(A_n)_{n \geq 0}$. $A_n, n \geq 0$ is defined as follows for all x:

$$A_0(x) = x+2$$

for $n \geq 0$

$$A_{n+1}(x) = A_n^x(2).//$$

Note: By the usual convention, for any function $\lambda x.h(x)$ h^0 is $\lambda x.x$. Hence $A_{n+1}(0) = 2$ for $n \geq 0$, thus $A_n(0) = 2$ for $n \geq 0$.

Lemma 1 $\lambda x.A_n(x) \in \mathcal{PR}$ for $n \geq 0$.

Proof Induction on n and closure of $\mathcal{PR}$ under pure iteration and substitution.//

Lemma 2   For all n,x, $A_n(x) > x+1$.

> *Proof* $n = 0$: For all x, $A_0(x) = x+2 > x+1$.
> Assume that for $n = k$: $A_k(x) > x + 1$ for all x.   (1)

> Let $n = k + 1$: For $x = 0$: $A_{k+1}(0) = 2 > 0+1$.

> Assume that for $x = m$: $A_{k+1}(m) > m + 1$

Let $x = m + 1$:

$$A_{k+1}(m+1) = A_k^{m+1}(2) = A_k(A_k^m(2)) = A_k(A_{k+1}(m))$$

By (1) (induction hypothesis on n)

$$A_k(A_{k+1}(m)) > A_{k+1}(m)+1$$

By induction hypothesis on x,

$$A_{k+1}(m) > m+1$$

Thus,

$$A_{k+1}(m+1) > m+2.//$$

Lemma 3 $\lambda k.A_n^k(x)$ is strictly increasing.

> *Proof* $A_n^{k+1}(x) = A_n(A_n^k(x)) > A_n^k(x)+1$ (Lemma 2).//

Lemma 4 $\lambda n.A_n(x+1)$ is strictly increasing.

Proof Let $n \geq 1$. $A_{n+1}(x+1) = A_n(A_{n+1}(x)) = A_{n-1}^{A_{n+1}(x)}(2) > A_{n-1}^{x+1}(2)$ by Lemmas 2 and 3. But $A_{n-1}^{x+1}(2) = A_n(x+1)$.
Finally, $A_1(x+1) = A_0^{x+1}(2) = 2(x+2) > x+3 = A_0(x+1).//$

Note: Since $A_n(0) = 2, n \geq 0, \lambda n.A_n(x)$ is increasing.

Lemma 5 $\lambda x.A_n(x)$ is strictly increasing.

Proof $A_{n+1}(x+1) = A_n(A_{n+1}(x)) > A_{n+1}(x)+1$ (Lemma 2).
Finally, $A_0(x) = x+2$, hence the result is true for all $n.//$

Definition 2   A relation $R(x)$ is true *almost everywhere* (a.e.) iff there is an x_0 such that $x > x_0$ implies $R(x).//$

Lemma 6   For any constant l, $A_{n+1}(x) > x + l$ a.e.

Proof $A_1(x) = A_0^x(2) = 2(x+1)$. For $x_0 = l-2$, we have that $x > x_0 \Rightarrow 2(x+1) > x+l$; i.e., $A_1(x) > x+l$.
Thus, $x > x_0 \Rightarrow A_{n+1}(x) > x+l$, since $A_{n+1}(x) \geq A_1(x)$ for all x ($A_{n+1}(x) = A_1(x)$ only if $n = 0$ or $x = 0$).$//$

Lemma 7 $A_{n+1}(x) > A_n^l(x)$ a.e.

Proof For $l = 1$, Lemma 7 follows from Lemma 4. Let $l > 1, n \geq 1$, and $x > l$.
By induction on l, it is easy to see that

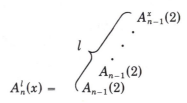

Thus,

$$A_{n+1}(x) = A_n^x(2) = \quad \underbrace{\begin{array}{l} A_{n-1}^2(2) \\ \cdot \\ \cdot \\ \cdot \\ A_{n-1}(2) \\ A_{n-1}(2) \end{array}}_{x}$$

By Lemma 3 it suffices to show that

$$x - l \left\{ \begin{array}{l} A_{n-1}^2(2) \\ \quad \cdot \\ \quad \cdot \\ \quad \cdot \\ A_{n-1}(2) \end{array} \right. > x \ \text{a.e.}$$

or

$$A_{n+1}(x - l) > x \ \text{a.e.}$$

which is true by Lemma 6.

Finally $A_1(x) = 2(x+1)$ and $A_0^l(x) = x + 2l$, thus Lemma 7 is true also for $n = 0.//$

Lemma 8   For $n \geq 0$, $A_{n+1}(x+y) \geq A_n^x(y)$ for all x, y.

Proof We know (Lemma 4) $A_{n+1}(y) \geq A_n(y)$, for $n \geq 0$, hence

$$A_n^y(2) \geq A_n(y) \tag{1}$$

By Lemma 5, $\lambda z.A_n^{x \dot- 1}(z)$ is increasing. Apply this function at both sides of (1):

$$A_n^{x \dot- 1 + y}(2) \geq A_n^x(y).$$

But $A_{n+1}(x+y) = A_n^{x+y}(2) \geq A_n^{x \dot- 1 + y}(2)$ by Lemma 3.//

Theorem 1 $\lambda \vec{x}.g(\vec{x}) \in \mathcal{PR}$ implies that there are m and k such that $g(\vec{x}) \leq A_m^k(\max(\vec{x}))$, for all $\vec{x}$.

Proof We shall use the characterization of $\mathcal{PR}$ given in Theorem 2.6.2. The proof is by induction with respect to $\mathcal{PR}$.

(1)   *Initial functions:* $x + 1 \leq A_0(x)$ and $x \dot- 1 \leq A_0(x)$.

(2)   The property (of A_n-majorization) *propagates* with *substitution:* Let $\lambda y\vec{x}.g(y,\vec{x})$ and $\lambda \vec{z}.h(\vec{z})$ in $\mathcal{PR}$ have the property. Set $f = \lambda \vec{z}\vec{x}.g(h(\vec{z}), \vec{x})$.

Let

$$g(y,\vec{x}) \leq A_m^k(\max(y,\vec{x}))$$

and

$$h(\vec{z}) \le A_n^l(\max(\vec{z}))$$

Then

$$g(h(\vec{z}), \vec{x}) \le A_m^k(\max(h(\vec{z}), \vec{x})) \le \text{(Lemma 5)}$$
$$A_m^k(\max(A_n^l(\max(\vec{z})), \vec{x})) \le \text{(Lemmas 2, 5)}$$
$$A_m^k(A_n^l(\max(\vec{z},\vec{x}))) \le \text{(Lemmas 4, 5)}$$
$$A_{\max(m,n)}^{k+l}(\max(\vec{z}, \vec{x}))$$

The verification that the property propagates with the remaining cases of substitution is left as an exercise.

(3) The property propagates with *bounded search*. Let $\lambda y \vec{x}.g(y,\vec{x}) \in \mathcal{PR}$ have the property. Then $(\mu y)_{\le z} g(y,\vec{x}) \le z + 1 < A_0(\max(z,\vec{x}))$.

Note: That g had the property proved irrelevant in this case.

Finally,

(4) the property propagates with *pure iteration*. So let $\lambda x.f(x) \in \mathcal{PR}$ and $f(x) \le A_m^k(x)$ for some k and m. By Lemma 5, $f^y(x) \le A_m^{ky}(x)$. By Lemma 8, $A_m^{ky}(x) \le A_{m+1}(ky+x)$. But $A_{m+1}(ky+x) \le A_{m+1}(2\max(ky,x)) \le A_{m} \circ_1 A_1(\max(ky,x)) \le A_{m+1}^2(\max(ky,x))$ [by Lemmas 4, 5]. Now, $\max(x,ky) \le k\max(x,y) \le A_1^l(\max(x,y))$ for some l (why?). Hence, $f^y(x) \le A_{m+1}^{l+2}(\max(x, y))$.//

Theorem 2 $\lambda xy.A_x(y) \in \mathcal{R} - \mathcal{PR}$.

Proof We show first that $\lambda xy.A_x(y) \notin \mathcal{PR}$. Let $\lambda xy.A_x(y) \in \mathcal{PR}$. Then also $\lambda x.A_x(x) \in \mathcal{PR}$. Hence, for some m and k, $A_x(x) \le A_m^k(x)$, hence (Lemma 7) $A_x(x) < A_{m+1}(x)$ a.e. By Lemma 4 $A_x(x) > A_{m+1}(x)$ a.e., hence $A_x(x) < A_x(x)$ a.e., a contradiction. Thus, $\lambda xy.A_x(y) \notin \mathcal{PR}$.

We now show that $\lambda xy.A_x(y) \in \mathcal{R}$. We have seen that $\lambda \vec{x}y.y = f(\vec{x}) \in \mathcal{R}_*$ implies $f \in \mathcal{R}$ (see note after Theorem 3.2.3).

Thus we shall proceed by showing $\lambda xyz.z = A_x(y) \in \mathcal{R}_*$. (Indeed, it is in $\mathcal{PR}_*$ and even "lower". See Chapter 13.)

To this end, let J, K, L be a particular primitive recursive pairing function and its first and second projections. For notational convenience we set $u_{i,j} = \exp(J(i, j), u)$ (for the definition of "exp", see Example 2.3.1). Clearly, $\lambda iju.u_{i,j} \in \mathcal{PR}$. Thus,

$$z = A_x(y) \equiv (\exists u)\{(\forall i)_{\le u}[u_{0,i} \ne 0 \Rightarrow u_{0,i} = i + 3] \;\&$$
$$(\forall j)_{<x}[u_{j+1,0} \ne 0 \Rightarrow u_{j+1,0} = 3] \;\&\; (\forall k)_{<x}(\forall l)_{\le u}[u_{k+1,l+1} \ne 0 \Rightarrow$$
$$(u_{k+1,l+1} = u_{k,u_{k+1,l} \dot- 1} \;\&\; u_{k+1,l} \ne 0)] \;\&\; u_{x,y} = z + 1\}.$$

Intuitively, u "codes" a "computation" for $z = A_x(y)$. $u_{i,j} \ne 0$ (i.e., $p_{J(i,j)}\,|\,u$) if $A_i(j)$ is needed as a stepping stone for the computation; in that case $u_{i,j} = A_i(j) + 1$. (This assertion can be easily proved by double induction with respect to i and j; see Problem 4.)

It is easy to see (Problem 4) that if $A_i(j)$ is needed for the computation of $A_x(y)$, then $A_i(j) \le A_x(y)$; thus for all i,j, $u_{i,j}$ is needed in the computation implies $u_{i,j} \le z + 1$.

Thus we can replace $(\exists u)$ in the definition of $z = A_x(y)$ by $(\exists u)_{\le \Pi_{\substack{i \le x \\ j \le z}} p_{J(i,j)}^{z+1}}$, where p_n is the nth prime. Clearly, $\lambda xyz.z = A_x(y) \in \mathcal{PR}_*.//$

Corollary   There exists a function $\lambda x.f(x)$ such that $f \in \mathcal{R} - \mathcal{PR}$ but $\lambda xy.y = f(x) \in \mathcal{PR}_*$.

   Proof   Take $f = \lambda x.A_x(x).//$

Are there predicates in $\mathcal{R}_* - \mathcal{PR}_*$? In view of the above, this is a reasonable question. We shall answer this affirmatively in §3.5.

§3.4   INDUCTIVELY DEFINABLE CLASSES OF FUNCTIONS

We say that a class of functions which is defined as the *closure* of a set of initial functions under a set of (functional) operations is called *inductively defined*. The question arises as to whether a class of functions not inductively defined can be redefined inductively. Most of this paragraph incorporates ideas of Grzegorczyk [4].

Definition 1   A class of functions $\mathcal{C}$ is *inductively definable* if there exists a set of functions $I \subset \mathcal{C}$ (the "initial functions") and a finite set of operators $\mathcal{O}$ such that $\mathcal{C}$ is the *closure* of I under the operators of $\mathcal{O}.//$

Definition 2   Let $\mathcal{C}$ be a class of functions; $\mathcal{C}^{(n)}$, $n \ge 1$, denotes the set of n-ary functions of $\mathcal{C}.//$

We shall see that $\mathcal{P}^{(1)}$ and $\mathcal{PR}^{(1)}$ are inductively definable. It will soon be apparent (§3.5) that inductive definability offers more than just the possibility of doing induction "with respect to the system".

In the following discussion, we shall assume a fixed (but otherwise unspecified) sequence $(I_n)_{n \geq 1}$ of primitive recursive coding functions, not necessarily onto, along with their projections $((\Pi_i^n)_{1 \leq i \leq n})_{n \geq 1}$. As in Definition 2.4.4, we shall write $\langle \vec{x}_n \rangle$ instead of $I_n(\vec{x}_n)$ (in particular $\langle x \rangle = x$).

Lemma 1 $f \in \mathcal{P}^{(1)}$ iff $f = \lambda x.\psi(\Pi_1^n(x), \ldots, \Pi_n^n(x))$ for some $\psi \in \mathcal{P}^{(n)}, n \geq 1$.

> ***Proof*** Let $\psi \in \mathcal{P}^{(n)}$. Then $f = \lambda x.\psi(\Pi_1^n(x), \ldots, \Pi_n^n(x)) \in \mathcal{P}^{(1)}$. Let $f \in \mathcal{P}^{(1)}$, then $\psi = \lambda x.f(\langle x \rangle) \in \mathcal{P}^{(1)}$ and $f = \lambda x.\psi(\Pi_1^1(x)).//$

We can now show that $\mathcal{P}^{(1)}$ is inductively definable.

Theorem 1   Let $\tilde{\mathcal{P}}$ be the closure of $\{\lambda x.x + 1, \lambda x.x \doteq 1, \lambda x.\langle x,x \rangle, \lambda x.\langle 0,x \rangle, \lambda x.\langle x,0 \rangle, \lambda x.K(x), \lambda x.L(x)\}$ under the following operations:

$$O_1: (f,g) \longmapsto f \circ g$$
$$O_2: (f,g) \longmapsto \lambda x.\langle f(x), g(x) \rangle$$
$$O_3: f \longmapsto \lambda x.f^{K(x)}(L(x))$$
$$O_4: f \longmapsto \lambda x.(\mu y) f(\langle y,x \rangle).$$

Then $\tilde{\mathcal{P}} = \mathcal{P}^{(1)}$.

> ***Proof*** The proof will be subdivided in a series of lemmas.

Lemma 2 $\mathcal{P} \subset \mathcal{P}^{(1)}$.

> ***Proof of Lemma 2***   Indeed,
>
> (i) $\tilde{\mathcal{P}} \subset \mathcal{P}$ (trivial)
>
> (ii)   All functions of $\tilde{\mathcal{P}}$ are unary (induction with respect to $\tilde{\mathcal{P}}$ proves both (i) and (ii)).// *Lemma 2*

Lemma 3   For $i = 1, \ldots, n, \lambda x.f_i(x) \in \tilde{\mathcal{P}}$ implies $\lambda x.\langle f_1(x), \ldots, f_n(x) \rangle \in \tilde{\mathcal{P}}$.

> ***Proof of Lemma 3*** For $n = 1, \langle f_1(x) \rangle = f_1(x)$; result immediate. If the result is true for $n = k$, consider the case $n = k + 1$:
>
> $$\langle f_1(x), \ldots, f_k(x), f_{k+1}(x) \rangle = \langle \langle f_1(x), \ldots, f_k(x) \rangle, f_{k+1}(x) \rangle.$$
>
> The result for $n = k + 1$ follows from the case $n = k$ and O_2.//*Lemma 3*

Lemma 4 $\mathcal{P}^{(1)} \subset \tilde{\mathcal{P}}$.

Proof of Lemma 4 By Lemma 1, we shall use induction with respect to $\mathcal{P}$. The property, (A), that we shall prove true for every function in $\mathcal{P}$, is: "$f(\Pi_1^n(x), \ldots, \Pi_n^n(x)) \in \tilde{\mathcal{P}}$, for all n-ary functions, $n \geq 1$".

(a) *Basis*. Trivially, all the initial functions of $\mathcal{P}$ (i.e., $\lambda x.x + 1$ and $\lambda x.x \dot- 1$) have the property (recall that $\Pi_1^1 = \lambda x.x$).

(b) *The property (A) propagates with substitution*:

Case 1. Let $\lambda \vec{x}_n z.f(\vec{x}_n, z)$ and $\lambda \vec{y}_m.g(\vec{y}_m)$ have the property. Consider $h = \lambda \vec{x}_n \vec{y}_m.f(\vec{x}_n, g(\vec{y}_m))$. Now, $h(\Pi_1^{n+m}(x), \ldots, \Pi_{n+m}^{n+m}(x)) = f(\Pi_1^{n+m}(x), \ldots, \Pi_n^{n+m}(x), g(\Pi_{n+1}^{n+m}(x), \ldots, \Pi_{n+m}^{n+m}(x)))$. By assumption, $\lambda x.g(\Pi_1^m(x), \ldots, \Pi_m^m(x)) \in \tilde{\mathcal{P}}$. Also, for $1 \leq i \leq m$, $\Pi_{n+i}^{n+m}(x) = L \circ K^{m-i}(x) \ [= K^{m-i}(x) \text{ if } n + i = 1]$, hence by O_1, $\Pi_{n+i}^{n+m} \in \tilde{\mathcal{P}}$.

Thus, (Lemma 3), $\lambda x.\langle \Pi_{n+1}^{n+m}(x), \ldots, \Pi_{n+m}^{n+m}(x) \rangle \in \tilde{\mathcal{P}}$ and hence, by O_1 and induction hypothesis, $\lambda x.g(\Pi_{n+1}^{n+m}(x), \ldots, \Pi_{n+m}^{n+m}(x)) = \lambda x.g(\Pi_1^m(\langle \Pi_{n+1}^{n+m}(x), \ldots, \Pi_{n+m}^{n+m}(x) \rangle), \ldots, \Pi_m^m(\langle \Pi_{n+1}^{n+m}(x), \ldots, \Pi_{n+m}^{n+m}(x) \rangle)) \in \tilde{\mathcal{P}}$. Set $t \overset{\text{def}}{=} \lambda x.g(\Pi_{n+1}^{n+m}(x), \ldots, \Pi_{n+m}^{n+m}(x))$. By an argument similar to the above, $\lambda x.f(\Pi_1^{n+1}(x), \ldots, \Pi_{n+1}^{n+1}(x)) \in \tilde{\mathcal{P}}$ (assumption on f) implies that $\lambda x.f(\Pi_1^{n+m}(x), \ldots, \Pi_n^{n+m}(x), t(x))$ is in $\tilde{\mathcal{P}}$.

Note: If $n = 0$ and $m = 1$, then $h = \lambda x.f(g(x))$ and $h(\Pi_1^1(x)) = h(x) = f(g(x)) = f(\Pi_1^1(g(\Pi_1^1(x))))$ and the result is trivial in this case, using O_1.

Case 2. Let $\lambda \vec{x}_n y.f(\vec{x}_n, y)$ have property (A). Show that $h = \lambda \vec{x}_n.f(\vec{x}_n, 0)$ also has the property. Now, if $n = 0$ the result is trivial, so let $n \geq 1$. $h(\Pi_1^n(x), \ldots, \Pi_n^n(x)) = f(\Pi_1^n(x), \ldots, \Pi_n^n(x), 0) = f(\Pi_1^{n+1}(\langle \Pi_1^n(x), \ldots, \Pi_n^n(x), 0 \rangle), \ldots, \Pi_{n+1}^{n+1}(\langle \Pi_1^n(x), \ldots, \Pi_n^n(x), 0 \rangle))$. By assumption on f and O_1, it suffices to show that $\lambda x.\langle \Pi_1^n(x), \ldots, \Pi_n^n(x), 0 \rangle = \lambda x.\langle \Pi_1^n(x), \ldots, \Pi_n^n(x), K(\langle 0, x \rangle) \rangle$ is in $\tilde{\mathcal{P}}$; this is true by Lemma 3, O_1 and the fact that $\Pi_1^n = K^{n-1}$, $\Pi_i^n = L \circ K^{n-i}$, $2 \leq i \leq n$.

Case 3. Let $\lambda xy.f(x, y)$ have property (A). Show that $h = \lambda x.f(x, x)$ also has the property.

Note: Without loss of generality, we consider only the case of two variables for the rule of identification of variables.

Now, $h(\Pi_1^1(x)) = h(x) = f(x, x) = f(\Pi_1^2(\langle x, x \rangle), \Pi_2^2(\langle x, x \rangle))$ and the result $(\lambda x.h(\Pi_1^1(x)) \in \tilde{\mathcal{P}})$ follows from the assumption on f, and O_1.

The remaining cases of substitution are left as an exercise.

(c) *The property (A) propagates with pure iteration*: Let $\lambda x.f(x)$ have the property. Show that $h = \lambda xy.f^y(x)$ has the property. By assumption, $f \in \tilde{\mathcal{P}}$ (why?).

Now,

$$h(\Pi_1^2(x), \Pi_2^2(x)) = f^{\Pi_2^2(x)}(\Pi_1^2(x)) = f^{K(\langle \Pi_2^2(x), \Pi_1^2(x) \rangle)}(L(\langle \Pi_2^2(x), \Pi_1^2(x) \rangle))$$

and the result follows, since $f \in \tilde{\mathcal{P}}$, by O_3, O_2, O_1 and the facts $\Pi_1^2 = K$, $\Pi_2^2 = L$.

(d) *The property (A) propagates with the μ-operation:* Let $\lambda \vec{x}_n y . f(\vec{x}_n, y)$ have property (A).

Consider $h = \lambda \vec{x}_n . (\mu y) f(\vec{x}_n, y)$. Set $k = \lambda x . f(\Pi_1^{n+1}(x), \ldots, \Pi_{n+1}^{n+1}(x)); k \in \tilde{\mathcal{P}}$ by assumption on f.

Since $\Pi_i^n \in \mathcal{P}$ for $i = 1, \ldots, n$, then, by O_1 and Lemma 3, $\lambda x . \langle \Pi_1^n(L(x)), \ldots, \Pi_n^n(L(x)), K(x) \rangle$ is in $\tilde{\mathcal{P}}$.

Set $w = \lambda x . k (\langle \Pi_1^n(L(x)), \ldots, \Pi_n^n(L(x)), K(x) \rangle)$. By O_1, $w \in \tilde{\mathcal{P}}$.

Clearly,

$$h(\Pi_1^n(x), \ldots, \Pi_n^n(x)) = (\mu y) f(\Pi_1^n(x), \ldots, \Pi_n^n(x), y) = (\mu y) w(\langle y, x \rangle),$$

thus h has property (A) by O_4.

We have proved by induction that $f \in \mathcal{P} \Rightarrow f$ has property (A). In particular every $f \in \mathcal{P}^{(1)}$ has property (A); that is, $f \in \mathcal{P}^{(1)} \Rightarrow f \in \tilde{\mathcal{P}}.//$ *Lemma 4*

Combining Lemmas 2 and 4, we obtain Theorem 1.// *Theorem 1.*

Corollary $\mathcal{P}^{(1)}$ is inductively definable.

Proof This is a restatement of Theorem 1 without the details.//

Theorem 2   Let $\tilde{\mathcal{PR}}$ be the closure of $\{\lambda x . x + 1, \lambda x . x \doteq 1, \lambda x . \langle x, x \rangle, \lambda x . \langle 0, x \rangle, \lambda x . \langle x, 0 \rangle, \lambda x . K(x), \lambda x . L(x)\}$ under the following operations:

$$O_1 \colon f, g \mapsto f \circ g$$

$$O_2 \colon f, g \mapsto \lambda x . \langle f(x), g(x) \rangle$$

$$O_3 \colon f \mapsto \lambda x . f^{K(x)}(L(x))$$

$$O_5 \colon f \mapsto \lambda x . (\mu y)_{\leq L(x)} f(\langle y, K(x) \rangle)$$

Then $\tilde{\mathcal{PR}} = \mathcal{PR}^{(1)}$.

Proof It is entirely analogous to that of Theorem 1. The only (trivial) difference is the treatment of O_5, instead of O_4 of Theorem 1. (See Problem 11.)//

§3.5 UNIVERSAL FUNCTIONS

This paragraph uses ideas of Grzegorczyk [4]. See also Péter ([9], §11)

Definition 1   Let $\mathcal{C}$ be a class of functions. The function $\lambda n \vec{x}_m . f(n, \vec{x}_m)$ is *universal* for $\mathcal{C}^{(m)}$ iff the mapping

$$\mathcal{N} \xrightarrow{n \mapsto \lambda \vec{x}_m . f(n, \vec{x}_m)} \mathcal{C}^{(m)}$$

is *onto* and *total*.//

Theorem 1   There exists a universal function $\lambda nx.\psi(x,x)$ for $\mathcal{P}^{(1)}$.

Proof Consider the function ψ defined as follows:

$$\psi(0,x) = x + 1$$
$$\psi(1,x) = x \dot- 1$$
$$\psi(2,x) = \langle x,x \rangle$$
$$\psi(3,x) = \langle 0,x \rangle$$
$$\psi(4,x) = \langle x,0 \rangle$$
$$\psi(5,x) = K(x)$$
$$\psi(6,x) = L(x)$$

and for $n \geq 6$

$$\psi(n+1, x) = \begin{cases} \psi(\Pi_1^3(n), \psi(\Pi_2^3(n), x)) & \text{if } \Pi_3^3(n) = 0 \\ \langle \psi(\Pi_1^3(n), x), \psi(\Pi_2^3(n), x) \rangle & \text{if } \Pi_3^3(n) = 1 \\ \psi^{K(x)}(\Pi_1^3(n), L(x)) \ (*) & \text{if } \Pi_3^3(n) = 2 \\ (\mu y)\psi(\Pi_1^3(n), \langle y,x \rangle) & \text{if } \Pi_3^3(n) > 2 \end{cases}$$

Note: If we had used $\tilde{\mu}$ instead of μ, the condition for the fourth case of the definition by cases above should have been "$\Pi_3^3(n) > 2$ & $\lambda xy.\psi(\Pi_1^3(n), \langle y,x \rangle)$ is *total*". The predicate in quotes is unfortunately nonrecursive.

We proceed to show that ψ is universal.

Observe that:

(1) The mapping $n \longmapsto \lambda x.\psi(n,x)$ is into $\mathcal{P}^{(1)}$ since it is easily shown by induction on n that for all n, $\lambda x.\psi(n,x)$ is partial recursive ($\in \mathcal{P}$) and, trivially, unary.

(2) To show that it is *onto*, we argue by induction with respect to $\mathcal{P}^{(1)}$ (recall Theorem 3.4.1).

Property (A) is: "f is in the *range* of the map $n \longmapsto \lambda x.\psi(n,x)$"

(a) *Basis.* Trivially, the initial functions of $\mathcal{P}^{(1)}$ have the property.

(b) *The property propagates with* O_1: Let $f = \lambda x.\psi(i,x)$ and $g = \lambda x.\psi(j,x)$. Then $f \circ g = \lambda x.\psi(\langle i,j,0 \rangle + 1, x)$.

(c) *The property propagates with* O_2: Let f and g be as above. Then $\lambda x.\langle f(x), g(x) \rangle = \lambda x.\psi(\langle i,j,1 \rangle + 1, x)$.

(*)If $h = \lambda x.\psi(\Pi_1^3(n), x)$, then $\psi^{K(x)}(\Pi_1^3(n), L(x)) = h^{K(x)}(L(x))$.

(d) *The property propagates with* O_3: Let f be as above. Then $\lambda x. f^{K(x)}(L(x)) = \lambda x. \psi(\langle i,0,2 \rangle + 1, x)$.

(e) *The property propagates with* O_4: Let f be as above. Then $\lambda x. (\mu y) f(\langle y,x \rangle) = \lambda x. \psi(\langle i,0,3 \rangle + 1, x). //$

Theorem 2   There exists a universal function $\lambda nx. \xi(n,x)$ for $\mathcal{PR}^{(1)}$.

Proof Consider the function ξ defined below:

$$\xi(0,x) = x + 1$$
$$\xi(1,x) = x \doteq 1$$
$$\xi(2,x) = \langle x,x \rangle$$
$$\xi(3,x) = \langle 0,x \rangle$$
$$\xi(4,x) = \langle x,0 \rangle$$
$$\xi(5,x) = K(x)$$
$$\xi(6,x) = L(x)$$

and for $n \geq 6$

$$
\xi(n+1, x) = \begin{cases}
\xi(\Pi_1^3(n), \xi(\Pi_2^3(n), x)) & \textbf{if } \Pi_3^3(n) = 0 \\
\langle \xi(\Pi_1^3(n), x), \xi(\Pi_2^3(n), x) \rangle & \textbf{if } \Pi_3^3(n) = 1 \\
\xi^{K(x)}(\Pi_1^3(n), L(x)) & \textbf{if } \Pi_3^3(n) = 2 \\
(\mu y)_{\leq L(x)} \xi(\Pi_1^3(n), \langle y, K(x) \rangle) & \textbf{if } \Pi_3^3(n) > 2
\end{cases}
$$

Entirely analogous with that of Theorem 1 is the remainder of the proof. (See Problem 15.)//

Corollary 1 ξ is total.

Proof For any n, $\lambda x. \xi(n,x) \in \mathcal{PR}.//$

Corollary 2 $\xi \notin \mathcal{PR}$.

Proof If $\xi \in \mathcal{PR}$, then $\lambda x. \xi(x,x) + 1 \in \mathcal{PR}$. Then, for some i, $\lambda x. \xi(i,x) = \lambda x. \xi(x,x) + 1$. Substituting i into x, we get a contradiction (recall that ξ is total!).//

Corollary 3 $\lambda xyn.y = \xi(n,x) \notin \mathcal{PR}_*$.

Proof If $\lambda xyn.y = \xi(n,x) \in \mathcal{PR}_*$, then $\lambda x.0 = \xi(x,x) \in \mathcal{PR}_*$ hence $\lambda x.0 \neq \xi(x,x) \in \mathcal{PR}_*$ (why?).

The characteristic function of $0 \neq \xi(x,x)$ is $\lambda x.1 \doteq \xi(x,x)$, which cannot be in $\mathcal{PR}$ (why?).//

Corollary 3 establishes an example of a function ξ such that its graph is not primitive recursive. We shall see in the next section that $\lambda xyn.y = \xi(n,x) \in \mathcal{R}_*$.

§3.6 EFFECTIVE ENUMERABILITY (PART I)

We have laid down enough groundwork toward the formalization of the notion "algorithmic function". As such we can take to mean partial recursive function. Further evidence for the reason behind such a position will be given in the next chapter. We shall neither adopt nor deny the position that $\mathcal{P}$ is a "complete formalism" for "algorithmic functions", however, we shall always rely on intuition to derive proofs even though, in the end, we shall always transform the informal arguments into formal ones.

Another important notion in Computability is that of the *effective* or *algorithmic* enumerability of a set (predicate). Intuitively, a set (predicate) is "effectively enumerable" iff there is an algorithm by which the xth element of the set can be produced. The following definition of *recursively enumerable* is the formal counterpart of effectively enumerable.

Definition 1. A predicate $P(\vec{x}_n)$ is *recursively enumerable*, or *re*, iff one of the following holds:

(a)   For all $\vec{x}_n$, $P(\vec{x}_n)$ is *false*.

(b)   There is an $f \in \mathcal{R}^{(1)}$ such that $f(\mathcal{N}) = \{\langle \vec{x}_n \rangle \mid P(\vec{x}_n)\}$, where $\langle \ \rangle$ is any (fixed) primitive recursive 1-1 coding $\mathcal{N}^n \to \mathcal{N}$.//

Note: The recursive function f "enumerates algorithmically" the set $\{\vec{x}_n \mid P(\vec{x}_n)\}$.

Example 1   The predicate $x \in \mathcal{N}$ is re, for if $f = \lambda x.x$, then $f(\mathcal{N}) = \{x \mid x \in \mathcal{N}\}.//$

Example 2   The predicate $x \equiv 0 \pmod 2$ is re, for if $f = \lambda x.2x$, then $f(\mathcal{N}) = \{x \mid x \equiv 0 \pmod 2\}.//$

Theorem 1   (*Projection theorem*). A predicate $P(\vec{x}_n)$ is re iff there exists a recursive predicate $R(y,\vec{x}_n)$ such that $P(\vec{x}_n) \equiv (\exists y)R(y,\vec{x}_n)$.

Proof (1) Let $P(\vec{x}_n)$ be re. If, for all $\vec{x}_n$, $P(\vec{x}_n)$ is false, then $P(\vec{x}_n)$ $\equiv (\exists y)y + \max(\vec{x}_n) + 1 = 0$; clearly, $\lambda y\vec{x}_n.y + \max(\vec{x}_n) + 1 = 0 \in \mathcal{R}_*$. Otherwise, there is an $f \in \mathcal{R}^{(1)}$ such that $f(\mathcal{N}) = \{\langle \vec{x}_n \rangle \,|\, P(\vec{x}_n)\}$. Thus, $P(\vec{x}_n)$ $\equiv (\exists y)(f(y) = \langle \vec{x}_n \rangle)$. Clearly, $\lambda y\vec{x}_n.f(y) = \langle \vec{x}_n \rangle \in \mathcal{R}_*$, since $\lambda \vec{x}_n.\langle \vec{x}_n \rangle \in \mathcal{PR}$.

(2) Let $P(\vec{x}_n) \equiv (\exists y)R(y,\vec{x}_n)$, where $R(y,\vec{x}_n) \in \mathcal{R}_*$. If $P(\vec{x}_n)$ is false for all $\vec{x}_n$, then $P(\vec{x}_n)$ is re. So let, $P(\vec{a}_n)$ be true. Define $f: \mathcal{N} \rightarrow \mathcal{N}$ by:

$$
f(z) = \begin{cases} \langle \Pi_2^{n+1}(z), \ldots, \Pi_{n+1}^{n+1}(z) \rangle & \text{if } R(\Pi_1^{n+1}(z), \ldots, \Pi_{n+1}^{n+1}(z)) \\ \langle \vec{a}_n \rangle & \text{if } \neg R(\Pi_1^{n+1}(z), \ldots, \Pi_{n+1}^{n+1}(z)) \end{cases}
$$

Since $\mathcal{R}$ is closed under definition by cases (Theorem 3.2.5), $f \in \mathcal{R}^{(1)}$. Clearly, $f(\mathcal{N}) \subset \{\langle \vec{x}_n \rangle \,|\, P(\vec{x}_n)\}$.
Let $P(\vec{b}_n)$ hold. Then there is a b such that $R(b,\vec{b}_n)$ is true.
Let $c = \langle b,\vec{b}_n \rangle$. Then $f(c) = \langle \vec{b}_n \rangle$, thus $f(\mathcal{N}) \supset \{\langle \vec{x}_n \rangle \,|\, P(\vec{x}_n)\}$.//

Note: The second part of the above proof essentially defines f by generating recursively all the $(n+1)$-tuples $(y,\vec{x}_n)$ and, for every such tuple that satisfies R, setting $f(i) = \langle \vec{x}_n \rangle$, where i is the position of $(y,\vec{x}_n)$ in the sequence. The projection theorem is important in providing tools for proving re-ness. Further, it characterizes re-ness *independently* of any particular coding function $\langle \vec{x}_n \rangle$.

Corollary 1   If $R(y,\vec{x}_n)$ is re, then $(\exists y)R(y,\vec{x}_n)$ is also re.

Proof By the projection theorem, there is a recursive predicate $P(z,y,\vec{x}_n)$ such that $R(y,\vec{x}_n) \equiv (\exists z)P(z,y,\vec{x}_n)$. Thus,

$$(\exists y)R(y,\vec{x}_n) \equiv (\exists y)(\exists z)P(z,y,\vec{x}_n)$$

$$= (\exists w)P(K(w),L(w),\vec{x}_n).$$

Since $P(K(w),L(w),\vec{x}_n) \in \mathcal{R}_*$, the result follows.//

Corollary 2   If $R(y,\vec{x}_n)$ is re, then $(\forall y)_{\leq z}R(y,\vec{x}_n)$ is re.

Proof As previously, let $P(w,y,\vec{x}_n) \in \mathcal{R}_*$ be such that $R(y,\vec{x}_n) \equiv (\exists w)P(w,y,\vec{x}_n)$. Then, $(\forall y)_{\leq z}(\exists w)P(w,y,\vec{x}_n)$ is, intuitively, the same as

$\exists\, w_0, w_1, \ldots, w_z$ such that

$$P(w_0,0,\vec{x}_n)\ \&\ P(w_1,1,\vec{x}_n)\ \&\ P(w_2,2,\vec{x}_n)\ \&\ \ldots\ \&P(w_z,z,\vec{x}_n)$$

Formally, it is equivalent to $(\exists w)(\forall i)_{\le z}P(\Pi_i^z(w),i,\vec{x}_n)$ where $\Pi_i^z(w)$ is a function of i, z, and w. We can use any primitive recursive coding scheme. For example, for any particular J, K, L pairing function and projections set:

$$\Pi_i^z(w) = \textbf{if } i = 0 \textbf{ then } K^z(w) \textbf{ else } L(K^{z \dot- i}(w))$$

Clearly, $\lambda izw.\Pi_i^z(w) \in \mathcal{PR}$ and $(\forall i)_{\le z}\, P(\Pi_i^z(w),\, i,\, \vec{x}_n) \in \mathcal{R}_*$. (Theorem 3.2.4). The result follows.//

Corollary 3   If $R(y,\vec{x}_n)$ is re, then $(\exists y)_{\le z}R(y,\vec{x}_n)$ is re.

Proof   Problem 17.//

Theorem 2   If $R(\vec{x}_n) \in \mathcal{R}_*$ then $R(\vec{x}_n)$ is re.

Proof $R(\vec{x}_n) \equiv (\exists y)(y=0\,\&c_R(\vec{x}_n)=y)$, where c_R is the characteristic function of R.//

Theorem 3 $R(\vec{x}_n) \in \mathcal{R}_*$ iff both $R(\vec{x}_n)$ and $\neg R(\vec{x}_n)$ are re.

Proof   (1) $R(\vec{x}_n) \in \mathcal{R}_*$ implies $\neg R(\vec{x}_n) \in \mathcal{R}_*$ and hence both $R(\vec{x}_n)$ and $\neg R(\vec{x}_n)$ are re.

(2) Let $R(\vec{x}_n)$ and $\neg R(\vec{x}_n)$ be re. If any of them is identically false, then clearly $R(\vec{x}_n) \in \mathcal{R}_*$. Next, let this not be the case.

Intuitively, we can have the following algorithm decide membership in R:

Given $\vec{a}_n$. Simultaneously and effectively list R and $\neg R$. Sooner or later $\vec{a}_n$ shall appear in one or other list. If it appears in that of R, then $R(\vec{a}_n)$ else $\neg R(\vec{a}_n)$.

Formally, let $f(\mathcal{N}) = \{\langle \vec{x}_n \rangle \mid R(\vec{x}_n)\}$ and $g(\mathcal{N}) = \{\langle \vec{x}_n \rangle \mid \neg R(\vec{x}_n)\}$, where f, g are in $\mathcal{R}$.
Define

$$h(\vec{x}_n) = (\mu y)(f(y) = \langle \vec{x}_n \rangle \vee g(y) = \langle \vec{x}_n \rangle)$$

Clearly, $h \in \mathcal{R}$ (it is total). Also, $R(\vec{x}_n) \equiv f(h(\vec{x}_n)) = \langle \vec{x}_n \rangle$.//

Note: h "generates" (a "directory" of) the two lists simultaneously.

§3.7 NORMAL FORM THEOREMS

Theorem 1   For every function $f \in \mathcal{P}^{(n)}$ there exist functions $u_f \in \mathcal{PR}^{(1)}$ and $t_f \in \mathcal{PR}^{(n+1)}$ such that $f = \lambda \vec{x}_n . u_f((\mu y) t_f(y, \vec{x}_n))$.

Proof (Following Grzegorczyk [4]). The proof is by induction with respect to $\mathcal{P}$. Further, without loss of generality (*) we assume that for any t_f-function, the equation $t_f(y, \vec{x}_n) = 0$ has *at most one* y-solution (given $\vec{x}_n$).
(a) *Basis.*

$$\text{For } s = \lambda x . x + 1 : u_s = \lambda x . x, t_s = \lambda xy . |y - (x + 1)|$$

$$\text{For } p = \lambda x . x \dot- 1 : u_p = \lambda x . x, t_p = \lambda xy . |y - (x \dot- 1)|$$

(b) *The property propagates with substitution.*
Case 1. Let $\lambda \vec{x}_n y . f(\vec{x}_n, y)$ and $\lambda \vec{y}_m . g(\vec{y}_m)$ have the property.
Then

$$
\begin{aligned}
f(\vec{x}_n, g(\vec{y}_m)) &= u_f((\mu y) t_f(y, \vec{x}_n, g(\vec{y}_m))) \\
&= u_f((\mu y) t_f(y, \vec{x}_n, u_g((\mu z) t_g(z, \vec{y}_m)))) \\
&= u_f \circ K((\mu w)[t_f(K(w), \vec{x}_n, u_g(L(w))) = 0 \ \& \ t_g(L(w), \vec{y}_m) = 0])
\end{aligned}
$$

by uniqueness of z-solution of $t_f(z, \vec{x}_n, y) = 0$ and y-solution of $t_g(y, \vec{y}_m) = 0$ (if such solutions exist). Thus, for $h = \lambda \vec{x}_n \vec{y}_m . f(\vec{x}_n, g(\vec{y}_m))$, $u_h = u_f \circ K$ and $t_h = \lambda w \vec{x}_n \vec{y}_m . t_f(K(w), \vec{x}_n, u_g(L(w))) + t_g(L(w), \vec{y}_m)$. The remaining cases of substitution are left as an exercise. (See Problem 19).
(c) *The property propagates with pure iteration*:
Let $\lambda x . f(x)$ have the property. Note that $z = f^y(x)$ holds iff there is a sequence $m_0, \ldots, m_y$ such that $m_0 = x, m_y = z$ and $(\forall i)_{<y} m_{i+1} = f(m_i)$.
    That is,

$$f^y(x) = \exp(y, (\mu m) [l(m) = y \ \& \ \exp(0, m) = x + 1 \ \&$$

$$(\forall i)_{<y}(\exp(i + 1, m) = f(\exp(i, m) \dot- 1) + 1)]) \dot- 1$$

where $l(m)$ = "the index of the largest prime that divides m" = $\pi((\mu y)_{\leq m}(\Pr(y) \ \& \ y \mid m \ \& \ (\forall i)_{\leq m} [i \mid m \ \& \ \Pr(i) \Rightarrow i \leq y])) \dot- 1$, which is in $\mathcal{PR}$ as a function of m. (See Example 2.3.1).

(*)Indeed, if some t_f does not have this property, then the characteristic function, $\tilde{t}_f$, of $t_f(y, \vec{x}) = 0 \ \& \ (\forall i)_{<y} t_f(i, \vec{x}_n) \neq 0$ has the property; clearly, $\tilde{t}_f \in \mathcal{PR}$ and $f = \lambda x_n . u_f((\mu y) \tilde{t}_f(y, \vec{x}_n))$.

By assumption,

$$f = \lambda x.u_f((\mu y)\, t_f(y,x)).$$

Thus,

$$f'(x) = \exp(y, (\mu m)[l(m) = y \,\&\, \exp(0,m) = x + 1 \,\&\, (\forall i)_{<y}(\exp(i + 1, m)$$
$$= u_f((\mu z)\, t_f(z, \exp(i,m) \div 1)) + 1)]) \div 1$$

To say, "$(\forall i)_{<y} \exp(i + 1, m) = u_f((\mu z)t_f(z, \exp(i, m) \div 1)) + 1$" amounts to saying that "$\exists z_0, \ldots, z_{y-1}$ such that $(\forall i)_{<y} \exp(i + 1, m) = u_f(z_i) + 1 \,\&\, t_f(z_i, \exp(i,m) \div 1) = 0$".

Note: Uniqueness of z-solution of $t_f(z,x) = 0$ was invoked here! Formally, the last statement in quotes amounts to

$$(\exists z)(l(z) + 1 = y \,\&\, ((\forall i)_{<y} p_i \,|\, z) \,\&\, (\forall i)_{<y} \exp(i + 1, m) =$$
$$u_f(\exp(i, z) \div 1) + 1 \,\&\, t_f(\exp(i,z) \div 1, \exp(i,m) \div 1) = 0.$$

Call the predicate following $(\exists z)$, above, $P(y,m,z)$. Since z codes the sequence $z_0, \ldots, z_{y-1}$, given y and x, the predicate $l(m) = y \,\&\, \exp(0,m) = x + 1 \,\&\, P(y,m,z)$ can be true for at most *one* pair (m,z).
Set

$$w = \langle m,z \rangle,$$

and

$$H(x,y,w) \equiv [l(K(w)) = y \,\&\, \exp(0,K(w)) = x + 1 \,\&\, P(y,K(w), L(w))].$$

The above remark means that given x and y, at most *one* w makes $H(x,y,w)$ true, hence

$$f'(x) = \exp(y,K((\mu w)H(x,y,w)))$$

Finally, setting $r = \langle y,w \rangle$, $u = \lambda r.\exp(K(r), K{\circ}L(r))$, and $T(x,y,r) \equiv H(x,K(r), L(r)) \,\&\, K(r) = y$, we obtain $f'(x) = u((\mu r)T(x,y,r))$.
(d) *The property propagates with μ-operator:* Let $f = \lambda y\vec{x}.u_f((\mu z)\, t_f(z,y,\vec{x}))$. Consider $h = \lambda \vec{x}.(\mu y)f(y,\vec{x}) = \lambda \vec{x}.(\mu y)u_f((\mu z)\, t_f(z,y,\vec{x}))$.
    To "compute" $h(\vec{x})$, for any given $\vec{x}$, we need find a sequence $z_0, \ldots, z_y$ such that $t_f(z_i,i,\vec{x}) = 0$ for all $i = 0, \ldots, y, u_f(z_y) = 0$ and $u_f(z_i) \neq 0, 0 \leq i < y$. Note that, consistent with the definition of "μ", we want $f(i,\vec{x})\downarrow$ (i.e., existence of z_i such that $t(z_i,i,\vec{x}) = 0$) for $0 \leq i < y$.

Thus, we are looking for a number z, coding the (*) z_i-sequence, such that

$$(\forall i)_{\leq l(z)}\,(p_i\,|\,z)\ \&\ (\forall i)_{\leq l(z)}\,t_f(\exp(i,z) \dot- 1,i,\vec{x}) = 0\ \&$$

$$(\forall i)_{<l(z)}\,u_f(\exp(i,z) \dot- 1) \neq 0\ \&\ u_f(\exp(l(z),z) \dot- 1) = 0.$$

Let us call the above predicate $P(z,\vec{x})$.

(1) $P(z,\vec{x}) \in \mathcal{PR}_*$.

(2) $y = h(\vec{x})$ iff $f(y,\vec{x}) = 0\ \&\ (\forall i)_{<y}\,(f(i,\vec{x})\!\downarrow\ \&\ f(i,\vec{x}) \neq 0)$
 iff $l(z) = y\ \&\ P(z,\vec{x})$.

By (2), if $h(\vec{x})\!\downarrow$, then $l(z) = h(\vec{x})\ \&\ P(z,\vec{x})$ for a unique z, hence, $h(\vec{x}) = l((\mu z)P(z,\vec{x}))$.//

Theorem 2   A partial function $\lambda \vec{x}_n.f(\vec{x}_n)$ is in $\mathcal{P}$ iff $\lambda y \vec{x}_n.y = f(\vec{x}_n)$ is re.

Proof (1) Let $f \in \mathcal{P}$.
Then for some $u_f \in \mathcal{PR}^{(1)}\ t_f \in \mathcal{PR}^{(n+1)}, f = \lambda \vec{x}_n.u_f((\mu z)t_f(z,\vec{x}_n))$.
Thus, $y = f(\vec{x}_n) \equiv (\exists z)(t_f(z,\vec{x}_n) = 0\ \&\ u_f(z) = y)$ which is re by the projection theorem.
(2) Let $y = f(\vec{x}_n)$ be re.
Intuitively, there exists an effective way to list all $(n+1)$-tuples $(\vec{x}_n,y)$ that satisfy $y = f(\vec{x}_n)$. Thus, to compute $f(\vec{a}_n)$, keep listing $(\vec{x}_n,y)$ such that $f(\vec{x}_n) = y$. If and when a tuple $(\vec{a}_n,b)$ is listed, answer b.
Formally, let $g \in \mathcal{R}^{(1)}$ be such that $g(\mathcal{N}) = \{\langle \vec{x}_n,y \rangle\,|\,y = f(\vec{x}_n)\}$.(†)
Then, $f = \lambda \vec{x}_n.L \circ g((\mu z)(K(g(z)) = \langle \vec{x}_n \rangle))$.//

Corollary   A total function $\lambda \vec{x}_n.f(\vec{x}_n)$ is in $\mathcal{R}$ iff $y = f(\vec{x}_n)$ is re.

We can now prove that there exists a universal function of $\mathcal{P}^{(1)}$ in $\mathcal{P}^{(2)}$ and a universal function of $\mathcal{PR}^{(1)}$ in $\mathcal{R}^{(2)}$.

Theorem 3   (*Universal Function Theorem*). The universal function ψ of $\mathcal{P}^{(1)}$ defined in Theorem 3.5.1 is in $\mathcal{P}^{(2)}$.

Proof By Theorem 2, it suffices to show that the predicate $y = \psi(n,x)$ is re.

(*)By uniqueness of z-solution for $t_f(z, y, \vec{x})$, given $\vec{x}$ and i, there is at most one z_i s.t. $t_f(z_i,i,\vec{x}) = 0$.
(†)$\{(\vec{x}_n, y)\,|\,y = f(\vec{x}_n)\} \neq \varnothing$ assumed, as otherwise $f = \lambda \vec{x}_n.(\mu y)y + 1 + \max(\vec{x}_n)$; hence, $f \in \mathcal{P}$.

The technique is familiar by now; we shall express $y = \psi(n,x)$ by "coding" its "computation" by a single number; i.e., we shall find a recursive predicate $R(u,n,x,y)$ such that $y = \psi(n,x) \equiv (\exists u)R(u,n,x,y)$ and then invoke the projection theorem. Intuitively, $R(u,n,x,y)$ holds iff u codes the computation that verifies $y = \psi(n,x)$.

As in Theorem 3.3.2, we set $u_x = \exp(x,u)$ and $u_{x,y} = \exp(\langle x,y \rangle, u)$. Clearly, $\lambda xy.u_x$ and $\lambda xyu.u_{x,y}$ are in $\mathcal{PR}$, and $x,y,u_x,u_{x,y}$ are $\leq u$.

Now,

$$y = \psi(n,x) \equiv (\exists u)((\forall i)_{\leq u}\,[(u_{0,i} > 0 \Rightarrow u_{0,i} = i + 2\,\&$$

$$(u_{1,i} > 0 \Rightarrow u_{1,i} = (i \dot{-} 1) + 1)\,\&\,(u_{2,i} > 0 \Rightarrow u_{2,i} = \langle i,i \rangle + 1)\,\&$$

$$(u_{3,i} > 0 \Rightarrow u_{3,i} = \langle 0,i \rangle + 1)\,\&\,(u_{4,i} > 0 \Rightarrow u_{4,i} = \langle i,0 \rangle + 1)\,\&$$

$$(u_{5,i} > 0 \Rightarrow u_{5,i} = K(i) + 1)\,\&\,(u_{6,i} > 0 \Rightarrow u_{6,i} = L(i) + 1)\,\&$$

$$(\forall i)_{<n}\,(\forall j)_{\leq u}\,(i \geq 6\,\&\,u_{i+1,j} > 0 \Rightarrow \{[\Pi_3^3(i) = 0\,\&$$

$$u_{\Pi_2^3(i),j} > 0\,\&\,u_{i+1,j} = u_{\Pi_1^3(i),u_{\Pi_2^3(i),j}\dot{-}1}]\,\vee\,[\Pi_3^3(i) = 1\,\&$$

$$u_{\Pi_1^3(i),j} > 0\,\&\,u_{\Pi_2^3(i),j} > 0\,\&\,u_{i+1,j} = \langle u_{\Pi_1^3(i),j} \dot{-} 1, u_{\Pi_2^3(i),j} \dot{-} 1 \rangle + 1]$$

$$\vee\,[\Pi_3^3(i) = 2\,\&\,(\exists w)_{\leq \Pi_{m \leq K(j)} p_m^u}\,(w_0 = L(j)\,\&\,w_{K(j)} + 1 = u_{i+1,j}\,\&$$

$$(\forall l)_{<K(j)}\,(w_{l+1} + 1 = u_{\Pi_1^3(i),w_l}))]\,\vee$$

$$[\Pi_3^3(i) > 2\,\&\,(\exists w)_{\leq u}\,(w + 1 = u_{i+1,j}\,\&\,u_{\Pi_1^3(i),\langle w,j \rangle} = 1\,\&$$

$$(\forall l)_{<w}\,(u_{\Pi_1^3(i),\langle l,j \rangle} > 1))]\})\,\&\,y + 1 = u_{n,x}.$$

The predicate quantified by $(\exists u)$ we shall call $R(u,n,x,y)$. Clearly, $R(u,n,x,y)$ $\in \mathcal{PR}_*$. Also, $R(u,n,x,y)$ is true iff there is a computation, coded by u, which verifies $y = \psi(n,x)$; indeed $u_{i,j} = \psi(i,j) + 1$ if $\psi(i,j)$ is a value of ψ needed in the evaluation of $\psi(n,x)$.//

Corollary 1   The universal function ξ of $\mathcal{PR}^{(1)}$ defined in Theorem 3.5.2 is recursive.

Proof As above, one shows that $y = \xi(n,x)$ is re. The result follows since ξ is total.//

Corollary 2 $\mathcal{R}_* - \mathcal{PR}_* \neq \varnothing$

Proof We know that $y = \xi(n,x) \notin \mathcal{PR}_*$ (Corollary 3 of Theorem 3.5.2.) By Corollary 1 above, $y = \xi(n,x) \in \mathcal{R}_*$.//

Corollary 3   There is an *onto* map $\phi: \mathcal{N} \to \mathcal{P}^{(1)}: i \mapsto \phi_i$ such that $\lambda xy.\phi_x(y) \in \mathcal{P}^{(2)}$.

Proof   Take $\lambda xy.\phi_x(y) \overset{\text{def}}{=} \lambda xy.\psi(x,y).//$

Definition 1   We shall reserve the symbol ϕ for the map $\phi: \mathcal{N} \to \mathcal{P}^{(1)}$ of Corollary 3 above. ϕ is called a *Gödel numbering* or *ϕ-indexing* of $\mathcal{P}^{(1)}$.//

Note: Theorem 3 above is intuitively pleasing. It says that our formalism is broad enough to include functions whose "algorithms" behave like "stored program computers". Specifically, i is (a code for) the "program" and x is the "data" in $\phi_i(x)$. Such a theorem should be expected to hold in any formalism which claims to be complete (with respect to the informal notion of algorithmic function).

Definition 2   Let $f \in \mathcal{P}^{(1)}$ and let i be such that $f = \phi_i$. Then i is called a *ϕ-index,* or simply *index,* or *Gödel number* for f.

Corollary 4   Each $f \in \mathcal{P}^{(1)}$ has (countably) infinitely many indices.

Proof   Each $f \in \mathcal{P}^{(1)}$ has infinitely many different inductive descriptions in the system $\mathcal{P}$; for example, if $i = \lambda x.x$, then $f, i\circ f, i\circ i\circ f, \ldots, i^k\circ f, \ldots$ all have different descriptions, thus they have different indices. However, they denote the same function.//

Note: Intuitively, Corollary 4 says that there are infinitely many different ways of "programming" a given function.

Lemma 1   Let $\lambda xy.\langle x,y \rangle, K, L$ be the pairing function with its projections which we used in Theorem 3.4.1. Then $f \in \mathcal{P}^{(n)}$ iff $f = \lambda \vec{x}_n.\phi_i(\langle \vec{x}_n \rangle)$ for some i.

Proof   Let $f \in \mathcal{P}^{(n)}$. Then $h = \lambda x.f(\Pi_1^n(x), \ldots, \Pi_n^n(x)) \in \mathcal{P}^{(1)}$; let $h = \phi_i$. Since $\lambda \vec{x}_n.\langle \vec{x}_n \rangle$ is 1-1 $(\Pi_1^n, \ldots, \Pi_n^n)\circ\langle \vec{x}_n \rangle = \vec{x}_n$; hence, $f = \lambda \vec{x}_n.h(\langle \vec{x}_n \rangle) = \lambda \vec{x}_n.\phi_i(\langle \vec{x}_n \rangle)$.
Conversely, $\lambda \vec{x}_n.\phi_i(\langle \vec{x}_n \rangle) \in \mathcal{P}^{(n)}$, trivially.//

Definition 3   Let $f \in \mathcal{P}^{(n)}$ and let i be such that $f = \lambda \vec{x}_n. \phi_i(\langle \vec{x}_n \rangle)$. Then i is called a *(ϕ-) index* (or *Gödel number*) of f. We shall use the notation $\phi_i^{(n)}$ for $\lambda \vec{x}_n.\phi_i(\langle \vec{x}_n \rangle)$. $\phi_i^{(1)}$ will be ϕ_i.//

Corollary   Each $f \in \mathcal{P}^{(n)}$ has (countably) infinitely many indices.

Theorem 4   (*Normal Form or Representation Theorem*). There is a function d in $\mathcal{PR}^{(1)}$ and for each $n > 0$ a predicate $T^{(n)}(z,\vec{x}_n,y) \in \mathcal{PR}_*$ such that for each $f \in \mathcal{P}^{(n)}$ there is an i such that $f = \lambda \vec{x}_n.d((\mu y) T^{(n)}(i,\vec{x}_n,y))$.

Proof Let u_ψ and t_ψ in $\mathcal{PR}^{(1)}$ $\mathcal{PR}^{(3)}$, respectively, be the functions such that, for all x, and y, $\phi_x(y) = u_\psi((\mu z)\, t_\psi(x,y,z))$. (See Theorem 1 and Corollary 3).

Define $d \overset{\text{def}}{=} u_\psi$ and, for $n > 0$,

$$T^{(n)}(z,\vec{x}_n,y) \equiv t_\psi(z, \langle \vec{x}_n \rangle, y) = 0$$

Clearly, $\lambda\vec{x}_n.d((\mu y)\,T^{(n)}(i,\vec{x}_n,y)) = \phi_i^{(n)}$. Let i be such that $f = \phi_i^{(n)}$.//

Note: Theorem 4 is due to Kleene. $T^{(n)}$ is known as the Kleene T-predicate. $T^{(n)}(z,\vec{x}_n,y)$ is true, intuitively, iff the "program" with code z and with input $\vec{x}_n$ has a "computation" characterized by y. d "decodes" the "code" of the "computation" y, and yields the output. Davis [3] shows that d and $T^{(n)}$ are primitive recursive by first defining Turing Machines (See Chapter 4), and then arithmetizing them. It can be shown that Davis' arithmetization actually puts d and $T^{(n)}$ in Grzegorczyk's 0th class. This is also true for our version of d and $T^{(n)}$, but this will only be apparent later (Chapter 13). Using still different techniques (Smullyan [14], Jones [5]), the Kleene predicate and the decoding function can be shown to belong to still simpler classes of functions and predicates.

One important consequence of the normal form theorem is that on one hand we can get any function of $\mathcal{P}^{(n)}$ from two initial functions (d and the characteristic function of $T^{(n)}$) by using *one* composition and *one* application of μ; on the other hand, nothing changes if instead of μ we use $\tilde{\mu}$ (why?).

Theorem 5   (*Basic theorem on re predicates*) $P(\vec{x}_n)$ is re iff for some i $P(\vec{x}_n)$ $\equiv \phi_i^{(n)}(\vec{x}_n)\!\downarrow$.

Proof (1) Let $P(\vec{x}_n) \equiv \phi_i^{(n)}(\vec{x}_n)\!\downarrow$. Then $P(\vec{x}_n) \equiv (\exists y)T^{(n)}(i,\vec{x}_n,y)$; result follows from the projection theorem.
(2) Let $P(\vec{x}_n)$ be re. Then, by the projection theorem, there is a $R(y,\vec{x}_n)$ $\in \mathcal{R}_*$ such that $P(\vec{x}_n) \equiv (\exists y)R(y,\vec{x}_n)$.
Let i be such that $\phi_i^{(n)} = \lambda\vec{x}_n.(\mu y)\,R(y,\vec{x}_n)$. Then $P(\vec{x}_n) \equiv \phi_i^{(n)}(\vec{x}_n)\!\downarrow$.//

Note: Intuitively, then, a predicate $P(\vec{x}_n)$ is re iff there is an effective procedure or algorithm (that for $\phi_i^{(n)}$) such that for any given $\vec{a}_n$, if $P(\vec{a}_n)$ it will verify it in finite time, but if $\neg P(\vec{a}_n)$ it will "run forever". Contrast this with recursive predicates.

Corollary 1   (*Normal Form Theorem*) $P(\vec{x}_n)$ is re iff there is an i such that $P(\vec{x}_n)$ $\equiv (\exists y)T^{(n)}(i,\vec{x}_n,y)$.

Definition 4   Let $P(x)$ be a (unary) re predicate, and let i be such that $P(x)$ $\equiv (\exists y)T(i,x,y)$. We call i a *re index* for $P(x)$. W_i denotes the set $P = \{x \mid P(x)\}$ in this case.//

Corollary 2   A unary predicate has infinitely many re indices.

> **Note:** A re index can be analogously defined for n-ary predicates, $n > 1$. Such practice will not, however, be necessary.

Example 1   The set $K = \{x \mid \phi_x(x)\!\downarrow\}$ is re. Indeed, $x \in K \equiv (\exists y)T(x,x,y).//$

Example 2   The set $\overline{K} = \{x \mid \phi_x(x)\!\uparrow\}$ is *not* re. Indeed, let $x \in \overline{K}$ be re. Then, for some i, $x \in \overline{K} \equiv (\exists y)T(i,x,y)$. But $x \in \overline{K}$ means $\neg(\exists y)T(x,x,y)$. Hence $\neg(\exists y)T(x,x,y) \equiv (\exists y)T(i,x,y)$. Substituting i for x we get a contradiction.$//$

Example 3   The set K is not recursive. If it were, then $\overline{K}$ should also be recursive, hence re. (See Theorem 3.6.3.)$//$

> **Note:** The sets K and $\overline{K}$ play an important role in the theory. We reserve the letters $K, \overline{K}$ for them. The context will not allow a confusion with the first projection, K, of a pairing function.

Example 4   The set $K_0 = \{(x,y) \mid \phi_x(y)\!\downarrow\}$ is not recursive. Indeed, if $K_0 \in \mathcal{R}_*$ then $c_{K_0} \in \mathcal{R}$. But then $c_K = \lambda x.c_{K_0}(x,x) \in \mathcal{R}$, a contradiction.$//$

> **Note:** The problem "$(x,y) \in K_0$" is the *halting problem*. It asks whether the program (with code) x with input y will eventually give an answer (i.e., will halt). We have just seen that there is no "algorithm" *formalizable in* $\mathcal{P}(*)$ which can solve the halting problem.

Example 5   The set K_0 is re. Indeed, $(x,y) \in K_0 \equiv (\exists z)T(x,y,z).//$

> **Note:** We have just seen that for the halting problem there exists a *partial* solution. That is, an algorithm exists such that, for any a, b, if $(a,b) \in K_0$ it will verify it, but if $(a,b) \notin K_0$ then it will run forever. Another way of viewing this is that there is an algorithm which effectively lists K_0; that is, given x it will produce the xth member of K_0 in finite time. The same comments apply to K. By contrast, for the problem $x \in \overline{K}$ there is not even a partial (algorithmic) solution.

Example 6   The set $\{x \mid \phi_x \in \mathcal{R}\}$ is *not* re. Indeed, if it were, there would exist a function $f \in \mathcal{R}^{(1)}$ such that $f(\mathcal{N}) = \{x \mid \phi_x \in \mathcal{R}\}$. Define $h \stackrel{\text{def}}{=} \lambda x.\phi_{f(x)}(x) + 1$. Clearly $h \in \mathcal{P}$; moreover, $h \in \mathcal{R}$ since it is total (why?). Thus, for some i, $h = \phi_{f(i)}$. Then, $h(i) = \phi_{f(i)}(i)$ [since $h = \phi_{f(i)}$], also $h = \phi_{f(i)}(i) + 1$ (by definition of h). Hence $\phi_{f(i)}(i) = \phi_{f(i)}(i) + 1$, a contradiction, since both sides are defined.$//$

Definition 5   If the set A is nonrecursive, then we shall call the problem $\vec{x} \in A$ *recursively unsolvable*, or just *unsolvable* or *undecidable*.$//$

(*)Whenever we say "there is" or "there is no algorithm" we always mean *formalizable in* $\mathcal{P}$, even if we don't mention the qualification explicitly.

Example 7 The *equivalence problem* of functions in $\mathcal{PR}$ is unsolvable; that is, there is no algorithm which when presented with (the numerical codes of) two inductive descriptions of functions in $\mathcal{PR}$ will decide whether the two descriptions define the same function or not.

Indeed, consider the case of functions $(\lambda y.c_T(x,x,y))_{x\geq0}$ and $\lambda y.1$. Let $A = \{x \mid \lambda y.c_T(x,x,y) = \lambda y.1\}$. If the equivalence problem for $\mathcal{PR}$ is solvable, so is the special case "$x \in A$".

Now $x \in A \equiv (\forall y)c_T(x,x,y) = 1 \equiv (\forall y)\neg T(x,x,y) \equiv \neg(\exists y)T(x,x,y) \equiv x \in \overline{K}$. Since $\overline{K}$ is not re, A is not either; in particular, it is not recursive.//

§3.8 THE *S-m-n* THEOREM

Consider a PL/1 program of the form

$$\left\{\begin{array}{l}\textbf{read } x \\ \textbf{read } y \\ \textbf{process} \\ \textbf{write } z,\end{array}\right.$$ where all variables and arithmetic is over the natural numbers.

This program defines a function $f = \lambda xy.z: \mathcal{N}^2 \to \mathcal{N}$. Assume that we want to obtain a new program from the above such that x is kept at the value a, thus defining $\lambda y.f(a,y)$.

This can be accomplished by simply replacing the instruction "**read** x" by "$x = a$". It is intuitively clear that this "parametrization" was achieved "mechanically". If our $\mathcal{P}$-formalism has any claim of completeness with respect to the notion "algorithmic function", such a mechanical parametrization should be possible in it.

We should be able to show that there is a *recursive* function $\lambda ix.\sigma(i,x)$ such that $\phi_{\sigma(i,x)} = \lambda y.\phi_i^{(2)}(x,y)$. To relate with the above, σ is the mechanical process, which given (the code of) the program i and the value of the "parameter", x, produces the new program (of code) $\sigma(i,x)$. The content of the *S-m-n* theorem is that such a $\sigma \in \mathcal{R}$ indeed exists. We shall utilize a sequence of lemmas; refer throughout to Theorem 3.5.1.

Lemma 1 There is an $h_1 \in \mathcal{R}^{(1)}$ such that $\phi_{h_1(y)} = \lambda x.x + y$.

Proof First, $\lambda x.x + 0 = \lambda x.(x + 1) \dot{-} 1 = \lambda x.\psi(1,\psi(0,x)) = \phi_{\langle 1,0,0 \rangle + 1}$. Hence, take $h_1(0) = \langle 1,0,0 \rangle + 1$.

$$\lambda x.x + y + 1 = \lambda x.x + 1 + y = \lambda x.\psi(h_1(y), \psi(0,x)) = \phi_{\langle h_1(y),0,0 \rangle + 1}.$$

Hence, take $h_1(y + 1) = \langle h_1(y),0,0 \rangle + 1$. Thus, $h_1 \in \mathcal{PR}^{(1)} \subset \mathcal{R}^{(1)}$ and $\phi_{h_1(y)} = \lambda x.x + y.$//

Lemma 2   There is an $h_2 \in \mathcal{R}^{(1)}$ such that $\phi_{h_2(y)} = \lambda x.y.$

Proof $\lambda x.y = \lambda x.K(\langle 0,x \rangle) + y.$
Now

$$K(\langle 0,x \rangle) = \psi(5,\psi(3,x)) = \phi_{\langle 5,3,0 \rangle + 1}(x)$$

Set

$$m = \langle 5,3,0 \rangle + 1$$

Then

$$K(\langle 0,x \rangle) + y = \psi(h_1(y), \psi(m,x)) = \phi_{\langle h_1(y),m,0 \rangle + 1}$$

Hence,

$$h_2 = \lambda y.\langle h_1(y),m,0 \rangle + 1$$

will do.//

Lemma 3   There is an $h_3 \in \mathcal{R}^{(1)}$ such that $\phi_{h_3(y)} = \lambda x.\langle y,x \rangle.$

Proof We have seen (Lemmas 1 & 2) that $\phi_{h_1(0)} = \lambda x.x$ and $\phi_{h_2(y)} = \lambda x.y.$
Thus,

$$\lambda x.\langle y,x \rangle = \lambda x.\langle \psi(h_2(y), x), \psi(h_1(0), x) \rangle = \phi_{\langle h_2(y),h_1(0),1 \rangle + 1}$$

Hence

$$h_3 = \lambda y.\langle h_2(y),h_1(0),1 \rangle + 1$$

will do.//

Theorem 1   (The *S-m-n* or *index* or *iteration* theorem). There exists a $\sigma \in \mathcal{R}^{(2)}$ such that $\phi_{\sigma(i,x)} = \lambda y.\phi_i^{(2)}(x,y).$

Proof Since $\phi_i^{(2)}(x,y) = \phi_i(\langle x,y\rangle)$,

$$\sigma = \langle i,h_3(x),0\rangle + 1$$

will do.//

Note: We have actually shown that $\sigma \in \mathcal{PR}^{(2)}$.

Corollary 1   There is a function $S_n^m \in \mathcal{PR}^{(m+1)}$ such that

$$\phi_{S_n^m}^{(n)}(i,\vec{x}_m) = \lambda\vec{y}_n.\phi_i^{(m+n)}(\vec{x}_m,\vec{y}_n)$$

Proof Set

$$h = \lambda uv.\langle\Pi_1^m(u),\ldots,\Pi_m^m(u),\Pi_1^n(v),\ldots,\Pi_n^n(v)\rangle$$

Since $h \in \mathcal{PR} \subset \mathcal{P}$, there is an e such that $h = \phi_e^{(2)}$.
Consider

$$w = \phi_i\circ\phi_e^{(2)}.$$

For all u and v,

$$w(u,v) = \phi_i(\phi_e^{(2)}(u,v)) = \phi_i(\phi_{\sigma(e,u)}(v))\ [\text{Theorem 1}] = \phi_{\langle i,\sigma(e,u),0\rangle+1}(v)$$

Thus, for all $(\vec{x}_m,\vec{y}_n)$,

$$w(\langle\vec{x}_m\rangle,\langle\vec{y}_n\rangle) = \phi_{\langle i,\sigma(e,\langle\vec{x}_m\rangle),0\rangle+1}(\langle\vec{y}_n\rangle) = \phi_{\langle i,s(e,\langle\vec{x}_m\rangle),0\rangle+1}^{(n)}(\vec{y}_n)$$

Note that

$$w(\langle\vec{x}_m\rangle,\langle\vec{y}_n\rangle) = \phi_i(\langle\vec{x}_m,\vec{y}_n\rangle) = \phi_i^{(m+n)}(\vec{x}_m,\vec{y}_n)$$

Thus,

$$S_n^m = \lambda i\vec{x}_m.\langle i,\sigma(e,\langle\vec{x}_m\rangle),0\rangle + 1$$

will do.
    Note that e is a constant, being a (fixed) index of a fixed function h.//

Corollary 2   Let $f \in \mathcal{P}^{(m+n)}$.
    Then there is an $h \in \mathcal{PR}^{(m)}$ such that $f = \lambda\vec{x}_m\vec{y}_n.\phi_{h(\vec{x}_m)}^{(n)}(\vec{y}_n)$.

Proof Let e be a fixed index for f; that is, $f = \phi_e^{(m+n)}$. $h = \lambda \vec{x}_m . S_n^m(e, \vec{x}_m)$ will do.//

Note: The *S-m-n* theorem will be mostly applied in its Corollary 2 version.

In this chapter, we have presented the foundations of a formal theory of effective computability getting also a first flavor of unsolvability results. We shall now digress from the further development of formal computability in order to look at the various other formalisms that were proposed as formalizations of the notions "algorithm" and "algorithmic function".

The deeper results about the set $\mathcal{P}$ will be presented from Chapter 7 onward.

§3.9 APPENDIX

The reader will often encounter in the literature (e.g., Davis [3], Kreider and Ritchie [7], Kleene [6]), the following alternative (number theoretic) definition of $\mathcal{P}$.

Definition 1 $\mathcal{P}_0$ is the closure of $\{\lambda xy.x + y, \lambda xy.x \doteq y, \lambda xy.xy\}$ under *substitution* and *$\tilde{\mu}$-operation* on *total* functions.//

We shall show that $\mathcal{P}_0 = \mathcal{P}$. Apart from inferring from this that $\mathcal{P}$ allows a fair amount of flexibility in its definition, some of the tools developed in the course of showing $\mathcal{P}_0 = \mathcal{P}$ will be useful later.

Lemma 1 $\mathcal{P}_0 \subset \mathcal{P}$.

Proof We know that the initial functions of $\mathcal{P}_0$ are in $\mathcal{P}$. Also, $\mathcal{P}$ is closed under *$\tilde{\mu}$-operation* applied on *total* functions, since on this type of functions μ and $\tilde{\mu}$ coincide. Finally, $\mathcal{P}$ is closed under *substitution* as well. $\mathcal{P}_0$ being the *smallest* set with these properties it follows that $\mathcal{P}_0 \subset \mathcal{P}$.//

In view of the normal form theorem (Theorem 3.7.4), to show the opposite inclusion, $\mathcal{P} \subset \mathcal{P}_0$, it suffices to show that $\mathcal{PR} \subset \mathcal{P}_0$. We embark on this task now.

Definition 2 $\mathcal{R}_0$ is $\{f \in \mathcal{P}_0 | f \text{ is } total\}$. $(\mathcal{R}_0)_*$ is the set of predicates with characteristic functions in $\mathcal{R}_0$.//

Lemma 2 $P(\vec{x}) \in (\mathcal{R}_0)_*$ iff for some $f \in \mathcal{R}_0, P(\vec{x}) \equiv f(\vec{x}) = 0$.

Proof See Theorem 2.2.4.//

Lemma 3 $(\mathcal{R}_0)_*$ is closed under the Boolean operations.

Proof See Theorem 2.2.5.//

Example 1 The relations $x \le y, x > y, x = y, x \ne y$ are in $(\mathcal{R}_0)_*$. Indeed, $x \le y \equiv x \dot- y = 0$. The rest follows from Lemmas 2 and 3.//

Lemma 4   If $P(\vec{x}_n) \in (\mathcal{R}_0)_*$ and $\lambda\vec{y}.f(\vec{y}) \in \mathcal{R}_0$, then $P(x_1, \ldots, x_{i-1}, f(\vec{y}), x_{i+1}, \ldots, x_n) \in (\mathcal{R}_0)_*$.

Proof Problem 33.//

Definition 3   Let $P(y,\vec{x})$ be a predicate.
Then $(\tilde{\mu}y)\, P(y,\vec{x})$ means $(\tilde{\mu}y)\, c_p(y,\vec{x})$.//

Lemma 5 $\mathcal{R}_0$ is closed under $(\mu y)_{\le z}$ and $(\mathcal{R}_0)_*$ is closed under $(\exists y)_{\le z}$ and $(\forall y)_{\le z}$.

Proof Let $\lambda y\vec{x}.f(y,\vec{x}) \in \mathcal{R}_0$.
Then $(\mu y)_{\le z} f(y,\vec{x}) = (\tilde{\mu}y)|z + 1 - y| \cdot f(y,\vec{x})$,
where $|z + 1 - y| = z + 1 \dot- y + y \dot- (z + 1)$. Also, let $P(y,\vec{x}) \in (\mathcal{R}_0)_*$.
Then $(\forall y)_{\le z} P(y,\vec{x}) \equiv (\mu y)_{\le z} (\neg P(y,\vec{x})) = z + 1.//$

Lemma 6 $\mathcal{R}_0$ is closed under definition by cases.

Proof See Theorem 2.2.8.//

Example 2 $\mathcal{R}_0$ contains pairing functions with their projections. Indeed, $J = \lambda xy.(x + y)^2 + x$ is 1-1 and clearly $J \in \mathcal{R}_0$. Furthermore, $K = \lambda z.(\mu x)_{\le z} ((\exists y)_{\le z} z = J(x,y))$ and $L = \lambda z.(\mu y)_{\le z} ((\exists x)_{\le z} z = J(x,y))$; thus $K, L \in \mathcal{R}_0$.//

In view of Example 2, to show that $\mathcal{PR} \subset \mathcal{R}_0$ it suffices to show that $\mathcal{R}_0$ is closed under pure iteration. (See Theorem 2.6.1.) The technique of showing that $f \in \mathcal{R}_0$, where $g \in \mathcal{R}_0$ and

$$f(0,y) = y$$

$$f(x + 1,y) = g(f(x,y))$$

is by now well known to the reader. We just have to "code" the computation of $f(x,y)$ by a single number. The problem lies in that the coding cannot be accomplished in this stage of development of $\mathcal{R}_0$-properties by using J, since the projection $\lambda inx.\Pi_i^n(x) = \lambda inx.$ **if** $i = 1$ **then** $K^{n \dot- 1}(x)$ **else** $L \circ K^{n \dot- i}(x)$ involves iteration already!

We shall use a different coding, a variant of which will be of further use in Chapter 5. (See Smullyan [14], Jones [5].)

Definition 4 An n-tuple $(a_n, a_{n+1}, \ldots, a_1, a_0)$ is a *dyadic* representation of the number $m \geq 1$ iff $m = \Sigma_{i=0}^n a_i 2^i$ and, for $i = 0, \ldots, n$, $a_i \in \{1,2\}$. It is customary to write $(a_n a_{n-1} \ldots a_1 a_0)_2 = m$ or simply $a_n a_{n-1} \ldots a_1 a_0 = m.//$

Exercise 1 Show that every $m \geq 1$ has a *unique* dyadic representation (*Hint.* For every $m \geq 1$ there are unique integers q and r such that $m = 2q + r$ and $0 < r \leq 2$).

Using the dyadic representation, a sequence of numbers $u_1, u_2, \ldots, u_l$ ($u_i \geq 0$ for $i = 1, \ldots l$) is coded as follows:(*) let $\overbrace{11 \ldots 1}^{m}$, $m \geq 0$, be the longest block of consecutive 1's appearing in any of the u_i's ($m = 0$ means that no u_i has the digit 1 in its dyadic representation). Let $\overline{w}$ denote the dyadic representation of w and let "$*$" denote string concatenation. Then code $u_1, \ldots, u_l$ by the number whose dyadic representation is

$$\overset{m+1}{\overbrace{211 \ldots 12}} * \overline{\overline{u_1}} * \overset{m+1}{\overbrace{21 \ldots 12}} * \overline{\overline{u_2}} * \overset{m+1}{\overbrace{21 \ldots 12}} * \ldots * \overset{m+1}{\overbrace{21 \ldots 12}} * \overline{\overline{u_l}} * \overset{m+1}{\overbrace{21 \ldots 12}}.$$

Denote this number by $\langle u_1, \ldots u_l \rangle$ as usual.

We shall now show that such a coding is expressible in $\mathcal{R}_0$.

Lemma 7. The predicates $x \mid y$, $\Omega(x) \equiv$ "x is a power of 2"(†) $y = 2^{|x|}$, where $|x|$ is the dyadic length of x, $z = x * y$(‡) are in $(\mathcal{R}_0)_*$.

Proof

(1) $x \mid y \equiv x \neq 0 \ \& \ (\exists z)_{\leq y} \ y = xz$

(2) $\Omega(x) \equiv x \geq 1 \ \& \ (\forall i)_{\leq x} \ (i \mid x \ \& \ i \neq 1 \Rightarrow 2 \mid i).$

(3)   Observe that $\overset{|x|}{\overbrace{1 \ldots 1}} \leq x \leq \overset{|x|}{\overbrace{2 \ldots 2}}$; that is $2^{|x|} - 1 \leq x \leq 2(2^{|x|} - 1)$, hence $2^{|x|} - 1 \leq x < 2^{|x|+1} - 1$ or $2^{|x|} \leq x + 1 < 2^{|x|+1}$.

(*)This coding is due to Quine [10]. See also Smullyan [14].

(†)See Smullyan [14].

(‡)In what follows, we shall not distinguish between a number and its dyadic string representation.

Thus $2^{|x|} = \max \{z \mid z \leq x + 1 \,\&\, \Omega(z)\}$. Hence, $y = 2^{|x|} \equiv \Omega(y) \,\&\, y \leq x + 1$ $\&\, (\forall i)_{\leq x+1} (\Omega(i) \Rightarrow i \leq y)$.

(4) $z = x*y \equiv (\exists w)_{\leq y+1} (\exists u)_{\leq z} (w = 2^{|y|} \,\&\, z = u + y \,\&\, u = xw).//$

Note: $0*y = y = y*0$ according to the scheme on the right of $\equiv$ in (4) above.

Definition 5   [1,14] xBy (resp. xEy, xPy) iff $y = x*u$ (resp. $y = u*x$, $y = u*x*v$) for some $u, v \geq 0$. tally(x) iff $\neg 2Px.//$

Lemma 8   The following predicates are in $(\mathcal{R}_0)_*$.

(5) xBy

(6) xEy

(7) xPy

(8)   tally(x)

(9) $y = $ maxtal$(x) \equiv y$ is the longest tally which is part of x.

Proof

(5) $xBy \equiv (\exists u)_{\leq y} \, y = x*u$

(6) $xEy \equiv (\exists u)_{\leq y} \, y = u*x$

(7) $xPy \equiv (\exists u,v)_{\leq y} \, y = u*v \,\&\, xEu$

(8)   tally$(x) \equiv \neg 2Px$

(9) $y = $ maxtal$(x) \equiv$ tally$(y) \,\&\, yPx \,\&\, \neg(y*1)Px$
      $\equiv$ tally$(y) \,\&\, yPx \,\&\, (\forall w)_{\leq x} [\text{tally}(w)$
      $\&\, wPx \Rightarrow wPy].//$

Definition 6   [1,14] "$(\exists y)_{Pz} \dots$" (resp. "$(\forall y)_{Pz} \dots$") denotes "$(\exists y)_{\leq z} \, yPz$ $\& \dots$" (resp. "$(\forall y)_{\leq z} \, yPz \Rightarrow \dots$").//

Important note: All the predicates of Lemma 8 are obtainable from $z = x*y$ and performance of explicit transformations (on predicates) Boolean operations and closure under $(\exists y)_{Pz}$ and $(\forall y)_{Pz}$. The *smallest* class of predicates thus obtainable is the class of *strictly rudimentary* predicates or *S-rudimentary* predicates (Bennett [1], Smullyan [14], Jones [5]), *base 2* (we get different classes for different bases [1]).

Exercise 2   Show that all predicates of Lemma 8 are *S*-rudimentary.

Exercise 3 Show that $y = x_1*x_2*\ldots*x_n$ is S-rudimentary.

Exercise 4 $(R_0)_*$ contains the S-rudimentary predicates.

Lemma 9 The following predicates are S-rudimentary, hence in $(R_0)_*$.

(10) $\text{code}(x) \equiv$ "x is $\langle u_1, \ldots, u_n \rangle$ for some sequence $u_1, \ldots, u_n$".

(11) $x \in y \equiv$ "$y = \langle u_1, \ldots, u_n \rangle$ for some sequence $u_1, \ldots, u_n$ and $x = u_i$ for some i".

(12) $\text{Next}(x,y,u) \equiv$ "$x \in u$ and $y \in u$ and y is the next term after x in the coded sequence".

(13) $\text{First}(x,u) \equiv$ "$x \in u$ and x is the *first* term of the coded sequence".

(14) $\text{Last}(x,u) \equiv$ "$x \in u$ and x is the *last* term of the coded sequence".

(15) $\lambda y \vec{u}_n . y = \langle u_1, \ldots, u_n \rangle$.

Proof (Essentially Smullyan [14].)

(10) $\text{code}(x) \equiv (\exists w)_{Px} (w = \text{maxtal}(x) \ \& \ (2*w*2)Bx \ \& \ (2*w*2)Ex$
$\& \ (2*w*2) \neq x \ \& \ \neg(2*w*2*w*2)Px \ \& \ (\exists u,v)_{Px} [(2*w*2*u*2*w*2)Px$
$\& \ w = v*1 \ \& \ vPu \ \& \ \neg wPu])$.

(11) $x \in y \equiv \text{code}(y) \ \& \ (\exists w)_{Py} (w = \text{maxtal}(y) \ \& \ (2*w*2*x*2*w*2)Py$
$\& \ \neg wPx)$.

(12) $\text{Next}(x,y,u) \equiv x \in u \ \& \ y \in u \ \&$
$(\exists w)_{Pu} (w = \text{maxtal}(u) \ \& \ (2*w*2*x*2*w*2*y*2*w*2)Pu)$.

(13) $\text{First}(x,u) \equiv x \in u \ \& \ (\exists w)_{Pu} (w = \text{maxtal}(u) \ \&$
$(2*w*2*x*2*w*2) Bu)$.

(14) $\text{Last}(x,u) \equiv x \in u \ \& \ (\exists w)_{Pu} (w = \text{maxtal}(u) \ \&$
$(2*w*2*x*2*w*2) Eu)$.

(15) $y = \langle u_1, \ldots, u_n \rangle \equiv \text{code}(y) \ \& \ (\exists w)_{Py} (w = \text{maxtal}(y) \ \& \ \neg wPu_1 \ \&$
$\ldots \& \ \neg wPu_n \ \& \ y = 2*w*2*u_1*2*w*2*u_2*\ldots*u_n*2*w*2)$.//

Definition 7 [14] We define K and L, the projections of $\lambda xy.\langle x,y \rangle$ by:

$$K(z) = \begin{cases} x \text{ if } z = \langle x,y \rangle \text{ for some } y \\ 0 \text{ otherwise} \end{cases}$$

$$L(z) = \begin{cases} y \text{ if } z = \langle x,y \rangle \text{ for some } x \\ 0 \text{ otherwise } // \end{cases}$$

Lemma 10 $y = K(x)$ and $y = L(x)$ are S-rudimentary.

    Proof $y = K(x) \equiv y = 0 \lor (\exists z)_{P_x} x = \langle y,z \rangle$.//

We are now, almost, ready to code "computations" using $\langle \ \rangle$. One thing we have to settle, is to define projections $\Pi_i^n(z)$ such that $y = \Pi_i^n(z)$ is in $(\mathcal{R}_0)_*$ as a predicate of i,n,y,z. The trick is [14] instead of coding $u_1, \ldots, u_n$ by $\langle u_1, \ldots, u_n \rangle$ to code it by $\langle \langle 1,u_1 \rangle, \ldots, \langle i,u_i \rangle, \ldots, \langle n,u_n \rangle \rangle$. This way, along with each u_i we also "store" its position in the code.

Lemma 11 There is a predicate $P(x,y,z,u)$ in $(\mathcal{R}_0)_*$ such that if $g \in \mathcal{R}_0^{(1)}$, then $z = g^y(x) \equiv (\exists u)P(x,y,z,u)$.

    Proof $z = g^y(x) \equiv (\exists u)(\text{code}(u) \ \& \ (\exists v,m)_{P_u}\{\text{First}(v,u) \ \& \ \text{Last}(m,u) \ \& \ K(v) = 1 \ \& \ K(m) = y + 1 \ \& \ L(v) = x \ \& \ L(m) = z\} \ \& \ (\forall i,j)_{P_u}\{\text{Next}(i,j,u) \Rightarrow K(i) + 1 = K(j) \ \& \ L(j) = g(L(i))\})$. $P(x,y,z,u)$ is the predicate, above, quantified by $(\exists u)$.//

    Note: It is clear that if $y = f(\vec{x}_n) \in (\mathcal{R}_0)_*$ then $\lambda \vec{x}_n.(\tilde{\mu}y)(y = f(\vec{x}_n)) \in \mathcal{R}_0$. Thus, for example $K,L \in \mathcal{R}_0$. We used this fact above.

Theorem 1 $\mathcal{PR} \subset \mathcal{R}_0$.

    Proof $g^y(x) = K((\tilde{\mu}v)P(x,y,K(v),L(v)))$. Thus $\mathcal{R}_0$ is closed under pure iteration. The presence of pairing functions, with their projections, in $\mathcal{R}_0$ yields the result.//

Theorem 2 $\mathcal{P} = \mathcal{P}_0$ and $\mathcal{R} = \mathcal{R}_0$.

    Proof By Theorem 1, the decoding function d and the Kleene predicate $T^{(n)}(z,\vec{x}_n,y)$ are in $\mathcal{R}_0,(\mathcal{R}_0)_*$, respectively, for all $n \geq 1$. Thus, for all i, $\phi_i^{(n)} = \lambda \vec{x}_n.d((\mu y)\,T^{(n)}(i,\vec{x}_n,y)) = \lambda \vec{x}_n.d((\tilde{\mu}y)\,T^{(n)}(i,\vec{x}_n,y)) \in \mathcal{P}_0$. That is, $\mathcal{P} \subset \mathcal{P}_0$. Combining with Lemma 1, we have $\mathcal{P} = \mathcal{P}_0$. $\mathcal{R} = \mathcal{R}_0$ is then trivial.//

    Note: For an alternative coding, using the Gödel β-function,(*) see Davis [3], Yasuhara [15], Kreider and Ritchie [7]. Using Gödel's β provides a shorter proof of Theorem 2, if we assume knowledge of number theory up to (and including) the Chinese Remainder Theorem.

    Our main reason of using the S-rudimentary relations, as implied earlier, is that in Chapter 5 we shall show the existence of an S-rudimentary T-predicate, a result we shall use in Chapter 13. This is the simplest class of relations we know that includes the T-predicate.

(*)$\beta \in \mathcal{PR}^{(3)}$ and for any sequence $u_1, \ldots, u_n$ there are numbers c, d such that for $i = 1, \ldots, n \ \beta(c,d,i) = u_i$.

PROBLEMS

1. Prove that if a class $\mathcal{C}$ of functions contains $\lambda xy.x + y$, $\lambda xy.x \div y$, $\lambda xy.xy$, and $\lambda x.\lfloor \sqrt{x} \rfloor$ and is closed under *substitution* and the operation [applied to *total* functions] $\lambda x.f(x) \mapsto \lambda x.f^{-1}(x)$ (where $f^{-1}(x)$ is shorthand for $\min\{y \mid f(y) = x\}$), then it is closed under $\tilde{\mu}$. (*Hint*: this problem is in the spirit of J. Robinson [11]. Let $\lambda xy.g(x,y) \in \mathcal{C}$ be total and consider $f = \lambda x.(\tilde{\mu}y)g(x,y)$. $\mathcal{C}$ contains pairing functions—e.g., $\lambda xy.((x+y)^2 + y)^2 + x$. Let K and L be its first and second projections. Then $f(x) = L((\tilde{\mu}z)[g(K(z),L(z)) = 0 \& K(z) = x])$. Fill in the rest, and extend to the case of the total function $\lambda \vec{x}_n y.g(\vec{x}_n, y) \in \mathcal{C}$, $n > 1$.)

2. Prove Theorem 3.2.3.

3. Prove that for $n \geq 0$, $A_n(x) < A_x(2)$ a.e. Conclude that

   (i) If $\lambda x.f(x) \in \mathcal{PR}$, then $f(x) < A_x(2)$ a.e.

   (ii) $\lambda x.A_x(2) \notin \mathcal{PR}$.

4. Fill in the missing details in the proof of Theorem 3.3.2.

5. Explain qualitatively why the Corollary of Theorem 3.3.2. should hold.

6. Show that it is impossible to obtain all of $\mathcal{PR}$ by starting with some *finite* set of initial functions and allowing *substitution* only as a functional operation. (*Hint*: Show that the closure of any finite set of $\mathcal{PR}$-functions under substitution contains functions strictly majorized by A_n, for some fixed n.)

7. Let g_1, g_2, and h be in $\mathcal{PR}$. Consider the schema of *unnested double recursion*

   $$(A) \begin{cases} f(0,y) = g_1(y) \\ f(x+1,0) = g_2(x) \\ f(x+1,y+1) = h(x,y,f(x+1,y),f(x,y+1)) \end{cases}$$

   which defines f.

   (a) Prove that $f(x,y) \leq A_n^k(x+y)$ for some appropriate n and k, and all x,y.

   (b) Using the method of proof employed for Theorem 3.3.2, show that $\lambda zxy.z = f(x,y) \in \mathcal{PR}_*$.

   (c) Conclude (by (a)) that $f \in \mathcal{PR}$.

(d) *Bypass* (a)–(c) and prove *directly* that $f \in \mathcal{PR}$. (*Hint for (d):* Observe that, in schema (A), the value $f(u,v)$ is computed from values $f(i,j)$ such that $i + j + 1 = u + v$. Use a "history" function $\lambda u . H(u)$ of such values, where

$$H(u) = \Pi_{i \leq u} p_i^{f(i,u-i)},$$

and show first that $H \in \mathcal{PR}$.)

8. Show that unnested double recursion *with* parameters does not lead outside $\mathcal{PR}$.

9. Prove that f, defined by the schema of unnested course-of-values double recursion below, is in $\mathcal{PR}$ if g_1, g_2, and h also are.

$$f(0,y) = g_1(y)$$
$$f(x+1,0) = g_2(x)$$
$$f(x+1,y+1) = h(x,y,H(x+y+1))$$

where

$$H = \lambda z . \Pi_{i \leq z} p_i^{\Pi_{j \leq i} p_j^{f(j,i-j)}}$$

is the "history" function of f. Prove, also, that presence of parameters does not change the conclusion of this problem.

10. (Péter [9]) If $\lambda x . h(x)$, $\lambda x y \vec{z}_k . g(x,y,\vec{z}_k)$ and for $i = 1, \ldots, k$, $\lambda x y . m_i(x,y)$ are in $\mathcal{PR}$, then so is the function f defined by the following "recursion with substitutions in the parameter x":

$$f(x,0) = h(x)$$
$$f(x,y+1) = g(x,y,f(m_1(x,y),y), \ldots, f(m_k(x,y),y))$$

(*Hint:* Show that

(a) For some p and k, $f(x,y) \leq A_p^k(x+y)$ for all x,y.

(b) $\lambda z x y . z = f(x,y) \in \mathcal{PR}_*$ (using the method of proof of Theorem 3.3.2.).

(c) Conclude, by (a), that $f \notin \mathcal{PR}$.)

   Note: Péter [9] proves the claim differently. (See Problem 14.)

11. Prove Theorem 3.4.2.

12. Prove that $\mathcal{PR}^{(1)}$ is the *closure* of

$$\{\lambda x.x+1, \lambda x.x \dot{-} \lfloor \sqrt{x} \rfloor^2, \lambda x.\lfloor \sqrt{x} \rfloor \}$$

under

$$\lambda x.f(x) \longmapsto \lambda x.f^x(0), \ (\lambda x.f(x), \lambda x.g(x)) \longmapsto \lambda x.f(g(x)),$$

$$(\lambda x.f(x), \lambda x.g(x)) \longmapsto \lambda x.f(x) + g(x).$$

Note: This is due to R. M. Robinson [12]. See also Péter [9]. The initial function $\lambda x.\lfloor \sqrt{x} \rfloor$ is not used in [9,12] (see next problem), but is introduced here for convenience.

(*Hint:* [9,12]. Show first that $\mathcal{PR}$ is the *closure* of $\{\lambda x.x+1, \lambda xy.x+y,$ $\lambda x.\lfloor \sqrt{x} \rfloor$, $\lambda x.x \dot{-} \lfloor \sqrt{x} \rfloor^2\}$ under *substitution* and *iteration at 0*, that is, $\lambda x.f(x) \longmapsto \lambda x.f^x(0))$. To this end, it suffices to show that the given initial functions and operations suffice to build the "special" pairing function $J = \lambda xy.((x+y)^2+y)^2+x$ and its projections $K = \lambda x.x \dot{-} \lfloor \sqrt{x} \rfloor^2$ (given) and $L = \lambda x.K(\lfloor \sqrt{x} \rfloor)$ on one hand, and that the above-mentioned closure, which we tentatively call $\mathcal{PR}'$, is closed under definition by cases. (Review Problem 2.25 here.)

Observe next that $\lambda x.k$, for any $k \in \mathcal{N}$, can be shown to be in $\mathcal{PR}'$ via a derivation involving unary functions only [e.g., if $s = \lambda x.x+1$ and $u = \lambda x.s^x(0)$, then $\lambda x.0 = \lambda x.u^x(0)$]. Once again, utilizing only unary functions, show that $\lambda x.x^2$ is in $\mathcal{PR}'$ [use iteration at 0 and $\lfloor \sqrt{x} \rfloor$]. $\lambda x.\lfloor x/2 \rfloor$ can be obtained by a simple simultaneous recursion which, using J *implicitly* as $G = \lambda x.((L(x)+K(x)+1)^2+K(x)+1)^2+L(x)$, is converted to an iteration at 0, showing $\lambda x.\lfloor x/2 \rfloor \in \mathcal{PR}'$ (once again, utilizing unary functions only and "$+$"). To show closure of $\mathcal{PR}'$ under definition by cases, it suffices to show that $\lambda xy.(1 \dot{-} x)y \in \mathcal{PR}'$. As subtasks, show first that $\lambda xy.xy \in \mathcal{PR}'$, by observing that $xy = ((x+y)^2-x^2-y^2)/2$, *and* $(x+y)^2 \ge x^2$, $(x+y)^2 - x^2 \ge y^2$. Thus, define a total function $\text{dif}(x,y)$ such that

$$\text{dif}(x,y) = \begin{cases} x - y \ \textbf{if } x \ge 0 \\ \text{whatever is most convenient } \textbf{otherwise} \end{cases}$$

[dif is obtained from K by substitution!]. Next, $\lambda x.1 \dot{-} x$ should be defined similarly, but independently of dif. (dif is *not* $\lambda xy.x \dot{-} y$.)

At the end of all this, $\mathcal{PR} \subset \mathcal{PR}'$ is concluded; hence, $\mathcal{PR} = \mathcal{PR}'$. Now argue that $\lambda xy.x+y$ can be eliminated in favor of,

$$(\lambda x.f(x), \lambda x.g(x)) \longmapsto \lambda x.f(x) + g(x), \text{ thus obtaining } \mathcal{PR}^{(1)}.)$$

13. Show that inclusion of $\lambda x.\lfloor \sqrt{x} \rfloor$ in the set of initial functions of the previous problem is superfluous.

14. (Péter's solution to Problem 10.) Prove that f, defined in Problem 10, is in $\mathcal{PR}$ if h, g, and m_i $(i=1,\ldots,k)$ are also. Constrain your method of solution along the following lines:

(a) Experiment with values such as $f(x,1),f(x,2)$, etc. to realize that $f(x,y)$ is obtained by certain "substitutions" (the number of substitutions being *variable* and depending on y) starting with initial functions h,g, and m_i.

(b) Show that there is a *universal* primitive recursive function $\lambda xy.\psi(x,y)$ for the class of functions obtained by the above substitutions, namely,

$$\psi(x,0)=x$$

$$\psi(x,y+1) = \begin{cases} h\circ\psi(x,\exp(1,y)) & \textbf{if } \exp(0,y)=0 \\[2mm] \textbf{for } i=1,\ldots,k: & \\[1mm] m_i(\psi(x,\exp(1,y)),\exp(2,y)) & \textbf{if } \exp(0,y)=i \\[2mm] g(x,\exp(1,y),\psi(x,\exp(2,y)),\ldots,\psi(x, \exp(k+1,y))) & \textbf{if } \exp(0,y)>k \end{cases}$$

(c) Show (if necessary, looking ahead at §3.8) that there is a $\lambda xy.p(x,y) \in \mathcal{PR}$ such that $\psi(\psi(x,s),t)=\psi(x,p(s,t))$ for all x,s,t. You can do this by defining p through primitive recursion. For example, $\psi(x,p(s,0))=\psi(\psi(x,s),0)=\psi(x,s)$. Thus, it suffices to set $p(s,0)=s$. Next find a (course-of-values) recurrence relation for $p(s,t+1)$.

(d) Finally, show that there is a function $e\in\mathcal{PR}^{(1)}$ such that $f(x,y)=\psi(x,e(y))$ for all x,y.

15. Complete the proof of Theorem 3.5.2.

16. Show that for all n, there is an m such that $A_n(x)<\xi(m,x)$, for all x, where ξ is that of Theorem 3.5.2 and A_n is the Ackermann function of §3.3. Conversely, for all n, there is an m such that $\xi(n,x)<A_m(x)$, for all x.

17. Prove Corollary 3 of Theorem 3.6.1.

18. Prove that if $R(y,\vec{z})$ is re and $\lambda\vec{x}.f(\vec{x})\in\mathcal{P}$, then $R(f(\vec{x}),\vec{z})$ is re. (*Hint:* $R(y,\vec{z})$ is re iff $R(y,\vec{z})=\mathrm{dom}(g)$ and $\lambda y\vec{z}.g(y,\vec{z})\in\mathcal{P}$, by the projection theorem.)

19. Fill in the missing details in the proof of Theorem 3.7.1.

20. Prove Theorem 3.7.2 (*only if* part) without relying on Theorem 3.7.1. (*Hint:* Use Problem 18.)

21. Show that Univ^m of Problem 2.36 is in $\mathcal{R}$.

22. Show that $\lambda zxy.z = \phi_x(y) \notin \mathcal{R}_*$ (*Hint:* If $\lambda zxy.z = \phi_x(y) \in \mathcal{R}_*$, then so is $\lambda x.0 = \phi_x(x)$. Conclude that $\psi = \lambda x.\textbf{if } 0 = \phi_x(x) \textbf{ then } 1 \textbf{ else } 0$ is both *in* and *not* in $\mathcal{P}$.)

23. *Argument:* $\lambda xy.x = y \in \mathcal{R}_*$; hence (Problem 18), $\lambda xyz.\phi_x(y) = z$ is re. $\lambda xy.x \neq y$ $\in \mathcal{R}_*$; hence (Problem 18), $\lambda xyz.\phi_x(y) \neq z$ is re. By Theorem 3.6.3, $\lambda xyz.\phi_x(y) = z \in \mathcal{R}_*$, contradicting Problem 22. What is wrong with the above "argument"?

24. Show that if $R(\vec{x})$ and $Q(\vec{y})$ are re, then so are $R(\vec{x}) \vee Q(\vec{y})$ and $R(\vec{x}) \& Q(\vec{y})$, whereas $\neg R(\vec{x})$ is not re in general.

25. Prove: A predicate $R(\vec{x})$ is re iff for some $\lambda \vec{x}.f(\vec{x})$ in $\mathcal{P}$, $R(\vec{x}) \equiv f(\vec{x}) = 0$.

26. Prove that $\{\langle x,y \rangle \mid \phi_x(y)\!\downarrow\}$, where $\lambda xy.\langle x,y \rangle$ is a recursive pairing function, is not recursive, but is re.

27. Prove that, given a primitive recursive pairing function $\lambda xy.\langle x,y \rangle$, there is a $\tau \in \mathcal{PR}^{(2)}$ such that $\xi(i,\langle x,y \rangle) = \xi(\tau(i,x),y)$ for all i,x,y, where ξ is that of Theorem 3.5.2.

28. Show that if $f \in \mathcal{PR}^{(m+n)}$, then there is an $h \in \mathcal{PR}^{(m)}$ such that $f = \lambda \vec{x}_m \vec{y}_n . \xi^{(n)}(h(\vec{x}_m),\vec{y}_n)$, where $\xi^{(n)}(i,\vec{y}_n)$ stands for $\xi(i,\langle \vec{y}_n \rangle)$.

29. Show that $\{x \mid 0 \in \text{dom}(\phi_x)\}$ is re.

30. Show that $\{x \mid 0 \in \text{range}(\phi_x)\}$ is re.

31. A real number x such that $0 \leq x < 1$ is *computable* iff there is a recursive function $f: \mathcal{N} \rightarrow \mathcal{N}$ such that $f(i)$ is the ith decimal digit of x, $i \geq 0$. Show that noncomputable real numbers exist. (*Hint:* Use a cardinality argument: The set of computable real numbers is enumerable, whereas $\{x \mid 0 \leq x < 1\}$ is not. Establish the claim made in this hint.)

32. (Rice) Prove: $A = \{x \mid \phi_x \in \mathcal{C}\}$ is recursive iff $\varnothing = \mathcal{C}$ or $\mathcal{C} = \mathcal{P}^{(1)}$. (*Hint:* (*only if* part) Argue by contradiction. Consider separately the cases (a) $\varnothing$ (the empty function) is in $\mathcal{C}$, and (b) $\varnothing \notin \mathcal{C}$. For case (a), use the *S-m-n* theorem to establish existence of an $h \in \mathcal{R}^{(1)}$ such that $\phi_{h(x)} = \varnothing$ iff $\phi_x(x)\!\uparrow$. Conclude that A is not re, hence A is not recursive. (b) is similar.)

33. Prove Lemma 3.9.4.

34. Let $\lambda x.f(x)$ be called (2-)*rudimentary* iff $\lambda xy.y = f(x)$ is rudimentary and $f(x) \leq x$ for all x.(*) Prove that for every function $\lambda \vec{x}.g(\vec{x})$ in $\mathcal{P}$, there are a rudimentary function $\lambda x.u_g(x)$ and a rudimentary predicate $P_g(w,\vec{x})$ such that $f = \lambda \vec{x}.u_g((\mu w)P_g(w,\vec{x}))$. Moreover, P_g can be chosen so that, for a given $\vec{x}$, $P_g(w,\vec{x})$ is true for at most *one* w. (*Hint:* See the proof of Theorem 3.7.1, and §3.9.)

(*)A more general definition is given in §5.1.

35. Show that there is a rudimentary $\lambda x.d(x)$ and a rudimentary $T(z,x,y)$ such that

(a) $f \in \mathcal{P}^{(1)}$ implies existence of i such that $f = \lambda x.d((\mu y)\,T(i,x,y))$

(b) $f \in \mathcal{P}^{(n)}$ implies existence of i such that $f = \lambda \vec{x}.d((\mu y)\,T(i,\langle \vec{x} \rangle,y))$, where $\langle \ \rangle$ is that of §3.9.

36. Prove that every nonempty re set is the range of a rudimentary function. (*Hint:* If S is re, then $x \in S \equiv (\exists y)\,T(i,x,y)$ for some i, where T is rudimentary. The x-components of a systematic list of pairs (x,y) such that $T(i,x,y)$ enumerate S with repetitions.)

37. (J. Robinson [11]) Show that $\mathcal{P}^{(1)}$ is the closure of $\{\lambda x.x+1, \lambda x.x \doteq \lfloor \sqrt{x} \rfloor^2\}$ under

$$(\lambda x.f(x), \lambda x.g(x)) \longmapsto \lambda x.f(x)+g(x) \tag{1}$$

$$(\lambda x.f(x), \lambda x.g(x)) \longmapsto \lambda x.f(g(x)) \tag{2}$$

$$\lambda x.f(x) \longmapsto \lambda x.f^{-1}(x) \text{ for } total\ f, \tag{3}$$

where $f^{-1}(x) \overset{\text{def}}{=} (\tilde{\mu}y)(f(y)=x)$. Prove that if, moreover, $f \mapsto f^{-1}$ is applied to *total, onto* functions only, then $\mathcal{R}^{(1)}$ is obtained. (*Hint:* [11]. Let $\mathcal{P}'$ be the closure of $\{\lambda x.x+1, \lambda xy.x+y, \lambda x.x \doteq \lfloor \sqrt{x} \rfloor^2\}$ under *substitution* and (3). Show that $\mathcal{P} \subset \mathcal{P}'$ by showing

(a) A certain *total* function dif, such that $\mathrm{dif}(x,y)=x-y$ if $x \geq y$, is in $\mathcal{P}'$

(b) $\lambda xy.x=y$ and $\lambda xy.x \leq y$ are in $\mathcal{P}'$

(c) $\lambda x.x^2$ is in $\mathcal{P}'$

(d) $\lambda x.1 \doteq x$ and $\lambda x.1 \doteq (1 \doteq x)$ are in $\mathcal{P}'$

(e) $\lambda x.\mathrm{rem}(x,2)$ and $\lambda x.\lfloor x/2 \rfloor$ are in $\mathcal{P}'$

(f) $\lambda xy.\,xy$ is in $\mathcal{P}'$ (this uses $\lfloor x/2 \rfloor$)

(g) $\lambda x.\lfloor \sqrt{x} \rfloor$ is in $\mathcal{P}'$

(h) $\mathcal{P}'$ is closed under $\tilde{\mu}$ applied to total functions (see Problem 1).

As examples, $\mathrm{dif} = \lambda xy.K(K^{-1}(2x+2y)+3x+y+4)$ will do, where $K = \lambda x.x \doteq \lfloor \sqrt{x} \rfloor^2$ (verify). Then $x^2 = \mathrm{dif}(K^{-1}(2x),2x)$ for all x. Also, $x = y \equiv \mathrm{dif}(x,y)+\mathrm{dif}(y,x)=0$. Once (a)-(h) have been established, a glance at §3.9 shows that dif, rather than $\lambda xy.x \doteq y$, would do equally well in the definition of $\mathcal{P}_0$. Thus $\mathcal{P}=\mathcal{P}_0 \subset \mathcal{P}'$; hence $\mathcal{P}=\mathcal{P}'$. Replacing $\lambda xy.x+y$ and substitution by the rules (1) and (2), we are done.)

REFERENCES

[1] Bennett, J. *On Spectra.* Ph.D. dissertation, Princeton University, 1962.

[2] Brainerd, W.S., and Landweber, L.H. *Theory of Computation.* New York: Wiley, 1974.

[3] Davis, M. *Computability and Unsolvability.* New York: McGraw-Hill, 1958.

[4] Grzegorczyk, A. "Some Classes of Recursive Functions". *Rozprawy Matematyczne* 4. Warsaw, 1953: 1-45.

[5] Jones, N.D. *Computability Theory; An Introduction.* New York: Academic Press, 1973.

[6] Kleene, S.C. *Introduction to Metamathematics.* Princeton, N.J.: Van Nostrand, 1950.

[7] Kreider, D.L., and Ritchie, R.W. *Notes on Recursive Function Theory.* Lecture notes for Mathematics 89 (Seminar in Logic). Dartmouth College, Winter term, 1965.

[8] Manna, Z. *Mathematical Theory of Computation.* New York: McGraw-Hill, 1974.

[9] Péter, R. *Recursive Functions.* New York: Academic Press, 1967.

[10] Quine, W.V. "Concatenation as a Basis for Arithmetic". *J. Symbolic Logic* 11 (1946): 105–114.

[11] Robinson, J. "General Recursive Functions". *Proceedings Amer. Math. Soc.* 1 (1950): 703–718.

[12] Robinson, R.M. "Primitive Recursive Functions". *Bulletin Amer. Math. Soc.* 53 (1947): 925-942.

[13] Rogers, H. *Theory of Recursive Functions and Effective Computability.* New York: McGraw-Hill, 1967.

[14] Smullyan, R. *Theory of Formal Systems.* Annals of Mathematics Studies, No. 47. Princeton, N.J.: Princeton University Press, 1961.

[15] Yasuhara, A. *Recursive Function Theory and Logic.* New York: Academic Press, 1971.

CHAPTER 4
Machines

In this chapter alternative (proposed) formalisms are presented for the intuitive notions of "algorithm" and "algorithmic function". Formally, this chapter, as well as Chapters 5 and 6, are not required for the continuation of our development of Recursion Theory.

However, since Chapters 2 and 3 introduced the notion of "algorithmic function" directly, without recourse to a formal counterpart for the notion of "algorithm", our intuitive grasp of both concepts will be considerably strengthened if we now discuss various formalisms for the notion of "algorithm". Moreover, machine formalisms are important in the study of the "Computational Complexity" of classes of functions in $\mathcal{P}$. However, it should be noted that a study of the complexity of $\mathcal{P}$-functions can also be based on purely number-theoretic considerations.

At the beginning of Chapter 3, we amended the proposed informal notion of algorithm, by allowing an algorithm to "compute forever" on some inputs. This informal notion has been formalized under various guises (e.g., Turing [15], Shepherdson and Sturgis [13], Post [11], Markov [9]). The key issue is what *type* of instructions to allow. A clear requirement exists: that the instructions should be chosen from the simplest possible repertoire so that, intuitively, no doubt can exist that they can be carried out by a "machine" (i.e., a "nonintelligent agent, that nevertheless is able to carry 'simple' instructions faithfully").

We shall present the Turing, Shepherdson/Sturgis, and Markov approaches in this chapter. The Post approach and some of its variants will be discussed in Chapter 6.

§4.1 TURING MACHINES

A *Turing Machine*, or TM in short, is an *abstract model* of a "computer". Like a computer, it can faithfully carry out simple instructions. To avoid technology-dependent limitations, it is defined so that it is, essentially, a computer with unbounded "memory". Thus, it never runs out of "storage" during a computation; correspondingly, no artificial limitation is put on the amount of "time"(*) it takes for the Turing Machine to complete a computation.

In the first instance, we shall introduce "fixed-program" Turing Machines which, unlike large real computers, can interpret a single fixed "program" rather than an unlimited variety of programs. Stored-program (or *universal*) TMs exist, as we shall soon discover. In the meanwhile, it is clear that (fixed-program) TMs can be thought of as formal programs or algorithms.

It is instructive to proceed informally at first and introduce formal definitions later.

Informally, a TM consists of:

(a)   An *infinite* two-way tape.

(b)   A *read/write tape-head*.

(c)   A "black-box", the *finite control*, which can be at any one of a (fixed) *finite* set of *internal states* (or simply, *states*).

Pictorially, a TM is often represented as in the figure below:

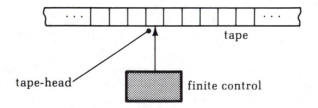

(*)The quoted terms such as "memory", "storage", "time", "program", etc. have formal counterparts, hence, the quotes.

The tape is subdivided into *squares*, each of which can hold a *single* symbol out of a *finite* set of admissible symbols associated with the TM (this is the *tape-alphabet*); the tape-head can scan only *one* square at a time. The TM tape corresponds to the memory of an actual computer. It can go *left* or *right* by *one* square at a time as instructed by the finite control(*); we actually find it more convenient to think of the tape-head moving, rather than the tape, in our informal discussions.

The head can *read* or *write* a symbol on the scanned square. Writing a symbol is assumed to *erase* first what was on the square previously. There is a distinguished alphabet symbol, the *blank* denoted B, which appears everywhere except a finite set of tape squares.

How does the machine operate?

Depending on

(a)   The currently scanned *symbol*

(b)   The current *state*

the machine will:

(i)   Write a symbol or leave the symbol unchanged on the scanned square;

(ii)   Enter a (possibly) new *state*;

(iii)   Move the head to the left or right or it will leave it stationary.

At the present, consistent with our informal definition of algorithm, as amended in Chapter 3, we shall require TMs to be *deterministic*; i.e., given the current symbol/state pair, a TM has a uniquely defined response. We shall eventually see that *nondeterministic* TMs do not have more computing power and are important notions in complexity (of computation) theory.

Given a set of consecutive symbols on the tape of a TM, we position the tape-head on the left-most *nonblank* symbol, "initialize" the machine and let it go.

Our convention is (following Davis [2]) that the machine will stop (or "halt", as we often say) iff at some instance *it is not specified* how to proceed, given the current symbol/state pair. At that time (when the machine has halted), whatever is on the tape (that is, the *largest* string of symbols which starts *and* ends with some *nonblank* symbol), is the *result* or *output* of the TM computation for the given input.

A question might now naturally arise in the reader's mind: How will the "TM operator" ever be sure that he has seen the *largest* string on tape delimited by nonblanks given the fact that the tape is infinite? Is he or she doomed to look at the tape forever and never be sure of the output?

(*)The finite control plays the role of the CPU of actual computers.

This problem arises due to the informality of our discussion above. In reality, a TM is an "agent" which produces a *sequence* of strings, the last string (if it exists) being the output of the computation which has as input the first string. At each step of the computation, the machine *finitely modifies* the current string to produce the next. Essentially, the tape referred to earlier is the "material" on which the machine writes the strings. Instead of an infinite tape, one can think then of an *extendible* but finite tape. The tape is always just long enough to hold the current string.

To be formal:

Definition 1   *A Turing Machine* (TM), *M*, is a 4-tuple $M = (A,K,q_0,\delta)$, where $A = \{s_0,s_1, \ldots , s_n\}$ is a *finite* set of tape symbols called the *tape-alphabet*, and $s_0 = B$ (the blank symbol).

$K = \{q_0,q_1, \ldots , q_m\}$ is a *finite* set of *states* (sometimes called the *state-alphabet*) of which q_0 is *distinguished*: It is the *start*-state.(*) (Intuitively, the machine must be "set" to q_0 before it starts; with this in mind it was indicated earlier that the machine be "initialized").

δ is the behavior or *transition function*, $\delta : A \times K \rightarrow A \times K \times \{L,R,S\}$. (It is similar to a "program" in that it specifies the sequence of operations of the machine.)//

Note: In the specification of δ above, *L* (resp. *R,S*) stand for "head move left" (resp. "right", "stay"). Also, requiring the δ relation to be single-valued is requiring TMs to be *deterministic*. Finally, note that $\delta(s_i,q_j) = (s_k,q_l,S)$ means that if the machine is in state q_j, and the head scans s_i, then it *replaces* s_i by s_k (of course, it may be $s_i = s_k$) enters state q_l (of course, q_l may be q_j) and the head does *not* move.

Since δ is a function on a finite set, it is itself a finite set and there are a number of ways it can be listed. For example, δ can be represented as a digraph (K,E), where K, the vertex set of the digraph, is the set of states of the TM, and the edge-set E is such that there is an edge (q_j,q_l) labeled (s_i,s_k,m) iff $\delta(s_i,q_j) = (s_k,q_l,m)$, where $m \in \{S,L,R\}$. (See Manna [8].)

Or, δ can be represented as a table (i.e., two-dimensional matrix), where each s_i-symbol labels exactly one row and each q_j-state labels exactly one column, so that $s_k q_l m$ (where $m \in \{S,L,R\}$) is in the matrix entry with coordinates (s_i,q_j) iff $\delta(s_i,q_j) = (s_k,q_l,m)$ (Trakhtenbrot [14]).

Also, δ can be specified as a set of *quintuples* (essentially Turing's own approach; Davis [2] is using a similar representation, but instead of quintuples he uses quadruples), where $q_j s_i s_k q_l m$ is in the set iff $\delta(s_i,q_j) = (s_k,q_l,m)$.

Finally, δ can be represented as a PL/1-like program which is a collection of instructions such as

---

(*)It is indicated that q_0 is distinguished by separately mentioning it in the 4-tuple (A,K,q_0,δ). Any name and/or subscript for the initial state may be used.

Q: **do**   **if** a **then** b; m; **goto** P;

             if a' **then** b'; m'; **goto** P';

        .

        .

        if $a^{(n)}$ **then** $b^{(n)}$; $m^{(n)}$; **goto** $P^{(n)}$;

   end

/* i.e., if the current symbol is a (resp. a', a'', ..., $a^{(n)}$), then replace it by b (resp. b', b'', ..., $b^{(n)}$), move the head according to m (resp. m', m'', ..., $m^{(n)}$), where $m^{(i)} \in \{S,L,R\}$, and goto P (resp. P', P'', ..., $P^{(n)}$) */

Clearly, for each state q_j, the statement "**if** s_i **then** s_k; m; **goto** q_l" is in the **do-end** group labeled q_j iff $\delta(s_i, q_j) = (s_k, q_l, m)$ (see, for example, Jones [6]).

Whenever we wish to specify a TM which does a particular task, we shall find it convenient to use either the graph or program approach. However, for the formal discussion of TM-computations in this and the next chapter we shall view δ as a set of quintuples.

We now proceed to precisely define TM-computations. First, let us rephrase Definition 1:

Definition 1'   A TM M *over* the (tape) alphabet $A = \{s_0, \ldots, s_n\}$, where $s_0 = B$, and with internal states $K = \{q_0, \ldots, q_l\}$ is a *finite* set of *quintuples* of the form $qabq'm$, where $\{q,q'\} \subset K$, $\{a,b\} \subset A$ and $m \in \{S,L,R\}$, such that no two quintuples begin with the *same* two symbols.//

Note: The last restriction essentially says that the relation $(a,q) \mapsto (b,q',m)$ is a *function*.

Definition 2   A *tape expression* (or simply *tape*), t, is a member of A^*.//

Definition 3   An *instantaneous description* (ID) is a string $t_1 q a t_2$, where $q \in K$, $a \in A$ and t_1, t_2 are tapes.//

Note: Intuitively, an ID is a "snapshot" of a computation. The tape contents are the *tape* $t_1 a t_2$, the current state is q and the scanned symbol is a.

Definition 4   Given two IDs α and β of a TM, M, we say that α *yields* β, in symbols $\alpha \longrightarrow \beta$ (or $\alpha \xrightarrow{M} \beta$ if we want to emphasize that we are refering to machine M) iff *one* of the following holds, where $\{t_1, t_2\} \subset A^*$ and $s_m \in A$:

(i) $q_i s_j s_k q_l S \in M$ and
    $\alpha = t_1 q_i s_j t_2$
    $\beta = t_1 q_l s_k t_2$

(ii) $q_i s_j s_k q_l R \in M$ and
either $\alpha = t_1 q_i s_j s_m t_2$
$\quad\quad \beta = t_1 s_k q_l s_m t_2$
or $\quad \alpha = t_1 q_i s_j$
$\quad\quad \beta = t_1 s_k q_l B$

(iii) $q_i s_j s_k q_l L \in M$ and
either $\alpha = t_1 s_m q_i s_j t_2$
$\quad\quad \beta = t_1 q_l s_m s_k t_2$
or $\quad \alpha = q_i s_j t_2$
$\quad\quad \beta = q_l B s_k t_2.//$

Note: The intuitive idea that the machine tape is extendible is incorporated in the "or" part of cases (ii) and (iii) of Definition 4.

Definition 5 α is a *final* ID iff $\alpha = t_1 q s t_2$ and M contains *no* quintuple starting with $qs.//$

Note: The above definition says that a TM halts, iff $\delta(s,q)\uparrow$, where (s,q) is the current symbol-state pair. Alternative conventions require the machine to halt, as soon as it enters any one of a distinguished subset of states called *halting* states (see Hopcroft and Ullman [5] for this approach as well as for a discussion of a large variety of variants of the TM model. For example, TMs with a one-way infinite tape, TMs with many tapes, TMs with "multi-track" tapes, TMs with multidimensional tapes, etc. See also Problems 1 through 6.)

Definition 6 A *computation* of M is a *finite* sequence of IDs $\alpha_1, \ldots, \alpha_k$, such that α_k is *final* and if $k > 1$ then $\alpha_i \xrightarrow{M} \alpha_{i+1}$ for $i = 1, \ldots, k - 1$, where α_1, the *initial* ID, has the form $q_0 t$, q_0 being the initial state and $t \in A^+$. For any i, $\alpha_i \rightarrow \alpha_{i+1}$ is one *step* of the computation. The *complexity* of the computation is $k.//$

Example 1 Consider the TM M over $\{B,\#\}$, where $M = \{q_0 B \# q_0 R\}$. It is clear that if $\alpha_1 = q_0 B$, then *no* computation exists with α_1 as initial ID. Indeed, $q_0 B \rightarrow \# q_0 B \rightarrow \#\# q_0 B \rightarrow \#\#\# q_0 B$, etc.$//$

Proposition 1 Let α_1 be an initial ID for TM M. Then there is *at most one* computation $\alpha_1, \ldots, \alpha_k$.

Proof Each α_i yields a unique α_{i+1} since δ is a function.$//$

Example 2 Construct a TM M over A, which when it receives as input (initial tape) $t \in (A - \{B\})^+$ it prints tBt and halts.

Here is the plan: (in pseudo-PL/1).

```
t ← tB
i ← 1; x ← ith symbol in t
do while (x ≠ B)
t ← tx
i ← i + 1
x ← ith symbol in t
end
```

We convert the above pseudo-PL/1 program to a TM program. Let $A = \{B, s_1, \ldots, s_n\}$. Note that for $t \leftarrow tB$ we do nothing (this is given by tape-extendibility property).

To do $t \leftarrow tx$, observe that, at some intermediate step of the TM computation the tape is $t_0 B t_1$, where t_1 is a prefix of t_0 and t_0 is the *original* t; indeed $t_0 = t_1 x t_2, \{t_1, t_2\} \subset (A - \{B\})^*$. To put x after $t_0 B t_1$ we must go over the first blank B until we find a *new* blank after t_1; also we must "remember" where x was so that the next position is processed next ("$i \leftarrow i + 1; x \leftarrow i$th symbol in t"). x's position is remembered by erasing x (replacing by B) as soon as "$x \leftarrow i$th symbol in t" is executed. Thus, the picture when we travel right-bound to carry x after current t is

$$t_1 B t_2 B t_1 B$$

↑ this is where x is to be copied
↑ original position of x ($t_0 = t_1 x t_2$)

Thus,

```
q₀:  do   if s₁ then B; R; goto q⁽¹⁾;    /* q⁽¹⁾ "remembers" to copy s₁ */
          if s₂ then B; R; goto q⁽²⁾;    /* q⁽²⁾ "remembers" to copy s₂ */
                            ·
                            ·
                            ·
          if sₙ then B; R; goto q⁽ⁿ⁾;    /* q⁽ⁿ⁾ "remembers" to copy sₙ */
     end
```

For $i = 1, \ldots, n$:

$q^{(i)}$:   **do** **if** B **then** B; R; **goto** $q_1^{(i)}$;   /* subscript "1" is a "switch". It indicates that the *first B* (right-bound travel) has been encountered. */

          if $s_j, j = 1, \ldots, n$ **then** s_j; R; **goto** $q^{(i)}$;   /* travel right over
   end   nonblanks */

For $i = 1, \ldots, n$:

$q_1^{(i)}$:   **do** **if** B **then** s_i; L; **goto** $r^{(i)}$;   /* Second B found. Print s_i, as remembered by $q^{(i)}$ & $q_1^{(i)}$; state $r^{(i)}$ now sends the head back left to the blanked position of s_i in t_0 */

          if $s_j, j = 1, \ldots, n$ **then** s_j; R; **goto** $q_1^{(i)}$;   /* travel right over nonblanks */
     end

For $i = 1, \ldots, n$:

$r^{(i)}$:   **do** **if** B **then** B; L; **goto** $r_1^{(i)}$;   /* The subscript "1" is a "switch". It indicates that the *first B* (left-bound travel) has been encountered. */

          if $s_j, j = 1, \ldots, n$ **then** s_j; L; **goto** $r^{(i)}$;   /* travel left over
   end   nonblanks */

For $i = 1, \ldots, n$:

$r_1^{(i)}$:   **do** **if** B **then** s_i; R; **goto** q_0;   /* Second B (left-bound) found. Restore s_i, as remembered by $q^{(i)}, q_1^{(i)}, r^{(i)}, r_1^{(i)}$, then go one square right to scan *next* symbol of t_0 and repeat the cycle */

          if $s_j, j = 1, \ldots, n$ **then** s_j; L; **goto** $r_1^{(i)}$;   /* travel left over
   end   nonblanks */

Note that no quintuple starts with $q_0 B$; this will stop the computation.//

It should be clear that it is not very easy to "program" a one-tape TM as its language is "assembly-like" rather than "high-level". This is why in computational complexity, whenever the TM is used as an abstract computer, a *multi-tape* TM is used.

Exercise 1   Informally, construct a 2-tape TM which does the same task as the TM in Example 2 above.

Example 3   Build a TM M which, when presented with an initial tape $x \in \{0,1\}^+$, computes $x + 1 \in \{0,1\}^+$ where x is interpreted as an integer in binary.
In pseudo-PL/1:

$i \leftarrow$ position of last symbol in x
do while $(x_i = 1)$ /* x_i denotes i th symbol of x */
   $x_i \leftarrow 0$
   $i \leftarrow i - 1$
end /* now i scans a 0 or a B. Make it 1 */
   $x_i \leftarrow 1$

In TM language:

q_0:   **do**   **if** 0 **then** 0; R; **goto** q_0;   /* go right skipping over
        if 1 **then** 1; R; **goto** q_0;   0s and 1s */
        if B **then** B; L; **goto** q_1;   /* Now last symbol of x is
                              scanned. q_1 "will perform
                              the addition" */

   end

q_1:   **do**   **if** 1 **then** 0; L; **goto** q_1;   /* This is "**do while** $(x_i = 1)$;
                            $x_i \leftarrow 0; i \leftarrow i - 1;$ **end**" */
        if 0 **then** 1; S; **goto** q_2;   /* No instructions are
        if B **then** 1; S; **goto** q_2;   labeled q_2, thus q_2
                              ends the computation */

   end

Note that the TM (tape) alphabet was $A = \{B,0,1\}$. In graph form, the above TM can be represented as

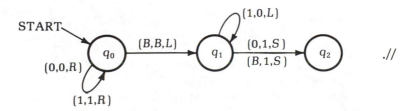

Example 4 Build a TM M which, when presented with an initial tape $x \in \{0,1\}^+$, computes $x \dot{-} 1 \in \{0,1\}^+$, where x is interpreted as an integer in binary:
In pseudo-PL/1:

> $i \leftarrow$ position of last symbol in x
>   **do while** $(x_i = 0)$
> $x_i \leftarrow 1$
> $i \leftarrow i - 1$
>   **end** /* Now i scans a 1 or B, the latter if, originally, $x \in \{0\}^+$. In the former case, change 1 to 0; in the latter turn B to 0 and erase all the 1s following to its right */
> **if** $x_i = 1$ **then** $x_i \leftarrow 0$ /* and now stop */
> **else do** $x_i \leftarrow 0$
> $i \leftarrow i + 1$
>   **do while** $(x_i = 1)$
> $x_i \leftarrow B$
> $i \leftarrow i + 1$
>   **end**
>   **end**

For a change, we now present the TM for the above PL/1-like program directly in graph form:

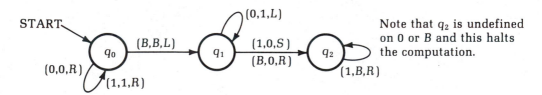

START

$(0,0,R)$

q_0 (B,B,L) q_1 $(0,1,L)$ $(1,0,S)$ $(B,0,R)$ q_2 $(1,B,R)$

$(1,1,R)$

Note that q_2 is undefined on 0 or B and this halts the computation.

Once again, the alphabet A is $\{B,0,1\}$.//

Example 5 Build a TM M which, when presented with an initial tape xBy, where $\{x,y\} \subset \{0,1\}^+$, computes $x + y \in \{0,1\}^+$, where x and y are interpreted as integers in binary:
Here is our PL/1-like plan:

> **do while** $(y \neq 0)$
> $y \leftarrow y \dot{-} 1$
> $x \leftarrow x + 1$
> **end** /* x holds the answer. y is now 0; i.e., the tape is $t0$, where $t = x_0 + y_0$ where x_0, y_0 are the *original* x and y */
> $y \leftarrow B$ /* Now the tape is $x_0 + y_0$ */

We may view $y \leftarrow y \div 1$ and $x \leftarrow x + 1$ as "calls" to the TM programs of Example 4 and Example 3, respectively. We shall incorporate them with appropriate amendments (to begin with, the procedures for $y \leftarrow y \div 1$ and $x \leftarrow x + 1$ should have different state-names since they are to be used in the same TM program; otherwise, havoc will occur).

Important note: Observe that y becomes 0 (in the repeated application of $y \leftarrow y \div 1$) as soon as (q_1, B) is the current state/symbol pair. At this stage, the picture would be

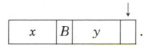

$$x_0 + y_0 \, B \, \underbrace{11 \ldots 1}$$

q_1 as many as binary
length of y_0.

We should *not* replace B by 0 (as we did in Example 4) as (see PL/1 program above) we would have to reset to B again! That is, instead of $(B,0,R)$ use (B,B,R) on edge (q_1, q_2).

Here is the next PL/1-like refinement:

$t \leftarrow x_0 B y_0$ /*t, the initial tape, is $x_0 B y_0$ */
$L: c \leftarrow$ last symbol in t /* to find it, find the *second* B from the beginning of t.
The picture is now:

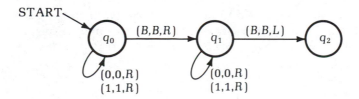

Note that x is, in general, the "partial sum", and y is the "balance" from original y_0 */
$y \leftarrow y \div 1$; **if** $y = 0$ /* i.e., machine scans B */ **then** clear the 1s to the right and **stop**;
$c \leftarrow$ last symbol in x;
$x \leftarrow x + 1$;
goto L

Now the TM built as a sequence of modules:

/* Find last symbol in t */

/* $y \leftarrow y \div 1$; **if** $y = 0$ **then** clear the 1s and **stop** */
/* *Important*: The machine is already scanning the right-most symbol of
y */

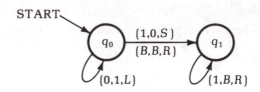

/* If $y \neq 0$, move to right-most position in x.

Picture before:

| x | B | y |

Picture after:

| x | | B | y |

*/

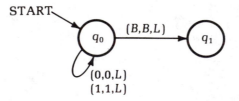

/* Do $x \leftarrow x + 1$, given that the head is in the right-most digit of x. Then
goto L (first module) */

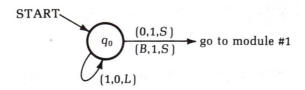

We now connect all these machines. Essentially, the "ending" state in each
becomes the start state of the next, while we are careful to shift the indices. In
connecting the second to the third machine, since q_1 (of the second) and q_0 (of
the third) have different behavior on the symbol 1, they cannot be identified.
Instead, we add a q_2 to second machine so that

and then identify q_2 with q_0 of the third machine.

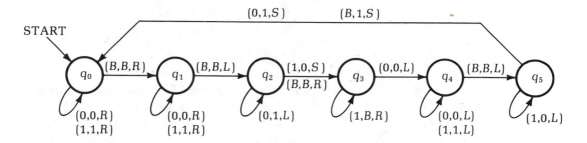

Note that $\delta(B,q_3)\uparrow$. This is the halting condition.//

Example 6 By enriching the alphabet of M and slightly amending the input/output requirements, we can shorten the TM-description that does "$x + y$ in binary".

Let the input be $x_0By_0\#$, where $\#$ is to serve as a right-most delimiter. As before, let an intermediate tape in the computation be called $xBy\#$. For output, we accept the form

$$\boxed{x + y} \boxed{B} \boxed{B} \quad \cdots \quad \boxed{B} \boxed{\#}$$

The following TM does the job:

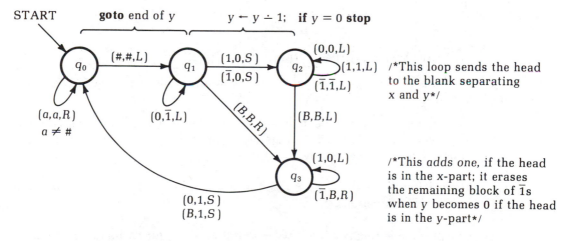

Note that the alphabet is $A = \{B,0,1,\bar{1},\#\}$. By using the extra symbol $\#$, we were able to reach the end of y using two states. By using $\bar{1}$ along with 1, we were able to use q_3 for two different tasks on input (essentially) 1. q_3 is undefined on # and the computation halts with final tape

$$\boxed{x_0 + y_0} \boxed{B} \quad \dots \quad \boxed{B}\boxed{\#}$$

$$\underbrace{\qquad\qquad}_{\text{length}(y_0) + 1}$$

.//

Let M be a TM over an alphabet A. M defines a function (often nontotal) from A^* to A^* in a natural way: (t_1,t_2) is a pair in the function iff there is a computation with initial tape t_1 and final tape t_2, where t_2 is taken to be the maximum tape delimited by non-blanks, or just a single B if the final is blank. Note that such a function can be interpreted as a function $\mathcal{N} \to \mathcal{N}$ since there is a 1-1 correspondence between $\mathcal{N}$ and A^*, where $\Lambda \in A^*$ corresponds to $0 \in \mathcal{N}$ and $x \in A^+$ corresponds to the number which has x as its $|A|$-adic expansion. (The elements of A are identified in some fixed way with the numbers 1,2, $\dots, |A|$.)

However, for work in computability, it is more convenient to take a somewhat different approach (see Davis [2]).

First of all, let $\alpha_1, \dots, \alpha_k$ be a computation of a TM M. Then, since α_k is the result of the computation that starts with α_1, we write $\alpha_k = \text{Res}_M(\alpha_1)$.

Further, for any string $s \in (A \cup K)^*$ let $[s]$ be the number of occurrences of the symbol "1" in s. If no 1's occur in s, then $[s] = 0$.

Definition 7   We say that $f:\mathcal{N}^n \to \mathcal{N}$ is *computed* by M iff, for all $\vec{x}_n \in \mathcal{N}^n$:

(a) $f(\vec{x}_n)\downarrow$ iff M has a computation with initial ID

$$\alpha_1 = q_0 \underbrace{11\dots 1}_{x_1+1} B\underbrace{11\dots 1}_{x_2+1}B \dots : B\underbrace{11\dots 1}_{x_n+1}.$$

(b)   Whenever $f(\vec{x}_n)\downarrow, f(\vec{x}_n) = [\text{Res}_M(\alpha_1)]$.

f is called *partially computable*. If it is total, we also say that it is *computable*.//

Note: $\overset{a}{\overbrace{SS\dots S}}$ will be denoted by S^a for convenience. Note that on input, a number a is denoted by 1^{a+1}. This avoids any ambiguity in the representation of 0, which is represented by 1.

Definition 8 $\mathcal{PC}$ (resp. $\mathcal{C}$) is the class of *partially computable* (resp. *computable*) functions.//

Our purpose is to show that $\mathcal{PC} = \mathcal{P}$ and $\mathcal{C} = \mathcal{R}$. Our earlier examples show how tedious it is to "program" (construct) TMs that do particular tasks. So even though it is possible to show directly that $\mathcal{PC}$ contains $\lambda x.x + 1, \lambda x.x \doteq 1$ and is closed under the same operations as $\mathcal{P}$ (this ensures $\mathcal{P} \subset \mathcal{PC}$), we shall instead first introduce a new type of machine which is easier to program. We shall then see that

(a) any function in $\mathcal{P}$ can be programmed in some such machine

(b) any such machine can be simulated by some TM. We conclude this paragraph with a few more examples.

Example 7   Show that $\lambda x.x + 1 \in \mathcal{C}$. Take $A = \{B,1\}$. The machine is

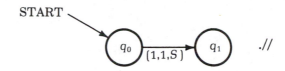

Example 8   Show that $\lambda xy.x + y \in \mathcal{C}$.
Take $A = \{B,1\}$.

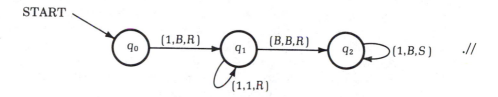

Exercise 2   Show that $\lambda x.2x \in \mathcal{C}$.
(*Hint:* If $x = 0$, then just erase the 1 representing 0. Else, erase a 1 to obtain x 1's on input. Then, using Example 1, double the number of 1's.)

§4.2 THE URM

The *Unbounded Register Machine*, in short URM, was proposed by Shepherdson and Sturgis [13] as an alternative formalism to that of the TM. As its "instruction set" is closer to "high-level" languages than that of the Turing Machine, the URM is easier "to program". Thus, some authors who prefer to introduce recursion theory through some machine-formalism have favored the URM over the TM (e.g., [1,3]). Indeed, as we shall see in this paragraph, to show that any function in $\mathcal{P}$ is URM-computable is very easy. Conversely, to show that any URM-computable function is in $\mathcal{P}$ is not difficult either.

The URM, as presented in this paragraph, is a fixed-program machine (but stored-program, or universal, URMs exist; see Chapter 5), thus we shall think of the URM as a programming formalism, since each machine corresponds to a unique program. Our approach, although equivalent to that of

Shepherdson and Sturgis [13], is not identical to it. It is suggested by the inductive definition of $\mathcal{P}$.

Intuitively, a URM is a pair (R,P) where R is a finite set of *registers*, and P a *program*. Each register can hold an arbitrary number from $\mathcal{N}$ (i.e., there is no bound on the register size as in "real" machines. Hence the term *unbounded* register machine). We use capital letters of the alphabet, with or without subscripts to denote registers. If X is a register, (X) denotes its contents (i.e., the number "stored" in X). A program is a finite sequence of instructions of the following forms only:

(a) $X = X + 1$

(b) $X = X \div 1$

(c)   **while** $X \neq 0$

(d)   **end**

where instructions (c) & (d) occur in matched pairs. Instructions *between* such a matched pair are to be repeated as long as $(X) \neq 0$; we understand that if $(X) = 0$ initially, then the instructions delimited by **while** $X \neq 0$ and matching **end** are skipped and whatever follows the **end** is executed next.

Normally, instructions of a program are executed sequentially, except that the instruction-pair **while-end** alters that sequence.

If the next instruction to execute is the **null** (nonexistent) instruction following the program, then the computation halts.

Example 1   The following program simulates (i.e., has the same effect as) the PL/1-like instruction $X = Y$, which, once it is executed, makes $(X) = (Y)$ and leaves the original (Y) invariant:

```
while X ≠ 0
   X = X ÷ 1
end /* now (X) = 0 */
while Z ≠ 0
   Z = Z ÷ 1
end /* now (Z) = 0 */
while Y ≠ 0
   X = X + 1
   Z = Z + 1
   Y = Y ÷ 1
end /* now, (X) = (Z) = original (Y); (Y) = 0 */
while Z ≠ 0
   Y = Y + 1
   Z = Z ÷ 1
end /* now (Y) is restored; (Z) = 0 */.//
```

We shall consistently take the point of view that if (R,P) is a URM, then a register is in R iff it is referenced in at least one instruction of P. Thus, we shall not refer to R explicitly in the sequel, as the R-information is included in P. We now present formal definitions:

Definition 1   Let A by the alphabet $\{X,1,\textbf{add},\textbf{sub},\textbf{while},\textbf{end},;\}$. A URM *register* is a string $X1^n(n \geq 0)$ over A.

   A URM is defined inductively as follows:

   (a)   **add** $X1^n$ is a URM for any $n \geq 0$

   (b)   **sub** $X1^n$ is a URM for any $n \geq 0$

   (c)   If P and Q are URMs, then $P;Q$ (and $Q;P$) is a URM

   (d)   If P is a URM, then **while** $X1^n;P;$ **end** is a URM for any $n \geq 0$.

   (e)   No string in A^* is a URM unless it can be shown to be one by a finite number of applications of rules (a) through (d).//

Definition 1 defines the syntax of URM programs or simply URMs. By **add** Z (resp. **sub** Z), where $Z = X1^n$ for some $n \geq 0$, we understand the instruction $Z = Z + 1$ (resp. $Z = Z \div 1$); i.e., increment (Z) by 1 (resp. decrement (Z) by 1 if $(Z) \neq 0$). By **while** Z, we understand "**while** $Z \neq 0$". These "semantics" are *intended* but by no means *present* in the above definition! We now postulate them formally, in a similar way as we did in the case of loop-programs.

   First, observe that by the above definition a URM is a *finite sequence* of instructions of the form **add** X, **sub** X, **while** X, and **end**; let the sequence be $(I_1; \ldots ;I_k)$. We say that i is the *instruction number* of I_i.

Definition 2   Let M be a URM with registers $R_1, \ldots , R_m$ and instruction sequence $(I_1; \ldots ;I_k)$. An *instantaneous description*, ID, of M is an $(m + 1)$-tuple $(r_1, \ldots , r_m;i) \in \mathcal{N}^m \times \{1,2, \ldots k,k + 1\}$.//

   Note: Intuitively, an ID is a snapshot of an M-computation, where $r_l = (R_l)$ for $l = 1, \ldots , m$ and i is the *instruction number* of the *current* instruction; the r_l-values are valid immediately *before* the execution of I_i.

Definition 3   Let $\alpha = (r_1, \ldots , r_m; i)$ and $\beta = (\tilde{r}_1, \ldots , \tilde{r}_m;\tilde{\imath})$ be two M-IDs where $M = (I_1; \ldots ;I_k)$. We say that α *yields* β, in symbols $\alpha \xrightarrow{M} \beta$, or simply $\alpha \rightarrow \beta$ if M is understood from the context, iff *one* of the following holds:

   (a) I_i is **add** R_j and
   $\tilde{r}_l = r_l, l \in \{1, \ldots , m\} - \{j\}$
   $\tilde{r}_j = r_j + 1$
   $\tilde{\imath} = i + 1$

(b) I_i is **sub** R_j and
$\tilde{r}_l = r_l, l \in \{1, \dots, m\} - \{j\}$
$\tilde{r}_j = r_j \div 1$
$\tilde{\imath} = i + 1$

(c) I_i is **while** R_j and
$\tilde{r}_l = r_l, l = 1, \dots, m$
$I_{\tilde{\imath}}$ is the **end**-instruction paired with **while** R_j

(d) I_i is **end** and

either $\begin{cases} \tilde{r}_l = r_l, l = 1, \dots, m \\ I_{\tilde{\imath}-1} \text{ is } \textbf{while } R_j, \text{ which matches } \textbf{end} \\ r_j \neq 0 \end{cases}$

or $\begin{cases} \tilde{r}_l = r_l, l = 1, \dots, m \\ \textbf{while } R_j \text{ matches } \textbf{end} \\ r_j = 0 \\ \tilde{\imath} = i + 1 \end{cases}$

(e) $i = k + 1$ and $\tilde{\imath} = k + 1$ and $\tilde{r}_l = r_l, l = 1, \dots, m.//$

Note: The yield-relation is similar to that for loop-programs, but simpler. Note that a **while**-instruction is an *unconditional jump* to the matching **end**, and an **end** is a *conditional jump* to either the instruction *following* the matching **while** or to the instruction following **end** itself, depending on whether the test register is $\neq 0$ or $= 0$, respectively. This makes URM behavior consistent with loop-program behavior. (See Definition 2.7.4; the purist may be inclined to rename "**while** X" "**repeat**", and the matching "**end**" "**until** X". In the interest of the above mentioned consistency we will resist this inclination.)

Rules (a) and (b) are straightforward. Rule (e) is similar to rule (e) of Definition 2.7.4. It is a technicality which ensures that the contents of the URM-registers, as well as the value of the i-component, are *total* functions of "time" (the usefulness of this will be appreciated in Chapter 13, where complexity of computation will be discussed).

An M-ID with $i = k + 1$, where $M = (I_1; \dots; I_k)$, is called *final*. An ID with $i = 1$ is called *initial*.

Definition 4   Let $\alpha_1 \dots, \alpha_t$ be a sequence of M-ID's such that

(1) α_1 is initial

(2) α_t is final

(3)   for $i = 1, \dots, t - 1 \; \alpha_i \xrightarrow[M]{} \alpha_{i+1}$

Then this sequence is called an *M-computation. t* is the *length* or *complexity* of the computation, an act such as $\alpha_i \rightarrow \alpha_{i+1}$ (for some i) being a *step* of the computation.//

Proposition 1   Let α_1 be an initial ID for a URM M. Then there is at most *one* computation starting with α_1.

$\quad\quad$ **Proof** For any α, there is a unique β such that $\alpha \xrightarrow[M]{} \beta$.//

Definition 5   Let M be a URM with registers $R_1, \ldots, R_l$. Let $X_1, \ldots, X_p$, $p \geq 1$, be a subset of the registers, where the X_i's are distinct, but Y may be one of them.

$\quad\quad$ Then $M_Y^{X_1, \ldots, X_p}$, or simply $M_Y^{\vec{X}_p}$, denotes the function f such that

$\quad\quad$ (1) $b = f(\vec{a}_p)$ iff an M-computation $\alpha_1, \ldots, \alpha_t$ exists, where $(X_i) = a_i, i = 1, \ldots, p$ and all other registers of α_1 hold a 0 and

$\quad\quad$ (2) $(Y) = b$ in α_t.

$\quad\quad$ If $\{X_1, \ldots, X_p\} = \{R_1, \ldots, R_l\}$, then instead of $M_Y^{\vec{X}_p}$ we may also write M_Y.//

Definition 6 $\mathcal{PU} = \{f | f = M_Y^{\vec{X}} \textit{ for some}$ URM M whose register set includes $\vec{X}$ and $Y\}$.

$\quad\quad$ $\mathcal{U}$ is the subset of *total* functions in $\mathcal{PU}$.//

We are ready to prove that $\mathcal{P} = \mathcal{PU}$ and $\mathcal{R} = \mathcal{U}$. We shall, however, digress to give an example.

Example 2   Let P be a loop-program. Then a URM M_p exists which simulates P.

$\quad\quad$ This claim means that the "*effect*" of any P-instruction is accomplished in M_p by some group of one or more instructions.

$\quad\quad$ More precisely,

$\quad\quad$ (a)   A 1-1 coding $C{:}\alpha^{(\mathrm{LP})} \rightarrow \alpha^{(\mathrm{URM})}$ exists which codes a loop-program ID ($\alpha^{(\mathrm{LP})}$) by a URM ID ($\alpha^{(\mathrm{URM})}$), and if $(\alpha_1, \ldots, \alpha_t)$ is a P-computation,

$\quad\quad$ (b)   An M_p-computation $(\beta_1, \ldots, \beta_m)$ exists, where $\beta_m = C(\alpha_t)$, and

$\quad\quad$ (c) $(C(\alpha_1), \ldots, C(\alpha_t))$ is a *subsequence* of $(\beta_1, \ldots, \beta_m)$.

$\quad\quad$ To this end, let $(X_1, \ldots, X_l; B_1, \ldots, B_k; I)$ be an ID-frame for P, where B_i corresponds to the P-instruction **Loop** X_{j_i}. (Note that X_{j_i}'s, $i = 1, \ldots, k$ need *not* be distinct, but B_i's *are* distinct! [Definition 2.7.4].)

$\quad\quad$ M_p is the *identical* sequence of instructions as P with the following exceptions:

$\quad\quad$ (a)   Each **Loop** X_{j_i} ($i = 1, \ldots, k$) is represented by the group of instructions $\quad \begin{cases} B_i = X_{j_i} \\ \textbf{while } B_i \end{cases}$

(b)   Each **end** instruction in P (where **Loop** X_{j_i} matches **end**) is represented by the group $\left\{ \begin{array}{l} \textbf{sub } B_i \\ \textbf{end} \end{array} \right.$

Note that instructions of the form $B_i = X_{j_i}$ are not primitive, but can be simulated. (Example 1.) A *single* auxiliary variable Z (different from all X_i's and all B_j's) can be used in M_p. Thus, the M_p-ID frame is $(Z, X_1, \ldots, X_l, B_1, \ldots, B_k; I)$.

Clearly, C: $(x_1, \ldots, x_l; b_1, \ldots, b_k; i) \rightarrow (0, x_1, \ldots, x_l, b_1, \ldots, b_k; \tilde{\imath})$ is an appropriate coding where $\tilde{\imath}$ is:

(1)   if i labels **add** X (resp. **sub** X, **null**) in P, then the instruction number of the corresponding **add** X (resp. **sub** X, **null**) in M_p

(2)   if i labels **end** in P, then the instruction number of the corresponding **sub** B_i.

(3)   if i labels **Loop** X_{j_i} in P (where B_i corresponds to X_{j_i}) then the instruction number of $B_i = X_{j_i}$.

The details are left to the reader to fill in, however, here is one of the cases to consider:

Let i_0 label **Loop** X_1 and i_1 label the matching **end** in P and assume for simplicity that B_1 corresponds to X_1.
Then

$$(x_1, \ldots, x_l; b_1, \ldots, b_k; i_0) \xrightarrow{\ P\ } (x_1, \ldots, x_l; x_1, b_2, \ldots, b_k; i_1)$$

and

$$(0, x_1, \ldots, x_l; b_1, \ldots, b_k; \tilde{\imath}_0) \xrightarrow{\ M_p\ } \cdots \xrightarrow{\ M_p\ } (0, x_1, \ldots, x_l; 0, b_2, \ldots, b_k; i_0^{(1)})$$

$$\xrightarrow{\ M_p\ } \cdots \xrightarrow{\ M_p\ } (x_1, 0, x_2, \ldots, x_l; x_1, b_2, \ldots, b_l; i_0^{(2)}) \xrightarrow{\ M_p\ } \cdots \xrightarrow{\ M_p\ }$$

$$(0, x_1, x_2, \ldots, x_l; x_1, b_2, \ldots, b_l; i_0^{(3)}) \xrightarrow{\ M_p\ } (0, x_1, \ldots, x_l; x_1, b_2, \ldots, b_l; \tilde{\imath}_1)$$

[Consult also with Example 1 and amend it appropriately for the needs of the present proof.] Note that instruction numbers $i_0^{(1)}$ and $i_0^{(2)}$ refer to instructions in the URM "subprogram" that accomplices $B_1 = X_1$. $i_0^{(3)}$ labels **while** B_1.//

By Example 2, URMs are at least as "powerful" a programming formalism as that for loop-programs. This result may be rephrased as $\mathcal{PR} \subset \mathcal{U}$. An alternative proof will be derived below as a side-effect of our proving that that $\mathcal{R} \subset \mathcal{U}$.

Lemma 1 $\mathcal{P} \subset \mathcal{PU}$ and $\mathcal{R} \subset \mathcal{U}$.

Proof It suffices to prove $P \subset PU$ as the rest follows trivially (why?). We shall employ induction with respect to P, to prove the property "$f \in PU$" for each $f \in P$.

(a)   *Basis.* $\lambda x.x + 1$ and $\lambda x.x \doteq 1$ are in PU. Indeed, $\lambda x.x + 1 = M_X^X$ and $\lambda x.x \doteq 1 = L_X^X$ where M is the program **add** X and L is the program **sub** X.

(b)   The property propagates with *substitution*: This claim is proved exactly as in Lemma 2.7.1.

(c)   The property propagates with *pure iteration*:

Let $\lambda x.f(x) = M_X^X$, where M is a URM and $f \in P$. We show $\lambda xy.f^y(x) \in PU$. Indeed, $\lambda xy.f^y(x) = Q_X^{X,Y}$, where Q is the program below:

$B = Y$ /* B is *not* a register of M */
while B
M /* this is $X \leftarrow f(X)$ */
$\left.\begin{array}{l} X_1 = 0 \\ \quad\vdots \\ \quad\vdots \\ X_m = 0 \end{array}\right\}$ /* zero all registers of M other than X in preparation for the next iteration */
sub B
end /* here X holds the result */

Note that $B = Y$ and $X_i = 0$ can be performed by groups of URM instructions.

(d)   The property propagates with *unbounded search*:

Let $\lambda xy.f(\vec{x},y) = M_Z^{\vec{x},Y}$, where M is a URM and $f \in P$. We show that $\lambda \vec{x}.(\mu y)f(\vec{x},y) \in PU$. Indeed, $\lambda \vec{x}.(\mu y)f(\vec{x},y) = Q_Y^{\vec{x}}$, where Q is the program below:

$Y = 0$
M
while Z
$\left.\begin{array}{l} R_1 = 0 \\ \quad\vdots \\ \quad\vdots \\ R_k = 0 \end{array}\right\}$ /* reset all registers of M other than $\vec{X}$ and Y to 0. Assume without loss of generality that M does not change any of $\vec{X}$ or Y */
add Y
M
end /* Y holds the result */   .//

Lemma 2 $\mathcal{PU} \subset \mathcal{P}$ and $\mathcal{U} \subset \mathcal{R}$.

Proof Once again, $\mathcal{U} \subset \mathcal{R}$ follows from $\mathcal{PU} \subset \mathcal{P}$. Thus, we concentrate in proving $\mathcal{PU} \subset \mathcal{P}$. We shall do induction with respect to the definition of URMs. As in the proof of Lemma 2.7.2, it will be convenient to prove the statement $M_Y \in \mathcal{P}$, for any URM M and any register Y of M (rather than proving $M_Y^{\vec{X}} \in \mathcal{P}$. This case follows from $M_Y \in \mathcal{P}$ by substituting 0's to "non-input" registers).

(a) *Basis.* Let $M = \textbf{add } X$ (resp. $\textbf{sub } X$). Then $M_X = \lambda X.X + 1$ (resp. $\lambda X.X \dot{-} 1$), hence $M_X \in \mathcal{P}$.

(b) The property (which is $M_X \in \mathcal{P}$) *propagates* with program *superposition* (or concatenation). See proof of Lemma 2.7.2.

(c) The property propagates with **while**-loop closure; i.e., prove that if M has the property (i.e., $M_X \in \mathcal{P}$ for *every* register X in M), then so does Q, where

$$
Q \left\{ \begin{array}{l} \textbf{while } Z \\ M \\ \textbf{end} \end{array} \right.
$$

We now show $Q_X \in \mathcal{P}$ for all registers X of Q.
Case 1. Z is *not* a register of M.
  Then, $Q_X = \lambda X.\, \textbf{if } (Z) = 0 \textbf{ then } (X) \textbf{ else } (\mu y)1(*)$, for any register X of Q (including Z); hence, $Q_X \in \mathcal{P}$. (**Note:** The induction hypothesis was not used in Case 1.)
Case 2. Z *is* a register of M. Let $Z, X_1, \ldots, X_k$ be all the registers of M (and hence of Q).
  By induction hypothesis, $M_R \in \mathcal{P}$ for all R in M. For convenience, we set $g_i = \lambda Z\vec{X}_k.M_{X_i}, i = 1, \ldots, k$ and $g_0 = \lambda Z\vec{X}_k.M_Z$.
  Now, for each i let $Q^{(i)}$ be the program $\underbrace{M;M;\ldots;M}_{i \text{ copies of } M}$. We set $f_0 = \lambda iZ\vec{X}_k.Q_Z^{(i)}$ and $f_m = \lambda iZ\vec{X}_k.Q_{X_m}^{(i)}, m = 1, \ldots, k$.
  Clearly, $\lambda Z\vec{X}_k.f_j (j = 0, \ldots, k)$ are in $\mathcal{P}$ (why?). We next see that $\lambda iZ\vec{X}_k.f_j$ are also in $\mathcal{P}$. (**Warning:** This is not obtained "automatically": that is, $\lambda x.f(x,y) \in C$, where C is a class of functions, does *not* necessarily imply $\lambda xy.f(x,y) \in C$. E.g., $\lambda y.A_x(y) \in \mathcal{PR}$ but $\lambda xy.A_x(y) \notin \mathcal{PR}$. Also, a $\lambda x.g(x,y) \in \mathcal{P}$ exists such that $\lambda xy.g(x,y) \notin \mathcal{P}$. See Problem 11.)

(*)By $(\mu y)1$, we understand $(\mu y)f(x,y)$, where $f = \lambda xy.1$. Clearly, $\lambda x.(\mu y)1 \uparrow$ for all x.

Note that

$$\begin{cases} f_0(0,Z,\vec{X}_k) = Z \\ f_j(0,Z,\vec{X}_k) = X_j, j = 1, \ldots, k \end{cases}$$

and

$$\begin{cases} f_j(i + 1,Z,\vec{X}_k) = g_j(f_0(i,Z,\vec{X}_k),f_1(i,Z,\vec{X}_k), \ldots, f_k(i,Z,\vec{X}_k)), \\ j = 0, \ldots, k \end{cases}$$

Since $\mathcal{P}$ contains pairing functions and is closed under pure iteration, it is closed under simultaneous primitive recursion. Thus, $g_j \in \mathcal{P}$, $j = 0, \ldots, k$ implies $f_j \in \mathcal{P}$, $j = 0, \ldots, k$.

Let $h = \lambda Z\vec{X}_k.(\mu i)f_0(i,Z,\vec{X}_k)$. Clearly, $h \in \mathcal{P}$, and

$$Q_{X_j} = \lambda Z\vec{X}_k. f_j(h(Z,\vec{X}_k), Z,\vec{X}_k), j = 1, \ldots, k$$
$$Q_Z = \lambda Z\vec{X}_k. f_0(h(Z,\vec{X}_k), Z,\vec{X}_k)$$

hence, $Q_R \in \mathcal{P}$ for any register in Q.

Note that Q_Z returns either 0 or is undefined.//

Note: In Chapter 13, we shall see an alternative proof of Lemma 2. We summarize Lemmas 1 and 2 in the important theorem below:

Theorem 1 $\mathcal{P} = \mathcal{PU}$ and $\mathcal{R} = \mathcal{U}$.

Apart from reinforcing our earlier claims that $\mathcal{P}$ (and $\mathcal{R}$) contain only "algorithmic" functions, Theorem 1 has provided an alternative formalism for $\mathcal{P}$. This will facilitate informal arguments in the sequel, as phrases such as " . . . perform y *steps* of the computation $\phi_x(z)$ . . . " now have an immediately formalizable meaning with the help of URMs.

The next paragraph begins the study of the relative strength (computing power) between URMs and TMs.

§4.3 URM vs. TM

We shall presently show that each URM M can be simulated by some TM. (The converse is also true, thus the two models have equal computing power. We shall prove this in §4.5.) In preparation for the TM simulation of

URMs, we shall make the URM model more "manageable". First we shall introduce the dreaded *goto*, next we shall see that two registers are enough for URMs to compute any $\mathcal{P}$-function.

Definition 1 (URMs with **goto**s.)
A URM+**goto** is a *finite sequence* of instructions of any of the forms below, separated by semicolons.

(a) L: **add** R

(b) L: **sub** R

(c) L: **if** $R = 0$ **then goto** M **else goto** N

(d) L: **stop**

subject to the constraints

(i) R is a register name of the same syntax as register names in §4.2

(ii) L, M, N are *integers* (we call them "labels")

(iii) If P is a program—that is, a sequence of instructions $(I_1; \ldots; I_k)$—then

(1) I_k is "k: **stop**" and the **stop** instruction does not appear anywhere else in P.

(2) The label of I_i is $i, i = 1, \ldots, k$

(3) **goto** M can appear in P *only* if $1 \leq M \leq k$.//

URMs originally appeared [13] in essentially the above form. We now present the *semantics* of URM+**goto**:

Definition 2 Let $P = (I_1; \ldots; I_k)$ be a URM+**goto**. Let $R_1, \ldots, R_m$ be all the registers of P. Then a P-ID, α, of P is an $(m + 1)$-tuple $(\alpha_1, \ldots, \alpha_m; i)$ in $\mathcal{N}^m \times \{1, \ldots, k\}$. If $\beta = (b_1, \ldots, b_m; j)$ is another P-ID, then α *yields* β, in symbols $\alpha \xrightarrow{P} \beta$, iff *one* of the following holds:

(a) i: **add** R_l is in P and
$j = i + 1$
$b_l = a_l + 1$
$b_q = a_q, q \in \{1, \ldots, k\} - \{l\}$

(b) i: **sub** R_l is in P and
$j = i + 1$
$b_l = a_l \dotdiv 1$
$b_q = a_q, q \in \{1, \ldots, k\} - \{l\}$

(c) i: **if** $R_l = 0$ **then goto** L **else goto** M is in P and

$\vec{a}_m = \vec{b}_m$ and

$$\text{either} \begin{cases} a_l = 0 \\ j = L \end{cases}$$

$$\text{or} \begin{cases} a_l \neq 0 \\ j = M \end{cases}$$

(d) $i = k$ (hence i: **stop** is the relevant instruction) and $j = k$ and $\vec{a}_m = \vec{b}_m$.//

Intuitively, "k: **stop**" plays the role of the "empty instruction following the program" we saw in §4.2. The other instructions got their natural interpretation in Definition 2.

Definition 3   Let $M = (I_1; \ldots ; I_k)$ be a URM+**goto**. Let $\alpha_1, \ldots, \alpha_t$ be a sequence of M-IDs such that

(1) $\alpha_1 = (\vec{a};1)$; i.e. is *initial* in the usual sense

(2) $\alpha_t = (\vec{b};k)$; i.e. is *final* in the usual sense, and

(3) $\alpha_i \xrightarrow{M} \alpha_{i+1}$ for $i = 1, \ldots, t - 1$.

Then $(\alpha_1, \ldots, \alpha_t)$ is an M-computation of length or *complexity t*. The act $\alpha_i \rightarrow \alpha_{i+1}$ (any i) is one *step* of the computation.//

The symbols $M_Y^{\vec{X}}$ and M_Y have an entirely similar definition as in §4.2 (Definition 4.2.5).

It should be clear now that a URM can be simulated by a URM+**goto**:

Proposition 1   Let M be a URM. Then a URM+**goto** exists which simulates M.

   Proof Let $M = I_1; \ldots ; I_k$, where $I_i(i = 1, \ldots, k)$ is a valid URM-instruction. First obtain the string $\overline{M} = 1:I_1; \ldots; i:I_i; \ldots; k:I_k; k + 1:$**stop**. Next obtain the string $\overline{\overline{M}}$ as follows: For $i = 1, \ldots, k$:

(a)   if I_i is **add** X (resp. **sub** X) leave it as is.

(b)   if I_i is **while** X where I_j is the matching **end** then replace I_i by
if $X = 0$ **then goto** j **else goto** j

(c)   if I_i is **end**, where I_j is **while** X which matches I_i, then replace I_i by
if $X = 0$ **then goto** $i + 1$ **else goto** $j + 1$.

   The remaining details are left to the reader. That is, show that an M-computation is also an $\overline{\overline{M}}$-computation. Clearly $\overline{\overline{M}}$ is a URM+**goto**.//

Corollary   Every function in $\mathcal{P}$ is some $M_Y^{\vec{X}}$ where M is a URM+**goto**.

Example 1   Let a be a constant > 0. Then $X = aX$ can be performed by a URM+**goto** by using only *one* additional register. Indeed, the following program does the job:

$\quad$ 1: **if** $Y = 0$ **then goto** 3 **else goto** 3

$\quad$ 2: $Y = Y \div 1$

$\quad$ 3: **if** $Y = 0$ **then goto** 4 **else goto** 2

$\quad$ 4: **if** $X = 0$ **then goto** $6 + a$ **else goto** $6 + a$

$\quad$ 5: $X = X \div 1$

$\quad$ 6: $Y = Y + 1$

$\quad$ 7: $Y = Y + 1$

$\qquad\qquad \vdots$ $\qquad\qquad$ /*this is $Y = Y + a$ */

$5 + a$: $Y = Y + 1$

$6 + a$: **if** $X = 0$ **then goto** $7 + a$ **else goto** 5

$7 + a$: **if** $Y = 0$ **then goto** $10 + a$ **else goto** $10 + a$

$8 + a$: $Y = Y \div 1$

$9 + a$: $X = X + 1$

$10 + a$: **if** $Y = 0$ **then goto** $11 + a$ **else goto** $8 + a$

$11 + a$: **stop**   .//

Example 2   Let $a > 0$. Then the instruction "$X =$ **if** $\mathrm{rem}(X,a) = 0$ **then** $\lfloor X/a \rfloor$ **else** /*no change*/ X" can be performed by a URM+**goto** by using only *one* additional register.

$\qquad$ Here is the plan in pseudo PL/1:

$\qquad\qquad Y = 0$

$\qquad B$: **if** $X = 0$ **goto** $L(0)$ /* case rem = 0 */

$\qquad\qquad X = X \div 1$

$\qquad\qquad$ **if** $X = 0$ **goto** $L(1)$ /* case rem = 1 */

$\qquad\qquad \vdots$

$\qquad\qquad X = X \div 1$

$\qquad\qquad$ **if** $X = 0$ **goto** $L(a - 1)$ /* case rem = $a - 1$ */

$\qquad\qquad X = X \div 1$

$\qquad\qquad Y = Y + 1$

$\qquad\qquad$ **goto** B

$$/\text{* now } (Y) = \left\lfloor \frac{\text{original } (X)}{a} \right\rfloor \text{*/}$$

$L(0)$: $X = Y$; **goto** L /* case rem $(X,a) = 0$ */
    /* rem$(X,a) \neq 0$; restore X */
$L(1)$: $Y = aY$
    $Y = Y + 1$
    $X = Y$
    goto L
    .
    .
    .
$L(i)$: $Y = aY$
    $Y = Y + i$
    $X = Y$
    goto L
    .
    .
    .
$L(a - 1)$: $Y = aY$
    $Y = Y + a - 1$
    $X = Y$
    L: **stop**

With the help of Example 1 it is straightforward to convert the above pseudo-PL/1 program to a URM+**goto** program. This task is left to the reader.//

Proposition 2   For any URM+**goto** machine M, a URM+**goto** machine $\overline{M}$ with only *two* registers exists which simulates M.(*)

Proof   Let $R_1, \ldots, R_m$ be the registers of M. If $m \leq 2$ the result is trivial. So let $m > 2$. $\overline{M}$ will have two registers, X and Y. At any step of the M-computation, $\overline{M}$ updates the register-value component of the appropriate M-ID and transfers to the appropriate next instruction.

To this end, $((R_1), \ldots, (R_m))$ is stored in X as $p_1^{(R_1)} \cdot p_2^{(R_2)} \cdot \ldots \cdot p_m^{(R_m)}$, where p_i is the *ith* prime ($p_1 = 2$). If $(\mathring{R}_1), \ldots, (\mathring{R}_m)$ are the initial contents of $R_1, \ldots, R_m$, then $\overline{M}$ is initialized with $(X) = 2^{(\mathring{R}_1)} \cdot 3^{(\mathring{R}_2)} \cdot \ldots \cdot p_m^{(\mathring{R}_m)}$ and $(Y) = 0$.

(*)In the sense of Example 4.2.2. If M has more than two registers, of which $R_1 \ldots$, R_n are input and R_1 output, $\overline{M}$ uses, say, X as input/output. For input convention, see the proof; the output is "read" off X as $\exp(1,(X))$, where $p_1 = 2$.

Let now $M = I_1; \ldots; I_k$, where each I_i is a valid URM+**goto** instruction. (Of course, I_k is k:**stop**.) $\overline{M}$ is $\overline{I}_1; \overline{I}_2; \ldots; \overline{I}_k$ where for $i = 1, \ldots, k$

(a)   if I_i is "i:**add** R_l" then
$\overline{I}_i$ is the instruction-group "$X = p_l X$" (with appropriate labels)

(b)   if I_i is "i:**sub** R_l" then
$\overline{I}_i$ is the instruction-group "$X = $ **if** $\mathrm{rem}(X, p_l) = 0$ **then** $\lfloor X/p_l \rfloor$ **else** X"
(with appropriate labels)

(c)   if I_i is "i:**stop**" then $\overline{I}_i$ is "p:**stop**" for appropriate p

(d)   if I_i is "i:**if** $R_l = 0$ **then goto** j **else goto** n" then $\overline{I}_i$ is the instruction-group "**if** $\mathrm{rem}(X, p_l) = 0$ **then goto** n_0 **else goto** j_0", where j_0 (resp. n_0) is the label of the first instruction of the $\overline{I}_j$-group (resp. $\overline{I}_n$-group) in $\overline{M}$.

The reader is requested once again to fill in the trivial details. For example, Examples 1 and 2 show how "$X = aX$" and "$X = $ **if** $\mathrm{rem}(X,a) = 0$ **then** $\lfloor X/a \rfloor$ **else** X" are done by a two-register URM+**goto**.

To incorporate these in $\overline{M}$, in general in several places with different a-values, the "**goto** '**stop**'" should be replaced by "**goto** '$next_a$'", where "$next_a$" in each implementation of $X = aX$ or $X = $ **if** $\mathrm{rem}(X,a) = 0$ **then** $\lfloor X/a \rfloor$ **else** X is the label of the *first* instruction following the subprogram $X = aX$(resp. $X = $ **if** $\mathrm{rem}(X,a) = 0$ **then** $\lfloor X/a \rfloor$ **else** X).

Finally, "**if** $\mathrm{rem}(X,a) = 0$ **then goto** n_0 **else goto** j_0" can be implemented by a trivial modification of the program in Example 2. Namely, in the $L(0)$ case use "$Y = aY; X = Y;$ **goto** n_0" (instead of "$X = Y;$**goto stop**"), whereas in the $L(i)$ case, $i = 1, \ldots, a - 1$, use **goto** j_0 instead of **goto stop**.//

Theorem 1   Given any two-register URM+**goto** M, a Turing Machine $\overline{M}$ exists which simulates M.

Proof   Informally first, the TM need only store the two registers in its tape, and each time it simulates an M-instruction to update the registers appropriately. For convenience, imagine first that $\overline{M}$ has *two* tapes and *two* heads, one for each tape:

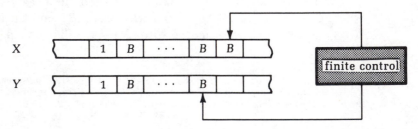

Say the upper tape implements register X, whereas the lower implements Y. This machine $\overline{M}$ need *not* write on its tapes. The *distance* of the head position from the single "1" on the (relevant) tape represents the *contents* of the (relevant) register. That is, $(X) = n$ iff the X-head is on the nth $(n \geq 0)$ blank to the right of 1; note that the head being on "1" means $(X) = 0$.

Assume a correspondence between M-labels and $\overline{M}$-states so that label i corresponds to state p_i.

Then the simulation is briefly described as follows, (where $R \in \{X, Y\}$):

M in label i		*$\overline{M}$ in state p_i*
"**add** R"	is simulated by	"Move the R-head one square right; enter p_{i+1}".
"**sub** R"	is simulated by	"If the R-head scans "1", then don't move, enter p_{i+1}; else move one square left, enter p_{i+1}".
"**stop**"	is simulated by	"Enter a state which is undefined for all inputs".
"**if** $R = 0$ **then goto** j **else goto** n"	is simulated by	"Do not move the heads; if the R-head scans "1", then enter p_j else enter p_n".

To formalize the previous two-tape machine within the one-tape model, we shall need to keep track of the heads of the two-tape model. Thus the pair $((X), (Y))$ is represented on *one*-tape format as

that is, the upper-track "1" (resp. lower-track "1") marks the position of the (simulated) first (resp. second) head of the two-tape model. "($,$)" marks the common (left) origin of the portion of the tape with "two tracks". Our convention is that $\$B^n1$ (upper or lower track) denotes $(R) = n$, $n \geq 0$ (where $R = X$ if upper track, $R = Y$ if lower).

Let now $M = I_1; I_2; \ldots ; I_k$, where I_k is "k:**stop**". $\overline{M}$ has as tape alphabet $A = \{B, (B,B), (B,1), (1,B), (1,1), (\$,\$)\}$. Among $\overline{M}$'s states, p_i $(i = 1, \ldots, k)$ correspond exactly to the labels i.

For $i = 1, \ldots, k$, if I_i is

(a) "i: **add** X" then $\overline{M}$ contains

$p_i a a p_i L$, $a \in \{B, (B,B), (B,1)\}$	/* go left searching for ($\$,\$$) or ($1,B$) */
$p_i (\$,\$) (\$,\$) p_i^{(r)} R$	/* if ($\$,\$$) found, go right to find
$p_i^{(r)} a a p_i^{(r)} R$, $a \in \{B, (B,B),(B,1)\}$	($1,B$) */
$p_i^{(r)} (1,B) (1,B) p_i S$	/* ($1,B$) found. Enter p_i state */
$p_i (1,B) (B,B) p_i^{(1)} R$	/*go right, after erasing upper-
$p_i (1,1) (B,1) p_i^{(1)} R$	track "1". Remember [switch
$p_i^{(1)} (B,B) (1,B) p_{i+1} S$	"(1)"] to print it in next square.
$p_i^{(1)} (B,1) (1,1) p_{i+1} S$	Then enter state p_{i+1} corres-
$p_i^{(1)} B (1,B) p_{i+1} S$	ponding to statement labeled
	$i + 1$ */

(a') "i: **add** Y", then $\overline{M}$ contains quintuples as in (a), but with the components of composite tape symbols interchanged.

(b) "i: **sub** X", then $\overline{M}$ contains

$p_i a a p_i L$, $a \in \{B, (B,B),(B,1)\}$
$p_i (\$,\$) (\$,\$) p_i^{(r)} R$
$p_i^{(r)} (1,B) (1,B) p_i S$
$p_i (1,B) (B,B) p_i^{(1)} L$
$p_i (1,1) (B,1) p_i^{(1)} L$
$p_i^{(1)} (B,B) (1,B) p_{i+1} S$
$p_i^{(1)} (B,1) (1,1) p_{i+1} S$
$p_i^{(1)} (\$,\$) (\$,\$) p_i^{(2)} R$
$p_i^{(2)} (B,B) (1,B) p_{i+1} S$
$p_i^{(2)} (B,1) (1,1) p_{i+1} S$

(b') "i: **sub** Y", then $\overline{M}$ contains quintuples as in (b), but with the components of composite tape symbols interchanged.

(c) "i: **stop**", then p_i in $\overline{M}$ is undefined on all input symbols.

(d) "i: **if** $X = 0$ **then goto** j **else goto** n", then $\overline{M}$ contains

$p_i a a p_i L$, $a \in \{B, (B,B),(B,1)\}$
$p_i (\$,\$) (\$,\$) p_i^{(r)} R$
$p_i^{(r)} (1,B) (1,B) p_i S$
$p_i (1,B) (1,B) p_i^{(1)} L$
$p_i (1,1) (1,1) p_i^{(1)} L$
$p_i^{(1)} (\$,\$) (\$,\$) p_j R$
$p_i^{(1)} (B,B) (B,B) p_n R$
$p_i^{(1)} (B,1) (B,1) p_n R$

(d') "i: **if** $Y = 0$ **then goto** j **else goto** n", then $\overline{M}$ contains quintuples as in (d) but with the components of each composite tape symbol reversed.

Finally, if the initial ID of M is $\alpha = (n,0;1)$, then the initial ID of $\overline{M}$ is $p_1(\$,\$)(B,1)(B,B)^{n-1}(1,B)$ if $n \geq 0$, $p_1(\$,\$)(1,1)$ if $n = 0.//$

Corollary 1   Any URM (resp. URM+**goto**) can be simulated(*) by some TM.

Corollary 2   Every function in $\mathcal{P}$ is TM-computable, *assuming some appropriate input-output conventions.*(*)

Corollary 3   Any function in $\mathcal{P}$ is computable(*) by a two-tape read-only TM with tape alphabet $\{B,1\}$.

Proof See the informal part of the proof at Theorem 1. This result is due to Minsky [10].//

We now wish to show that $\mathcal{P} \subset \mathcal{PC}$. Note that this result is not an immediate consequence of Corollary 2 above, as $\mathcal{PC}$ assumes an input/output scheme which is *incompatible* with the way the TM of Theorem 1 works.

Lemma 1   There is a TM M with $A = \{B,0,1,\$\}$ such that

$$[\text{Res}_M(q_0 1^{x_1+1} B 1^{x_2+1} \ldots B 1^{x_n+1})] = 2^{2^{x_1+1} + 2^{x_1+x_2+3} + \ldots + 2^{x_1+ \ldots + x_i+2i-1} + \ldots + 2^{\Sigma_{i-1}^n x_i + 2n-1}}$$

Proof First, observe that $2^{x_1+1} + 2^{x_1+x_2+3} + \ldots + 2^{x_1 + \cdots + x_n+2n-1}$ has binary representation $(10^{x_n+1} 10^{x_{n-1}+1} 1 \ldots 10^{x_1+1})_2$.
This is the *reverse* of $1^{x_1+1} B 1^{x_2+1} B \ldots B 1^{x_n+1} B$ after "1" has been replaced by "0" and "B" by "1".
Clearly,

$$2^{(10^{x_n+1}1 \ldots 10^{x_1+1})_2} = (10^{(10^{x_n+1}1 \ldots 10^{x_1+1})_2})_2;$$

Our machine M will end up with this number in *unary* (a tally of 1's) in its final tape.
     M is:

(*)The "operator" of the simulating TM must be prepared to live (for a while) with the cumbersome input/output conventions of the proof of Proposition 2 as they are reflected in the TM. This situation will be remedied, starting with Lemma 1 below.

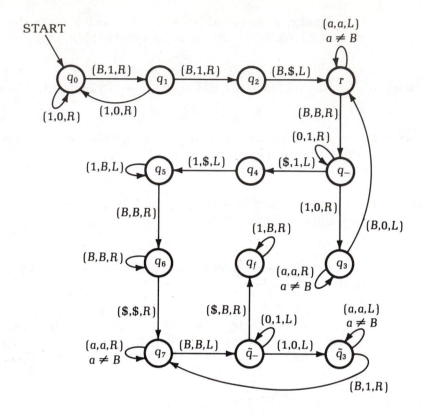

Comments

The ID at the *first* invocation of r is:

$0^{x_1+1}10^{x_2+1}1 \ldots 10^{x_n+1}1r\$$.

The loop (r, q_-, q_3, r)
repeatedly subtracts 1 from
the *reverse* binary number to the
left of "$\$$", and concatenates
a "0" *right* of "$\$$". The
process ends when the binary
number becomes 0.

Departing from q_5, the situation is

$$q_5 B^{\Sigma_{i=1}^n x_i + 2n} \$ 10^{(10^{x_n+1}1 \ldots 10^{x_1+1})_2}$$

The remaining states convert the binary string in the above ID, interpreted as
a *forward* binary number, to a tally of 1's of length equal to that number.

The final ID is

$$1^{(10^{(10^{(10^{x_n+1}1\ldots 10^{x_1+1})}2)}2)}2}BBB^{(10^{x_n+1}1\ldots 10^{x_1+1})_2}q_fB$$

Note that the loop $(q_7,\tilde{q}_-,\tilde{q}_3,q_7)$ is similar to (r,q_-,q_3,r), but processes a *forward* binary number.//

Lemma 2   There is a TM C which has initial state named q_f, its other states, all different from those of M of Lemma 1, are p_1, p_2, p_3, and $\text{Res}_C(1^{m+1}B^{n+1}q_fB)$ $= p_3(\$,\$)(B,1)(B,B)^m(1,B)$, where the alphabet of C is $A=\{B,1,(\$,\$),(B,B),$ $(B,1),(1,B)\}$.

Proof C is:

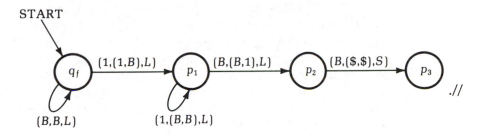

Theorem 2 $\mathcal{P} \subset \mathcal{PC}$.

Proof Let $f \in \mathcal{P}^{(n)}, n \geq 1$.
Consider, for $n \geq 1$, the codings

$$c_n = \lambda\vec{x}_n \cdot 2^{x_1+1} + 2^{x_1+x_2+3} + \ldots + 2^{\Sigma_{i-1}^n x_i + 2n-1}$$

It is easy to see that c_n is indeed 1-1. As $c_n \in \mathcal{P}\,\mathcal{R}$, Π_i^n, its ith projection is in $\mathcal{PR}$ as well, since

$$z = c_n(\vec{x}_n) \Rightarrow x_i \leq z, i = 1, \ldots, n$$

Consider

$$h = \lambda z.f(\Pi_1^n(z), \ldots, \Pi_n^n(z)).$$

Clearly, $h \in \mathcal{P}^{(1)}$ and hence a URM+**goto**, M, exists such that h is computed by M. Let $R_1, \ldots, R_m$ be M's registers.
Without loss of generality (why?), we assume:

(a) $h = M_{R_1}^{R_1}$

(b)   The final ID of M has $(R_2) = (R_3) = \ldots = (R_m) = 0$.

Clearly, M, with input $2^{x_1+1} + 2^{x_1+x_2+3} + \ldots + 2^{\sum_{i-1}^{n} x_i + 2n - 1}$ will output $f(\vec{x}_n)$(*) in R_1 and 0 in R_j, $j = 2, \ldots, m$. Let $\tilde{M}$ be a two-register URM+**goto** simulating M, so that $(X) = 2^{(R_1)} \cdot 3^{(R_2)} \cdot \ldots \cdot p_m^{(R_m)}$ and $(Y) = 0$, initially and finally.

Let $\overline{M}$ be the TM of Theorem 1, which simulates $\tilde{M}$.

For an *initial* $\tilde{M}$-ID of the form $(2^{2^{x_1+1} + \ldots + 2^{\sum_{i-1}^{n} x_i + 2n - 1}}, 0, 1)$

(1)   the corresponding *initial* $\overline{M}$-ID is

$$p_1(\$,\$)(B,1)(B,B)^{2^{2^{x_1+1} + \ldots + 2^{\sum_{i-1}^{n} x_i + 2n - 1}} - 1}(1,B).$$

(2)   the *final* $\tilde{M}$-ID (if it exists) is $(2^{f(\vec{x}_n)}, 0, k)$, where k:**stop** is an instruction of $\tilde{M}$.

(3)   the corresponding *final* $\overline{M}$-tape is $(\$,\$)(B,1)(B,B)^{2^{f(\vec{x}_n)} - 1}(1,B)$.

The TM $M_0 = M \cup C \cup \overline{M}_{+2}$ [where M, C are as in Lemmas 1 and 2, $\overline{M}_{+2}$ is the $\overline{M}$-machine (where *each* p_i, $p_i^{(1)}$, $p_i^{(2)}$, $p_i^{(r)}$, has been replaced by p_{i+2}, $p_{i+2}^{(1)}$, $p_{i+2}^{(2)}$, $p_{i+2}^{(r)}$, respectively), A_0, the alphabet of M_0, is the union of the M, C, and $\overline{M}$ alphabets, and q_0 is the initial state of M_0] when started with $q_0 1^{x_1+1} B \ldots B 1^{x_n+1}$ has as final tape $(\$,\$)(B,1)(B,B)^{2^{f(\vec{x}_n)} - 1}(1,B)$, whenever it halts.

Let, without loss of generality, $p_k(\$,\$)(B,1)(B,B)^{2^{f(\vec{x}_n)} - 1}(1,B)$ be the final ID of such a computation, where k is the maximum index of all p_i's in M_0.

Consider the machine M_1 below:

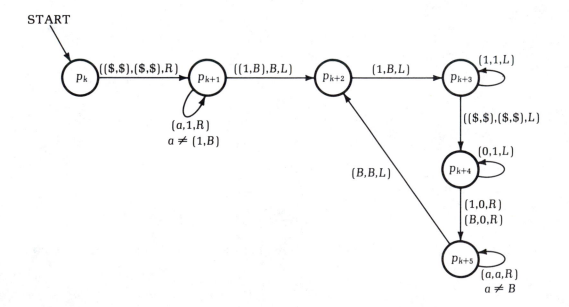

(*)If $f(\vec{x}_n)\downarrow$; if $f(\vec{x}_n)\uparrow$, then M will not halt.

It is left as an exercise (See Problem 15) to show that $M_0 \cup M_1$ (initial state q_0) has the behavior

$$\mathrm{Res}_{M_0 \cup M_1}\,(q_0 1^{x_1+1}B \ldots B1^{x_n+1}) = 01^{f(\vec{x}_n)}p_{k+2}(\$,\$).//$$

Note: The reader may have observed that $01^{f(\vec{x}_n)}$ is the "dual" of $10^{f(\vec{x}_n)}$, which represents $2^{f(\vec{x}_n)}$ in binary.

Corollary 1   Each $f \in P$ is (partially) computable by some TM which has as tape alphabet $\{B,0,1,\$,(B,B),(B,1),(1,B),(1,1),(\$,\$)\}$ and its final IDs are of the form $01^n q(\$,\$)$, where q is a state.

Corollary 2 $\mathcal{R} \subset \mathcal{C}$.

Proof $\mathcal{R} \subset \mathcal{P} \subset \mathcal{PC}$. But $\mathcal{R}$ contains total functions, hence $\mathcal{R} \subset \mathcal{C}.//$

§4.4 MARKOV ALGORITHMS

The last formalism we present in this chapter is that due to Markov [9]. His Normal Algorithms (or as they are presently known, Markov Algorithms) are more akin to TMs rather than URMs as they are "string-processing" rather than "number-processing" devices. The machine hosting such an algorithm need have exactly one register (or tape) which initially holds the input string. The instructions modify this string, and if the process halts, the last string in the register is the result.

Markov algorithms are a special case of "grammar-like" formalisms (e.g., Post, Thue—see Chapter 6) but unlike them, Markov algorithms are deterministic.

Definition 1   A *Markov algorithm*, or *MA*, *M* over an alphabet *A*, is a finite *sequence*(*) of pairs (x,y), where $x \in A^*$, $y \in A^*$. These pairs are the *instructions* of $M.//$

Informally, given a string w as input, we scan M, from the beginning to end, until we find the *first* instruction (if any) whose first component, say x, is a substring of w.

Let $w = uxv$, where $ux = u_1 x v_1$ implies $v_1 = \Lambda$ (the empty string). Then change w to uyv, where y is the second component of the instruction we are executing. Rename uyv "w" and repeat the process. Stop iff *no* instruction is applicable.

(*)Unlike TMs which are *sets* of instructions, Markov algorithms are *sequences*. The order of the instructions is important.

Example 1   Let $M = (\Lambda,1)$ be over $A = \{1\}$. Then, starting with Λ, M produces the infinite sequence Λ, 1, 11, 111, ... that is, M does *not* halt. Note that Λ matches *any* string as its left-most substring.//

Example 2   Let $M = (1B,11)$ over $A = \{B,1\}$. Then M "adds one" and halts if started with $1^{x+1}B$.//

Definition 2   Let $M = (I_1, \ldots, I_k)$ be a Markov algorithm over A, where I_i is its ith instruction (x_i, y_i), $i = 1, \ldots, k$.

An M-ID is any string $w \in A^*$. Let w and u be two M-IDs. Then w *yields* u, in symbols $w \xrightarrow{M} u$, iff the following is true:

$$(\exists i)_{\leq k}(w = \alpha x_i \beta \,\&\, u = \alpha y_i \beta \,\&$$

$$(\forall v)_{P\alpha x_i}(\forall z)_{P\alpha x_i}(\alpha x_i = v x_i z \Rightarrow z = \Lambda)(*) \,\&\, (\forall j)_{<i} \neg x_j P w)$$

An M-*computation* is a sequence of M-IDs $(w_1, \ldots, w_t)$ such that

(a) $(\forall i)_{1 \leq i \leq k} \neg x_i P w_t$

(b) $(\forall i)_{<t} w_i \longrightarrow w_{i+1}$

An ID w_t with property (a) is called *final*. *One* execution of the yield operation is one *step* of the computation. t is the *complexity* of the computation.//

Note: It should be clear that if an instruction (Λ,y) is present in a Markov algorithm M, then, according to Definition 2, M does not halt on any input. Also, if (x_i,y_i) and (x_j,y_j) are in M, $i < j$ and $x_i = x_j$, then the (x_j,y_j) instruction is never applicable. The following is immediate:

Proposition 1   If α_1 is an initial ID of a Markov algorithm, then there is at most one computation starting with α_1.

Definition 3   Let M be a Markov algorithm over an alphabet containing $\{B,\#,1\}$. We say that $f:\mathcal{N}^n \to \mathcal{N}$ is M-computable, iff

(a) $f(\vec{x}_n)\!\downarrow$ iff there is an M-computation starting with $\#1^{x_1+1}B1^{x_2+1}B \ldots B1^{x_n+1}$

(b)   Whenever $f(\vec{x}_n)\!\downarrow$, if w is the final ID of the above mentioned M-computation, then $[w] = f(\vec{x}_n)$, where, as in §4.1, $[w]$ = number of 1's in w.//

Example 3 $\lambda x.x \doteq 1$ is Markov-computable. Indeed, let the algorithm be $M = ((\#1,B))$, over $\{B,1,\#\}$.

Then $\#1^{x+1} \to B1^x$ is a computation.//

(*)Refer to notation introduced in the Appendix to Chapter 3.

Definition 4 $\mathcal{PM}$ denotes the class of all functions $f:\mathcal{N}^n \to \mathcal{N}$ that are Markov-computable according to Definition 3. $\mathcal{M}$ is the subclass of $\mathcal{PM}$ which contains only *total* functions.//

In the next section, we shall complete the proof that all the formalisms of this chapter are equivalent. So far, we know that $\mathcal{P} = \mathcal{PU} \subset \mathcal{PC}$.

§4.5 MA vs. TM vs. URM

Lemma 1 $\mathcal{PC} \subset \mathcal{PM}$.

Proof Let $f \in \mathcal{PC}^{(n)}$. This means that a TM M exists such that $[\mathrm{Res}_M(q_0 1^{x_1+1}B1^{x_2+1}B\ldots B1^{x_n+1})] = f(\vec{x}_n)$, for all $\vec{x}_n$. Let A be M's tape-alphabet and K be M's state-alphabet (without loss of generality, $\# \notin A \cup K$). We now construct a Markov algorithm $\overline{M}$ as follows:
The alphabet for $\overline{M}$ is $\{\#\}\cup A\cup K$.
The instructions of $\overline{M}$ are:

(1) $(\#,q_0)$

(2) if $q_i s_l s_j q_m L \in M$, then (and only then)
$(aq_i s_l, q_m as_j)$, for all $a \in A$, are in $\overline{M}$ *followed* by $(q_i s_l, q_m Bs_j)$.

(3) if $q_i s_l s_j q_m R \in M$, then (and only then)
$(q_i s_l a, s_j q_m a)$, for all $a \in A$, are in $\overline{M}$ *followed* by $(q_i s_l, s_j q_m B)$

(4) if $q_i s_l s_j q_m S \in M$, then (and only then)
$(q_i s_l, q_m s_j) \in \overline{M}$.

Note: As $\overline{M}$ is defined, the *order* of instructions is immaterial, except for the cases mentioned. Clearly,

$$\#1^{x_1+1}B\ldots B1^{x_n+1} \xrightarrow{\overline{M}} q_0 1^{x_1+1}B\ldots B1^{x_n+1} \text{ and from now}$$

on any $\overline{M}$-yield operation is identical to an M-yield operation.//

Corollary 1 $\mathcal{PU} \subset \mathcal{PM}$.

Corollary 2 $\mathcal{C} \subset \mathcal{M}$.

At this point, we can pause and let the arithmetization of the next chapter prove $\mathcal{PM} \subset \mathcal{P}$.

However, the arithmetization is used mainly to prove more powerful theorems such as the universal function and S-m-n theorems. Actually, any formalism (machine or not) is useful up to the point where one can prove the above theorems. From then on, everything in Recursion Theory is derivable from these two theorems. (This is well known. For example, Kleene [7] has introduced a formalism [see Chapter 10] which essentially postulates both the S-m-n and universal function theorems. See also Hennie [4]).

Because of this, we shall prove the rather minor statement that $\mathcal{PM} \subset \mathcal{P}$ in this paragraph.

Lemma 2 $\mathcal{PM} \subset \mathcal{PU}$.

Proof Let $M = (I_1, \ldots, I_k)$ be a Markov algorithm over $A = \{s_1, \ldots, s_m\}$, where $I_i = (x_i, y_i)$, $i = 1, \ldots, k$. M operates on strings over A. The simulating URM $\overline{M}$ will operate on numbers in m-adic notation; these numbers are in 1-1 correspondence with A^+ and, as usual, we let $\Lambda \in A^*$ correspond with $0 \in \mathcal{N}$. Indeed, A may be identified with $\Sigma = \{1, \ldots, m\}$, where s_i is identified with i, $i = 1, \ldots, m$.

$\overline{M}$ will have a register W which holds the current ID of M (i.e., an element of A^*) in coded form; that is, as an element of Σ^*. r denotes (R) for any register R.

$\overline{M}$ will also have other registers for scratchwork. In pseudo-PL/1, $\overline{M}$ is:

cycle: $U = w + 1$;
$\left\{\begin{array}{l} \textbf{do } i = 1 \textbf{ to } k \textbf{ while } (u = w + 1); \\ U = (\mu u)_{\leq w}(u \| x_i)Bw; \\ \textbf{end;} \end{array}\right.$ /* "$\|$" denotes concatenation. "B" is the
$\textbf{if } u = w + 1 \textbf{ then stop}$ "begins" relation of §3.9 */
$\textbf{else do}$ /* "$u = w + 1$" means
$\left\{\begin{array}{l} \qquad V = (\mu v)_{\leq w}w = u\|x_i\|v; \\ \qquad W = u\|y_i\|v; \\ \qquad \textbf{goto cycle} \\ \textbf{end} \end{array}\right.$ that the search for a u such that $(u \| x_i)Bw$ was unsuccessful */.

Note that the operations on the righthand sides of the assignment statements above are primitive recursive. Thus, the above program can be expanded to a URM program since $\mathcal{PR} \subset \mathcal{P} = \mathcal{PU}$. (The purist will indeed observe that the above can be readily converted to a loop-program with the addition of a *single* **goto**.) The details are left to the reader.//

Corollary   Any functions in $\mathcal{P}$ can be computed by some loop-program possibly with the help of a *single* **goto**.

Theorem 1 $\mathcal{P} = \mathcal{PU} = \mathcal{PC} = \mathcal{PM}$ and $\mathcal{R} = \mathcal{U} = \mathcal{C} = \mathcal{M}$.

PROBLEMS

1. Give a formal definition of a TM which halts (regardless of scanned symbol) iff it enters any one of a set of distinguished states (*final* or *halting* states). Show that for every TM M, defined as in §4.1, there exists a TM M', with the same tape alphabet that halts by final state, such that, for each pair of tapes (t_0, t_1), M with t_0 as initial tape yields t_1 as final tape iff M' does so as well.

2. A TM with a "2-way infinite tape" is what Definitions 4.1.1-4.1.6 formalize. (Tapes are extendible both to the left and to the right. See Definition 4.1.4.) A TM with a "1-way infinite tape" is one with a distinguished symbol, #, in its tape alphabet A, such that:

   (a) the initial tape is always of the form $\#t$, where $t \in (A - \{\#\})^*$

   (b) The TM never erases or writes the symbol #

   (c) The TM never moves left of #.

   (1) Formalize the 1-way infinite tape TM model. (See Definition 4.1.4.)

   (2) Show that for each 2-way infinite tape TM M over A ($\# \notin A$), there is a 1-way infinite tape TM M' over A' such that $\{\#\} \cup A \subset A'$ and for all t_0, t_1 in A^+, M with initial tape t_0 yields t_1 iff M' with initial tape $\#t_0$ yields $\#t_1$.

   (*Hint:* By adding to A symbols which are elements of $A \times A$, build an M' which behaves like M, but on a "folded" 2-way infinite tape, i.e., a 1-way infinite tape.)

3. (*Multitrack Turing Machines*). Intuitively, a k-track TM has its tape subdivided in k parallel tracks. The tape head has k "sections" and at each given moment it scans k squares, one per track, aligned perpendicularly to the tape. Formally, a k-track TM has a set of states Q, and k tape alphabets A_i ($i = 1, \ldots, k$), one for each track. (In practice, all A_i being the same does not restrict generality.) The transition function δ is $\delta : A_1 \times A_2 \times \ldots \times A_k \times Q \to A_1 \times \ldots \times A_k \times Q \times \{L, R, S\}$. It is immediately clear that (under some appropriate input/output conventions) for any $k > 1$ and any k-track machine M, a 1-track machine M' can be found over alphabet $A_1 \times \ldots \times A_k$ which "does what M does". Formalize and prove this claim.

4. (*Multitape Turing Machines*). Intuitively, a k-tape TM has k tapes and k heads, one per tape, which can move independently of each other. Formally, a k-tape machine has k tape alphabets $A_1, \ldots, A_k$, one per tape, and a state set Q.

The transition function is $\delta: A_1 \times \ldots \times A_k \times Q \to A_1 \times \ldots \times A_k \times Q \times \{L,R,S\}^k$. Assume, for example, that input is given in tape 1 and output is read from tape 2. Show that under reasonable input/output coding schemes, any k-tape TM ($k \geq 2$) can be "simulated" by a 1-tape (1-track) TM. (*Hint:* Reduce this to Problem 3, by showing that a k-tape TM can be simulated by a 1-tape $2k$-track TM. Each tape is stored in two tracks, one track for remembering the tape contents, and one for remembering the corresponding head's position.)

5. (*TM transducer*). A TM transducer is a special case of a multitape TM. It has one *read-only* input tape, one *write-only* output tape, and one or more *work-tapes*. The input is presented in the read-only tape and is delimited by two special symbols (say, $ at left and ¢ at right). The read-only head cannot go left of $(resp. right of ¢). The write-only head goes only to the right and only *after* it prints a nonblank symbol. The work heads operate on the work-tapes in the usual manner. Formalize the transducer model and show that it is equivalent to the 1-tape, 1-track model.

6. Formalize TM models with one head, but with a two-dimensional tape subdivided into infinitely many squares forming an infinite rectangular grid. Consider 1-, 2-, 3-, and 4-quadrant models. Show that none of these models is more powerful than the 1-tape, 1-track model.

   Note: Problems 1-6 are to be solved by direct simulations in this context. The reader will also realize, as soon as he reads Chapter 5, that alternative proofs for problems 1-6 can be provided through arithmetizations in $\mathcal{P}$, and will thus conclude that the $\mathcal{P}$ (and hence the simple TM) formalism subsumes those suggested in the problems.

7. (TM according to Davis [2].) Show that a standard TM restricted to moving (*left* or *right*) *only* if it has *not* changed the scanned symbol is as powerful as the unrestricted TM standard model.

8. Redo examples 4.1.2, 4.1.3, 4.1.4, and 4.1.5 using multitape TMs.

9. Show, relying directly on the Definitions in §4.1, that

   (a) $\mathcal{PC}$ is closed under substitution

   (b) $\mathcal{PC}$ is closed under unbounded search

   (c) $\mathcal{PC}$ is closed under primitive recursion.

10. Fill in the missing details in Example 4.2.2.

11. Show that there are total number-theoretic functions not in $\mathcal{R}$. Conclude that there is a $\lambda xy.g(x,y) \notin \mathcal{P}$ such that $\lambda x.g(x,y) \in \mathcal{P}$, for all y.

12. Construct URMs (with and without **goto**) which compute $\lambda xy.x + y$, $\lambda xy.xy$, and $\lambda xy.x^y$.

13. Complete the proof of Proposition 4.3.1.

14. Fill in the missing details in the proof of Proposition 4.3.2.

15. Fill in the missing details in the proof of Theorem 4.3.2.

16. Restrict the transducer model of Problem 5 so that the TM has exactly *two* 1-way work tapes, each restricted so that the work-head can move left *only* if it has first printed a blank symbol. (Such operation makes each work tape behave like a *stack*. This is why the transducer of the present problem is known as a *2-stack machine*.)

   Formalize the 2-stack machine and prove that it has computing power equivalent to the standard TM. (*Hint:* The square scanned in each work tape is the corresponding top-stack square. Show that the standard TM tape can be mimicked by the two stacks, intuitively "glued" at their tops.)

   Note: The content of Problem 16 is due to Minsky.

17. Show by direct construction that a *universal* TM U exists in the following sense: U has, of course, a fixed tape alphabet A. The only inputs it "likes" are of the form $t_1 B t_2$, where t_1 is a coding over $A - \{B\}$ of some TM M and $t_2 \in (A - \{B\})^+$ is to be viewed as input for M. U is to simulate M on input t_2. If the input is not of the required form, U loops forever.

18. Repeat Problem 17, where now $t_1 B t_2$ is acceptable as a (TM, input)-pair iff the TM part is coded by the *maximum*-length TM coding that *begins* t_1, the rest of t_1 being a "don't care" string in $(A - \{B\})^*$.

19. Repeat Problem 17 for URMs.

20. Repeat Problem 17 for MAs.

21. *Without* relying on the unsolvability of the *halting problem*, already proved in Chapter 3, show that *there is no TM M* such that whenever it receives as input a pair (t, t'), where t is a TM-coding (say in the sense of Problem 18) and t' an input for t, it halts and prints 1 if the TM (coded by) t with input t' halts; otherwise it halts and prints 2 if the TM t with input t' does not halt.

22. (Busy Beaver Problem of Rado [12]). Consider the set of all TMs over a fixed alphabet A. Call this set $\mathcal{A}$. Define b by $b = \lambda x.$(the *maximum* number of 1's printed by an x-state TM in $\mathcal{A}$ starting with blank tape). Prove:

   (a) b is total

   (b) For all $f \in \mathcal{R}^{(1)}$, $f(x) < b(x)$ for all but finitely many x

   (c) $b \notin \mathcal{R}$.

   Note: We prove this in Chapter 11, using the second recursion theorem of Kleene. We seek here a solution that does not rely on the recursion theorem.

23. Complete the proof of Lemma 4.5.2, paying attention in particular to input/ output conventions as set out in the appropriate definitions.

24. Define carefully the *Loop-Program* + **goto** formalism suggested in the proof of Lemma 4.5.2, and prove the Corollary of the aforesaid Lemma.

REFERENCES

[1] Cutland, N.J. *Computability; an Introduction to Recursive Function Theory.* Cambridge: Cambridge University Press, 1980.

[2] Davis, M. *Computability and Unsolvability.* New York: McGraw-Hill, 1958.

[3] Diller, J. *Rekursionstheorie.* Institut für mathematische Logik und Grundlagenforschung, Westfalische Wilhelms-Universität, Münster, 1976.

[4] Hennie, F. *Introduction to Computability.* Reading, Mass.: Addison-Wesley, 1977.

[5] Hopcroft, J.E., and Ullman, J.D. *Introduction to Automata Theory, Languages, and Computation.* Reading, Mass.: Addison-Wesley, 1979.

[6] Jones, N.D. *Computability Theory; an Introduction.* New York: Academic Press, 1973.

[7] Kleene, S.C. "Recursive Functionals and Quantifiers of Finite Types, I". *Transactions of the Amer. Math. Soc.* 91 (1959):1–52.

[8] Manna, Z. *Mathematical Theory of Computation.* New York: McGraw-Hill, 1974.

[9] Markov, A.A. *Theory of Algorithms.* Transl. Amer. Math. Soc. Series 2, 15 (1960).

[10] Minsky, M.L. *Computation; Finite and Infinite Machines.* Englewood Cliffs, N.J.: Prentice-Hall, 1967.

[11] Post, E.L. "Formal Reductions of the General Combinatorial Decision Problem". *Amer. Journal of Math.* 65 (1943):197–215.

[12] Rado, T. "On Non-computable Functions". *Bell System Tech. Journal* 41 (1962):877–884.

[13] Shepherdson, J.C., and Sturgis, H.E. "Computability of Recursive Functions". *Journal of the ACM* 10 (1963):217–255.

[14] Trakhtenbrot, B.A. *Algorithms and Automatic Computing Machines.* Topics in Mathematics series, Lexington, Mass.: D.C. Heath and Co., 1963.

[15] Turing, A.M. "On Computable Numbers, with an Application to the Entscheidungsproblem". *Proc. of the London Math. Soc., series 2,* 42 (1936): 230–265 and 43 (1937): 544–546. (Also in Davis, M., ed. *The Undecidable.* Hewlett, N.Y.: Raven Press, 1965:115–154.)

CHAPTER 5

Arithmetization

In this chapter, a different proof than the one in Chapter 3 is given for the universal function and *S-m-n* theorems for $\mathcal{P}$.

For the universal function theorem, TM's U_n, will be built. When U_n receives an input $(x, \vec{y}_n)$, $n \geq 1$, it interprets x as the code of a TM X and then simulates the action of X on input $\vec{y}_n$ (if x cannot be recognized as a code, then U_n runs forever). In other words, U_n acts as a "stored program computer", and the function $\lambda x \vec{y}_n . \psi_n(x, \vec{y}_n)$ which it computes is *universal* for $\mathcal{PC}^{(n)}$ (that is, $\mathcal{P}^{(n)}$), since

(1)   for any $x \in \mathcal{N}$, $\lambda \vec{y}_n . \psi_n(x, \vec{y}_n) \in \mathcal{P}^{(n)}$ [if x is not a code, then $\lambda \vec{y}_n . \psi_n(x, \vec{y}_n) = \varnothing \in \mathcal{P}^{(n)}$].

(2)   for any $f \in \mathcal{P}^{(n)}$, if X is a TM which computes f and if $x \in \mathcal{N}$ is a code for it, then $\lambda \vec{y}_n . \psi_n(x, \vec{y}_n) = f$.

As for the *S-m-n* theorem, an "algorithmic procedure" must be found such that for each two-input TM X (which computes the function $\lambda \vec{y}_2 . \psi_2(x, y_1, y_2)$, by the above remarks) and each value of y_1, say a, it produces the description of a one-input TM $M_{x,a}$ which with input y_2 produces exactly the same output as X with input (a,y_2). The "algorithmic procedure" we are seeking will be formalized by a function $\sigma \in \mathcal{R}^{(2)}$ such that, for all x,y_1 and $y_2, \psi_2(x,y_1,y_2) = \psi_1(\sigma(x,y_1), y_2)$.

It should be clear that to carry out this plan we must at least find a way to code TM's by numbers. This is where "arithmetization" enters the picture. This is not difficult to do. Recall that a TM is a set of strings (quintuples) and hence it can also easily be thought of as a *string* itself (by concatenating the quintuples in some order). The results of Chapter 4 (in particular, Corollary 1 of Theorem 4.3.2) show that a *single fixed finite* alphabet is good enough to express *any* TM over it. In other words, a TM is a string over this fixed alphabet; we may interpret each string as a number (strictly speaking as the m-adic representation of a number, where m is the number of elements of the alphabet).

§5.1 RUDIMENTARY PREDICATES AND FUNCTIONS

In this chapter we revisit the rudimentary predicates which we briefly saw in §3.9. Rudimentary predicates were introduced by Smullyan [7] and studied further by Bennett [1].

First of all, we shall expand our horizon by talking in terms of m-adic representations in general, rather than dyadic as in §3.9.

Definition 1   Let $m > 1$. Let $A = \{1,2,\ldots,m\}$ and $x = a_1a_2\ldots a_n \in A^+$ ($a_i \in A$, $i = 1,\ldots,n$). Then q has as *m-adic* representation the string x iff $q = \Sigma_{i=1}^n a_i m^i$.//

Note: The number 0 has as *m-adic* representation Λ, the empty string. It is an easy matter to show that each $q \in \mathcal{N}$ has a unique m-adic representation. We shall *not* distinguish between a number and its m-adic representation. By $x *_m y$ we mean m-adic concatenation. If m is fixed throughout a discussion, we write simply $x*y$. From now on, all our numbers are expressed in *m-adic* for an unspecified, but fixed, m. For x,y in $\mathcal{N}$, xBy means $(\exists u)y=x*u$, xEy means $(\exists u)y=u*x$, and xPy means $(\exists u,v)y=u*x*v$. For each $i \in A$ ($A = \{1,\ldots,m\}$) $\text{tally}_i(x)$ means that $x \in \{i\}^+$; $\text{tally}(x)$ means that $x \in \bigcup_{i=1}^n \{i\}^+$. We shall find it convenient to write $(\exists y)_{Pz}[\ldots]$ instead of $(\exists y)(yPz \ \& \ [\ldots])$. Also, we write $(\forall y)_{Pz}[\ldots]$ instead of $(\forall y)(yPz \Rightarrow [\ldots])$.

Definition 2   (Smullyan [7])

The class of *m-rudimentary* (or *rudimentary* base m) predicates is the *closure* of $\{z = x*_m y\}$ under *explicit transformations* (*), *Boolean operations* and $(\exists y)_{\leq z}, (\forall y)_{\leq z}.//$

Definition 3   (Bennett [1], Smullyan [7])

The class of *strictly m-rudimentary* (or *S-rudimentary*(†) base m) predicates is the *closure* of $\{z = x*_m y\}$ under *explicit transformations, Boolean operations* and $(\exists y)_{Pz}, (\forall y)_{Pz}.//$

Note: In Definition 3, each occurrence of "$\dots P \dots$" is implying m-adic notation for both of P's arguments. Bennett [1] shows that the m-rudimentary predicates are independent of m whereas the S-rudimentary predicates are not.

Further, note that $(\exists y)_{Pz}$ implies a simpler "search" for y than $(\exists y)_{\leq z}$ does. In the former case, we try at most about $\log_m z$ different y's (that is, as many as the m-adic length of z), whereas in the latter case we try at most $z + 1$ different y's.

Lemma 1   The predicate xPy is rudimentary and S-rudimentary.

> ***Proof*** $xPy \equiv (\exists u)_{\leq y}(\exists v)_{\leq y} \, y = u*x*v \equiv (\exists u)_{Py}(\exists v)_{Py} \, y = u*x*v.//$

Note: In the proof above, we tacitly relied on the trivial fact that $\lambda \vec{x}_n y . y = x_1 * x_2 * \dots * x_n$ is both rudimentary and S-rudimentary. (See Problem 2.)

Proposition 1   The class of rudimentary predicates contains the class of S-rudimentary predicates.

> ***Proof*** Observe that $(\exists y)_{Pz}[\dots] \equiv (\exists y)_{\leq z}(yPz \, \& \, [\dots]).//$

Lemma 2   The predicates $\lambda x.\mathrm{Pow}_m(x) \equiv$ "x is a power of m", $\lambda xy.y = m^{|x|}$, where $|x|$ is the m-adic length of x, and $\lambda xyz.z = x*y$ are in $\mathcal{PR}_*$.

(*)See Definition 1.2.2.

(†)This term is used by Smullyan [7] and Jones [6].

Proof (1) Let $p_{i_1}, \ldots, p_{i_k}$ be the *distinct* primes occurring in the prime number decomposition of m, with exponents $a_{i_1}, \ldots, a_{i_k}$ (all ≥ 1), respectively.

Then $\mathrm{Pow}_m(x) \equiv a_{i_1} \mid \exp(i_1, x) \& \ldots \& a_{i_k} \mid \exp(i_k, x) \&$

$$\left\lfloor \frac{\exp(i_1, x)}{a_{i_1}} \right\rfloor = \left\lfloor \frac{\exp(i_2, x)}{a_{i_2}} \right\rfloor \& \left\lfloor \frac{\exp(i_2, x)}{a_{i_2}} \right\rfloor = \left\lfloor \frac{\exp(i_3, x)}{a_{i_3}} \right\rfloor$$

$$\& \ldots \& \left\lfloor \frac{\exp(i_{k-1}, x)}{a_{i_{k-1}}} \right\rfloor = \left\lfloor \frac{\exp(i_k, x)}{a_{i_k}} \right\rfloor \& (\forall y)_{\leq x}(y \mid x \& \Pr(y)$$

$$\Rightarrow (y = p_{i_1} \vee y = p_{i_2} \vee \ldots \vee y = p_{i_k})).$$

(2) Observe that $\overbrace{1 \ldots 1}^{|z|} \leq z \leq \overbrace{m \ldots m}^{|z|}$, hence

$$\frac{m^{|z|} - 1}{m - 1} \leq z \leq m \frac{m^{|z|} - 1}{m - 1}.$$

Thus, $m^{|z|} \leq (m - 1) z + 1 \leq m^{|z|} - m + 1$ and finally $m^{|z|} \leq (m - 1) z + 1 < m^{|z|+1}$. Hence, $y = m^{|x|} \equiv \mathrm{Pow}_m(y) \& y \leq (m - 1) x + 1 \& my > (m - 1) x + 1$

(3) $z = x*y \equiv z = xm^{|y|} + y.//$

Proposition 2 Every S-rudimentary (resp. rudimentary) predicate is primitive recursive.

Proof By Lemma 2 and Proposition 1.//

Lemma 3 The predicates below are S-rudimentary.

(1) xBy

(2) xEy

(3) xPy

(4) $x = y$

(5) $\mathrm{tally}_i(x)$, for $i = 1, 2, \ldots, m$. (x is a tally of i's)

(6) $y = \mathrm{maxtal}_i(x)$ (y is the maximal length tally of i's which is part of x)

(7) $D(x) \equiv x$ is a digit

(8) $\mathrm{tally}(x) \equiv x$ is some tally of the same (unspecified) digit

Proof

(1) $\quad xBy \equiv (\exists z)_{Py}\, y = x*z$

(2) $\quad xEy \equiv (\exists z)_{Py}\, y = z*x$

(3) $\quad xPy \equiv (\exists z)_{Py}\, (\exists w)_{Py}\, y = z*x*w$

(4) $\quad x = y$ is obtained from $x = y*z$ by explicit transformation $(z \leftarrow 0)$.

(5) $\quad \text{tally}_i(x) \equiv x \neq 0\ \&\ (\forall y)_{Px}\, (y \neq 0 \Rightarrow iPy)$

(6) $\quad y = \text{maxtal}_i(x) \equiv \text{tally}_i(y)\ \&\ yPx\ \&\ (\forall z)_{Px}(\text{tally}_i(z) \Rightarrow zPy)$

(7) $\quad D(x) \equiv x = 1 \vee x = 2 \vee \ldots \vee x = m$

(8) $\quad \text{tally}(x) \equiv \text{tally}_1(x) \vee \text{tally}_2(x) \vee \ldots \vee \text{tally}_m(x).//$

Lemma 4   The following predicates are rudimentary:

(1) $\quad x \leq y$

(2) $\quad |x| < |y|$

(3) $\quad |x| \leq |y|$

(4) $\quad |x| = |y|$

Proof

(1) $\quad x \leq y \equiv (\exists z)_{\leq y}\, z = x$

(2) $\quad |x| < |y| \equiv (\exists z)_{\leq y}\, \text{tally}_1(z)\ \&\ \neg (z \leq x)$

(3) $\quad |x| \leq |y| \equiv \neg (|y| < |x|)$

(4) $\quad |x| = |y| \equiv |x| \leq |y|\ \&\ |y| \leq |x|.//$

The above lemmas provide us with enough tools to carry out the bulk of our arithmetization. However, for the "decoding" function ("d" of Theorem 3.7.4), as well as for converting the $q_0 1^{x_1+1} B 1^{x_2+1} \ldots B 1^{x_n+1}$ input format into the format required by the arithmetization, we shall need the fact that $\lambda xy.y = |x|$ is rudimentary.

To this end, we devote the balance of this paragraph, following Bennett [1] closely. (We only deviate in the proof that rudimentary predicates are closed under substitution by "concatenation polynomials".)

Lemma 5   The predicate $\lambda \vec{x}_n \vec{y}_m . x_1 * \ldots * x_n = y_1 * \ldots * y_m$ is strictly rudimentary (hence also rudimentary, by Proposition 1).

Proof [1] Induction on n (m is arbitrary).

$n = 1$: $x_1 = y_1* \ldots *y_m$ is strictly rudimentary (see note following Lemma 1).

$n = k$: Assume $x_1* \ldots *x_k = y_1* \ldots *y_m$ is S-rudimentary for arbitrary m.

Case $n = k + 1$: $x_1* \ldots *x_k*x_{k+1} = y_1* \ldots *y_m$ iff some initial part of $y_1* \ldots *y_m$ is "responsible" for $x_1* \ldots *x_k$ and the balance for x_{k+1}.

Hence,

$$x_1* \ldots *x_k*x_{k+1} = y_1* \ldots *y_m \equiv (\exists z,z')_{Py_1}\, y_1 = z*z'\, \&$$

$$x_1* \ldots *x_k = z\, \&\, x_{k+1} = z'*y_2* \ldots *y_m \lor \ldots$$

$$\lor (\exists z,z')_{Py_i}\, y_i = z*z'\, \&\, x_1* \ldots *x_k = y_1* \ldots *y_{i-1}*z\, \&\, x_{k+1}$$

$$= z'*y_{i+1}* \ldots *y_m \lor \ldots \lor (\exists z,z')_{Py_m}\, y_m = z*z'\, \&\, x_1* \ldots *x_k$$

$$= y_1* \ldots *y_{m-1}*z\, \&\, x_{k+1} = z'.//$$

Lemma 6 $|x| + 1 = |y|$ is rudimentary.

Proof $|x| + 1 = |y| \equiv (\exists d)_{Py}\, (\exists u)_{Py}\, y = u*d\, \&\, |x| = |u|\, \&\, D(d).//$

Lemma 7   The class of rudimentary predicates is closed under substitution of terms like $x*d$, where $D(d)$.

Proof Induction with respect to the definition of the class.

(1) *Basis.* "$z = x*y$ is still rudimentary after substituting $u*d$, where $D(d)$, for any of x,y,z". The statement in quotes is true as a special case of Lemma 5.

(2) Property propagates with $\&$ and $\neg$. For example, if the rudimentary predicates $R(\vec{x}_n)$ and $Q(\vec{y}_m)$ have the property (that substitution of $u*d$ for any of x_i or y_j leaves them rudimentary), then, clearly, so do $R(\vec{x}_n)\&Q(\vec{y}_m)$ and $\neg R(\vec{x}_n)$.

(3) Property propagates with explicit transformation; i.e., "if the rudimentary predicate $R(\vec{x}_n)$ has the property, then so does any explicit transform $Q(\vec{y}_m)$ of $R(\vec{x}_n)$".

Indeed, let $Q(\vec{y}_m) \equiv R(\vec{t}_n)$ where each t_i is either some y_j or constant.

Let $u*d$, where $D(d)$, be substituted for y_p.

Case 1. No t_i is y_p. Then $Q(y_1, \ldots, y_{p-1}, u*d, y_{p+1}, \ldots, y_m) \equiv R(\vec{t}_n)$, hence Q as amended is rudimentary (since R is).

Case 2. $t_{i_1}, \ldots, t_{i_l}$ all are the same symbol as y_p. Then,

$$Q(y_1, \ldots, y_{p-1}, u*d, y_{p-1}, \ldots, y_m) \equiv R(t_1, \ldots, t_{i_1} - 1, u*d,$$

$$t_{i_1+1}, \ldots, t_{i_2-1}, u*d, t_{i_2+1}, \ldots, t_{i_l-1}, u*d, t_{i_l+1}, \ldots, t_n).$$

Hence, Q as amended is rudimentary since R is (after substitution) by induction hypothesis.

(4) Property propagates with $(\exists y)_{\leq}$.

So let $P(y,\vec{z})$ have the property. Consider $(\exists y)_{\leq x}P(y,\vec{z})$ and substitute $u*d$ for x, where $D(d)$.

Now $(\exists y)_{\leq u*d}P(y,\vec{z}) \equiv (\exists y_1)_{\leq u}(\exists y_2)_{\leq m}D(y_2) \;\&\; (y_2 \leq d \lor y_1 < u) \;\&\; P(y_1*y_2,\vec{z})$. The predicate to the right of "$\equiv$" is rudimentary by induction hypothesis.//

Note: Recall that we are using m-adic notation.

Definition 4   [1] A term such as $x_1*\ldots*x_n$ $(n \geq 1)$, where each x_i is a variable or a constant, is a *concatenation polynomial* of length n.//

For example, x, $x*x$, $x*y$, $x*2$, $1*3*y$ are concatention polynomials.

Proposition 3   [1] Rudimentary predicates are closed under substitution by concatenation polynomials.

Proof   By induction with respect to the polynomial length n.

Basis $(n = 1)$. The result is trivial as substitution of a length-one polynomial is a special case of explicit transformation.

Assume that, for $n \leq k$ $(k \geq 1)$, rudimentary predicates are closed under substitution by a concatenation polynomial of length n.

Case $n = k + 1$. We are to show that rudimentary predicates are closed under substitution by a concatenation polynomial of length $k + 1$.

We do induction with respect to the definition of rudimentary predicates.

Clearly, the basis (substitution in $z = x*y$) is satisfied for substitutions of polynomials of *any* length (Lemma 5) and again the property (of "staying rudimentary after substitution of a concatenation polynomial for a variable") propagates with $\&$, $\neg$, and explicit transformations. (Note that so far the induction hypothesis on n was not needed).

Finally, show that the property propagates with $(\exists y)_{\leq x}$; i.e., if substitution of $x_1*\ldots*x_{k+1}$ for any variable of the rudimentary predicate $R(y,\vec{z})$ leads to a rudimentary predicate, then this is also the case for the rudimentary predicate $(\exists y)_{\leq x}R(y,\vec{z})$.

Clearly, the interesting case is substitution in x (why?), which we now consider:

$$(\exists y)_{\leq x_1*\ldots*x_{k+1}} R(y,\vec{z}) \equiv (\exists y_1)_{\leq x_1*\ldots*x_k} (\exists y_2)_{\leq x_{k+1}} R(y_1*y_2,\vec{z})$$
$$\lor (\exists y_1)_{\leq x_1*\ldots*x_k} (\exists d)_{\leq m} (\exists y_2)_{\leq x_{k+1}} D(d) \;\&\; y_1 < x_1*\ldots*x_k$$
$$\&\; |y_2| + 1 = |x_{k+1}| \;\&\; (\exists u)_{\leq y_1*d} R(u*y_2,\vec{z}).$$

By the induction hypothesis on n, all the predicates to the right of $(\exists)$-symbols are rudimentary. Again, $(\exists y_1)_{\leq x_1 * \ldots * x_k}$, by induction hypothesis on n, and $(\exists u)_{\leq y_1 * d}$ (Lemma 7) do not lead outside the rudimentary class. We are done.//

Note: (a) A number y is less than or equal to a number $z * w$ if either the last (right-most) $|w|$ digits of y form a number $>w$ but the left-most digits (if any) a number $<z$, or if $y = y_1 * y_2$ where both $y_1 \leq z$ and $y_2 \leq w$. This observation was used in the splitting of $(\exists y)_{\leq x_1 * \ldots * x_{k+1}}$.

(b) The use of Lemma 7 in order to have $(\exists u)_{\leq y_1 * d} R(u * y_1, \vec{z})$ rather than $R(y_1 * d * y_2, \vec{z})$ was essential in our induction from $n \leq k$ to $n = k + 1$, because the *least* value of $k + 1$ is 2, and thus without Lemma 7 we would not be able to infer that $R(y_1 * d * y_2, \vec{z})$ is rudimentary, and our induction would not progress from $n = 1$ to $n = 2$ [remember, we assumed that $R(y, \vec{z})$ is staying rudimentary after substitution of polynomials of length $k + 1$ for variables. Thus, if $n = k = 1$, the assumption is for substitution of polynomials of length $k + 1 = 2$].

Corollary $x_1 * \ldots * x_k \leq y_1 * \ldots * y_m$ is rudimentary.

> **Proof** $x_1 * \ldots * x_k \leq y_1 * \ldots * y_m \equiv (\exists z)_{\leq y_1 * \ldots * y_m} z = x_1 * \ldots * x_k.//$

Note: Bennett proves the corollary without the help of Proposition 3; indeed, he uses the result stated in the corollary for the proof of Proposition 3.

Also it is true that the *strictly* rudimentary predicates are closed under substitution of concatenation polynomials [1]. As we do not need this result, we leave it as an interesting exercise. (See Problem 3).

Definition 5   (Jones [6]) A function $\lambda \vec{x}.f(\vec{x})$ is strictly rudimentary iff $\lambda \vec{x} y.y = f(\vec{x})$ is strictly rudimentary and for some fixed i, $f(\vec{x})Px_i$ for all $\vec{x}$.//

Example 1 $f = \lambda x.x$ is strictly rudimentary, since $\lambda xy.y = x$ is, and $f(x)Px$ is true for all x.//

Example 2   Although $z = x * y$ is strictly rudimentary, $\lambda xy.x * y$ is not, as $(x * y)$ is not a part of x for all x, y nor is it a part of y for all x, y.//

Example 3 $\lambda x.x * 1$ is not strictly rudimentary since $\neg(\exists x)(x * 1)Px$. On the other hand, the predicate $y = x * 1$ *is* strictly rudimentary.//

Example 4 $\lambda x.0$ is, but $\lambda x.1$ is not (!) strictly rudimentary. Indeed, both $y = 0$ and $y = 1$ are strictly rudimentary. But whereas $0Px$ for all x, "$1Px$" is *not* true for all x. For example, if $x = 2$, $\neg 1P2$.//

Even though the strictly rudimentary predicates form a very restricted class and the corresponding functions are even more restricted, as seen by the examples above, they are powerful enough to be used in a derivation of a Kleene-normal form theorem for partial recursive functions, *provided* we are prepared to adjust input/output conventions around the restrictions imposed on us by these predicates and functions. If we want the input/output convention to be flexible, then we need to use rudimentary functions (see below) and predicates.

Definition 6   (Smullyan, Bennett [7, 1]) A function $\lambda \vec{x}_n.f(\vec{x}_n)$ is rudimentary iff $\lambda \vec{x}_n y.y = f(\vec{x}_n)$ is rudimentary, and $f(\vec{x}_n)$ is bounded for all $\vec{x}_n$ by some concatenation polynomial.//

Note: Trivially, every *strictly* rudimentary function is also rudimentary.

Example 5 All the functions discussed in Examples 1-4 are rudimentary.//
     We now embark on showing that $\lambda x.|x|$ is rudimentary. As $|x| \le x$ for all x, the nontrivial subtask is to show that $\lambda xy.y = |x|$ is rudimentary. This will follow from a general result concerning functions obtained through "recursion on notation". We first present a convenient lemma:

Lemma 8   [1] The rudimentary predicates are closed under substitution of a variable by a rudimentary function.   This statement is also true if the term "rudimentary" is replaced by "strictly rudimentary" everywhere.

     Proof (a) *rudimentary case:* Let $R(x,\vec{z})$ be rudimentary and $\lambda \vec{y}_l.f(\vec{y}_l)$ be rudimentary. Let $u_1 * \ldots * u_k$ be a concatenation polynomial. (Each u_i is some y_j or a constant), which bounds $f(\vec{y}_l)$ for all $\vec{y}_l$.
     Then

$$R(f(\vec{y}_l), \vec{z}) \equiv (\exists x)_{\le u_1 * \ldots * u_k} x = f(\vec{y}_l) \& R(x,\vec{z})$$

     (b) *strictly rudimentary case:* Let $R(x,\vec{z})$ be strictly rudimentary and let $\lambda \vec{y}_l.f(\vec{y}_l)$ be also strictly rudimentary so that, without loss of generality, $f(\vec{y}_l)Py_1$ for all $\vec{y}_l$. Then

$$R(f(\vec{y}_l), \vec{z}) \equiv (\exists x)_{Py_1} x = f(\vec{y}_l) \& R(x,\vec{z}).//$$

Definition 7   (Bennett, Cobham [1, 2]) Given functions $\lambda \vec{y}_k.h(\vec{y}_k)$ and $\lambda x\vec{y}_k dz.g(x,\vec{y}_k,d,z)$, where $D(d)$, the function $\lambda x\vec{y}_k.f(x,\vec{y}_k)$ is obtained from them by *recursion on (m-adic) notation* iff for all $(x,\vec{y}_k)$ the f-values are defined according to the schema below:

$$\begin{cases} f(0,\vec{y}_k) = h(\vec{y}_k) \\ \text{for } i = 1,2,\ldots, m \;\; f(x*i,\vec{y}_k) = g(x,\vec{y}_k,i,f(x,\vec{y}_k)).// \end{cases}$$

Note: The above schema is, strictly speaking, *right* recursion on notation as from the f-value on x we obtain next the $x*i$-value. Clearly, ordinary primitive recursion is a special case of the above, when $m = 1$. Moreover, $\mathcal{PR}$ is closed under recursion on notation for any $m > 1$. (See Problem 4.)

The important question that arises in connection with the schema of recursion on notation is, assuming that both h and g are rudimentary, what is a sufficient condition that f is also rudimentary?

Bennett [1] has shown that $|f(x,\vec{y}_k)|^2 \le a + b \cdot \max(|x|, |y_i|_{i=1,\ldots,k})$ is sufficient.

We now embark on presenting his proof here. First, observe that the condition above ensures that f is bounded by a concatenation polynomial (why?), so that the main task is to show that $\lambda x\vec{y}_k z.z = f(x,\vec{y}_k)$ is rudimentary, given the bounding condition and the assumptions on h and g.

As usual, one deals with this by finding a predicate, which in this case we want it to be rudimentary, $R(u,x,\vec{y}_k,z)$, which is true iff u "codes" a computation which verifies "$z = f(x,\vec{y}_k)$". Because of the recursion schema (Definition 7), this computation must have one "aside computation" (or "step") for each digit of x. Therefore, the length of u will be $\le \{|x| \cdot$ (length of longest aside computation)$\}$, which is $\le |x| \cdot \sqrt{a + b \cdot \max(|x|, |y_i|_{i=1,\ldots,k})}$, as during the aside computations we will have to write down the subresults $f(q,\vec{y}_k)$ for all q such that qBx.

This makes the estimated *value* of u too big (larger than any concatenation polynomial), so that in $z = f(x,\vec{y}_k) \equiv (\exists u)R(u,x,\vec{y}_k,z)$ the "$(\exists u)$" cannot be bounded in a way which makes the right hand side rudimentary.

Bennett suggested that x be partitioned in consecutive parts $x_1, \ldots, x_l$, $l \le \sqrt{|x|}$ (roughly speaking) each part x_i having length $\le \sqrt{|x|}$.

Then instead of coding the computation of f at $x_1*x_2*\ldots*x_l$ we code the l subcomputations of f from 0 to x_1 from x_1 to $x_1*x_2, \ldots$, from $x_1*\ldots*x_{l-1}$ to $x_1*\ldots*x_l$, and then, somehow, put the subcomputations together to achieve the effect of $R(u,x,\vec{y}_k,z)$.

Note that each subcomputation has length $O(|x|),(*)$ roughly speaking, since it has $O(\sqrt{|x|})$ steps (each of size $O(\sqrt{|x|})$). We need then a predicate $Q(u,x,v,\vec{y}_k,w,w')$ which is true iff u codes a (sub)computation of f from the initial value $w = f(x,\vec{y}_k)$ to the final value $w' = f(x*v,\vec{y}_k)$.

(*)*Big-O notation:* if f and g are functions $\mathcal{N} \to \mathcal{N}$, then $f(n) = O(g(n))$ stands for $(\exists c,k)(\forall n)(n > k \Rightarrow f(n) \le c \cdot g(n))$.

Then $z = f(x, \vec{y}_k)$ is true iff

$$(\exists x_1, \ldots, x_l)_{\substack{|x_i| \le \sqrt{|x|} \\ x = x_1 * \ldots * x_l}} (\exists w_0, \ldots, w_l)(\exists u) Q(u, 0, x_1, \vec{y}_k, w_0, w_1) \, \& \, w_0$$

$$= h(\vec{y}) \, \& \, [(\forall i)_{1 < i \le l}(\exists u) Q(u, x_1 * \ldots * x_{i-1}, x_i, \vec{y}_k, w_{i-1}, w_i)] \, \& \, w_l = z$$

Note that each u in the occurrences of "$(\exists u)$" has length $O(|x|)$, and hence can be bounded by a concatenation polynomial (clearly, $|u| \le c|x| \Rightarrow u \le 1 * \underbrace{x * x * \ldots * x}_{c}$). Also the x_i, and w_i sequences can be coded each with total length $O(|x|)$ (for the x_i obviously; for the w_i due to the f-bound).

As the only source of concern toward an implementation of the above idea is the necessity to have along with the sequence $x_1, x_2, \ldots, x_l$ the sequence $x_1, x_1 * x_2, x_1 * x_2 * x_3, \ldots, x_1 * \ldots * x_l$. This latter sequence is too long for our purpose $[O(|x|^{3/2})]$. Bennett has suggested a sequence coding scheme which, unlike the ones we have seen so far (including that of §3.9), "stores" $x_1, \ldots, x_l$ contiguously in a word (number) x so that $x = x_1 * \ldots * x_l$, and uses an auxiliary, let us say, "index"-word t such that $|t| = |x|$ to access the individual x_i's, given x. This is done by choosing t so that, seen from left to right, it consists of l tallies $t_1, \ldots, t_l$ such that no two consecutive tallies t_i, t_{i+1} are formed by the same digit, and for all i, $|x_1 * \ldots * x_i| = |t_1 * \ldots * t_i|$ (therefore, also, $|x_i| = |t_i|$). Clearly, the sequence $x_1, x_1 * x_2, \ldots, x_1 * \ldots * x_l$ is also represented by the pair (x, t)!

This coding scheme is entirely analogous to the sequential allocation of variable length memory blocks (the x_i sequence) in a computer; t gives the "addressing scheme". To avoid the ambiguity arising from storing 0's, and also whenever we wish to access more than one coded sequence through the *same* index word t, we often "pad" each stored element of a sequence with a tally.

For example, in the figure above, u and w are sequence codes and t is a common index number (word). To t_i correspond u_i and w_i and the actual number "stored" in the "u_i-block" (resp. w_i-block) is retrieved by removing the *longest left-most tally* of u_i (resp. w_i) (Bennett [1]). How much padding one uses in any case (*if at all*) depends on the total length of u (resp. w) we can tolerate; also padding should *not* be used if both $u_1, \ldots, u_i, \ldots$ and $u_1, u_1 * u_2, u_1 * u_2 * u_3, \ldots$ must be easily retrievable from u.

We now give the technical details (see also Bennett [1]).

Lemma 9   The following predicates are rudimentary.

(a) $\text{Cons}(s,t) \equiv$ "s begins t and is a sequence of one or more consecutive maximal tallies of t".

(b) $\text{Init}(u,w;t,s) \equiv$ "$|u|=|t|$ and u is indexed by t and $\text{Cons}(s,t)$ and $|w|=|s|$ and w begins u".

(c) $\text{Term}(u,v;t,s) \equiv |u|=|t|$ and u is indexed by t and $\text{Cons}(s,t)$ and v, where vPu, corresponds in position and length to the last tally of s".

(d) $\text{Rid}(x,y) \equiv$ "y is obtained from x after getting rid of the left-most maximum tally of x".

Proof

(a) $\text{Cons}(s,t) \equiv s \neq 0 \ \& \ t \neq 0 \ \& \ sBt \ \& \ (\forall d)_{Ps} \ [dEs \ \& \ D(d) \Rightarrow \neg (s*d)Bt]$

(b)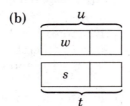

We refer to the figure at left.

$\text{Init}(u,w;t,s) \equiv |u|=|t| \ \& \ \text{Cons}(s,t) \ \& \ wBu \ \& \ |w|=|s|$

(c)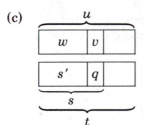

We refer to the figure at left.

$$\text{Term}(u,v;t,s) \equiv (\exists w)_{Pu} \ (\exists s',q)_{Pt} \ \text{Cons}(s',t) \ \& \ \text{Cons}(s,t)$$
$$\& \ s = s'*q \ \& \ \text{tally}(q) \ \& \ \text{Init}(u,w*v;t,s) \ \& \ |v|=|q|$$

(**Note:** Recall that concatenation polynomials may be substituted for variables in a rudimentary relation.)

(d) $\text{Rid}(x,y) \equiv x \neq 0 \ \& \ (\exists w)_{Px} \ [\text{tally}(w) \ \& \ x = w*y \ \& \ \text{Cons}(w,x)]. //$

In what follows, it will be convenient to rephrase the predicates in (b), (c) and (d) in function form. We use the notation of Bennett:

Define the functions SQ_0, SQ, and MK by:

$$SQ_0(u;t,s) = \begin{cases} w \text{ if } \mathrm{Init}(u,w;t,s) \\ 0 \text{ otherwise} \end{cases}$$

$$SQ(u;t,s) = \begin{cases} v \text{ if } \mathrm{Term}(u,v;t,s) \\ 0 \text{ otherwise} \end{cases}$$

$$MK(x) = \begin{cases} y \text{ if } \mathrm{Rid}(x,y) \\ 0 \text{ otherwise} \end{cases}$$

Clearly, SQ_0, SQ, and MK are rudimentary (why?). We are now fully equipped to prove the following:

Theorem 1   (Bennett [1]) If $\lambda x \vec{y}.f(x,\vec{y}_k)$ is obtained from rudimentary functions $\lambda \vec{y}_k.h(\vec{y}_k)$ and $\lambda x \vec{y}_k iz.g(x,\vec{y}_k,i,z)$, $i = 1, \ldots, m$ by the schema of recursion on notation (Definition 7) and if $|f(x,\vec{y}_k)|^2 \leq a + b \cdot \max(|x|,|y_j|_{j=1,\ldots,k})$, then f is rudimentary.

Proof (Bennett [1]) As earlier discussed, it suffices to show that $\lambda x \vec{y}_k z.z = f(x,\vec{y}_k)$ is rudimentary.

We first show that the Q-predicate of the earlier discussion is rudimentary. Now that we have crystalized our ideas about our coding scheme for sequences, the computation will be coded, instead of simply by "u", by a triple "p,q,t", where t is the common index word for p and q.

The Q-predicate now becomes $Q(p,q,t,x,v,\vec{y}_k,w,w')$, and it represents a computation of f from f-value "w" to f-value "w'''" and from x-value "x" to x-value "$x*v$", where for each i in the index range, $MK(q_i) = v_i$ (where $v_i Bv$) and $MK(p_i) = f(x*v_i,\vec{y}_k)$. We assume that each stored value $(v_i, f(x*v_i,\vec{y}_k))$ is padded by a tally of length ≥ 1 leading to q_i and p_i such that $|q_i| = |p_i|$ for all i.

Finally,

$Q(p,q,t,x,v,\vec{y}_k,w,w') \equiv |p| = |q| \,\&\, |q| = |t| \,\&\, (\exists s)_{Pt} [\mathrm{tally}(s) \,\&\, \mathrm{Cons}(s,t)$

$\&\, MK(SQ(p;t,s)) = w \,\&\, MK(SQ(q;t,s)) = 0] \,\&$

$MK(SQ(p;t,t)) = w' \,\&\, MK(SQ(q;t,t)) = v \,\&$

$(\forall s,s')_{Pt} [\mathrm{Cons}(s,t) \,\&\, \mathrm{Cons}(s*s',t) \,\&\, \mathrm{tally}(s') \Rightarrow$

$(\exists d)_{\leq m} \{D(d) \,\&\, MK(SQ(q;t,s*s')) = MK(SQ(q;t,s))*d \,\&$

$MK(SQ(p;t,s*s')) = g(x*MK(SQ(q;t,s)),\vec{y}_k,d,MK(SQ(p;t,s)))\}]$

Clearly, Q is rudimentary.

Now in a computation of $z = f(x, \vec{y}_k)$ either $\vec{y}_k$ is "too big", so that $|x|^2 \le a + b \cdot \max(|x|, |y_i|_{i=1,\ldots,k})$, or it is not, so that $|x|^2 > a + b \cdot \max(|x|, |y_i|_{i=1,\ldots,k})$.

In the first case, we have $|x| \le \lfloor \sqrt{a + b \cdot \max(|x|, |y_i|)} \rfloor$ and $|f(l, \vec{y}_k)| \le \lfloor \sqrt{a + b \cdot \max(|x|, |y_i|)} \rfloor$ for all l such that lBx. Due to padding, and the fact that there are $|x| + 1$ steps in the computation, we get the bound $|p| \le (\lfloor \sqrt{a + b \cdot \max(|x|, |y_i|)} \rfloor + 1)^2 \le 3(a + b \cdot \max(|x|, |y_i|)) + 1$; thus

$$p \le \underbrace{11 \ldots 1}_{3a+2} * \underbrace{x * \ldots * x}_{3b} * \underbrace{y_1 * \ldots * y_1}_{3b} * \ldots * \underbrace{y_k * \ldots * y_k}_{3b} .$$

Let us call the righthand side concatenation polynomial $J(x, \vec{y}_k)$.

Hence, in this case, $z = f(x, \vec{y}_k)$ amounts to

$$(\exists p, q, t)_{\le J(x, \vec{y}_k)} \, Q(p, q, t, 0, x, \vec{y}_k, h(\vec{y}_k), z) \tag{1}$$

In the second case, $(|x|^2 > a + b \cdot \max(|x|, |y_i|_{i=1,\ldots,k}))$ let $x = x_1 * x_2 * \ldots * x_l$, where

$$|x_1| = \ldots = |x_{l-1}| = \lfloor \sqrt{a + b \cdot \max(|x|, |y_i|)} \rfloor + 1 \text{ and}$$

$$\lfloor \sqrt{a + b \cdot \max(|x|, |y_i|)} \rfloor + 1 \le |x_l| < 2(\lfloor \sqrt{a + b \cdot \max(|x|, |y_i|)} \rfloor + 1)$$

In accordance with the discussion preceding Theorem 1, let $w_1, \ldots, w_l$ be the f-values at the end of the subcomputations from 0 to x_1, from x_1 to $x_1 * x_2$, $\ldots$, from $x_1 * \ldots * x_{l-1}$ to $x_1 * \ldots * x_l$, respectively.

Since for

$$h = 1, \ldots, l, |w_h| \le \lfloor \sqrt{a + b \cdot \max(|x_1 * \ldots * x_h|, |y_i|_{i=1,\ldots,k})} \rfloor$$

$$< \lfloor \sqrt{a + b \cdot \max(|x|, |y_i|)} \rfloor + 1$$

it follows that in coding $w_1, \ldots, w_l$ (by u) and $x_1, \ldots, x_l$ (by $x = x_1 * x_2 * \ldots * x_l$) with common index t we pad the w_i's into u_i's, but we do *not* pad the x_i's. (Thus, $|u| = |t| = |x|$.) Each subcomputation (for w_h, $h = 1, \ldots, l$) has *length* bounded by

$$\{2(\lfloor \sqrt{a + b \cdot \max(|x|, |y_i|)} \rfloor + 1)\} \cdot \{\lfloor \sqrt{a + b \cdot \max(|x|, |y_i|)} \rfloor + 1\}$$

$$\le 6(a + b \cdot \max(|x|, |y_i|)) + 2$$

hence the numerical *value* of the subcomputation is

$$\le \underbrace{11 \ldots 1}_{6a+3} * \underbrace{x * \ldots * x}_{6b} * \ldots * \underbrace{y_k * \ldots * y_k}_{6b} .$$

Let us call this last concatenation polynomial $I(x,\vec{y}_k)$. Thus, in this case, $z = f(x,\vec{y}_k)$ amounts to

$$(\exists u,t)_{\leq 1*x} [\,|t| = |u| \,\&\, |u| = |x| \,\&\, z = MK(SQ(u;t,t)) \,\&$$

$$(\exists s)_{Pt} \{\text{tally}(s) \,\&\, \text{Cons}(s,t) \,\&\, (\exists p,q,\tau)_{\leq I(x,\vec{y}_k)} Q(p,q,\tau,0,SQ(x;t,s),\vec{y}_k,h(\vec{y}_k),$$

$$MK(SQ(u;t,s)))\} \,\&\, (\forall s,s')_{Pt} \{\text{Cons}(s,t) \,\&\, \text{Cons}(s*s',t) \,\&\, \text{tally}(s')$$

$$\Rightarrow (\exists p,q,\tau)_{\leq I(x,\vec{y}_k)} Q(p,q,\tau,SQ_0(x;t,s),SQ(x;t,s*s'),\vec{y}_k,MK(SQ(u;t,s)),$$

$$MK(SQ(u;t,s*s')))\}] \tag{2}$$

Clearly, $z = f(x,\vec{y}_k)$ is the disjunction of the predicates (1) and (2), hence it is rudimentary.//

Note: Bennett has shown that a slightly more general recursion schema, with the same boundedness assumption, still does not lead outside the set of rudimentary functions. (See Problem 6).

Example 6 $\lambda x.x + 1$ is rudimentary. Indeed, observe first that $x + 1 \leq m*x$, hence $x + 1$ is bounded by some concatenation polynomial.
Next,

$$y = x + 1 \equiv [x = 0 \,\&\, y = 1] \vee [\text{tally}_m(x) \,\&$$

$$(\exists u)_{Py} \{y = 1*u \,\&\, |u| = |x| \,\&\, \text{tally}(y)] \vee [(\exists w,d,v)_{Px}(\exists \delta,u)_{Py} \{x = w*d*v$$

$$\&\, y = w*\delta*u \,\&\, \text{tally}_m(v) \,\&\, \text{tally}_1(u) \,\&\, D(d) \,\&\, D(\delta)$$

$$\&\, d \neq m \,\&\, (d = 1 \,\&\, \delta = 2 \vee d = 2 \,\&\, \delta = 3 \vee \ldots \vee d = m - 1 \,\&\, \delta = m)\}].$$

The predicate to the right of "$\equiv$" is certainly rudimentary.//
Finally, our goal:

Proposition 4 $\lambda x.|x|$ is rudimentary.

Proof First$|\,|x|\,| = O(\log_m(|x|)) \leq \sqrt{a + b|x|}$ for appropriate a and b. (See Problem 7.)
Next,

$$\begin{cases} |0| = 0 \\ |x*d| = |x| + 1, \text{ for } d = 1, \ldots, m \end{cases}$$

defines $\lambda x.|x|$ by recursion on notation, starting from the rudimentary functions $\lambda x.0$ and $\lambda x.x + 1$.
Therefore, $\lambda x.|x|$ is rudimentary.//

§5.2 UNIVERSAL FUNCTION THEOREM

We now proceed to obtain a rudimentary Kleene predicate for Turing Machines and through it to derive the universal function theorem. The method, as already indicated, is by arithmetization of TM computations.

We know (Corollary 1 of Theorem 4.3.2) that TMs over a *fixed* and finite tape alphabet are sufficient to compute any function in $\mathcal{P}$.

Let this finite alphabet be $A = \{1, B, s_2, \ldots\}$. In talking about Turing Machines some other symbols are important: The state symbol q (this is a generator for any state name: q_0 is "q", q_1 is "qq", and q_i is "$\overbrace{qq \ \ldots. \ q}^{i+1}$"), the STAY, LEFT, RIGHT symbols "S", "L", "R", and finally a separator symbol ";" used to separate consecutive quintuples of a TM as well as consecutive IDs of a TM computation.

Thus, both TMs and computations are certain strings over $\tilde{A}$, where $\tilde{A}$ is (order of elements fixed throughout this discussion) $\{1, B, ;, q, S, L, R, s_2, \ldots, s_l\}$. To be exact, a TM is a string of the form ;quintuple$_1$;quintuple$_2$;...;quintuple$_k$; where there are k distinct quintuples (their order is immaterial, but nevertheless kept fixed in any given discussion). A computation is a string ;ID$_1$;ID$_2$;...;ID$_l$; such that ID$_i \rightarrow$ ID$_{i+1}$ for all i. Note that the separator ";" is *not* a symbol of any of the substrings "quintuple$_j$" or "ID$_i$".

Finally, let $|\tilde{A}| = m$ and identify $\tilde{A}$'s elements $1, B, ;$ etc., with the numbers $1, 2, 3$, etc—that is, with their *position number*. Thus, any string over $\tilde{A}$ naturally corresponds to a number x. Throughout the remaining discussion $|\tilde{A}| = m > 1$ is fixed. We present a sequence of predicates which "talk about" TM's. Each predicate is seen to be *strictly m-rudimentary*.

(1)   tape$(x) \equiv x \in \{1, 2, 8, \ldots, m\}^*$ (i.e., x is a *tape* string).

$$\text{tape}(x) \equiv \neg\,(3Px \lor 4Px \lor 5Px \lor 6Px \lor 7Px)$$

(2)   symbol$(x) \equiv x \in \{1, 2, 8, \ldots, m\}$ (i.e., x is a *tape* symbol)

$$\text{symbol}(x) \equiv x = 1 \lor x = 2 \lor x = 8 \lor x = 9 \lor \ldots \lor x = m$$

(3)   state$(x) \equiv \text{tally}_4(x)$ (i.e., x denotes a state, therefore it is a string of q's)

(4)   ID$(x) \equiv x$ is an ID, i.e., of the form $tqas$ where t and s are tapes, q is a state and a is a symbol.

$$\text{ID}(x) \equiv (\exists t, q, a, s)_{Px}\ x = t*q*a*s \ \&\ \text{tape}(t) \ \&\ \text{state}(q) \ \&$$

$$\text{symbol}(a) \ \&\ \text{tape}(s)$$

(5) $\mathrm{quint}(x) \equiv x$ is a quintuple string.

$\mathrm{quint}(x) \equiv (\exists\,q,a,b,p,r)_{Px}\ \mathrm{state}(q)\ \&\ \mathrm{symbol}(a)\ \&\ \mathrm{symbol}(b)\ \&$

$\mathrm{state}(p)\ \&\ (r = 5 \vee r = 6 \vee r = 7)\ \&\ x = q*a*b*p*r$

(6) $\mathrm{func}(x,y) \equiv x$ and y code quintuples which are identical *if* they start with the same state-symbol pair (i.e., the TM transition relation is a function)

$\mathrm{func}(x,y) \equiv \mathrm{quint}(x)\ \&\ \mathrm{quint}(y)\ \&\ (\forall q,a)_{Px}\ [\mathrm{state}(q)\ \&\ \mathrm{symbol}(a)$

$\&\ (q*a)Bx\ \&\ (q*a)By \Rightarrow x = y]$

(7) $\mathrm{TM}(x) \equiv x$ is a string denoting a TM.

$\mathrm{TM}(x) \equiv \neg 33Px\ \&\ 3Bx\ \&\ 3Ex\ \&\ (\forall y)_{Px}\ [((3*y*3)Px\ \&\ \neg 3Py)$

$\Rightarrow \mathrm{quint}(y)]\ \&\ (\forall y,z)_{Px}\ [\mathrm{quint}(y)\ \&\ \mathrm{quint}(z)\ \&$

$(3*y*3)Px\ \&\ (3*z*3)Px \Rightarrow \mathrm{func}(y,z)](*)$

(8) The "yield" operation $X \longrightarrow_Z Y$ for IDs X and Y of TM Z is captured by the yield-relation "yield(z,x,y)", where z is the string denoting Z, x denotes X and y denotes Y.
We express "yield" as a disjunction of "yield$_1$", "yield$_2$", "yield$_3$" (see also Davis [3]) according to the case (see Definition 4.1.4).
Case 1. $qabpS \in Z\ \&\ X = tqas\ \&\ Y = tpbs$.

$\mathrm{yield}_1(z,x,y) \equiv \mathrm{TM}(z)\ \&\ (\exists w,q,a,b,p)_{Pz}\ (\exists t,s)_{Px}\ \{\mathrm{tape}(t)$

$\&\ \mathrm{tape}(s)\ \&\ \mathrm{state}(q)\ \&\ \mathrm{state}(p)\ \&\ \mathrm{symbol}(a)\ \&\ \mathrm{symbol}(b)\ \&$

$x = t*q*a*s\ \&\ y = t*p*b*s\ \&\ w = q*a*b*p*5\ \&\ (3*w*3)Pz\}$

Case 2. $qabpL \in Z\ \&\ (x = tcqas\ \&\ Y = tpcbs\ \vee$
$\qquad\qquad\quad X = qas\ \&\ Y = pBbs)$.

$\mathrm{yield}_2(z,x,y) \equiv \mathrm{TM}(z)\ \&\ (\exists w,q,a,b,p)_{Pz}\ \{\mathrm{state}(q)\ \&\ \mathrm{state}(p)\ \&$

$\mathrm{symbol}(a)\ \&\ \mathrm{symbol}(b)\ \&\ w = q*a*b*p*6\ \&\ (3*w*3)Pz\ \&$

(*)Recall that "3" corresponds to ";". We could have used ";Bx", etc., instead of "$3Bx$", etc. Our indirect approach (3 for ; etc.) is only to *emphasize* that our predicates are on numbers, which would, of course, not be contradicted even if we used ";Bx", etc. (After all, "three" was "3" for the Arabs, "III" for the Romans, and "γ" for the ancient Greeks; it could very well be ";" for us.)

$$[(\exists c,t,s)_{Px} \ (\text{symbol}(c) \ \& \ \text{tape}(t) \ \& \ \text{tape}(s) \ \& \ x = t*c*q*a*s \ \& \ y$$

$$= t*p*c*b*s) \vee (\exists s)_{Px} \ (\text{tape}(s) \ \& \ x = q*a*s \ \& \ y = p*2*b*s)]\}$$

Case 3. $qabpR \in Z \ \& \ (X = tqacs \ \& \ Y = tbpcs \vee$
$$X = tqa \ \& \ Y = tbpB).$$

$\text{yield}_3(z,x,y) \equiv \text{TM}(z) \ \& \ (\exists w,q,a,b,p)_{Pz}\{\text{state}(q)$

$\& \ \text{symbol}(a) \ \& \ \text{symbol}(b) \ \& \ \text{state}(p) \ \& \ w = q*a*b*p*7 \ \&$

$(3*w*3)Pz \ \& \ [(\exists c,t,s)_{Px} \ (\text{symbol}(c) \ \& \ x = t*q*a*c*s \ \&$

$y = t*b*p*c*s \ \& \ \text{tape}(t) \ \& \ \text{tape}(s)) \vee (\exists t)_{Px} \ (x = t*q*a \ \& \ y$

$= t*b*p*2 \ \& \ \text{tape} \ (t))]\}.$

(9) $\text{InitID}(x) \equiv x$ is an initial ID; i.e., one of the form $q_0 t$ where $t \neq 0$ is a tape.

$$\text{InitID}(x) \equiv \text{ID}(x) \ \& \ x \neq 4 \ \& \ 4Bx \ \& \ \neg 44Bx$$

(10) $\text{Fin}(z,x) \equiv x$ is a *final* ID of TM z.

$\text{Fin}(z,x) \equiv \text{TM}(z) \ \& \ (\exists t,q,a,s)_{Px} \ [\text{tape}(t) \ \& \ \text{state}(q)$

$\& \ \text{symbol}(a) \ \& \ \text{tape}(s) \ \& \ x = t*q*a*s \ \&$

$(\forall w)_{Pz} \ (\text{quint}(w) \ \& \ (3*w*3)Pz \Rightarrow \neg (q*a)Bw)]$

(11)   A computation of Z is a sequence $I_1, \ldots, I_k$ of IDs such that

$$I_1 \text{ is initial}$$
$$I_k \text{ is final}$$
$$I_i \rightarrow {}_z I_{i+1}, i = 1, \ldots, k-1$$

We "code" a computation by $;I_1;I_2;\ldots;I_k;$
$\text{Comp}(z,x,y) \equiv y$ is a computation of TM Z starting with ID x. Thus,

$\text{Comp}(z,x,y) \equiv \text{TM}(z) \ \& \ \text{InitID}(x) \ \& \ 3By \ \& \ 3Ey \ \& \ \neg 33Py$

$\& \ (\forall w)_{Py} \ [((3*w*3)Py \ \& \ \neg 3Py) \Rightarrow \text{ID}(w)] \ \&$

$(3*x*3)By \ \& \ (\forall u,w)_{Py} \ [\{(3*u*3*w*3)Py \ \& \ \neg (3Pu \vee 3Pw)\}$

$\Rightarrow \text{yield}(z,u,w)] \ \& \ (\exists w)_{Py} \ ((3*w*3)Ey \ \& \ \neg 3Pw \ \& \ \text{Fin}(z,w))$

We have proved:

Theorem 1   There is a *strictly* rudimentary predicate $\text{Comp}(z,x,y)$ which is true iff the TM (with code) z when started with initial ID (with code) x has a computation (with code) y.

Note: A string z which codes a TM Z is called *a Gödel number* of Z. In symbols $gn(Z) = z$. This "gn" relation is not a function, as $gn(Z)$ depends on the particular ordering of the quintuples of Z.

Theorem 1, adapted for nondeterministic TMs (see next chapter) will be useful in Chapter 13. Note that "Comp" is essentially a Kleene-predicate for TM computations. We now conclude this paragraph by obtaining normal form theorems for $\mathcal{PC}$-*functions* (recall that $\mathcal{P} = \mathcal{PC} = \mathcal{PU} = \mathcal{PM}$ by Theorem 4.5.1).

We obtain first a normal-form theorem involving strictly rudimentary predicates and functions.

Theorem 2   There exists a strictly rudimentary function, "dec", of one variable, and a strictly rudimentary predicate $\text{Comp}(z,x,y)$ such that for each TM Z,

$$[\text{Res}_Z(q_0 1^{x_1+1}B1^{x_2+1}B\ldots B1^{x_n+1})] = [\text{dec}((\mu y)\text{Comp}(z,41^{x_1+1}2\ldots 21^{x_n+1}, y))]$$

where 4 is q_0, 2 is B, $z = gn(Z)$, and arithmetic is on m-adic notation.

Note: $[s]$ = number of 1's in the string s, as in §4.1 The string s, as all strings in this section, is over $\tilde{A}$ earlier introduced, hence it is (the m-adic notation of) a number in $\mathcal{N}$.

Proof Recall (Corollary 1, Theorem 4.3.2) that final IDs of TMs that compute functions are of the form $01^n q(\$,\$)$, $n \ge 0$.

Define $y = \text{dec}(x)$ by

$$(\exists z)_{Px}\,(3*z*3)Ex\ \&\ \text{ID}(z)\ \&\ y = \text{maxtal}_1(z)$$

Clearly, $\lambda xy.y = \text{dec}(x)$ is strictly rudimentary. Define $\lambda x.\text{dec}(x)$ by

$$\text{if } (\exists y)\ y = \text{dec}(x)\ \textbf{then } y\ \textbf{else } 0$$

Clearly, $\text{dec}(x)Px$; hence, "dec" is strictly rudimentary. The statement of the theorem follows.//

Note: (a) A different "dec" defined by $\lambda x.$"last ID of computation x" would be sufficient for Theorem 2 and would still be strictly rudimentary. The particular way we defined "dec" will be useful in the proof of Theorem 3 below.

(b) $\lambda s.[s]$ cannot be strictly rudimentary (why?), thus we cannot put the Theorem 2 statement in the form "$\psi(\vec{x}_n) = \text{dec}((\mu y)\text{Comp}(z,\langle\vec{x}_n\rangle,y))$" for strictly rudimentary dec, where $\psi \in \mathcal{P}^{(n)}$ and $\langle\vec{x}_n\rangle = 41^{x_1+1}21^{x_2+1}2\ldots21^{x_n+1}$.

It should be appreciated that changing the "output convention" of TMs so that results, instead of in "unary", are in dyadic or in some other basis $< |\tilde{A}|$ does not change the situation. Still "dec" has to convert, say, the string over $\{1,2\}$ to a number (*) (that is, string over $\tilde{A} \supsetneq \{1,2\}$) and this cannot be done by a strictly rudimentary function.

A way out of this difficulty is to change our input/output conventions and view TMs as string processors (Jones [6]). That is, an n-Tm defines a function from $\{1,2,4\}^*$ to $\{1\}^*$ which is defined *only* on strings of the form $41^{x_1+1}2\ldots21^{x_n+1}$, where $x_1 \geq 0, \ldots, x_n \geq 0$. The output of such a function is given by "$\text{dec}((\mu y)\text{Comp}(z,41^{x_1+1}2\ldots21^{x_n+1},y))$" (Jones [6]), which is a normal form theorem for computable functions from $\{1,2,4\}^*$ to $\{1\}^*$.

By allowing more powerful predicates and functions, we now obtain a normal form theorem for $\mathcal{P}$.

Theorem 3   (*Normal Form Theorem for $\mathcal{P}$*) There exists a rudimentary function $\lambda x.d(x)$ and for $n \geq 1$ a rudimentary predicate $T^{(n)}(z,\vec{x},y)$ such that for all $\psi \in \mathcal{P}^{(n)}$ a z (depending on ψ) exists such that $\psi = \lambda\vec{x}_n.d((\mu y)T^{(n)}(x,\vec{x}_n,y))$.

Note: This theorem is identical to Theorem 3.7.4 except for the sharper characterization of d and $T^{(n)}$. Thus (Proposition 5.1.2) Theorem 3 implies Theorem 3.7.4.

Proof First, set $d = \lambda x.|\text{dec}(x)|$ (recall that $|\;|$ is m-adic length, where $m = |\tilde{A}|$ is the fixed base of our number representation). Next,

$$T^{(n)}(z,\vec{x}_n,y) \equiv (\exists x)_{Py}(\text{Comp}(z,x,y) \;\&$$

$$(\exists t_1,\ldots,t_n)_{Px}\; x = 4*t_1*2*t_2*2*\ldots*2*t_n \;\& \;\text{tally}_1(t_1) \;\&$$

$$\ldots \;\& \;\text{tally}_1(t_n) \;\& \;|t_1| = x_1 + 1 \;\& \;|t_2| = x_2 + 1 \;\& \ldots\& \;|t_n| = x_n + 1$$

Clearly, $\lambda z\vec{x}_n y.T^{(n)}(z,\vec{x}_n,y)$ is rudimentary and if $\lambda\vec{x}_n.\psi(\vec{x}_n)$ is computable by a TM with $gn(Z) = i$, then $\psi = \lambda\vec{x}_n.d((\mu y)T^{(n)}(i,\vec{x}_n,y)).//$

Theorem 4   (*The Universal Function Theorem for $\mathcal{P}^{(1)}$*) There is a function $\psi \in \mathcal{P}^{(2)}$ such that $i \mapsto \lambda x.\psi(i,x)$ is *onto* from $\mathcal{N}$ to $\mathcal{P}^{(1)}$.

Note: This is Theorem 3.7.3.

(*)Recall that $\mathcal{P}$ has been defined as a set of number theoretic functions.

Proof Take $\psi = \lambda ix.d((\mu y)T^{(1)}(i,x,y)).//$

Note: As in Chapter 3, we write T instead of $T^{(1)}$. For $n \geq 1$, $T^{(n)}$ is the Kleene predicate of order n.

Since $\psi \in \mathcal{P} = \mathcal{PC}$, let U_1 be a particular TM which computes ψ. Then U_1 is a "stored program" (or "universal") TM for $\mathcal{PC}^{(1)}$: Given a TM description i (this is the "program") for some function $f \in \mathcal{PC}^{(1)}$ and an input x (the "data"), U_1 comes up with $f(x)$ as output (this is $\psi(i,x)$).

Also note that for each $n > 0$ a universal TM (function) for $\mathcal{PC}^{(n)}$ exists: Indeed, U_n computes the function $\lambda z \vec{x}_n.d((\mu y)\ T^{(n)}(z,\vec{x}_n,y))$.

As in Chapter 3, the function $i \mapsto \lambda x.\psi(i,x)$ is denoted by ϕ; ϕ_i is $\lambda x.\psi(i,x)$. We call ϕ a *Gödel numbering* of $\mathcal{P}^{(1)}$ and i a *Gödel number* of f just in case $\phi_i = f$. Note that this chapter's Gödel numbering is *not* identical to the one of Chapter 3 (Definition 3.7.1); for example, $\phi_0 = \lambda x.x + 1$ (Chapter 3), thus according to the approach of Chapter 3 "0" is a Gödel number of $\lambda x.x + 1$. With the present approach this is *not* the case, as for no TM Z is it possible that $gn(Z) = 0$; thus $\phi_0 = \lambda x.d((\mu y)T(0,x,y))$ is the *empty* function since $T(0,x,y)$ is *false* for all x,y.

Yet the two ϕ's are "abstractly identical", thus we continue using the same symbol for both. By "abstractly identical" we mean the following: Both Gödel numberings lead to a universal function and an *S-m-n* (see §5.3) theorem. These two theorems along with some simple assumptions (a set of initial functions, including the **if-then-else** function, and closure under *substitution*) uniquely characterize $\mathcal{P}$ (this observation is due to Kleene. See also Hinman [5] and Hennie [4]. We shall establish this result in Chapter 10).

Further, note that it is still true for the new Gödel numberings that each $f \in \mathcal{P}^{(1)}$ has infinitely many distinct Gödel numbers (see Corollary 4 of Theorem 3.7.3) since one can easily see that there are infinitely many different TMs which compute the same function f (just add redundant quintuples).

On notation, we set for $n \geq 1$ $\phi_i^{(n)} = \lambda \vec{x}_n.d((\mu y)T^{(n)}(i,\vec{x}_n,y))$, where $\phi_i^{(1)} = \phi_i$.

We finally note that the universal function theorem can be put in the form "There is a u_1(*) such that for each $f \in \mathcal{P}^{(1)}$ an x (depending on f) exists so that $f = \lambda y.\phi_{u_1}^{(2)}(x,y)$" or "There is a u_1 such that $\phi_{u_1}^{(2)} = \lambda xy.\phi_x(y)$". This can also be expressed in terms of the Gödel numbering ϕ without involving $\phi^{(2)}$.

"For any given primitive recursive pairing function $(x,y) \mapsto \langle x,y \rangle$ an i_0 exists such that the map $i \mapsto \lambda x.\phi_{i_0}(\langle i,x \rangle): \mathcal{N} \to \mathcal{P}^{(1)}$ is onto". [Indeed, let K and L be the projections of $\langle\ \rangle$. Let i_0 be some Gödel number of $\lambda z.d((\mu t)\ T(K(z),L(z),t))$. Then $\phi_x(y) = \phi_{i_0}(\langle x,y \rangle)$ for all x,y].

Note that $i_0 \neq u_1$ in general (however, an i_1 exists such that $\phi_{i_1}^{(2)} = \lambda xy.\phi_{i_1}(\langle x,y \rangle)$; this can be seen using the "recursion theorem" of Chapter 11).

(*)$u_1 = gn(U_1)$, where U_1 is a *universal* TM for $\mathcal{P}^{(1)}$.

§5.3 THE *S-m-n* THEOREM

We prove now the *S-m-n* theorem (see Theorem 3.8.1) relying solely on the tools developed in the present chapter. Intuitively, we seek a primitive recursive procedure σ which for any $(1 + m)$-input TM I (of Gödel number i) and any x constructs an m-input TM $\sigma(i,x)$ such that for all $\vec{y}_m$,

$$\phi^{(m)}_{\sigma(i,x)}(\vec{y}_m) = \phi^{(1+m)}_i(x,\vec{y}_m)$$

The plan is:

(a) Given TM I, x, and $\vec{y}_m$,

(b) Construct a TM Z_x, which starting with $\vec{y}_m$ on tape halts with $(x,\vec{y}_m)$ on tape (i.e., $1^{x+1}B1^{y_1+1}B\ldots B1^{y_m+1}$)

(c) Now TM I takes over, on input $(x,\vec{y}_m)$

(d) Compute the Gödel number of the composite TM $(Z_x \cup I)(*)$; this Gödel number is $\sigma(i,x)$.

Proposition 1   For any $m > 0$, a function $\sigma \in \mathcal{PR}^{(2)}$ exists (σ depends on m) such that for all $i,x,\vec{y}_m$,

$$\phi^{(m)}_{\sigma(i,x)}(\vec{y}_m) = \phi^{(1+m)}_i(x,\vec{y}_m)$$

Proof [Note: Machine I, such that $gn(I) = i$, computes $\phi^{(1+m)}_i$]. First construct Z_x of the preceding the proposition discussion:

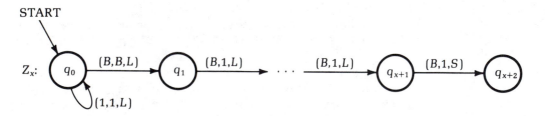

The behavior of Z_x is

$$q_0 1^{y_1+1}B\ldots B1^{y_m+1} \to \ldots \to q_{x+2}1^{x+1}B1^{y_1+1}B\ldots B1^{y_m+1}$$

(*)Care must be taken not to mix up the states of Z_x and I.

Now, machine I takes over, but each q_i of I is renamed to q_{i+x+2} (thus q_0 of I is q_{x+2}). $Z_x \cup I$ is appropriate for $\phi^{(m)}$ at the left hand side of the proposition's stated equality. $\sigma(i,x)$ is a Gödel number of $Z_x \cup I$ which we now compute:

(a) *Find a Gödel number of Z_x:*

$$Z_x = \{q_011q_0L, q_0BBq_1L, (q_iB1q_{i+1}L)_{i=1,\ldots,x}, q_{x+1}B1q_{x+2}S\}$$

Consider the functions QS, $\tilde{Z}$, and Z defined as follows:

$$\begin{cases} QS(0) = 0 \\ QS(s+1) = QS(s)*4 \qquad \text{[Note: ``4'' stands for (or ``is'') ``q''].}\end{cases}$$

Clearly, $QS \in \mathcal{PR}$ and $QS(x) = \underbrace{44\ldots 4}_{x}$; i.e., $Q(x+1)$ denotes "q_x".

$$\begin{cases} \tilde{Z}(0) = 3411463422446 \text{ (in m-adic!)} \\ \tilde{Z}(i+1) = Z(i)*3*QS(i+2)*21*QS(i+3)*6\end{cases}$$

Clearly, $\tilde{Z} \in \mathcal{PR}$ and $\tilde{Z}(i)$ is the (number denoted by the) string "$;q_011q_0L; q_0BBq_1L; q_1B1q_2L; \ldots; q_iB1q_{i+1}L$" (Note the absence of "$;$" at the right end). Then set

$$Z(x) \stackrel{\text{def}}{=} \tilde{Z}(x)*3*QS(x+2)*21*QS(x+3)*5$$

It follows that, $Z \in \mathcal{PR}$ and "$Z(x)$;" is a Gödel number of Z_x.

Finally, define for all x, z

$\text{disp}(x,z) \stackrel{\text{def}}{=}$ "number obtained from z after *each* tally of 4's in z is lengthened by x 4's"

We see that "disp" is definable by right recursion on notation from primitive recursive functions; hence, it is primitive recursive (See Problem 4):

$$\begin{cases} \text{disp}(x,0) = 0 \\ \text{for } d = 1, \ldots, m: \text{disp}(x, z*d) = \textbf{if } d \neq 4 \text{ \& } 4Ez \textbf{ then } \text{disp}(x,z)*QS(x)*d \\ \qquad\qquad\qquad\qquad\qquad\qquad \textbf{else } \text{disp}(x,z)*d\end{cases}$$

It follows that $\sigma(i,x) = Z(x)*\text{disp}(x+2, i).//$

Theorem 1   (The S-m-n or *index* or *iteration* theorem)   For each m,n a primitive recursive function S_m^n exists such that for all $i, \vec{x}_n, \vec{y}_m$

$$\phi_{S_m^n(i,\vec{x}_n)}^{(m)}(\vec{y}_m) = \phi_i^{(n+m)}(\vec{x}_n, \vec{y}_m)$$

Proof Induction on n:

$n = 1$: This is Proposition 1.

Assume that for $n = k$, $S_m^n \in \mathcal{PR}^{(1+n)}$ exists satisfying the equality stated in the theorem.

Consider the case $n = k + 1$ $(k \geq 1)$:

By Proposition 1,

$$\phi_i^{(n+m)}(\vec{x}_n, \vec{y}_m) =$$

$$\phi_{\sigma(i,x_1)}^{(n-1+m)}(x_2, \ldots, x_n, \vec{y}_m) = \text{(induction hypothesis on } n\,)$$

$$\phi_{S_m^{n-1}(\sigma(i,x_1),x_2,\ldots,x_n)}^{(m)}(\vec{y}_m)$$

Let us set

$$S_m^n = \lambda i \vec{x}_n . S_m^{n-1}(\sigma(i,x_1), x_2, \ldots, x_n)$$

Since $S_m^{n-1} \in \mathcal{PR}$, we are done.//

Corollary 1   Let $\lambda xy.\langle x,y \rangle$ be some primitive recursive pairing function. Then an $h \in \mathcal{PR}^{(2)}$ exists such that for all i, x and $y, \phi_{h(i,x)} = \lambda y.\phi_i(\langle x,y \rangle)$.

Note: This is the S-m-n theorem given in terms of ϕ (without) intervention of $\phi^{(m)}$, $m > 1$). This is the form in which it was proved in Chapter 3 (Theorem 3.8.1).

Proof Now $\lambda ixy.\phi_i(\langle x,y \rangle) \in \mathcal{P}^{(3)}$ since $\phi_i(\langle x,y \rangle) = d((\mu t)\, T(i, \langle x,y \rangle, t))$.

Let z be such that $\phi_z^{(3)} = \lambda ixy.\phi_i(\langle x,y \rangle)$. By Theorem 1, there is an $h \in \mathcal{PR}$ such that $\phi_z^{(3)}(i,x,y) = \phi_{h(i,x)}(y)$ (recall, z is fixed).//

Corollary 2   Let $f \in \mathcal{P}^{(m+n)}$. Then an $h \in \mathcal{PR}^{(m)}$ exists such that $f = \lambda \vec{x}_n . \phi_{h(\vec{y}_m)}^{(n)}(\vec{x}_n)$.

Proof $f \in \mathcal{P}^{(m+n)}$ implies existence of z such that $\phi_z^{(m+n)} = f$. Now apply Theorem 1, observing that z is fixed.//

§5.4 CONSTRUCTIVE ARITHMETIC PREDICATES

For the purpose of presenting Gödel's incompleteness theorem (Chapter 9), and also for use in our discussion of the Arithmetical Hierarchy (Chapter 10), we now introduce the Constructive Arithmetic predicates (CA), (Smullyan [7], Bennett [1]) and show that for each $m > 1$ the class of the m-rudimentary predicates, Rud_m, is included in CA. Our proof is different from that in Bennett [1].

Definition 1 The class of the Constructive Arithmetic predicates, CA, is the closure of $\{\lambda zxy.z = x + y, \lambda zxy.z = xy\}$(*) under *explicit transformation, Boolean operations* $(\neg, \&)$, and *bounded quantification* $((\exists y)_{\leq z}, (\forall y)_{\leq z})$.//

Example 1 The predicate $\lambda xy.x \mid y$ ("x divides y") is in CA. Indeed,

$$x \mid y \equiv (\exists z)_{\leq y} (y = xz).//$$

Example 2 The predicates $x \leq y$, $x < y$, $x = y$, $x \neq y$ are in CA. Indeed,

$$x \leq y \equiv (\exists z)_{\leq y} x + z = y.$$

Then

$$x < y \equiv \neg(y \leq x), \quad x = y \equiv x \leq y \& y \leq x, \quad x \neq y \equiv \neg(x = y).//$$

Example 3 $\lambda xyz.z = x \mathbin{\dot-} y \in \text{CA}$. Indeed,

$$z = x \mathbin{\dot-} y \equiv z = 0 \& x < y \vee x = y + z.//$$

Example 4 The predicates $\lambda xyz.\lfloor x/y \rfloor = z$ and $\lambda xyz.z = \text{rem}(x,y)$ (where $\text{rem}(x,y)$ = "remainder of the division x/y") are in CA. Indeed, $z = \lfloor x/y \rfloor \equiv z = 0 \& y = 0 \vee (zy \leq x \& zy + y > x)$ where we adopt the convention that $\lfloor x/0 \rfloor = 0$. Note that

$$zy \leq x \equiv (\exists w)_{\leq x} w = zy, \text{ and } zy + y > x \equiv \neg(x \geq zy + y)$$
$$\equiv \neg((\exists w)_{\leq x} w = zy + y) \equiv \neg((\exists w)_{\leq x} (\exists u)_{\leq w} (w = u + y \& u = zy))$$

(*)By "xy", multiplication of x by y is implied.

Finally,

$$z = \mathrm{rem}\,(x,y) \equiv z = x \dot- \lfloor x/y \rfloor\, y$$
$$\equiv (\exists u,w)_{\leq x}(z = x \dot- u\; \&\; u = wy\; \&\; w = \lfloor x/y \rfloor\,).//$$

Example 5 $\lambda xyzw.w = \mathrm{rem}\,(x + y,z) \in \mathrm{CA}$. Indeed, observe first that

$$\mathrm{rem}\,(x + y,z) = \mathrm{rem}\,(\mathrm{rem}\,(x,z) + \mathrm{rem}\,(y,z),z)$$
$$= \mathbf{if}\ \mathrm{rem}\,(x,z) + \mathrm{rem}\,(y,z) < z\ \mathbf{then}\ \mathrm{rem}\,(x,z) + \mathrm{rem}\,(y,z)\ \mathbf{else}$$
$$\mathrm{rem}\,(x,z) + \mathrm{rem}\,(y,z) \dot- z$$

Thus,
$$w = \mathrm{rem}\,(x + y,z) \equiv (\exists u)_{\leq x}\,(\exists v)_{\leq y}\,(u = \mathrm{rem}\,(x,z)\; \&\; v$$
$$= \mathrm{rem}\,(y,z)\; \&\; [u + v < z\; \&\; w = v + u \vee u + v \geq z\; \&\; w + z = u + v]$$

The observations that follow complete the argument:

(a) $u + v < z \equiv (\exists q)_{\leq z}\,(q < z\; \&\; q = u + v)$

(b) $u + v \geq z \equiv \neg(u + v < z)$

(c) $w + z = u + v \equiv (\exists q)_{\leq z}\,z = u + q\; \&\; v = w + q$. [Recall that $u < z$ and $v < z$, so that *in the context of this example* there are no more cases to consider. (See Problem 8)].//

In what follows, $|x|_m$ denotes m-adic length.

Lemma 1   The following predicates are in the CA, if p is prime.

(a) $\lambda x.\Omega(p,x) \equiv$ "x is a power of p"

(b) $\lambda y.y = p^{|x|_p}$

(c) $\lambda xyz.z = x *_p y$

Proof [**Note:** The assumption that p is prime is essential in the following *proof;* however, the statements are true for *any* number p (see Lemma 5, below).]

(a) $\Omega(p,x) \equiv x = 1 \vee (\forall i)_{\leq x}\,(i \mid x\; \&\; i > 1 \Rightarrow p \mid i)$

(b)   Set

$$l = |x|_p$$

Then,

$$\frac{p^{l-1}}{p-1} \le x < \frac{p^{l+1}-1}{p-1}.$$

hence,

$$p^l \le x(p-1) + 1 < p^{l+1}.$$

Thus,

$$y = p^{|x|_p} \equiv \Omega(p,y) \;\&\; y \le x(p-1) + 1 \;\&\; x(p-1) + 1 < yp.$$

We already know from the previous examples that $\lambda xy.y \le x(p-1) + 1$ $\in$ CA. Also, $x(p-1) + 2 \; \le yp$ is an instance of $\lambda \vec{x}_n \vec{y}_p.x_1 + \cdots + x_n \le y_1 + \cdots + y_p$, which is in CA [induction on n and the observation: $x_1 + \cdots + x_{n+1}$ $\le y_1 + \cdots + y_p \equiv (\exists i)_{1 \le i \le p}(\exists z,w)_{\le y} x_1 + \cdots + x_n \le y_1 + \cdots + y_{i-1} + z$ $\& x_{n+1} \le w + y_{i+1} + \cdots + y_p \;\&\; y_i = z + w]$.

(c) $z = x *_p y \equiv z = x p^{|y|_p} + y \equiv (\exists w,u)_{\le z} (z = w + y \;\&\; u = p^{|y|_p} \;\&\; w = xu).//$

Lemma 2   For any prime p, $\mathrm{Rud}_p \subset$ CA.

Proof By Lemma 1, the initial predicate of Rud_p, namely $\lambda xyz.z = x *_p y$, is in CA. Further, Rud_p is closed under the same operations as CA.//

Lemma 3   For any prime p, $\lambda xy.y = |x|_p \in$ CA.

Proof $\lambda xy.y = |x|_p \in \mathrm{Rud}_p \subset$ CA.//

Lemma 4   For any prime p, $\lambda xy.y = \lfloor \log_p x \rfloor \in$ CA.

Note: We arbitrarily define $\lfloor \log_p 0 \rfloor = 0$.

Proof Let $l = \lfloor \log_p (\lfloor x/(p-1) \rfloor (p-1) + 1) \rfloor$. That is,

$$p^l \le \left\lfloor \frac{x}{p-1} \right\rfloor (p-1) + 1 < p^{l+1}$$

hence

$$\frac{p^l - 1}{p-1} \le \left\lfloor \frac{x}{p-1} \right\rfloor < \frac{p^{l+1}-1}{p-1}$$

hence

$$\left\|\left\lfloor \frac{x}{p-1} \right\rfloor\right\|_p = l$$

For $p = 2$ then,

$$y = \lfloor \log_2 x \rfloor \ \equiv\ y = |x \div 1|_2 \equiv x < 1 \ \&\ y = 0 \lor (\exists z)_{\leq x}\,[x = z + 1\ \&\ y = |z|_2]$$

and the result follows by Lemma 3. Let next $p > 2$.
Now

$$\left\lfloor \frac{x}{p-1} \right\rfloor (p-1) + 1 = x \div \mathrm{rem}\,(x, p-1) + 1$$

and this leads to

$$y = \lfloor \log_p x \rfloor \ \equiv\ x < p\ \&\ y = 0 \lor (\exists u)_{\leq x + p}\,(\exists r)_{\leq p}\Big(x + r = u + 1\ \&$$

$$r = \mathrm{rem}\,(u, p-1)\ \&\ y = \left\|\left\lfloor \frac{u}{p-1} \right\rfloor\right\|_p\Big) \equiv x < p\ \&\ y = 0 \lor$$

$$(\exists u)_{\leq x}\,(\exists w)_{\leq p}\,(\exists r)_{\leq p}\Big(x + r = u + w + 1\ \&\ r = \mathrm{rem}\,(u + w, p-1)\ \&$$

$$y = \left\|\left\lfloor \frac{u + w}{p-1} \right\rfloor\right\|_p\Big)$$

To see that this last predicate is in CA, observe that *in the present context:*

(a) $\quad x + r = u + w + 1 \equiv (\exists l)_{\leq x}\ x = u + l\ \&\ w + 1 = r + l$

$\qquad\qquad \equiv (\exists l)_{\leq x}\ x = u + l\ \&\ [(\exists q)_{\leq r}\,(r = q + 1\ \&\ w = q + l) \lor$

$\qquad\qquad (\exists q)_{\leq l}\,(l = q + 1\ \&\ w = q + r)]$

(b) $\ y = \left\lfloor \dfrac{u + w}{m} \right\rfloor \equiv (\exists z, t)_{\leq y}\,(\exists r_1, r_2)_{\leq m}\Big(z = \left\lfloor \dfrac{u}{m} \right\rfloor \&\ t = \left\lfloor \dfrac{w}{m} \right\rfloor \&\ r_1 = \mathrm{rem}\,(u, m)$

$\qquad \&\ r_2 = \mathrm{rem}\,(w, m)\ \&\ [y = z + t\ \&\ r_1 + r_2 < m \lor$

$\qquad y = z + t + 1\ \&\ r_1 + r_2 \geq m]\Big)$

(c) $\ y = \left\|\left\lfloor \dfrac{u + w}{p-1} \right\rfloor\right\|_p \equiv (\exists z)_{\leq x}\Big(z = \left\lfloor \dfrac{u + w}{p-1} \right\rfloor \&\ y = |z|_p\Big) \lor x \leq 1\ \&\ y = 0.//$

Lemma 5

The following predicates are in CA.

(a) $\lambda xy.y = \overline{\exp}(p,x) \equiv$ "y is the exponent of the prime p in the prime decomposition of x"

(b) $\Pr(x) \equiv$ "x is prime"

(c) $\lambda x.\Omega(m,x) \equiv$ "x is a power of m" (*any* $m > 1$)

(d) $\lambda xy.y = m^{\lfloor x \rfloor_m}$ (*any* $m > 1$)

(e) $\lambda xyz.z = x *_m y$ (*any* $m > 1$)

Proof

(a) $y = \overline{\exp}(p,x) \equiv p^y | x \,\&\, \neg p^{y+1} | x$

$\equiv (\exists z)_{\leq x} (\Omega(p,z) \,\&\, y = \lfloor \log_p z \rfloor \,\&\, z | x \,\&\, \neg(\exists w)_{\leq x}\{w = zp \,\&\, w | x\})$

(b) $\Pr(x) \equiv x > 1 \,\&\, (\forall i)_{\leq x} (i | x \Rightarrow i = 1 \lor i = x)$

(c) See proof of Lemma 5.1.2. Observe that

$$\left\lfloor \frac{\overline{\exp}(p_i,x)}{a_i} \right\rfloor = \left\lfloor \frac{\overline{\exp}(p_j,x)}{a_j} \right\rfloor \text{ amounts to}$$

$$(\exists u_i, u_j, w)_{\leq x} \; u_i = \overline{\exp}(p_i,x) \,\&\,$$

$$u_j = \overline{\exp}(p_j,x) \,\&\, w = \left\lfloor \frac{u_i}{a_i} \right\rfloor \,\&\, w = \left\lfloor \frac{u_j}{a_j} \right\rfloor$$

(d) See proof of Lemma 1

(e) See proof of Lemma 1.//

Theorem 1

$\mathrm{Rud}_m \subset \mathrm{CA}$ for all $m > 1$.

Proof By Lemma 5.//

The importance of Theorem 1 is in showing that the behavior of Turing Machines can be described by a Kleene predicate built upon $z = x + y$ and $z = xy$.

Similarly, any re set is the projection of such an "arithmetic" predicate (why?).

It is interesting to note that $\mathrm{Rud}_m = \mathrm{CA}$ for all $m > 1$. This result, originally proved by Bennett, we shall not need, and we do not prove here. The interested reader is asked to prove it in the problems.

PROBLEMS

1. Given $m > 1$, prove that every integer $n \geq 1$ has a unique m-adic representation. (*Hint:* Show that there are unique q and r, where $1 \leq r \leq m$ and $n = qm + r$. Then use course-of-values induction with respect to n.)

2. Show that for any $n \geq 1$, $\lambda \vec{x}_n y.y = x_1 * \ldots * x_n$ is S-rudimentary and rudimentary. ($*$ is m-adic concatenation.) (*Hint:* Induction on n.)

3. Show that the strictly m-rudimentary predicates ($m > 1$) are closed under substitution by concatenation polynomials.

4. Show that $\mathcal{PR}$ is closed under recursion on (m-adic) notation. (*Hint:* Such recursion is actually a [number-theoretic] course-of-values recursion. Indeed, $x * i = mx + i$, where $i \in \{1, 2, \ldots, m\}$.)

5. Show that if

$$|f(x, \vec{y}_k)|^2 \leq a + b \cdot \max(|x|, |y_i|_{i=1,\ldots,k}),$$

where $|\ldots|$ is m-adic length, then f is bounded by a concatenation polynomial (in m-adic notation).

6. (Bennett [1]) (A restricted form of bounded multiple recursion on s-adic notation.) Let $\lambda \vec{x}_n \vec{y}_m . f(\vec{x}_n, \vec{y}_m)$ be defined for all $\vec{x}_n, \vec{y}_m$ by the following schema:

$$f(0, \ldots, 0, \vec{y}_m) = h(0, \ldots, 0, \vec{y}_m)$$

for $\vec{\imath}_n \in \{1, \ldots, s\}^n$, $f(x_1 * i_1, \ldots, x_n * i_n, \vec{y}_m) = g(\vec{x}_n, \vec{y}_m, \vec{\imath}_n, f(\vec{x}_n, \vec{y}_m))$ **if**

$$|x_1| = |x_2| = \ldots = |x_n| \ (|\ldots| \text{ is } s\text{-adic length})$$

$$h(x_1 * i_1, \ldots, x_n * i_n, \vec{y}_m) \textbf{ otherwise.}$$

$$|f(\vec{x}_n, \vec{y}_m)|^2 \leq a + b \cdot |\max(\vec{x}_n, \vec{y}_m)|, \text{ where } a \text{ and } b \text{ are constants.}$$

Show that if h and g are s-rudimentary, then so is f.

7. Fill in the missing details in the proof of Proposition 5.1.4.

8. Fill in the missing details in Example 5.4.5.

9. Show that if $R(y, \vec{x})$ is m-rudimentary, then $\lambda \vec{x} z.(\mu y)_{\leq z} R(y, \vec{x})$ is an m-rudimentary function (and this is true whether unsuccessful search returns 0 or $z + 1$).

10. (Bennett [1]) Show that $\lambda xyz.z = x + y$ is m-rudimentary for any $m > 1$. (*Hint: $z = x + y$* iff for any $n > 0$, the nth m-adic digit of z from the right end is the one obtained by adding the corresponding position digits of x and y, taking into consideration the appropriate carry. To complete the picture, define by bounded recursion on notation the digit and carry as functions of x, y, and w, where $|w| = n$.)

> **Note:** It is remarked in [1] that R. Ritchie had also obtained, independently from Bennett, a proof of the above claim.

11. (Bennett [1]) Show that $\lambda xyz.z = x \cdot y$ is m-rudimentary for any $m > 1$. (*Hint:* [1] Think first of each of x and y as polynomials in, say, t, with integer coefficients x_i, y_i between 1 and m (inclusive). Thus,

$$x = \Sigma_{i=0}^{|x| \dot{-} 1} x_i t^i,$$

$$y = \Sigma_{i=0}^{|y| \dot{-} 1} y_i t^i,$$

and

$$xy = \Sigma_{j=0}^{|x| + |y| \dot{-} 2} t^j (\Sigma_{i+k=j} x_i \cdot y_k)$$

Now, in general, the "convolution"

$$\Sigma_{i+k=j} x_i \cdot y_k$$

is not a digit (i.e., it is $>m$). Inductively, reduce the convolution, for each $j = 0,1, \ldots, |x| + |y| \dot{-} 2$, into a *digit* and a *carry* by a formula such as

$$\text{carry}(j + 1,x,y) * \text{digit}(j + 1,x,y) = \text{carry}(j,x,y) + \Sigma_{i+k=j+1} x_i \cdot y_k.$$

Show that

(1) $\lambda jxy.\text{carry}(j,x,y)$, $\lambda jxy.\text{digit}(j,x,y)$, and $\lambda jxy.\Sigma_{i+k=j} x_i \cdot y_k$ are m-rudimentary. [That $\lambda ix.x_i$ is m-rudimentary, where it is understood that $x_i = 0$ if $i = 0$ or $i > |x|$, was already implicit in the proof of Problem 10.]

(2) $x \cdot y = \text{carry}(|x| + |y| \dot{-} 2,x,y) * \text{digit}(|x| + |y| \dot{-} 2,x,y) * \ldots * \text{digit}(0,x,y)$, and continue from there.)

12. (Bennett [1]) Show that for $m \geq 2$, $\text{Rud}_m = \text{CA}$.

13. (Bennett [1]) Show that $\lambda xyz.z = x^y$ is m-rudimentary, for $m > 1$. (*Hint:* Reduce the problem to the case $1 \leq y \leq m$ by looking at the m-adic representation of y. You will also need Bennett's coding scheme, as presented in §5.1, and the observation that to compute $m,m^2,m^3, \ldots, m^i$ one need compute $v_1, v_2, \ldots, v_i$, where $v_1 = m$ and $v_{j+1} = v_j \cdot m$.)

REFERENCES

[1] Bennett, J. *On Spectra*. Ph.D. dissertation, Princeton University, 1962.

[2] Cobham, A. "The Intrinsic Computational Difficulty of Functions". *Proceedings of the 1964 International Congress for Logic, Methodology and Philosophy of Science*. Edited by Y. Bar-Hillel. Amsterdam:North-Holland, 1964:24–30.

[3] Davis, M. *Computability and Unsolvability*. New York: McGraw-Hill, 1958.

[4] Hennie, F. *Introduction to Computability*. Reading, Mass.: Addison-Wesley, 1977.

[5] Hinman, P.G. *Recursion-Theoretic Hierarchies*. Heidelberg: Springer-Verlag, 1978.

[6] Jones, N.D. *Computability Theory; an Introduction*. New York: Academic Press, 1973.

[7] Smullyan, R. *Theory of Formal Systems*. Annals of Mathematics Studies, No. 47. Princeton, N.J.: Princeton University Press, 1961.

CHAPTER 6

Nondeterminism

In this chapter, we conclude our study of alternative formulations of the notions "algorithm" and "algorithmic" by considering a number of nondeterministic formalisms. Loosely speaking, a nondeterministic formalism makes mathematically precise the notion of a (computer) program which may contain one (or more) "instruction" such that after its execution there is a choice of two or more next instructions. (Note that an ordinary *test*-instruction as found in usual programming languages, after its execution leads to a unique successor instruction.)

Nondeterministic formalisms are important in the theory of Formal Languages, in Computational Complexity (where, at the present state of the art, there are problems which we can solve more efficiently nondeterministically than we can deterministically; see Chapter 13) and also in the art of programming, where certain problems which involve exhaustive searches with backtracking can be conveniently approached through a nondeterministic program, which abstracts from the backtracking details (Floyd [9]). Furthermore, the reader is familiar with at least one nondeterministic, but nevertheless algorithmic, process—that of proving a mathematical theorem. The "instructions" here are the rules of inference. The nondeterministic element is "to what part of the so far generated proof should I apply a new 'instruction', and what type of instruction should I apply?" (*)

It is intuitively apparent that given an input $\vec{x}_n$ to a nondeterministic algorithm A, during execution of A it may be that some choices of next instruction lead to an "infinite loop" (no halting), whereas among those choices that lead to halting (if such choices exist) not all are producing the same output. This non-uniqueness of output suggests that we do not view nondeterministic algorithms as function computing devices but rather as predicate accepting devices:

Given an n-input nondeterministic device A, a predicate $R_A(\vec{x}_n)$ is associated to it by

$R_A(\vec{x}_n)$ iff there is a (halting) computation of A with input $\vec{x}_n$.

$R_A(\vec{x}_n)$ is the predicate "accepted" by device A. We can now say that the (partial) function $\lambda\vec{x}_n.f(\vec{x}_n)$ is "computable" by a nondeterministic device, just in case some such device B accepts $\lambda y\vec{x}_n.y = f(\vec{x}_n)$. The intuition behind this suggestion is obvious (see also Theorem 3.7.2):

Given $\vec{x}_n$ as input, the computation of $f(\vec{x}_n)$ proceeds as follows:

(1) $i \leftarrow 1$

(2) *perform i steps* of computation of device B for *each* of the first i $(n + 1)$-tuples $(b,\vec{a}_n)$(†) trying in turn *all* possible choices of next instruction, whenever choice is available.

(3) **if** during (2), for some c, $(c,\vec{x}_n)$ leads B to a halting configuration **then** stop everything and output c; **else** set $i \leftarrow i + 1$ and **goto** (2).

Starting with the next section, we discuss various ways of formalizing nondeterminism.

§6.1 NONDETERMINISTIC TURING MACHINES

The definition of a nondeterministic Turing Machine is identical to the deterministic counterpart (Definition 4.1.1), with the exception that instead of a transition *function* δ, we now have a transition *relation* Δ.

(*)For example, in the context of Propositional Calculus, there are two types of instructions: *Modus Ponens* (i.e. deriving B from A and $A \rightarrow B$) and *substitution*.

(†)We already know that there is an algorithm for generating all such tuples.

Definition 1   A *Nondeterministic Turing Machine* (NTM),*M*, is a 4-tuple $M = (A,K,q_0,\Delta)$, where $A = \{s_0,s_1, \ldots, s_n\}$ is a *finite* set of tape symbols called the *tape alphabet*, and $s_0 = B$ (the blank symbol).

$K = \{q_0,q_1, \ldots, q_m\}$ is a *finite* set of *states* (sometimes called the *state alphabet*) of which q_0 is *distinguished*: it is the *start*-state.(*)

Δ is the *transition relation*: $\Delta \subset (A \times K) \times (A \times K \times \{L,R,S\})$. Δ is the "hardwired" program of *M*. *L, R, S* indicate movement (of the head) *left, right, stay*, respectively.//

Note: Clearly, Definition 1 has Definition 4.1.1 as a special case. Thus, *every* TM may be thought of as an NTM.

The various ways of representing δ, already discussed in §4.1, are also valid ways of representing Δ. We shall use the graph model in examples and the quintuple model in the formal discussions. Thus,

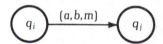

is part of *M*'s graph representation iff $\Delta(a,q_i;b,q_j,m)$ holds for *M* (where $m \in \{L,R,S\}$). Similarly, the quintuple q_iabq_jm is in *M*'s quintuple representation iff $\Delta(a,q_i;b,q_j,m)$ holds for *M*.

As in §4.1, let us rephrase Definition 1 before defining NTM-computations formally:

Definition 1′   A NTM *M over* the (tape) alphabet $A = \{s_0, \ldots, s_n\}$, where $s_0 = B$, and with internal states $K = \{q_0, \ldots, q_l\}$ is a *finite* set of *quintuples* of the form $qabq'm$, where $\{q,q'\} \subset K, \{a,b\} \subset A$ and $m \in \{S,L,R\}$.//

Note: In the nondeterministic model, we do not require anymore (but we do not prohibit either!) that no two distinct quintuples start with the same state-tape symbol pair (see Definition 4.1.1′).

The reader is referred to Definitions 4.1.2–5.

If α and β are IDs of NTM *M*, then $\alpha \xrightarrow{M} \beta$ is defined as in Definition 4.1.4.

Definition 2   A *computation* of NTM *M* is a *finite* sequence of IDs $\alpha_1, \ldots, \alpha_k$, such that α_k is *final* and if $k > 1$ then $\alpha_i \xrightarrow{M} \alpha_{i+1}$ for $i = 1, \ldots, k - 1$, where α_1, the *initial* ID, has the form q_0t, q_0 being the initial state and $t \in A^+$.

For any i, $\alpha_i \rightarrow \alpha_{i+1}$ is one *step* of the computation. The *complexity* of the computation is k.//

(*)The special role of q_0 is indicated by its separate listing in *M*; we may use any name and/or subscript for the initial state.

Definition 3   Let M be an NTM over the (tape) alphabet A. Let $w \in A^+$. Then M *accepts* w iff there exists an M-computation with initial ID $q_0 w$.

M accepts $S \subset A^+$ iff $w \in S \Leftrightarrow M$ accepts w.//

Note: By Definition 3, w not accepted by M means that there is *no* computation of M with input w; i.e., any choice strategy we may try for next instructions leads to an infinite loop (no halting). On the other hand, for acceptable w at least one computation is guaranteed to exist by definition (of course, some "bad" choices of next instruction might still lead to an infinite loop (see Example 1 below) but they are irrelevant to w's acceptability).

Definition 4   Let M be an NTM over A, where $\{1,B\} \subset A$. Then we say that M accepts $\vec{x}_n (\in \mathcal{N}^n)$ iff it accepts $1^{x_1+1} B 1^{x_2+1} B \ldots B 1^{x_n+1}$. M accepts $R(\vec{x}_n)$ iff $R(\vec{x}_n)$ is equivalent to "M accepts $\vec{x}_n$". The (partial) function $\lambda \vec{x}_n . f(\vec{x}_n)$ is (partial) computable by M iff M accepts $\lambda \vec{x}_n y . y = f(\vec{x}_n)$. We call $\mathcal{PC}^{ND}$ (resp. $\mathcal{C}^{ND}$) the set of all partial (resp. total) functions computable by some nondeterministic TM.//

Example 1   The predicate $\lambda x . "x$ is composite" is accepted by some NTM M.
Indeed, consider first the following NTM N:

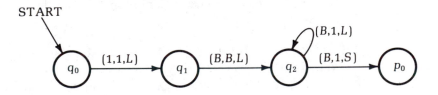

It is easy to verify that N has a computation $q_0 1^{x+1} \to \ldots \to p_0 1^{y+1} B 1^{x+1}$ for arbitrary $y \geq 0$.

Also note that should the choice $q_2 B 1 p_0 S$ be avoided, then N is in an "infinite loop".

Now the predicate $y \geq 2$ & $y < x$ & $y | x$ is primitive recursive. Let $\lambda x y . c(x,y)$ be its characteristic function.

By Corollary 1, Theorem 4.3.2, there is a (deterministic) TM L such that its states are $p_0, p_1, \ldots, p_k$ for some k, where $\delta((\$,\$),p_k) \uparrow$ and p_0 is initial, and moreover

$$c(x,y) = 0 \Rightarrow p_0 1^{y+1} B 1^{x+1} \xrightarrow[L]{} \ldots \xrightarrow[L]{} 0 q_k (\$,\$)$$

$$c(x,y) = 1 \Rightarrow p_0 1^{y+1} B 1^{x+1} \xrightarrow[L]{} \ldots \xrightarrow[L]{} 01 q_k (\$,\$)$$

Modify L by adding

$$q_k(\$,\$)(\$,\$)q_{k+1}L$$

$$q_{k+1}11q_kR$$

Call the new machine $\overline{L}$.

Clearly, $M = NU\overline{L}$ will do, for if x is composite then there is a number $2 \leq y < x$ such that $y \mid x$. For such a number, there is an N-computation $q_0 1^{x+1}$ $\xrightarrow{N} \cdots \xrightarrow{N} p_0 1^{y+1} B 1^{x+1}$. Now $\overline{L}$ takes over and halts with the ID $q_{k+1}0(\$,\$)$.

Note that, for any $y \geq 0$ such that $\neg(2 \leq y < x \;\&\; y \mid x)$, $\overline{L}$ does *not* halt with input $1^{y+1}B1^{x+1}$, thus there is *no* M-computation with input $x(= 1^{x+1})$ in this case. In other words, M does not accept anything but composite numbers.

Also note that if x is composite, there are two types of "bad choice" of next instruction of M which do not lead to a (terminating) M-computation. Namely, one type of bad choice is always opting for q_2B1q_2L given the alternative q_2B1p_0S. The other is producing the "wrong y"; i.e., one which violates $2 \leq y < x \;\&\; y \mid x$. Finally, if, for example, $x = 6$ then either of $y = 2$ or $y = 3$ lead to an M-computation.//

Note: It is customary to abbreviate detailed constructions such as those in Example 1 by saying informally: " ... given input x, M first *guesses*(*) an appropriate y and then proceeds to verify $2 \leq y < x \;\&\; y \mid x$; if no such y exists, then M is in an infinite loop."

It is emphasized that a statement such as " ... given $\vec{x}$, the NTM T guesses a y such that $R(y,\vec{x})$ and then proceeds to verify $R(y,\vec{x})$; if $\neg(\exists y)R(y,\vec{x})$, then T loops forever (or is in an infinite loop)" is a colorful way of saying " ... given $\vec{x}$, *if* $(\exists y)R(y,\vec{x})$, then for each particular y_0 such that $R(y_0,\vec{x})$ there is a T-computation with initial tape $\vec{x}$, which constructs $(y_0,\vec{x})$ and then verifies $R(y_0,\vec{x})$; *if* $\neg(\exists y)R(y,\vec{x})$, then there is *no* computation with initial tape $\vec{x}$ ".

We may view T as the "set of tools" we need in proving a theorem of the form $(\exists y)R(y,\vec{x})$ *constructively* (rather than by contradiction, where one would start with "Let $\neg(\exists y)R(y,\vec{x})$").

These tools enable us to

(1)   *write down* any y, in particular one that satisfies $R(y,\vec{x})$.

(2)   Verify mechanically that $R(y_0,\vec{x})$ just in case y_0 is appropriate.

The machine T *does not* discover the *right* y_0 by itself but is able to write down any y *we* guide it to. The main job of the machine is to do the verification of $R(y_0,\vec{x})$. (In some of the literature, the reader is being misled to believe that

(*)Sometimes we also say that " ... *M nondeterministically generates y ...*"

the NTM T "tries all the y's in parallel"(*); a look back at Example 1 shows that nothing is further from fact.)

Returning to Example 1, note that the main job of the machine M is to verify $2 \leq y < x$ & $y \mid x$, whereas for a deterministic machine for the same job it appears (†) that it must essentially try all y's in the range $2 \leq y \leq \lfloor \sqrt{x} \rfloor$. It thus appears(†) that nondeterministic computations for the same predicate are shorter ("faster") than corresponding deterministic computations.

Theorem 1   Every re predicate is accepted by some NTM M.

Proof Let $R(\vec{x})$ be re. Then for some function $f \in \mathcal{P}$ it is $R(\vec{x}) \equiv f(\vec{x})\downarrow$. Let M be a (deterministic) TM which computes f (we know that $\mathcal{P} = \mathcal{PC}$; Theorem 4.5.1). Since M may be viewed as nondeterministic, we are done.//

Note: $\lambda x.$ "x is composite" is in $\mathcal{PR}_*$, since its negation (that x is prime) we already know is in $\mathcal{PR}_*$ (See Example 2.3.1(g)). Since every predicate of $\mathcal{PR}_*$ is re, Theorem 1 is a shorter presentation of Example 1. By the latter we simply meant to provoke the discussion in the note following it.

Corollary $\mathcal{PC} \subset \mathcal{PC}^{\mathrm{ND}}$ and $\mathcal{C} \subset \mathcal{C}^{\mathrm{ND}}$.

Proof $\lambda \vec{x}_n.f(\vec{x}_n) \in \mathcal{PC} \Rightarrow f \in \mathcal{P}$. Hence, $\lambda \vec{x}_n y.y = f(\vec{x}_n)$ is re (Theorem 3.7.2). Therefore $\lambda \vec{x}_n y.y = f(\vec{x}_n)$ is accepted by some NTM. Hence (Definition 4), $f \in \mathcal{PC}^{\mathrm{ND}}$. Next, $g \in \mathcal{C} \Rightarrow g \in \mathcal{PC} \Rightarrow g \in \mathcal{PC}^{\mathrm{ND}}$. Since g is total, $g \in \mathcal{C}^{\mathrm{ND}}$.//

We have already mentioned in the first footnote below that NTMs are not more powerful (i.e., they do not accept non-re sets) than deterministic TMs.

We shall prove this relying on the following proposition and the results of Chapter 5.

Proposition 1   For any NTM M over the tape alphabet A, which accepts $S \subset \{0,1\}^+$ there is an NTM N over the tape alphabet $\{0,1,B\}$ which accepts S.

(*)It is true, however, that for each NTM T a *deterministic* TM $\tilde{T}$ exists which accepts the same predicate as T. Intuitively, $\tilde{T}$ may be constructed to simulate $i + 1$ steps ($i = 0,1,2,\dots$) of all possible subcomputations of T that are involved with the verification of $R(y,\vec{x})$ for $y = 0,1,\dots,i$ (dovetailing; see also the informal discussion at the beginning of this chapter). It appears then that $\tilde{T}$ (but *not* T!) "tries all y's in parallel".

(†)In the present state of Complexity Theory we cannot say more than "it appears", as we do not know whether predicates can be recognized by DTMs as "fast" as by NTMs. In particular, even though it is conceivable that $\lambda x.$"x is composite" could be tested (deterministically) without having to try all y's, $2 \leq y \leq \lfloor \sqrt{x} \rfloor$, as possible divisors, such a "fast" algorithm is not known. We shall expand on the theme of this footnote in Chapter 13.

Proof If $A \subset \{0,1,B\}$, then we are done. So let $A = \{B,0,1,s_2, \ldots, s_k\}$, $k \geq 2$.

Intuitively, each of $B,0,1,s_2, \ldots, s_k$ will be coded by fixed length strings over $\{0,1,B\}$. First, without loss of generality, let $k = 2^l - 1$ for some $l > 1$ (if not, add some redundant symbols to A until k attains the required form).

Code now 0 by $\underbrace{0 \ldots 0}_{l}$ ($= 0^l$), 1 by $0^{l-1}1$, B by B^l, and s_i by i in binary notation (for example, s_k is 1^l). Note that l is a constant associated with the machine M.

N is defined as the union of two submachines. The first (L) converts the input w, where $w = a_1 \ldots a_n$ and $a_i \in \{0,1\}$ for $i = 1, \ldots, n$ to $\tilde{w} = \text{code}(a_1)\text{code}(a_2) \ldots \text{code}(a_n)$. The second (P) starts with tape $\tilde{w}$ and treats blocks of l symbols as single symbols over A. Its action is to simulate M step by step.

The details:

L is the following machine.

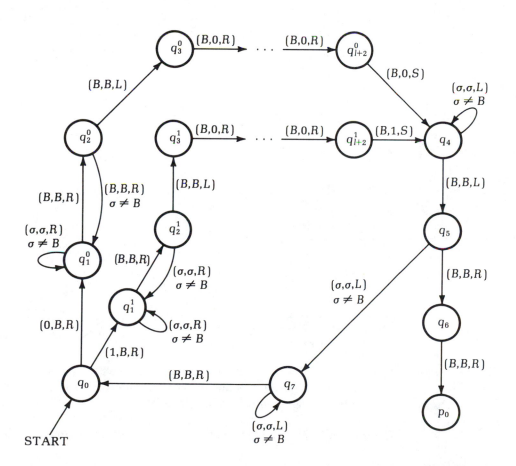

The reader can verify that L behaves as follows:

$$q_0 a_1 \ldots a_n \xrightarrow[L]{} \cdots \xrightarrow[L]{} q_5 BB \mathrm{code}(a_1) \ldots \mathrm{code}(a_n)$$

$$\xrightarrow[L]{} q_6 B \mathrm{code}(a_1) \ldots \mathrm{code}(a_n) \xrightarrow[L]{} p_0 \, \mathrm{code}(a_1) \ldots \mathrm{code}(a_n)$$

For the rest of the proof, we assume that $p_0, \ldots, p_m$ are the states of M where p_0 is the initial state.

P simulates M by first recognizing which symbol of A is represented by the length-l string over $\{0,1,B\}$ whose left-most symbol is at the currently scanned tape square. It then changes this string to represent the change dictated by M, and the head moves l squares left or right or it stays at the beginning of the changed string.

For simplicity, we set $s_{-1} = B$, $s_0 = 0$ and $s_1 = 1$.

Case 1. $p_i s_j s_h p_n R$ is in M. Let $\mathrm{code}(s_j) = \sigma_1 \ldots \sigma_l$ and $\mathrm{code}(s_h) = \tau_1 \ldots \tau_l$ where σ_t, τ_t are in $\{0,1,B\}$, $t = 1,2,\ldots,l$. Then P contains

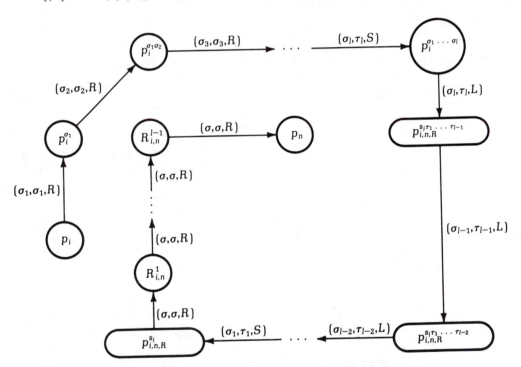

The state $p_i^{\sigma_1 \cdots \sigma_l}$ "says" that the simulated machine M, while at state p_i, "saw" an s_j. The remaining edges and nodes implement the response $s_h p_n R$, which may be one of many in the nondeterministic machine M.

The states $R_{i,n}^j$ cause P to travel right. j counts how many times to shift, i

indicates that the operation originated at state p_i, n indicates the target state p_n. Similar comment for states $L_{i,n}^j$; the only difference: shifting left.
Case 2. $p_i s_j s_h p_n L$ is in M.
Case 3. $p_i s_j s_h p_n S$ is in M.
    Both these cases are left to the reader. The first is similar to Case 1 [use $L_{i,n}^j$ instead of $R_{i,n}^j$ and (σ,σ,L) instead of (σ,σ,R)], the second is even simpler.
    Now $N = L \cup P$ with start state q_0.//

Corollary   Any predicate $R(\vec{x}_n)$ which is accepted by some NTM M, is accepted by some NTM N over the alphabet $\{0,1,B\}$.

   Proof Let A be the alphabet of M. If $A \subset \{0,1,B\}$ we are done.
    In the contrary case, N will be the union of two machines $\tilde{L}$ and P, where P is the heart of the simulation and is defined as in the proof of Proposition 1.
    The job of $\tilde{L}$ is to receive an input such as $1^{x_1+1}B1^{x_2+1}B \ldots B1^{x_n+1}$, to convert it first in the form $(0^{l-1}1)^{x_1+1}B^l(0^{l-1}1)^{x_2+1}B^l \ldots B^l(0^{l-1}1)^{x_n+1}$, and then move the head to the left-most 0 entering the initial state of P, p_0.
    We shall define $\tilde{L}$ as the union of machines L_1, L, and L_2 (L is the same as in the proof of Proposition 1, for an appropriate constant l).
    L_1:/*Changes separator blanks to 0*/

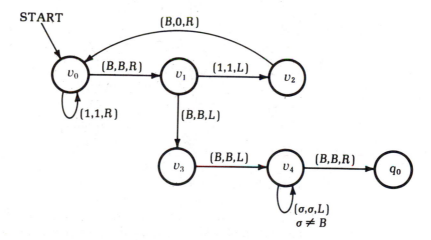

This machine behaves as follows:

$$v_0 1^{x_1+1}B \ldots B1^{x_n+1} \xrightarrow[L_1]{} \cdots \xrightarrow[L_1]{} 1^{x_1+1}0\ldots01^{x_n+1}Bv_1B$$

$$\xrightarrow[L_1]{} 1^{x_1+1}0\ldots01^{x_n+1}v_3B \xrightarrow[L_1]{} 1^{x_1+1}0\ldots01^{x_n}v_41$$

$$\xrightarrow[L_1]{} \cdots \xrightarrow[L_1]{} v_4B1^{x_1+1}0\ldots01^{x_n+1} \xrightarrow[L_1]{} q_01^{x_1+1}0\ldots01^{x_n+1}$$

Thus, $L_1 \cup L$ (start-state v_0) behaves as

$$v_0 1^{x_1+1} B \ldots B 1^{x_n+1} \xrightarrow[L_1 \cup L]{} \cdots \xrightarrow[L_1 \cup L]{} \tilde{p}_0 (0^{l-1}1)^{x_1+1} 0^l \ldots 0^l (0^{l-1}1)^{x_n+1}$$

Note that instead of p_0 we use $\tilde{p}_0$ in L (no other changes) because we are not ready for machine P (with start-state p_0) to take over. Using L_2 (below), we first change the separator 0^l to B^l:

L_2:/* processing is from right to left. 0^l is easily spotted as, unlike $0^{l-1}1$, it ends with 0*/

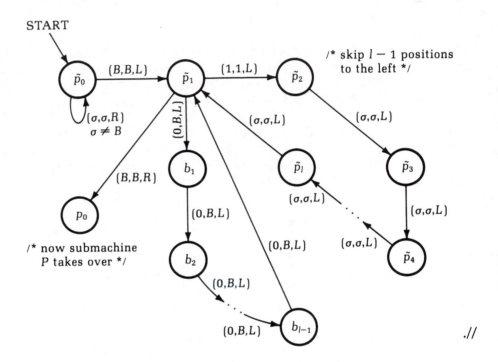

Note: By Proposition 1 and its corollary, unless otherwise specified, from now on without loss of generality every NTM considered is over the alphabet $\{0,1,B\}$. Thus, the results of Chapter 5 (we based our derivation there on the consideration of TMs over a fixed tape alphabet) can be applied here. We prove first:

Theorem 2   There is a *strictly* rudimentary predicate $\mathrm{Comp}^{\mathrm{ND}}(z,x,y)$ which is equivalent to the statement "The (possibly nondeterministic) Turing Machine (with code) z when started with initial ID (with code) x has a computation (with code) y". Throughout, the coding of Chapter 5 is implied.

Note: This is the Theorem 5.2.1 adapted for NTMs.

Proof The proof is the same as that of Theorem 5.2.1, except for the fact that the predicate "func(x,y)" [item (6) in the proof of Theorem 5.2.1] is not needed for NTMs; indeed, in the present case, we have

$$\text{TM}(x) \equiv \neg 33Px \,\&\, 3Bx \,\&\, 3Ex \,\&$$

$$(\forall y)_{Px}[((3*y*3)Px \,\&\, \neg 3Py) \Rightarrow \text{quint}(y)]$$

[Compare with item (7) of the proof of Theorem 5.2.1.]//

Theorem 3   For every $n \geq 1$, there is a *rudimentary* predicate $T_n^{\text{ND}}(z,\vec{x}_n,y)$ which is equivalent to the statement "The NTM (with code) z accepts $\vec{x}_n (\in \mathcal{N}^n)$ through a computation (with code) y".

Proof (See proof of Theorem 5.2.3)

$$T_n^{\text{ND}}(z,\vec{x}_n,y) \equiv (\exists x)_{Py}\text{Comp}^{\text{ND}}(z,x,y) \,\&\, (\exists t_1, \ldots, t_n)_{Px}\, x =$$

$$4*t_1*2*t_2*2* \ldots *2*t_n \,\&\, \text{tally}_1(t_1) \,\&\, \ldots \,\&\, \text{tally}_1(t_n) \,\&$$

$$|t_1| = x_1 + 1 \,\&\, |t_2| = x_2 + 1 \,\&\, \ldots \,\&\, |t_n| = x_n + 1.//$$

Corollary 1 $R(\vec{x}_n)$ is accepted by some NTM iff $R(\vec{x}_n)$ is re.

Proof The *if* part is Theorem 1.
For the *only if* part, let $R(\vec{x}_n)$ we accepted by an NTM with code z. Then $R(\vec{x}_n)$ $\equiv (\exists y)T_n^{\text{ND}}(z,\vec{x}_n,y)$.
Hence, $R(\vec{x}_n)$ is re by the projection theorem.//

Note: The reader is reminded, in connection with the previous proof, that rudimentary predicates are primitive recursive, therefore also recursive.

Corollary 2 $\mathcal{PC} = \mathcal{PC}^{\text{ND}}, \mathcal{C} = \mathcal{C}^{\text{ND}}$.
We conclude this section with a useful proposition, that NTM computations can be normalized so that the final tape is always (independently of the machine used) the same.

Proposition 2   Let $S \subset A^+$ be accepted by an NTM M over the tape alphabet A. Then there is an NTM N over $A \cup \{\vdash, \dashv\}$, where $\{\vdash, \dashv\} \cap A = \varnothing$, which accepts S and whose final ID is $\vdash \tilde{q} \dashv$, where $\tilde{q}$ is some distinguished state.

Proof Intuitively, the initial tape w is modified into $\vdash w \dashv$ by N. Then N behaves exactly as M but whenever tape extension to the left (resp. right) is needed by M, then N moves $\vdash$ (resp. $\dashv$) one place to the left (resp. right) to make room for a new symbol from A to be written on tape. In the end of a computation, the tape is cleared to contain only $\vdash \dashv$ (these symbols facilitate cleaning up in the presence of blanks embedded in the tape. Such blanks could be avoided (see Problem 2) but if no modification of the machine is undertaken, they will in general exist (see Example 4.1.8)).

N is $N_1 \cup N_2 \cup N_3$ and has q_0 as initial state, where

N_1:START

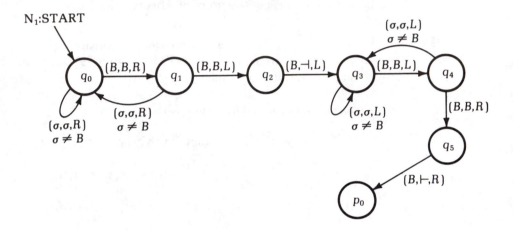

Note: In building N_1, we worked with the reasonable assumption that the only role of B on input is as a one-symbol separator of arguments in the vector input case.

N_2 is essentially M (we assumed that the state alphabet of M is $\{p_0, p_1, \ldots, p_k\}$), but for every state p_i we add the paths

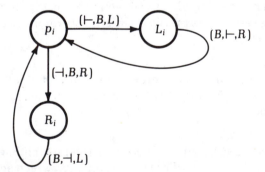

/* if, in state p_i, $\vdash$ (resp. $\dashv$) is scanned, then M would like to see a B. Thus, $\vdash$ (resp. $\dashv$) is shifted one square left (resp. right.*/

Now, without loss of generality, we assume that there is exactly one state, say p_f, of M such that an M-ID is final iff it involves p_f [indeed, if that were not the case, add a new state p_f to M such that for no $a \in A$ does a quintuple start with $p_f a$. Then, for all p_i, add

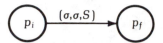

to M for all $\sigma \in A$ such that no quintuple starts with $p_i \sigma$].

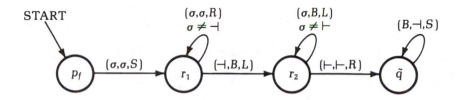

§6.2 NONDETERMINISTIC MARKOV ALGORITHMS

Definition 1 A *Nondeterministic Markov Algorithm*, or NMA, M over an alphabet A is a finite *set* of pairs (x,y), where $x \in A^*$, $y \in A^*$. These pairs are the *instructions* of M.//

Note first that an NMA is an unordered set of instructions, unlike the MA. Informally, given a string w as input to an NMA M, we may replace *any* (nondeterminism!) substring x of w by y just in case (x,y) is some instruction (again contrast with the orderly fashion this is done in the case of an MA). We thus obtain a new string w' ($w' = uyv$ whereas $w = uxv$ for some $\{u,v\} \subset A^*$).

We may continue this process as long as we please, every time transforming the most recently obtained string.

If we manage to obtain eventually some string $w^{(n)}$ such that no instruction is applicable, then we just finished an NMA computation.

Example 1 Let $M = \{(ab,a) \ (ab,abab)\}$ over $A = \{a,b\}$. Then if $w = abab$ is the input, we may form the computation $abab$, aba, aa (also $abab$, aab, aa) by using always the instruction (ab,a).

Other computations are possible: e.g., $abab$, $ababab$, $aabab$, $aaab$, aaa.

Clearly, as in the case of NTMs, due to the possible multitude of outputs for a given input, we shall consider NMAs as accepting devices.

A final comment: if we opt in using exclusively the instruction $(ab,abab)$ then our "computation", with input $abab$, will not terminate as $(ab,abab)$ is always applicable.//

Definition 2   Let M be an NMA over A.

An M-ID is any string $w \in A^*$. Let w and u be two M-IDs. Then w *yields* u, in symbols $w \xrightarrow{M} u$ iff the following is true:

$$(\exists x,y) [(x,y) \in M \& (\exists \alpha,\beta) (\{\alpha,\beta\} \subset A^* \& w = \alpha x \beta \& u = \alpha y \beta)]$$

An *M-computation* is a sequence of M-IDs $(w_1, \ldots, w_t)$ such that

   (a) $(\forall x,y) [(x,y) \in M \Rightarrow \neg x P w_t]$

   (b) $(\forall i)_{<t} \, w_i \xrightarrow{M} w_{i+1}$

An ID w_t with property (a) is called *final. One* execution of the yield operation is one *step* of the computation. t is the *complexity* of the computation.//

Note: As in the case of MAs, if (Λ,y) is an instruction of an NMA N, then there is no N-computation for any input (no ID is final since $\Lambda P w$ for all w).

Definition 3   Let M be an NMA over A, and $w \in A^+$. w is *accepted* by M iff there is an M-computation $(w,w_1, \ldots, w_t)$. $S \subset A^+$ is *accepted* by M iff the statement "$w \in S$" is equivalent to "w is accepted by M".

Moreover, let $\{\#,B,1\} \subset A$. Then $\vec{x}_n \in \mathcal{N}^n$ is accepted by M iff $\#1^{x_1+1}B \ldots B1^{x_n+1}\#$ is accepted by M. $R \subset \mathcal{N}^n$ is accepted by M iff $R(\vec{x}_n)$ is equivalent to "$\vec{x}_n$ is accepted by M".//

Definition 4 $\mathcal{PM}^{\mathrm{ND}}$ is the set of partial functions $\lambda \vec{x}_n.f(\vec{x}_n)$ such that $\lambda y \vec{x}_n.y = f(\vec{x}_n)$ is accepted by some NMA. $\mathcal{M}^{\mathrm{ND}}$ is the subset of $\mathcal{PM}^{\mathrm{ND}}$ consisting of total functions.//

It is perhaps apparent that the string transformation power of NMAs is enough to simulate the NTM yield operation and therefore NTM computations. The following theorem states just that.

Theorem 1   If $R(\vec{x}_n)$ is re, then $R(\vec{x}_n)$ is accepted by some NMA M.

Proof(*) By theorem 6.1.1, there is an NTM N which accepts $R(\vec{x}_n)$.

(*)The essentials of this proof are due to Post [21]. See also Davis [7]. In both instances, deterministic TMs were simulated.

Let A be the tape alphabet and K the state alphabet of N. Without loss of generality, let $\{\$,\#\} \cap (A \cup K) = \varnothing$. M is an NMA over $\{\$,\#\} \cup A \cup K$ whose instructions are as follows:

(1) $(\#1,\$q_0 1)$

(2) if $q_i s_l s_j q_m L \in N$, then (and only then)
$(a q_i s_l, q_m a s_j) \in M$ for all $a \in A$.
Also $(\$q_i s_l, \$q_m B s_j) \in M$ ($\$$ identifies the left end of N-IDs which M is to generate in its computations)

(3) if $q_i s_l s_j q_m R \in N$, then (and only then)
$(q_i s_l a, s_j q_m a) \in M$ for all $a \in A$.
Also $(q_i s_l \#, s_j q_m B\#) \in M$ (# identifies the right end of N-IDs which M is to generate in its computations)

(4) if $q_i s_l s_j q_m S \in N$, then (and only then)
$(q_i s_l, q_m s_j) \in M$.

Starting with the initial M-ID $\#1^{x_1+1}B \ldots B1^{x_n+1}\#$ only instruction (1) is applicable to yield $\$q_0 1^{x_1+1}B \ldots B1^{x_n+1}\#$. From now on only instructions of types (2)-(4) are applicable; indeed the reader can easily verify that from this point on $w \xrightarrow{M} u$ iff $w = \$\tilde{w}\#, u = \$\tilde{u}\#$, and $\tilde{w}, \tilde{u}$ are N-IDs such that $\tilde{w} \xrightarrow{N} \tilde{u}$. Hence, $\#1^{x_1+1}B \ldots B1^{x_n+1}\#$ is M-accepted iff $1^{x_1+1}B \ldots B1^{x_n+1}$ is N-accepted.//

Corollary $\mathcal{P} \subset \mathcal{PM}^{\text{ND}}, \mathcal{R} \subset \mathcal{M}^{\text{ND}}$.

String transformation systems such as NMAs have been studied by Thue at the beginning of this century. They are widely referred to as Thue-like systems in the literature. More specifically, following Davis [7], we define:

Definition 5   A *semi-Thue* system S over the alphabet A is a pair (w_0,M), where $w_0 \in A^*$ is the *axiom* of S and M is an NMA over A.
   A *proof* in S is a sequence $w_0,w_1, \ldots, w_n (n \geq 0)$ where (if $n > 0$) $w_i \xrightarrow{M} w_{i+1}$ in the sense of Definition 2, for $i = 0, \ldots, n - 1$. Note that w_n is *not* required to be final.
   w_n is called a *theorem* of S, in symbols $\vdash_S w_n$. $T_S \subset A^*$ is the set of theorems of S.//

Note: By Definition 5, each of $w_0, \ldots, w_{n-1}$ in a proof $w_0,w_1, \ldots, w_n$ is a theorem of S, since omitting all w_i's, $i = m_0 + 1, \ldots, n$, leaves a proof $w_0, \ldots, w_{m_0}$, implying that w_{m_0} is a theorem.

Proofs and theorems in semi-Thue systems are over-simplified counterparts of proofs in mathematical theories. It should be observed that a semi-Thue system is in some sense an NMA "in reverse". Whereas the NMA is an *acceptor*, the semi-Thue system is a *generator*.

Theorem 2   If $R(\vec{x}_n)$ is re, then for some semi-Thue system S,

$$R(\vec{x}_n) \text{ iff } \vdash_S \$q_0 1^{x_1+1} B \dots B 1^{x_n+1} \#$$

Proof Let N be an NTM which accepts $R(\vec{x}_n)$. Without loss of generality (Proposition 6.1.2), the only final ID possible for N is $\vdash q_f \dashv$. Let M be its associated NMA according to Theorem 1. Next, let $\tilde{M}$ be obtained from M by dropping the instruction (1).

S is $(\$\vdash q_f \dashv \#, \tilde{M}^{-1})$ where $(x,y) \in \tilde{M}^{-1}$ iff $(y,x) \in \tilde{M}$.

Indeed, it is clear that $w \xrightarrow[\tilde{M}]{} u$ iff $u \xrightarrow[\tilde{M}^{-1}]{} w$, hence the statement (Theorem 1)

$R(\vec{x}_n)$ if and only if $\$q_0 1^{x_1+1} B \dots B 1^{x_n+1} \#$ is accepted by $\tilde{M}$,

translates to

$R(\vec{x}_n)$ iff $\vdash_S \$q_0 1^{x_1+1} B \dots B 1^{x_n+1} \#$

since an N-computation ends with $\vdash q_f \dashv$, and therefore an $\tilde{M}$-computation ends with $\$\vdash q_f \dashv \#.//$

Corollary   If $R(\vec{x}_n)$ is re, then for some semi-Thue system T,

$$R(\vec{x}_n) \text{ iff } \vdash_T 1^{x_1+1} B \dots B 1^{x_n+1}.$$

Proof (Davis [7]) The idea is to modify S of Theorem 2 so that $\$q_0$ and $\#$ are removed from the S-theorem $\$q_0 1^{x_1+1} B \dots B 1^{x_n+1} \#$.

To obtain T, add to $\tilde{M}^{-1}$ the instructions

$$\left. \begin{array}{c} (\$q_0, r) \\[2mm] (r1, 1r) \\[2mm] (rB, Br) \\[2mm] (r\#, \Lambda) \end{array} \right\} \quad (I)$$

Clearly then, $R(\vec{x}_n) \underset{\text{(Th.2)}}{\Rightarrow} \vdash_S \$q_0 1^{x_1+1} B \dots B 1^{x_n+1} \# \Rightarrow \vdash_T \$q_0 1^{x_1+1} B \dots B 1^{x_n+1} \#$

(The last implication since T has the same axiom as S and in the proof of $\$q_0 1^{x_1+1} B \dots B 1^{x_n+1} \#$ only $\tilde{M}^{-1}$-instructions are involved.) It is clear now that by using the added instructions (I) we can continue the proof as follows:

$$\$q_0 1^{x_1+1} B \dots B 1^{x_n+1} \# \xrightarrow[T]{} r 1^{x_1+1} B \dots B 1^{x_n+1} \# \xrightarrow[T]{} \dots \xrightarrow[T]{}$$

$$1^{x_1+1} r B \dots B 1^{x_n+1} \# \xrightarrow[T]{} 1^{x_1+1} B r 1^{x_2+1} \dots B 1^{x_n+1} \# \xrightarrow[T]{} \dots \xrightarrow[T]{}$$

$$1^{x_1+1} B \dots B 1^{x_n+1} r \# \xrightarrow[T]{} 1^{x_1+1} B \dots B 1^{x_n+1}, \text{ thus } \vdash_T 1^{x_1+1} B \dots B 1^{x_n+1}$$

Conversely, let $\vdash_T 1^{x_1+1}B \ldots B1^{x_n+1}$; show that $R(\vec{x}_n)$. There is a T-proof $w_0, w_1,$ $\ldots, w_l$, where $w_0 = \$\vdash q_f \dashv \#$ (the axiom) and $w_l = 1^{x_1+1}B \ldots B1^{x_n+1}$.

Clearly, only S-instuctions are applicable to w_0, and no S-instruction is applicable to w_l (due to the absence of a q-symbol in w_l). So let t be the least index i such that w_i contains no q-symbol. It follows that the proof is

$$w_0 \xrightarrow{S} w_1 \xrightarrow{S} \cdots \xrightarrow{S} w_{t-1} \xrightarrow{T} w_t \xrightarrow{T} \cdots \xrightarrow{T} w_l \,(*)$$

Moreover (induction on t),

$$w_{t-1} = \$uqav\#, \text{ where } u \in A^*, v \in A^*, a \in A$$

Since w_t contains no q-symbol, $w_{t-1} \xrightarrow{T} w_t$ must involve the first (I)-instruction, thus $w_{t-1} = \$q_0 1v\#$ and $w_t = r1v\#$. Since the only effect of (I)-instructions is to walk r to the right until r is adjacent to $\#$ ($r\#$) with subsequent action the erasure of $r\#$, then, since $w_l = 1^{x_1+1}B \ldots B1^{x_n+1}$, it must be that $w_{t-1} = \$q_0 1^{x_1+1}B \ldots B1^{x_n+1}\#$.

But then, due to $\vdash_S w_{t-1}$, $R(\vec{x}_n)$ holds by Theorem 2.//

Definition 6 A *Thue* system over A is a semi-Thue system $S = (w_0, M)$ such that the relation(†)M is *symmetric*.//

Since every Thue system is a semi-Thue system as well (but not conversely, as nonsymmetric M's clearly exist), the notions of *proof, theorem*, etc., are defined as in the case of semi-Thue systems. In particular, M is an NMA.

Theorem 3(‡) For any re $R(\vec{x}_n)$, a Thue system $S = (w_0, L)$ exists such that $R(\vec{x}_n)$ iff $\vdash_S \$q_0 1^{x_1+1}B \ldots B1^{x_n+1}\#$.

Proof Let N be a *deterministic* TM such that $R(\vec{x}_n)$ iff N with input $1^{x_1+1}B \ldots B1^{x_n+1}$ halts. By Proposition 6.1.2 we may assume that the only final ID possible for N is $\vdash q_f \dashv$.

Let $w_0 = \$\vdash q_f \dashv \#$ and $L = \tilde{M} \cup \tilde{M}^{-1}$ (where $\tilde{M}, \tilde{M}^{-1}$ are those in the proof of Theorem 2).

Let next $R(\vec{x}_n)$. Then (Theorem 2) $\vdash_S \$q_0 1^{x_1+1}B \ldots B1^{x_n+1}\#$ by using only $\tilde{M}^{-1}$ instructions in the proof.

(*)If at some point t, w_t contains no q-symbol, then no S-instruction is applicable. This is true for all subsequent steps of the proof due to the fact that instructions (I) do not reintroduce q-symbols.

(†)M is, of course, a binary relation on A^*.

(‡)The essence of this theorem and of its proof are due to Post [21]. See also Davis [7].

Conversely, let $\vdash_S \$q_0 1^{x_1+1} B \ldots B 1^{x_n+1} \#$. Thus, there is a proof $w_0, w_1, \ldots,$ w_l where $w_l = \$q_0 1^{x_1+1} B \ldots B 1^{x_n+1} \#$. Clearly, no $\tilde{M}$-instruction is applicable to w_0 (no such instruction has q_f as part of its left side). So let t be the largest $t \leq l$ such that $w_0, w_1, \ldots, w_t$ involves only $\tilde{M}^{-1}$-instructions.

If $t = l$, then $w_0, \ldots, w_l$ is a proof in the system $(w_0, \tilde{M}^{-1})$; hence, by Theorem 2, $R(\vec{x}_n)$ holds.

Assume next that $t < l$.
Then

$$w_{t-1} \rightarrow w_t \text{ using } \tilde{M}^{-1}$$

and

$$w_t \rightarrow w_{t+1} \text{ using } \tilde{M}$$

Clearly (induction on i), each w_i $(i \leq l)$ is of the form $\$uqav\#$, where $uqav$ is an N-ID. Thus, we have $w_t \rightarrow w_{t+1}$ and $w_t \rightarrow w_{t-1}$ using $\tilde{M}$, hence some N-ID $uqav$ (where $w_t = \$uqav\#$) has two distinct(*) successor IDs, contradicting the fact that N is deterministic. The case $t < l$ being rejected, we are done.//

We leave it until §6.5 to prove appropriate converses of Theorems 1, 2, and 3.

§6.3 POST SYSTEMS

The type of instructions involved in (semi-) Thue systems is a special case of the *canonical productions* of Post [22]. A system equipped with such instructions is known as a *Post canonical system*. Thus (semi-) Thue systems are (as far as their definition is concerned) special cases of the Post canonical systems(†).

Of particular interest, due to its simple instruction-set, is still another special case of the canonical systems formalism, called *normal system* by Post [22]. It is known [22] that the normal systems have the same set-descriptive power as the canonical systems; both are capable of generating any re set, but neither can generate non-re sets.

We shall continue on our previous pattern of presentation, introducing normal systems first as *acceptors*, calling them *Nondeterministic Post Algo-*

(*)If $w_{t+1} = w_{t-1}$, then the proof could be shortened to $w_0, \ldots, w_{t-1}, w_{t+2}, w_{t+3}, \ldots, w_l$. This observation allows the implicit assumption that no w_i's are repeated in $w_0, \ldots,$ w_l.

(†)In terms of descriptive power, though, it turns out that (semi-) Thue systems are equivalent to canonical systems.

rithms (NPA), and next as *generators*, calling them *Post systems*. In this choice of name we are influenced by Scott [25], who apparently was the first to give the name *Post Machine* to a TM-like *deterministic* device with circular tape(*), which was suggested by Arbib [2] in the course of a proof that "monogenic normal systems" (essentially a deterministic version of the NPA) are as powerful as TMs. Post machines also appear (but not by name) in Shepherdson and Sturgis [26] as "string processing" variants of the URM.

Definition 1   A *Nondeterministic Post Algorithm*, or NPA, M, over A is a finite set of pairs (x,y), where $x \in A^*$, $y \in A^*$. These pairs are the *instructions* of M.//

Superficially, there is no difference between Definition 1 and Definition 6.2.1. However, there is a difference in the way instructions are applied: Given an input w to a NPA, we may replace any substring x (nondeterminism) of w, which *begins* w, by Λ and concatenate to the *end* of the resulting string the string y, just in case (x,y) is an instruction of the NPA.

We thus obtain a new string w' ($w' = uy$ whereas $w = xu$ for some $u \in A^*$). We may continue this process as long as we please, every time transforming the most recently obtained string. If we eventually manage to obtain a string $w^{(n)}$ such that no instruction is applicable, then we just finished an NPA computation.

Example 1   Consider the NPA over $\{a,b\}$ whose only instruction is (a,a).

Then every string which contains a "b" leads to a computation, whereas no string over $\{a\}$ leads to a computation.//

Definition 2   Let M be an NPA over A. An M-ID is any string $w \in A^*$. Let w and u be two M-IDs. Then w *yields* u, in symbols $w \xrightarrow{M} u$, iff the following is true:

$$(\exists x,y) [(x,y) \in M \mathbin{\&} (\exists \alpha) (\alpha \in A^* \mathbin{\&} w = x\alpha \mathbin{\&} u = \alpha y)]$$

An *M-computation* is a sequence of M-IDs $(w_1, \ldots, w_t)$ such that

(a) $(\forall x,y) [(x,y) \in M \Rightarrow \neg x B w_t]$

(b) $(\forall i)_{<t}\, w_i \xrightarrow{M} w_{i+1}$

An ID w_t with property (a) is called *final*. One execution of the yield operation is one *step* of the computation. t is the *complexity* of the computation.//

(*)This device, under the same name was subsequently used by others; e.g., Manna [19].

Note: As for MAs and NMAs, if (Λ,y) is an instruction of an NPA N then there is no N-computation for any input.

Also observe that an NPA treats each ID as a *queue*; i.e., a string of symbols where all *deletions* occur in one end and all *additions* at the other end.

Theorem 1 below will essentially say that using a *single* queue one can compute every function in $\mathcal{P}$ and accept every re predicate. A similar result is false for the other very popular and simple data structure, the *stack*. However, with *two* stacks we can do the job. (See Problems 51 and 4.16.)

Definition 3   Let M be an NPA over A, and $w \in A^*$. w is *accepted* by M iff there is an M-computation $(w,w_1, \ldots, w_t)$. $S \subset A^*$ is *accepted by M* iff the statement "$w \in S$" is equivalent to "w is accepted by M".

Moreover, let $\{\#,B,1\} \subset A$. Then $\vec{x}_n \in \mathcal{N}^n$ is accepted by M iff $\#1^{x_1+1}B \ldots B1^{x_n+1}$ is accepted by M. $R \subset \mathcal{N}^n$ is accepted by M iff $R(\vec{x}_n)$ is equivalent to "$\vec{x}_n$ is accepted by M".//

Note: In some sense, due to the nature of NPA instructions, the left and right ends of an NPA-ID are "adjacent"; this is the reason we employ only one delimiter $\#$.

Definition 4   A (partial) function $\lambda \vec{x}_n.f(\vec{x}_n)$ is NPA-computable iff $\lambda y \vec{x}_n.y = f(\vec{x}_n)$ is accepted by some NPA.//

Theorem 1   If $R(\vec{x}_n)$ is re, then it is accepted by some NPA M.

Proof By Theorem 6.1.1, there is an NTM N which accepts $R(\vec{x}_n)$. By Proposition 6.1.2, it is legitimate to assume that there is a state q_f such that every N-ID involving it is final, and moreover no ID involving other states can be final.

Let A, K be the tape and state alphabets of N respectively, where $\{\$,\#\} \cap (A \cup K) = \varnothing$.

The alphabet of the NPA M will be $\{\$,\#\} \cup A \cup K$. The instruction set is:

(1) $(\#1, \$q_0 1)$

(2) (σ,σ) for all $\sigma \in \{\$\} \cup A \cup (K - \{q_f\})$

(3)   For every $q_i s_l s_j q_m R \in N$, the following instructions are in M:

$(q_i s_l a, s_j q_m a)$ for all $a \in A$

$(q_i s_l \$, s_j q_m B\$)$ (this captures NTM's tape extendibility
to the right)

(4) For every $q_i s_l s_j q_m L \in N$, the following are instructions of M:

$(aq_i s_l, q_m as_j)$ for all $a \in A$

$(\$q_i s_l, \$q_m Bs_j)$ (this captures the NTM's tape extendibility to the left)

(5) For every $q_i s_l s_j q_m S \in N$, the following instruction is in M:

$(q_i s_l, q_m s_j)$

Let us verify that the defined NPA does the job: First, let $R(\vec{x}_n)$. Then there is an N-computation $I_0, I_1, \ldots, I_t$ where $I_0 = q_0 1^{x_1+1} B \ldots B1^{x_n+1}$ and $I_t = uq_f av$ for some $\{u, v\} \subset A^*, a \in A$. (**Note:** The full strength of Proposition 6.1.2, that I_t may be assumed to have the form $\vdash q_f \dashv$, is not needed in the present proof).

Let now $\tilde{I}$ denote $qav\$u$ whenever I is $uqav$, an N-ID.

We can easily see, by considering cases, that $I \xrightarrow[N]{} J$ implies $\tilde{I} \xrightarrow[M]{} \ldots \xrightarrow[M]{} \tilde{J}$. For example, if $I = qav$, $J = q'Bbv$ and $qabq'L \in N$ then $\tilde{I} = qav\$ \xrightarrow[(2)]{} \ldots \xrightarrow[(2)]{} \$qav \xrightarrow[(4)]{} v\$q'Bb \xrightarrow[(2)]{} \ldots \xrightarrow[(2)]{} q'Bbv\$ = \tilde{J}$.

Similarly, if $I = ucqav$, $J = uq'cbv$ and $qabq'L \in N$, then

$$\tilde{I} = qav\$uc \xrightarrow[(2)]{} \ldots \xrightarrow[(2)]{} cqav\$u \xrightarrow[(4)]{} v\$uq'cb \xrightarrow[(2)]{} \ldots \xrightarrow[(2)]{} q'cbv\$u = \tilde{J}$$

The remaining cases are left to the reader. We conclude that we have an M-computation

$$\#1^{x_1+1}B \ldots B1^{x_n+1} \xrightarrow[(1)]{} 1^{x_1}B \ldots B1^{x_n+1}\$q_0 1 \xrightarrow[(2)]{} \ldots \xrightarrow[(2)]{}$$

$$q_0 1^{x_1+1}B \ldots B1^{x_n+1}\$ = \tilde{I}_0 \xrightarrow[M]{} \ldots \xrightarrow[M]{} \tilde{I}_1 \xrightarrow[M]{} \ldots \xrightarrow[M]{} \tilde{I}_2 \xrightarrow[M]{} \ldots \xrightarrow[M]{} \tilde{I}_t$$

[Note that $\tilde{I}_t = q_f au\$v$ for some $a \in A$, $\{u, v\} \subset A^*$. No (2)-instruction is applicable to it neither any of the instructions (3)-(5). (1) is not applicable to any M-ID except the initial].

Conversely, let $w_0, w_1, \ldots, w_l$ be an M-computation, where $w_0 = \#1^{x_1+1}B \ldots B1^{x_n+1}$. It is immediate that $w_1 = 1^{x_1}B \ldots B1^{x_n+1}\$q_0 1$. An easy induction on $l(\geq 1)$ shows that each w_i $(i = 1, \ldots, l)$ contains exactly one occurrence of $\$$, and exactly one occurrence of a symbol in K, each other symbol being in A. Hence, each w_i is a rotation(*) of a string $\$I_i$ where I_i is an N-ID.

Next, for any $i = 0, \ldots, l - 1$, whenever $w_i \xrightarrow[M]{} w_{i+1}$ it is either the case

(*)We say that u is a rotation of v just in case $u = xy$ and $v = yx$ for some strings x and y.

that w_i and w_{i+1} are rotations of the same string $\$I$ where I is an N-ID or there exist distinct N-IDs I and J such that w_i is a rotation of $\$I$ and w_{i+1} a rotation of $\$J$.

Considering the relevant cases we show that $I \xrightarrow{N} J$. It is clear that in $w_i \to w_{i+1}$, instructions (1) and (2) are not responsible. Therefore, one of (3)-(4) was applied.

Case 1. (i) $(qab, cq'b)$ was applied, where $\{a,b,c\} \subset A$. Then $qacq'R \in N$. It follows that $w_i = qabv\$u$ and $w_{i+1} = v\$ucq'b$; hence $I = uqabv$ and $J = ucq'bv$ and clearly $I \xrightarrow{N} J$ via $qacq'R$.

(ii) $(qa\$, cq'B\$)$ was applied. Then $qacq'R \in N$. It follows that $w_i = qa\$u$ and $w_{i+1} = ucq'B\$$; hence $I = uqa$ and $J = ucq'B$ and clearly $I \xrightarrow{N} J$ via $qacq'R$.

Case 2. (i) $(bqa, q'bc)$ was applied, where $\{a,b,c\} \subset A$. Then $qacq'L \in N$. It follows that $w_i = bqav\$u$ and $w_{i+1} = v\$uq'bc$; hence, $I = ubqav$ and $J = uq'bcv$ and clearly $I \xrightarrow{N} J$ via $qacq'L$.

(ii) $(\$qa, \$q'Bc)$ was applied. Then $qacq'L \in N$. It follows that $w_i = \$qav$ and $w_{i+1} = v\$q'Bc$; hence $I = qav$ and $J = q'Bcv$ and clearly $I \xrightarrow{N} J$ via $qacq'L$.

Case 3. $(qa, q'b)$ was applied. This case is left to the reader.

It follows that we may associate with each w_i, $i = 1, \ldots, l$ an N-ID I_i such that w_i is a rotation of $\$I_i$ where $I_1 = q_0 1^{x_1+1}B \ldots B1^{x_n+1}$, $I_l = uq_f av$ for some $\{u,v\} \subset A^*$, $a \in A(*)$ and for $i = 1, \ldots, l-1$ either $I_i = I_{i+1}$ or $I_i \xrightarrow{N} I_{i+1}$. Picking up the subsequence of distinct I_i's we get an N-computation starting with $q_0 1^{x_1+1}B \ldots B1^{x_n+1}$. Hence [due to the relationship between N and $R(\vec{x}_n)$], $R(\vec{x}_n)$ holds.//

Corollary   Every function in $\mathcal{P}$ is NPA-computable.

As in §6.2, we now consider the generator aspect of NPAs.

Definition 5   A *Post system*(†) S over an alphabet A is a pair (w_0, M), where M is an NPA; w_0 is the *axiom* of S. A sequence $w_0, w_1, \ldots, w_n$ $(n \geq 0)$, where (*if* $n > 0$) $w_i \xrightarrow{M} w_{i+1}$ for $i = 0, \ldots, n-1$ is a *proof* in S. w_n is a *theorem* of S, in symbols $\vdash_S w_n$. $T_S \subset A^*$ is the set of theorems of S.//

Note: As in Definition 6.2.5, each of $w_0, w_1, \ldots, w_{n-1}$ in a proof $w_0, \ldots, w_n$ is a theorem of S.

Theorem 2   If $R(x)$ is re, then for some Post system S, $R(x)$ iff $\vdash_S 1^{x+1}$.

Proof The proof can be modeled around the proofs of Theorem 6.2.2 and its corollary. To have some variety of methods, we will present a different proof.

(*)The only form possible for a final M-ID is $q_f u$, $u \in (A \cup \{\$\})^*$.

(†)Called *normal system* by Post. The statement of Theorem 2 is due to him [24].

Now "$R(x)$ re" means that there is a function f in $\mathcal{P}^{(1)}$ such that $R(x)$ $\equiv f(x)\!\downarrow$. Intuitively, the function g defined by

$$g(x) = \begin{cases} x + 1 & \text{if } f(x)\!\downarrow \\ \uparrow & \textbf{otherwise} \end{cases}$$

is "computable" [given x, use an f-program to compute $f(x)$. If this program ever halts, output $x + 1$] and $R(x) \equiv g(x)\!\downarrow$. Formally, $g(x) = x + 1 + 0\cdot f(x)$ for all x. There is then a deterministic TM N such that a computation $q_0 1^{x+1}$ $\xrightarrow[N]{} \cdots \xrightarrow[N]{} 01^{x+1}q_f(\$,\$)$ exists iff $R(x)$. [Without loss of generality q_f is the only state of N such that $\delta(a,q_f)\!\uparrow$ for all a in N's alphabet, whereas $\delta(a,q)\!\downarrow$ for all such a if $q \neq q_f$.] The alphabet of S is as in Theorem 1, extended with the new symbol T. Let the axiom be T. The Post system S we construct has in its M-part instructions (1)-(5) as in the proof of Theorem 1. In addition, it has the instructions

(6) $\quad \begin{cases} (T,T1) \\ (T,\#1) \end{cases}$

(7) $\quad (q_f(\$,\$)\$0, \Lambda)$ [Note that $(\$,\$) \neq \$$].

Since T is the axiom, a proof is a sequence $T,w_1, \ldots, w_n$ where $T \xrightarrow[M]{} w_1$ and $w_i \xrightarrow[M]{} w_{i+1}$, $i = 1, \ldots, n - 1$. Now instructions (2) and (6) can "nondeterministically generate" a rotation w_t ($t \geq 1$) of $\#1^{x+1}$ for any $x \geq 0$. Clearly, there is a proof of "1^{x+1}" if $R(x)$, because then there is an M-computation $w_t,w_{t+1}, \ldots, w_l$ $= q_f(\$,\$)\$01^{x+1}$, where $l < n$, by Theorem 1. Subsequently, $w_l \xrightarrow[(7)]{} 1^{x+1}$. Conversely, if $\vdash_S 1^{x+1}$ through a proof $T,w_1,w_2, \ldots, w_n$ $(w_n = 1^{x+1})$, then there is a $y \geq 0$ and a largest $t \geq 1$ such that w_t is a rotation of $\#1^{y+1}$ [inevitable due to (6) and (2)]. As the only instruction applicable now is (1), w_{t+1} is a rotation of $\$q_0 1^{y+1}$. w_i cannot be stripped of the presence of a q-symbol unless one applies (7). It must be then that there is an s $(1 \leq s < n)$ such that $w_s = q_f(\$,\$)\$01^{y+1}$ ($y + 1$ is the output of the TM N, which M simulates). From now on, only instructions (7) (once) and (2) are applicable to yield 1^{y+1}. Hence, $y = x$. But then the M-computation $w_{t+1}, \ldots, w_s$ implies $R(x)$.//

Corollary   If $R(\vec{x}_n)$ is re, then for some Post system S it is

$$R(\vec{x}_n) \text{ iff } \vdash_S 1^{x_1+1}B \ldots B1^{x_n+1}.$$

Proof   One may either use a proof similar to those of Theorem 6.2.2 and its corollary, being careful about the fact that the inverse of an NPA(*) is not,

(*)An NPA is, of course, a binary relation M on A^*, with some appropriate interpretation (that of Definition 2).

strictly speaking (but essentially it is), an NPA since instructions of the inverse make deletions/additions at exactly the opposite ends of a string than the NPA does.

Alternatively, one may first prove that $R(\vec{x}_n)$ is re iff for some deterministic TM N, N with input $1^{x_1+1}B \ldots B1^{x_n+1}$ halts iff $R(\vec{x}_n)$, and when it halts $1^{x_1+1}B \ldots B1^{x_n+1}$ is the final tape. Then the method of proof of Theorem 2 is applicable. The reader is encouraged to try both methods.//

Appropriate converses of Theorems 1 and 2 are valid. (See §6.5.)

The (semi-) Thue systems of the previous section as well as the Post (normal) systems of the present section are special cases of the Post Canonical Systems (in short PCS) studied by Post [22].

Briefly, a PCS over an alphabet A is a finite set of *axioms* along with a finite set of *productions* of the form $(a_1u_1a_2 \ldots u_{n-1}a_n, b_1v_1b_2 \ldots b_{m-1}v_{m-1}b_m)$, where a_i ($i = 1, \ldots, n$) and b_i ($i = 1, \ldots, m$) are in A^* and u_i ($i = 1, \ldots, n - 1$), v_i ($i = 1, \ldots, m - 1$) are *production variables* out of an auxiliary alphabet V ($V \cap A = \varnothing$). It is required that each v_i is the same as some u_j. Given w_1, w_2 in A^* and a PCS M, then $w_1 \xrightarrow{M} w_2$ just in case $w_1 = a_1\bar{u}_1 \ldots a_{n-1}\bar{u}_{n-1}a_n$ and $w_2 = b_1\bar{v}_1 \ldots b_{m-1}\bar{v}_{m-1}b_m$ for some production $(a_1u_1 \ldots a_{n-1}u_{n-1}a_n, b_1v_1 \ldots b_{m-1}v_{m-1}b_m)$ and $\bar{u}_i, \bar{v}_j$ in A^*.

Note: u_i, v_j are *variables* over A^*; $\bar{u}_i, \bar{v}_j$ are *strings* in A^*.

Multiple premises are allowed, thus productions such as $(a_1u_1a_2 \ldots a_{n-1}u_{n-1}a_n, a_1'u_1' \ldots u_{n'-1}'a_{n'}'; b_1v_1 \ldots v_{m-1}b_m)$ are possible. In this case, we will have $(w_1, w_2) \to w_3$ if $w_1 = a_1\bar{u}_1 \ldots a_{n-1}\bar{u}_{n-1}a_n$, $w_2 = a_1'\bar{u}_1' \ldots a_{n'-1}'\bar{u}_{n'-1}' a_{n'}'$ and $w_3 = b_1\bar{v}_1 \ldots b_{m-1}\bar{v}_{m-1}b_m$ for some strings $\bar{u}_i, \bar{u}_j', \bar{v}_l$ in A^*.

Example 2 Semi-Thue instructions (a,b) [where $\{a,b\} \subset A^*$, for some alphabet A] take the PCS production form (xay, xby) where x, y are production variables. Normal productions (a,b) take the PCS production form (ax, xb) where x is a production variable. Thus, the claim earlier made that PCS subsumes normal and Thue-like systems is seen to be substantiated.//

Example 3 The PCS S over $A = \{1\}$ with single axiom 1 and production $(x, x11)$ generates the set of odd positive integers in unary notation. That is, $\vdash_S 1^n (n \geq 1)$ iff $n = 2k - 1$ for some $k \geq 1$. The PCS T over $A = \{1\}$ with axiom 1 and production (x, xx) generates exactly the strings over A of the form 1^{2^k}, $k \geq 0$. In both cases the productions have the required format for PCS.//

Example 4 Consider the PCS S over $A = \{1, \$\}$ with axiom $\$1$ and single production $(x\$y, y\$xy)$. Then it can be shown that $\vdash_S w$ iff $w = 1^m\$1^n$, where m and n

$(m < n)$ are consecutive Fibonacci numbers. To this end, we have to prove two things:

(1)   For any Fibonacci numbers F_n and F_{n+1}, $\vdash_S 1^{F_n}\$1^{F_{n+1}}$.
Induction on n:

(i) $n = 0$. Then $F_0 = 0$, $F_1 = 1$, and indeed $\vdash_S \$1$ (this is the axiom).(*)

(ii)   Assume that $1^{F_k}\$1^{F_{k+1}}$ is provable in S. Then (apply the production once again) $\vdash_S 1^{F_{k+1}}\$1^{F_k}1^{F_{k+1}}$, that is, $\vdash_S 1^{F_{k+1}}\$1^{F_{k+2}}$. The induction is complete.

(2)   Let $\vdash_S w$. Prove that $w = 1^{F_n}\$1^{F_{n+1}}$ for some $n \geq 0$. We apply *induction with respect to the system* which is the set of strings provable in S (see Chapter 2 for the first occurrence of this principle, in connection with the system $\mathcal{PR}$). This system is the *smallest* set of strings containing $\$1$ and closed under the application of the production $(x\$y, y\$xy)$. (The reader will observe that our induction, viewed differently, is with respect to the length of proofs in S.)

(i)   Now if w is the axiom, it does "have the right form", namely, $1^{F_0}\$1^{F_1}$.

(ii)   Let w "have the right form". Clearly, the property of "having the right form" is propagated by application of the production. Hence the set of strings that "have the right form" is closed under application of $(x\$y, y\$xy)$ and contains the axiom. It follows that it includes $\{w \mid \vdash_S w\}$, as the latter is the *smallest* that includes the axiom and is closed under the same operation.//

Example 5   Add to the PCS of Example 4 the production $(x\$y, x)$. Then $\vdash_S w$ and $\$$ is not part of w iff $w = 1^{F_k}$ for some $k \geq 0$. (The verification is left to the reader.)//

Example 6   Consider the PCS S over $A = \{a, b, c, \#, \$\}$ with axiom $a\#b\$c$ and production $(x\#y\$z, xa\#yb\$zc)$. It can be shown that $\vdash_S w$ iff $w = a^n\#b^n\$c^n$ for some $n \geq 1$.

If we add the production $(x\#y\$z, xyz)$ to S, then $\vdash_S w$ & $w \in \{a, b, c\}^*$ iff $w = a^n b^n c^n$ for some $n \geq 1$.//

Example 7   Consider the PCS S over $A = \{a, b\}$ with axiom Λ and productions (x, axa) and (x, bxb). Then $\vdash_S w$ iff w is a *palindrome* of even length; that is, for some x_i, $i = 1, \ldots, n$, where $x_i \in \{a, b\}$, $w = x_1 x_2 \ldots x_{n-1} x_n x_n x_{n-1} \ldots x_2 x_1$.//

Example 8   Consider the PCS S over $A = \{a, b, \#\}$ axiom $\#$ and productions $(x\#, xa\#)$, $(x\#, xb\#)$, $(x\#, xx)$. Then $\vdash_S w$ & $w \in \{a, b\}^*$ iff $w = xx$ for some $x \in \{a, b\}^*$.//

(*)We employ the convention $1^0 = \Lambda$.

We conclude this section with the formal definition of Smullyan's [27] elegant variant of PCS.

Definition 6   [27] Consider the (finite) alphabets A, V, Π. The symbols of V are the *variables*. The symbols of Π are the *predicates*; each predicate is associated with a unique integer ≥ 1, its *degree*.

A string over $A \cup V$ is a *term*.

A string Pw is an *atomic formula* (in short, *af*) iff P is a predicate of degree n and w is a string of n terms separated by $n - 1$ *commas*(*)(,).

A *well-formed-formula* (in short, *wff*) is defined recursively by:

(i)   Every *af* is a *wff*.

(ii)   If W is *af* and U is *wff*, then $W \rightarrow U$(*) is a *wff*.//

Note: Intuitively, an *af* Pw states a "property P" about w (this is formalized below). "$\rightarrow$" is intuitively the "implies" symbol. By definition of wff, "$\rightarrow$" is "right associative" (or "evaluates from right to left") since only *after* U, which involves all the "$\rightarrow$" symbols except the left-most, has been checked to be a wff (and presumably "evaluated") we apply the left-most "$\rightarrow$". Thus $W_1 \rightarrow W_2 \rightarrow W_3$, where each of W_1, W_2, W_3 are *af*'s, intuitively means "**if** W_1 **then if** W_2 **then** W_3", or "W_1 & W_2 implies W_3" or "$W_2 \rightarrow W_3$ provided W_1". Of course, this note gives motivational discussion. The intuitive interpretation of *wff*'s and *af*'s is *not* to be employed in formal proofs.

Definition 7   [27] An *Elementary Formal System* (in short, EFS) over A is a 4-tuple $S = (\Phi, A, V, \Pi)$, where Φ is a finite set of *wff*'s, called the *axioms*, and A, V, Π are as in Definition 6.//

Definition 8   [27] A *proof in S* is a finite sequence $F_1, F_2, \ldots, F_k$ of wffs such that for each $i = 1, \ldots, k$, $F_i \in \Phi$, or F_i is obtained from some F_j ($j < i$) by *substitution* of *each* occurrence of some variable in F_j by a string in A^*, or there is an *af* W such that $F_m = W$ and $F_j = W \rightarrow F_i$ for some $m < i, j < i$.

Note: $F_m = W$ is equality of strings; same with $F_j = W \rightarrow F_i$.

A wff W is a *theorem* of S, in symbols $\vdash_S W$, iff W is the last formula of a proof in S.//

Note: The rule of going from *af* A and *wff* $A \rightarrow B$ to *wff* B is called *modus ponens* and is a basic instrument in mathematical proofs. The premise is an *af* rather than a *wff* due to right associativity of "$\rightarrow$".(†) The notion of proof in an EFS is much closer to the notion of mathematical proof than proofs in Thue-like systems are.

(*)The symbols "," and "$\rightarrow$" are not in $A \cup V \cup \Pi$.

(†)A different "definition" of modus ponens, although possible, would be counterintuitive.

EFS simulate the action of PCS in a straightforward manner. For example, if S is a PCS over A with axioms $A_1, \ldots, A_k$ and productions $P_1, \ldots, P_m$, then the simulating EFS has the same alphabet A, and same set of variables V as the PCS. Π contains a single predicate of *degree* 1, let us call it T.

Axioms of the EFS are $TA_1, \ldots, TA_k$ and for each $i = 1, \ldots, m$, if P_i is a production (in general with multiple premises) $(x_1^i, \ldots, x_{n_i}^i ; y^i)$ then and only then the EFS contains the axiom $Tx_1^i \to Tx_2^i \to \ldots \to Tx_{n_i}^i \to Ty^i$. [Recall that, intuitively, due to right-associativity of "$\to$", the previous string of implications stands for "$Tx_1^i \& \ldots \& Tx_{n_i}^i \to Ty^i$"]. It is left to the reader to verify that $\vdash_S w$ iff $\vdash Tw$ in the EFS. (See Problem 8.)

Example 9 Let $A = \{1\}$ and let an EFS over A have the only axiom $Px,111$, where P is a predicate of degree 2 and x a variable.

Clearly, $\vdash Px,111$ for all $x \in A^*$. In some sense, then, this EFS "computes" the function $\lambda x.3.//$

Definition 9 [27] Let S be an EFS over A. Then a relation(*) $W \subset (A^*)^k$ is *formally representable over A in S* iff for some k-degree predicate T of S

$$\vec{x}_k \in W \quad \text{iff} \quad \vdash_S T\vec{x}_k$$

We say that T *represents* W. We say that $U \subset (A^*)^k$ is *formally representable over A* or simply *formally representable* if the EFS concerned or, respectively, both the EFS and the alphabet concerned are either understood from the context or we wish to leave them unspecified.$//$

Definition 10 Let $R \subset \mathcal{N}^n$ $(n \geq 1)$. Then $R(\vec{x}_n)$ is formally representable in *unary notation* iff $\{(1^{x_1+1}, \ldots, 1^{x_n+1}) \mid R(\vec{x}_n)\}$ is formally representable.

It is representable in *m-adic notation* $(m \geq 1)$ iff $\{(x_1^{(m)}, \ldots, x_n^{(m)}) \mid R(\vec{x}_n)\}$ is formally representable, where $x^{(m)}$ in general is the m-adic notation of $x \in \mathcal{N}.//$

Note: In *1-adic*, 0 is Λ and n is 1^n. In *unary*, $n \geq 0$ is 1^{n+1}.

Example 10 Let $A = \{a,b,c\}$. Form an EFS over A with axioms (1) Pa, (2) Pb, (3) $Px \to Pxa$, (4) $Px \to Pxb$. It is easy to verify that $\{a,b\}^+$ is formally representable in this system, through P.

An alternative EFS over A [27] which represents $\{a,b\}^+$ as well, is obtained by the previous by replacing axioms (3) and (4) by (3′) $Px \to Py \to Pxy$ (intuitively, "if x and y are in $\{a,b\}^+$, then so is xy").$//$

(*)Whenever our discussions involve EFS, we shall call subsets of $(A^*)^k$ or N^k *relations* rather than predicates so to avoid confusion with the *predicates* of the EFS.

Example 11   The set $\{a,b,c\}^+ - \{a,b\}^+$ is also formally representable. Indeed, take as axioms (1) Pc, (2) $Px \rightarrow Pyxz$ [27] (intuitively, "w is in $\{a,b,c\}^+ - \{a,b\}^+$ iff it is just c or if it is of the form yxz where $x \in \{a,b,c\}^+ - \{a,b\}^+$ and $\{y,z\} \subset \{a,b,c\}^*$").//

Note: Representable relations which have representable complements are called *solvable* by Smullyan. Since representable relations fully correspond to re relations (to be shown shortly), solvable relations correspond to recursive ones.

Example 12   Consider the EFS S over $A = \{a,b,c,\#,\$\}$ with axioms (1) $Pa\#b\$c$, (2) $Px\#y\$z \rightarrow Pxa\#yb\$zc$, (3) $Px\#y\$z \rightarrow Pxyz$.

Then $\vdash_S Pw$ iff $w = a^n b^n c^n$ or $w = a^n \# b^n \$ c^n$ ($n \geq 1$). (Compare with Example 6.)

If (3) is replaced by $Px\#y\$z \rightarrow Txyz$, where T is a new (1-degree) predicate, then T represents $\{a^n b^n c^n \,|\, n \geq 1\}$.//

Theorem 3   If $R \subset \mathcal{N}^n$ is re, then for some EFS S, $R(\vec{x}_n)$ iff $\vdash_S P1^{x_1+1}B \ldots B1^{x_n+1}$, where P (of degree 1) is in S.

Proof   Let T be some semi-Thue system over an alphabet A (such that $\{1,B\} \subset A$) with the property $R(\vec{x}_n)$ iff $\vdash_T 1^{x_1+1}B \ldots B1^{x_n+1}$ (Corollary of Theorem 6.2.2).

The appropriate EFS S over A has a single 1-degree predicate P, and axioms

(1) Pw_0, where w_0 is the axiom of T,

(2)   For each production (a,b) of T ($\{a,b\} \subset A^*$), $Pxay \rightarrow Pxby$ is an axiom of S.

It is trivial to verify that $\vdash_T w$ iff $\vdash_S Pw$. Hence, $R(\vec{x}_n)$ iff $\vdash_T 1^{x_1+1}B \ldots B1^{x_n+1}$ iff $\vdash_S P1^{x_1+1}B \ldots B1^{x_n+1}$.//

Corollary   If $R \subset \mathcal{N}^n$ is re, then $R(\vec{x}_n)$ is formally representable in unary notation.

Proof   Add an n-degree predicate Q and a 1-degree predicate N to S of the previous proof. Add the axioms:

$$\left.\begin{array}{l} N1 \\[1em] Nx \rightarrow Nx1 \end{array}\right\} \text{Intuitively, this says that } N \text{ represents nonempty tallies of 1's.}$$

Add the axiom:

$$Nx_1 \rightarrow Nx_2 \rightarrow \ldots \rightarrow Nx_n \rightarrow Px_1Bx_2B \ldots Bx_n \rightarrow Qx_1,x_2, \ldots, x_n$$

It is clear that in the new EFS, and for *any* nonempty tallies $x_1, \ldots, x_n$,

$$\vdash Px_1Bx_2B \ldots Bx_n \to Qx_1, \ldots, x_n$$

Then

$$\vdash Qx_1, \ldots, x_n \text{ iff } \vdash Px_1Bx_2B \ldots Bx_n \text{ iff } \vdash_T 1^{|x_1|}B \ldots B1^{|x_n|}$$
$$\text{iff } R(|x_1| - 1, \ldots, |x_n| - 1),$$

where $|s|$ = length of s, as usual.//

The question naturally arises (*), whether re predicates are formally representable even when the integers are represented in m-adic ($m > 1$). This can be answered in the affirmative, proving seemingly (but not really, in view of §6.5) more (Theorem 4 below).

To this end, and for the balance of this section, let integers be represented in m-adic ($m > 1$) for some fixed m. We define a function *code*: $\mathcal{N} \to \mathcal{N}$ by *code* $= \lambda x.$"the number, which in m-adic notation, is a sequence of x 1's". Of course, *code* can be thought of as a function $A^* \to A^*$, where $A = \{1, 2, \ldots, m\}$. *code* can be defined using (right) recursion on notation:

$$code(0) = 0$$
$$\text{for } i = 1, \ldots, m \quad code(x*i) = \overbrace{code(x)*code(x)*\ldots*code}^{m}(x) * 1^i$$

where "$*$" denotes concatenation.

Lemma 1 $code(x) = y$ is formally representable over the alphabet $A = \{1, \ldots, m\}$.

Proof $code(x) = y$ is represented by the 2-degree predicate C in the EFS over A with axioms

$$C\Lambda, \Lambda$$
$$Cx, y \to Cx1, y^m1$$
$$Cx, y \to Cx2, y^m1^2$$
$$\vdots$$
$$Cx, y \to Cxm, y^m1^m$$

(*)This question also arises in the context of Thue-related and Post-related systems. Due to the equivalence of these formalisms with EFS (to be shown in §6.5), we postponed asking this question until now.

The reader will verify that $code(x) = y$ iff $\vdash Cx,y.//$

$code$ converts x to 1^x. For unary notation, we need $x \longmapsto 1^{x+1}$. Therefore:

Lemma 2 $code(x)1 = y$ is formally representable over the alphabet $A = \{1, \ldots, m\}$.

> **Proof** Add the axiom $Cx,y \to \hat{C}x,y1$ to the EFS of Lemma 1, where $\hat{C}$ is an additional 2-degree predicate.
> Then $code(x)1 = y$ iff $\vdash \hat{C}x,y.//$

Theorem 4   If $R(\vec{x}_n)$ is formally representable over A in *unary notation*, then it is formally representable in *m-adic notation* for any $m > 1$.

> **Proof** If A does not already include $\{1, \ldots, m\}$, then extend A to $\overline{A} = A \cup \{1, \ldots, m\}$.
> First, observe (Smullyan [27]) that $R(\vec{x}_n)$ is representable (in unary) over $\overline{A}$; indeed, add the new predicate Q to system S, and the axioms

$$(1) \quad Q\Lambda \quad (\Lambda \text{ is the empty string})$$

$$(2) \quad Qx \to Qxa \text{ for each } a \in A.$$

Clearly, Q represents A^* over $\overline{A}$.

Next, replace each axiom X of S, which involves variables $x_1, \ldots, x_n$ by the axiom $Qx_1 \to Qx_2 \to \ldots \to Qx_n \to X$ [27].

Intuitively, this way we force "useful substitutions" during a proof in the new system over $\overline{A}$ to involve only strings in A^*.

It can be easily seen now that $R(\vec{x}_n)$ is representable (in unary) over $\overline{A}$ in the new system.

Let now

$$R^{(1)} = \{(1^{x_1+1}, \ldots, 1^{x_n+1}) \mid R(\vec{x}_n)\}$$

and

$$R^{(m)} = \{(x_1^{(m)}, \ldots, x_n^{(m)}) \mid R(\vec{x}_n)\}$$

We have

$$(x_1^{(m)}, \ldots, x_n^{(m)}) \in R^{(m)} \text{ iff } (code(x_1^{(m)})1, \ldots, code(x_n^{(m)})1) \in R^{(1)}.$$

Add the axioms of Lemmas 1 and 2 to the system (without loss of generality, we assume that neither C nor $\hat{C}$ were in the predicate set of the original system

S). Add a new predicate of degree n to the system. Call it F. Add the axiom below, where $\boldsymbol{R}$ represents R in unary over $\overline{A}$:

$$\hat{C}x_1, z_1 \rightarrow \ldots \rightarrow \hat{C}x_n, z_n \rightarrow \boldsymbol{R}z_1, \ldots, z_n \rightarrow Fx_1, \ldots, x_n.$$

Clearly, F represents $R^{(m)}$, i.e.,

$$R(\vec{x}_n) \text{ iff } \vdash Fx_1^{(m)}, \ldots, x_n^{(m)}.//$$

Corollary 1   If $R(\vec{x}_n)$ is formally representable over A in m-adic notation $(m > 1)$, then it is formally representable over A in unary notation.

   Proof   The reader is asked to adapt the proof of Theorem 4.//

Corollary 2 $R(\vec{x}_n)$ is formally representable in m-adic notation iff it is so in n-adic notation for any $m \geq 1$ and $n \geq 1$.

Corollary 3   If $R(\vec{x}_n)$ is re then it is formally representable in m-adic for any $m \geq 1$.

   An interesting and stronger result is that every formally representable relation is so representable over $A = \{1\}$ (Smullyan [27]). A seemingly weaker statement than that, namely that re relations are so representable, is left as an exercise (Problems 13-17). That this is also true for formally representable relations follows from the results of §6.5.

§6.4 SYSTEMS OF EQUATIONS

   In Chapter 2, we saw functions on $\mathcal{N}^n$ defined by equations such as

$$(1) \quad \left\{ \begin{array}{l} h(0,\vec{x}) \overset{\text{def}}{=} g(\vec{x}) \\ h(y+1,\vec{x}) \overset{\text{def}}{=} f(y,\vec{x},h(y,\vec{x})) \end{array} \right\}, \text{ where } g \text{ and } f \text{ are given,}$$

or

$$(2) \quad h(\vec{x}) \overset{\text{def}}{=} f(g_1(\vec{x}), \ldots, g_n(\vec{x})), \text{ where } f \text{ and the } g_i\text{'s are given.}$$

   It was noted early by Ackermann [1] and others that there were other equational schemas which defined functions impossible to be defined by schemas (1) and (2) (see, for example, §3.3, Ackermann's function). Herbrand

and Gödel (in that chronological order [12,11]) were the first to suggest very general equational schemas which included (1) and (2) as special cases. The functions definable by these schemas were called *general recursive* as opposed to just *recursive* for those definable by schemas (1) and (2). (The "modern" name for the latter is, of course, *primitive recursive* and is due to Kleene.)

Next, Kleene [15], following essentially Gödel, formalized systems of equations as strings of symbols over finite alphabets subject to simple string transformations. A "concrete" function on $\mathcal{N}^n$ was defined to be *general recursive* by Kleene just in case it was in some sense representable (see below) in some particular abstract system of equations.

We point out, before we proceed with the details, that whereas Kleene represented integers in his systems of equations by strings such as $0 \underbrace{{}'\,'\cdots\,'}_{n}$

(this represents $n \in \mathcal{N}$; $'$ is a successor operator. The symbol 0 stands for "zero" of $\mathcal{N}$) we shall allow arbitrary alphabets $A = \{1, 2, \ldots, m\}$ and represent integers over A though their m-adic notation (in this we are similar to Jones [14]).

Definition 1   Let $A = \{1, 2, \ldots, m\}$ be a fixed given alphabet.

Given are also a finite alphabet V of *variables* and a finite alphabet F of *function symbols*. Each function symbol has associated with it a number $n \geq 1$, its *degree*.

The set of *terms* T over $A \cup V \cup F$ is the *smallest* set which contains $\{\Lambda\} \cup V$ and satisfies

(1) $t \in T \Rightarrow ta \in T$ for all $a \in A$

(2) $t_1, \ldots, t_n$ in T and f of degree n in F implies $f(w) \in T$ where w is the string composed of t_i, $i = 1, \ldots, n$, from left to right, separated by the symbol ",".

An equation over $A \cup V \cup F$ is a string $t = s$ where $\{t, s\} \subset T$ and "=" is a new symbol.//

Note: Intuitively, "=" stands for "equals", but in Definition 1 it has no meaning; it is just some uninterpreted symbol. In any case, it does *not* indicate that t and s are identical strings.

Example 1   Let $A = \{1, 2\}$, $V = \{x, y\}$, $F = \{f, g\}$, where f has degree 2 and g degree 3.

Then Λ, 11, 12, 212, $x2$, $f(x2, 121)2$, $f(1, g(122, y211, x)22)$ are terms. $f(f(x, y), g(1, 2, x1)) = 11$ is an equation.//

Example 2   Let $A = \{1\}$, $V = \{x\}$, $F = \{p\}$, where p has degree 1.
Then

$$p(0) = 0$$

$$p(x1) = x$$

are equations over $A \cup V \cup F$. Intuitively, p "defines" the predecessor function $\lambda x . x \doteq 1.//$

Definition 2 [15] A *system of equations* S over $A \cup V \cup F$ is a finite *set* of equations $S = \{E_1, \ldots, E_n\}$, each E_i over $A \cup V \cup F$.

A *deduction* in S is a finite *sequence* of equations $e_1, e_2, \ldots, e_p$ such that for each $i = 1, \ldots, p$ one of the following holds:

(1) $e_i \in S$

(2) e_i is obtained from some $e_j, j < i$, by substituting *each* occurrence of some variable of e_j with the same string $u \in A^*$

(3) e_i is obtained from two equations e_j and e_k ($j < i$ and $k < i$) as follows:

e_j is $f(t_1, \ldots, t_k) = t_{k+1}$ where each t_i, $i = 1, \ldots, k + 1$, is in A^*, and e_i results from e_k by substituting a *single* occurrence of the term $f(t_1, \ldots, t_k)$ in e_k by t_{k+1}.

We write $\vdash_S e_p$, and say that e_p is *deduced* in $S.//$

Note: In many discussions, one considers several systems of equations with the same alphabet A, even though the V and F alphabets may be different. Because of this, we shall usually say a "system of equations over A" rather than "over $A \cup V \cup F$".

Definition 3   Let $R(\vec{x}_n, y)$ be a relation on the integers. R is *deducible* in some system of equations S over A, where without loss of generality $A = \{1, \ldots, m\}$ and $m \geq 1$, iff for some n-degree function symbol r of S it is true that $R(\vec{x}_n, y)$ iff $\vdash_S r(\vec{x}_n^{(m)}) = y^{(m)}$, where, in general, $a^{(m)}$ is the m-adic notation for $a \in \mathcal{N}$.

We say that r *represents* R. A function $g : \mathcal{N}^n \rightarrow \mathcal{N}$ is *general recursive over A* (in S) iff $\lambda \vec{x}_n y . g(\vec{x}_n) = y$ is deducible (in S) over A. If $\breve{g}$ represents $\lambda \vec{x}_n y . g(\vec{x}_n) = y$, we also say that it represents $g.//$

Note: Just because the symbols in F are called *function symbols*, it does *not* follow that each deducible relation $R(\vec{x}_n, y)$ is a function $\vec{x}_n \mapsto y$.

Example 3   Let $A = \{1\}$, $V = \{x\}$, $F = \{f\}$, where f has degree 1.

Consider the system $S = \{f(x) = 1, f(111) = 11\}$. Then both $\vdash f(111) = 11$ [(1) in Definition 2] and $\vdash f(111) = 1$, the latter through the deduction $(f(x) = 1, f(111) = 1)$.

Thus, if $R(x, y)$ is *defined* by "$R(x, y)$ iff $\vdash f(1^x) = 1^y$" then as R is not single-valued it is not a function.

It turns out that the problem of whether an arbitrary function symbol of an arbitrary system S represents a function is recursively unsolvable. (See

Problem 18.) In this respect, perhaps, the system of equations formalism is a bit awkward compared to the $\mathcal{P}$-formalism of Chapter 3.//

Example 4 Let $\lambda x.h(x)$ and $\lambda yxzi.g(i,y,x,z)$ be general recursive over $A = \{1,2,\ldots,m\}$, the first in the system H the second in the system G. Assume without loss of generality that H and G have disjoint function symbol alphabets F_H, F_G and that i,x, and y are different symbols. Consider $H \cup G$ over $A \cup V_H \cup V_G \cup F_H \cup F_G$.

Let $\tilde{h}$ represent h in H and $\tilde{g}$ represent g in G. Add the equations below, where $\tilde{r}$ is a *new* function letter:

$$\tilde{r}(\Lambda,x) = \tilde{h}(x)$$

$$\tilde{r}(yi,x) = \tilde{g}(i,y,x,\tilde{r}(y,x)), i = 1,\ldots,m$$

Suppose we want the "value" of $\tilde{r}(31,22)$. Assume that $\vdash_H \tilde{h}(22) = 12$ and $\vdash_G \tilde{g}(3,\Lambda,22,12) = 1$, $\vdash_G \tilde{g}(1,3,22,1) = 1112$. Then (a) $\vdash_{H\cup G} \tilde{h}(22) = 12$, (b) $\vdash_{H\cup G} \tilde{g}(3,\Lambda,22,12) = 1$ and (c) $\vdash_{H\cup G} \tilde{g}(1,3,22,1) = 1112$. Continue these three $(H \cup G)$-deductions by

$\tilde{r}(\Lambda,x) = \tilde{h}(x), \tilde{r}(\Lambda,22) = \tilde{h}(22)$, [using (a)]

$\tilde{r}(\Lambda,22) = 12, \tilde{r}(y3,x) = \tilde{g}(3,y,x,\tilde{r}(y,x)), \tilde{r}(3,x) = \tilde{g}(3,\Lambda,x,\tilde{r}(\Lambda,x))$,

$\tilde{r}(3,22) = \tilde{g}(3,\Lambda,22,\tilde{r}(\Lambda,22)), \tilde{r}(3,22) = \tilde{g}(3,\Lambda,22,12)$, [by (b)]

$\tilde{r}(3,22) = 1, \tilde{r}(y1,x) = \tilde{g}(1,y,x,\tilde{r}(y,x)), \tilde{r}(31,x) = \tilde{g}(1,3,x,\tilde{r}(3,x))$,

$\tilde{r}(31,22) = \tilde{g}(1,3,22,\tilde{r}(3,22)), \tilde{r}(31,22) = \tilde{g}(1,3,22,1)$, [by (c)]

$\tilde{r}(31, 22) = 1112$

Thus, the "value" of $\tilde{r}$ at $(31,22)$ has been "calculated". It is a trivial matter to prove, by induction on the m-adic length of y, that the unique function $r: (A^*)^2 \rightarrow A^*$ defined by

$$r(\Lambda,x) = h(x)$$

$$r(y*i,x) = g(i,y,x,r(y,x)), i = 1,\ldots,m$$

("$*$" denotes m-adic concatenation) is represented by $\tilde{r}$ in our system over A; hence, it is general recursive (viewed as a function on the integers).//

We now show that every function in $\mathcal{P}$ is general recursive over $A = \{1\}$.

Lemma 1 $\lambda x.x + 1$ and $\lambda x.x \doteq 1$ are general recursive over $A = \{1\}$.

Proof For $\lambda x . x + 1$ take $V = \{x\}$, $F = \{s\}$, where s has degree 1. Take $s(x) = x1$ as the only equation. Then s represents $\lambda x . x + 1$. The case for $\lambda x . x \dot{-} 1$ was the subject of Example 2.//

Lemma 2   The set of general recursive functions over $\{1\}$ is closed under substitution.

Proof There are several cases to consider. We illustrate one and leave the rest to the reader. (See Problem 19).

Let $\lambda \vec{x}_n . f(\vec{x}_n)$ and $\lambda \vec{y}_m . g(\vec{y}_m, z)$ be general recursive.

Consider $\lambda \vec{x}_n \vec{y}_m . g(\vec{y}_m, f(\vec{x}_n))$. We show it is general recursive:

Let f be represented by $\tilde{f}$ in system S_f over $\{1\}$ and g be represented by $\tilde{g}$ in system S_g. Consider the system $S' = S_f \cup S_g$ where, without loss of generality, S_f and S_g do not share any function symbols.

Add the equation $\tilde{h}(\vec{x}_n, \vec{y}_m) = \tilde{g}(\vec{y}_m, \tilde{f}(\vec{x}_n))$, where $\tilde{h}$ is a new function symbol and the x_i and y_j are all distinct, to obtain system S.

Clearly, $\lambda \vec{x}_n \vec{y}_m . g(\vec{y}_m, f(\vec{x}_n))$ is represented over $\{1\}$ in S by the symbol $\tilde{h}$.//

Lemma 3   The set of general recursive functions over $\{1\}$ is closed under pure iteration.

Proof Consider $\lambda x . g(x)$. Let $\tilde{g}$ be a function symbol, in some system S_g over $\{1\}$, which represents g. Add a new variable y and new function symbol $\tilde{f}$ (degree 2) to the system, along with the equations

$$\tilde{f}(\Lambda, y) = y$$
$$\tilde{f}(x1, y) = \tilde{g}(\tilde{f}(x, y))$$

Then $\tilde{f}$ represents $\lambda x y . g^x(y)$.//

Lemma 4   The set of general recursive functions over $\{1\}$ is closed under unbounded search.

Proof Consider $\lambda y \vec{x}_n . g(y, \vec{x}_n)$. Let g be represented by $\tilde{g}$ in some system S_g over $\{1\}$. In order to enlarge this system so that it represents $\lambda \vec{x}_n . (\mu y) g(y, \vec{x}_n)$, observe first that

$$(\mu y) g(y, \vec{x}_n) = h(0, \vec{x}_n), \text{ where}$$
$$h(y, \vec{x}_n) = \textbf{if } g(y, \vec{x}_n) = 0 \textbf{ then } y \textbf{ else } h(y + 1, \vec{x}_n)$$
$$= sw(g(y, \vec{x}_n), y, h(y + 1, \vec{x}_n)),$$

where *sw* is the "switch" or **if-then-else** function $sw = \lambda xyz.$ **if** $x = 0$ **then** y **else** z.

A primitive recursion easily defines *sw* by

$$\left\{ \begin{array}{c} sw(0,y,z) = y \\ sw(x+1,y,z) = z \end{array} \right\}$$

Hence, add the function symbol $\tilde{sr}$ (for "search"; degree n), $\tilde{h}$ (degree $n+1$), $\tilde{sw}$ (degree 3) to S_g.

Add the equations

$$\tilde{sw}(\Lambda,y,z) = y$$
$$\tilde{sw}(x1,y,z) = z$$
$$\tilde{h}(y,\vec{x}_n) = \tilde{sw}(\tilde{g}(y,\vec{x}_n),y,\tilde{h}(y1,\vec{x}_n))$$
$$\tilde{sr}(\vec{x}_n) = \tilde{h}(\Lambda,\vec{x}_n)$$

where the variables $y, z, \vec{x}_n$ are distinct symbols.
Then $\tilde{sr}$ represents $\lambda \vec{x}_n.(\mu y)g(y,\vec{x}_n)$ over $\{1\}$.//

Theorem 1   Every function in $\mathcal{P}$ is general recursive over the alphabet $A = \{1\}$.

   Proof The initial functions of $\mathcal{P}$ are general recursive (Lemma 1), and the set of general recursive functions is closed under the same operations as $\mathcal{P}$ (Lemmas 2,3,4). $\mathcal{P}$ being the smallest such set, the result follows.//

   We have seen a number of formalisms in §6.1–§6.4, each of which can do at least as well as TMs (deterministic or not, as the two are equivalent) in that they all can accept (generate, represent) any re set or, equivalently, can "calculate" any function in $\mathcal{P}$. In the next section we show that all the formalisms (including TMs) are equivalent.

§6.5 EQUIVALENCE OF THE FORMALISMS IN §6.1–§6.4

   We first show that every relation $R(\vec{x}_n,y)$ $(n \geq 1)$ which is deducible in some system of equations is formally representable in some EFS. Also, the EFS is shown to subsume Thue-like systems as well as Post systems (hence it also subsumes the TM formalism). Next, each relation $R(\vec{x}_n)$ which is formally represented over A in some EFS S is shown to be re. This closes the circle.

Lemma 1

Let S be a system of equations over A and let r be a function symbol in S of degree n. Then $\vdash_S r(t_1, \ldots, t_n) = t_{n+1}$, where $\{t_1, \ldots, t_{n+1}\} \subset A^*$, iff there is a sequence $e_0, e_1, \ldots, e_q$ where e_q is $r(t_1, \ldots, t_n) = t_{n+1}$ and for $i = 0, \ldots, q$, e_i is one of the following:

(1) e_i is obtained from some equation $\tilde{e}$ of S by substituting elements of A^* in each variable of $\tilde{e}$.

(2)   There are e_k and e_j ($k < i$, $j < i$) in the sequence, where e_k is $f(t_1, \ldots, t_l) = t_{l+1}$ ($\{t_1, \ldots, t_{l+1}\} \subset A^*$), and e_i is obtained from e_j by substituting *one* occurrence of $f(t_1, \ldots, t_l)$ in e_j by t_{l+1}.

Proof *if* part. Each e_i in the sequence, which satisfies (1) above, is replaced by $\tilde{e}_1, \ldots, \tilde{e}_{m_i}$, where $\tilde{e}_1$ is the corresponding equation $\tilde{e}$, $\tilde{e}_{m_i}$ is e_i and each $\tilde{e}_j$ is obtained from $\tilde{e}_{j-1}$ ($j = 2, \ldots, m_i$) by application of rule (2) of Definition 6.4.2. This way the sequence $e_0, \ldots, e_q$ is transformed into an S-deduction.

only if part. Let $\tilde{e}_0, \ldots, \tilde{e}_l$ be an S-deduction. We show that for any *instance* of $\tilde{e}_l$, e (i.e., equation obtained from $\tilde{e}_l$ by substituting all variables of $\tilde{e}_l$ by strings in A^*) there is a sequence $e_0, e_1, \ldots, e_q$, where e_q is e and (1) and (2) of the lemma statement are satisfied.

We prove this by induction on l:

$l = 1$: Then $\tilde{e}_l$ is an equation of S. Any instance of $\tilde{e}_l$ by itself satisfies (1) and (2).

Assume the claim true for $l \leq k$.

$l = k + 1$: We consider cases according to Definition 6.4.2:

Case 1. $\tilde{e}_l$ is an equation of S. Then any instance of $\tilde{e}_l$ by itself satisfies (1) and (2).

Case 2. $\tilde{e}_l$ is obtained from $\tilde{e}_j$ ($j < l$) by substituting some variable, say x, of $\tilde{e}_j$ by some t in A^*. Now, $\tilde{e}_0, \ldots, \tilde{e}_j$ is a deduction of length $< l$. Moreover, any instance of $\tilde{e}_l$ is an instance of $\tilde{e}_j$. By induction hypothesis, we are done.

Case 3. $\tilde{e}_l$ is obtained from $\tilde{e}_j$ ($j < l$) by substituting *one* occurrence of some $f(t_1, \ldots, t_l)$ of $\tilde{e}_j$ ($t_i \in A^*$, $i = 1, \ldots, l$) by t_{l+1}, where $t_{l+1} \in A^*$ and $f(t_1, \ldots, t_l) = t_{l+1}$ is $\tilde{e}_k$ for some $k < l$.

By induction hypothesis, there is a sequence $e_0, \ldots, e_h$ (e_h is $\tilde{e}_k$, as $\tilde{e}_k$ is variable-free) satisfying (1) and (2) of the lemma. Now any instance of $\tilde{e}_l$, e, induces a uniquely corresponding instance of $\tilde{e}_j$, e'. By induction hypothesis, there is a sequence $e_{h+1}, \ldots, e_p$ (e_p is e') satisfying (1) and (2) of the lemma. Then $e_0, \ldots, e_h, e_{h+1}, \ldots, e_p, e$ is what we want.//

Theorem 1   Let $R(\vec{x}_n, y)$ $(n \geq 1)$ be deducible over A in some system of equations E. Then R is formally representable in some EFS.

Proof Let $F = \{f_1, \ldots, f_n\}$ be the set of function symbols of E, $V = \{x_1, \ldots, x_m\}$ be the set of variables of E. Without loss of generality, we assume that $A = \{1, 2, \ldots, m\}$ $(m \geq 1)$ and that f_1 is of degree n and represents R (Definition 6.4.3).

We construct an EFS S_E as follows:

(1)   Its alphabet is $A \cup F \cup \{(,), =, ;\}$

(2)   Its variables-alphabet is $V \cup \{\xi_1, \ldots, \xi_k\}$, where $\{\xi_1, \ldots, \xi_k\} \cap V = \varnothing$ and k is large enough to accommodate (4.4) below.

(3)   It has two predicates P and Q, both of degree 1, and a predicate ***R*** of degree $n + 1$.

(4)   Axioms of S_E:

(4.1) $P\Lambda$ (Λ is the empty string)

(4.2) $Px \rightarrow Pxa$, for all $a \in A$
(Clearly, P represents A^* in S_E.)

(4.3)   Let $\{e_1, \ldots, e_q\}$ be the set of equations in E. For each e_i, which has, say, variables $u_1, \ldots, u_{m_i}$ ($u_j \in V$ for $j = 1, \ldots, m_i$) add the axiom

$$Pu_1 \rightarrow Pu_2 \rightarrow \ldots \rightarrow Pu_{m_i} \rightarrow Q\hat{e}_i$$

where $\hat{e}_i$ is e_i with the only difference being that every occurrence of "," in e_i has been replaced by ";".(*)
(Intuitively, this axiom says that the only "useful" substitutions for variables in $\hat{e}_i$ are by strings in A^*. Indeed, for no other substitutions is it possible to derive $\vdash_{S_E} Q\hat{e}_i$).

(4.4)   For each equation e_i of E, consider the largest set of equations S_i such that $e \in S_i$ iff either e is e_i or e contains at least one F-symbol and for some equation e_j of S_i there is a term $f(t_1, \ldots, t_h)$ in e_j and a variable $u \in V \cup \{\xi_1, \ldots, \xi_k\}$, not in e_j, such that e is obtained from e_j by replacing *one* occurrence of $f(t_1, \ldots, t_h)$ by u.
Under these circumstances, for each e and e_j of S_i $(i = 1, \ldots, q)$ so related add the axiom

$$Pu \rightarrow Pu_1 \rightarrow \ldots \rightarrow Pu_m \rightarrow Qf(t_1; \ldots; t_h) = u \rightarrow Q\hat{e}_j \rightarrow Q\hat{e},$$

where $u, u_1, \ldots, u_m$ are all the variables which occur in e_j and e.

(*)This is done since "," has a different role in S_E, namely to separate the "arguments" of ***R***.

It is now clear, using Lemma 1, that $\vdash_E e$, where e is variable-free, iff $\vdash_{S_E} Q\hat{e}$.

(4.5) Add the axiom $Qf_1(u_1; \ldots; u_n) = u_{n+1} \to Ru_1, \ldots, u_n, u_{n+1}$, where $u_1, \ldots, u_{n+1}$ are variables.

By the above remark,

$\vdash_E f_1(a_1^{(m)}, \ldots, a_n^{(m)}) = a_{n+1}^{(m)}$ iff

$\vdash_{S_E} Qf_1(a_1^{(m)}; \ldots; a_n^{(m)}) = a_{n+1}^{(m)}$ iff $\vdash_{S_E} Ra_1^{(m)}, \ldots, a_n^{(m)}, a_{n+1}^{(m)}$

(It is clear that $\vdash_{S_E} Rt_1, \ldots, t_n, t_{n+1}$ only if $t_i \in A^*$, $i = 1, \ldots, n + 1$). Thus, $R(\vec{x}_n, y)$ is represented in S_E by R.//

Theorem 2   Let S be a semi-Thue system over A. Then T_S is formally representable over A.

Proof See the proof of Theorem 6.3.3.//

Corollary   Let S be a Thue system over A. Then T_S is formally representable over A.

Theorem 3   Let S be a Post system over A. Then T_S is formally representable over A.

Proof Let w_0 be the axiom of S. We construct an EFS T over A, with a single 1-degree predicate P.

Its axioms are

(1) Pw_0

(2) $Pax \to Pxb$ for each instruction (a, b) of S.

It is straightforward to verify that $\vdash_S w$ iff $\vdash_T Pw$; that is, $w \in T_S$ iff $\vdash_T Pw$.//

In preparation for showing that every formally representable relation is re, we present the following lemma.

Lemma 2   Let S be an EFS over A. If w is a variable-free wff in S, then $\vdash_S w$ iff there is a sequence $w_0, w_1, \ldots, w_n$, where w is w_n and for $i = 0, \ldots, n$ w_i is either (1) an instance of an axiom(*) of S or (2) for some w_k, w_j ($k < i, j < i$) w_k is $w_j \to w_i$ and w_j is an af.

Proof Entirely analogous to that of Lemma 1 and is left as an exercise.//

(*)An *instance* of a wff involving the variables $x_1, \ldots, x_m$ is obtained by substituting each occurrence of x_i ($i = 1, \ldots, m$) by a term t_i in A^*.

Note: It is not difficult to see that each w_i in the statement of Lemma 2 is necessarily variable-free.

Let now S be an EFS over $A = \{a_1, \ldots, a_q\}$, with $V = \{x_1, \ldots, x_l\}$ as the set of variables and $\Pi = \{R_1^{(n_1)}, \ldots, R_m^{(n_m)}\}$ as the set of predicates. (The superscript n_i of R_i indicates its degree.) We shall arithmetize the behavior of S. To this end, observe that an instance of a wff is a string over $A \cup \Pi \cup \{,\} \cup \{\rightarrow\}$.

A proof of a variable-free wff w_n can be modified to be (Lemma 2) a sequence of strings $w_0, \ldots, w_n$ over $A \cup \Pi \cup \{,\} \cup \{\rightarrow\}$. By adding the new symbol ";" as a separator between w_i's, we can view a (modified à la Lemma 2) proof as a string "$;w_0;w_1; \ldots ;w_n;$" over $A \cup \Pi \cup \{,\} \cup \{\rightarrow\} \cup \{;\}$. Let us call this last set $\hat{A}$.

Then instances of afs and wffs, as well as proofs in the sense of Lemma 2, are (naturally interpreted to be) integers in $|\hat{A}|$-adic notation.(*) We shall use the above-mentioned symbols of $\hat{A}$, unchanged, to stand as (names of) $|\hat{A}|$-adic digits. In what follows, "$*$" denotes $|\hat{A}|$-adic concatenation.

We next define a sequence of useful predicates, each of which is easily shown to be *strictly $|\hat{A}|$-rudimentary* (see the righthand side of the equivalence "$\equiv$" inside [ ]-brackets).

(1) $A(x) \equiv$ "$x \in A^*$"

$$[A(x) \equiv \neg((;Px) \vee (\rightarrow Px) \vee (,Px) \vee (R_1^{(n_1)}Px) \vee \ldots \vee (R_m^{(n_m)} Px))]$$

(2) $af(x) \equiv$ "x is an *instance* of an *af*"

$$[af(x) \equiv (\exists t_1, \ldots, t_{n_1})_{Px} (x = R_1^{(n_1)}*t_1*,*\ldots*,*t_{n_1} \&$$

$$A(t_1) \& \ldots \& A(t_{n_1})) \vee \ldots \vee (\exists t_1, \ldots, t_{n_m})_{Px}$$

$$(x = R_m^{(n_m)}*t_1*,*\ldots*,*t_{n_m} \& A(t_1) \& \ldots \& A(t_{n_m}))]$$

(3) $a_i(x) \equiv$ "x is an *instance* of the ith axiom of S", where the axioms are $a_1, \ldots, a_h$.
[Noting that a_i is a string over $\hat{A} \cup V$, let $u_1, \ldots, u_{m_i}$ be all the elements of V occurring in a_i. That is, let $a_i = t_1\xi_1 t_2\xi_2 \ldots \xi_l t_{l+1}$, where ξ_i $(i = 1, \ldots, l)$ is some u_j $(j = 1, \ldots, m_i)$, and $t_i \in (\hat{A})^*$ $(i = 1, \ldots, l + 1)$ are fixed strings. Then $a_i(x) \equiv (\exists u_1, \ldots, u_{m_i})_{Px} (x = t_1*\xi_1*t_2*\ldots*\xi_l*t_{l+1} \& A(u_1) \& \ldots \& A(u_{m_i}))$.
E.g. if a_i is $Ru \rightarrow Rv \rightarrow Ruv$, then $a_i(x)$ is

$$(\exists u,v)_{Px} (x = R*u*\rightarrow R*v*\rightarrow R*u*v \& A(u) \& A(v))].$$

(4) $a(x) \equiv$ "x is an *instance* of some axiom"

$$[a(x) \equiv a_1(x) \vee a_2(x) \vee \ldots \vee a_h(x)]$$

(*)An unspecified (but fixed) ordering of $\hat{A}$ is assumed.

(5) $Pf(x) \equiv$ "x is a proof in the sense of Lemma 2"

$[Pf(x) \equiv ;Bx \ \& \ ;Ex \ \& \ \neg \, (;;)Px \ \& \ (\forall u,v)_{Px} \, ((;u;v;)Bx$

$\& \ \neg \, ;Pv \Rightarrow a(v) \vee (\exists y,w)_{Pu} \, \{(;y;)Pu \ \& \ (;w;)Pu \ \&$

$\neg ;Py \ \& \ \neg \, ;Pw \ \& \ af(y) \ \& \ w = y* \!\longrightarrow\! *v\})]$

Theorem 4   If $R \subset \mathcal{N}^n$ is formally representable in some EFS S over A, then R is re.

Proof Without loss of generality (Corollary 1 of Theorem 6.3.4), we assume that $R(\vec{x}_n)$ is formally represented in *unary*, say by predicate $\mathcal{R}$ of degree n, that is,

$$R(\vec{x}_n) \text{ iff } \vdash_S \mathcal{R}1^{x_1+1}, \ldots, 1^{x_n+1}.$$

Let us adopt the "arithmetization" of S developed prior to Theorem 4.
    Then

$R(\vec{x}_n) \equiv (\exists y) \, (Pf(y) \ \& \ (\exists t, t_1, \ldots, t_n)_{Py} \, \{t = \mathcal{R}*t_1*,*\ldots*,* t_n \ \&$

$(;t;)Ey \ \& \ \text{tally}_1(t_1) \ \& \ \ldots \ \& \ \text{tally}_1(t_n) \ \& \ |t_1| = x_1 + 1 \ \& \ |t_2| = x_2 + 1 \ \& \ \ldots \ \&$

$|t_n| = x_n + 1\})$

where $\lambda x. |x|$ is the $|\hat{A}|$-adic length function; also, all concatenations ("$*$") are in $|\hat{A}|$-adic.
    Since the predicate to the right of $(\exists y)$ is rudimentary (hence, recursive) $R(\vec{x}_n)$ is re by the projection theorem.//

Note: The basis of proof of Theorem 4 is "arithmetization" of EFS S. In this respect, the similarity to Smullyan's proof of the theorem is obvious. On the other hand, Smullyan uses the coding scheme of §3.9 and finds it necessary to treat *pure* EFS rather than (full) EFS (after proving that the two are equivalent formalisms. See problem 21).
    Also, even though it is easy to arithmetize all EFS S (and have the code, or Gödel number, of S as a parameter in the discussion), since we do not need to reprove universality results for a third time we do not pursue this avenue (the interested reader can do so. See Problem 22).
    We can now state:

Theorem 5   The NTM, semi-Thue, Thue, Post, Herbrand-Gödel-Kleene, and Smullyan formalisms are all equivalent to the $\mathcal{P}$-formalism of Chapter 3.

We have left the NMA and NPA formalisms out of this discussion. The reader can easily show (Problem 23) that the NTM formalism subsumes these two, through direct simulation. Alternatively, we can accomplish this by easy arithmetizations.

Theorem 6   Let $R \subset \mathcal{N}^n$ be accepted by an NMA M over A, where $\{\#, B, 1\} \subset A$. (See Definition 6.2.3.) Then R is re.

Proof The assumption says that $R(\vec{x}_n)$ iff there is an M-computation starting with $\#1^{x_1+1}B \ldots B1^{x_n+1}\#$.

Let ";" be a symbol not in A, to use as separator of consecutive IDs of an M-computation. Consider $\hat{A} = A \cup \{;\}$. An M-computation can be thought of as a string over $\hat{A}$. Moreover, IDs and computations can be thought of as $(|\hat{A}|$-adic notations of) integers.

First, we observe that the following predicates are *strictly* $|\hat{A}|$-rudimentary (all "arithmetic" below is done in $|\hat{A}|$-adic notation).

(1) $\mathrm{ID}(x) \equiv$ "x is an M-ID" $[\mathrm{ID}(x) \equiv \neg;Px]$

(2) $\mathrm{yield}_i(x,y) \equiv$ "ID x yields ID y according to

   M-instruction number i, where M has the

   instructions $\{(a_i, b_i)\}_{i-1,\ldots,m}$"

   $[\mathrm{yield}_i(x,y) \equiv \mathrm{ID}(x) \ \& \ \mathrm{ID}(y) \ \&$

   $(\exists u,v)_{Px} \ (x = u*a_i*v \ \& \ y = u*b_i*v)$

   see Definition 6.2.2]

(3) $\mathrm{yield}(x,y) \equiv$ "ID x yields ID y according to

   some M-instruction"

   $[\mathrm{yield}(x,y) \equiv \mathrm{yield}_1(x,y) \vee \ldots \vee \mathrm{yield}_m(x,y)]$

(4) $\mathrm{Fin}(x) \equiv$ "x is a final ID"

   $[\mathrm{Fin}(x) \equiv \mathrm{ID}(x) \ \& \ \neg (a_1Px \vee a_2Px \vee \ldots \vee a_nPx)]$

(5) $\mathrm{Comp}(y) \equiv$ "y is a computation of M"

   $[;By \ \& \ ;Ey \ \& \ (\forall u,v)_{Py} \ ((;u;v;)Py \ \& \ \neg;Pu$

   $\& \ \neg;Pv \Rightarrow \mathrm{yield}(u,v)) \ \& \ (\exists v)_{Py} \ \{(;v;)Ey \ \& \ \mathrm{Fin}(v)\}]$

Next, observe that

$$R(\vec{x}_n) \equiv (\exists y) \ \{\mathrm{Comp}(y) \ \& \ (\exists x)_{Py} \ ((;x;)By \ \& \ \mathrm{ID}(x) \ \&$$

$$(\exists t_1, \ldots, t_n)_{Px} \ [x = \#*t_1*B*\ldots*B*t_n*\# \ \& \ \mathrm{tally}_1(t_1) \ \& \ldots \&$$

$$\mathrm{tally}_1(t_n) \ \& \ |t_1| = x_1 + 1 \ \& \ldots \& \ |t_n| = x_n + 1])\}$$

Since the predicate to the right of $(\exists y)$ is clearly rudimentary, hence recursive, $R(\vec{x}_n)$ is re by the projection theorem.//

Theorem 7 Let $R \subset \mathcal{N}^n$ be accepted by an NPA M over A, where $\{\#, B, 1\} \subset A$. (See Definition 6.3.3). Then R is re.

Proof Entirely analogous to that of Theorem 6.//

§6.6 WORD PROBLEMS

By *word problems*, we understand questions of the form "is such and such a word transformable to such and such a word, given a set of transformation rules?" This question can be made precise in the context of NMAs and NPAs, semi-Thue, Thue and Post systems.

Definition 1 Given an NMA over A (or NPA, semi-Thue system, Thue system, Post system) and two words w and u in A^*, we say that w is *transformable* into u iff either (a) $w = u$ or (b) there is a sequence of words in A^*, $s_0, s_1, \ldots, s_n$ $(n \geq 1)$ where $w = s_0, u = s_n$ and for $i = 0, \ldots, n - 1$ $s_i \rightarrow s_{i+1}$ in the NMA (or NPA, semi-Thue system, Thue system, Post system). We shall use the symbol $w \rightarrow * u$ just in case w is transformable to u.//

Definition 2 The *word problem* for NMAs (or NPAs, semi-Thue systems, Thue systems, Post systems) is the problem: "Given an NMA M (resp. NPA, etc.) and two arbitrary words w and u is it true that $w \rightarrow * u$?"//

Definition 3 The *decision problem* for semi-Thue, Thue, and Post systems as well as for EFS is "given such a system and an arbitrary word w, is it true that $\vdash w$ in the system?"//

For any given alphabet A, words in A^* are naturally interpreted as numbers in $|A|$-adic notation (an unspecified, but fixed in any given discussion, ordering of A is assumed). In the case of an EFS over A, with Π as the predicate set, a wff w is viewed as a number written in $|B|$-adic notation, where $B = A \cup \Pi \cup \{,\} \cup \{\rightarrow\}$.

Thus, whenever we are talking about such and such a word-problem (resp. decision-problem) being or not being *recursively solvable*, we are referring to the recursiveness or not recursiveness of sets such as $\{(w, u) \in \mathcal{N}^2 \,|\, w^{(|A|)} \rightarrow * u^{(|A|)}\}$ (resp. $\{w \in \mathcal{N} \,|\, \vdash w^{(|B|)}\}$).

Clearly, recursive solvability as defined above is tantamount to saying that the problem is "solvable by a TM", in the sense that the TM, given input

$x \in A^*$, $y \in A^*$ (resp. $x \in B^*$) will reply *yes* just in case $x \longrightarrow * y$ (resp. $\vdash x$); it will reply *no* otherwise.

Theorem 1   There is an NMA M which has a recursively unsolvable word-problem.

> *Proof* We know that $K = \{x \,|\, \phi_x(x)\!\downarrow\}$ is re but not recursive (Examples 3.7.1 and 3.7.3).
>
> By Theorem 6.2.1, there is an NMA over some alphabet A with the behavior $\#1^{x+1}\# \longrightarrow * \$\vdash q_f\dashv \#$ iff $x \in K$.(*)
>
> Define $S(w)$ by $w^{(|A|)} \xrightarrow[M]{} * \$\vdash q_f\dashv\#$; we claim that $S \notin \mathcal{R}_*$.
>
> Let instead $S \in \mathcal{R}_*$.
>
> Define $g \colon \mathcal{N} \to \mathcal{N}$ by $g(x)^{(|A|)} = \#1^{x+1}\#$ for all x. Clearly, $g \in \mathcal{PR}$, hence $S(g(x)) \in \mathcal{R}_*$. But $S(g(x)) \equiv x \in K$; contradiction.
>
> Finally, $G = \{(w,u) \in \mathcal{N}^2 \,|\, w^{(|A|)} \longrightarrow * u^{(|A|)}\} \notin \mathcal{R}_*$ since S is an explicit transform of G.//

Corollary 1   The set $G = \{(M,w,u) \in \mathcal{N}^3 \,|\, M^{(|A \cup \{,\} \cup \{(),;\}|)}$ is an NMA(†) over A and $w^{(|A|)} \longrightarrow * u^{(|A|)}$ according to this NMA$\}$ is not recursive.

> *Proof* If $G \in \mathcal{R}_*$ then so is S of Theorem 1, since the latter is an explicit transform of the former.//

Note: This is the "general" word-problem of NMAs, where the NMA itself is a parameter of the problem.

Corollary 2   The decision problem of semi-Thue systems is recursively unsolvable.

> *Proof* See Corollary of Theorem 6.2.2.//

Corollary 3   The decision problem for Thue systems is recursively unsolvable.

> *Proof* By Theorem 6.2.3, there is a Thue system S over an appropriate A so that $x \in K$ iff $\vdash_S \$q_0 1^{x+1}\#$. Define $f \colon \mathcal{N} \to \mathcal{N}$ by $f(x)^{(|A|)} = \$q_0 1^{x+1}\#$ for all x. It is easily seen that $f \in \mathcal{PR}$.
>
> Hence, if $C = \{w \in \mathcal{N} \,|\, \vdash_S w^{(|A|)}\} \in \mathcal{R}_*$, then so is $C(f(w))$. But the latter is equivalent to $w \in K$.

(*)The simulated NTM is assumed to satisfy Proposition 6.1.2.

(†)An NMA over A can be viewed as a string $(a_1,b_1); \ldots ; (a_n,b_n)$ over $A \cup \{,\} \cup \{(),;\}$, where $(a_i,b_i) \in (A^*)^2$, $i = 1, \ldots, n$.

Finally, $C \notin \mathcal{R}_*$ implies the (general) decision problem being unsolvable.//

Note: Intuitively, what we have done in the proof of Theorem 1 and Corollary 3 was to say "If I had an algorithmic(*) solution for problem A, then, since I could effectively transform any instance of problem B to an instance of problem A (this was the role of the functions g and f), this would result to an algorithmic solution for problem B.

But I know that there is no algorithmic solution for problem B (this is "$x \in K$" in the proof of Theorem 1 and Corollary 3); hence A has no algorithmic solution either."

This is the technique of "reducing problem B to problem A". Reducibility among "problems" will be discussed in Chapter 9.

Corollary 3 provides an interesting unsolvability result in Algebra. We shall discuss this briefly now.

A *semigroup* A is a pair $(A, \circ)$, where A is some set (the *underlying* or *carrier* set) and $\circ$ is a *total* function $\circ : A \times A \rightarrow A$. Moreover, $\circ$ satisfies for all a, b, c in A the relation

$$\circ(a, \circ(b, c)) = \circ(\circ(a, b), c) \text{ ("associativity" of } \circ)$$

or, in traditional "infix" notation [where $\circ(x, y)$ is denoted by $x \circ y$],

$$a \circ (b \circ c) = (a \circ b) \circ c$$

Clearly, associativity allows us to write "$a_1 \circ a_2 \circ \ldots \circ a_n$" unambiguously without brackets. Let now $B \subset A$ and let $\circ | B$ have its range in B (we say that B is *closed* under $\circ$). Then clearly $\circ | B$ is associative; we call $(B, \circ | B)$ a *subsemigroup* of A. Usually, one writes $\circ$ instead of $\circ | B$.

Next let $C \subset A$ be a subset of A. By gen(C) we denote the *smallest* subsemigroup of A containing C [that is, gen$(C) = \bigcap_{B \text{ subsemigroup of } A} B$]. We call gen$(C)$ the semigroup *generated* by C. It is easy to verify that gen(C) has as underlying set the set $\bigcup_{i=1}^{\infty} C_i$, where $C_1 = C$ and C_i $(i > 1)$ is $\{a_1 \circ a_2 \circ \ldots \circ a_i | a_j \in C, j = 1, \ldots, i\}$.

Following the usual practice of "implied-times notation" and treating $\circ$ as an abstract "times" operation, we see that gen(C) is "structurally identical"(†) to $(C^+, *)$, where "$*$" is concatenation of strings. Thus, in what follows, we shall view gen(C) as $(C^+, *)$. (Note that we are not bound by the original A anymore.)

(*)"Algorithmic solution" is a term we use here because of its intuitive connotations. We use it as a *synonym* for (e.g.) "TM obtainable solution" and certainly we do *not* imply that there is a *provable* equivalence between the two terms.

(†)"Isomorphic" is the technical term, but we shall not get too technical in this brief discussion. The interested reader may consult, for example, Kurosh [17].

Next, we introduce a finite set of pairs $\{(x_1,y_1), \ldots, (x_n,y_n)\}$ where $\{x_i,y_i\} \subset C^+$, $i = 1, \ldots, n$, which we call *relations* on $\text{gen}(C)$. Intuitively, we may view (x_i,y_i) as saying "$x_i = y_i$".

Using these relations one may transform an element a of C^+ to some other element of C^+, b, by repeatedly applying the rule "replace products by equal products" (where any $x \in C^+$ we call a "product" of its component elements from C). More technically, whereas "$=$" on $\text{gen}(C)$ is simply equality of strings in C^+, for $\text{gen}(C)$ *equipped* with the relations $R = \{(x_1,y_1), \ldots, (x_n,y_n)\}$ we define: $a \doteq b$ on $(\text{gen}(C),R)$ iff there is a sequence $a_0,a_1,\ldots,a_l$ ($l \geq 0$) of elements of C^+, where $a = a_0$, $b = a_l$ and (if $l > 0$) for $i = 0,1,\ldots,l-1$ either $a_i = ux_mv$ and $a_{i+1} = uy_mv$ or $a_i = uy_mv$ and $a_{i+1} = ux_mv$ for some $\{u,v\} \subset C^*$ and $1 \leq m \leq n$.

We shall call *word problem* of the semigroup $(\text{gen}(C),R)$ the problem "$a \doteq b$?" for arbitrary $\{a,b\} \subset C^+$. We can immediately state:

Corollary 4   There is some semigroup, generated by a finite set and equipped with relations, which has an unsolvable word problem.

Proof Consider the Thue system S over A of Corollary 3. Let R be the set of instructions of this Thue system.

Consider the semigroup $G = (\text{gen}(A),R)$ (carrier set A^+, operation concatenation).

Clearly, if $\{(a,b) \in \mathcal{N}^2 \,|\, a^{(|A|)} \doteq b^{(|A|)}$ in $G\}$ is recursive then so is $\{a \in \mathcal{N} \,|\, a^{(|A|)} \doteq \$ \vdash q_f \dashv \#$ in $G\}$ as the latter is an explicit transform of the former (for an explanation of $\$ \vdash q_f \dashv \#$, see Theorem 6.2.3). But this last set is $\{a \in \mathcal{N} \,|\, \vdash_S a^{(|A|)}\}$.//

Note: The unsolvability of the word problem for semigroups was proved by Post [21] and independently by Markov [20].

Theorem 2   The word problem for NPAs is recursively unsolvable.

Proof Problem 25.//

Corollary   The decision problem for Post systems is recursively unsolvable.

Proof Problem 26.//

Theorem 3   The decision problem for EFS is recursively unsolvable.

Proof Let S be an EFS over A which represents K in unary notation (say, by the predicate P of degree 1); that is, $x \in K$ iff $\vdash_S P1^{x+1}$.

Set $W(w) \equiv w \in \mathcal{N}$ & $\vdash_S w^{(|B|)}$, where $B = A \cup \Pi \cup \{\rightarrow\} \cup \{,\}$. If $W(w)$

$\in \mathcal{R}_*$, then so is $W(h(x))$, where $h: \mathcal{N} \to \mathcal{N}$ is the primitive recursive function defined by $h(x)^{(|B|)} = P1^{x+1}$ for all x in $\mathcal{N}$. But $W(h(x)) \equiv x \in K$, hence $W(w) \notin \mathcal{R}_*$.

It follows that the (general) decision problem for EFS is also unsolvable; i.e., $\{(S,w) \in \mathcal{N}^2 \mid S^{(|B \cup \{;\}|)}$ is an EFS over A and $\vdash w^{(|B|)}$ in this EFS$\} \notin \mathcal{R}_*.//$

§6.7 THE POST CORRESPONDENCE PROBLEM

The following problem, posed and proved recursively unsolvable by Post [23] has useful applications in Formal Language theory, where it is "reduced" to certain problems thus showing them to be recursively unsolvable.

Definition 1   The *Post Correspondence Problem* (PCP):

Given an alphabet A and a set of ordered pairs $R = \{(a_1, b_1), \ldots, (a_n, b_n)\}$, where $\{a_i, b_i\} \subset A^*, i = 1, \ldots, n$.

The PCP of R is the question of whether or not there is a sequence $(i_1, \ldots, i_k)$, where $1 \leq i_j \leq n$ for $j = 1, \ldots, k$, such that $a_{i_1} a_{i_2} \ldots a_{i_k} = b_{i_1} b_{i_2} \ldots b_{i_k}$. A sequence such as $(i_1, \ldots, i_k)$ is a *solution* of the PCP of $R.//$

Example 1   Let $R = \{(a_i, b_i)\}_{i=1,\ldots,n}$ over A, where $|a_i| < |b_i|$ for $i = 1, \ldots, n$. Since, clearly, $|a_{i_1} \ldots a_{i_k}| < |b_{i_1} \ldots b_{i_k}|$ for any sequence $(i_1, \ldots, i_k)$, the PCP of R has no solution. It is solvable though as a problem (!) in the sense that given R we answered its PCP negatively.$//$

Example 2   Let $R = (a_i, b_i)_{i=1,\ldots,n}$ over $A = \{a\}$.

Then $a_{i_1} \ldots a_{i_k} = b_{i_1} \ldots b_{i_k}$ for some $(i_1, \ldots, i_k)$ iff there is a solution $(x_1, \ldots, x_n) \in \mathcal{N}^n$ to the equation $x_1|a_1| + x_2|a_2| + \ldots + x_n|a_n| = x_1|b_1| + x_2|b_2| + \ldots + x_n|b_n|$, such that $x_1 + x_2 + \ldots + x_n > 0$. This last *yes/no* problem being solvable by well-known algorithmic methods (see, for example, LeVeque [18]), we infer that the PCP problem over a one-symbol alphabet is "algorithmically solvable"; i.e., there is, say, a TM which when presented with R as input it eventually halts and prints *yes* if the PCP of R has a solution, it prints *no* otherwise.$//$

Definition 2   Let A be a given alphabet. The *PCP class over A* is the set

$$C_A = \{(x,y) \in ((A \cup \{\mathcal{c}\})^*)^2 \mid \mathcal{c} \notin A \ \& \ x = a_1 \mathcal{c} \ldots \mathcal{c} a_n \ \&$$

$$y = b_1 \mathcal{c} \ldots \mathcal{c} b_n \ \& \ R = \{(a_i, b_i)\}_{i=1,\ldots,n} \subset (A^*)^2 \ \&$$

$$\text{the PCP of } R \text{ has a solution}\}$$

We view C_A as a subset of $\mathcal{N}^2$, where integers are represented in $|A \cup \{\mathfrak{c}\}|$-adic notation. We then say that the *PCP over A* is *recursively solvable* (resp. *unsolvable*) whenever $C_A \in \mathcal{R}_*$ (resp. $C_A \notin \mathcal{R}_*$).//

We have seen that the PCP over A is recursively solvable (i.e., $C_A \in \mathcal{R}_*$) if $|A| = 1$. We shall see that $C_A \notin \mathcal{R}_*$ if $|A| > 1$.

Lemma 1   For any re set S, there is an NMA M such that

$$x \in S \text{ iff } \#1^{x+1}\# \xrightarrow[M]{} * \Lambda.$$

Proof To the NMA of the proof of Theorem 6.2.1, add the instruction $(\$ \vdash q_f \dashv \#, \Lambda)$, assuming that the simulated NTM satisfies Proposition 6.1.2.//

Lemma 2   The special word problem "$w \rightarrow * \Lambda$" for NMAs is recursively unsolvable.

Proof Amend the proof of Theorem 6.6.1, using Lemma 1.//

Lemma 3   Let $1 \le m \le n$. Then the function $f_{m,n} \colon \mathcal{N} \rightarrow \mathcal{N}$ defined by $(f_{m,n}(x))^{(n)} = x^{(m)}$ for all x [in English: "x and $f_{m,n}(x)$, when the former is represented in m-adic and the latter in n-adic, are identical strings over $\{1, \ldots, m\}$"] is primitive recursive, regardless of the particular orderings chosen for the alphabets of digits $\{1, \ldots, m\}$ and $\{1, \ldots, n\}$.

Proof Let for $i = 1, \ldots, m$ $p(i)$ be i's position (value) in the alphabet $\{1, \ldots, n\}$ according to the adopted lexicographic ordering of $\{1, \ldots, n\}$ (without loss of generality, the lexicographic ordering of the alphabet $\{1, \ldots, m\}$ is $1 < 2 < \ldots < m$).
Then $f_{m,n}$ is defined for all $x \ge 0$ by the following course-of-values recursion:

$$\text{for } i = 1, \ldots, m \begin{cases} f_{m,n}(0) = 0 \\ f_{m,n}(mx + i) = nf_{m,n}(x) + p(i).// \end{cases}$$

Example 3   Let $n = m = 2$, but let the lexicographic ordering in the $\{1, \ldots, n\}$ alphabet be $2 < 1$ (that is, $p(1) = 2$, $p(2) = 1$).

Then

$$f_{2,2}(0) = 0$$

$$f_{2,2}(2x + 1) = 2f_{2,2}(x) + 2$$

$$f_{2,2}(2x + 2) = 2f_{2,2}(x) + 1$$

For example, $4^{(m)}$ is "12". On the other hand, "12" in the n-system represents $2 \cdot 2 + 1 = 5$ since "1" has value (lexicographic position) 2 and "2" has value 1. The reader can readily verify that indeed $f_{2,2}(4) = 5$.//

Theorem 1   Let M be an NMA over A, and let $w \in A^{+}$. Then there is a set of pairs $R \subset ((A \cup \{\perp, \uparrow\})^{*})^{2}$, where $\{\perp, \uparrow\} \cap A = \varnothing$, such that the PCP of R has a solution iff $w \xrightarrow[M]{} * \Lambda$.

Proof   Let $\{(a_1, b_1), \ldots, (a_n, b_n)\}$ be the set of instructions of M, where $\{a_i, b_i\} \subset A^{*}$ for $i = 1, \ldots, n$. R contains exactly the following pairs:

(1) $(\uparrow \perp \sigma_1 \perp \ldots \perp \sigma_k \perp \uparrow, \uparrow \perp)$, where $w = \sigma_1 \sigma_2 \ldots \sigma_k$ and $\sigma_i \in A$ for $i = 1, \ldots, k$.

(2)   For each (a_i, b_i), $i = 1, \ldots, n$, if $a_i = t_1 t_2 \ldots t_l$ and $b_i = s_1 s_2 \ldots s_q \neq \Lambda$, where each t_j and s_m are in A,

   (a) $(\perp s_1 \perp s_2 \perp \ldots \perp s_q, t_1 \perp t_2 \ldots \perp t_l \perp)$ and

   (b) $(\perp s_1 \perp s_2 \perp \ldots \perp s_q \perp \uparrow, t_1 \perp t_2 \perp \ldots \perp t_l \perp \uparrow \perp)$.
   If a_i is as above, but $b_i = \Lambda$, then

   (c) $(\perp \uparrow, t_1 \perp t_2 \perp \ldots \perp t_l \perp \uparrow \perp \uparrow)$.

(3)   For each $\sigma \in A$, $(\perp \sigma, \sigma \perp)$ and $(\perp \sigma \perp \uparrow, \sigma \perp \uparrow \perp)$.

For convenience, let us call $\{(x_i, y_i)\}_{i \in I}$ the set of pairs R so constructed. Let now $w_0 \xrightarrow[M]{} w_1 \xrightarrow[M]{} \ldots \xrightarrow[M]{} w_n$, where $w_0 = w$ and $w_n = \Lambda$. This computation can also be represented as a string over $A \cup \{\perp, \uparrow\}$, namely $\uparrow \hat{w}_0 \uparrow \hat{w}_1 \uparrow \ldots \uparrow \hat{w}_n \uparrow$, where for any u, $\hat{u}$ is defined inductively on $A \cup \{\perp, \uparrow\}$ by $\hat{\Lambda} = \perp$ and if $a \in A$ and $u \in A^{*}$ then $(\hat{ua}) = \hat{u}a \perp$.

Under these circumstances, the PCP of R is easily seen to have a solution $(i_1, \ldots, i_m)$ such that $x_{i_1} \ldots x_{i_m} = \uparrow \hat{w}_0 \uparrow \hat{w}_1 \uparrow \ldots \uparrow \hat{w}_n \uparrow$.

Conversely, let there be a solution $(i_1, \ldots, i_m)$. Now the only pairs (of R) that start (resp. end) with the same symbol are (1) [resp. (2) (c)].

Hence, $x_{i_1} x_{i_2} \ldots x_{i_m} = \uparrow \perp \sigma_1 \perp \ldots \perp \sigma_k \perp \uparrow w_1 \uparrow w_2 \uparrow \ldots \uparrow w_h \uparrow$ for some h, and $w_i \in (A \cup \{\perp\})^{*}$, $i = 1, \ldots, h$.

Set $w_0 = \perp \sigma_1 \perp \ldots \perp \sigma_k \perp$. We show, by induction on l, that if $w_i \neq \Lambda$ for $i = 0, \ldots, l + 1$, then for some $\{a, b\} \subset A^{*}$, $w_l = \hat{a}$, $w_{l+1} = \hat{b}$, $a \xrightarrow[M]{} * b$, and moreover for some $u \neq \Lambda$, either $w_{l+1} = \perp$ and $(\perp \uparrow, u \uparrow \perp \uparrow) \in R$, or $w_l = \perp u$ and

$w_{l+1}\uparrow$ corresponds to $u\uparrow\perp$ in a partial solution $(i_1, \ldots, i_q)$ of $(i_1, \ldots, i_m)$.(*)
We shall call x-*part* (resp. y-*part*) of the partial solution the string $x_{i_1} \ldots x_{i_q}$
(resp. $y_{i_1} \ldots y_{i_q}$).
$l = 0$: In this case, necessarily $w_{l+1} = w_1 \neq \Lambda$ (why?).

The x-part:
$$\uparrow \underbrace{\perp \sigma_1 \perp \ldots \perp \sigma_k \perp}_{w_0} \uparrow w_1 \uparrow$$

The y-part: $\uparrow \perp \phi$, where ϕ corresponds to $w_1 \uparrow$

Clearly, $\sigma_1 \perp \ldots \perp \sigma_k \perp \uparrow$ begins ϕ. If an instance of the third case of pairs of type
(2) has been applied, then it must be $(\perp\uparrow, \sigma_1\perp \ldots \perp\sigma_k\perp\uparrow\perp\uparrow) \in R$. Thus,
$\sigma_1 \ldots \sigma_k \xrightarrow{M} \Lambda$, $w_1 = \perp$, $(\perp\uparrow, u\uparrow\perp\uparrow) \in R$, where $w_0 = \perp u$.

Otherwise, a mix of pairs (2) and (3) [excluding case (c) of (2)] has been
used, hence,

(1) ϕ ends with $\uparrow\perp$; i.e., $\phi = u\uparrow\perp$ $(u \neq \Lambda)$, since "$\uparrow$" of $w_1\uparrow$ forces an
instance of (2)(b) or second case of (3) to be used. So $w_1\uparrow$ corresponds to
$u\uparrow\perp$.

(2)   If $\hat{a} = w_0$ and $\hat{b} = w_1$, then clearly $a \xrightarrow{M} * b$ (Note that we do not use
"$\rightarrow$", since it may be $w_0 = w_1$; i.e., only pairs (3) were used, or pairs (2)
were used more than once).

Let, for $l \leq k$, what we set out to prove be true.

$l = k + 1$: x-part: $\uparrow w_0 \uparrow \ldots \uparrow w_{l-1} \uparrow w_l \uparrow w_{l+1} \uparrow$

y-part: $\uparrow w_0 \uparrow \ldots \uparrow w_{l-1} \uparrow \perp \phi$

Since $w_i \neq \Lambda$ for $i = 0, \ldots, l + 1$, $w_l \neq \perp$ (otherwise, by induction hypothesis,
for some $u \neq \Lambda$ $(\perp\uparrow, u\uparrow\perp\uparrow) \in R$. Hence, $w_{l+1} = \Lambda$, since only pair (1) can
continue to the right of $w_l\uparrow$).

Hence, for some $u \neq \Lambda$, $w_l\uparrow$ corresponds to $u\uparrow\perp$ (where $w_{l-1} = \perp u$).
Clearly then, ϕ corresponds to $w_{l+1}\uparrow$ and the argument concludes as in the case
$l = 0$.

Now, for some l, w_l is $\perp$ (certainly w_h is); let l_0 be the smallest such l.
Hence for $i = 0, \ldots, l$ $w_i \neq \Lambda$ and the above induction results in $\sigma_1 \ldots \sigma_k$
$\rightarrow *\Lambda$.//

Note: A similar construction for R (but based on "Post Machines" rather
than NMAs) is given in Scott [25] and Manna [19].

(*)a corresponds to b in a partial solution $(i_1, \ldots, i_q)$ iff (1) $(i_1, \ldots, i_q)$ is an initial
segment of $(i_1, \ldots, i_m)$ and (2) For some $p \leq q$, $a = x_{i_p} x_{i_{p+1}} \ldots x_{i_q}$ and
$b = y_{i_p} y_{i_{p+1}} \ldots y_{i_q}$.

Corollary 1   There is an alphabet A, such that $C_A \notin \mathcal{R}_*$.

Proof Let M be an NMA over B such that the predicate $S(x) \overset{\text{def}}{\equiv} x \in \mathcal{N}$ & $x^{(|B|)} \xrightarrow[M]{} * \Lambda$ is not in $\mathcal{R}_*$ (Lemma 2).

Define $T(x)$ by $x \in \mathcal{N}$ & $x^{(|A|)} \xrightarrow[M]{} * \Lambda$, where $A = B \cup \{\bot, \uparrow, \mathcal{c}\}$ and $B \cap \{\bot, \uparrow, \mathcal{c}\} = \varnothing$.

Clearly, $S(x) \equiv T(f_{|B|,|A|}(x))$, hence $T(x) \notin \mathcal{R}_*$ (Lemma 3). Now if $C_A(x,y) \in \mathcal{R}_*$ then so is $C(x)$ obtained from $C_A(x,y)$ by substituting for y the number whose $|A|$-adic notation is $y_1 \mathcal{c} y_2 \mathcal{c} \ldots \mathcal{c} y_n$, where R is as in Theorem 1 and (x_1, y_1) is pair (1).

Define: $h: \mathcal{N} \to \mathcal{N}$ for all x by

$$h(0) = \bot$$

for $i = 1, \ldots, |A|$ $h(x*i) = h(x)*i*\bot$, where concatenation "$*$" is in $|A|$-adic.

Clearly, $h \in \mathcal{PR}$. Set $g = \lambda x. \uparrow * h(x) * \uparrow * c$, where $c^{(|A|)} = \mathcal{c} x_2 \mathcal{c} \ldots \mathcal{c} x_n$.

Now $C(x) \in \mathcal{R}_* \Rightarrow C(g(x)) \in \mathcal{R}_*$. But $C(g(x))$ & $x^{(|A|)} \in B^* \equiv T(x)(*)$; a contradiction.//

Note: Informally, what we did in the above proof was to say "If a TM N solves (yes/no) the problem "$(x,y) \in C_A$?" then the special word problem for an NMA M over A is solved as follows: Given x (possible input for the NMA), a TM L builds R (of Theorem 1). Then, using N as a "subroutine", it decides the PCP of R, hence also the question $x \xrightarrow[M]{} * \Lambda$".

Now that the reader has seen a number of formal reduction arguments we shall feel free, whenever convenient, to present reduction arguments informally and leave the formal details as an easy exercise.

Corollary 2 $C_A \notin \mathcal{R}_*$ for any A with $|A| > 1$.

Proof It suffices to show this for $|A| = 2$.

Let $\{a_1, \ldots, a_q\}$ be the elements of an alphabet B such that $C_B \notin \mathcal{R}_*$. Let $A = \{a, b\}$. Code each a_i of B by $\underbrace{aa \ldots a}_{i}b$. The result follows.//

§6.8 FORMAL LANGUAGES

Given an alphabet A, a *Language* over A is any set $L \subset A^*$. Naturally we are interested in languages definable by finitary means (TMs, Thue systems, Post systems, to name a few examples, are certainly "finitary means").

(*)$x^{(|A|)} \in B^* \equiv \neg((\bot Px) \vee (\uparrow Px) \vee (\mathcal{c} Px))$, hence it is in $\mathcal{R}_*$.

The linguist Noam Chomsky suggested the following classification of sets definable by semi-Thue systems [5]. Such sets are widely referred to as *Formal Languages.*

Central to the classification is the notion of *grammar*; a grammar is essentially a semi-Thue system.

Definition 1   A *grammar* G is a 4-tuple (V_N, V_T, P, S), where V_N is a finite set, the set of *nonterminals* or *variables,* V_T is a finite set, the set of *terminals.* It is required that $V_N \cap V_T = \varnothing$. $S \in V_N$ is a distinguished symbol, the *start*-symbol. P is a set of semi-Thue instructions over $V_N \cup V_T$, usually called *productions* (or *rewriting rules*) in this context.

Thus, G is a semi-Thue system (S, P) over $V_N \cup V_T$, where S is the axiom and P the associated NMA in the sense of Definition 6.2.5.

The *(Formal) Language defined* or *generated* by G is denoted by $L(G)$ and is $\{w \in V_T^* \mid \vdash_G w\}.//$

Note: Usually, if $(x, y) \in P$ one writes "$x \to y$". "$\Rightarrow$" is the symbol for the yield relation in this context; "$\Rightarrow_+$" (resp. "$\Rightarrow_*$") denotes *one* (resp. *zero*) or *more* applications of "$\Rightarrow$". Thus, $\vdash_G w$ is synonymous to $S \Rightarrow_* w$; if w is understood to be in V_T^* then $S \Rightarrow_+ w$ is the same as $\vdash_G w$. In the context of grammars, a sequence $\sigma_0 \Rightarrow \sigma_1 \Rightarrow \ldots \Rightarrow \sigma_n$ is a *derivation* of σ_n which starts with σ_0. Note that a derivation that starts with S is a proof in the semi-Thue system (S, P).

Definition 2   [5] A *type 0 grammar* is one that satisfies Definition 1, with no other restrictions except those already mentioned in Definition 1.

A grammar is of *type 1* iff for every $(x, y) \in P, |x| \le |y|$, where $|\ |$ is the length function on $(V_N \cup V_T)^*$.

A grammar is of *type 2* (also known as *context free grammar*—in short, CFG) iff for *every* $(x, y) \in P, |x| \le |y|$ and $x \in V_N$.(*)

A grammar is of *type 3* (or *regular*) iff for every $(x, y) \in P$, $x \in V_N$ and $y = aw$, where $a \in V_T$ and $w \in V_N \cup \{\Lambda\}$.

A *language* $L \subset A^*$ is of *type i* $(i = 0, 1, 2, 3)$ iff there is a type i grammar G such that $L = L(G)$. Two grammars G and G' are *equivalent* iff $L(G) = L(G').//$

Note: It is immediate from the above definition that a grammar (language) of type i is also of type $(i - 1)$ for $i = 1, 2, 3$.

The type 1 grammars are also known as *context sensitive* (CSG), since every type 1 grammar is equivalent to one whose productions are of the form $\alpha A \beta \to \alpha x \beta$, where $A \in V_N, \{\alpha, \beta\} \subset (V_N \cup V_T)^*$ and $x \in (V_N \cup V_T)^+$ (Problem 28.) The name derives from the fact that in order to "rewrite" A by x, the "context" has to be α (left) and β (right). Now, the term *context free* is clearly

---

(*)The restriction $x \in V_N$ can be replaced by $|x| = 1$; that is, $x \in V_N \cup V_T$. (See Problem 27.)

derived from the fact that in CFGs a production $A \to x$ is applicable to a string $w \in (V_N \cup V_T)^+$ just in case A is part of w; the "context" around A is immaterial.

As a matter of notational convenience we shall use the symbols CSL and CFL for the class of all *context sensitive* (i.e., type 1) languages and *context free* languages, respectively. "Type 0 L" will denote the class of all *type 0 languages* and REGL the class of all *regular languages*.

Our first comment in this note says that REGL $\subset$ CFL $\subset$ CSL $\subset$ Type 0 L. In fact, the inclusions are proper. We refer to the set of these four classes of languages as the "Chomsky hierarchy".

Example 1 Let $V_T = \{0,1\}$, $V_N = \{\langle \text{NUMBER} \rangle\, (*)\}$ and let the productions be

$$\langle \text{NUMBER} \rangle \to 0,\ \langle \text{NUMBER} \rangle \to 1,\ \langle \text{NUMBER} \rangle \to 0 \langle \text{NUMBER} \rangle,$$
$$\langle \text{NUMBER} \rangle \to 1 \langle \text{NUMBER} \rangle$$

where $\langle \text{NUMBER} \rangle$ is the start symbol.

In BNF, one frequently uses the *metasymbol* " | " to combine productions with the same lefthand side. So, " | " separates alternatives and is usually read as "or". We call it metasymbol since $| \notin V_N \cup V_T$. It is used in discussions *about* CFGs. Also, "::=" is often used in place of "$\to$" in BNF notation.

The grammar we just defined is clearly type 3, and it generates V_T^+. It is compactly written in BNF as

$$\langle \text{NUMBER} \rangle ::= 0\,|\,1\,|\,0\langle \text{NUMBER} \rangle\,|\,1\langle \text{NUMBER} \rangle$$

We can compactify the notation further: Introduce a new nonterminal, $\langle \text{BIT} \rangle$. Amend the grammar to:

$$\langle \text{BIT} \rangle ::= 0\,|\,1$$

$$\langle \text{NUMBER} \rangle ::= \langle \text{BIT} \rangle\,|\,\langle \text{BIT} \rangle \langle \text{NUMBER} \rangle$$

(One can appreciate this compactification more, if a V_T such as, e.g., $\{0,1,2,3,4,5,6,7,8,9\}$ is imagined.)

Clearly, the new grammar also generates V_T^+. However, even though it is a CFG, it is *not* regular.//

Example 2 Let $G = (V_N, V_T, P, S)$, where $V_T = \{a,b\}$, $V_N = \{S\}$ and P contains $S \to aSb$ and $S \to ab$.

It is easy to verify that $L(G) = \{a^n b^n \,|\, n \geq 1\}$, where x^n is $\overbrace{xx \ldots x}^{n}$, as usual. This verification is left as an exercise (*Hint:* (1) Show $L(G) \subset \{a^n b^n \,|\, n \geq 1\}$ by

(*)In BNF (Backus Naur- or Backus Normal-Form) notation for CFG's, we are free to use any multiple-symbol string as a nonterminal; in return, every nonterminal is enclosed in $\langle\ \rangle$-brackets.

induction on the length of the derivation of a string in $L(G)$; (2) show $\{a^n b^n \mid n \geq 1\} \subset L(G)$ by induction on n). Clearly (by Definition 3), $\{a^n b^n \mid n \geq 1\}$ is in CFL. One can show that it is *not* regular, i.e., that there is *no* regular grammar G' such that $L(G') = \{a^n b^n \mid n \geq 1\}$.

This is the usual language L one employs to show that CFL $-$ REGL $\neq \emptyset$.//

Example 3   The language $L = \{a^n b^n c^n \mid n \geq 1\}$ is the standard example which establishes CSL $-$ CFL $\neq \emptyset$. We show $L \in$ CSL in this example and delegate $L \notin$ CFL to the problems (Problem 40).

To this end, take $V_T = \{a,b,c\}$, $V_N = \{S,B,C\}$, where S is the *start* symbol. The productions are:

$$S \to aBC$$
$$S \to aSBC$$
$$CB \to BC$$
$$aB \to ab$$
$$bB \to bb$$
$$bC \to bc$$
$$cC \to cc$$

Call the grammar so constructed G. It is clearly of type 1. Now (Example 2), $S \Rightarrow_{+} a^n (BC)^n$, for any $n \geq 1$, using the first two productions. The third production separates B's from C's, putting the former before the latter.

Replacement of B's by b's is proceeding from left to right (productions 4 and 5). Similarly with C's and c's (productions 6 and 7). The relation $L \subset L(G)$ is obvious; the details for the case $L(G) \subset L$ are left to the reader.//

Exercise 4   Consider the grammar with $V_T = \{a,b,c\}$, $V_N = \{S\}$ and productions:

$$S \to abc$$
$$S \to Sabc$$
$$ab \to ba$$
$$ba \to ab$$
$$ca \to ac$$
$$ac \to ca$$
$$bc \to cb$$
$$cb \to bc$$

Observe that this is a CSG. Also note that in some productions the lefthand side contains no nonterminals; this is not forbidden by Definition 2. It is easy to see that this grammar generates strings which contain the same number of a's, b's, and c's.//

Theorem 1   For every re predicate $R(\vec{x}_n)$ on $\mathcal{N}^n$ there is a type 0 grammar G, with $V_T = \{1,B\}$ such that

$$L(G) = \{1^{x_1+1}B \ldots B1^{x_n+1} \mid R(\vec{x}_n)\}$$

Proof Refer to the proof of the corollary to Theorem 6.2.2. Add a new instruction (production) $Q \rightarrow \$ \vdash q_f \dashv \#$, where Q is a new symbol, the start symbol. All the symbols in the aforementioned proof, except for 1 and B, are in V_N.//

Corollary   A set $S \subset \mathcal{N}$ is re iff $\{1^{x+1} \mid x \in S\}$ is a type 0 language.

Proof The *only if* follows from Theorem 1. The *if* follows from Theorem 6.5.5.//

Theorem 2   Every language L over A in CSL is a primitive recursive set (where each $x \in L$ is viewed as the $|A|$-adic notation of some number $w \in \mathcal{N}$).

Proof Let G be a CSG which generates L, where $V_T = A$.

Using the technique of proof employed in Theorem 6.5.4, we can prove that there is a strictly $|B|$-rudimentary(*) predicate $\mathrm{Deriv}(y)$, which holds iff the $|B|$-adic notation of y is a G-derivation $x_0 \Rightarrow x_1 \Rightarrow \ldots \Rightarrow x_n$, where $x_0 = S$ (the start symbol) and $x_i \in (V_N \cup V_T)^+$ for $i = 1, \ldots, n$. B is an extension of $V_N \cup V_T$ which also contains "$\Rightarrow$".

Thus, $x^{(|B|)} \in L(\dagger) \equiv (\exists y) \, [\mathrm{Deriv}(y) \, \& \, \neg (\Rightarrow Px) \, \& \, (\Rightarrow x)Ey](\ddagger)$. Now let y code a *shortest* derivation $x_0 \Rightarrow \ldots \Rightarrow x_n$, where $x_0 = S$ and $x_n = x^{(|B|)}$. Thus, no two x_i's are the same. Since G is type 1, we have $|x_i| \leq |x_{i+1}|$ for $i = 0, \ldots, n - 1$.

Next observe that $|y|_B \leq (n + 1)|x|_B + n$, where $||_B$ is the $|B|$-adic length on $\mathcal{N}$.

(*)**Warning:** In this proof, $||$ is used in two different ways. (1) $|B|$ is number of elements of the set B; (2) $|x_i|$ is the *length* of the string $x_i \in (V_N \cup V_T)^+$.

(†)As in other parts of this chapter, $x^{(m)}$, $(m \geq 1)$, is the m-adic representation of $x \in \mathcal{N}$.

(‡)In "$(\Rightarrow x) Ey$," concatenation of "$\Rightarrow$" and "x" (in $|B|$-adic) is implied, but our usual concatenation symbol "$*$" was judged potentially confusing in view of the meaning attached to "$\Rightarrow *$".

Also,

$$n + 1 \leq |B|^{|x|_B} + |B|^{|x|_B - 1} + \ldots + |B| \leq |B| \cdot x$$

hence

$$|y|_B \leq |B| \cdot |x|_B \cdot x + n$$

and therefore

$$y \leq |B|^{|B| \cdot |x|_B \cdot x + n + 1},$$

Set

$$g = \lambda x. |B|^{|B| \cdot |x|_B \cdot x + n + 1}$$

Clearly, $g \in \mathcal{PR}$ and hence

$$x^{(|B|)} \in L \equiv (\exists y)_{\leq g(x)} [\text{Deriv}(y) \ \& \ \neg(\Rightarrow Px) \ \& \ (\Rightarrow x)Ey] \text{ is in } \mathcal{PR}_*$$

To conclude, set

$$L_A(x) \stackrel{\text{def}}{\equiv} x^{(|A|)} \in L$$

and

$$L_B(x) \stackrel{\text{def}}{\equiv} x^{(|B|)} \in L$$

Clearly,

$$L_A(x) \equiv L_B(f_{|A|,|B|}(x))$$

where $f_{|A|,|B|}$ is that of Lemma 6.7.3. It follows that $L_A(x) \in \mathcal{PR}_*.//$

Note: Once the class of *Elementary* functions and predicates has been defined (see Chapter 13), it will be obvious that the above proof establishes each language in CSL to be Elementary. It follows that there are primitive recursive languages(‡) which are not in CSL. (For the time being, we at least saw that there are *recursive* languages not in CSL). In some sense then, there is a big jump from CSL to type 0 languages.

Corollary   (Type 0 L) − CSL ≠ ∅.

(‡)In Chapter 13 it will be seen that not every function (predicate) in $\mathcal{PR}(\mathcal{PR}_*)$ is Elementary.

We next develop machinery to show CSL − CFL ≠ ∅. Let G be a CFG and let $\sigma_0 \Rightarrow \sigma_1 \Rightarrow \ldots \Rightarrow \sigma_n$ be a G-derivation, where $\sigma_0 \in V_N$ and $\sigma_i \in (V_N \cup V_T)^+$, for $i = 0, 1, \ldots, n$.

A *chain* (associated with the given derivation) is a sequence $(c_{i_0}, c_{i_1}, \ldots, c_{i_q})$, $0 \leq q \leq n$, such that

(1) $c_{i_0} = \sigma_0$.

(2) $c_{i_k} P \sigma_{i_k}$ for $k = 1, \ldots, q$.

(3) For $k = 0, \ldots, q - 1$, the G-derivation above uses a production $x \rightarrow c_{i_{k+1}}$, for some $x \in V_N$ such that $x P c_{i_k}$.

The *length* of a chain is the number of its elements. A *maximal* chain is one which is not a proper subsequence of another chain. The *depth* of the derivation $\sigma_0 \Rightarrow \ldots \Rightarrow \sigma_n$ is the length of a *longest* maximal chain.

Example 5   Consider the grammar $G = (V_N, V_T, P, S)$, where $V_N = \{S\}$, $V_T = \{a, b\}$ and $P = \{S \rightarrow aSb, S \rightarrow ab\}$.

Consider the derivation $S \Rightarrow aSb \Rightarrow aaSbb \Rightarrow aaaSbbb$. Then (S), (S, aSb), (S, aSb, aSb) and (S, aSb, aSb, aSb) are all (associated) chains. Only the last is maximal. The *depth* of this derivation equals its *length*; but this is not always the case.

For example, let $G = (V_N, V_T, P, E)$, where $V_N = \{E\}$, $V_T = \{i, (,), +, \times\}$ and P consists of $E \rightarrow E + E \mid E \times E \mid (E) \mid i$. Consider the derivation $E \Rightarrow E + E \Rightarrow E \times E + E \Rightarrow i \times E + E \Rightarrow i \times i + E \Rightarrow i \times i + i$. $(E, E + E, i)$ is a chain. $(E, E + E, E \times E, i)$ is a maximal chain. It is easily seen to be longest. Thus, the above derivation has *depth* 4 but *length* 6.//

Lemma 1   Consider a CFG G. Let $\sigma_0 \Rightarrow \ldots \Rightarrow \sigma_n$ be a G-derivation where $\sigma_0 \in V_N$ and $|\sigma_n| \geq m^l$, where $m = \max\{|y| \mid (\exists x)(x \rightarrow y \text{ is a } G\text{-production})\}$. Then the depth of the derivation is at least $l + 1$.

Proof   By induction of l, show that, if the derivation has depth $l + 1$, then $|\sigma_n| \leq m^l$ (see Problem 39).//

Theorem 3   (The "*uvwxy* theorem" or "*pumping Lemma for CFL*") If $L \subset \Sigma^*$ is in CFL, then there is a constant p (depending on L) such that $z \in L$ and $|z| \geq p$ implies that $z = uvwxy$ for some $\{u, v, w, x, y\} \subset \Sigma^*$ such that $|v| + |x| \neq 0$, $|vwx| \leq p$ and $uv^i wx^i y \in L$ for $i \geq 0$.(*)

Proof   Let G be some CFG such that $L = L(G)$, where $V_T = \Sigma$. Let $k = |V_N|$ and $m = \max\{|y| \mid (\exists x)(x \rightarrow y \text{ is a } G\text{-production})\}$.

(*)By definition, $x^0 = \Lambda$ for any $x \in \Sigma^*$.

Next, let $\sigma_0 \Rightarrow \ldots \Rightarrow \sigma_n$ be a *shortest* derivation of $z\,(=\sigma_n)$, where $\sigma_0 = S$ (the start symbol) and $|z| \geq m^{k+1}$.

Let $(c_{i_0}, c_{i_1}, \ldots, c_{i_q})$ be a *longest maximal* chain. Then $q \geq k + 1$ (Lemma 1) and $c_{i_q} P z$ (otherwise, the chain would not be maximal).

Let $(A_{i_{q-k-1}}, A_{i_{q-k}}, \ldots, A_{i_{q-1}})$ be the sequence of nonterminals such that $A_{i_j} P c_{i_j}$ and $A_{i_j} \rightarrow c_{i_{j+1}}$ is the production used in the above derivation, for $j = q - k - 1, q - k, \ldots, q - 1$. It must be that for some s, t such that $q - k - 1 \leq s < t \leq q - 1$, $A_{i_s} = A_{i_t}\,(=A$, say).

Then $S \Rightarrow_+ \sigma_{i_s} \Rightarrow_+ \sigma_{i_t} \Rightarrow_+ z$, where $\sigma_{i_s} = \phi A_{i_s} \psi = \phi A \psi$ $\sigma_{i_t} = \tilde{\phi} \phi_1 A_{i_t} \psi_1 \tilde{\psi} = \tilde{\phi} \phi_1 A \psi_1 \tilde{\psi}$ and ϕ, A, ψ are transformed into $\tilde{\phi}, \phi_1 A \psi_1, \tilde{\psi}$ respectively in the course of the derivation ($\phi_1 A \psi_1 = \tilde{\phi}_1 c_{i_t} \tilde{\psi}_1$, where $\tilde{\phi}_1$ begins ϕ_1 and $\tilde{\psi}_1$ ends ψ_1).

Now, $|\phi_1| + |\psi_1| \neq 0$, otherwise the derivation of z would not be *shortest*.

Next, let $z = uvwxy$, where u, v, w, x, y are derived from ϕ (hence, $\tilde{\phi}$), ϕ_1, $A_{i_t}\,(=A]$, ψ_1, ψ (hence, $\tilde{\psi}$) respectively in the course of the derivation.

It follows:

(1) $|v| + |x| \neq 0$ (since $|\phi_1| + |\psi_1| \neq 0$)

(2) $S \Rightarrow_+ \phi A \psi \Rightarrow_+ uwy$ (case $i = 0$)

(3) $S \Rightarrow_+ \phi A \psi \Rightarrow_+ \tilde{\phi} \phi_1^i A \psi_1^i \tilde{\psi} \Rightarrow_+ uv^i w x^i y$ (case $i > 0$)

(4) $A_{i_s} \Rightarrow_+ \phi_1 A_{i_t} \psi_1 \Rightarrow_+ vwx$ has *depth* at most $k + 2$(*); hence (Lemma 1), $|vwx| \leq m^{k+1}$.

Take $p \overset{\text{def}}{=} m^{k+1}$.//

Corollary   CSL – CFL $\neq \varnothing$.

Proof $\{a^n b^n c^n \mid n \geq 1\} \notin$ CFL (see Problem 40).//

Example 6 $\{a^n \mid n \in \mathcal{N}$ is a prime$\} \notin$ CFL. For if it is in CFL, then by the "*uvwxy* theorem" there is a p such that $n > p$ implies $a^n = uvwxy$, so that $|v| + |x| \neq 0$ and, for all $i \geq 0$, $|u| + |w| + |y| + i\,(|v| + |x|)$ is prime.

Hence, $|u| + |w| + |y|$ is prime, therefore $|u| + |w| + |y| > 1$. But then, for $i = |u| + |w| + |y|$, $|u| + |w| + |y| + i\,(|v| + |x|)$ is composite. This contradiction establishes our original claim.//

Theorem 4   (*"Pumping Lemma for REGL"*) If $L \subset \Sigma^*$ is regular, then there is a constant p (depending on L) such that $z \in L$ and $|z| \geq p$ implies $z = uvw$ for $\{u, v, w\} \subset \Sigma^*$, where $0 < |v| < p$ and $uv^i w \in L$ for all $i \geq 0$.

(*)$(A_{i_s}, c_{i_{s+1}}, \ldots, c_{i_q})$ is a *longest* maximal chain of that derivation; if not, a longer maximal chain starting with A_{i_s} could be used to construct a longer maximal chain $(c_{i_0}, \ldots, c_{i_s}, \bar{c}_{i_{s+1}}, \ldots, \bar{c}_{i_h})$.

Proof Let G be a regular grammar such that $L = L(G)$, where $V_T = \Sigma$. Let $p = |V_N| + 1$.

Consider a derivation $\sigma_0 \Rightarrow \ldots \Rightarrow \sigma_n$, where $\sigma_0 = S$ (the start symbol) and $\sigma_n = z$.

Clearly, each σ_i ($i = 0, \ldots, n - 1$) has the form xA, where $x \in \Sigma^*$ and $A \in V_N$; moreover $|\sigma_{i+1}| = |\sigma_i| + 1$ for $i = 0, \ldots, n - 2$. $|\sigma_n| = |\sigma_{n-1}|$, thus $|\sigma_i| = i + 1, i = 0, \ldots, n - 1, |\sigma_n| = n$.

Let $|z| \geq p$. Then $n \geq |V_N| + 1$, hence for some i,j such that $0 \leq i < j \leq p - 1$ ($\leq n - 1$) σ_i and σ_j contain the same nonterminal, say A.

That is,

$$S \Rightarrow_+ \underbrace{uA}_{\sigma_i} \Rightarrow \underbrace{uvA}_{\sigma_j} \Rightarrow_+ uvw = z$$

where A (of σ_i) derives vA, and A (of σ_j) derives w in the course of the derivation. The nature of the regular productions guarantees $|v| \neq 0$; also $|v| \leq |uv| < |\sigma_j| = j + 1 \leq p$. Hence

(1) $0 < |v| < p$ (also $|uv| < p$)

(2) $S \Rightarrow_+ uA \Rightarrow_+ uw$ (case $i = 0$)

(3) $S \Rightarrow_+ uA \Rightarrow_+ uv^iA \Rightarrow_+ uv^iw$ (case $i > 0$).//

Corollary CFL − REGL $\neq \varnothing$.

Proof $\{a^n b^n \mid n \geq 1\} \notin$ REGL. Indeed, if this language were regular, then if $n \geq p$ (p as in Theorem 4) $a^n b^n = uvw, |v| \neq 0, |uv| < n$ (hence, $uv \in \{a\}^+$), and uv^iw has the form $a^k b^k$ for all $i \geq 0$. In particular ($i = 0$) uw is of that form, but on the other hand $uw = a^k b^n$, $k < n$; a contradiction.//

Note: The observation that $|uv| < n$ is convenient but not essential. The reader will find an alternative proof of the corollary not based on that observation.

We conclude this chapter with another unsolvability result.

Example 7 Consider again the grammar $G = (V_N, V_T, P, E)$ where $V_N = \{E\}$, $V_T = \{i, (,), +, \times\}$ and P contains $E \to E + E \mid E \times E \mid (E) \mid i$. It is easy to see that $i + i \times i \in L(G)$. Indeed,

$$E \Rightarrow E \times E \Rightarrow E + E \times E \Rightarrow i + E \times E \Rightarrow i + i \times E \Rightarrow i + i \times i$$

also

$$E \Rightarrow E + E \Rightarrow i + E \Rightarrow i + E \times E \Rightarrow i + i \times E \Rightarrow i + i \times i.$$

This grammar is a skeletal version of the way many programming languages define the "arithmetic expression".

Both derivations above are *leftmost,* which is what type of derivation a *top-down compiler*(*) would attempt to generate in practice. By *leftmost,* we mean that at each step of the derivation the *leftmost* nonterminal of the current string is being replaced.

Note that the two (leftmost) derivations above are *different* (as sequences of strings). This annoys us since, intuitively, the first suggests that "+" has higher *precedence* than "×" (since you get to do "×" in "$E \times E$" *after* you have "recognized" that the "$i + i$" part is an "E"); the second suggests that "×" has higher precedence than "+" (since E is a "sum of E's" it says, multiplications imbedded in an E are done first). This *ambiguity* is harmful, for example, in the context of a compiler.

Naturally, we become interested in the question "Given a CFG G, is there an algorithm to test it for possible ambiguities?" The answer is no, if by "algorithm" we understand one expressible in any of the formalisms already presented in this book.//

Definition 3   Let G be a CFG and $\sigma_0 \Rightarrow \ldots \Rightarrow \sigma_n$ be a G-derivation. This derivation is *leftmost* iff at each step $\sigma_i \Rightarrow \sigma_{i+1}$, the *leftmost* nonterminal of σ_i was rewritten.

G is *ambiguous* iff there is a string $z \in L(G)$ and two *distinct* leftmost G-derivations of z.//

Theorem 5   The problem "is G an ambiguous CFG?" is recursively unsolvable.

Note: The argument below will be quasi-formal. It can easily be extended to a completely formal "reduction argument" modeled around those found in §6.6 and §6.7.

Proof(†) Let instead the ambiguity problem be solvable; we shall show that the PCP then is also solvable.

Let $R = \{(a_1, b_1), \ldots, (a_n, b_n)\} \subset (\Sigma^*)^2$, where Σ is any alphabet with at least two elements. Let $t_1, \ldots, t_n$ be n distinct symbols different from all a_i and $b_i, i = 1, \ldots, n$. Moreover $\{t_1, \ldots, t_n\} \cap \Sigma = \varnothing$.

(*)A *compiler* is a program which translates programs written in some programming language into equivalent programs written in "machine" language.

(†)This proof appears in many places; e.g., [3,13,28]. Various proofs have been independently discovered by Cantor [4], Floyd [10], and Chomsky and Schutzenberger [6].

Consider the CFG G with $V_N = \{S,A,B\}$, $V_T = \Sigma \cup \{t_1, \ldots, t_n\}$, start symbol S and productions

$$S \to A$$

$$S \to B$$

$$\left.\begin{array}{l} A \to a_i t_i \\ A \to a_i A t_i \end{array}\right\} \text{ for all } i = 1, \ldots, n$$

$$\left.\begin{array}{l} B \to b_i t_i \\ B \to b_i B t_i \end{array}\right\} \text{ for all } i = 1, \ldots, n$$

Clearly, A generates strings of the form $a_{k_1} \ldots a_{k_l} t_{k_l} \ldots t_{k_1}$ and B generates strings of the form $b_{q_1} \ldots b_{q_k} t_{q_k} \ldots t_{q_1}$. $L(G)$ is the totality of all such strings.

Let the PCP of R have a solution $(i_1 \ldots i_p)$. Then $a_{i_1} \ldots a_{i_p} = b_{i_1} \ldots b_{i_p}$, hence also $a_{i_1} \ldots a_{i_p} t_{i_p} \ldots t_{i_1} = b_{i_1} \ldots b_{i_p} t_{i_p} \ldots t_{i_1}$. This last string has two distinct leftmost derivations in G, one starting with $S \Rightarrow A$ the other with $S \Rightarrow B$.

Conversely, let $z \in L(G)$ have two distinct leftmost derivations in G. The only way this is possible is that one derivation starts with $S \Rightarrow A$, the other with $S \Rightarrow B$, since once A (resp. B) appears in σ_i in a derivation $\sigma_0 \Rightarrow \ldots \Rightarrow \sigma_i \Rightarrow \ldots$ there is a unique way to proceed, each σ_j $(j \geq i)$ being of the form xAy (resp. xBy) where $\{x,y\} \subset V_T^+$.

Let now $z = x t_{i_1} \ldots t_{i_p}$, where x contains no t_i-symbols. Since $S \Rightarrow A \Rightarrow_+ z$ can only produce $z = a_{i_p} \ldots a_{i_1} t_{i_1} \ldots t_{i_p}$ and $S \Rightarrow B \Rightarrow_+ z$ can only produce $z = b_{i_p} \ldots b_{i_1} t_{i_1} \ldots t_{i_p}$, it follows that $(i_p, i_{p-1}, \ldots, i_1)$ is a solution of the PCP of R. [So is $(i_1, \ldots, i_p)$.]

Thus, G is ambiguous iff the PCP of R has a solution; therefore, a recursive solution of the ambiguity problem implies a recursive solution of the PCP.//

PROBLEMS

1. Show by direct simulation, that for every NTM M, there exists a deterministic TM N which accepts the same predicate as M. (*Hint:* N systematically tries all possible computation attempts of M, for given input w, and accepts the input iff M has at least one (succesful) computation with input w.)

2. Show that for every TM M (deterministic or not), there is a TM N which accepts the same predicate as M, and moreover, except possibly for the initial ID, all IDs of an N-computation involve tapes with no embedded blanks.

3. Give a formal definition of a *deterministic* Post algorithm, appropriately amending Definitions 6.3.1 and 6.3.2. Show that for any TM M, there is a deterministic Post algorithm, which accepts the same predicate as M.

4. Prove Theorem 6.3.2 with a proof similar to that used for Theorem 6.2.2 and its Corollary.

5. Prove the Corollary of Theorem 6.3.2 two ways:

    (a) Using the method of proof of Theorem 6.3.2, and

    (b) Using the method of proof of Theorem 6.2.2 and its Corollary.

6. Show directly from the definition that sets generated by PCS are closed with respect to *union* and *intersection*.

7. Define a PCS S over an alphabet containing 1, such that $\vdash_S w$ and $w \in \{1\}^+$ iff the length of w is prime.

8. Prove that EFS are at least as powerful as PCS.

9. Find an EFS S in which $\{a^m b^n c^k \mid m = n \text{ or } n = k\}$ is representable.

10. Prove Corollary 1 of Theorem 6.3.4.

11. Prove Corollary 2 of Theorem 6.3.4.

12. Prove Corollary 3 of Theorem 6.3.4.

13. Prove that the set of relations formally representable over A is closed under

    (a) Union

    (b) Intersection

    (c) Explicit transformation

    (d) $(\exists y)$.

14. Show that $\lambda xyz.z = x + y$, $\lambda xyz.z = xy$, $\lambda xy.x \le y$ and their negations are formally representable over any alphabet A in m-adic notation ($1 \le m \le |A|$). (*Hint:* It suffices to deal with $m = 1$, $\{1\} \subset A$.)

15. Show that the set of relations formally representable over A is closed under

(a) $(\exists y)_{\le z}$

(b) $(\forall y)_{\le z}$.

(*Hint:* (b) Let $R(y,\vec{x})$ be formally representable over A. Let r be the characteristic function of R. Define s by

$$s(0,\vec{x}) = 0$$
$$s(y + 1,\vec{x}) = \textbf{if } s(y,\vec{x}) = 0 \text{ \& } r(y,\vec{x}) = 0$$

$$\textbf{then } 0$$

$$\textbf{else } 1$$

Then s is the characteristic function of $(\forall y)_{<z} R(y,\vec{x})$. Imitate now the above construction in an appropriate EFS. The case for $(\forall y)_{\le z}$ follows from that for $(\forall y)_{<z}$.)

16. Show that every constructive arithmetic predicate is formally representable over $A = \{1\}$. (*Hint.* Show, by induction with respect to CA, that if a CA predicate R is formally representable over $A = \{1\}$, then so is $\neg R$. The other cases of operations under which CA is closed are covered in Problems 13 and 15.)

17. Prove that every re relation is formally representable over $A = \{1\}$.

18. Prove that the problem of whether an arbitrary function symbol f of an arbitrary system of equations S represents a function (i.e., single-valued relation) is recursively unsolvable. (*Hint:* (a) Arithmetize the formalism of systems of equations, using, e.g., prime power coding or the techniques of Chapter 5. (b) Show that if the problem at hand is solvable, then so is a problem of the form $\lambda x.W_x \in \mathcal{C}$, where $\mathcal{C}$ is a class of re sets. (c) Adapt Rice's Theorem (Problem 3.32) to show that $\{x \mid W_x \in \mathcal{C}\}$ is recursive iff $\mathcal{C} = \emptyset$ or $\mathcal{C} = $ all re sets.)

19. Fill in the details in the proof of Lemma 6.4.2.

20. Prove Lemma 6.5.2.

21. [27] A *pure term* (in an EFS) over A is either a *variable* or an *element of A*. A *pure EFS* is one which contains the axiom Cx,y,xy (which contains the impure term xy), but *all* the other axioms involve *pure* terms *only*. Show that every relation formally representable over A is also formally representable over some *pure* EFS over A.

22. "Arithmetize" the EFS over A, and show that a *"universal"* EFS exists.

23. Show by direct simulation arguments that the NTM formalism is at least as powerful as the NMA and NPA formalisms.

24. The word problem of NMAs (NPAs, semi-Thue, Thue, Post systems) with *one* instruction is recursively solvable.

25. Show that the word problem for NPAs is recursively unsolvable.

26. Prove that the decision problem for Post systems is recursively unsolvable.

27. Show that a grammar with productions $x \longrightarrow y$ restricted so that $|x| = 1$, $|x| \leq |y|$ is equivalent to a CFG.

28. Show that each type 1 grammar is equivalent to one with productions of the form $\alpha A\beta \longrightarrow \alpha x\beta$, where $A \in V_N$, $\{\alpha,\beta\} \subset (V_N \cup V_T)^*$, and $x \in (V_N \cup V_T)^+$.

29. Show that a grammar whose productions have the forms $A \longrightarrow a$, $A \longrightarrow aB$, $A \longrightarrow B$, where $\{A,B\} \subset V_N$, $a \in V_T$, is equivalent to a regular grammar.

30. Show that the set of regular languages over Σ is closed under *union, intersection,* and *complementation* (i.e., if L is regular, then so is $\Sigma^+ - L$).

    Note: As defined in the text, languages of types 1, 2, 3 cannot contain the empty string, Λ. It is convenient to bend the rules and allow Λ via the rule $S \longrightarrow \Lambda$, *provided* S (the *start* symbol) does not occur in the righthand side of any production.

31. Prove that any grammar of type 1, 2, or 3 is equivalent to one of the same type where the start symbol does not appear on the righthand side of any production.

32. Extend grammars of types 1, 2, 3 so that a production $S \longrightarrow \Lambda$ is allowed, provided S (the start symbol) does not appear on the righthand side of any production. Prove that if $L \subset \Sigma^*$ is of type 1, 2, 3, then so is $L \cup \{\Lambda\}$ and $L - \{\Lambda\}$.

33. Revisit Problem 30, where now complements are taken from Σ^*.

34. Let $A \subset \Sigma^*$, $B \subset \Sigma^*$. $A \cdot B$ denotes $\{xy \mid x \in A \ \& \ y \in B\}$, where "$xy$" implies concatenation of x and y. Show that if A and B are regular, then so is $A \cdot B$.

    Note: We can make the same claim for the other types of languages, where the proof is much easier.

35. Let $A \subset \Sigma^*$ be regular. Prove that A^* is also regular. (*Hint:* Experiment first with the case $\Lambda \notin A$, trying to show that A^+ is regular. To this end, productions

such as $P \to a$ which end a derivation should be supplemented by the alternative $P \to aS$, where S is the start symbol.)

36. Show that every finite $A \subset \Sigma^*$ is regular.

37. *Regular expressions* (Kleene [16]). Regular expressions are notations, over a given alphabet Σ, aimed at describing, in a finite way as grammars do, regular languages over Σ. The goal of the present and next problem is to show that regular expressions are indeed adequate for the job.
Definition The set of *regular expressions* over Σ is the *smallest* set of strings over $\Sigma \cup \{\cdot, *, \cup, \emptyset, \Lambda\}$ such that

(1) $\emptyset$, Λ, and any $a \in \Sigma$ is a regular expression

(2) If α and β are regular expressions, then so are $\cup\alpha\beta$, $\cdot\alpha\beta$, and $*\alpha$
End of Definition.

The *sets* defined by regular expressions are, inductively, as follows:

$\emptyset$ defines $\emptyset$
Λ defines $\{\Lambda\}$
$a(\in \Sigma)$ defines $\{a\}$
If α and β define A and B respectively, then $\cup\alpha\beta$ defines $A \cup B$
$\cdot\alpha\beta$ defines $A \cdot B$ (see Problem 34)
$*\alpha$ defines the "Kleene closure" A^*.
Prove that every regular *expression* over Σ defines a regular *language* over Σ. (*Hint.* This is just a rephrasing of Problems 30(33), 34, 35, 36.)

   Note: In view of the fact that the definition of regular expressions does not involve *intersection* and *complementation,* and in view of Problem 30(33) the following is striking.

38. [16] Prove: Every regular language L over Σ is definable by some regular expression over Σ. (*Hint:* Let $L(G) = L$, where G is regular. Let $V_N = \{S_1, \ldots, S_n\}$ and S_{n+1} a new nonterminal, *not* in V_N already. Let S_1 be the start symbol. As a matter of notational uniformity, change each production $S_j \to a$ into $S_j \to aS_{n+1}$, thus making

$$L(G) = \{x \in V_T^* \mid S_1 \Rightarrow *xS_{n+1}\}.$$

Define the sets R_{ij}^k for i, j in $\{1, \ldots, n + 1\}$ and $0 \le k \le n$ by

$R_{ij}^k = \{x \in V_T^* \mid S_i \Rightarrow *xS_j$, where in all the productions $S_m \to aS_\ell$ (*except the* last) involved in the above derivation, $\ell \le k\}$.
   Clearly, $L = R_{1(n+1)}^n$. Note that the R_{ij}^k can be given by the following

inductive definition, and conclude that some regular expression defines $R_{1(n+1)}^n$.

$$R_{ij}^0 = \{a \in V_T \cup \{\Lambda\} \mid S_i \rightarrow aS_j\}, \text{ for all } i, j$$

$$R_{ij}^{k+1} = R_{i(k+1)}^k \cdot (R_{(k+1)(k+1)}^k)^* \cdot R_{(k+1)j}^k \cup R_{ij}^k, \text{ for all } i, j \text{ and } 0 \le k \le n.)$$

39. Prove Lemma 6.8.1.

40. Prove that $\{a^n b^n c^n \mid n \ge 1\} \notin \text{CFL}$.

41. Let G be a regular grammar, where $|V_N| = n$. Show that

(i) $L(G) \ne \varnothing$ iff $(\exists x)(x \in L(G) \,\&\, |x| \le n)$

(ii) $L(G)$ is infinite iff $(\exists x)(x \in L(G) \,\&\, n < |x| < 2n + 2)$.

Conclude that the *emptiness* and *finiteness* problems for regular languages are solvable. (*Hint:* Use Theorem 6.8.4.)

42. Prove that the equivalence problem of

(i) regular grammars

(ii) regular expressions

is solvable. (*Hint:* Use Problems 41 and 30(33). For (ii), observe that given a regular expression, a regular grammar defining the same regular language can be found algorithmically.)

43. (A restricted form of Post's correspondence problem.) Given two lists of words over Σ, $L_1 = \{x_1, \ldots, x_m\}$ and $L_2 = \{y_1, \ldots, y_m\}$, the problem is to detect if there are two sequences $i_1, \ldots, i_n$ and $j_1, \ldots, j_k$ of indices in the range $\{1, \ldots, m\}$ such that

$$x_{i_1} x_{i_2} \ldots x_{i_n} = y_{j_1} y_{j_2} \ldots y_{j_k}. \tag{1}$$

Prove that this problem is solvable. (*Hint.* Condition (1) is satisfied iff $L_1^+ \cap L_2^+ \ne \varnothing$.)

44. Let G be a CFG and let p be the associated constant as in Theorem 6.8.3. Then prove

(i) $L(G) \ne \varnothing$ iff $(\exists x)(x \in L(G) \,\&\, |x| < p)$

(ii) $L(G)$ is infinite iff $(\exists x)(x \in L(G) \,\&\, p \le |x| < 2p)$

Conclude that the *emptiness* and *finiteness* problems for CFLs are solvable.

45. Show that the problem "$L(G_1) \cap L(G_2) = \varnothing$?", where G_1 and G_2 are CFGs, is recursively unsolvable. (*Hint:* See the proof of Theorem 6.8.5.) Conclude that CFL is *not* closed under intersection.

46. Give a simple proof that CFL is not closed under intersection by finding two context-free languages whose intersection is $\{a^n b^n c^n \mid n \geq 1\}$.

47. A *nondeterministic pushdown automaton* (in short NPDA, or simply PDA, since nondeterminism is normally understood unless otherwise stated) is a two-tape nondeterministic TM restricted as follows:

(1) It has a *read-only* input tape, whose tape head can either *stay* or go *right*.

(2) It has a work-tape which works like a stack—i.e., the head moves left *only* *after* it prints a B (blank) in the currently scanned square.

(3) Each tape has its own alphabet (this does not preclude that the two alphabets have a nonempty intersection), and the input does not contain blanks.

(4) The set of states is partitioned into *scan* and *stack* states. If in a *scan* state, the machine's next move, namely, state update and input head's move to the right, depends only on the *input* symbol (it is *undefined* if the input symbol is blank); the *stack* (worktape) is *not* updated.

If the PDA is in a *stack* state, then the input symbol scanned is irrelevant, and the input head does not move. The stack is updated.

(5) There is another partition of Q, viz., into *final* (or *accepting*) states and nonfinal states.

(6) An ID is a tuple (q, ϕ, γ), where q is the "present" state and ϕ is the *initial* part of the input already *scanned*. (Initially it is Λ, empty, since the input-head is positioned at the first input symbol and no move has taken place yet.) Clearly, then, the input head is at the *next symbol after* ϕ on the input tape. $\gamma = \#x$, where $\#$ is a special symbol of the stack alphabet S (called bottom of stack marker) and $x \in S^*$, or $\gamma = \Lambda$. The stack head is on the last symbol of γ, if $\gamma \neq \Lambda$.

(7) An ID is final if q is final. It is *initial* if $q = q_0$ (the initial state), $\phi = \Lambda$, and $\gamma = \#$.

(8) (q, ϕ, γ) yields (q', ϕ', γ') (in symbols, $(q, \phi, \gamma) \rightarrow (q', \phi', \gamma')$) iff one of the following holds:

(a) q is a scan state, and

$\phi' = \phi a$

$\gamma' = \gamma$

and the PDA contains the instruction (in pseudo-PL/1)
"q:**if** input-symbol $= a$ **then** [move input-head right and **goto** q']"
(b) q is a *stack* state, and

(i) $\begin{cases} \phi' = \phi \\ \gamma'b = \gamma \ (b \in S) \text{ and the PDA contains} \\ \text{"}q\text{: \textbf{if} stack-symbol} = b \textbf{ then } [\text{replace } b \text{ by blank, move(*)} \\ \text{left, \textbf{goto} } q'\text{]"} \end{cases}$

or (ii) $\begin{cases} \phi' = \phi \\ \gamma' = xa \ (x \in S^*, a \in S) \\ \gamma = xc \ (x \in S^*, c \in S) \text{ and the PDA contains} \\ \text{"}q\text{: \textbf{if} stack-symbol} = c \textbf{ then } [\text{replace } c \text{ by } a, \text{ stay, \textbf{goto} } q'\text{]"} \end{cases}$

or (iii) $\begin{cases} \phi' = \phi \\ \gamma' = xaB \ (x \in S^*, a \in S, B = \text{blank}) \\ \gamma = xc \ (x \in S^*, c \in S) \text{ and the PDA contains} \\ \text{"}q\text{: \textbf{if} stack-symbol} = c \textbf{ then } [\text{replace } c \text{ by } a, \text{ move right, \textbf{goto} } q'\text{]"} \end{cases}$

(9) An input w is *accepted* iff there is a computation $(q_0, \Lambda, \#) \to \dots \to (q, w, \gamma)$, where $\gamma \in S^*$ and q is *final*. We say that w is accepted by *final state*.

(9) An alternative definition of acceptance is by *empty stack*: w is accepted by empty stack iff there is a computation $(q_0, \Lambda, \#) \to \dots \to (q, w, \Lambda)$.

(10) A set L of strings is *accepted* by some PDA M iff [$w \in L$ iff w is PDA acceptable (under 9 or 9')].

    Prove that a set of strings L is PDA acceptable by final state iff it is so by empty stack. (*Hint:* If L is acceptable by final state by M, then M' can be found to simulate M and accept by empty stack. Stack moves will empty the stack once the whole input is being scanned. Ensure that inadvertent erasure of M's stack does not fool M' into accepting necessarily. The idea is similar for the converse simulation.)

48. Show that any CFL L is PDA acceptable. (*Hint:* Use empty stack acceptance. Let $L = L(G)$, where G is CFG with start symbol N. The PDA starts with stack moves, which put N in the stack initially. From then on, it "guesses" the applicable productions and simulates them using the stack. For example, if, reading the stack from *right to left*, we see $A\gamma\#$ where A is nonterminal, then A is replaced by the righthand side t of an applicable production $A \to t$. If we see

(*)Under (b), movement refers to the stack head.

$a\gamma\#$, where a is a terminal, we remove a and (hopefully) scan an a in the input.)

49. Show that a grammar whose productions are either context free or of the form $A \to \Lambda$ (empty productions) is equivalent to a context-free grammar. (*Hint:* If $A \to \alpha B\beta$ is a production of the original and $B \to \Lambda$ is also a production, drop the latter and add $A \to \alpha\beta$. See also Problems 31, 32.)

50. Show that any set of strings L which is PDA acceptable is in CFL. (*Hint:* Let M be a PDA which accepts L by empty stack. Build a CFG G [with empty productions] as follows: The set of terminals of G is the input alphabet of M. The nonterminals of G are Σ (the start symbol), which is not a symbol associated with M, and all triples (q,A,p), where A is a stack symbol of M and q, p are states. The G-productions are:

(i) $\Sigma \to (q_0,\#,p)$ for all p, where q_0 is the start state, and $\#$ the bottom stack symbol.

(ii) $(q,A,p) \to \Lambda$ for all q,A,q' such that M contains
"q: **if** stack-symbol $= A$ **then** [erase A; move left; **goto** p]"

(iii) $(q,A,p) \to (q_1,C,p)$ for all q,A,C,q_1,p such that M contains
"q: **if** stack-symbol $= A$ **then** [replace A by C; stay; **goto** q_1]"

(iv) $(q,A,p) \to (q_1,B,q_2)(q_2,C,p)$ for all q,A,p,q_1,q_2,C such that M contains
"q: **if** stack-symbol $= A$ **then** [replace A by C; move right; **goto** q_1]"
(**Note.** B is the blank symbol)

(v) $(q,A,p) \to a(q_1,A,p)$ for all q,A,p,a,q_1 such that M contains
"q:**if** input-symbol$=a$ **then**[move right;**goto** q_1].

   Show that $(q,A,p) \Rightarrow *w$ iff $(q,z,\gamma A) \to \ldots \to (p,zw,\gamma)$ for *any* z and γ (possibly empty).)

51. Prove that PDAs are strictly less powerful than NTMs.

52. A *nondeterministic finite automaton* (NFA) is a one-tape NTM restricted as follows:

(1) The input contains no blanks.

(2) The head moves *right* only(*) and *cannot* write.

(3) The NFA halts as soon as the head confronts the blank symbol immediately right of the input (i.e., no move is defined with blank input).

(4) The input is accepted if the automaton has halted while in a *final* (or *accepting*) state. A set of strings L is accepted by an NFA M iff $x \in L \iff x$ is accepted by M.

(*)This is for convenience only. "2-way" NFAs are not any more powerful than the NFA model above. See, for example, Hopcroft and Ullman [13].

Formalize the suggested model above, and show that L is NFA acceptable iff it is a regular language. (*Hint:* Given a regular grammar $G = (V_N, V_T, P, S)$, where S is the start symbol, build an NFA which has V_T as its tape alphabet and $V_N \cup \{\#\}$ as its set of states, where $\# \notin V_N \cup V_T$ is to serve as the *final* state. For any production $A \to aB$ (resp. $A \to a$), the transition relation. $\Delta: (V_N \cup \{\#\}) \times V_T \to V_N \cup \{\#\}$ contains $((A, a), B)$ (resp. $((A, a), \#)$). S is the start state for the NFA. Prove that $x \in L(G)$ iff x is acceptable by the constructed NFA. Also, show how, starting with an NFA, we can obtain an equivalent regular grammar, and prove the correctness of your construction.)

53. Show that if L is context free over A and R is regular over A, then $L \cap R$ is context free over A. (In words, "CFL's are closed under intersection by regular languages".) (*Hint:* Let M be a PDA accepting L by *final state*, and N an NFA accepting R. Construct a PDA P, which has N incorporated in its "control"— i.e., P's states are pairs (q, q'), where q is an N-state and q' is an M-state. A state (q, q') is final in P iff both q and q' are final in the respective machines. A state (q, q') is a *stack* (resp. *scan*) state iff q' is a *stack* (resp. *scan*) state in M. During stack moves, the first component of a state does not change. A scan-move "(q, q'): **scan** a;**goto** $(\bar{q}, \bar{q}')$" is in P just in case M contains the scan move "q':**scan** a;**goto** $\bar{q}'$" and N contains the move "q:**scan** a;**goto** $\bar{q}$".)

54. (Another simplified version of Post's correspondence problem). Let two lists of nonempty strings over A, $(x_1, \ldots, x_n)$ and $(y_1, \ldots, y_n)$, be given. Show that the problem of whether there are two *equal-length* sequences of indices $i_1, \ldots, i_m$ and $j_1, \ldots, j_m$, both in the range $\{1, \ldots, n\}$, such that

$$x_{i_1} x_{i_2} \ldots x_{i_m} = y_{j_1} y_{j_2} \ldots y_{j_m}$$

is *recursively solvable*. (*Hint:* Let a and b be two distinct symbols not in A. For each $x_i = x_i^1 x_i^2 \ldots x_i^{k_i}$ (resp. $y_j = y_j^1 y_j^2 \ldots y_j^{\ell_j}$), where x_i^p (resp. y_j^q) are in A, consider the regular expression

$$a^* b a^* x_i^1 a^* x_i^2 a^* \ldots a^* x_i^{k_i} (\dagger)$$

(resp. $b^* a b^* y_j^1 b^* y_j^2 b^* \ldots b^* y_j^{\ell_j}$). Call the regular *language* represented R_{x_i} (resp. R_{y_j}). Prove that

(1) There is a solution

$$x_{i_1} \ldots x_{i_m} = y_{j_1} y_{j_2} \ldots y_{j_m} \text{ iff } \varnothing \neq \left(\bigcup_{i=1}^n R_{x_i} \right)^+ \cap \left(\bigcup_{j=1}^n R_{y_j} \right)^+ \cap C,$$

($\dagger$)Here we used "infix notation for regular expressions"; i.e. $(a)^*$ instead of $^*a$, $(a)(b)$ rather than $\cdot ab$. Further, letting * have higher precedence than concatenation $[(a)(b)]$ we have omitted brackets without ambiguity.

where $C = \{x \in (A \cup \{a,b\})^* \mid x$ has an equal number of a's and b's$\}$

(2) C is in CFL

(3) Observing that the proofs of Problems 48, 50, 53 were constructive and that

$$(\bigcup_{i=1}^{n} R_{x_i})^+ \cap (\bigcup_{j=1}^{n} R_{y_j})^+$$

is regular, the problem reduces to the emptiness of CFLs [Problem 44]).

REFERENCES

[1] Ackermann, W. "Zum Hilbertschen Aufbau der reellen Zahlen". *Math. Annalen* 99 (1928): 118–133.

[2] Arbib, M.A.: "Monogenic Normal Systems are Universal". *J. Australian Math. Soc.* 3 (1963): 301–306.

[3] Brainerd, W.S. and Landweber, L.H. *Theory of Computation.* New York: Wiley, 1974.

[4] Cantor, D.C. "On the Ambiguity Problem of Backus Systems". *Journal of the ACM* 9 (1962): 477–479.

[5] Chomsky, N. "On Certain Formal Properties of Grammars". *Information and Control* 2 (1959): 137–167.

[6] Chomsky, N. and Schutzenberger, M.P. "The Algebraic Theory of Context-free Languages". In *Computer Programming and Formal Systems,* edited by Braffort, P., and Hirschberg, D. Amsterdam: North Holland, 1963.

[7] Davis, M. *Computability and Unsolvability.* New York: McGraw-Hill, 1958.

[8] Davis, M. *The Undecidable.* Hewlett, New York: Raven Press, 1965.

[9] Floyd, R.W. "Nondeterministic algorithms". *Journal of the ACM* 14 (1967): 636–644.

[10] Floyd, R.W. "On Ambiguity in Phrase Structure Languages". *Comm. ACM* 5(10) (1962a): 526–534.

[11] Gödel, K. "On Undecidable Propositions of Formal Mathematical Systems". (1934). In Davis [8]: 39–73.

[12] Herbrand, J. "Sur la non-contradiction de l'arithmetique". *J. für die reine und angewandte Math.* 166 (1932): 1–8.

[13] Hopcroft, J.E. and Ullman, J.D. *Introduction to Automata Theory, Languages, and Computation.* Reading, Mass.: Addison-Wesley, 1979.

[14] Jones, N.D. *Computability Theory; an Introduction.* New York: Academic Press, 1973.

[15] Kleene, S.C. "General Recursive Functions of Natural Numbers". *Math. Annalen* 112 (1936): 727–742. Also in Davis [8]: 237–252.

[16] Kleene, S.C. "Representation of Events in Nerve Nets and Finite Automata". In *Automata Studies*, edited by Shannon, C.E., and McCarthy, J. Princeton, N.J.: Princeton University Press, 1956: 3–42.

[17] Kurosh, A.G. *Lectures on General Algebra*. New York: Chelsea Publishing Co., 1963.

[18] LeVeque, W.J. *Topics in Number Theory, Vol. 1*. Reading, Mass.: Addison-Wesley, 1956.

[19] Manna, Z. *Mathematical Theory of Computation*. New York: McGraw-Hill, 1974.

[20] Markov, A.A. "On the Impossibility of Certain Algorithms in the Theory of Associative Systems". English translation in *Comptes rendus de l'academie des sciences de l'U.R.S.S.*, n.s. 55 (1947): 583–586.

[21] Post, E. "Recursive Unsolvability of a Problem of Thue". *J. Symb. Logic* 12 (1947): 1–11. Also in Davis [8], pp. 293–303.

[22] Post, E. "Formal Reductions of the General Combinatorial Decision Problem". *Amer. J. Math.* 65 (1943): 197–215.

[23] Post, E. "A Variant of a Recursively Unsolvable Problem". *Bulletin Amer. Math. Soc.* 52 (4) (1946): 264–268.

[24] Post, E. "Recursively Enumerable Sets of Positive Integers and their Decision Problems". *Bulletin Amer. Math. Soc.* 50 (1944): 284–316. Also in Davis [8]: 305–337.

[25] Scott, D. "Some Definitional Suggestions for Automata Theory". *J. of Computer and System Sciences* 1 (1967): 187–212.

[26] Shepherdson, J.C., and Sturgis, H.E. "Computability of Recursive Functions". *Journal of ACM* 10 (1963): 217–255.

[27] Smullyan, R. *Theory of Formal Systems*. Annals of Mathematics Studies, No. 47. Princeton, N.J.: Princeton University Press, 1961.

[28] Yasuhara, A. *Recursive Function Theory and Logic*. New York: Academic Press, 1971.

CHAPTER 7

Recursiveness and Unsolvability (Part II)

§7.1 CHURCH'S THESIS

After the digression of Chapters 4, 5, and 6, we now resume our study of Computability (Recursion Theory). As already mentioned at the beginning of Chapter 2, the purpose of this theory is to formalize the intuitive notions of "algorithm" and "algorithmic", or "computable", function on one hand, and study the properties of algorithms (and algorithmic functions) on the other. In particular, the relative "difficulty" with which computable functions can be computed concerns computer scientists (Chapters 12 and 13). Logicians are also interested in a classification of uncomputable (nonrecursive) functions as to their relative degree of "uncomputability" (Chapter 10).

Strong motivation for the development of formal recursion theory was Hilbert's belief that for any formal mathematical theory there must exist a "method" (or "algorithm") by which *the provability or not of a formula of the theory can be decided* (this was Hilbert's *Entscheidungsproblem*, or decision problem).

Naturally, one would like to know what *precisely* was meant by "algorithm", more so after Gödel's incompleteness results and Church's related undecidability results (Chapter 9). Vigorous activity by Kleene, Church,(*) Turing, Post, Markov, and others gave rise to a big variety of formalisms, each attempting to describe the "class of algorithmic functions".

Many of these formalisms we saw in Chapters 3, 4 and 6, where we also saw that they are equivalent.

There seems then to be something quite "natural" of the class P, since so many researchers independently came up with "seemingly dissimilar" formalisms, which at the end proved to be equivalent to it. Moreover, *at the present state of our understanding of computational processes,* no obvious way of defining an "algorithmic function" outside P is known. Thus, P so far appears to be a "maximal" class of "computable" functions.

Such empirical evidence led to the formulation of the belief of Church (known as "Church's thesis"), that, essentially, *the intuitive notions of "algorithmic function" and "algorithm" coincide respectively with "partial recursive function" and (say) "Turing Machine".* In other words, Church maintains that *any algorithmic procedure, described informally, can be formalized as a Turing Machine* (it is worth observing that once a "procedure" has been formally presented in any of the known formalisms, then a "method" exists to translate this presentation in any of the other formalisms. Certainly this can be seen for the formalisms discussed in this book).

Church's thesis is widely accepted; in fact, very often in the literature one finds informal descriptions of computable functions, which are eventually ratified by an appeal to Church's thesis.

There have also been a number of dissenters, notably Kalmár [3] and Péter [4]. That there is room for disagreement lies in the fact that Church's thesis is not provable; not *before* the notion of "algorithm" has been formalized (†) anyway!

Thus, we shall view Church's thesis as a philosophical position with which one may agree or disagree, and shall not involve it in our proofs.

Those that may disagree with Church's thesis may argue that (1) The various formalisms (of P) known to us are not *that* dissimilar after all: they all are, essentially, *string manipulation systems,* where in each case *very simple rules* are employed in transforming one string to the next. Thus, the independent discovery and eventual convergence of these formalisms to P, one may choose not to attribute to the alleged "maximality" of P but rather to the way mathematical knowledge is transmitted which has the effect that, generally,

---

(*)"λ-Calculus", due to Church, is still another formalism equivalent to the P-formalism. For an introduction, see [2].

(†)A *formal proof,* of course, is a manipulation of formally defined objects. In this case, even though P *is* defined formally, the "class of algorithmic functions" is not.

mathematicians of any given era approach problems in similar ways.(*) (2) It is not unreasonable to allow the possibility that mathematicians in the future will come up with new more powerful concepts of "method" or "computation" than the ones presently held. Such changes in mathematical thought are not unknown. E.g. for about 2000 years it was impossible to escape from the frame of mind according to which *from a point outside a line l exactly one parallel to l can be drawn.*

Only recently *infinitesimals,* until then regarded as void (self-contradictory) notions, found their place on the real line [5]. Another example, from physics, is Einstein's extension of Newtonian mechanics.

In the end, in our opinion, the problem is not so much whether you agree with Church's thesis or not, but how to apply it reliably (assuming you agree with it). We saw that it states: *for any informal algorithm, a TM equivalent to it exists.* But how are we to *agree* that the "informal algorithm" is being *given?* How are we to *recognize* it in the absence of *formal rules* telling us what is and what is not an "algorithm"? It is too demanding to expect of the beginning student of Recursion Theory to feel confident about this question.

On the other hand, the value of informal techniques cannot be overemphasized. A certain amount of "handwaving" can considerably shorten tedious proofs and make their essential features stand out. Moreover, one usually *discovers* proofs (and theorems) by first thinking informally, assuming some familiarity with the subject.

Most of our proofs are given both informally and formally. In those cases that formal arguments are omitted (due to their length or in order to serve as exercises), the implication invariably is: "A formal proof/construction is known to exist; dear reader, be prepared to construct one upon request."

In the sequel, we shall often say "algorithm" or "algorithmic" function meaning (say) URM and $\mathcal{P}$-function respectively; we shall simply use the terms as *synonyms* without any deeper philosophical implications.

Let us next remember, before we proceed, the following basic facts about $\mathcal{P}$ (these were proved in Chapter 3 for $\mathcal{P}$ and again in Chapter 5 for $\mathcal{PC}$).

(1)   There is an *onto* function $i \rightarrow \phi_i \colon \mathcal{N} \rightarrow \mathcal{P}^{(1)}$.

(2)   There is a function $\sigma \in \mathcal{R}^{(2)}$, such that $\phi_i^{(2)}(x,y) = \phi_{\sigma(i,x)}(y)$ for all i,x,y (index theorem)

(2')   If $\psi \in \mathcal{P}^{(n+m)}$, then there is a function $h \in \mathcal{R}^{(n)}$ such that $\psi(\vec{x}_n, \vec{y}_m) = \phi_{h(\vec{x}_n)}(\vec{y}_m)$ for all $\vec{x}_n, \vec{y}_m$.

(*)We may cite as examples the independent and almost simultaneous discoveries of (a) hyperbolic geometry by Bolyai and Lobachevski, (b) solution to Post's problem by Muchnik and Friedberg, (c) the unsolvability of the semigroup word-problem by Post and Markov.

(3) For any pairing function $J\colon \mathcal{N}^2 \to \mathcal{N}$ in $\mathcal{R}^{(2)}$, there is an i_0 such that $\phi_{i_0}(J(x,y)) = \phi_x(y)$ for all x,y (*universal function theorem;* also given in the form $(\exists z_0)\phi_{z_0} = \lambda xy.\phi_x(y)$, or yet $(\exists \psi)\psi \in \mathcal{P}^{(2)}$ and $\psi = \lambda xy.\phi_x(y)$).

(4) (Kleene Normal Form) There are $u \in \mathcal{PR}^{(1)}$ and $t \in \mathcal{PR}^{(3)}$ such that $\phi_i^{(n)} = \lambda \vec{x}_n.u((\mu y)t(i,\langle \vec{x}_n \rangle,y))$ for all i [for a definition of $\langle \vec{x}_n \rangle$, see Definition 2.4.4].

Very often, in the course of some informal proof we shall say things such as "... perform z steps of the computation of program i...".

By "perform program (or algorithm) i", we mean "compute $\phi_i^{(n)}$ using fact (4)", where n is understood from the context. The notion of "step" has been defined in the Turing, URM, and Markov formalisms. But how does it relate to the basic facts (1)–(4) above? In the case of TMs, it is clear from the results of Chapter 5 that y in $T^{(n)}(i,\vec{x}_n,y)$ [the Kleene predicate] is an increasing function of the number of steps of the computation of TM i (on input $\vec{x}_n$). Thus, y is a "measure" of the number of steps in this case. This is intuitively true in general,(*) regardless of how the Kleene predicate $t(i,\langle \vec{x}_n \rangle,y) = 0$ was derived. Indeed, for the computation of the function ϕ_i from the formula $u((\mu y)t(i,\langle \vec{x}_n \rangle,y))$, u is simply "overhead" as it does not depend on i.

What can greatly influence the speed of the computation of $\phi_i(\vec{x}_n)$ is how long the search for the first y such that $t(i,\langle \vec{x}_n \rangle,y) = 0$ will take; that is, $(\mu y)t(i,\langle \vec{x}_n \rangle,y)$ measures in some sense the efficiency of "program i" for the computation of ϕ_i. In the end, "program", "step", etc. are for guiding our intuition. These notions are not explicitly present in formal proofs.

§7.2 DIAGONALIZATION

In this section, we show that certain predicates and functions are not recursive. Those proofs which rely on diagonalization are carefully structured to look similar to Cantor's diagonal proof of the non-enumerability of the real numbers. We first revisit the "halting problem". Although its unsolvability (nonrecursiveness) was shown in Examples 3.7.2, 3.7.3, 3.7.4 by essentially a diagonal argument (Example 3.7.2; one always suspects a diagonal argument, whenever a predicate of the form $\neg Q(x,x)$ is involved) its relationship to Cantor diagonalization was not perhaps clear.

Theorem 1   The "halting problem" is (recursively) unsolvable (i.e., the predicate $\lambda xy.\phi_x(y)\downarrow$ is not in $\mathcal{R}_*$).

(*)In Chapter 12, we shall see that $\lambda x.(\mu y)t(i,x,y)$ satisfies Blum's axioms of abstract complexity measure for ϕ_i.

Proof As in Example 3.7.4, we observe that it suffices to prove that $\phi_x(x)\!\downarrow \notin \mathcal{R}_*$. We proceed informally first; we shall argue by contradiction.

So let $\phi_x(x)\!\downarrow \in \mathcal{R}_*$. Consider the infinite matrix below:

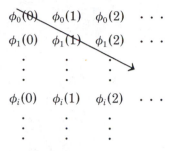

$$
\begin{array}{cccc}
\phi_0(0) & \phi_0(1) & \phi_0(2) & \cdots \\
\phi_1(0) & \phi_1(1) & \phi_1(2) & \cdots \\
\vdots & \vdots & \vdots & \\
\phi_i(0) & \phi_i(1) & \phi_i(2) & \cdots \\
\vdots & \vdots & \vdots &
\end{array}
$$

Utilizing the diagonal entries $\{\phi_i(i)\}_{i\geq 0}$, we shall build a function f which is (*under the assumption* $\phi_x(x)\!\downarrow\in\mathcal{R}_*$) both *in* $\mathcal{P}$ and *not* in $\mathcal{P}$; this contradiction will establish $\phi_x(x)\!\downarrow \notin \mathcal{R}_*$.

Define f to be *different* from any ϕ_i (it suffices to make $f(i) \neq \phi_i(i)$ for all i)

$$
\text{for all } x \text{ let } f(x) =
\begin{cases}
\downarrow \text{ if } \phi_x(x)\!\uparrow \\
\uparrow \text{ if } \phi_x(x)\!\downarrow
\end{cases}
$$

f is not *defined* yet; let us agree to set $f(x) = 6$, whenever $f(x)\!\downarrow$ is required. (**Note:** Other choices for $f(x)\!\downarrow$ are, of course, possible. For example, $f(x) = 0$ or $f(x) = 10^{350000}$ or $f(x) = x$ or $f(x) = x^{x^x}$, etc. As the reader may have seen elsewhere, the choice "$f(x) = 0$" with no explanation, we want to make the point that there is nothing sacred about "0" in the course of this proof).

Thus $f(x) = $ **if** $\phi_x(x)\!\uparrow$ **then** 6 **else** $\uparrow$. On the assumption that $\phi_x(x)\!\downarrow$ $\in \mathcal{R}_*$, f is intuitively "computable" (given x, test whether the xth program with input x halts or not. If it does not, stop everything and output 6; otherwise loop forever); one suspects it is in $\mathcal{P}$.(*) Indeed, $f(x) = $ **if** $\chi_H(x) = 1$ **then** 6 **else** $(\mu y)(y + 1)$, where χ_H, the characteristic function of $\phi_x(x)\!\downarrow$, is by assumption in $\mathcal{R}$.

But $f \in \mathcal{P}$ implies $f = \phi_i$ for some i. This cannot be (hence $f \notin \mathcal{P}$!) since

$$
\phi_i(i)\!\downarrow \underset{(f=\phi_i)}{\text{ iff }} f(i)\!\downarrow \underset{(\text{def. of } f)}{\text{ iff }} \phi_i(i)\!\uparrow .//
$$

Corollary 1   There is a function $\psi \in \mathcal{P}^{(2)}$ for which $(\tilde{\mu}y)\psi(y,x) \notin \mathcal{P}$.(†)

(*)An application of Church's thesis here would consist of saying: "By Church's thesis, $f \in \mathcal{P}$".

(†)For the definition of $\tilde{\mu}$, see Definition 3.1.2.

Proof Intuitively, ψ is defined so that in the loop

$$y \leftarrow 0$$
$$\textbf{do while } (\psi(y,x) \neq 0)$$
$$\qquad y \leftarrow y + 1$$
$$\textbf{end}$$

advancement of y ($y \leftarrow y + 1$) is blocked due to $\psi(y,x)\uparrow$.(*)

Let

$$\psi(y,x) = \begin{cases} 0 & \textbf{if } y = 0 \ \& \ \phi_x(x)\downarrow \ \vee \ y = 1 \\ \uparrow & \textbf{otherwise} \end{cases}$$

Now $\psi \in \mathcal{P}$, for

$$\psi(y,x) = \textbf{if } y = 0 \textbf{ then } 0 \cdot \phi_x(x) \textbf{ else if } y = 1 \textbf{ then } 0 \textbf{ else } (\mu y)(y + 1)$$

Next, $0 = (\tilde{\mu}y)\psi(y,x) \Rightarrow \phi_x(x)\downarrow \Rightarrow \psi(0,x) = 0 \Rightarrow 0 = (\tilde{\mu}y)\psi(y,x)$; that is, $0 = (\tilde{\mu}y)\psi(y,x)$ iff $\phi_x(x)\downarrow$. On the other hand, $\phi_x(x)\uparrow \Rightarrow 1 = (\tilde{\mu}y)\psi(y,x) \Rightarrow \phi_x(x)\uparrow$; that is, $1 = (\tilde{\mu}y)\psi(y,x)$ iff $\phi_x(x)\uparrow$. Thus, $\lambda x.(\tilde{\mu}y)\psi(y,x)$ is the characteristic function of the set $K = \{x \mid \phi_x(x)\downarrow\}$, hence it cannot be in $\mathcal{P}$.//

Corollary 2 The problem "the xth TM halts with input y" is unsolvable. (This is the "halting problem of TMs").

Proof This is the problem $\phi_x(y)\downarrow$, where the map $x \mapsto \phi_x : \mathcal{N} \to \mathcal{P}^{(1)}$ has been obtained by the arithmetization of Turing Machines. (See Chapter 5).//

Corollary 3 Consider the family of TMs over the alphabet A (which includes $\{1, B\}$). For any fixed symbol $t \in A$, the problem "A TM in the family ever prints t" is unsolvable.

Proof Observe that this problem is reducible to the halting problem for TMs. Indeed, for any TM M a TM N can be constructed such that M on input x halts iff N on input x eventually prints a t. [To this end, modify M, if necessary, so that it never prints a t. Next, for any state symbol pair (q,a) of M

(*)This in itself does not guarantee that $f = \lambda x.(\tilde{\mu}y)\psi(y,x) \notin \mathcal{P}$ since a method, more "ingenious" than the simple **do-while** above, might still exist and compute f. However, this simple requirement is sufficient motivation for a careful definition of ψ.

such that $\delta(q,a)\uparrow$ add the quintuple $qatq'S$ where q' is a *new* state, not already in the state alphabet of M. The machine so constructed is N.]

If we could (recursively) solve the problem of Corollary 3, then we could also solve the problem of Corollary 2.//

Theorem 2   The problem of whether an arbitrary function in P for an arbitrary input attains a given (fixed) value is unsolvable.

Proof It suffices to prove that

$$\lambda xy.\phi_x(y) = z \notin \mathcal{R}_* \ (z \text{ is the given fixed value})$$

We argue by contradiction. So let $\lambda xy.\phi_x(y) = z \in \mathcal{R}_*$. Consider the infinite matrix below. As in the proof of Theorem 1, we shall build a g which is both *in* P and *not* in P,

diagonal values considered, to construct g.

Define g to differ from all ϕ_i:

$$g(y) = \begin{cases} z + 1 \ \textbf{if } z = \phi_y(y) \text{ /* thus } g \neq \phi_y \text{ in this case */} \\ z \qquad \textbf{if } z \neq \phi_y(y) \text{ /* thus } g \neq \phi_y \text{ in this case as well */} \end{cases}$$

Clearly, $g \notin P$, for otherwise $g = \phi_i$ for some i, hence

$$z + 1 = g(i) \Longleftrightarrow z + 1 = \phi_i(i) \text{ /* since } g = \phi_i \text{ */}$$
$$\Longleftrightarrow z = \phi_i(i) \text{ /* definition of } g \text{ */}.$$

On the other hand, the characteristic function χ of $\lambda xy.\phi_x(y) = z$ is recursive, hence

$$\lambda y.\textbf{if } \chi(y,y) = 0 \textbf{ then } z + 1 \textbf{ else } z, \text{ which is } g, \text{ is in } P.//$$

Corollary $\lambda xyz.\phi_x(y) = z \notin \mathcal{R}_*$.

Proof $\lambda xy.\phi_x(y) = z$ is an explicit transform of $\lambda xyz.\phi_x(y) = z$ so if the latter is in $\mathcal{R}_*$, so is the former.//

Note: It is interesting to note that $\lambda xyz.\phi_x(y) = z$ [hence, $\lambda xy.\phi_x(y) = z$] is re. Indeed, $\phi_x(y) = z \equiv (\exists w)(t(x,y,w) = 0 \,\&\, u(w) = z)$. The contention follows from the projection theorem (Theorem 3.6.1).

Theorem 3   There is a function $f \in \mathcal{P}$ which is not extendible to a function in $\mathcal{R}$.

Proof We shall obtain by diagonalization an $f \in \mathcal{P}$ such that whenever $\phi_x(x)\downarrow$ then $f(x)\downarrow$ and $f(x) \neq \phi_x(x)$. Pretend for a while that we already have such an f and let $g \in \mathcal{R}$ such that $f(x)\downarrow \Rightarrow f(x) = g(x)$ (that is, $f \subset g$).

$g \in \mathcal{R} \Rightarrow g \in \mathcal{P}$, hence for some $i, g = \phi_i$. Then $g(i) = \phi_i(i)$ (since $g = \phi_i$). Also $g(i) = f(i)$ ($\phi_i(i)[= g(i)]\downarrow$); hence, $g(i) \neq \phi_i(i)$. This contradiction proves that *if* f exists, then no $g \in \mathcal{R}$ such that $f \subset g$ exists.

Define f by

$$f(x) = \begin{cases} \phi_x(x) + 1 \text{ if } \phi_x(x)\downarrow \\ \qquad\uparrow \qquad \text{if } \phi_x(x)\uparrow \end{cases}$$

f is "algorithmic" [Given x, run program x on input x. If it ever halts, add 1 to the computed value and halt everything.] Indeed, $f \in \mathcal{P}$ since $f = \lambda x.\phi_x(x) + 1$ and $\lambda x.\phi_x(x) \in \mathcal{P}$ by the universal function theorem [alternatively, $\phi_x(x) = u((\mu y)t(x,x,y))$].//

Note: The function $\lambda x.\phi_x(x)$ is not extendible to a recursive function h either, since otherwise it would follow that $\lambda x.h(x) + 1 \in \mathcal{R} - \mathcal{P}$! (why?)

Theorem 4   The problem of testing whether an arbitrary function in $\mathcal{P}$ is also in $\mathcal{R}$ is unsolvable.

Note: This we already know, since the predicate $\phi_x \in \mathcal{R}$ is not re (Example 3.7.6). We give a direct proof here.

Proof Let instead the predicate $\phi_x \in \mathcal{R}$ be recursive.

We first obtain a subsequence $(\tilde\phi_i)_{i\geq0}$ of $(\phi_i)_{i\geq0}$, which contains all the total functions:

For each i, put ϕ_i in the subsequence iff $\phi_i \in \mathcal{R}$. Since this last predicate (of i) is assumed recursive, we can effectively (= algorithmically) build the sequence $(\tilde\phi_i)_{i\geq0}$.

Since each $\tilde\phi_i$ is *total*, the standard Cantor argument proves that $f = \lambda x.\tilde\phi_x(x) + 1(*)$ is *not* an $\tilde\phi_j$ (for any j).

On the other hand, f, besides being total, is computable. [Given x, obtain

(*)Note that $f = \lambda x.\tilde\phi_x(x) + 1$ is different from each $\tilde\phi_i$ since (total functions!) $f(i) \neq \tilde\phi_i(i)$ for all i.

$\tilde{\phi}_x$ by building a long enough initial segment of $(\phi_i)_{i\geq0}$ and building the list $(\tilde{\phi}_i)_{i\geq0}$ in parallel. Next, run $\tilde{\phi}_x$ on x. When it halts, add 1 to the result and stop.] This is a contradiction.

Formally, first let m be such that $\phi_m \in \mathcal{R}$ (recall $\mathcal{R} \neq \varnothing$; e.g., $\lambda x.x \in \mathcal{R}$). Then define h by $h(x) = $ **if** $\phi_x \in \mathcal{R}$ **then** x **else** m.(*) Since "$\phi_x \in \mathcal{R}$" is recursive, $h \in \mathcal{R}$. Moreover, $h(\mathcal{N}) = \{x \mid \phi_x \in \mathcal{R}\}$

Thus, $f = \lambda x.\phi_{h(x)}(x) + 1$, and by the universal function theorem $f \in \mathcal{R}.//$

Note: At the beginning of Chapter 3, it was said that if we were to avoid an obvious diagonalization which produced an algorithmic function not in our formal system we had to allow nontotal functions in the formalism. Part of the price to pay for this is that the question "is f total" is recursively unsolvable in the system (Theorem 4).

The diagonal method we saw in the previous proofs (as well as the instances of it which we saw in Example 3.7.2, Theorem 2.7.3, and in §1.5) is, in general, a technique for constructing (defining) a set R which lies *outside* a given family of subsets $(R_i)_{i\in I}$ of some set $\mathcal{S}$. R is $\{i \in \mathcal{S} \mid i \notin R_i\}$. This more "general" diagonalization is essentially the same as that of Cantor's. Indeed, if χ_i is the characteristic function of R_i, then R_i is essentially the family $(\chi_i(a))_{a\in\mathcal{S}}$. If χ is the characteristic function of R, clearly $\chi(a) = 1 \dot- \chi_a(a)$, $a \in \mathcal{S}$. In the case $\mathcal{S} = \mathcal{N}, I = \mathcal{N}, (\chi_i(a))_{a\in\mathcal{S}}$ is $\chi_i(0),\chi_i(1), \ldots, \chi_i(i), \ldots$ the ith row of the infinite matrix $(\chi_i)_{i\geq0}$. χ is the infinite sequence obtained from $\chi_0(0),\chi_1(1), \ldots, \chi_i(i), \ldots$ (the diagonal of the matrix) by interchanging 0 with 1 everywhere.

In the case of diagonalizing over a set of functions $(f_i)_{i\geq0}$ the general technique is to form f so that $f(x) \neq f_x(x)$ for all x (hence, $f \neq f_x$ for all x). For *total* functions f_i, this can be accomplished by all sorts of ways; e.g., $f = \lambda x.f_x(x) + 1$ or $\lambda x.1 \dot- f_x(x)$ or $\lambda x.f_x(x) + 9$, etc. If the f_i's are not necessarily total, one way of obtaining f is by "$f(x)\downarrow$ iff $f_x(x)\uparrow$" (this we used in the proof of Theorem 1). In the end, practice is what is needed for a good mastering of diagonal arguments.

We next give two applications of the diagonal technique where functions with certain required properties are constructed.

Theorem 5   For any $g \in \mathcal{R}^{(1)}$, there is an $f \in \mathcal{R}^{(1)}$ such that for any i satisfying $f = \phi_i$, the program i needs more than $g(x)$ steps to compute $f(x)$ for all $x \geq i$.

Note: Recall our discussion at the end of §7.1 for the proper interpretation of the imprecise terms used in the theorem statement. Another, still looser way, of stating the theorem is that "There are arbitrarily hard to compute recursive functions".

(*)"**else** m" guarantees that h is total.

Proof Consider once again the infinite matrix

$$
\begin{array}{ccccc}
\phi_0(0) & \phi_0(1) & \cdots & \phi_0(i) & \cdots \\
\phi_1(0) & \phi_1(1) & \cdots & \phi_1(i) & \cdots \\
\vdots & & & & \\
\phi_i(0) & \phi_i(1) & \cdots & \phi_i(i) & \cdots \\
\vdots & & & &
\end{array}
$$

We construct an $f \in \mathcal{R}$ such that for each $x, f(x) \neq \phi_i(x)$ if $\phi_i(x)\downarrow$ in $\leq g(x)$ steps *and* $i \leq x$. In other words, for each x we inspect all $\phi_i(x)$, $i \leq x$, rather than just the diagonal value $\phi_x(x)$, in order to define $f(x)$.

Set $I_x = \{i \mid i \leq x \,\&\, \phi_i(x)\downarrow$ in $\leq g(x)$ steps$\}$. Then

$$
f(x) = \begin{cases}
\Sigma_{i \in I_x} \phi_i(x) + 1 \text{ if } I_x \neq \varnothing \\
1(*) \text{ otherwise}
\end{cases}
$$

Intuitively, f is computable.

Setting temporarily aside the formal proof that $f \in \mathcal{R}$, we see that once that is settled, there must be an i (in fact, infinitely many; Corollary 4 of Theorem 3.7.3) such that $f = \phi_i$.

For any such "program" i, let $x \geq i$. Then, if $\phi_i(x)\downarrow$ in $\leq g(x)$ steps, it follows that

$$
I_x \neq \varnothing \text{ and } f(x) = \Sigma_{k \in I_x} \phi_k(x) + 1 > \phi_i(x) = f(x);
$$

a contradiction.

For the formal details, observe

(1) $i \in I_x$ iff $i \leq x \,\&\, (\exists y)_{\leq g(x)} \, t(i,x,y) = 0$. Hence, $\lambda ix.i \in I_x$ is in $\mathcal{R}_*$.

(2) Let χ be the characteristic function of $\lambda ix.i \in I_x$. Define first $h = \lambda ix.$**if** $\chi(i,x) = 0$ **then** $\phi_i(x)$ **else** 0. Clearly, $h \in \mathcal{R}$ and $f(x) = 1 + \Sigma_{i \leq x} h(i,x)$ for all $x.//$

Note: (1) $\lambda x.1 + \Sigma_{i \leq x}(1 \div \chi(i,x))\phi_i(x)$ is not necessarily total, as for some $i \leq x$ it may be $\phi_i(x)\uparrow$. This is why we utilized h.

(*)Instead of 1 ($\lambda x.1$ really), we can let f be equal to any fixed recursive function in this case.

(2) Theorem 5 is true even if we restrict the range of f in $\{0,1\}$. It then states that there are arbitrarily hard to compute *predicates* in $\mathcal{R}_*$. The proof is a bit more involved, since $f(x) \neq \phi_i(x)$ cannot be guaranteed for all $i \in I_x$. The "trick" is to make $f(x) \neq \phi_i(x)$ for *some* $i \in I_x$ (usually the smallest) which has not been *earlier used* to set $f(y) \neq \phi_i(y)$ ($y < x$); then i is marked *used* (or *cancelled*). (This idea is traceable at least as far back as Blum [1].)

Theorem 6   For any $g \in \mathcal{R}^{(1)}$ there is a 0-1 valued function $f \in \mathcal{R}^{(1)}$ which, for any i such that $f = \phi_i$, needs more than $g(x)$ steps for its computation on program i for all but a finite number of x values.

    Proof   Define

$$f(x) = \begin{cases} 1 \div \phi_i(x) \text{ if } i \text{ is the } smallest \text{ } uncancelled \text{ } j \text{ in } I_x; \text{ } cancel \text{ } i \\ 1(*) \qquad \text{ if no uncancelled } j \text{ exists in } I_x \end{cases}$$

Clearly, range$(f) \subset \{0,1\}$, and f is total. It is also intuitively computable. [Given x, compute (inductively) $f(0), \ldots, f(x-1)$ to obtain the list of cancelled indices. If no uncancelled index exists in I_x,(†) then output 1; otherwise, output $1 \div \phi_i(x)$, where i is the smallest uncancelled index in I_x; cancel i].

    Assume for a minute that $f \in \mathcal{P}$ has been shown. Let then i_0 be *any* index such that $f = \phi_{i_0}$. If now there is an infinite sequence $x_1 < x_2 < \ldots$, where $i_0 \leq x_1$, such that $\phi_{i_0}(x_i) \downarrow$ in $\leq g(x_i)$ steps ($i = 1,2,\ldots$), then there is another infinite sequence $j_1 < j_2 < \ldots < i_0$ such that j_i is the smallest uncancelled index $< i_0$ in I_{x_i} ($i = 1,2,\ldots$). [Note that if such a $j_i < i_0$ does not exist, $\phi_{i_0}(x_i) \downarrow$ in $\leq g(x_i)$ steps suggests that we set $f(x_i) = 1 \div \phi_{i_0}(x_i)$ and cancel i_0; this contradicts $f = \phi_{i_0}$]. Since $j_1 < j_2 < \ldots < i_0$ is impossible for infinitely many j_i, it follows that for some $m \geq i_0$, $x \geq m \Rightarrow \phi_{i_0}(x) \downarrow$ in $> g(x)$ steps.

    We now turn to formalizing the definition of f and show that $f \in \mathcal{R}$:

    Define first the function *cancelled* by

cancelled$(x) = p_0 \cdot \Pi_{i>1} p_i^{(i\text{th cancelled index}) + 1}$, where all cancelled indices up to (and including) the definition of $f(x)$ are accounted for (p_i is the ith prime; e.g., $p_0 = 2, p_1 = 3$, etc.).

    Define h by

$$h(x,z) = \textbf{if } (\exists y)_{\leq x+1}([(\forall i)_{<\text{length}(z)} i > 0 \Rightarrow \exp(i,z) \neq y + 1] \,\&$$

$$\chi(y, x+1) = 0) \textbf{ then } 1 + (\mu y)_{\leq x+1}([(\forall i)_{<\text{length}(z)} i > 0 \Rightarrow \exp(i,z) \neq y + 1] \,\&$$

$$\chi(y, x+1) = 0) \textbf{ else } 0$$

(*)Instead of $\lambda x.1$, we can use also $\lambda x.0$.

(†)Recall that the membership in I_x is "algorithmically" decidable, since $\lambda ix.i \in I_x$ $\in \mathcal{R}_*$. Also, $\phi_i(x) \downarrow$, since indeed it does in $\leq g(x)$ steps ($i \in I_x$).

Clearly, $h \in \mathcal{R}$.(*)

Thus,

$$\begin{cases} \text{cancelled}(0) = \textbf{if } \chi(0,0) = 0 \textbf{ then } p_0 p_1 \textbf{ else } p_0 \\ \text{cancelled}(x + 1) = \text{cancelled}(x) \cdot p_{\text{length(cancelled}(x))}^{h(x, \text{cancelled}(x))} \end{cases}$$

Finally,

$$f(x) = \begin{cases} \textbf{if } x = 0 \quad \textbf{then if } \chi(0,0) \textbf{ then } 1 \dot{-} \phi_0(0) \textbf{ else } 1 \\ \textbf{else} \quad \begin{cases} \textbf{if } h(x \dot{-} 1, \text{cancelled}(x \dot{-} 1)) > 0 \\ \textbf{then } 1 \dot{-} \phi_{h(x \dot{-} 1, \text{cancelled}(x \dot{-} 1)) \dot{-} 1}(x) \\ \textbf{else } 1 \end{cases} \end{cases}$$

Clearly, $f \in \mathcal{R}$.//

Note: It is important to emphasize that the functions f of Theorems 5 and 6 require more than $g(x)$ steps for almost all inputs, *regardless* of the program i_0 (such that $\phi_{i_0} = f$) chosen, no matter how "clever" it may be. This is clear because our argument was in terms of the arbitrary i_0, such that $\phi_{i_0} = f$.

Corollary   There is no $g \in \mathcal{R}^{(1)}$ such that $f \in \mathcal{R}^{(1)}$ implies $f = \lambda x . u((\mu y)_{\leq g(x)} t(i, x, y))$ for some i.

Proof Indeed (Theorem 5 or 6) an $f \in \mathcal{R}^{(1)}$ exists such that for *all i* such that $\phi_i = f$, $f(x) = u((\mu y)_{\leq g(x)} t(i, x, y))$ is false for all but finitely many values of x.//

§7.3 UNSOLVABILITY AND NON RE-NESS PROOFS VIA THE *S-m-n* THEOREM

The main result of this section is Rice's theorem which states roughly that "any nontrivial problem about programs is recursively unsolvable". By *trivial* problem about programs we understand one for which *every* program (or *no* program) is a solution. We prefix Rice's theorem with a number of

(*)χ is the characteristic function of $i \in I_x$. The functions *exp* and *length* are primitive recursive, where length$(x) = (\mu y)_{\leq x} [p_y | x \,\&\, (\forall i)_{\leq x} \{p_i | x \Rightarrow i \leq y\}] + 1$. (See Example 2.3.1.)

ad-hoc unsolvability and non re-ness results. The general method used in proving that $x \in S$ is *not recursive* (resp. *not re*) is in finding, with the help of the *S-m-n* theorem, a function $h \in \mathcal{R}^{(1)}$ such that $h(x) \in S$ iff $x \in L$, where L is some nonrecursive or non re set, respectively. This says that "if an algorithm or algorithmic enumeration exists for S, then this must be the case for L". This is an instance of the reducibility technique we saw in Chapter 6; problem L is reduced to problem S. In practice, L is usually taken to be K (to prove nonrecursiveness) or $\overline{K}$ (to prove non re-ness).

Theorem 1   The following sets are not recursive.

$$A = \{x \mid \phi_x \text{ is a constant}\}$$
$$B = \{(x,y) \mid y \in \text{range}(\phi_x)\}$$
$$C = \{(x,y,z) \mid \phi_x(y) = z\}$$
$$D = \{x \mid \text{dom}(\phi_x) \text{ is infinite}\}$$
$$E = \{x \mid \text{range}(\phi_x) \text{ is infinite}\}$$
$$F = \{x \mid \text{dom}(\phi_x) \text{ is finite}\}$$
$$G = \{x \mid \text{range}(\phi_x) \text{ is finite}\}$$
$$H = \{x \mid \text{dom}(\phi_x) = \varnothing\}$$
$$I = \{x \mid \text{range}(\phi_x) = \varnothing\}$$
$$J = \{x \mid \phi_x \in \mathcal{R}\} = \{x \mid \phi_x \text{ is total}\}.$$

Note: Nonrecursiveness of C and J have been proved in §7.2 as well.

Proof Define ψ by

$$\psi(x,y) = \begin{cases} 0 \text{ if } \phi_x(x)\downarrow \\ \uparrow \text{ otherwise} \end{cases}$$

Clearly, $\psi \in \mathcal{P}$ ($\psi = \lambda xy.0 \cdot \phi_x(x)$, where $\lambda x.\phi_x(x) \in \mathcal{P}$ by the universal function theorem).

By the *S-m-n* theorem, there is an $h \in \mathcal{R}^{(1)}$ such that $\psi(x,y) = \phi_{h(x)}(y)$ for all x,y. Clearly, $\phi_{h(x)} = \lambda x.0$ iff $\phi_x(x)\downarrow$. Another way of stating this is: $\phi_{h(x)}$ is totally undefined iff $\phi_x(x)\uparrow$.

For A: $h(x) \in A$ iff $x \in K$. Hence, $A \notin \mathcal{R}_*$ (otherwise, $h(x) \in A$ is in $\mathcal{R}_*$, i.e., $K \in \mathcal{R}_*$).

For B: $(h(x),0) \in B$ iff $x \in K$. Hence, $B \notin \mathcal{R}_*$ (otherwise, $(h(x),0) \in B$ is in $\mathcal{R}_*$; that is, $K \in \mathcal{R}_*$).

For C: $(h(x),x,0) \in C$ iff $x \in K$. Hence, $C \notin \mathcal{R}_*$.

For D: $h(x) \in D$ iff $x \in K$. Hence, $D \notin \mathcal{R}_*$.

For F: $h(x) \in F$ iff $x \in \overline{K}$. Hence, $F \notin \mathcal{R}_*$ (of course, $\overline{K} \notin \mathcal{R}_*$). [One can also say that $F \notin \mathcal{R}_*$ because its complement D is not in $\mathcal{R}_*$].

For H: $h(x) \in H$ iff $x \in \overline{K}$. Hence, $H \notin \mathcal{R}_*$.

For I: $h(x) \in I$ iff $x \in \overline{K}$. Hence, $I \notin \mathcal{R}_*$.

For J: $h(x) \in J$ iff $x \in K$. Hence, $J \notin \mathcal{R}_*$.

For E and G we utilize a different function χ:

$$\chi(x,y) = \begin{cases} y \text{ if } \phi_x(x)\downarrow \\ \uparrow \text{ otherwise} \end{cases}$$

$\chi \in \mathcal{P}$, for $\chi = \lambda xy.y + 0 \cdot \phi_x(x)$. Then, there is an $f \in \mathcal{R}^{(1)}$ such that $\chi = \lambda xy.\phi_{f(x)}(y)$. Clearly, $\phi_{f(x)} = \lambda y.y$ iff $\phi_x(x)\downarrow$, i.e., $\phi_{f(x)}$ is totally undefined iff $\phi_x(x)\uparrow$.

It follows

For E: $f(x) \in E$ iff $x \in K$. Hence, $E \notin \mathcal{R}_*$.

For G: $f(x) \in G$ iff $x \in \overline{K}$. Hence, $G \notin \mathcal{R}_*$.//

Note: It is incorrect to state that $h(x) \in G$ iff $x \in \overline{K}$, for $h(x) \in G$ iff $\phi_{h(x)}$ has finite range. This may very well be when $\phi_{h(x)} = \lambda y.0$; but $x \in K$ in this case. It can also be when $\phi_{h(x)}$ is totally undefined. But then $x \in \overline{K}$. This explains the need for a "new" ψ (χ in the proof) and a new h (f in the proof) to deal with E and G.

Lemma 1   If $\lambda y\vec{x}_n.R(y,\vec{x}_n)$ is re and $\lambda \vec{y}_m.f(\vec{y}_m) \in \mathcal{R}$, then $\lambda \vec{y}_m\vec{x}_n.R(f(\vec{y}_m),\vec{x}_n)$ is re.

Proof By the projection theorem, $R(y,\vec{x}_n) \equiv (\exists z)Q(z,y,\vec{x}_n)$ for some $Q \in \mathcal{R}_*$.
But $\lambda z\vec{y}_m\vec{x}_n.Q(z,f(\vec{y}_m),\vec{x}_n) \in \mathcal{R}_*$ and $R(f(\vec{y}_m),\vec{x}_n) \equiv (\exists z)Q(z,f(\vec{y}_m),\vec{x}_n)$.//

Theorem 2   All the sets, except B and C, defined in Theorem 1 are not re.

Proof re-ness can be proved by direct application of the definition (Definition 3.6.1) but usually more conveniently by application of the projection theorem.

re-ness of C has already been established in §7.2. As for B, $(x,y) \in B \equiv (\exists z)\phi_x(z) = y$. re-ness of B follows from that of C by Corollary 1, of Theorem 3.6.1.

As for non re-ness of the remaining sets, first, non re-ness of F,G,H,I follows from the proof of Theorem 1 and Lemma 1. That J is not re is already known (Example 3.7.6). We now turn to the cases A,D,E.

For A we need a $\sigma \in \mathcal{R}^{(1)}$ such that $\sigma(x) \in A$ iff $x \in \overline{K}$. We choose a σ so that $\phi_{\sigma(x)} = \lambda y.0$ iff $\phi_x(x)\uparrow$. One is tempted to set

$$
\tau(x,y) = \begin{cases} 0 \text{ if } \phi_x(x)\uparrow \\ \uparrow \text{ otherwise} \end{cases}
$$

Should $\tau \in \mathcal{P}$, existence of the required σ would follow from the S-m-n theorem. Unfortunately, $\tau \notin \mathcal{P}$ [if $\tau \in \mathcal{P}$, then $\lambda x.\tau(x,x) \in \mathcal{P}$, requiring $\text{dom}(\lambda x.\tau(x,x))$ to be re; this cannot be since $\text{dom}(\lambda x.\tau(x,x)) = \overline{K}$]. τ is what we want but is not computable; we define then a computable "approximation" of it:

$$
\tilde{\tau}(x,y) = \begin{cases} 0 \text{ if } \phi_x(x) \text{ does } not \text{ converge in} \le y \text{ steps} \\ \uparrow \text{ otherwise} \end{cases}
$$

Intuitively, $\tilde{\tau}$ is computable as the reader can readily verify. We postpone proof of $\tilde{\tau} \in \mathcal{P}$ for a while.

$\tilde{\tau} \in \mathcal{P}$ implies $\exists \sigma \in \mathcal{R}^{(1)}$ such that $\tilde{\tau} = \lambda xy.\phi_{\sigma(x)}(y)$.

Now,

$$\phi_x(x)\uparrow \Rightarrow (\forall y)\phi_x(x) \text{ does } not \text{ converge in} \le y \text{ steps}$$

$$\Rightarrow (\forall y)\tilde{\tau}(x,y) = 0 \Rightarrow \phi_{\sigma(x)} = \lambda y.0$$

Also,

$$\phi_x(x)\downarrow \Rightarrow (\exists y_0)[(\forall y)y \ge y_0 \Rightarrow \phi_x(x)\downarrow \text{ in} \le y \text{ steps}]$$

$$\Rightarrow (\forall y)[y \ge y_0 \Rightarrow \phi_{\sigma(x)}(y)\uparrow] \Rightarrow \phi_{\sigma(x)} \text{ } not$$

$$\text{a constant } [\phi_{\sigma(x)}(y) = 0 \text{ for } y < y_0]$$

Thus, $\phi_x(x)\uparrow \Rightarrow \sigma(x) \in A$. $\sigma(x) \in A \Rightarrow \phi_{\sigma(x)}$ is a constant; the only possibility is that $\phi_{\sigma(x)} = \lambda y.0$, hence $(\forall y)[\phi_x(x)$ does not converge in $\le y$ steps]; that is, $\phi_x(x)\uparrow$.

This establishes $\sigma(x) \in A$ iff $x \in \overline{K}$; hence, A is not re (Lemma 1). [Note that $\tilde{\tau}(x,y) = 0$ if $\phi_x(x)\uparrow$, but when $\phi_x(x)\downarrow$ is $almost \ everywhere$ undefined rather than everywhere. In this sense, we meant that $\tilde{\tau}$ approximates τ].

For D: $\sigma(x) \in D$ iff $x \in \overline{K}$; hence, D is not re (Lemma 1). For E we utilize a $k \in \mathcal{R}^{(1)}$ such that $\text{range}(\phi_{k(x)})$ is infinite iff $\phi_x(x)\uparrow$.

To this end, let

$$\tilde{\tilde{\tau}}(x,y) = \begin{cases} y \text{ if } \phi_x(x) \text{ does } not \text{ converge in } \leq y \text{ steps} \\ \uparrow \textbf{ otherwise} \end{cases}$$

By S-m-n theorem (and an argument as that for $\tilde{\tau}$), a $k \in \mathcal{R}^{(1)}$ exists such that $\phi_{k(x)} = \lambda y.y$ iff $x \in \overline{K}$; hence, $k(x) \in E$ iff $x \in \overline{K}$; it follows that E is not re.

Finally, $\tilde{\tau} \in \mathcal{P}$ and $\tilde{\tilde{\tau}} \in \mathcal{P}$ for

$$\tilde{\tau} = \lambda xy. \textbf{ if } (\forall z)_{\leq y} \, t(x,x,z) \neq 0 \textbf{ then } 0 \textbf{ else } (\mu y)(y + 1)$$

$$\tilde{\tilde{\tau}} = \lambda xy. \textbf{ if } (\forall z)_{\leq y} \, t(x,x,z) \neq 0 \textbf{ then } y \textbf{ else } (\mu y)(y + 1).//$$

Note: The sets D and E are examples of non re sets whose complements (F and G, respectively) are also non re. The complement $\overline{J}$ of J is non re as well [$h(x) \in \overline{J}$ iff $x \in \overline{K}$, where h is that in the proof of Theorem 1].

The above two theorems involved *complete index sets*—i.e., sets of the form $\{x \,|\, \phi_x \in \mathcal{C}\}$—where $\mathcal{C}$ is a class of $\mathcal{P}$-functions. Intuitively, a complete index set is a set of *all programs* (x), whose associated (i.e., computed by x) functions (ϕ_x) have a common property (belonging to $\mathcal{C}$).

Sets of the form $\{x \,|\, W_x \in \mathcal{C}\}$ are special cases of complete index sets, since $W_x \in \mathcal{C}$ iff $\text{dom}(\phi_x) \in \mathcal{C}$ therefore $\{x \,|\, W_x \in \mathcal{C}\} = \{x \,|\, \phi_x \in \tilde{\mathcal{C}}\}$, where $\tilde{\mathcal{C}} = \{\psi \in \mathcal{P} \,|\, \text{dom}(\psi) \in \mathcal{C}\}$. In particular, D of Theorem 1 is $\{x \,|\, W_x \text{ is infinite}\}$, $H = \{x \,|\, W_x = \varnothing\}$, $J = \{x \,|\, W_x = \mathcal{N}\}$.

Lemma 2   Let $\mathcal{P}_\mathcal{C} = \{x \,|\, \phi_x \in \mathcal{C}\}$. If $\mathcal{P}_\mathcal{C}$ is re, then for all $\psi \in \mathcal{C}$ and χ in $\mathcal{P}$ such that $\psi \subset \chi$ it must be that $\chi \in \mathcal{C}$.

Proof   We argue by contradiction. So let there be m_0 and m_1 such that $\phi_{m_0} \in \mathcal{C}, \phi_{m_0} \subset \phi_{m_1}$ but $\phi_{m_1} \notin \mathcal{C}$ [we shall contradict re-ness of $\mathcal{P}_\mathcal{C}$].

The plan is to find an $h \in \mathcal{R}^{(1)}$ such that $h(x) \in \mathcal{P}_\mathcal{C}$ iff $x \in \overline{K}$. A first attempt to accomplish this is to define h through the S-m-n theorem from the (hopefully!) computable function ψ:

$$\psi(x,y) = \begin{cases} \phi_{m_0}(y) \textbf{ if } \phi_x(x)\uparrow \\ \phi_{m_1}(y) \textbf{ if } \phi_x(x)\downarrow \end{cases}$$

As ψ is given, it is not easy to see that it is computable. We next try

$$\tilde{\psi}(x,y) = \begin{cases} \phi_{m_0}(y) \textbf{ if } \phi_{m_0}(y)\downarrow \textit{ before } \phi_x(x)\downarrow (*) \\ \phi_{m_1}(y) \textbf{ otherwise} \end{cases}$$

(*)That is, $\phi_{m_0}(y)\uparrow \& \phi_x(x)\uparrow \vee (\exists z)[\phi_{m_0}(y)\downarrow$ in $\leq z$ steps $\& \phi_x(x)$ does *not* converge in $< z$ steps].

Intuitively, $\tilde{\psi}$ is computable. [Given x and y, start program m_0 on y, and program x on x. If the former converges *first*, then output $\phi_{m_0}(y)$ and stop. Otherwise, discard the computation of m_0 and start program m_1 on y; if it ever halts, output $\phi_{m_1}(y)$ and stop.] Let us postpone formal proof that $\tilde{\psi} \in \mathcal{P}$ for a while.

By S-m-n (assuming $\tilde{\psi} \in \mathcal{P}$), there is an $h \in \mathcal{R}^{(1)}$ such that

$$\tilde{\psi} = \lambda xy.\phi_{h(x)}(y).$$

Let $\phi_x(x)\uparrow$. Then $\phi_{h(x)} = \phi_{m_0}$ as it is $\phi_{m_0}(y)\downarrow$ *before* $\phi_x(x)\downarrow$.

Let $\phi_x(x)\downarrow$. Let y be given.

Case 1. $\phi_{m_0}(y)\downarrow$ *before* $\phi_x(x)\downarrow$. Then $\phi_{h(x)}(y) = \phi_{m_0}(y) = \phi_{m_1}(y)$ (since $\phi_{m_0} \subset \phi_{m_1}$).

Case 2. $\phi_x(x)\downarrow$ *before* $\phi_{m_0}(y)\downarrow$.(*) Then $\phi_{h(x)}(y) = \phi_{m_1}(y)$. That is, $\phi_x(x)\downarrow \Rightarrow \phi_{h(x)} = \phi_{m_1}$. This coupled with the case "$\phi_x(x)\uparrow$" gives $\phi_{h(x)} = \phi_{m_0}$ iff $x \in \overline{K}$. [Note that this establishes, incidentally, that $\psi = \tilde{\psi}$; hence, $\psi \in \mathcal{P}$ since $\tilde{\psi} \in \mathcal{P}$ (see below).] That is, $h(x) \in \mathcal{P}_{\mathcal{C}}$ iff $x \in \overline{K}$, thus (Lemma 1) $\mathcal{P}_{\mathcal{C}}$ is not re. This contradiction establishes the lemma.

We now settle the claim $\tilde{\psi} \in \mathcal{P}$:

Define $g = \lambda xy.(\mu z)(t(m_0,y,z)\cdot t(x,x,z))$ [this is the "minimum number of steps" among the two computations $\phi_{m_0}(y)$ and $\phi_x(x)$]. Clearly, $g \in \mathcal{P}$ and

$$\tilde{\psi} = \lambda xy.\textbf{if } t(m_0,y,g(x,y)) = 0 \textbf{ then } \phi_{m_0}(y) \textbf{ else } \phi_{m_1}(y).//$$

The corresponding statement for $\{x \mid W_x \in \mathcal{C}\}$ is

Corollary   Let $\mathcal{P}_{\mathcal{C}} = \{x \mid W_x \in \mathcal{C}\}$ be re. Then for all $A \in \mathcal{C}$, if $A \subset B$ and B is re, it must be that $B \in \mathcal{C}$.

Proof $\mathcal{P}_{\mathcal{C}} = \{x \mid \phi_x \in \tilde{\mathcal{C}}\}$, where $\tilde{\mathcal{C}} = \{\psi \in \mathcal{P} \mid \text{dom}(\psi) \in \mathcal{C}\}$. Let $A \in \mathcal{C}$, B be re and $A \subset B$. Let ψ and χ in $\mathcal{P}$ be such that $\text{dom}(\psi) = A$ and $\text{dom}(\chi) = B$. We may assume that $\psi(x) = 0$ iff $\psi(x)\downarrow$, and same for χ. [Let $f \in \mathcal{P}$ be such that $\text{dom}(f) = A$. Then $\psi = \lambda x.0\cdot f(x) \in \mathcal{P}$.] But $\psi \in \tilde{\mathcal{C}}$, hence Lemma 2 and $\psi \subset \chi$ imply $\chi \in \tilde{\mathcal{C}}$, that is $B = \text{dom}(\chi) \in \mathcal{C}.//$

Theorem 3   (Rice) *A complete index set is recursive iff it is either $\varnothing$ or $\mathcal{N}$.*

Proof (1) *if* part: This is trivial, as both $\varnothing$ and $\mathcal{N}$ are recursive.

(2) *only if* part: Let $\mathcal{P}_{\mathcal{C}} = \{x \mid \phi_x \in \mathcal{C}\}$ be recursive.

Case (i). $\psi = \lambda x.(\mu y)(y + 1)$, the empty function, is in $\mathcal{C}$. Then $\mathcal{C} = \mathcal{P}$, since $\mathcal{P}_{\mathcal{C}}$ is re and, by Lemma 2, any χ such that $\psi \subset \chi$ must be in $\mathcal{C}$. But $(\forall \chi)\, \chi \in \mathcal{P} \Rightarrow \psi \subset \chi$.

It follows that $\mathcal{P}_{\mathcal{C}} = \mathcal{N}$.

(*)This condition understood as $\neg(\phi_{m_0}(y)\downarrow$ *before* $\phi_x(x)\downarrow)$.

Case (ii). $\psi \notin \mathcal{C}$. Clearly, $\mathcal{N} - \mathcal{P}_\mathcal{C} = \{x \mid \phi_x \notin \mathcal{C}\} = \{x \mid \phi_x \in \mathcal{P} - \mathcal{C}\}$ and $\psi \in \mathcal{P} - \mathcal{C}$. According to *Case (i)*, $\mathcal{N} - \mathcal{P}_\mathcal{C} = \mathcal{N}$, hence, $\mathcal{P}_\mathcal{C} = \varnothing$.//

Corollary $\{x \mid W_x \in \mathcal{C}\}$ is recursive iff it is either $\varnothing$ or $\mathcal{N}$.

Proof $\{x \mid W_x \in \mathcal{C}\}$ is a complete set of indices.//

Note: As earlier noted, Rice's theorem says that the *only* recursive "complete sets of programs" (ϕ-indices) for $\mathcal{P}$-functions are $\varnothing$ and $\mathcal{N}$.(*) It is crucial that (1) *complete* sets of indices are considered: For example, $\{x \mid \phi_x = \lambda x.0\}$ is nonrecursive. [It is neither $\varnothing$ nor $\mathcal{N}$.) However, if we do not take *all* "programs" (ϕ-indices) of $\lambda x.0$, but instead take only a *finite* number of them, then this set of programs is certainly recursive (even primitive recursive; indeed, S-rudimentary.] (2) The fact that $\mathcal{P}$ contains nontotal functions plays a role as well in the validity of Rice's theorem. If instead we considered a universal function (enumeration function) $i \mapsto \psi_i \colon \mathcal{N} \to \mathcal{R}$ (of course, this enumeration cannot be *onto*; why?), say, the enumeration ξ of $\mathcal{PR}$ (Theorem 3.5.2), then there are nontrivial properties of functions whose corresponding complete index sets *are* recursive. For example, the set $S = \{x \mid \xi(x,0) = 0\}$ is recursive, even though it is neither $\varnothing$ nor $\mathcal{N}$ [e.g. $\lambda x.0 \in S$ but $\lambda x.1 \notin S$].

Rice's theorem (and its corollary) is a powerful tool for proving nonrecursiveness of complete sets of indices. Lemma 2 (and its corollary) serve well in proving non re-ness. In the next chapter, we shall see a generalization of Lemma 2 (and its corollary) in an "iff-version".

We conclude by citing a number of direct interesting applications.

(1) $\{(x,y) \mid \phi_x = \phi_y\}$ is not recursive. This is "the equivalence problem of programs (x and y)". Indeed, if it where solvable, then so would be $\{x \mid \phi_x = \phi_{y_0}\}$, where y_0 is a particular index of $\lambda x.(\mu y)(y + 1)$. As $\varnothing \neq \{x \mid \phi_x = \phi_{y_0}\} \neq \mathcal{N}$ obviously, this last complete index set is not recursive.

(2) $\{(x,y) \mid \phi_x = \phi_y\}$ is not re. Indeed, if it were, then so would be $\{x \mid \phi_x = \phi_{y_0}\}$. This cannot be since any nontrivial extension ψ of a ϕ_x such that $\phi_x = \phi_{y_0}$ (for example, $\psi = \lambda x.0$) does not satisfy $\psi = \phi_{y_0}$.

(3) Nonrecursiveness of $\{x \mid W_x = \mathcal{N}\} = \{x \mid \phi_x \in \mathcal{R}\}$ is an easy corollary of Rice's theorem. We know that this set is not re, but this is not provable by Lemma 2 (or its corollary).

(4) Clearly, the set $\{x \mid W_x \in \mathcal{R}_*\}$ is not recursive. It is also true that it is not re. Indeed, $\varnothing \in \mathcal{R}_*$ but $\varnothing \subset K$, yet $K \notin \mathcal{R}_*$. The result follows from the corollary of Lemma 2. It is extremely interesting to note (see Theorem 8.4.3) that there is a *re* subset S of $\{x \mid W_x \in \mathcal{R}_*\}$ such that for any $A \in \mathcal{R}_*$ there is a $y \in S$ such that $A = W_y$. Of course, $S \neq \{x \mid W_x \in \mathcal{R}_*\}$. Any recursive enumeration of S provides an enumeration of *all* recursive sets.

(*)Of course, quite nontrivial recursive sets which are not complete sets of indices exist. For example, recall that $\mathcal{PR}_* \subset \mathcal{R}_*$ and we already know that $\mathcal{PR}_*$ contains more than just $\varnothing$ and $\mathcal{N}$.

PROBLEMS

1. Prove that if $f \in \mathcal{R}^{(1)}$, then for some $g \in \mathcal{R}^{(2)}$, $f = \lambda x.(\mu y)g(y,x)$.

2. Let $\tilde{\phi}:\mathcal{N} \to \mathcal{P}^{(1)}$ be any *onto* map satisfying the *S-m-n* and universal function theorems. Show that for any $f \in \mathcal{P}^{(1)}$, $A_f = \{x \mid \tilde{\phi}_x = f\}$ is not recursive. (*Hint:* You may observe that Rice's theorem holds for any $\tilde{\phi}$ like the above.)

3. (Rogers) An onto map $\tilde{\phi}:\mathcal{N} \to \mathcal{P}^{(1)}$ is an *acceptable numbering* if (i) and (ii) below hold, where $\phi:\mathcal{N} \to \mathcal{P}^{(1)}$ is, say, the "standard" Gödel numbering of $\mathcal{P}^{(1)}$ obtained in Chapter 5.

   (i) There is an $f \in \mathcal{R}^{(1)}$ such that $\phi \circ f = \tilde{\phi}$

   (ii) There is a $g \in \mathcal{R}^{(1)}$ such that $\tilde{\phi} \circ g = \phi$

   Note: Intuitively, we can go back and forth between ϕ and $\tilde{\phi}$ *effectively*, but not necessarily in a 1-1 fashion.

   Prove that

   (1) (i) holds iff $\tilde{\phi}$ satisfies the universal function theorem
   (i.e., $\lambda xy.\tilde{\phi}_x(y) \in \mathcal{P}^{(2)}$)

   (2) (ii) holds iff $\tilde{\phi}$ satisfies a version of the *S-m-n* theorem, namely, for any $h \in \mathcal{P}^{(2)}$, there is a $\sigma \in \mathcal{R}^{(1)}$ such that $\tilde{\phi}_{\sigma(x)}(y) = h(x,y)$ for all x,y.

   (3) If $\tilde{\phi}$ is acceptable, then for any $f \in \mathcal{P}^{(1)}$, $\{x \mid \tilde{\phi}_x = f\}$ is infinite.

4. Is $\lambda x.x \in \mathrm{dom}(\phi_y)$ recursive? (*Hint:* Consider cases according to y.)

5. Is $\{x \mid \phi_x \in \mathcal{PR}\}$ recursive? re?

6. Is $\{x \mid \phi_x \notin \mathcal{PR}\}$ recursive? re?

7. Is $\{x \mid \phi_x$ is *not* 1-1$\}$ re? How about its complement?

8. Is $\{x \mid \phi_x$ is *not* onto$\}$ re? How about its complement?

9. (Rice-Myhill-Shapiro) Let $A = \{x \mid \phi_x \in \mathcal{C}\}$ be re, where $\mathcal{C} \subset \mathcal{P}^{(1)}$. Prove that $f \in \mathcal{C}$ iff there is a *finite* function(*) $\xi \in \mathcal{C}$ such that $\xi \subset f$. (*Hint:* Let A be re, and assume that there is an $f \in \mathcal{C}$ such that $\xi \subset f$ & ξ finite $\Rightarrow \xi \notin \mathcal{C}$. Next, define an $h \in \mathcal{R}^{(1)}$ by *S-m-n* such that $\phi_{h(x)}$ is *finite* and $\phi_{h(x)} \subset f$ if $x \in K$, $\phi_{h(x)} = f$ otherwise. This would make $\overline{K}$ re. (Why?) *Further hint:* To obtain h, show that the function informally computed as follows is in $\mathcal{P}$:

(*)I.e., a function with finite domain.

Given x and y, start computing $\phi_x(x)$. If $\phi_x(x)\downarrow$ in $\leq y$ steps, then compute forever (no output). If $\phi_x(x)$ does *not* converge in $\leq y$ steps, then discard the $\phi_x(x)$ computation, and compute and output (iff $f(y)\downarrow$) the number $f(y)$. This settles the *only if* part. The *if* part follows from Lemma 7.3.2.)

10. ("Vector" version of the Rice-Myhill-Shapiro theorem)
Let $A = \{\vec{x}_n \mid (\phi_{x_1}, \ldots, \phi_{x_n}) \in \mathcal{C}\}$ be re, where $\mathcal{C}$ is a subset of

$$\underbrace{\mathcal{P}^{(1)} \times \ldots \times \mathcal{P}^{(1)}}_{n \text{ times}}$$

Show that
(1) $(\psi_1, \ldots, \psi_n) \in \mathcal{C}, \psi_i \subset \chi_i$ and $\chi_i \in \mathcal{P}^{(1)}$ $(i = 1, \ldots, n)$ implies $(\chi_1, \ldots, \chi_n) \in \mathcal{C}$.

(2) $(\psi_1, \ldots, \psi_n) \in \mathcal{C}$ iff for some finite functions $\xi_1, \ldots, \xi_n, \xi_i \subset \psi_i$ $(i = 1, \ldots, n)$ and $(\xi_1, \ldots, \xi_n) \in \mathcal{C}$.

11. ("Vector" version of Rice's Theorem). Prove that the set A of Problem 10 is recursive iff $\varnothing = A$ or $\mathcal{N}^n = A$.

12. Is the theorem of Rice applicable to $\{x \mid \phi_x(x)\downarrow\}$? To $\{x \mid \phi_x(x)\uparrow\}$? To $\{x \mid \phi_x = \phi_{f(x)}\}$, where $f \in \mathcal{R}^{(1)}$? What can you say about re-ness or recursiveness of the last set?

13. $\{x \mid \phi_x(x) = c\}$ is re but not recursive (c is a constant). Is Rice's Theorem applicable here?

14. Two sets A and B are *recursively inseparable* iff the following hold:

(1) $A \cap B = \varnothing$

(2) There is no recursive set C such that $A \subset C$ and $B \subset \mathcal{N} - C$.
Prove that $A = \{x \mid \phi_x(x) = 0\}$ and $B = \{x \mid \phi_x(x) = 1\}$ are recursively inseparable. (*Hint*: Clearly, $A \cap B = \varnothing$. Let there be a $C \in \mathcal{R}_*$ such that $A \subset C$ and $B \subset \mathcal{N} - C$. Let i be a ϕ-index of the characteristic function of $\mathcal{N} - C$. Argue that $i \notin C \cup (\mathcal{N} - C)$, thus deriving a contradiction.)

REFERENCES

[1] Blum, M. "A Machine-independent Theory of the Complexity of Recursive Functions". *Journal of the ACM* 14 (1967):322-336.

[2] Church, A. "An Unsolvable Problem of Elementary Number Theory". *Amer. J. Math.* 58 (1936): 345–363. (Also in Davis, M. *The Undecidable.* Hewlett, New York: Raven Press, 1965:89–107.)

[3] Kalmár, L. "An Argument Against the Plausibility of Church's Thesis". *Constructivity in Mathematics.* Proceedings of the Colloquium held at Amsterdam (1957):72–80.

[4] Péter, R. *Recursive Functions.* New York: Academic Press, 1967.

[5] Robinson, A. *Nonstandard Analysis.* Amsterdam: North-Holland, 1966.

CHAPTER 8

Effective Enumerability (Part II)

The notion of *recursive enumerability* (effective enumerability in intuitive terms) and its elementary consequences have been introduced in §3.6 and §3.7. A powerful related result was established in §7.3 (Lemma 2 and its corollary). Before we proceed with the further study of re predicates, the reader is advised to quickly review §3.6 (especially Definition 1, Theorem 1 and its corollaries, and Theorem 2) and §3.7 (especially Theorem 2 and corollary, Theorem 5 and its corollaries, and Definition 4).

§8.1 ENUMERATIONS OF RE SETS

Theorem 1 $P(\vec{x}_n)$ is re and nonempty iff there is a rudimentary function(*) $f\colon \mathcal{N} \to \mathcal{N}$ such that $f(\mathcal{N}) = \{\langle \vec{x}_n \rangle \mid P(\vec{x}_n)\}$, where $\lambda \vec{x}_n.\langle \vec{x}_n \rangle$ is defined from $\lambda xy.\langle x,y \rangle$ of Section 3.9 according to Definition 2.4.3.

(*)We mean "m-rudimentary function" for some m. Since the classes of m- and n-rudimentary functions coincide ($m \neq n$) as shown by Bennett [1], the unspecified m is immaterial.

Proof To begin with, $\lambda \vec{x}_n . \langle \vec{x}_n \rangle$ is rudimentary for any n. Indeed, $y = \langle \vec{x}_n \rangle \equiv (\exists z)_{\leq p(\vec{x}_{n-1})} (y = \langle z, x_n \rangle \;\&\; z = \langle \vec{x}_{n-1} \rangle)$, where p is a concatenation polynomial in the $\vec{x}_{n-1}$ (Definition 5.1.4) and we have inductively assumed that $\lambda \vec{x}_{n-1} . \langle \vec{x}_{n-1} \rangle$ is rudimentary. To complete the argument, (1) the induction basis states that $\lambda x . x$ is rudimentary (true), (2) $\langle \vec{x}_n \rangle$ is bounded by some concatenation polynomial in $\vec{x}_n$ (it is itself a concatenation polynomial) and, (3) $\lambda xyz . z = \langle x, y \rangle$ is rudimentary (Lemma 3.9.9. Note that its proof is valid for the m-rudimentary case, $m > 2$).

Next, let $f(\mathcal{N}) = \{ \langle \vec{x}_n \rangle \mid P(\vec{x}_n) \}$ with f and $\langle \; \rangle$ as in the theorem. Since f and $\lambda \vec{x}_n . \langle \vec{x}_n \rangle$ are clearly in $\mathcal{R}$, $P(\vec{x}_n)$ is re (and obviously nonempty since f is total) by Definition 3.6.1.

Finally, let $P(\vec{x}_n)$ be nonempty and re. Thus, (i) For some $\vec{a}_n \in \mathcal{N}^n$, $P(\vec{a}_n)$, (ii) For some z, $P(\vec{x}_n) \equiv (\exists y) T^{(n)}(z, \vec{x}_n, y)$, where $T^{(n)}$ is rudimentary (Theorem 5.2.3).

The required f is defined as in the proof of the projection theorem (Theorem 3.6.1).

$$
f(w) = \begin{cases} \langle \Pi_1^{n+1}(w), \ldots \Pi_n^{n+1}(w) \rangle \textbf{ if } T^{(n)}(z, \Pi_1^{n+1}(w), \ldots, \Pi_{n+1}^{n+1}(w)) \\[2mm] \langle \vec{a}_n \rangle \textbf{ otherwise} \end{cases}
$$

[Note that (1) rudimentary functions are closed under definition by cases, (2) $\lambda xy . y = \Pi_i^{n+1}(x)$ is rudimentary, since $y = \Pi_i^{n+1}(x) \equiv (\exists x_1, \ldots, x_{i-1}, x_{i+1}, \ldots, x_{n+1})_{\leq x} \; x = \langle x_1, \ldots, x_{i-1}, y, x_{i+1}, \ldots, x_{n+1} \rangle$ and (3) $\lambda x . \Pi_i^{n+1}(x)$ is rudimentary, by (2), since it is also bounded by x].//

Corollary   Theorem 1 holds, when "rudimentary" is everywhere substituted by "primitive recursive".

Note: The proof of Theorem 1 was done by a "dovetailing" computation (as in the proof of Theorem 3.6.1 and in Example 2 in the preamble of Chapter 3). The principle of dovetailing is, essentially, to proceed in the computation by "cycles": In the ith cycle, the first i inputs of $\phi_z^{(n)}$ (where the n-tuple inputs are enumerated by some standard way) are used, each for i steps of the computation of z. For each such computation that may halt the corresponding input is being output (by f). In the proofs of Theorem 3.6.1 and 1 however, instead of subdividing the computation in "cycles", we enumerate, in standard form, $(n + 1)$-tuples $(\vec{x}_n, i)$ and try $\phi_z^{(n)}(\vec{x}_n)$ for i steps;(*) we output (through f) $\vec{x}_n$ iff $\phi_z^{(n)}(\vec{x}_n) \downarrow$ in i steps.

The above corollary is the usual form in which Theorem 1 appears in the literature. Grzegorczyk [3] has shown that $\mathcal{E}^0$-functions (to be defined in Chapter 13) are sufficient to enumerate any re set. This follows the same proof

(*)"Steps" is used in the sense described at the end of §7.1.

as in Theorem 1 using a Kleene-predicate in $\mathcal{E}^0$ (such is the $T^{(n)}$ of Theorem 3.7.4, but this will not be obvious until we introduce $\mathcal{E}^0$ in Chapter 13).

At the present state of knowledge, it is not known whether the inclusion $\text{Rud} \subset \mathcal{E}^0_*$ is proper or not.(*)

Thus, on one hand, rudimentary functions $\lambda\vec{x}.f(\vec{x})$ are "simpler" than $\mathcal{E}^0$ functions, perhaps strictly, in the sense that their associated predicates ($y = f(\vec{x})$) are a fortiori in $\mathcal{E}^0_*$, on the other hand the $\mathcal{E}^0$-functions are of smaller *size* than rudimentary functions. (It is known [3] that $\lambda\vec{x}_n.f(\vec{x}_n) \in \mathcal{E}^0$ implies that $f(\vec{x}_n) \le x_i + k$ for all $\vec{x}_n$ and some fixed i and k depending on f.)

Theorem 2 $P(\vec{x}_n)$ is re iff for some $f \in \mathcal{P}^{(1)}$ and $\lambda\vec{x}_n.\langle\vec{x}_n\rangle$ in $\mathcal{PR}, f(\mathcal{N}) = \{\langle\vec{x}_n\rangle \mid P(\vec{x}_n)\}$.

   Proof *only if* part. Let $P(\vec{x}_n) \equiv (\exists y)T^{(n)}(i,\vec{x}_n,y)$ for some i, where $T^{(n)} \in \mathcal{PR}_*$.

Then f is given by

$$f(w) = \begin{cases} \langle\Pi_1^{n+1}(w),\ldots,\Pi_n^{n+1}(w)\rangle \text{ if } T^{(n)}(i,\Pi_1^{n+1}(w),\ldots,\Pi_{n+1}^{n+1}(w)) \\ \uparrow \textbf{ otherwise} \end{cases}$$

That is,

$f = \lambda w.$ **if** $T^{(n)}(i,\Pi_1^{n+1}(w),\ldots,\Pi_{n+1}^{n+1}(w))$ **then** $\langle\Pi_1^{n+1}(w),\ldots,\Pi_n^{n+1}(w)\rangle$

   else $(\mu y)(y+1)$

if part. Let $f \in \mathcal{P}^{(1)}$ and $f(\mathcal{N}) = \{\langle\vec{x}_n\rangle \mid P(\vec{x}_n)\}$. Then $P(\vec{x}_n) \equiv (\exists y)f(y) = \langle\vec{x}_n\rangle$ Now $\lambda xy.x = f(y)$ is re (Theorem 3.7.2) and hence $\lambda y\vec{x}_n.f(y) = \langle\vec{x}_n\rangle$ is re; thus $P(\vec{x}_n)$ is re by Corollary 1 of the projection theorem (3.6.1).//

Corollary 1   There is an $h \in \mathcal{R}^{(1)}$ such that $\text{dom}(\phi_x) = \text{range}(\phi_{h(x)})$; i.e., $W_x = \text{range}(\phi_{h(x)})$.

   Proof $z \in W_x (z \in \text{dom}(\phi_x))$ is equivalent to the predicate $(\exists y)T(x,z,y)$.

The f-function is defined as in the above proof, except that now it is defined to depend on x as well as z. That is,

$$f = \lambda xz. \text{ if } T(x,K(z),L(z)) \text{ then } K(z) \text{ else } (\mu y)(y+1)$$

where $(K,L)\circ\langle x,y\rangle = (x,y)$ for all x,y.

(*)On the other hand, the improper inclusion can be seen to hold as soon as $\mathcal{E}^0$ and its simple properties are discussed.

Since $f \in \mathcal{P}$, there is an $h \in \mathcal{R}^{(1)}$ such that $f = \lambda xy.\phi_{h(x)}(y)$. For fixed x, $\phi_{h(x)}$ is f of the previous proof.//

Note: An alternative proof, not relying on the normal form theorem, is to define (as in Rogers [6], for example)

$$f(x,y) = \begin{cases} y \text{ if } \phi_x(y)\downarrow \\ \uparrow \text{ otherwise} \end{cases}$$

that is, $f = \lambda xy.y + 0 \cdot \phi_x(y) \in \mathcal{P}$, hence $f = \lambda xy.\phi_{m(x)}(y)$ for some $m \in \mathcal{R}$ and, clearly, $\text{range}(\phi_{m(x)}) = \text{dom}(\phi_x)$.

Corollary 2   There is a $\sigma \in \mathcal{R}^{(1)}$ such that $\text{range}(\phi_x) = \text{dom}(\phi_{\sigma(x)})$ [that is, $\text{range}(\phi_x) = W_{\sigma(x)}$].

Proof $y \in \text{range}(\phi_x) \equiv (\exists z)(y = \phi_x(z)) \equiv (\exists z)(\exists w)[T(x,z,w) \& d(w) = y]$, where d is the decoding or output function of the normal form theorem (3.7.4).

It follows that $y \in \text{range}(\phi_x) \equiv (\exists m)[T(x,K(m),L(m)) \& d(L(m)) = y]$. Let $\psi = \lambda xy.(\mu m)[T(x,K(m),L(m)) \& d(L(m)) = y]$. Clearly, $y \in \text{range}(\phi_x) \equiv \psi(x,y)\downarrow \equiv \phi_{\sigma(x)}(y)\downarrow$ where $\sigma \in \mathcal{R}$ was obtained by applying the S-m-n theorem to ψ.

Thus $\text{range}(\phi_x) = \text{dom}(\phi_{\sigma(x)})$.//

Note: The proof of Corollary 2 is a formalization of another "dovetailing" computation:

We are given a program x (for ϕ_x) and we want to write another program, using x [the new program is $\sigma(x)$], which will halt for input y iff program x will output y for *some* input z; the output of the program $\sigma(x)$ is unimportant (z, of course, is unknown initially, if it exists).

The details of the informal counterpart of the above proof are found in Example 2 (preample section of Chapter 3). Essentially, all computations $\phi_x(z)$ are tried ($z = 0,1,\ldots$) in the search of a z_0 such that $\phi_x(z_0) = y$ (if such a z_0 exists), but in a manner that we shall not be "trapped" in an "infinite loop" attempting to compute some $\phi_x(z)$, where $\phi_x(z)\uparrow$ and $z < z_0$. That is, each z is tried an unbounded number of times; the ith time it is tried by doing *exactly i* steps of the $\phi_x(z)$ computation. Formally, this amounts to looking at $T(x,z,i)$ for all (z,i), or looking at $T(x,K(m),L(m))$ for all m, until T becomes "true".

Corollary 1 can be sharpened:

Corollary 3   There is a $\tau \in \mathcal{R}^{(1)}$ such that $W_x = \text{range}(\phi_{\tau(x)})$, and $W_x \neq \varnothing \Rightarrow \phi_{\tau(x)} \in \mathcal{R}$.

Proof Refer to the proof of Theorem 1. We want to make "$\langle \vec{a}_n \rangle$" a $\mathcal{P}$-function of the set's re-index so that f can be viewed as a function of both w and z, to allow an application of the S-m-n theorem.

To this end, we define a "selector" function $g \in \mathcal{P}^{(1)}$ such that $W_x \neq \varnothing$ $\Rightarrow g(x) \in W_x$, $W_x = \varnothing \Rightarrow g(x)\uparrow$. $g(x)$ will be the z-component of the first(*) pair (z,w) such that $\phi_x(z)\downarrow$ in w steps.

$g(x) = K((\mu m) T(x, K(m), L(m)))$. Clearly, g has the required properties.
Next, let f be

$$f = \lambda xz. \text{ if } T(x, K(z), L(z)) \text{ then } K(z) \text{ else } g(x).$$

By S-m-n, $f = \lambda xy.\phi_{\tau(x)}(y)$ and $W_x \neq \varnothing \Rightarrow \phi_{\tau(x)} \in \mathcal{R}.//$

We have seen (Theorem 2) that ranges of partial recursive functions are re; hence, whenever $f(\mathcal{N}) = A$, where $f \in \mathcal{P}$, then a $k \in \mathcal{R}$ exists such that $k(\mathcal{N}) = A$, provided $A \neq \varnothing$. It is intuitively clear that if we know f (through a program P_f), then we can construct a program P_k for k which has P_f as a "subroutine". (Indeed, P_k will dovetail P_f computations to obtain as many items of A as we like.) This is formalized below; m is the "method" by which from P_f (that is, x below) we construct P_k (that is, $m(x)$).

Corollary 4   There is an $m \in \mathcal{R}^{(1)}$ such that $\text{range}(\phi_x) = \text{range}(\phi_{m(x)})$ and $\text{range}(\phi_x) \neq \varnothing \Rightarrow \phi_{m(x)} \in \mathcal{R}$.

Proof By Corollary 2, for some $\sigma \in \mathcal{R}^{(1)}$, $\text{range}(\phi_x) = W_{\sigma(x)}$.
By Corollary 3, for some $\tau \in \mathcal{R}^{(1)}$, $W_{\sigma(x)} = \text{range}(\phi_{\tau\circ\sigma(x)})$ and $W_{\sigma(x)} \neq \varnothing \Rightarrow \phi_{\tau\circ\sigma(x)} \in \mathcal{R}$. Thus, $m \overset{\text{def}}{=} \tau \circ \sigma$ will do.//

The enumerations we employed so far are, in general, enumerations with repetitions. For example, in the proof of Theorem 1 and Corollary 3 of Theorem 2 a certain fixed item of W_x is being enumerated infinitely often if $W_x \neq \mathcal{N}$ (why?). Also, regardless of choice of enumeration, a total function can enumerate *finite* sets only with repetitions. We will observe that re-ness is an algorithmic (effective) counterpart of the notion of "at most enumerable" of set theory (Definition 1.5.2). In this latter case, a set A is at most enumerable iff there is a 1-1 and onto function $f: \mathcal{N} \rightarrow A$(†) (of course, f will not be total necessarily, for example, when A is finite).
There is an effective counterpart of this in a very strong sense: Not only for any re set there is a way to "effectively enumerate" it without repetitions, but this enumerating program can be *constructed*, once the re set is given. [The re set is *given* either as a W_x or as $\text{range}(\phi_x)$].

Corollary 5   There are functions p and q in $\mathcal{R}^{(1)}$ such that

(1) $W_x = \text{range}(\phi_{p(x)})$, $\phi_{p(x)}$ is 1-1 and, if nontotal, $\text{dom}(\phi_{p(x)}) = \{i \in \mathcal{N} \mid i < n\}$ for some n.

(*)"First" in some effective enumeration of pairs.
(†)This can be easily deduced from Corollary 1 of Theorem 1.5.1.

(2) $\text{range}(\phi_x) = \text{range}(\phi_{q(x)})$, $\phi_{q(x)}$ is 1-1 and, if nontotal, $\text{dom}(\phi_{q(x)})$ $= \{i \in \mathcal{N} \mid i < l\}$ for some l.

Proof (1) We adapt the proof of Corollary 3 to obtain an f such that $\lambda z.f(x,z)$ is 1-1:

$$f(x,0) = g(x) \text{ (} g \text{ as in the proof of Corollary 3)}$$

$$f(x,z + 1) = K((\mu m)[T(x,K(m),L(m)) \text{ \& } K(m) \notin \{f(x,0),\ldots,f(x,z)\}])$$

Since $\mathcal{P}$ is closed under course-of-values recursion, $\lambda xz.f(x,z) \in \mathcal{P}$, and for some $p \in \mathcal{R}$, $f = \lambda xz.\phi_{p(x)}(z)$. Clearly $\phi_{p(x)}$ is what we want.

(2) By Corollary 2, $\text{range}(\phi_x) = W_{\sigma(x)}$, where $\sigma \in \mathcal{R}^{(1)}$. By part (1), $W_{\sigma(x)}$ $= \text{range}(\phi_{p \circ \sigma(x)})$. $q = p \circ \sigma$ will do.//

Note: Theorem 2 and its corollaries appear as such or as exercises in the literature (for example, in Rogers [6], Kreider and Ritchie [4], Brainerd and Landweber [2]). The basis of their proofs were dovetailing arguments. Frequently, using the above results spares us from devising more dovetailing arguments.

Example 1 If $A \subset \mathcal{N}$ is re and $f \in \mathcal{P}^{(1)}$, then $f^{-1}(A)$ is re. This can be established from first principles via a dovetailing argument. [For example, enumerate A; for each x $\in A$ obtained so far, at the ith step of the enumeration, do i steps of each of the computations $f(0),\ldots,f(i)$. List under $f^{-1}(A)$ those j's in $\{0,\ldots,i\}$ that have already been found to satisfy $f(j) = x$.

Another way of seeing this is to list systematically triples (x,y,z) and put x in the $f^{-1}(A)$-list just in case $f(x) = y$ & $y \in A$ is verifiable in $\leq z$ steps]. Formally, we can prove that $r \in \mathcal{R}^{(2)}$ exists such that $W_{r(x,y)} = \phi_x^{-1}(W_y)$.

Indeed, $z \in \phi_x^{-1}(W_y)$ iff $\phi_x(z) \in W_y$ iff $\phi_y \circ \phi_x(z)\!\downarrow$; that is, $\phi_x^{-1}(W_y) = \text{dom}(\phi_y \circ \phi_x)$. By the universal function theorem, $\lambda xyz.\phi_y \circ \phi_x(z)$ is in $\mathcal{P}$, and by $S\text{-}m\text{-}n$ this last function is $\lambda xyz.\phi_{r(x,y)}(z)$ for some $r \in \mathcal{R}^{(2)}$, thus $\text{dom}(\phi_y \circ \phi_x) = W_{r(x,y)}.$//

§8.2 ENUMERATIONS OF RECURSIVE SETS

Example 1 As motivation for the next theorem let us recall the "standard" sequential search of an ordered finite sequence of integers. More specifically, let $f(0),\ldots,$ $f(n)$ be the sequence (stored in ascending order(*) in a PL/1-like "array" f)

(*)In the sense that $f(i) \leq f(i + 1)$, $i = 0,\ldots,n - 1$.

and let us try to solve the question

$$\text{``}(\exists i)(0 \le i \le n \ \& \ f(i) = m)?\text{''} \tag{1}$$

This is usually done as follows in pseudo-PL/1.

```
i ← 0
do while (f(i) < m & i < n)
i ← i + 1
end
if f(i) = m then output "m is in f"
else output "m is not in f"      /* Note: if f(i) overshoots m, the
                                    search terminates unsuccessfully */
```

Of course, f is a partial recursive function $\{0, \ldots, n\} \to \mathcal{N}$. Let next $f \colon \mathcal{N} \to \mathcal{N}$ be an increasing(*) recursive function. The question corresponding to (1) above becomes

$$\text{``}m \in \text{range}(f)?\text{''}$$

The solution is, in a first approximation, obtained from the above by dropping the condition "$i < n$".

This will always work for f such that range(f) is infinite, otherwise, say, when each $y \in \text{range}(f)$ is $<m$, the amended algorithm is incorrect (does not terminate).

Returning to the case that range(f) is infinite, the above amendment can be easily formalized (say, via a URM program) to show that range$(f) \in \mathcal{R}_*$. Note that in such a proof, $f \in \mathcal{R}$ is used to derive that $\lambda im.f(i) < m$ and $\lambda im.f(i) = m$ are in $\mathcal{R}_*$ and hence their characteristic functions are URM computable. Of course, when range(f) is finite, it is also recursive.//

Theorem 1 $A \in \mathcal{R}_*$ and $A \neq \varnothing$ iff A is the range of an increasing recursive function.

Proof *if* part. Let $A = \text{range}(f), f \in \mathcal{R}$ and f be increasing. Then, $A \neq \varnothing$ (since f is total).

Case 1. A is finite. Then $A \in \mathcal{R}_*$.

Case 2. A is infinite. Then $A \in \mathcal{R}_*$ by the argument of Example 1. (However, to be consistent with the plan we set out at the end of §7.1, we also show $A \in \mathcal{R}_*$ with no recourse to a specific formalism:

$$m \in \text{range}(f) \equiv m = f((\mu i)(f(i) \ge m))$$

(*)In the sense $f(i) \le f(i + 1)$ for all i.

Note that $g = \lambda m.(\mu i)(f(i) \geq m)$ is in $\mathcal{P}$, since $\lambda im.f(i) \geq m \in \mathcal{R}_*. g \in \mathcal{R}$ as well, since it is total due to the fact that $range(f)$ is infinite. Thus, $range(f)$—that is, A—is in $\mathcal{R}_*$).

only if part. Let $A \in \mathcal{R}_*$ and $A \neq \emptyset$.

Case 1. A is finite, say $A = \{f_0, f_1, \ldots, f_k\}$, where $f_i < f_{i+1}$ for $i = 0, \ldots, k-1$. Clearly, f defined by

$$f(x) = \begin{cases} \textbf{if } x = 0 \textbf{ then } f_0 \\ \textbf{if } x = 1 \textbf{ then } f_1 \\ \cdot \\ \cdot \\ \cdot \\ \textbf{if } x \geq k \textbf{ then } f_k \end{cases}$$

is recursive, increasing, and $A = range(f)$.

Case 2. A is infinite. $A \in \mathcal{R}_*$ means that c_A, the characteristic function of A, is in $\mathcal{R}$.

Define f by

$$f(0) = (\mu y)c_A(y) \text{ [that is, smallest } y \text{ in } A]$$

$$f(x + 1) = (\mu y)(c_A(y) = 0 \text{ \& } y > f(x))$$

Since $\mathcal{P}$ is closed under primitive recursion, $f \in \mathcal{P}$. Since A is infinite, f is total, hence $f \in \mathcal{R}$. Finally, $range(f) = A$, otherwise for some $m \in A$ there is an x such that $f(x) < m < f(x + 1)$, contradicting the definition of $f(x + 1).//$

Corollary 1 $A \in \mathcal{R}_*$ and A infinite iff $A = range(f)$, $f \in \mathcal{R}$ and f is strictly increasing.(*)

Proof *if* part. As in the proof of Theorem 1. Since f is strictly increasing, only the case $A (= range(f))$ infinite obtains.

only if part. Same as case 2 of the corresponding part of the proof of Theorem 1. Clearly, the f defined there is *strictly* increasing.//

Corollary 2   Let A be re and infinite. Then it contains an infinite recursive set.

Proof Let $A = range(f), f \in \mathcal{R}$. We pick an increasing subsequence g of f (possible since A is infinite, hence unbounded) to be the required recursive set:

$$g(0) = f(0)$$

$$g(x + 1) = f((\mu y)(f(y) > g(x)))$$

(*)That is, $f(i) < f(i + 1)$ for all i.

Clearly, $g \in \mathcal{R}$ and g strictly increasing. By Corollary 1, range(g) $(\subset A)$ is recursive and infinite.//

Note: If A is "given" in Corollary 2, then the required recursive subset can be "found".

Corollary 3   There is an $r \in \mathcal{R}^{(1)}$, such that if W_x is infinite, then $W_{r(x)} \subset W_x$, $W_{r(x)} \in \mathcal{R}_*$ and $W_{r(x)}$ is infinite.

> **Proof** By Corollary 3 of Theorem 8.1.2, there is a $\tau \in \mathcal{R}^{(1)}$ such that W_x = range$(\phi_{\tau(x)})$ and $W_x \neq \emptyset \Rightarrow \phi_{\tau(x)} \in \mathcal{R}$. Define g as follows
>
> $$g(x,0) = \phi_{\tau(x)}(0)$$
>
> $$g(x,y+1) = \phi_{\tau(x)}((\mu z)(\phi_{\tau(x)}(z) > g(x,y))).$$

Clearly, $g \in \mathcal{P}^{(2)}$; thus, by S-m-n, $g = \lambda xy.\phi_{h(x)}(y)$ for some $h \in \mathcal{R}^{(1)}$. By Corollary 2, range$(\phi_{h(x)})$ is recursive and infinite if W_x is infinite.

By Corollary 2 of Theorem 8.1.2, there is a $\sigma \in \mathcal{R}^{(1)}$ such that range(ϕ_x) = $W_{\sigma(x)}$. The required r is $\sigma \circ h$.//

Note: We do not only know that r exists, but we can *construct* r, as a review of the proofs of the corollaries mentioned in the above proof will clearly indicate. This is not always the case: Sometimes we can prove that an algorithm *exists,* but we do not know how to construct it [for example, if W_x is finite, then a recursive function f which enumerates it in increasing order exists.(*) We *cannot,* however, given a procedure to construct (an index of) f, since given (finite) W_x there is no algorithm (function in $\mathcal{P}$) which will let us know what *all* the elements of W_x are. See Corollary 2, Theorem 8.3.3. Such proofs are called *nonconstructive*].

Corollary 4   There is an $\eta \in \mathcal{R}^{(1)}$ such that for all x

(1)   range$(\phi_{\eta(x)}) \subset$ range(ϕ_x)

(2)   range$(\phi_{\eta(x)}) \in \mathcal{R}_*$

(3) ϕ_x strictly increasing(†) $\Rightarrow \phi_x = \phi_{h(x)}$

> **Proof** Let ψ be defined by
>
> $$\psi(x,0) = \phi_x(0)$$
>
> $$\psi(x,y+1) = \phi_x((\mu z)(\phi_x(z) > \psi(x,y)))$$

(*)Proof of Theorem 1, case 1 of *only if* part.
(†)$f \in \mathcal{P}$ is strictly increasing if for $\{x,y\} \subset$ dom(f), $x < y \Rightarrow f(x) < f(y)$.

Clearly $\psi \in \mathcal{P}$, hence for some $\eta \in \mathcal{R}^{(1)}$, $\psi = \lambda xy.\phi_{\eta(x)}(y)$.

(1)   is clearly satisfied by $\phi_{\eta(x)}$.

(2) $\phi_{\eta(x)}$ is strictly increasing. If range($\phi_{\eta(x)}$) is finite, then the result is obvious, otherwise it follows from Corollary 1.

(3)   When ϕ_x is strictly increasing then clearly (induction on y) $\phi_x(y)\downarrow$ $\Rightarrow \phi_{\eta(x)}(y)\downarrow$ and $\phi_x(y) = \phi_{\eta(x)}(y).//$

§8.3 NAMES OF RECURSIVE SETS

Since every recursive set is also re, one way of naming recursive sets is through their re-indices. With this notation, a recursive set W_x is "given" through its enumeration algorithm, x.(*) Recursive sets, by definition, have recursive characteristic functions, thus they can also be named (or given) by programs of such functions.

Definition 1   A *characteristic index* of a recursive set $S \subset \mathcal{N}$ is a ϕ-index of its characteristic function.//

Theorem 1   If a recursive set is given by a *characteristic* index, a re index for it can be computed. (More formally, there is an $h \in \mathcal{R}^{(1)}$ such that if ϕ_x is a 0-1 total function then $W_{h(x)} = \{y \mid \phi_x(y) = 0\}$).

Proof Define ψ by

$$\psi(x,y) = \begin{cases} 0 \text{ if } \phi_x(y) = 0 \\ \uparrow \text{ otherwise} \end{cases}$$

That is, $\psi(x,y) = $ **if** $\phi_x(y) = 0$ **then** 0 **else** $(\mu z)(z + 1)$. Clearly, $\psi \in \mathcal{P}$ and, by S-m-n, there is an $h \in \mathcal{R}^{(1)}$ such that $\psi = \lambda xy.\phi_{h(x)}(y)$. Thus, if x is a characteristic index of a set, then $h(x)$ is a re index of the same set.//

Note: Theorem 1 says that *if* x is a characteristic index, then $h(x)$ is a re index of the same set. Of course, by Rice's theorem, given x we cannot test algorithmically whether ϕ_x is a 0-1 recursive function or not. Thus the

---

(*)For recursive sets S in $\mathcal{N}^n$, a re index of S is, of course, an i such that $\vec{x}_n \in S$ iff $\phi_i^{(n)}(\vec{x}_n)\downarrow$. Without loss of generality, we shall present only the case $n = 1$. Results are translatable to the case $n > 1$ with the help of coding functions $\lambda \vec{x}.\langle \vec{x} \rangle$ in $\mathcal{R}$.

"output" $h(x)$ is "useful" only if we know beforehand that x is a characteristic index.(*)

Moreover, the converse procedure, by which we transform re indices to characteristic indices is not algorithmic. Certainly we do not hope to have a $\psi \in P$ such that if W_x is recursive then $\psi(x)\downarrow$ and $\psi(x)$ is a characteristic index, otherwise $\psi(x)\uparrow$ (such a ψ makes the set $\{x \mid W_x \in \mathcal{R}_*\}$ re; we have seen already that it is not, in (4) under the concluding remarks of Chapter 7). The following theorem says that *even a* ψ which is *allowed to converge* on the complement of $\{x \mid W_x \in \mathcal{R}_*\}$ cannot exist such that it is in P and $W_x \in \mathcal{R}_* \Rightarrow \psi(x)$ is a characteristic index of W_x (such a ψ would be less informative than the previous, since $\psi(x)\downarrow$ does not mean necessarily that $\psi(x)$ is a characteristic index of W_x: W_x might not be recursive!).

Theorem 2   There is no ψ in P with the property $W_x \in \mathcal{R}_* \Rightarrow \psi(x)\downarrow$ and $\phi_{\psi(x)}$ is the characteristic function of W_x.

Proof (Essentially that in Rogers [6].) We argue by contradiction. So let $\psi \in P$ with the required property exist.

Let first $h \in \mathcal{R}^{(1)}$ be such that $\phi_{h(x)} = \lambda y.0$ if $x \in K$, $\phi_{h(x)} = \lambda y.(\mu y)(y + 1)$ if $x \in \overline{K}$ (see proof of Theorem 7.3.1).

In other words,

$$W_{h(x)} = \begin{cases} \mathcal{N} \text{ if } x \in K \\ \varnothing \text{ if } x \in \overline{K} \end{cases}$$

Since $W_{h(x)} \in \mathcal{R}_*$ for all x,
$\psi \circ h \in \mathcal{R}$ and

$\phi_{\psi \circ h(x)} = \lambda y.0$ if $x \in K$ ($\phi_{\psi \circ h(x)}$ is characteristic function of $\mathcal{N}$)

$\phi_{\psi \circ h(x)} = \lambda y.1$ if $x \in \overline{K}$ ($\phi_{\psi \circ h(x)}$ is characteristic function of $\varnothing$)

In other words, $1 \in \text{range}(\phi_{\psi \circ h(x)})$ iff $x \in \overline{K}$; i.e., $(\exists y)(\exists z)(T(\psi \circ h(x), z, y)$ & $d(y) = 1) \equiv x \in \overline{K}$. The lefthand side is re by the projection theorem, the righthand side is not; a contradiction.//

A subclass of recursive sets is that of *finite* sets. Besides *re* and *characteristic* indices, a third type of naming is possible for finite sets. If D

(*)We cannot expect to have a $\chi \in P$ such that if x is a characteristic index then $\chi(x)\downarrow$ and is a re-index, $\chi(x)\uparrow$ otherwise. Such a function would make the set $\{x \mid \phi_x$ is 0-1 and total$\}$ re, say equal to range(f), $f \in \mathcal{R}$. The function $\lambda x.1 \dot{-} \phi_{f(x)}(x)$ would provide a contradiction.

$= \{x_0, \ldots, x_{n-1}\}$, then $\langle x_0, \ldots, x_{n-1}, n \rangle$ (using any recursive 1-1 coding $\langle \ \rangle$) can serve as a *name* of D. In particular, one can use $\Pi_{i<n} \, p_i^{x_i+1}$, where p_i is the ith prime ($p_0 = 2, p_1 = 3$, etc.), or $x_0^{(2)} \, 21^{m+1} \, 2x_1^{(2)} \, 21^{m+1} \, 2 \ldots 21^{m+1} \, 2x_{n-1}^{(2)}$, where $x_i^{(2)}$ is x_i in dyadic and 1^m is the longest tally of 1's in any of the x_i's. Note that in the specific codings we mentioned, n (the cardinality of D), is implicitly coded in the code.

We shall agree to use 0 as the code for $\varnothing$ under any scheme whatsoever.

In what follows we shall use a fixed but unspecified coding of finite sets.

Definition 2 A *canonical index* of a finite set D is a number u such that if $D = \{x_0, \ldots, x_{n-1}\}$ then $u = \langle x_0, \ldots, x_{n-1}, n \rangle$ under the selected (fixed) recursive 1-1 coding scheme $\langle \ \rangle$. We write $D = D_u.//$

Note: There is a different (in general) u for each of the $n!$ permutations of D.

Example 1 The following coding is suggested in Rogers [6] for assigning a *unique* number to a finite *set*:

If $D = \{x_1, \ldots, x_n\}$ then $u = 2^{x_1} + \ldots + 2^{x_n}$. Clearly, $\lambda \vec{x}_n . 2^{x_1} + \ldots + 2^{x_n}$ is *not* 1-1. u in binary has 1's exactly at positions x_i ($i = 1, \ldots, n$) from the right, where the right-most position is the 0th. Also u is independent of the particular permutation of $D.//$

The next question we consider is whether transformations between the various indices are effective.

First, it is easy to see intuitively that given a canonical index of a (finite) set, a characteristic index can be computed, therefore also a re index (by Theorem 1) can be computed (the reader is asked to present a formal argument to this effect—see Problem 8).

The converse process, going from a characteristic index to a canonical index, is not effective. Indeed, let $\psi \in \mathcal{P}$ exist such that ϕ_x is a characteristic function of a finite set $S \Rightarrow \psi(x) \downarrow$ and $S = D_{\psi(x)}$; if ϕ_x is not a characteristic function of a finite set then $\psi(x) \uparrow$.

This implies that the set $A = \{x \mid \phi_x$ is 0-1, total, and its range contains finitely many 0's$\}$ is re, since it equals $\mathrm{dom}(\psi)$. We shall derive a contradiction:

Let χ be defined by

$$\chi(x,y) = \begin{cases} 1 & \textbf{if } \phi_x(x) \text{ does } not \text{ converge in } \leq y \text{ steps} \\ 0 & \textbf{otherwise} \end{cases}$$

Clearly, $\chi = \lambda xy.$ **if** $(\forall i)_{\leq y} \, \neg T(x,x,i)$ **then** 1 **else** 0 and $\chi \in \mathcal{P}$ (indeed it is in $\mathcal{PR}$), hence for some $h \in \mathcal{R}^{(1)}$, $\chi = \lambda xy.\phi_{h(x)}(y)$.

Now,

$$\phi_x(x)\uparrow \; \Rightarrow \; \phi_{h(x)} = \lambda y.1 \; \Rightarrow \; \text{range}(\phi_{h(x)}) \text{ has } \textit{finitely} \text{ many 0's}$$

$$\phi_x(x)\downarrow \; \Rightarrow \; \text{range}(\phi_{h(x)}) \text{ has } \textit{infinitely} \text{ many 0's}$$

Thus, $h(x) \in A \Leftrightarrow \phi_x(x)\uparrow$, hence $\phi_x(x)\uparrow$ is re as well (contradiction)!

Next, suppose the "information yield" of ψ is weakened, so that $x \in A \Rightarrow \psi(x)\downarrow$ and $\phi_x^{-1}(0) = D_{\psi(x)}$, but where $x \notin A$ does *not* mean $\psi(x)\uparrow$ (thus, it must(*) be that $\psi(x)\downarrow$, for some $x \notin A$. But then $\psi(x)$ cannot be interpereted as a canonical index. A not being re we cannot in general know how to take a $\psi(x)$-output unless we already know that $x \in$ or $\notin A$).

Such a ψ is not available either:

Theorem 3   There is no $\psi \in \mathcal{P}$ such that ϕ_x is the characteristic function of a finite set S implies $\psi(x)\downarrow$ and $D_{\psi(x)} = S$.

Proof We adapt the proof non re-ness of A from our previous discussion. There, $\phi_x(x)\uparrow \Rightarrow \phi_{h(x)}$ is the sequence $\{\underbrace{1,1,1,\ldots}_{\text{all 1's}}\}$ (characteristic function of $\varnothing$) whereas $\phi_x(x)\downarrow \Rightarrow \phi_{h(x)}$ is a sequence $\{\underbrace{1, 1, \ldots,}_{\text{all 1's}} 1,\underbrace{0, 0, \ldots}_{\text{all 0's}}\}$, that is, a characteristic function of an *infinite* set. [Note that since $\phi_x(x)\downarrow$ in y_0 steps, then $\phi_x(x)\downarrow$ in $\leq y$ steps for *all* $y > y_0$. Thus we remove the inequality from the definition of χ, to make $\phi_{h(x)}$ a characteristic function of a finite set for all x]. So let $\psi \in \mathcal{P}$ with the required property exist.

Let

$$\chi(x,y) = \begin{cases} 1 & \textbf{if } \phi_x(x) \text{ does not converge in } \textit{exactly } y \text{ steps} \\ 0 & \textbf{if } \phi_x(x) \text{ converges in } \textit{exactly } y \text{ steps} \end{cases}$$

Of course, $\chi \in \mathcal{PR}$ [$\chi = \lambda xy.$ **if** $t(x,x,y) \neq 0$ **then** 1 **else** 0, where without loss of generality (see Theorem 3.7.1) $t(x,x,y) = 0$ has at most one y-solution, given x]; hence, for some $h \in \mathcal{R}^{(1)}$, $\chi = \lambda xy.\phi_{h(x)}(y)$.

Then

$$\phi_x(x)\uparrow \; \Rightarrow \; \phi_{h(x)} = \lambda y.1 \text{ (characteristic function of } \varnothing)$$

$$\Rightarrow \psi{\circ}h(x)\downarrow \; \& \; \psi{\circ}h(x) = 0$$

$$\phi_x(x)\downarrow \; \Rightarrow \; \phi_{h(x)}(y) = 0 \text{ for exactly } \textit{one } y_0 \text{ (characteristic}$$

$$\text{function of } \{y_0\}) \Rightarrow \psi{\circ}h(x)\downarrow \; \& \; \psi{\circ}h(x) \neq 0.$$

(*)"must" since A is not re.

Thus, $\psi \circ h \in \mathcal{R}$ and $x \in \overline{K}$ iff $\psi \circ h(x) = 0$; that is, $\overline{K}$ is recursive; a contradiction.//

Corollary 1   There is no $\theta \in \mathcal{P}$ such that if ϕ_x is the characteristic function of a finite set S, then $\theta(x)\downarrow$ and $\theta(x)$ is the cardinality of S.

Proof If such a θ exists, then so does the ψ of Theorem 3. We give an informal argument and leave the formal details to the reader.

/* Program to compute ψ, such that *if* ϕ_x is the characteristic function of a finite set S then $\psi(x)\downarrow$ and $S = D_{\psi(x)}$ */

(0)   Given x

(1) $z \leftarrow \theta(x)$

(2)   **if** $z = 0$ **then output** 0 and **halt**

(3) $y \leftarrow 2^{(\mu t)\phi_x(t)+1}$ /* we use prime power coding for convenience */

(4)   **do** $i = 1\ to\ z - 1$
$$y \leftarrow y \cdot p_i^{1 + (\mu t)[\phi_x(t) = 0 \& (\forall j)_{<\text{length}(y)}\{t + 1 > \exp(j,y)\}]}$$
end

(5)   **output** y and **halt**   /* of course, if ϕ_x is *not* the characteristic function of a finite set, at best this program outputs a nonsensical result and halts, at worst it does not terminate. "Infinite looping" can occur in lines (1), (3) and in the do-loop (4). */ //

Note: Although the above proof is instructive, a more direct proof based on that of Theorem 3 is possible. (See Problem 9).

Corollary 2   There is no $\eta \in \mathcal{P}$ such that W_x finite implies $\eta(x)\downarrow$ and $W_x = D_{\eta(x)}$.

Proof Using such an $\eta \in \mathcal{P}$, we could obtain the ψ of Theorem 3 (in $\mathcal{P}$):

Let $h \in \mathcal{R}^{(1)}$ such that if ϕ_x is a 0-1 total function, then $W_{h(x)} = \phi_x^{-1}(0)$ (Theorem 1).

Now, if $\phi_x^{-1}(0)$ is finite, then $\eta \circ h(x)\downarrow$ and $\phi_x^{-1}(0) = D_{\eta \circ h(x)}$. Take $\psi \overset{\text{def}}{=} \eta \circ h$.//

§8.4 ENUMERATIONS OF CLASSES OF RE SETS

Let $\mathcal{C}$ be a class of re sets. The sets of this class can be enumerated (of course, $\mathcal{C}$ is enumerable; why?) in various ways:

First, by viewing each set in $\mathcal{C}$ as a W_x we can enumerate $\mathcal{C}$ by enumerating the "names" (re-indices) of the sets in $\mathcal{C}$.

Second, if we know that $\mathcal{C} \subset \mathcal{R}_*$, then we can enumerate $\mathcal{C}$ by enumerating characteristic functions of sets in $\mathcal{C}$. These functions are of course "given" (and enumerated) by their ϕ-indices.

Third, if each set in $\mathcal{C}$ is finite, then another method of enumeration is enumerating canonical indices of sets in $\mathcal{C}$.

Of course, we are interested in effective (algorithmic) enumerations. This leads to the following definitions (Rice).

Definition 1   A class $\mathcal{C}$ of *re sets* is a *recursively enumerable class* (re class) iff there is a re set S such that $A \in \mathcal{C} \equiv (\exists x)(x \in S \ \& \ W_x = A).//$

Note: The above definition can be rephrased: "$\mathcal{C}$ is a *re class* iff there is an $f \in \mathcal{R}^{(1)}$ such that $A \in \mathcal{C} \equiv (\exists x)(x \in \mathcal{N} \ \& \ W_{f(x)} = A)$".

Definition 2   A class of $\mathcal{C}$ of *recursive sets* is a *characteristically enumerable class* iff there is a re set S such that

(1) $x \in S \Rightarrow \phi_x \in \mathcal{R}$ and range$(\phi_x) \subset \{0,1\}$

(2) $A \in \mathcal{C}$ iff $(\exists x)(x \in S \ \& \ A = \phi_x^{-1}(0)).//$

Note: A rephrasing of Definition 2 is: "$\mathcal{C}$ is characteristically enumerable iff there is an $f \in \mathcal{R}^{(1)}$ such that

(1) $x \in \mathcal{N} \Rightarrow \phi_{f(x)} \in \mathcal{R}$ and range$(\phi_{f(x)}) \subset \{0,1\}$

(2) $A \in \mathcal{C}$ iff $(\exists x)(x \in \mathcal{N} \ \& \ A = \phi_{f(x)}^{-1}(0))$".

Definition 3   A class $\mathcal{C}$ of *finite* sets is *canonically enumerable* iff there is a re set S such that $A \in \mathcal{C} \equiv (\exists x)(x \in S \ \& \ A = D_x).//$

Note: Alternative phrasing: "$\mathcal{C}$ is canonically enumerable iff there is an $f \in \mathcal{R}^{(1)}$ such that $A \in \mathcal{C} \equiv (\exists x)(x \in \mathcal{N} \ \& \ A = D_{f(x)})$".

Definition 4   A class $\mathcal{C}$ is a *completely recursively enumerable class* (completely re class) iff $S = \{x \mid W_x \in \mathcal{C}\}$ is re.$//$

Note: The S of Definition 1 contains at least one (but *not* necessarily all) re index of each A in $\mathcal{C}$. By contrast, the S of Definition 4 contains *all* such re indices. It is a complete index set.

Theorem 1 $\mathcal{R}_*$ is not characteristically enumerable.

Proof If it were, then for some $f \in \mathcal{R}^{(1)}$, such that range$(\phi_{f(x)}) \subset \{0,1\}$ and $\phi_{f(x)} \in \mathcal{R}$ for all x, $A \in \mathcal{R}_*$ iff $\phi_{f(x)}^{-1}(0) = A$.

Consider $\psi = \lambda x.1 \div \phi_{f(x)}(x)$.

(1) $\psi \in \mathcal{R}$.

(2)   range$(\psi) \subset \{0,1\}$; hence, $\psi^{-1}(0) \in \mathcal{R}_*$; hence, $\psi^{-1}(0) = \phi_{f(x)}^{-1}(0)$ for some x.
But $x \in \psi^{-1}(0)$ iff $x \notin \phi_{f(x)}^{-1}(0).//$

Theorem 2 $\mathcal{R}_*$ is not a completely re class.

Proof See (4) of the concluding remarks in Chapter 7.//

Interestingly, we have the following result:

Theorem 3 $\mathcal{R}_*$ is a re class.

Proof We start with the sequence ϕ_x, $x = 0,1,\ldots$. Each re set is equal to range(ϕ_x) for some x. Thus range$(\phi_x)_{x=0,1,\ldots}$ is the (effective) sequence of all re sets.
By Corollary 4 of Theorem 8.2.1, there is an $\eta \in \mathcal{R}^{(1)}$ such that

(1)   range$(\phi_{\eta(x)}) \subset$ range(ϕ_x)

(2)   range$(\phi_{\eta(x)}) \in \mathcal{R}_*$

(3) ϕ_x strictly increasing $\Rightarrow \phi_x = \phi_{\eta(x)}$.

Now every infinite recursive set is equal to range$(\phi_{\eta(x)})$ (some x) by (3) and Corollary 1 of Theorem 8.2.1. This is also true for finite recursive sets (see proof of Theorem 8.2.1, case 1 of *only if* part. Set there $f(x)\uparrow$ if $x > k$ rather than f_k).
     Thus $\mathcal{R}_* = \{$range$(\phi_{\eta(x)}) \mid x \in \mathcal{N}\}$
By Corollary 2 of Theorem 8.1.2, there is a $\sigma \in \mathcal{R}^{(1)}$ such that range$(\phi_{\eta(x)})$ $= W_{\sigma \circ \eta(x)}$, hence
     $A \in \mathcal{R}_*$ iff $(\exists x)(x \in \mathcal{N} \,\&\, A = W_{\sigma \circ \eta(x)}).//$

Example 1 It is possible to have a recursively enumerable class $\mathcal{D}$ of finite sets which is not characteristically enumerable (Rice).
     Consider $\mathcal{D} = \{D \mid (\exists x)(x \in \overline{K} \,\&\, D = \{x\}) \vee (\exists x)(x \in K \,\&\, D = \{x, x+1\})\}$.
Define f as follows:

$$f(0,x) = x$$

$$f(y+1,x) = \textbf{if } \phi_x(x)\downarrow \textbf{ then } x+1 \textbf{ else } (\mu z)(z+1)$$

$$= x + 1 + 0 \cdot \phi_x(x)$$

Clearly, $f \in P$; hence, there is an $h \in \mathcal{R}^{(1)}$ such that $f(y,x) = \phi_{h(x)}(y)$ for all x,y. Let $\sigma \in \mathcal{R}^{(1)}$ such that range$(\phi_x) = W_{\sigma(x)}$ (Corollary 2 of Theorem 8.1.2). Then range$(\phi_{h(x)}) = W_{\sigma \circ h(x)}$, and

$$ W_{\sigma \circ h(x)} = \begin{cases} \{x\} & \text{if } x \in \overline{K} \\ \{x, x+1\} & \text{if } x \in K \end{cases} $$

Thus, $D \in \mathcal{D}$ iff $(\exists x)(D = W_{\sigma \circ h(x)})$, which means that $\mathcal{D}$ is a re class.

Let $\mathcal{D}$ also be characteristically enumerable. That means a $\tau \in \mathcal{R}^{(1)}$ exists such that

(1) $\phi_{\tau(x)}$ is 0-1 and total for all x,

(2) $D \in \mathcal{D}$ iff $(\exists x)(D = \phi_{\tau(x)}^{-1}(0))$.

But then $x \in \overline{K}$ iff $(\exists y)(\phi_{\tau(y)}(x) = 0 \,\&\, \phi_{\tau(y)}(x+1) = 1)$; hence $\overline{K}$ is re. This contradiction establishes that $\mathcal{D}$ is not characteristically enumerable.//

Note: It follows immediately that $\mathcal{D}$ is not canonically enumerable (otherwise it would be characteristically enumerable). Characteristically enumerable classes of finite sets which are not canonically enumerable exist. (See Problem 19).

§8.5 A CHARACTERIZATION OF COMPLETELY RE CLASSES

Theorem 1   (Rice, Shapiro, McNaughton, Myhill [5]). A class $\mathcal{C}$ of re sets is *completely re* iff there is a canonically enumerable class $\mathcal{D}$ of finite sets such that $A \in \mathcal{C} \equiv (\exists D)(D \in \mathcal{D} \,\&\, D \subset A)$.

Note: Rice called the class $\mathcal{D}$ a "key array" of $\mathcal{C}$.

Proof *if* part. Let $\mathcal{C}$ be a class of re sets such that $A \in \mathcal{C} \equiv (\exists D)(D \in \mathcal{D} \,\&\, D \subset A)$, where $\mathcal{D}$ is canonically enumerable; say $\psi \in \mathcal{R}$ enumerates canonical indices of all sets in $\mathcal{D}$.

Let $\mathcal{P}_{\mathcal{C}} = \{x \mid W_x \in \mathcal{C}\}$.

$$ x \in \mathcal{P}_{\mathcal{C}} \equiv W_x \in \mathcal{C} \equiv (\exists y) D_{\psi(y)} \subset W_x $$

For convenience, let finite sets be prime power coded. Then $D_{\psi(y)} \subset W_x \equiv (\forall z)_{<\text{length}(\psi(y))} [\exp(z, \psi(y)) \doteq 1 \in W_x]$.

Now $\lambda xyz.\exp(z,\psi(y)) \dot- 1 \in W_x$ is re since it is equivalent to $\lambda xyz.(\exists w)T(x,\exp(z,\psi(y)) \dot- 1,w)$. Thus, $\lambda xy.D_{\psi(y)} \subset W_x$ is re (Corollary 2 of Theorem 3.6.1) and hence $\mathcal{P}_{\mathcal{C}}$ is re (Corollary 1 of Theorem 3.6.1).

only if part. Let $\mathcal{P}_{\mathcal{C}} = \{x \mid W_x \in \mathcal{C}\}$ be re.

Note that $A \in \mathcal{C} \equiv (\exists D)(D \in \mathcal{D} \,\&\, D \subset A)$ [which we want to prove under our assumption] implies that $\mathcal{D} \subset \mathcal{C}$ (since $D \subset D$). Thus, our first task is to show

$$(1) \qquad\qquad A \in \mathcal{C} \Rightarrow (\exists D)(D \in \mathcal{C} \,\&\, D \subset A).$$

Let instead there be an $A \in \mathcal{C}$ such that D finite and $D \subset A \Rightarrow D \notin \mathcal{C}$.

Plan: Find $h \in \mathcal{R}^{(1)}$ such that

$$W_{h(x)} = \begin{cases} A \text{ is } \phi_x(x)\uparrow \\[2mm] D \text{ (where } D \subset A \text{ and } D \text{ finite) if } \phi_x(x)\downarrow \end{cases}$$

Since no such D is in $\mathcal{C}$ (the assumption we want to contradict), $h(x) \in \mathcal{P}_{\mathcal{C}}$ iff $\phi_x(x)\uparrow$, proving $\overline{K}$ re (contradiction). This establishes (1).

The details: Let $A = W_m$. Define ψ by

$$\psi(x,y) = \begin{cases} \phi_m(y) \text{ if } \phi_x(x) \text{ does } not \text{ converge in } \leq y \text{ steps} \\[2mm] \uparrow \textbf{ otherwise} \end{cases}$$

$\psi \in \mathcal{P}$ $[\psi = \lambda xy.$ **if** $(\forall i)_{\leq y} \neg T(x,x,i)$ **then** $\phi_m(y)$ **else** $(\mu z)(z + 1)]$; hence, for some $h \in \mathcal{R}^{(1)}$, $\psi = \lambda xy.\phi_{h(x)}(y)$. Clearly,

$$\phi_x(x)\uparrow \;\Rightarrow\; \phi_{h(x)} = \phi_m \;\Rightarrow\; W_{h(x)} = A.$$

$$\phi_x(x)\downarrow \;\Rightarrow\; (\exists y_0)(\phi_x(x)\downarrow \text{ in } y_0 \text{ steps})$$

$$\Rightarrow (\exists y_0)\mathrm{dom}(\phi_{h(x)}) \subset \{0,1,\ldots,y_0 \dot- 1\}$$

Now, $\mathrm{dom}(\phi_{h(x)}) \subset A$ and $\mathrm{dom}(\phi_{h(x)})$ is finite in this case. This is the $h \in \mathcal{R}^{(1)}$ we want. (1), and the earlier observation that $\mathcal{D} \subset \mathcal{C}$, suggest that $\mathcal{D}$ be defined as $\{D \mid D$ is finite $\&\ D \in \mathcal{C}\}$. For such a $\mathcal{D}$, clearly $A \in \mathcal{C} \Rightarrow (\exists D)(D \in \mathcal{D} \,\&\, D \subset A)$ by (1). Also, $(\exists D)(D \in \mathcal{D} \,\&\, D \subset A) \Rightarrow A \in \mathcal{C}$ by the corollary of Lemma 7.3.2 The proof will be complete if $\mathcal{D}$ is shown to be canonically enumerable. To this end, we know that there is a $\sigma \in \mathcal{R}^{(1)}$ such that $D_x = W_{\sigma(x)}$. The set $\mathcal{S}$

$= \{\sigma(x) \mid W_{\sigma(x)} \in \mathcal{C}\}$ is re ($y \in \mathcal{S}$ iff $\sigma(x) \in \mathcal{P}_{\mathcal{C}}$) and $D \in \mathcal{D}$ iff $(\exists y)(y \in \mathcal{S}$ & $D = D_y).//$

Example 1 $\{x \mid \phi_x \in \mathcal{R}\}$ is not re as we already know. That is, $\{x \mid W_x = \mathcal{N}\}$ is not re. We prove this again, this time relying on Theorem 1. Indeed, if it is re, then there is a finite set D such that $D \subset W_x$ and $D = \mathcal{N}$; a contradiction. [Note that Lemma 7.3.2 is not applicable in this case].//

PROBLEMS

1. (*Reduction property* of the class of re predicates) Prove that for any re predicates $R(\vec{x})$ and $S(\vec{x})$, there are re predicates $R'(\vec{x})$ and $S'(\vec{x})$ such that

   (i) $R' \subset R$ and $S' \subset S$

   (ii) $R' \cap S' = \varnothing$

   (iii) $R' \cup S' = R \cup S$

   (*Hint*: Express R and S as $(\exists y)\overline{R}(y,\vec{x})$ and $(\exists y)\overline{S}(y,\vec{x})$, where $\overline{R}$ and $\overline{S}$ are recursive. Define R' and S' as projections of appropriate Boolean combinations of $\overline{R}$ and $\overline{S}$. In the term "Boolean" here, we include the bounded quantification operations.)

2. (*Selection Theorem*) Prove that for each $m \geq 1$, there is a partial recursive "selection" function $\mathrm{Sel}^{(m)}$ such that

   (i) $(\exists y)\phi_i^{(m)}(y,\vec{x}_m)\!\downarrow \equiv \mathrm{Sel}^{(m)}(i,\vec{x}_m)\!\downarrow$

   (ii) $(\exists y)\phi_i^{(m)}(y,\vec{x}_m)\!\downarrow \Rightarrow \phi_i^{(m)}(\mathrm{Sel}^{(m)}(i,\vec{x}_m),\vec{x}_m)\!\downarrow$

   where $T^{(m)}$ is the Kleene-predicate (say, of Chapter 5).

3. Show that for each $m \geq 1$ and re predicate $P(\vec{x},y)$, there is a function $\mathrm{Sel}_P \in \mathcal{P}$ such that

   $$(\exists y)P(\vec{x},y) \equiv \mathrm{Sel}_P(\vec{x})\!\downarrow \equiv P(\vec{x}, \mathrm{Sel}_P(\vec{x})).$$

4. (*Definition by "positive cases"*, i.e., when the "otherwise" case holds, you let the "under definition" function be undefined)
   f is defined by *positive cases* from functions $g_i \in \mathcal{P}^{(m)}$ $(i = 1, \ldots, k)$ and re predicates $R_i(\vec{x}_m)$ $(i = 1, \ldots, k)$ iff

   $$f(\vec{x}_m) = \begin{cases} g_1(\vec{x}_m) \text{ if } R_1(\vec{x}_m) \\ \vdots \qquad \vdots \quad \vdots \\ g_k(\vec{x}_m) \text{ if } R_k(\vec{x}_m) \\ \uparrow \qquad \textbf{otherwise} \end{cases}$$

where the R_i's are mutually exclusive. Show that under these circumstances, $f \in \mathcal{P}$. (*Hint:* The proof is *not* as straightforward as when the R_i's are *recursive*. Intuitively, you want to dovetail computations for all the R_i's simultaneously, and the first convergent such computation (say, for R_j) should trigger the start of the computation of the corresponding function, g_j. Formally, you may use Problem 3.)

5. Let us call *co-re n*-ary predicates, sets $S \subset \mathcal{N}^n$ such that $\mathcal{N}^n - S$ is *re*. Prove that the *co-re n*-ary predicates have the *separation property*, i.e., given any two such predicates P and Q such that $P \cap Q = \varnothing$, there is a recursive predicate S(*), such that $P \subset S \subset \mathcal{N}^n - Q$. (We say that S *separates* P and Q.) (*Hint:* Use Problem 1.)

6. Prove: There is an $h \in \mathcal{R}^{(2)}$ such that $W_{h(x,y)} = \phi_x(W_y)$.

7. Is every infinite recursive subset of $\mathcal{N}$ the range of a strictly increasing *primitive* recursive function?

8. Show that there are recursive functions f and g such that whenever x is a *canonical* index of a finite set (under some particular coding), then $f(x)$ (resp. $g(x)$) is a *characteristic* (resp. *re*) index of the set.

9. Give a proof of Corollary 1 of Theorem 8.3.3, modelled around the proof of the theorem.

10. *Argument:* Theorem 8.4.3 states that $\mathcal{R}_*$ is an *re* class. Let, then, $f(0), f(1), \ldots$ be a recursive enumeration of some indices of *all* recursive sets. Construct the "diagonal" set $S = \{x \mid x \notin W_{f(x)}\}$. Since f and $W_{f(x)}$ are recursive, S is recursive. Thus, for some i, $S = W_{f(i)}$. Then

    (1) $i \in S$ iff $i \in W_{f(i)}$ [since $S = W_{f(i)}$] and

    (2) $i \in S$ iff $i \notin W_{f(i)}$ [Definition of S]

    We have just contradicted Theorem 8.4.3! What is wrong with this argument?

11. We call a class $\mathcal{C}$ of partial recursive functions a *re class* iff there is an $f \in \mathcal{R}^{(1)}$ such that $\psi \in \mathcal{C} \iff (\exists x)\psi = \phi_{f(x)}$.
    Is $\mathcal{R}$ a *re* class of functions?

12. Is $\mathcal{P}\mathcal{R}$ a *re* class? (*Hint:* A Gödel numbering of $\mathcal{P}$ can be based on a Loop-program formalism augmented with the **goto**-instruction.)

13. Is $\mathcal{P} - \mathcal{R}$ a *re* class of functions? (*Hint:* Consider an effective listing of pairs of integers $(x_0, y_0), (x_1, y_1), \ldots$ For each (x_i, y_i), add to "program" with "code" x_i an instruction which makes it "loop forever" with input y_i. The effective listing of programs so obtained computes all the functions in $\mathcal{P} - \mathcal{R}$.)

(*)Thus both S and $\neg S$ are *co-re*.

14. Let $\mathcal{C}$ be a re subclass of $\mathcal{R}$. Show that $\mathcal{P}-\mathcal{C}$ is a re class. (*Hint*: Modify each "program" x into a program $f(x)$ ($f \in \mathcal{R}$) incapable of computing a function of $\mathcal{C}$. Make sure that you do not exclude functions of $\mathcal{P}-\mathcal{C}$ from the class of $f(x)$-computable functions. *Further hints*: Let g enumerate ϕ-indices of all the functions in $\mathcal{C}$. If $\phi_{f(x)}$ is nontotal, then there's nothing to worry about. If it is total, then ensure that $\phi_{f(x)} \neq \phi_{g(y)}$ for all y. To this end, for each y and x,

(1) $f(x)$ computes $\phi_x(y)$

(2) If $\phi_x(y)\!\downarrow$, then $f(x)$ won't output $\phi_x(y)$ *unless*, via a dovetailing subcomputation, it finds a z such that $\phi_x(z) \neq \phi_{g(y)}(z)$. To ensure that no functions from $\mathcal{P} - \mathcal{R}$ are omitted, incorporate the scheme of Problem 13.)

15. Prove that $\mathcal{P}-\mathcal{PR}$ is a re class.

16. Prove that for no re subclass $\mathcal{C}$ of $\mathcal{R}$ is $\mathcal{R}-\mathcal{C}$ a re class.

17. Disprove Problem 14, where $\mathcal{C}$ is now a re subclass of $\mathcal{P}$.

18. Show that the class of finite (and hence the class of infinite) functions in $\mathcal{P}$ is re.

19. Consider the class of finite sets,

$$\mathcal{C} = \{D \,|\, (\exists x)(D = \{x\} \ \& \ x \in \overline{K})$$
$$\vee \ (\exists x)(\exists y)(D = \{x, x + y + 1\} \ \& \ \phi_x(x)\!\downarrow \text{ in exactly y steps})\}.$$

Show that $\mathcal{C}$ is characteristically, but not canonically, enumerable.

20. Prove the Rice-Shapiro-McNaughton-Myhill theorem (Theorem 8.5.1) for *functions*:
First, a *finite* function $\theta = \{(x_0, y_0), \ldots, (x_n, y_n)\}$ is coded as

$$\Pi_{i \leq n} \, p_{x_i}^{y_i + 1}.$$

Denote this code by $[\theta]$. We can now state that, whenever $\mathcal{C} \subset \mathcal{P}$, $\{x \,|\, \phi_x \in \mathcal{C}\}$ is re *iff* there is a re set $\mathcal{F}$ such that $\psi \in \mathcal{C} \equiv (\exists \theta)(\theta \subset \psi \ \& \ \theta \text{ finite } \& \ [\theta] \in \mathcal{F})$.

21. State and prove Problem 20 for sets such as $\{\vec{x}_n \,|\, (\phi_{x_1}, \ldots, \phi_{x_n}) \in \mathcal{C}\}$, where $\mathcal{C} \subset \mathcal{P}^{(n)}$.

REFERENCES

[1] Bennett, J. *On Spectra.* Ph.D. dissertation, Princeton University, 1962.

[2] Brainerd, W.S., and Landweber, L.H. *Theory of Computation.* New York: Wiley, 1974.

[3] Grzegorczyk, A. "Some Classes of Recursive Functions". *Rozprawy Matematyczne* 4. Warsaw, (1953):1–45.

[4] Kreider, D.L., and Ritchie, R.W. *Notes on Recursive Function Theory.* Lecture notes for Mathematics 89 (Seminar in Logic). Dartmouth College, Winter term, 1965.

[5] Rice, H.G. "On Completely Recursively Enumerable Classes and their Key Arrays". *J. of Symbolic Logic* 21 (1956): 304–308.

[6] Rogers, H. *Theory of Recursive Functions and Effective Computability.* New York: McGraw-Hill, 1967.

CHAPTER 9

Reducibility, Creativity, Productiveness

In §7.3, among other things, we have seen repeated use of the technique by which a set (problem) S is proved non recursive (resp. non re) by *reducing* some known non recursive (resp. non re) set T to S. This reduction was always effected with the help of some recursive function f (usually depending on S and T) in the sense that $x \in T$ was equivalent to $f(x) \in S$. Besides providing proofs of nonrecursiveness or non re-ness, reducibility helps to establish a classification between problems as to the degree of their unsolvability that is, if T is reducible to S, but not vice versa, then a solution for S furnishes a solution for T, but not vice versa. In this case, intuitively, S is "more unsolvable" than T. On the other hand, if S is reducible to T as well, then it is natural to say that S and T are of the same degree of unsolvability.

Example 1 The set K is reducible to $\{x \mid \phi_x \in \mathcal{R}\}$. Indeed we have seen this in the proof of Theorem 7.3.1, where it was shown that for some $h \in \mathcal{R}^{(1)}$, $x \in K$ iff $h(x) \in \{x \mid \phi_x \in \mathcal{R}\}$.//

325

Example 2   The set D is reducible to J (these sets are defined in Theorem 7.3.1).

Indeed, $D = \{x \mid W_x \text{ is infinite}\}$ and $J = \{x \mid \phi_x \in \mathcal{R}\}$.

By Corollary 5 of Theorem 8.1.2, there is a $p \in \mathcal{R}^{(1)}$ such that $W_x = \text{range}(\phi_{p(x)})$ and if $\phi_{p(x)} \notin \mathcal{R}$ then W_x is finite (since then $\text{dom}(\phi_{p(x)}) = \{i \in \mathcal{N} \mid i < n\}$ for some n).

It follows that $x \in D$ iff $p(x) \in J$, hence D is reducible to J.

In this connection, it is interesting to observe that $\{x \mid W_x \text{ is finite}\}$ is reducible to $\{x \mid \phi_x \notin \mathcal{R}\}$ via p.//

Example 3   The set J is reducible to D (both as in Example 2).

Indeed, let ψ be defined by

$$\psi(x,y) = \begin{cases} 0 \text{ if } \phi_x(z)\downarrow \text{ for all } z \leq y \\ \uparrow \text{ otherwise} \end{cases}$$

Intuitively, ψ is computable, since given x,y then all the computations $\phi_x(0), \ldots, \phi_x(y)$ are dovetailed either forever, or until *all* halt, in which case 0 is being output and the process terminates.

Assuming then for a minute that $\psi \in \mathcal{P}$ has been proved, there is an $f \in \mathcal{R}^{(1)}$ such that $\psi = \lambda xy.\phi_{f(x)}(y)$. Clearly, $x \in J \Rightarrow \phi_x(z)\downarrow$ for all $z \leq y$ and all y; hence $\phi_{f(x)} = \lambda y.0$, that is $f(x) \in D$. Conversely, $f(x) \in D \Rightarrow \phi_{f(x)}(y)\downarrow$ for infinitely many y, hence $\phi_x(z)\downarrow$ for *all* z [Consider an increasing sequence $y_1 < y_2 < \ldots$ such that $\phi_{f(x)}(y_i)\downarrow$ for all $i = 1,2,\ldots$]. That is $f(x) \in J$. Thus, $x \in D$ iff $f(x) \in J$.

Finally, to settle $\psi \in \mathcal{P}$:

$$z = \psi(x,y) \equiv z = 0 \ \& \ (\forall i)_{\leq y}(\exists w)T(x,i,w)$$

The righthand side of $\equiv$ is re (by Corollary 2, Theorem 3.6.1), and $\psi \in \mathcal{P}$ by Theorem 3.7.2.//

Example 4 K is *not* reducible to $\overline{K}$. For let, for some $f \in \mathcal{R}^{(1)}$, $x \in K$ iff $f(x) \in \overline{K}$. That is, $x \notin K$ iff $f(x) \notin \overline{K}$; that is, $x \in \overline{K}$ iff $f(x) \in K$. This makes $\overline{K}$ re; impossible.//

Example 5 $\{x \mid \phi_x \in \mathcal{R}\}$ is not reducible to K. Indeed, if it were this would make $\{x \mid \phi_x \in \mathcal{R}\}$ re.//

Example 6   Every re set A is reducible to K. Indeed, let ψ be defined by

$$\psi(x,y) = \begin{cases} 0 \text{ if } x \in A \\ \uparrow \text{ otherwise} \end{cases}$$

ψ is intuitively computable. [Given x, y; enumerate A. If x ever shows up in the enumeration then output 0 and stop.] Formally, let $A = W_e$. Then $\psi = \lambda xy.0 \cdot \phi_e(x)$, thus $\psi \in \mathcal{P}$. Let $h \in \mathcal{R}^{(1)}$ such that $\psi = \lambda xy.\phi_{h(x)}(y)$. Then $\phi_{h(x)} = \lambda y.0$ iff $x \in A$. $\phi_{h(x)}$ is empty iff $x \notin A$. Thus,

$$x \in A \Rightarrow \phi_{h(x)} \in \mathcal{R} \Rightarrow \phi_{h(x)}(h(x))\downarrow \Rightarrow h(x) \in K.$$

$$x \notin A \Rightarrow W_{h(x)} = \varnothing \Rightarrow h(x) \notin W_{h(x)} \Rightarrow h(x) \notin K.$$

That is, $x \in A$ iff $h(x) \in K$.//

Example 7   We revisit the proof of Theorem 7.3.2, where $\{x \mid \phi_x$ is constant$\}$ is shown non re.

A function $\sigma \in \mathcal{R}^{(1)}$ is defined there, where

$$x \notin K \Rightarrow \phi_{\sigma(x)} = \lambda y.0 \Rightarrow W_{\sigma(x)} \text{ infinite}$$

$$x \in K \Rightarrow (\exists y_0)[(\forall y)y \geq y_0 \Rightarrow \phi_{\sigma(x)}(y)\uparrow] \Rightarrow W_{\sigma(x)} \text{ finite}$$

That is, $x \in K$ iff $W_{\sigma(x)}$ is finite, in other words, K is reducible (via σ) to $\{x \mid W_x$ is finite$\}$. This latter set being non re (Theorem 7.3.2), it is not reducible to K.//

We now proceed to put in some order various important notions implicit in the previous examples.

§9.1 STRONG REDUCIBILITY

Definition 1   Let $A \subset \mathcal{N}$ and $B \subset \mathcal{N}$.
A is *1-1 reducible* to B, in symbols $A \leq_1 B$, if for some 1-1 function $f \in \mathcal{R}$ the predicates $x \in A$ and $f(x) \in B$ are equivalent.//
1-1 reducibility is usually referred to as *1-reducibility*.

Definition 2   Let $A \subset \mathcal{N}$ and $B \subset \mathcal{N}$. Then A is *many-one reducible* to B, in symbols $A \leq_m B$, if for some $f \in \mathcal{R}$ the predicates $x \in A$ and $f(x) \in B$ are equivalent.//

Note: Many-one reducibility is usually referred to as *m-reducibility*. Both these reducibilities, due to Post, imply other types of reducibilities (such as truth-table and Turing reducibilities) thus they are sometimes called *strong* reducibilities (Shapiro).

Proposition 1 $A \leq_1 B$ (resp. $A \leq_m B$) iff for some 1-1 $f \in \mathcal{R}$ (resp. some unrestricted $f \in \mathcal{R}$) any of the following equivalent statements holds

(i) $A = f^{-1}(B)$.

(ii) $\chi_A = \chi_B \circ f$, where χ_A (resp. χ_B) is the characteristic function of A (resp. B).

(iii) $f(A) \subset B$ & $f(\overline{A}) \subset \overline{B}$ (where $\overline{S} = \mathcal{N} - S$, for any $S \subset \mathcal{N}$).

Proof Easy exercise.//

Proposition 2   (i) $\leq_1$ and $\leq_m$ are reflexive and transitive

(ii) $\leq_1 \underset{\neq}{\subseteq} \leq_m$

(iii) $A \leq_1 B \Rightarrow \overline{A} \leq_1 \overline{B}$ and $A \leq_m B \Rightarrow \overline{A} \leq_m \overline{B}$

(iv) $A \leq_m B$ & $B \in \mathcal{R}_* \Rightarrow A \in \mathcal{R}_*$

(v) $A \leq_m B$ & B re $\Rightarrow A$ re

(vi)   There are recursive sets A and B which are incomparable with respect to $\leq_m$ (and hence with respect to $\leq_1$).

(vii)   There are nonrecursive sets A and B which are incomparable with respect to $\leq_m$ (and hence with respect to $\leq_1$).

(viii)   There is a re set A such that $A \nleq_m \overline{A}$ (and hence $A \nleq_1 \overline{A}$).

(ix)   If $A \in \mathcal{R}_*$ and $\emptyset \underset{\neq}{\subseteq} A \underset{\neq}{\subseteq} \mathcal{N}$, then $A \leq_m \overline{A}$.

Proof (i) $A \leq_1 B \leq_1 C$ means $x \in A \Leftrightarrow f(x) \in B$ and $x \in B \Leftrightarrow h(x) \in C$ where f and h are 1-1 recursive. Clearly, $x \in A \Leftrightarrow h \circ f(x) \in C$ and $h \circ f \in \mathcal{R}$ is 1-1.

For reflexivity, $x \in A \Leftrightarrow h(x) \in A$, where $h = \lambda x.x$. The case for $\leq_m$ is similar.

(ii) $A \leq_1 B$ means $x \in A \Leftrightarrow f(x) \in B$ for some 1-1 $f \in \mathcal{R}$. Ignoring 1-1-ness of f we get $A \leq_m B$. Thus, the (binary) relation $\leq_1$ is a subset of $\leq_m$. Let now $A = \mathcal{N} - \{0\}$ and $B = \{0\}$.

Define f by $f(x) = $ **if** $x \neq 0$ **then** 0 **else** 1. Clearly, $f \in \mathcal{R}$ and $x \in A$ iff $f(x) \in B$, thus $A \leq_m B$. On the other hand, if $h \in \mathcal{R}$ is 1-1 and $x \in A \Leftrightarrow h(x) \in B$ then $A = h^{-1}(B)$, which is impossible since A is infinite and B is finite. Thus, $\leq_1 \neq \leq_m$.

(iii) $A \leq_1 B$ means [Proposition 1,(iii)] that for some 1-1 $f \in \mathcal{R}, f(A) \subset B$ & $f(\overline{A}) \subset \overline{B}$. The symmetry of this condition implies $\overline{A} \leq_1 \overline{B}$. Similarly for $\leq_m$.

(iv) Follows from Proposition 1(ii). [This fact has already been established and used in the reducibility proofs of §7.3].

(v) By Lemma 7.3.1, or, alternatively by Proposition 1(i) and Example 8.1.1.

(vi) Take $A = \emptyset$ and $B = \mathcal{N}$.

(vii) Take $A = K$ and $B = \overline{K}$. We know that $K \not\leq_m \overline{K}$. (Example 4 in the preamble to this chapter.) $\overline{K} \not\leq_m K$ as well, otherwise $\overline{K}$ would be re ($\overline{K} = f^{-1}(K)$, where f effects the reducibility $\overline{K} \leq_m K$).

(viii) Take $A = K$.

(ix) Let $a \in A$ and $\overline{a} \in \overline{A}$. Then $f = \lambda x.$ **if** $x \in A$ **then** $\overline{a}$ **else** a is recursive and $A \leq_m \overline{A}$ via f.//

Note: It should be obvious that Proposition 2(iv) and (v) is valid for 1-reducibility as well.

Definition 3

$$\equiv_1 \;\overset{\text{def}}{=}\; \leq_1 \cap \leq_1^{-1}.$$

$$\equiv_m \;\overset{\text{def}}{=}\; \leq_m \cap \leq_m^{-1}.$$

where R^{-1}, as usual, is the converse of R, that is, $\{(x,y) \mid (y,x) \in R\}$.//

Proposition 3 $\equiv_1$ and $\equiv_m$ are equivalence relations.

> *Proof* (We consider the case of $\equiv_1$; that of $\equiv_m$ is similar).
>
> (1) *Reflexivity*. Since $\leq_1$ is reflexive, so is $\leq_1^{-1}$ and hence $\leq_1 \cap \leq_1^{-1}$.
>
> (2) *Symmetry*. $A \equiv_1 B \Rightarrow A \leq_1 B \;\&\; B \leq_1 A \Rightarrow B \leq_1 A \;\&\; A \leq_1 B \Rightarrow B \equiv_1 A$.
>
> (3) *Transitivity*. Since $\leq_1$ is transitive, so is $\leq_1^{-1}$ and hence $\leq_1 \cap \leq_1^{-1}$.//

Definition 4   The equivalence classes of $\equiv_1$ are called *1-degrees* and the equivalence classes of $\equiv_m$ the *m-degrees*.

We denote by $\deg_1(A)$ [resp. $\deg_m(A)$] the equivalence class of $\equiv_1$ (resp. $\equiv_m$) determined by A.

We define a relation R_1 (resp. R_m) on the set of 1-degrees (resp. m-degrees) by

$$x R_1 y \text{ iff } A \leq_1 B \text{ and } A \in x, B \in y$$

$$x R_m y \text{ iff } A \leq_m B \text{ and } A \in x, B \in y.//$$

Proposition 4   (1) R_1 and R_m are well-defined; i.e., they do not depend on the representative sets A and B used in Definition 4.

(2) R_1 and R_m are partial orders.

Proof (1) (case of R_1; the other is similar.) Let xR_1y due to $A \leq_1 B$ where $A \in x$, $B \in y$. Let $C \in x$ and $D \in y$. It is $A \equiv_1 C$ and $D \equiv_1 B$. In particular, $C \leq_1 A$ and $B \leq_1 D$. Thus, $C \leq_1 D$ by transitivity of $\leq_1$, which is consistent with xR_1y.

(2) (case of R_1; the other is similar). Reflexivity and transitivity are inherited from $\leq_1$. Antisymmetry: Let xR_1y & yR_1x. Then $A \leq_1 B$ for some $A \in x$, $B \in y$ (because of xR_1y) and $B \leq_1 A$ [because of yR_1x and part (1)]. It follows that $A \equiv_1 B$; hence (see discussion following Example 1.3.6) $x = y$.//

Note: We shall denote (hopefully unambiguously) R_1 by $\leq_1$ and R_m by $\leq_m$; the context should imply whether they act on sets or on 1- or m-degrees.

By Proposition 2 (iv) [resp. (v)], it follows that if a 1-(m-)degree contains a recursive (resp. re) set then all its sets are recursive (resp. re). Such degrees are called *recursive* (resp. *re*) 1-(m-)*degrees*. Note in this connection that $\deg_1(K)$ [also $\deg_m(K)$] is an *re* but *not recursive* 1-(m-)degree.

By Proposition 2(vi) [resp. (vii)], there are recursive but incomparable (resp. nonrecursive but incomparable) 1-(m-)degrees. Thus, $\leq_1$ ($\leq_m$) is not a linear ordering of 1-(m-)degrees.

The question whether there are incomparable re 1-(m-)degrees. suggests itself; also whether there are incomparable non-re 1-(m-)degrees. We shall deal with these questions later.

Example 1 Let x and y be any two m-degrees. There is an m-degree z which is the least upper bound of x and y (i.e., $x \leq_m z$, $y \leq_m z$ and for any m-degree c such that $x \leq_m c$ and $y \leq_m c$ it is $z \leq_m c$).

Let $A \in x$ and $B \in y$. Consider Z (sometimes termed the *join* of A and B) defined by $Z = \{2m \mid m \in A\} \cup \{2m + 1 \mid m \in B\}$.

Next, $A \leq_m Z$ via $\lambda m.2m$ and $B \leq_m Z$ via $\lambda m.2m + 1$, that is $x \leq_m z$ and $y \leq_m z$, where we have set $z = \deg_m(Z)$.

Now let $x \leq_m c$ and $y \leq_m c$ and let $C \in c$. Thus, for some f and g in $\mathcal{R}$, $x \in A$ iff $f(x) \in C$ and $x \in B$ iff $g(x) \in C$.

Define h by

$$h(x) = \textbf{if } 2 \mid x \textbf{ then } f\left(\left\lfloor \frac{x}{2} \right\rfloor\right) \textbf{ else } g\left(\left\lfloor \frac{x \doteq 1}{2} \right\rfloor\right)$$

It follows that $Z \leq_m C$ via h; hence, $z \leq_m c$. It is worth observing that if x and y are re m-degrees then so is z, since Z is then re. It can be shown (this result is due to Young, see Rogers [18], §10.3) that there are two 1-degrees which have no least upper bound with respect to $\leq_1$.//

§9.2 COMPLETENESS

We next give a name to the property exhibited by K in Example 6 of this chapter's preamble.

Definition 1 A *recursively enumerable* set S is *1-complete* (resp. *m-complete*) iff

$$\text{for all re sets } T, T \leq_1 S \text{ (resp. } T \leq_m S).//$$

Note: Intuitively, a complete set S has "maximum complexity" among *re sets*. It is clear that if a 1-(m-)degree contains a complete set, then all its sets are complete (why?).

Theorem 1 K is m-complete.

Proof The proof was given in the aforementioned Example 6.//

Example 1 Let $\lambda xy.\langle x,y\rangle$ be a fixed but unspecified recursive pairing function with recursive projections K and L. Let $K_0 = \{\langle x,y\rangle \mid \phi_x(y)\downarrow\}$, the halting problem set. Its counterpart in the context of the Post system formalism has been shown to be 1-complete by Post [16].
    Indeed, first K_0 is re, for $z \in K_0 \equiv (\exists w)T(K(x),L(x),w)$.
    Next, let A be re and let e be one of its re-indices (that is, $A = W_e$).
    Then $x \in W_e$ iff $\phi_e(x)\downarrow$ iff $\langle e,x\rangle \in K_0$. That is $W_e \leq_1 K_0$ via $\lambda x.\langle e,x\rangle.//$

Theorem 2 K is 1-complete.

Proof For convenience, we assume that the Gödel numbering $i \mapsto \phi_i$ is that of Chapter 3.
    An inspection of the proof of Theorem 3.8.1 (S-m-n theorem) and of its Corollary 1 shows that the S-m-n functions $\lambda i\vec{x}_m.S_n^m(i,\vec{x}_m)$ are 1-1 for any m and n.
    Thus, the function h, which effects the reducibility $A \leq_m K$ in Example 6 of the preamble to this chapter is 1-1, hence $A \leq_1 K.//$

Note: It is now clear that all the reducibilities which were constructed in §7.3 and in the examples of the preamble to this chapter with the help of the S-m-n theorem can be viewed as 1-reducibilities.

Corollary $K \equiv_1 K_0$.

Proof Both K and K_0 are 1-complete. Hence, $K \leq_1 K_0$ and $K_0 \leq_1 K$; i.e., $K \equiv_1 K_0$.//

We shall eventually see that 1-completeness and m-completeness are equivalent and that there are re sets which are not m-complete.

§9.3 RECURSIVE ISOMORPHISM

In studying Recursion Theory one studies certain properties of sets and functions (for example, re-ness, m-completeness, recursiveness etc.). It is desirable to strip sets (functions) of their inessential (for the theory) properties so that "similar" sets (functions) can be identified with each other for the purposes of the theory.

This approach is not peculiar to Recursion Theory. In Euclidean Geometry, any two figures such that one is obtained from the other by a *displacement* are considered equivalent (similar, indistinguishable). In Topology, any two subspaces of a space are indistinguishable if they are *homeomorphic*.

Thus, in a branch of mathematics one studies a particular *space* (set) and its subsets, the space being equipped with a set of similarity transformations.(*) These transformations form a *group* under composition,(†) and two sets A and B in the theory are similar (indistinguishable) iff $f(A) = B$ where f is a similarity transformation. Another way of putting this, is that in this theory the "essential" properties of sets (functions) are those that remain invariant under the similarity transformations. Thus in Euclidean Geometry, being a circle is essential, in Projective Geometry though it is not, as the admitted similarity transformations (projective collineations) are not circle-preserving.

The space-transformation approach is due to Klein. In the case of Recursion Theory we are not worried when the elements of the sets we study undergo a systematic "change of name", this change being effected algorithmically. Thus the relevant similarity transformations are *1-1* and *onto* recursive functions $f: \mathcal{N} \to \mathcal{N}$.

Definition 1 A *recursive permutation* $f: \mathcal{N} \to \mathcal{N}$ is a *1-1* and *onto* recursive function.

Two sets A and B in $\mathcal{N}$ are *recursively isomorphic* iff $f(A) = B$ for some recursive permutation f. We write $A \sim B$.

Two sets A and B in $\mathcal{N}^k$ are recursively isomorphic iff $g(A) = B$, where $g = \lambda x.(\underbrace{f(x), \ldots, f(x)}_{k})$ and f is a recursive permutation. We write $A \sim B$.

(*)A transformation of the space is a *1-1, onto, total* map from the space to itself.

(†)A structure $G = (S, \circ)$ is a *group* under $\circ$, iff (1) It is a *semigroup* under $\circ$ (see §6.6). (2) There is an *identity* e in S; i.e., $(\forall a) [a \in S \Rightarrow a \circ e = e \circ a = a]$. (3) The equations $a \circ x = b$ and $x \circ a = b$ have x-solutions in S for any choice of a,b in S.

Two (partial) functions ψ and χ are recursively isomorphic iff $\psi \sim \chi$, where ψ and χ are viewed as sets in $\mathcal{N} \times \mathcal{N}$.//

Proposition 1   The set of recursive permutations $\mathcal{G}$ is a group under composition.

> **Proof** That it is a semigroup is clear as
> (1) for any $f, g \in \mathcal{G}$, $f \circ g$ is clearly 1-1 and onto, and recursive.
> (2) $f \circ (g \circ h) = (f \circ g) \circ h$ is true for any functions. Next, $e = \lambda x.x$ is in $\mathcal{G}$ and $e \circ f = f \circ e = f$ for any $f \in \mathcal{G}$. Finally, since f^{-1} is a 1-1 and onto function, clearly the equation $f \circ x = h$ has the x solution $f^{-1} \circ h$ and $x \circ f = h$ has the x-solution $h \circ f^{-1}$. Finally, $f^{-1} \in \mathcal{R}$ since $f^{-1} = \lambda x.(\mu y)(x = f(y))$.//

Proposition 2 $\sim$ is an equivalence relation.

> **Proof** (We restrict attention to $\sim$ acting on $\mathcal{P}(\mathcal{N})$.)
>
> (1)   *reflexivity*: $A \sim A$ via $\lambda x.x$.
>
> (2)   *symmetry*: $A \sim B$ means $f(A) = B$, where $f \in \mathcal{G}$. But then $f^{-1} \in \mathcal{G}$ (Proposition 1) and $f^{-1}(B) = A$, that is $B \sim A$.
>
> (3)   *transitivity*: Let $A \sim B$ and $B \sim C$. Then $f(A) = B$ and $g(B) = C$, where f, g are in $\mathcal{G}$. But then $g \circ f(A) = C$ and $g \circ f \in \mathcal{G}$, that is $A \sim C$.//

Note: By Proposition 2, any two sets (functions) related via $\sim$ will be viewed as indistinguishable for the purposes of Recursion Theory. Thus, this theory, essentially studies the equivalence classes of $\sim$, or, in other words, studies those properties of sets and functions which remain invariant under recursive permutations. For example, recursiveness, re-ness, m-completeness are such invariants.

On the other hand, the property that a set contains 0 is not a recursive invariant, since, clearly, some $f \in \mathcal{G}$ exists which maps the set with the property to one without the property.

Also, the property of belonging to $\mathcal{PR}_*$ is not a recursive invariant. Indeed, let $A \in \mathcal{PR}_*$ be infinite with an infinite complement $\overline{A}$. Let $B \subset \overline{A}$ be in $\mathcal{R}_* - \mathcal{PR}_*$; sets such as A and B can be easily found. (See Problem 11.) By Corollary 1 of Theorem 8.2.1 there are 1-1 (indeed, strictly increasing) recursive functions f and g such that $f(\mathcal{N}) = A$ and $g(\mathcal{N}) = B$. Set $C = \mathcal{N} - (A \cup B)$ and $h = g \circ f^{-1}$, where, of course, f^{-1} is defined only on A. Now $C \in 0_*$ and $h \in \mathcal{P}$ [since $h = \lambda x.g((\mu z)(f(z) = x))$]. Clearly, $h \mid A$ is 1-1 and onto B. Define j by

$$
j(x) = \begin{cases} \textbf{if } x \in A \textbf{ then } h(x) \\ \textbf{if } x \in C \textbf{ then } x \\ \textbf{if } x \in B \textbf{ then } h^{-1}(x) \qquad \text{/* i.e., } (\mu y)(h(y) = x) \text{ */} \end{cases}
$$

Clearly $j \in \mathcal{G}$, thus $A \sim B$, but $A \in \mathcal{PR}_*$ whereas $B \notin \mathcal{PR}_*$.

Thus, $\mathcal{PR}_*$ is not important in the general Recursion Theory. It is however important in its own right; study of $\mathcal{PR}_*$ is the study of invariants of a subgroup of $\mathcal{G}$ the same way Euclidean Geometry is the study of the invariants of a subgroup of the group of projective collineations.

Example 1 $\psi \sim \chi$, where ψ and χ are (possibly partial) functions iff $\psi = f^{-1} \circ \chi \circ f$ for some $f \in \mathcal{G}$.

Indeed, $\psi \sim \chi$ iff $(f,f)(\psi) = \chi (*)$ for some $f \in \mathcal{G}$.

Now $(f,f)(\psi) = \chi$ iff $\psi(x) = y \Longleftrightarrow \chi(f(x)) = f(y)$ iff $\psi(x) = y \Longleftrightarrow f^{-1} \circ \chi \circ f(x) = y$ iff $\psi = f^{-1} \circ \chi \circ f.//$

We now present a theorem, due to Myhill, which is an algorithmic analogue of the Schröder-Bernstein theorem (Theorem 1.5.2).

It shows that if there are 1-1 recursive functions f and g such that f sends A and $\overline{A}$ into B and $\overline{B}$, respectively, and g sends B and $\overline{B}$ into A and $\overline{A}$, respectively, then $A \sim B$. The proof is less involved than that of Theorem 1.5.2 since A and B are in $\mathcal{P}(\mathcal{N})$ and $\mathrm{dom}(f) = \mathrm{dom}(g) = \mathcal{N}$. The behavior of f and g on $\overline{A}$ and $\overline{B}$ also helps.

Theorem 1   (Myhill) $A \equiv_1 B$ iff $A \sim B$.

Proof *if* part. $A \sim B$ iff for some $f \in \mathcal{G}, f(A) = B$. Thus, $A = f^{-1}(B)$; hence, $A \leq_1 B$ and $B = (f^{-1})^{-1}(A)$, hence $B \leq_1 A$.

only if part. Let $A \leq_1 B$ via f and $B \leq_1 A$ via g. The idea is to form an *effective list* of pairs $h = \{(x_i,y_i)\}_{i \geq 0}$ such that

(1) $h:\mathcal{N} \to \mathcal{N}$ is a 1-1 correspondence (i.e., total, 1-1 and onto)

(2) $h(A) = B$.

Observe that if h (as a relation $\lambda xy.h(x) = y$) is an effective list (i.e., a re relation) then, as a function, it is in $\mathcal{R}$, since $h \in \mathcal{P}$ (Theorem 3.7.2) and it is total.

To meet (1) one simply arranges two copies of $\mathcal{N}$ in two lists $\{x_i\}_{i \geq 0}$ and $\{y_i\}_{i \geq 0}$; to this end one alternates between selecting x_i or y_i as the "independent choice".(†) Once such an x_i (resp. y_i) is chosen to simply satisfy $x_i \notin \{x_0, \ldots, x_{i-1}\}$ (resp. $y_i \notin \{y_0, \ldots, y_{i-1}\}$), y_i (resp. x_i) is selected as a "dependent choice" to satisfy $y_i \notin \{y_0, \ldots, y_{i-1}\}$ (resp. $x_i \notin \{x_0, \ldots, x_{i-1}\}$).

To meet (2), each time y_i is a dependent choice, it is selected as follows:

Consider the sequence $(x_{j_0}, x_{j_1}, \ldots)$ such that $x_{j_0} = x_i$ and $f(x_{j_l}) = y_{j_{l+1}} \in \{y_0, \ldots, y_{i-1}\}$ for $l \geq 0$. The numbers $y_{j_1}, y_{j_2}, \ldots, y_{j_l}$ are distinct for any $l > 1$.

(*)$(f,f) = \lambda x.(f(x),f(x))$.

(†)This alternation ensures $\bigcup_{i \geq 0} \{x_i\} = \bigcup_{i \geq 0} \{y_i\} = \mathcal{N}$; i.e., totalness and ontoness of h.

[For $l = 2$, $y_{j_1} = f(x_i)$ and $y_{j_2} = f(x_{j_1})$. Since $x_{j_1} \in \{x_0, \ldots, x_{i-1}\}$ and $x_i \notin \{x_0, \ldots, x_{i-1}\}$, $y_{j_1} \neq y_{j_2}$ by 1-1-ness of f. Next, let $y_{j_1}, \ldots, y_{j_l}$ be distinct. It follows that $x_{j_1}, \ldots, x_{j_l}$ are distinct [requirement (1) is being maintained as we "grow" the x_i and y_i lists] and $x_{j_0} = x_i \notin \{x_{j_1}, \ldots, x_{j_l}\} \subset \{x_0, \ldots, x_{i-1}\}$.

Thus $y_{j_1}, \ldots, y_{j_l}, y_{j_{l+1}}$ are distinct by 1-1-ness of f.] Hence, for some smallest m, $y_{j_m} \notin \{y_0, \ldots, y_{i-1}\}$. Take y_i to be y_{j_m}. Assume now that $x_l \in A$ iff $y_l \in B$ for $l = 0,1, \ldots, i - 1$. The above construction guarantees that $x_i \in A$ iff $y_i \in B$; for $x_i \in A$ iff $x_{j_0} \in A$ iff $y_{j_1} \in B$ iff $x_{j_1} \in A$ iff $y_{j_2} \in B$ iff $\ldots$ iff $y_{j_m} \in B$ iff $y_i \in B$.

To choose dependent x_i's the above construction is repeated with the roles of x_l and y_l interchanged and using g instead of f. Setting $x_0 = 0$ and $y_0 = f(0)$, we have, by induction, that $x_i \in A$ iff $y_i \in B$ for all i. This is equivalent to requirement (2).

The formal construction of h:

Define $\chi: \mathcal{N} \to \mathcal{N}$ and $\psi: \mathcal{N} \to \mathcal{N}$ so that for all i, $\chi(i) = x_i$ and $\psi(i) = y_i$.

(1) $\chi(0) = 0$ $\psi(0) = f(0)$

(2)   For $t > 0$ (independent choices of x_i and y_i)

$$\chi(2t) = (\mu x)[(\forall i)_{\leq 2t \dot- 1} \, x \neq \chi(i)]$$

$$\psi(2t + 1) = (\mu y)[(\forall i)_{\leq 2t} \, y \neq \psi(i)]$$

(3)   For $t > 0$ (dependent choices of x_i and y_i), set

$$p = (\mu x)[\exp(0,x) = \chi(2t) + 1 \,\&$$
$$\{(\forall l)_{<\text{length}(x) \dot- 1}(\exists j)_{\leq 2t \dot- 1} \, (f(\exp(l,x) \dot- 1) = \psi(j)$$
$$\&\ \exp(l + 1,x) = \chi(j) + 1)\} \,\&$$
$$\{(\forall i)_{\leq 2t \dot- 1} f(\exp(\text{length}(x) \dot- 1,x) \dot- 1) \neq \psi(i)\}]$$

$$q = (\mu y)[\exp(0,y) = \psi(2t + 1) + 1 \,\&$$
$$\{(\forall l)_{<\text{length}(y) \dot- 1}(\exists j)_{\leq 2t} \, (g(\exp(l,y) \dot- 1) = \chi(j)$$
$$\&\ \exp(l + 1,y) = \psi(j) + 1)\} \,\&$$
$$\{(\forall i)_{\leq 2t} g(\exp(\text{length}(y) \dot- 1, y) \dot- 1) \neq \chi(i)\}].$$

Then,

$$\psi(2t) = f(\exp(\text{length}(p) \dot- 1, p) \dot- 1)$$

$$\chi(2t + 1) = g(\exp(\text{length}(q) \dot- 1, q) \dot- 1)$$

Note: x codes the sequence $x_{j_0}, \ldots, x_{j_m}$, where $x_{j_0} = x_{2t}$, $f(x_{j_l}) = y_{j_{l+1}}$ $\in \{y_0, \ldots, y_{2t \doteq 1}\}$ for $l \leq m - 1$ and $f(x_{j_m}) \notin \{y_0, \ldots, y_{2t \doteq 1}\}$, according to the scheme $\Pi_{l \leq m} p_l^{x_{j_{l+1}}}$. Similar comment for y. This note applies to (3) only.

Since the occurrence of $\chi(2t)$ [resp. $\psi(2t + 1)$] in (3) above can be avoided using (2), the above is an ordinary course-of-values simultaneous recursion, thus χ and ψ are in $\mathcal{R}$. Finally, $h(x) = y$ iff $(\exists t)(x = \chi(t)$ & $y = \psi(t))$, thus $\lambda xy.h(x) = y$ is re by the projection theorem, and $h \in \mathcal{R}$ as per our earlier remarks.//

Note: We note three consequences of Theorem 1:

(1) $K \sim K_0$, since $K \equiv_1 K_0$ (Corollary of Theorem 9.2.2).

(2) $\{x \mid \phi_x \in \mathcal{R}\} \sim \{x \mid W_x$ is infinite$\}$, due to the note following Theorem 9.2.2 and Examples 2 and 3 presented in the preamble to this chapter.

(3)   Theorem 1 can be restated as $\sim = \equiv_1$. This gives special significance to the 1-degrees as the "natural objects" of study of Recursion Theory. Further, $\equiv_1 \subset \equiv_m$ implies $\sim \subset \equiv_m$.

This has as consequence that each $\equiv_m$-equivalence class (i.e., m-degree) is the union of one or more isomorphism types (i.e., $\sim$-equivalence classes).

§9.4 CYLINDERS

In what follows, $J \in \mathcal{R}^{(2)}$ is a fixed but unspecified *onto* pairing function with projections K and L in $\mathcal{R}^{(1)}$.

Definition 1   A set $A \subset \mathcal{N}$ is a *cylinder* iff $A \sim J(B \times \mathcal{N})$ for some set $B \subset \mathcal{N}$.//

Note: Geometrically, $B \times \mathcal{N} = \{(x,y) \mid x \in B$ & $y \in \mathcal{N}\}$ is a cylinder with base B in the two-dimensional integer lattice $\mathcal{N} \times \mathcal{N}$; hence, the term. We shall see that the definition is independent of the choice of J [Theorem 1(iii) below]. Also, let $J = \lambda xy.2^y(2x + 1) \doteq 1$ and consider $J(\{0\} \times \mathcal{N})$. Next, let $f \stackrel{\text{def}}{=} \lambda x.$ **if** x even **then** $x + 1$ **else** $x \doteq 1$. Clearly, f is a recursive permutation and $f(J(\{0\} \times \mathcal{N}))$, consisting only of even numbers and 1, cannot *equal* a set $J(B \times \mathcal{N})$ for any $B \subset \mathcal{N}$. Thus, to give recursive invariance to the concept of cylinder "$\sim$", instead of "$=$", is used in the above definition.

Theorem 1   (i) $A \leq_1 J(A \times \mathcal{N})$

(ii) $J(A \times \mathcal{N}) \leq_m A$

(iii) A is a cylinder iff $(\forall X)(X \leq_m A \Rightarrow X \leq_1 A)$

(iv) $A \leq_m B$ iff $J(A \times \mathcal{N}) \leq_1 J(B \times \mathcal{N})$.

Proof

(i) $x \in A$ iff $J(x,x) \in J(A \times \mathcal{N})$ and $\lambda x.J(x,x)$ is 1-1.

(ii) $x \in J(A \times \mathcal{N})$ iff $K(x) \in A$ and $K \in \mathcal{R}^{(1)}$.

(iii)   *if* part. By assumption and (ii) $J(A \times \mathcal{N}) \leq_1 A$. By (i) and Theorem 9.3.1 $A \sim J(A \times \mathcal{N})$.
only if part. Let $A \sim J(B \times \mathcal{N})$ for some B.
Let $X \leq_m A$ via f; that is, $X = f^{-1}(g^{-1}(J(B \times \mathcal{N})))$, where g is a recursive permutation under which $g(A) = J(B \times \mathcal{N})$. Thus
$X \leq_m J(B \times \mathcal{N})$ via $g{\circ}f$, and hence
$X \leq_m B$ via $K{\circ}g{\circ}f$. Clearly,
$X \leq_1 A$ via $\lambda x.g^{-1}{\circ}J(K{\circ}g{\circ}f(x),x)$, which is, of course, 1-1.

(iv)   *if* part. Let $J(A \times \mathcal{N}) \leq_1 J(B \times \mathcal{N})$.
Hence $A \leq_1 J(A \times \mathcal{N}) \leq_1 J(B \times \mathcal{N}) \leq_m B$ and the result follows from $\leq_1 \subset \leq_m$ and transitivity of $\leq_m$
only if part. Let $A \leq_m B$. Then
$J(A \times \mathcal{N}) \leq_m A \leq_m B \leq_1 J(B \times \mathcal{N})$ hence, as in the *if* part,
$J(A \times \mathcal{N}) \leq_m J(B \times \mathcal{N})$. But $J(B \times \mathcal{N})$ is a cylinder, hence
$J(A \times \mathcal{N}) \leq_1 J(B \times \mathcal{N})$ by (iii).//

Corollary 1   The definition of cylinder is independent of the choice of J.

Proof By (iii) above.//

Corollary 2 A is a cylinder iff $J(A \times \mathcal{N}) \leq_1 A$ iff $A \sim J(A \times \mathcal{N})$.

Proof A cylinder $\Rightarrow J(A \times \mathcal{N}) \leq_1 A$ [by (ii), using (iii)] $\Rightarrow A \sim J(A \times \mathcal{N})$ [by (i) and Theorem 9.3.1] $\Rightarrow A$ is a cylinder (by Definition 1).//

Note: $J(A \times \mathcal{N})$ is the *cylindrification* of A. Within recursive isomorphism, a cylinder stays invariant if cylindrified (Corollary 2). Also, let $A \in \deg_m(B)$. Then $A \leq_m B \leq_1 J(B \times \mathcal{N})$; hence, $A \leq_1 J(B \times \mathcal{N})$ [Theorem 1, (iii)].

In other words, each m-degree x contains a maximum (with respect to $\leq_1$) 1-degree (or isomorphism type) which can be represented by a cylindrification of any set in x.

The next theorem is yet another characterization of cylinders which does not involve J.

Theorem 2 A is a cylinder iff for some $g \in \mathcal{R}^{(1)}$ and all x

(a) $\varnothing \neq D_x \subset A \Rightarrow g(x) \in A - D_x$

(b) $\varnothing \neq D_x \subset \overline{A} \Rightarrow g(x) \in \overline{A} - D_x$, where D_x is a finite set with prime-power code x (see §8.3).

Proof *if* part. Let the conditions (a) and (b) of the Theorem hold. We show A to be a cylinder relying on Theorem 1(iii). So let $X \leq_m A$ via $f \in \mathcal{R}^{(1)}$. Using f and g, we obtain a 1-1 reduction function h:
Define h by $h(0) = f(0)$. Clearly, $0 \in X$ iff $h(0) \in A$. To define $h(n + 1)$, we must meet two requirements:

(1) $n + 1 \in X$ iff $h(n + 1) \in A$

(2) $h(n + 1) \notin \{h(0), \ldots, h(n)\}$.

To satisfy (1), set $h(n + 1) = f(n + 1)$. If (2) is not satisfied, change $h(n + 1)$ to $g(2^{f(n+1)+1})$. [Note that $D_{2^{f(n+1)+1}} = \{f(n + 1)\}$]. Observe that [by (a) and (b)] (1) is still satisfied. If (2) is not satisfied, change $h(n + 1)$ to $g(2^{f(n+1)+1} \cdot 3^{g(2^{f(n+1)+1})+1})$ [Note that $D_{2^{f(n+1)+1} \cdot 3^{g(2^{f(n+1)+1})+1}} = \{f(n + 1), g(2^{f(n+1)+1})\}$.] Observe that (1) is still satisfied, since either $\{f(n + 1), g(2^{f(n+1)+1})\}$ is in A or in $\overline{A}$. If (2) is still not satisfied, add to the finite set $\{f(n + 1), g(2^{f(n+1)+1})\}$ the entry $g(2^{f(n+1)+1} \cdot 3^{g(2^{f(n+1)+1})+1})$ and change $h(n + 1)$ to the outcome of g applied to the prime-power code of the new finite set. Eventually (2) will be satisfied since the entries $f(n + 1), g(2^{f(n+1)+1}), g(2^{f(n+1)+1} \cdot 3^{g(2^{f(n+1)+1})+1}), \ldots$ are distinct.

Formally, the above sequence of entries is given by the function $\lambda kn.\text{seq}(k,n)$ such that

$$\text{seq}(0,n) = f(n + 1)$$
$$\text{seq}(k + 1,n) = g(\Pi_{i \leq k} \, p_i^{\text{seq}(i,n)+1})$$

$\text{seq} \in \mathcal{R}$, since $\mathcal{R}$ is closed under course-of-values recursion. Thus, $h(n + 1) = \text{seq}((\mu y)_{\leq n+1}((\forall x)_{\leq n} h(x) \neq \text{seq}(y,n)),n)$. h being 1-1 and recursive, and since it satisfies (1) and (2) for all n we have $X \leq_1 A$ via h which establishes A to be a cylinder.

only if part. Let A be a cylinder. By Corollary 2 of Theorem 1, $f(A) = J(A \times \mathcal{N})$, where f is a recursive permutation.
Define $g \overset{\text{def}}{=} \lambda x.f^{-1} \circ J(K \circ f(\exp(0,x) \doteq 1), L \circ f(\exp(0,x) \doteq 1) + 1)$. Now if $\varnothing \neq D_x \subset A$, then $f(D_x) \subset f(A) = J(A \times \mathcal{N})$, hence $K \circ f(\exp(0,x) \doteq 1) \in A$ and thus $g(x) \in A$. On the other hand, $g(x) \in D_x \Rightarrow f \circ g(x) \in f(D_x) \Rightarrow (\exists i)_{<\text{length}(x)} f(\exp(i,x) \doteq 1) = J(K \circ f(\exp(0,x) \doteq 1), L \circ f(\exp(0,x) \doteq 1) + 1)$, which is impossible since it implies $L \circ f(\exp(0,x) \doteq 1) = L \circ f(\exp(0,x) \doteq 1) + 1$. Similarly, $\varnothing \neq D_x \subset \overline{A} \Rightarrow g(x) \in \overline{A} - D_x.//$

We shall see in Chapter 11 that 1-completeness and m-completeness coincide. Then by Theorem 1(iii) each m-complete set is a (recursively enumerable) cylinder, therefore it satisfies the conditions of Theorem 2. In the Problems (12-15) hints are given for a direct proof that (a) and (b) of Theorem 2 are satisfied by m-complete sets, proving m-complete sets to be (re) cylinders and hence 1-complete.

§9.5 CREATIVE AND PRODUCTIVE SETS

Sets such a $\overline{K}$ are non-re in a constructive sense. Indeed, for any "claim" that $W_x \subset \overline{K} \Rightarrow W_x = \overline{K}$, x provides a "counterexample" since $x \in \overline{K} - W_x$ [$x \in W_x$ means $x \in K$, hence it is inconsistent with $W_x \subset \overline{K}$. Thus, $x \notin W_x$; hence also $x \in \overline{K}$, thus $x \in \overline{K} - W_x$]. Thus, the function $h = \lambda x.x$ "constructs" counterexamples of the form $h(x) \in \overline{K} - W_x$.

Sets such as $\overline{K}$ have been termed *productive* (Dekker) and the functions such as h productive functions.

Definition 1 A is *productive* with *productive function* $f \in \mathcal{R}$ iff $(\forall x)(W_x \subset A \Rightarrow f(x) \in A - W_x).//$

Definition 2 A is *creative* iff

    (1) A is re

and   (2) $\overline{A}$ is productive.//

Note: Creative sets (studied by Post) are important in the study of *formal mathematical systems*. Post used a recursive productive function as in Definition 1. It turns out (Chapter 11) that the notion is not broadened if we allow partial recursive productive functions.

Example 1 K is creative. Indeed, it is re and $\overline{K}$ is productive with $\lambda x.x$ as productive function.//

Example 2 The notion of productiveness (and creativity) are recursive invariants. Indeed, let A be productive with productive function h and let $g(A) = B$ where g is a recursive permutation.

Now $W_x \subset B$ iff $g^{-1}(W_x) \subset A$. There is a recursive function f such that $W_{f(x)} = g^{-1}(W_x)$ for all x (Example 1 in §8.1).

Thus, $W_x \subset B$ iff $W_{f(x)} \subset A$. Hence, $h \circ f(x) \in A - W_{f(x)} = g^{-1}(B) - g^{-1}(W_x)$ $= g^{-1}(B - W_x)$; i.e., $g \circ h \circ f(x) \in B - W_x$; that is, B is productive with productive function $g \circ h \circ f$.

The result for creativity follows by the fact that re-ness is preserved by recursive permutations.//

Example 3   A productive set S is *not* re. Indeed, if $W_x = S$ for some x, $h(x) \in S - W_x$ provides a counterexample, where h is a productive function of S. Similarly, a creative set is *not* recursive (its complement is not *re*).//

Theorem 1   If A is productive and $A \leq_m B$, then B is productive.

   Proof   Say $A \leq_m B$ via g and let h be a productive function for A. It follows that $A = g^{-1}(B)$ and the rest of the proof is as in Example 2 above.//

Corollary 1   If A is creative and $A \leq_m B$, then $\overline{B}$ is productive.

   Proof $\overline{A}$ is productive and $\overline{A} \leq_m \overline{B}$.//

Corollary 2   If A is m-complete then it is creative.

   Proof m-completeness implies $K \leq A$. Now K creative implies $\overline{A}$ productive. But A is re.//

   In Chapter 11, we prove a converse of Corollary 2.

Example 4   The sets $\{x \mid \phi_x \text{ is constant}\}$, $\{x \mid \phi_x \in \mathcal{R}\}$ and $\{x \mid W_x \text{ is infinite}\}$ are productive.
   Indeed, from the proof of Theorem 7.3.2 we know that $\overline{K} \leq_m \{x \mid \phi_x \text{ is constant}\}$.
   In the same place we learn that $\overline{K} \leq_m \{x \mid W_x \text{ is infinite}\}$.
   Finally, the recursive invariance of productiveness and the fact that $\{x \mid \phi_x \in \mathcal{R}\} \sim \{x \mid W_x \text{ is infinite}\}$, proves that $\{x \mid \phi_x \in \mathcal{R}\}$ is also productive.
   Are all productive sets isomorphic? The answer is *no* via the counterexample $\{x \mid \phi_x \in \mathcal{R}\}$ and $\{x \mid \phi_x \notin \mathcal{R}\}$.
   The latter is productive (observe the note following the proof of Theorem 7.3.2), but we shall see in Chapter 11 that these two sets are incomparable with respect to $\leq_m$ (and, hence, $\leq_1$) thus they cannot be isomorphic.//

Example 5   Let B be re and $A \cap B$ productive. We show A is productive. Let f be a productive function of $A \cap B$ and let $W_x \subset A$. Hence $W_x \cap B \subset A \cap B$. There is an $h \in \mathcal{R}^{(1)}$ such that $W_{h(x)} = W_x \cap B$ for all x. (See Problem 27). Thus, $f \circ h(x) \in A \cap B - W_{h(x)} = A \cap B - W_x \cap B = A \cap B - W_x \subset A - W_x$. Thus $f \circ h$ is a productive function for A.//

Theorem 2   For each productive set A and each $W_x \subset A$ there is an infinite re subset of

A, disjoint to W_x. Moreover, given x, an re index for the aforementioned set can be effectively found.

Proof   Let h be a productive function for A. Build an infinite sequence (obviously without repetitions) as follows:

$$z_0 = h(x) \in A - W_x$$

$$z_1 = h(x_1) \in A - W_{x_1}, \text{ where } W_{x_1} = W_x \cup \{z_0\}$$

$$z_2 = h(x_2) \in A - W_{x_2}, \text{ where } W_{x_2} = W_{x_1} \cup \{z_1\} = W_x \cup \{z_0, z_1\}$$

$$\vdots$$

More formally, let first $f \in \mathcal{R}^{(2)}$ be such that $W_{f(x,y)} = W_x \cup W_y$. [The "program" $f(x,y)$ enumerates W_x and W_y taking turns; i.e., dovetailing. More precisely, define $\psi(x,y,z)$ for all x,y,z by $(\mu w)(T(x,z,w) \vee T(y,z,w))$. Clearly, $\psi \in \mathcal{P}$; hence, $\psi = \lambda xyz.\phi_{f(x,y)}(z)$ for some f; $\text{dom}(\phi_{f(x,y)}) = W_x \cup W_y$.] Define a function $g \in \mathcal{R}^{(2)}$ by the requirements

$$W_{g(x,0)} = W_x$$

$$W_{g(x,y+1)} = W_{g(x,y)} \cup \{h \circ g(x,y)\}$$

To establish that the g above exists, observe that for some $m \in \mathcal{R}^{(1)}$, $W_{m(x)} = \{x\}$. [Let $\psi \overset{\text{def}}{=} \lambda xy.\text{ if } y = x \text{ then } 0 \text{ else } (\mu z)(z + 1)$. Then, for some $m \in \mathcal{R}^{(1)}$, $\psi(x,y) = \phi_{m(x)}(y)$ for all x,y and $\text{dom}(\phi_{m(x)}) = \{x\}$].
Thus, simply set

$$g(x,0) = x$$

$$g(x,y + 1) = f(g(x,y), m \circ h \circ g(x,y)).$$

Clearly, $g \in \mathcal{R}^{(2)}$ and $\text{range}(\lambda y.h \circ g(x,y))$ is re, infinite, disjoint to W_x and equal to $\{z_0, z_1, z_2, \ldots\}$.//

Note: We state for emphasis, that $g(x,y)$ is defined for *all* x even if $W_x \notin A$.

Corollary 1   A productive set has an infinite re subset.

Proof   Take $W_x = \varnothing$, for some appropriate x.//

Corollary 2   A productive set has an infinite recursive subset.

   Proof Corollary 1, and Corollary 2 of Theorem 8.2.1.//

Corollary 3   A productive set A has a 1-1 productive function.

   Proof Let h be a productive function of A and let g be that of the proof of Theorem 2. We specify a 1-1 productive function p as follows:

   (1) $p(0) = h(0)$

For $p(x + 1)$ we consider two cases:

   (a) $h(x + 1) \notin \{p(0), \dots, p(x)\}$

   (b) $h(x + 1) \in \{p(0), \dots, p(x)\}$

In case (a) take $p(x + 1) = h(x + 1)$.
In case (b) there are two subcases:

   (ba) $W_{x+1} \subset A$

   (bb) $W_{x+1} \not\subset A$

(ba) implies $\{h \circ g(x + 1, y)\}_{0 \leq y \leq x+1}$ *are distinct.* Let us call this condition $C(x)$. In case (bb), it may be that $C(x)$ holds, but also $\neg C(x)$ may hold.
   So, in case (b) we set $p(x + 1)$ either equal to the first $h \circ g(x + 1, y)$ which is not in $\{p(0), \dots, p(x)\}$ if $C(x)$ holds, else we put it equal to the first integer not in $\{p(0), \dots, p(x)\}$. Thus,

   (2) $p(x + 1) = $ **if** $h(x + 1) \notin \{p(0), \dots, p(x)\}$ **then** $h(x + 1)$
else if $C(x)$ **then** $h \circ g(x + 1,(\mu y)[h \circ g(x + 1,y) \notin \{p(0), \dots, p(x)\}])$
else $(\mu y)[y \notin \{p(0), \dots, p(x)\}]$

Clearly, p is 1-1 and recursive, since $\mathcal{R}$ is closed under course-of-values recursion, and $\lambda x.C(x) \in \mathcal{R}_*$. Further, p is a productive function for A since whenever $p(x)$ is chosen to be different from $h(x)$ *and* $W_x \subset A$, $p(x)$ is nevertheless a $h \circ g(x,y)$, hence in $A - W_x$.//

Corollary 4   With the assumptions of Corollary 3, range(h) $\subset$ range(p).

   Proof For any x, $h(x) \in$ range(p). Indeed, if $x = 0$ this is so since $h(0) = p(0)$.
   If $x \neq 0$, this is so by the definition of $p(x)$ in case (a) (proof of Corollary 3).//

Corollary 5   A productive set A has a productive function which is a recursive permutation.

Proof It suffices to show that if h is productive for A then an *onto* productive function q can be found (result then follows from Corollaries 3 and 4).

We must meet two requirements:

(1) q is productive.

(2) q is onto.

To meet (1), we set $q(x) = h(x)$ when $W_x \subset A$. [**Note:** in the end, it will be $q(x) = h(x)$ for a *recursive* superset of $\{x \mid W_x \subset A\}$, as this set is not recursive.] To meet (2), without spoiling (1), we ensure that q covers that part of $\mathcal{N}$ left uncovered by h, by acting on a *recursive* subset of $\{x \mid W_x \not\subset A\}$ [**Note:** our preoccupation with *recursive* supersets and subsets is so that the definition of q, by cases from h, leads to a recursive function].

So let us start with a recursive set of re-indices of some set S that $S \not\subset A$. Conveniently, S is chosen to be $\mathcal{N}$ [if $\mathcal{N} \subset A$ then $A = \mathcal{N}$ and hence A is not productive].

The set of indices of $\mathcal{N}$ is $\{x \mid \phi_x \in \mathcal{R}\}$, which, because it is productive (Example 4), in turn contains an *infinite recursive* set I (Corollary 2). Let $g(\mathcal{N}) = I$, where $g \in \mathcal{R}$ is strictly increasing (Corollary 1 of Theorem 8.2.1).

Define

$$q = \lambda x.\ \textbf{if } x \in I \textbf{ then } (g^{-1} \mid I)(x)$$

$$\textbf{else } h(x)$$

That is, $q = \lambda x.\ \textbf{if } x \in I \textbf{ then } (\mu y)(g(y) = x) \textbf{ else } h(x)$. Since $I \in \mathcal{R}_*$ and $g \in \mathcal{R}$, $q \in \mathcal{R}$. Since $g^{-1} \mid I$ is onto $\mathcal{N}$, q is onto $\mathcal{N}$. Finally, $W_x \subset A \Rightarrow x \notin I \Rightarrow q(x) = h(x) \Rightarrow q(x) \in A - W_x.//$

§9.6 GÖDEL'S INCOMPLETENESS THEOREM

Hilbert believed that in order to avoid logical vicious circles, matters such as the consistency(*) of logical and mathematical theories should be tackled with the most elementary *finitary* means so that these methods, being so close to intuitive acceptance, would not have to be in turn subject to scrutiny as to their logical consistency. Hilbert's "program" was therefore to develop such finitary means so that the consistency of all known mathematical (logical) systems could be proved (or disproved, as the case may be). He further

(*)That is, freedom from contradiction.

believed that there must be a *method* (algorithm) by which the *decision problem* (*Entscheidungsproblem*) of an axiomatic mathematical (logical) theory could be solved. The decision problem is, in this context, the provability or non-provability of a given formula which is written according to the admitted syntax rules of the theory.

Gödel's incompleteness results [8] pointed at the fruitlessness of both of Hilbert's beliefs. First, Gödel showed that if a mathematical theory is just rich enough so that Hilbert's finitary methods are expressible in the theory then the consistency of the theory (if it *is* consistent!) cannot be proved within the theory, and hence cannot be proved by the *finitary means* either!

A corollary of Gödel's result is that reasonably rich theories have a recursively unsolvable decision problem.(*) Thus, if by *method* we mean a *recursive procedure*, then for such theories no *method* exists to settle their *Entscheidungsproblem*; this shatters the second of Hilbert's beliefs.

In this section we give a brief account of Gödel's incompleteness results. We rely on methods of Recursion Theory (unlike Gödel) which at the same time both shorten and, hopefully, make more approachable the proofs.

Let us proceed informally at first. We shall call *Elementary Arithmetic* the set of *formulas* we can build by utilizing subscripted or unsubscripted lower-case letters (e.g., $a, b, \ldots, x, y, z \ldots, x_1, a_1, \ldots$) as (names of) variables over $\mathcal{N}$, along with the symbols $+, \times$ (times), $=, \neg, \Rightarrow, \vee, \&, (,), \forall, \exists, 0, 1, 2, 3, \ldots$ (an arbitrary integer $n \in \mathcal{N}$, will be denoted by $\tilde{n}$ to avoid confusion with the variable named n). Of course *not every* string utilizing these symbols is a formula of Elementary Arithmetic. The rules for constructing syntactically correct ("well-formed") formulas are well known from our experience with elementary high school algebra and will be left unspecified here.

For convenience, we shall write $F(x_1, \ldots, x_n)$ or $F(\vec{x}_n)$ to denote some (perhaps unspecified) formula where $\vec{x}_n$ are the only *free* variables (that is, not acted upon by $\exists$ or $\forall$). For example, $E(x)$ has x only as a free variable, F has none, $E(\tilde{x})$ has none ($\tilde{x}$ is some unspecified but fixed integer) and $(\forall t)E(t)$ has none. Formulas with *no* free variables are called *sentences*. *Truth* of a sentence (although it can be defined rigorously following Tarski) will be taken in the intuitive sense. (For example, $(\forall t)E(t)$ is true iff $E(0), E(1), \ldots, E(\tilde{t}), \ldots$ are *all* true, where $\tilde{t}$ ranges over all of $\mathcal{N}$, whereas $(\exists t)E(t)$ is true just in case, for some $\tilde{x} \in \mathcal{N}, E(\tilde{x})$ is true).

Specifically, $(\forall t)(\exists x)(t = 2x)$ is *false* since it "states" that all integers (t) are even, whereas $(\exists p)(\forall x)[(\exists t)(p = xt) \Rightarrow x = 1 \vee x = p]$ is *true* since it states that prime numbers (p) exist.

(*)That undecidable mathematical theories exist was first explicitly stated and proved by Church [4] and is known as Church's undecidability result. It should be noted that at the time Gödel wrote his paper, Recursion Theory was not yet developed. Indeed, Gödel relied on primitive recursive functions to prove his results.

We shall call *Elementary Number Theory*, or ENT, the set of *true* sentences of Elementary Arithmetic.

The notion of *proof, provable formula* (and *sentence*) we shall leave undefined, assuming the usual (syntactic) notion of proof. In this connection, let us argue, informally, that the set of all *provable formulas* (as well as *provable sentences*) is recursively enumerable.(*)

Indeed, list *effectively* all *proofs* by listing lexicographically for each $l = 1, 2, \ldots$ all the proofs (if any) of *length* l, where a proof $F_1, F_2, \ldots, F_n$ is thought of as a string $F_1 * F_2 * \ldots * F_n$ where F_i's are formulas and $*$ a *new* symbol not in the initial alphabet.

Simultaneously build two more lists, $\mathscr{F}$ and $\mathscr{S}$: Every time a proof has been added in the proof-list, put its last formula in the F-list (effective list of formulas). If this formula is a sentence, then put it also in the S-list (effective list of sentences).

An axiomatization of Elementary Arithmetic is *sound* if *no false* sentence is provable, within the axiomatization. We now state and "prove" a version of Gödel's *First Incompleteness Theorem*.

Theorem 1   (Gödel [8]) If an axiomatization of Elementary Arithmetic is sound, then it contains some sentence such that neither it nor its negation are provable.(†)

Proof Both $\phi_x(x)\downarrow$ and $\phi_x(x)\uparrow$ are statements which "can be made" (this *does not* mean that they are going to be *true* or even *provable* for every $x \in \mathcal{N}$!) within Elementary Arithmetic, since $\phi_x(x)\downarrow$ iff $(\exists y)T(x,x,y)$ and $\phi_x(x)\uparrow$ iff $\neg(\exists y)T(x,x,y)$. By the result of §5.4, T can be chosen to be constructive arithmetic hence obviously "stateable" in Elementary Arithmetic; then this is also the case for $(\exists y)T(x,x,y)$ and its negation.

Consider the function $g = \lambda x.\mathbf{if}\ \phi_x(x)\downarrow\ \mathbf{then}\ 0\ \mathbf{else}\ 1$ (this is the characteristic function of K, of course). We claim that under the assumption of soundness, for some $\tilde{x}$ that makes $\phi_{\tilde{x}}(\tilde{x})\uparrow$ true, $\phi_{\tilde{x}}(\tilde{x})\uparrow$ is nevertheless *unprovable* (in the system as given).

By contradiction, let $\phi_{\tilde{x}}(\tilde{x})\uparrow$ be provable for *all* $\tilde{x} \in \mathcal{N}$ that make $\phi_{\tilde{x}}(\tilde{x})\uparrow$ true.

(*)It is clear that the *alphabet* over which Elementary Arithmetic is being built is finite, since the integers are going to be represented (say) in 10-adic notation. Thus formulas can be thought of as numbers in m-adic notation, where m is the cardinality of the alphabet. Thus, re-ness, recursiveness, etc., of a set of formulas is interpreted as an attribute of the corresponding set of (m-adic notations of) numbers.

(†)Provable within the axiomatization, of course, as given. Existence of sentences S such that neither S nor $\neg S$ is provable implies that *true* sentences which are *non-provable* exist (it cannot be that *both* S and $\neg S$ are *false*!) Thus, the system (Elementary Arithmetic) is called *incomplete* as it does not "capture" all true sentences.

Then g is recursive, since to compute g, given $\tilde{x}$, we compute $\phi_{\tilde{x}}(\tilde{x})$ and at the same time enumerate $\mathcal{S}$ by dovetailing.

If $\phi_{\tilde{x}}(\tilde{x})\downarrow$ eventually, then we stop and output 0. ($\phi_{\tilde{x}}(\tilde{x})\uparrow$, or more precisely $\neg(\exists y)T(\tilde{x},\tilde{x},y)$, being *false* in this case, is *not* in the $\mathcal{S}$-list [by soundness], thus our decision to halt the computation is justified.)

If $\phi_{\tilde{x}}(\tilde{x})\uparrow$ is true on the other hand, we stop as soon as we locate $\neg(\exists y)T(\tilde{x},\tilde{x},y)$ in the $\mathcal{S}$-list, outputting 1 in this case [again by soundness, $\phi_{\tilde{x}}(\tilde{x})\downarrow$ cannot occur in this case]. (This procedure can easily be formalized as soon as a recursive function f, which has as range all the provable sentences, viewed as integers in m-adic notation, is given).

But $g \in \mathcal{R}$ contradicts the nonrecursiveness of K, thus for some $\tilde{x} \in \mathcal{N}$, $\phi_{\tilde{x}}(\tilde{x})\uparrow$ is *true* but *unprovable*. Since then $\phi_{\tilde{x}}(\tilde{x})\downarrow$ is *false*, also $\phi_{\tilde{x}}(\tilde{x})\downarrow$ is unprovable, by soundness.//

Note: We have scattered around the statement of Theorem 1 and its proof qualifications of the form "in the system as given" whenever such and such is provable or unprovable was claimed. The origin of such caution comes from the fact that sometimes *augmenting* an axiomatic system by *adding* more axioms one may make it *complete* (in other words, an axiomatic system receptive of such treatment was incomplete due to "too few" axioms originally being given. Sound and complete axiomatic systems exist. See, for example, Wilder [22], p. 36). Can Elementary Arithmetic be made complete by adding axioms, under the reasonable assumption that it (was and) stays sound? The answer must be clearly "no", as in the above (informal) proof exactly *what axioms were already present* did not play any role.

Thus, for example, if we add to the already existing axioms the axiom $\phi_{\tilde{x}}(\tilde{x})\uparrow$ (we cannot add $\phi_{\tilde{x}}(\tilde{x})\downarrow$ if we insist on soundness, since then $\phi_{\tilde{x}}(\tilde{x})\downarrow$ would become provable in the *new* system, nevertheless being *false*!) some other $\tilde{x}_1$ will exist (according to the previous proof) such that both formulas $\phi_{\tilde{x}_1}(\tilde{x}_1)\downarrow$ and $\phi_{\tilde{x}_1}(\tilde{x}_1)\uparrow$ are unprovable in the *new* system while the latter sentence is *true*. Continuing like this, it is clear that for every augmentation of the axiom set by adding a *true* sentence of the form $\phi_{\tilde{x}_i}(\tilde{x}_i)\uparrow$ to it, the new axiomatization obtained for Elementary Arithmetic is still incomplete (if sound).

No matter what subset of the set of *true* sentences of the form $\phi_{\tilde{x}}(\tilde{x})\uparrow$ we adjoin to the set of axioms, under the sole proviso that the set of axioms remains "recognizable" (that is, *recursive*), the proof of Theorem 1 still insists that for some $x \in \mathcal{N}$, $\phi_{\tilde{x}}(\tilde{x})\uparrow$ is true but unprovable in the augmented axiomatization of Elementary Arithmetic. [Incidentally, as a corollary, it is not possible then to adjoin *all* of $\{\phi_{\tilde{x}}(\tilde{x})\uparrow \mid \phi_{\tilde{x}}(\tilde{x})\uparrow$ is *true*$\}$ to the set of axioms since then all such sentences would be provable]. This phenomenon exhibited by sound axiomatizations of Elementary Arithmetic, according to which the system cannot be *made* complete, is referred to as the Gödel *incompletableness* phenomenon.

Gödel's original proof of Theorem 1 was obtained by a "direct diagonalization" as opposed to indirect (via the halting problem) employed in the above proof. By first arithmetizing the axiomatization of Elementary Arithmetic with the help of primitive recursive functions (prime power coding) he next showed that the statement *"I am not a theorem"* is a sentence of Elementary Arithmetic.(*) In a sound arithmetic such a sentence must be unprovable (otherwise, its provability implies its falsehood, hence unsoundness of the system!), hence it must be *true*, since it "states" just that! But then its negation is not provable either (by soundness).

Corollary   ENT cannot be axiomatized.(†)

Proof Any axiomatization of ENT is sound (why?) and contains all true sentences of the form $\phi_{\tilde{x}}(\tilde{x}){\uparrow}$ as theorems.//

Note: Theorem 1 is equivalent to the above corollary. Indeed, if a sound axiomatization of Elementary Arithmetic is given then some sentence in ENT must be unprovable, else this would be an axiomatization of ENT.

The above corollary can be sharpened:

Theorem 2   ENT is a productive set, hence (not being re) is not axiomatizable.

Proof Consider the following sets of sentences:

$$A = \{\phi_{\tilde{x}}(\tilde{x}){\uparrow} \mid x \in \mathcal{N}\}$$

$$\tilde{K} = \{\phi_{\tilde{x}}(\tilde{x}){\uparrow} \mid \phi_{\tilde{x}}(\tilde{x}){\uparrow} \text{ is } true\}$$

ENT

Now, A is clearly recursive, where, of course, $\phi_{\tilde{x}}(\tilde{x}){\uparrow}$ is shorthand for $\neg(\exists y)T(\tilde{x},\tilde{x},y)$ and the latter is viewed as the m-adic notation of an integer.

$\tilde{K}$ is productive. Indeed, $\tilde{K} = \{\phi_{\tilde{x}}(\tilde{x}){\uparrow} \mid x \in \overline{K}\}$. Consider the (obviously) recursive function $f\colon \mathcal{N} \to \mathcal{N}$ defined by $f(x) = z$ iff z in m-adic is the string $\neg(\exists y)T(\tilde{x},\tilde{x},y)$ [of course, $T(\tilde{x},\tilde{x},y)$ is shorthand. T is not a member of the

---

(*)He first assigned *numerical codes* to each formula, sentence and proof, in an *effective way* (this is the origin of "arithmetizations" and of the term *Gödel number*). Then he produced
(1) A "substitution function" f in $\mathcal{PR}^{(2)}$ (fully analogous to an S-m-n function) such that $f(x,y)$ is what becomes of the Gödel number of formula (with Gödel number) x after a (fixed) *variable* z has been substituted by the *number* y.
(2) A predicate $P(z,y)$ in $\mathcal{PR}_*$, which says that y is a Gödel number of a proof of the formula (with Gödel number) z. Thus, "I am not a theorem" is captured by $\neg(\exists y)P(f(\tilde{x},\tilde{x}),y)$, where $\tilde{x}$ = Gödel number of $\neg(\exists y)P(f(z,z),y)$.
(†)An axiomatization of ENT has as theorems exactly all the sentences of ENT.

alphabet of Elementary Arithmetic, but it is a string over it], where $\tilde{x}$ is x in 10-adic.

Thus, $\overline{K} \leq_m \tilde{K}$ via f, and the claim follows from Theorem 9.5.1. Finally, $\tilde{K} = \text{ENT} \cap A$, thus ENT is productive by Example 9.5.5.//

Corollary The set of *false* sentences of Elementary Arithmetic is also productive, thus it is not axiomatizable either.

Proof Let

$$B = \{\phi_{\tilde{x}}(\tilde{x})\!\downarrow \mid x \in \mathcal{N}\}$$

$$C = \{\phi_{\tilde{x}}(\tilde{x})\!\downarrow \mid \phi_{\tilde{x}}(\tilde{x})\!\uparrow \text{ is } \textit{true}\}$$

Now, $\phi_{\tilde{x}}(\tilde{x})\!\downarrow$ is shorthand for $(\exists y)T(\tilde{x},\tilde{x},y)$. Also, $C = \{\phi_{\tilde{x}}(\tilde{x})\!\downarrow \mid x \in \overline{K}\}$; thus, $\overline{K} \leq_m C$, hence C is productive.

Finally {false sentences}$\cap B = C$ and the result follows from Example 9.5.5.//

We shall next present a more formal version of Gödel's incompleteness theorem. The soundness requirement will be replaced by that of "ω-consistency" which is defined independently of the notion of *truth*.

First we define formal mathematical systems:

Definition 1 A *Formal Mathematical System* (or *Axiomatic Mathematical System*), in short FMS consists of

(1)   A *finite* alphabet of symbols Σ.

(2)   A *recursive* subset $\mathcal{F}$ of Σ^*, called *formulas*.

(3)   A *recursive* subset $\mathcal{A}$ of $\mathcal{F}$, called *axioms*.

(4)   A *finite* set of *recursive* relations on Σ^*, none of them unary, called *rules of inference.*//

Note: The set Σ^* is naturally identified with $\mathcal{N}$ under the 1-1 correspondence which assigns to $n \in \mathcal{N}$ its $|\Sigma|$-adic representation (assuming a fixed but unspecified ordering of Σ). Thus, the notions of recursive, re, etc., in connection with relations (sets) on Σ^*, as well as functions on Σ^*, have their usual number-theoretic meaning. For example, $\mathcal{A}$ recursive means that for some recursive $A \subset \mathcal{N}$, $v \in A$ iff the $|\Sigma|$-adic notation of v is in $\mathcal{A}$. $\mathcal{F}$ and $\mathcal{A}$ are required to be recursive so that, intuitively, we can *recognize* a formula or axiom when we see one. Similarly, the rules of inference are required to be recursive so that, intuitively, they can be "effectively" applied.

Smullyan [19] (following Post) gives a more general definition, requiring only that the set of theorems of the system (to be defined below) is re, relaxing all recursiveness requirements. The above definition is similar to that in Davis [5]. [In Davis, rule (2) in the definition is dropped, but we find it convenient].

Definition 2   Let $\mathcal{S} = (\Sigma, \mathcal{F}, \mathcal{A}, \mathcal{I})$ be a formal mathematical system, where $\mathcal{I} = \{R_i^{(n_i)}(\vec{x}_{n_i})\}_{i=1, \ldots, m}$ is the set of rules of inference, n_i denoting the *rank* ("arity") of $R_i^{(n_i)}(\vec{x}_{n_i})$, and $\vec{x}_{n_i} \in (\Sigma^*)^{n_i}$.

A *sequence* $y_1, y_2, \ldots, y_k$ of elements in $\mathcal{F}$ is a *proof in $\mathcal{S}$* (or simply *proof*, if $\mathcal{S}$ is understood from the context) iff each y_i $(i = 1, \ldots, k)$ is either an *axiom*, or is obtained from *previous* y_j's (i.e., $j < i$) by application of some rule of inference, say $R_q^{(n_q)}(\vec{x}_{n_q})$, in the sense that $R_q^{(n_q)}(y_i, y_{j_1}, \ldots, y_{j_{n_q-1}})$ holds, where $\{j_1, \ldots, j_{n_q-1}\} \subset \{1, 2, 3, \ldots, i-1\}$.

y_k will then be called a *theorem of $\mathcal{S}$*, in symbols $\vdash_{\mathcal{S}} y_k$.

The set of theorems of $\mathcal{S}$ is denoted by $T_{\mathcal{S}}$.//

Note: Clearly, $\mathcal{A} \subset T_{\mathcal{S}}$.

Theorem 3   For any FMS $\mathcal{S}$, $T_{\mathcal{S}}$ is re.

Proof (This is a simpler version of the proof of Theorem 6.5.4.)

Let Σ be the alphabet of $\mathcal{S}$. Consider $\hat{\Sigma} = \{;\} \cup \Sigma$, where $; \notin \Sigma$ and where $a <\ ;$ is (to fix ideas) the order relationship of $;$ with all $a \in \Sigma$. All arithmetic in this proof is done in $|\hat{\Sigma}|$-adic notation.

An $\mathcal{S}$-proof $y_1, y_2, \ldots, y_k$ will be "coded" by the number $;y_1;y_2; \ldots ;y_k;$

Let $R_1^{(n_l)}, \ldots, R_m^{(n_m)}$ be the rules of inference of $\mathcal{S}$. Let $Pf(y)$ be true iff the $|\hat{\Sigma}|$-adic notation of y "codes" an $\mathcal{S}$-proof.

Then,

$$Pf(y) \equiv\ ;By\ \&\ ;Ey\ \&\ (\forall u,v)_{Py}((u;v;)By\ \&\ \neg(;Pv)$$

$$\Rightarrow [v \in \mathcal{A} \vee (\exists t_1, \ldots, t_{n_1-1})_{Pu}\{\neg(;Pt_1 \vee \ldots \vee ;Pt_{n_1-1})\ \&$$

$$(;t_1;)Pu\ \&\ \ldots\ \&\ (;t_{n_1-1};)Pu\ \&\ t_1 \in \mathcal{F}\ \&\ \ldots\ \&\ t_{n_1-1} \in \mathcal{F}\ \&\ v \in \mathcal{F}\ \&$$

$$R_1^{(n_1)}(v, t_1, \ldots, t_{n_1-1})\} \vee \ldots \vee (\exists t_1, \ldots, t_{n_m-1})_{Pu}\{\neg(;Pt_1$$

$$\vee \ldots \vee ;Pt_{n_m-1})\ \&\ (;t_1;)Pu\ \&\ \ldots\ \&\ (;t_{n_m-1};)Pu\ \&$$

$$t_1 \in \mathcal{F}\ \&\ \ldots\ \&\ t_{n_m-1} \in \mathcal{F}\ \&\ v \in \mathcal{F}\ \&\ R_m^{(n_m)}(v, t_1, \ldots, t_{n_m-1})\}])$$

Note that: (1) it is conceivable that y_i could be Λ [empty] thus, unlike the proof of Theorem 6.5.4, we do not require $\neg(;;)Py$ in the definition of $Pf(y)$.

(2) $(u;v;)$, $(;t_1;)$, etc., used in the definition of $Pf(y)$ are more accurately written $(u*;*v*;)$ $(;*t_1*;)$, etc., where $*$ denotes $|\hat{\Sigma}|$-adic concatenation. We slightly abused notation for a cleaner visual effect.

(3) The predicates $v \in \mathcal{A}$, $t \in \mathcal{F}$, $R_i^{(n_i)}(\vec{x}_{n_i})$ are recursive (in $|\Sigma|$-adic notation) according to Definition 1. More precisely, there are recursive predicates $A(v)$, $F(t)$, $P_i^{(n_i)}(\vec{x}_{n_i})$ such that

$$A(v) \text{ iff } v^{(|\Sigma|)} \in \mathcal{A}$$

$$F(t) \text{ iff } t^{(|\Sigma|)} \in \mathcal{F}$$

$$P_i^{(n_i)}(\vec{x}_{n_i}) \text{ iff } R_i^{(n_i)}(\vec{x}_{n_i}^{(|\Sigma|)})$$

where, generally, $x^{(m)}$ denotes the m-adic notation of $x \in \mathcal{N}$.

The predicates $v \in \mathcal{A}$, $t \in \mathcal{F}$, $R_i^{(n_i)}(\vec{x}_{n_i})$ in the definition of of $Pf(y)$ are abuse of notation, once again, and stand for $v^{(|\hat{\Sigma}|)} \in \mathcal{A}$, $t^{(|\hat{\Sigma}|)} \in \mathcal{F}$, and $R_i^{(n_i)}(\vec{x}_{n_i}^{(|\hat{\Sigma}|)})$, that is,

$$(\exists s)_{\leq v}[f_{|\Sigma|,|\hat{\Sigma}|}(s) = v \ \& \ A(s)], \quad (\exists s)_{\leq t}[f_{|\Sigma|,|\hat{\Sigma}|}(s) = t \ \& \ F(t)]$$

and

$$(\exists \vec{s}_{n_i})_{\leq \max(\vec{x}_{n_i})}[f_{|\Sigma|,|\hat{\Sigma}|}(s_1) = x_1 \ \& \ \ldots \ \& \ f_{|\Sigma|,|\hat{\Sigma}|}(s_{n_i}) = x_{n_i} \ \& \ P_i^{(n_i)}(\vec{s}_{n_i})]$$

where $f_{m,n}$ is that of Lemma 6.7.3; thus, they are recursive (as predicates on $\mathcal{N}$).

Thus, $Pf(y)$ is recursive.

Finally, let $T \stackrel{\text{def}}{=} \{x \mid (\exists y)[Pf(y) \ \& \ \neg;Px \ \& \ (;x;)Ey]\}$. Then T is obviously re (projection theorem) and so is $T_{\mathcal{S}} = f_{|\Sigma|,|\hat{\Sigma}|}^{-1}(T)$ (Example 8.1.1).//

We will next appreciate the generality of Definitions 1 and 2 by considering some examples.

Example 1 An EFS (Φ, A, V, Π) (Definitions 6.3.6–6.3.8) is an instance of an FMS. Indeed, referring to Definitions 1, 2 and 6.3.6-6.3.8:

(1) Take $\Sigma = A \cup V \cup \Pi \cup \{,\} \cup \{\rightarrow\}$.

(2) $\mathcal{F}$ is the set of *wff*'s of the EFS. Clearly, $\mathcal{F}$ is recursive.

(3) $\mathcal{A}$ is Φ, and is finite, hence recursive.

(4) There are two rules of inference, $\text{Sub}(v,w)$ and $\text{MP}(v,w,z)$. $\text{Sub}(v,w)$ holds iff wff v is obtained by substitution of each occurrence of some variable in wff w by some string in A^*. $\text{MP}(v,w,z)$ holds iff w is an *af*, z is $w \rightarrow v$ and v is a *wff*.//

Example 2   *Propositional Calculus.*
Here

(1) $\Sigma = \{p,1,\neg,\vee\}$

(2) $\mathcal{F}$ is the *smallest* subset of Σ^* satisfying:

(2.1) $\bigcup_{n\geq 1}\{p1^n\} \subset \mathcal{F}$, where $1^n = \overbrace{1\ldots 1}^{n}$ as usual.

[Intuitively, $p1, p11, p111$, etc., *are* the so-called *propositional variables*. The term "variable" suggests the existence of the term "constant". This is not the case here. Propositional variables, informally, stand for (or vary over the set of) [mathematical] statements, or "propositions". Thus, for *any* statement s, "(*not s*) *or s*" is (classically) "true" and the statement in quotes is mirrored in Propositional Calculus by any string of the form $\vee\neg p1^n p1^n$ ($n \geq 1$); intuitively, the "propositional variable" $p1^n$ stands for the statement s.

In a formal proof within the calculus, only strings such as $p1^n$ can be variables. Usually though, one uses *"metasymbols"* for convenience, i.e., symbols usually employed in *talking about* (rather than operating *within*) the calculus. Thus, x,v,p,p',p_{12},y, etc., may, hopefully without confusion, be used as *names* of variables.]

(2.2)   If $F \in \mathcal{F}$, then $\neg F \in \mathcal{F}$

(2.3)   If $\{F,G\} \subset \mathcal{F}$, then $\vee FG \in \mathcal{F}$

[For simplicity, we avoid the use of brackets, following the example of Bourbaki [2], by using *prefix* notation in writing the formulas. Prefix (or Polish) notation, according to which "operators" (here $\neg,\vee$) are put ahead of their operands, originated with the Polish logician Lukasiewicz, hence the term "Polish notation". This notation, incidentally, is popular in computer science, e.g., in the context of compiler-writing, where the "reverse" Polish notation is more common (that is, "operand, operand, operator").]

Examples of elements in $\mathcal{F}$: $p1, p111, \vee p1p1, \vee\neg p111p111$.

As usual, we conveniently use metasymbols (preferably short ones!) as *names* of formulas. Thus, if x stands for $p111$, then $\vee\neg xx$ stands for the last example above. If necessary, we invent another *name*, say y, for $\vee\neg xx$.

Prefix has been used to obtain a convenient definition of $\mathcal{F}$. In practice, we are more comfortable with *infix* (i.e., "operand, operator, operand") notation, and we shall informally use the latter, with the understanding that, if $x \in \mathcal{F}$ then the *infix form of* x, $\hat{x}$, is defined as:

(a)   If x is a variable, then $\hat{x} = x$

(b)   If $y = \neg x$, then $\hat{y} = \neg(\hat{x})$

(c)   If $y = \vee xz$, then $\hat{y} = (\hat{x}) \vee (\hat{z})$.

Moreover, we introduce the symbol $\Rightarrow$ to mean $\vee\neg$. Brackets can be reduced in numbers if we agree never to put variables in brackets, and that $\neg$ has strongest priority, $\vee$ coming second and $\Rightarrow$ last; all of $\neg,\vee,\Rightarrow$ are *right-associative*. Informally, $\Rightarrow$ is the (classical) implication symbol, $\vee$ is *or* and $\neg$ is *not*.

Clearly, $\mathcal{F}$ is recursive.

(3) $\mathcal{A}$ is infinite and consists of *all* strings in $\mathcal{F}$ which have the *forms* (we use infix notation for convenience; the reader will have no trouble putting these strings in the prefix notation).

(3.1) $((\hat{X})\vee(\hat{X}))\Rightarrow(\hat{X})$

(3.2) $(\hat{X})\Rightarrow((\hat{X})\vee(\hat{Y}))$

(3.3) $((\hat{X})\vee(\hat{Y}))\Rightarrow((\hat{Y})\vee(\hat{X}))$

(3.4) $((\hat{X})\Rightarrow(\hat{Y}))\Rightarrow(((\hat{Z})\vee(\hat{X}))\Rightarrow((\hat{Z})\vee(\hat{Y})))$

where X,Y,Z stand for any item in $\mathcal{F}$ [they are (meta)variables over $\mathcal{F}$].

Clearly, $\mathcal{A}$ is recursive.(*)

For example, an instance of (3.1) is $(x\ \vee\ y)\vee(x\ \vee\ y)\Rightarrow x\vee y$ where x and y are propositional variables, and the redundant brackets have been eliminated.

(4)   There is only one rule of inference (*Modus Ponens*): $\mathrm{MP}(X,Y,Z)$ holds, iff Z is $\vee\neg YX$ for any items Y,X in $\mathcal{F}$. [Intuitively, we infer $\hat{X}$ just in case we got both $\hat{Y}$ and $(\hat{Y})\Rightarrow(\hat{X})$.]//

Note: Rules (3.1)–(3.4) form *axiom schemata*. Substituting formulas for the metavariables one gets *axioms*.

Example 2   (continued). *Transitivity of implication.*

For any F,G,H in $\mathcal{F}$, if $\Rightarrow FG$ and $\Rightarrow GH$ are theorems of Propositional Calculus, then so is $\Rightarrow FH$.

Indeed, let $\dots,\Rightarrow FG$, and $\dots,\Rightarrow GH$ be proofs. We then have a proof $\dots,\Rightarrow FG,\dots,\Rightarrow GH,((\hat{G})\Rightarrow(\hat{H}))\Rightarrow((\neg(\hat{F})\vee(\hat{G}))\Rightarrow(\neg(\hat{F})\vee(\hat{H})))$ [where we used 3.4 here in *infix*, for convenience, and have set $\hat{G},\hat{H},\neg(\hat{F})$ instead of the metavariables $\hat{X},\hat{Y},\hat{Z}$ respectively], $((\hat{F})\Rightarrow(\hat{G}))\Rightarrow((\hat{F})\Rightarrow(\hat{H}))$ [by MP and definition of "$\Rightarrow$"], $(\hat{F})\Rightarrow(\hat{H})$. Thus, $(\hat{F})\Rightarrow(\hat{H})$; that is, strictly speaking, $\Rightarrow FH$ is a theorem.

Note that our proof here is a *metaproof* and we established a *metatheorem*, since we did not prove a theorem *of* Propositional Calculus (which is some string in $\mathcal{F}$) but a theorem *about* the behavior of proofs in Propositional Calculus.//

(*)This is a subset of those axioms (schemata) proposed in *Principia Mathematica* [21]. The axiom (schema) $((\hat{X})\vee((\hat{Y})\vee(\hat{Z})))\Rightarrow(((\hat{X})\vee(\hat{Y}))\vee(\hat{Z}))$ present there was shown provable from the others by Bernays [1].

Example 2   (concluded). We observe that the semantic model of Propositional Calculus— i.e., the one based in the notion of *truth*—satisfies De Morgan's laws, in particular $A \& B$ has the same "truth values" as $\neg(\neg A \vee \neg B)$ for any formulas A and B.

This motivates the *definition* of "&" in the formal system by: For all X and Y in $\mathcal{F}$, the formula $\neg \vee \neg X \neg Y$ is denoted by $\& XY$. In the context of *infix* notation, the priority of & is, traditionally, taken between those of $\neg$ and $\vee$.

Further, $\Leftrightarrow XY$ is shorthand for $\& \Rightarrow XY \Rightarrow YX$. The priority of $\Leftrightarrow$ is taken as the weakest. When $\Leftrightarrow XY$ (or, in infix, $\hat{X} \Leftrightarrow \hat{Y}$) is provable, then we say that X and Y are *demonstrably equivalent*.

It can be shown that any formula is demonstrably equivalent to itself; that is, $\Leftrightarrow XX$ is provable for any formula X.//

Example 3   As we have already stated, propositional variables stand for [mathematical] statements. However, if we must do any interesting (formal) mathematics at all, we should be able to have an FMS which involves some of the inner details of mathematical statements; we cannot get too far by simply manipulating *"names"* of such statements.

To this end, a richer FMS, known as *Lower Predicate Calculus*(*) (or *First-Order* Predicate Calculus) *with equality* is introduced in this example. We shall refer to it as LPC.

Here

(1) $\Sigma = \{a, \square, *, \tau, 1, ;, \neg, \vee, =\}$

(2)   To define $\mathcal{F}$, define

> (2.1)   Variables: $\{a1^n a\}_{n \geq 1}$ [As metasymbols we shall use low case letters with or without subscripts/primes. For example, a, b, y, x', x_1, w''_{52}, etc.]

> (2.2)   *Substitution slots*: $\{*1^n *\}_{n \geq 1}$ [As metasymbols, $*_n$ will stand for $*1^n *$, $n \geq 1$]

> (2.3)   A *formula-term schema*, in short *ft-schema* of *rank* n $(n \geq 1)$, is a nonempty string over $(\Sigma - \{*\}) \cup \{*1*, *11*, \ldots, *1^n *\}$, which has each of $*1^i *$, $i = 1, \ldots, n$ as substrings.
>
> If F is an ft-schema of rank n, and $s_1, \ldots, s_n$ are any strings over $\Sigma - \{*\}$, then $F(s_1, \ldots, s_n)$ or $F(\vec{s}_n)$ denotes the string obtained from F by substituting each occurrence of $*1^i *$ by s_i, $i = 1, \ldots, n$. [With the help of ft-schemata formulas and terms will be defined. It should be noted that the restriction that strings substituted into "substitution slots" do not contain '*', ensures that regardless of the order of substitution the result will be the same. Also, to

(*)As introduced, it should be more appropriately called an *applied* LPC (Kleene) or *first-order theory* (Tarski) since *predicate variables* are not included in its arsenal of notation.

substitute a string s over $\Sigma - \{*\}$ for x in $F(x, \ldots)$, where F does not contain x, is to form $F(s, \ldots)$. For example, if F is $*1*a$, $F(x)$ is xa, $F(axb)$ is $axba$.

In metanotation, if we are particularly interested in s_3, among the s_i's in $F(\vec{s}_n)$, we shall write $F(\ldots s_3 \ldots)$ or even $F(s_3)$. The rank of the ft-schema F in this case will either be understood from the context, or will be unimportant.]

(2.4)   We now inductively define *formulas* and *terms* simultaneously:

(a)   Each variable is a *term*.

(b)   For any formula F and any variable v, if v is *not* part of $F(*)$ then τF is a term. If v *does* occur in F, so that, for some ft-schema G of rank 1 which does *not* contain v,F is the string $G(v)$, then $\tau \hat{p}_1 ; \hat{p}_2 ; \ldots ; \hat{p}_k G(\square)$ is a term, where there are exactly k occurrences of $\square$ in $G(\square)$ at positions $p_1, \ldots, p_k$ symbols from the beginning of $G(\square)$, and $\hat{p}_i = 1^{p_i}$ for $i = 1, \ldots, k$, and $p_1 < p_2 < \ldots < p_k$.

In both cases the term is denoted by the metasymbol $(\tau v)F$. [Once again, we follow Bourbaki [2]. We may think of the $\hat{p}_i$'s as representing "pointers" from τ to the $\square$-symbols introduced as above. Thus, if F is $=xy$ then $(\tau v)F$ is $\tau=xy$. If F is $\lor=vy=zv$ then $(\tau v)F$ is $\tau 111;1111111\lor=\square y=z\square$ or, in the

notation of Bourbaki, $\tau\overline{\lor=\square y=z\square}$.

It is clear that $(\tau v)F$ does *not* contain v. Thus v is an *apparent* or *dummy* variable, the same way t is in $\int_a^b f(t)dt$. It is called a *bound* variable. Variables actually *present* in a formula (term) F are called *free*. Finally, note that if the variable u does not occur in the formula $G(v)$, and if the ft-schema G does not contain v, then $(\tau v)G(v)$ and $(\tau u)G(u)$ are identical strings(†).]

(c)   *Nothing* else is a term unless it can be proved to be so by a finite number of applications of (a) and (b).

(d) $=st$ is a *formula*, for any *terms* s and t.

(*)Variables are defined differently in LPC than in Propositional Calculus, since it would be inconvenient (in the LPC case) if part of a variable can also be a variable.

(†)There is a weakness in the *metanotation* "$(\tau v)F(v)$". Namely, $(\tau v)(\tau v)G(v,v)$ is ambiguous. It may mean $(\tau v)(\tau u)G(u,v)$, or $(\tau v)(\tau u)G(v,u)$, or $(\tau v)(\tau u)G(v,v)$, or $(\tau v)(\tau u)G(u,u)$. One should use judgment in choosing names (metasymbols) for "bound" variables.

(e)   If F and G are *formulas*, then so are $\neg F$ and $\vee FG$.

(f)   *Nothing* else is a formula, unless it can be proved to be so by a finite number of applications of (d) and (e).

$\mathcal{F}$ is the set of all formulas so defined; clearly, it is recursive.

Note that $*$ is not part of any formula or term (Problem 30). τ can be thought of, *intuitively*(*), as a (non-effective) *selector function*, which selects some item $(\tau x)F(x)$ among those that have "property" $F(a)$ [where $F(a)$ is a formula such that a is not part of the ft-schema F], if such items exist, it is undefined otherwise. It is not surprising then, that Bourbaki-style Set Theory is powerful enough to contain the "Axiom of Choice" as a theorem. τ is essentially Hilbert's ϵ-symbol; the interested reader may consult Hilbert and Bernays [11].

A *formula* $F(\vec{a}_n)$, intuitively, "defines" the set of all n-tuples that have "property-F". We usually call $F(\vec{a}_n)$ a predicate of rank n, just in case F is an ft-schema of rank n which contains none of the variables a_i, $i = 1, \ldots, n$.

$(\exists x)F(x)$ is introduced as shorthand for $F((\tau x)F(x))$ [Intuitively, "$(\exists x)F(x)$" states that there is some term verifying $F(a)$ since this is what $F((\tau x)F(x))$ says. This is in accordance with the usual interpretation of "$(\exists x)F(x)$". $\exists$ is the *existential* quantifier.]

$(\forall x)F(x)$ is shorthand for $\neg(\exists x)\neg F(x)$ as usual. [It says, intuitively, that $F(a)$ is true for all terms a, since "there is no term s to verify the opposite, namely $\neg F(s)$". $\forall$ is the *universal* quantifier.]

(3) $\mathcal{A}$ is infinite and consists of the axiom schemata (3.1)–(3.4) of Propositional Calculus and additionally the following schemata:

(3.5) $\Rightarrow F(s)F((\tau x)F(x))$ for all terms s and formulas $F(a) \in \mathcal{F}$ such that the ft-schema F does not contain the variable a. [In infix, this is $F(s) \Rightarrow (\exists x)F(x)$, assuming $F(a)$ and s are in infix. Intuitively, this axiom schema says that if some term s has property F, then the selector τ *will* select some term satisfying $F(a)$.]

(3.6)   *Axiom schemata for equality:* [Note that "$=$" of Σ is an *undefined* symbol, intended to capture "equality of terms". This is accomplished by the following axioms. It is important *not* to confuse this "$=$" with the metasymbol "$=$". In particular, $t = s$, where t and s are terms *does not* say that t and s are identical strings. This is why, repeatedly, we said things such as $(\tau v)F$ *is* [the string] $\tau{=}xy$ for example, rather than writing $(\tau v)F = \tau{=}xy$.]

(*)"Intuitively" cannot be overemphasized. These comments cannot be made precise outside the context of some concrete *interpretation* of the purely syntactic ("meaningless") system that LPC is.

(i)   For every variable x, terms s and t, and ft-schema F not containing x, and such that $F(x) \in \mathcal{F}$,

$$(s = t) \Rightarrow (F(s) \Rightarrow F(t));$$

where everything is in infix for convenience. [In Bourbaki, $F(s) \Leftrightarrow F(t)$ is used instead. The present formulation is equivalent to it. The intuitive grounds for this axiom schema is that "if two items s and t are equal then if one satisfies $F(a)$ then so does the other, no matter what property F describes."]

(ii)   For every variable x, and F,G in $\mathcal{F}$

$$((\forall x)(F \Leftrightarrow G)) \Rightarrow ((\tau x)F = (\tau x)G)$$

where everything is in infix for convenience. [Intuitively, if F and G are just alternative *names* for the same property, or, synonymously, set, then τ should select the same item in this set regardless of the name used.]

Clearly, $\mathcal{A}$ is recursive.

(4)   The only rule of inference is Modus Ponens, which is clearly recursive.//

Example 3   (continued).

If $\vdash_{\text{LPC}} X \Rightarrow Y$ and $\vdash_{\text{LPC}} Y \Rightarrow Z$, then $\vdash_{\text{LPC}} X \Rightarrow Z$ for any $\{X,Y,Z\} \subset \mathcal{F}$. The proof is as in the case of Propositional Calculus. We refer to this property of $\Rightarrow$ as *transitivity*.

We obtain next $\vdash_{\text{LPC}} X \Rightarrow X$. Indeed, by (3.1) and (3.2)

$\vdash_{\text{LPC}} X \Rightarrow (X \vee X)$ and $\vdash_{\text{LPC}} X \vee X \Rightarrow X(*)$. The result follows. It follows immediately that also $\vdash_{\text{LPC}} \neg X \vee X$ (definition of $\Rightarrow$), and hence $\vdash_{\text{LPC}} X \vee \neg X$, by (3.3), thus also $\vdash_{\text{LPC}} \neg X \vee \neg\neg X$, i.e., $\vdash_{\text{LPC}} X \Rightarrow \neg\neg X$. Next, $\vdash_{\text{LPC}} X \Rightarrow (Y \vee X)$ is established, which gives "symmetry" to (3.2): This is immediate, by transitivity of $\Rightarrow$, and by (3.2), (3.3). As a result, $\vdash_{\text{LPC}} X \Rightarrow (Y \Rightarrow X)$ [using $\neg Y$ instead of Y in $\vdash_{\text{LPC}} X \Rightarrow (Y \vee X)$] and hence the rule:

If $\vdash_{\text{LPC}} X$, then $\vdash_{\text{LPC}} Y \Rightarrow X$ for *any* $Y \in \mathcal{F}$. Another important theorem is $\vdash_{\text{LPC}} (X \Rightarrow Y) \Rightarrow (\neg Y \Rightarrow \neg X)$ [Intuitively, "if X implies Y then the opposite of Y implies the opposite of X"].

Arguing in reverse, we want to show that $\vdash_{\text{LPC}} (\neg X \vee Y) \Rightarrow (\neg\neg Y \vee \neg X)$. Now [(3.4)] $\vdash_{\text{LPC}} (Y \Rightarrow \neg\neg Y) \Rightarrow ((\neg X \vee Y) \Rightarrow (\neg X \vee \neg\neg Y))$ and since $\vdash_{\text{LPC}} Y \Rightarrow \neg\neg Y$ we get $\vdash_{\text{LPC}} (\neg X \vee Y) \Rightarrow (\neg X \vee \neg\neg Y)$.

The result follows by (3.3) and transitivity of $\Rightarrow$. Finally, we establish

---

(*)Whenever the intention is fairly evident, we shall always use fewer brackets than is formally necessary. In this case, it should be $(X) \vee (X) \Rightarrow (X)$, since, e.g., X may contain $\Leftrightarrow$.

that, if $\vdash_{\mathrm{LPC}} (X \Rightarrow Y)$ then $\vdash_{\mathrm{LPC}} (Y \Rightarrow Z) \Rightarrow (X \Rightarrow Z)$. Indeed, by the last result above, and modus ponens we get $\vdash_{\mathrm{LPC}} \neg Y \Rightarrow \neg X$.

Hence, $\vdash_{\mathrm{LPC}} (Z \vee \neg Y) \Rightarrow (Z \vee \neg X)$ by (3.4) and modus ponens. By (3.3) we have $\vdash_{\mathrm{LPC}} (\neg Y \vee Z) \Rightarrow (Z \vee \neg Y)$ and $\vdash_{\mathrm{LPC}} (Z \vee \neg X) \Rightarrow (\neg X \vee Z)$ and the result follows by transitivity of $\Rightarrow$.//

Example 3   (continued). We establish here an important metalogical rule known as the *Deduction Theorem*.

Its proof can become quite complicated (e.g., Kneebone [13]), due to tedious details arising from *substitution* of terms for variables, the distinction between bound and free variables, and the so-called *rules for quantifiers*(*), which are necessary additional rules of inference in some versions of the LPC. The Bourbaki approach, involving *axiom schemata* rather than *axioms* dispenses with substitution; moreover, bound variables do not really exist. The introduction of quantifiers via τ dispenses with the *rules for quantifiers*.

The Deduction Theorem: "To prove $X \Rightarrow Y$ in LPC, it suffices to prove Y in the FMS obtained from LPC by simply adding X as an axiom" [X is an *explicit* axiom now, *not a schema*, since it is a *fixed*, but unspecified, formula.]

Indeed, let $Y_1, \ldots, Y_m$ (where Y_m is Y) be a proof of Y *in the augmented* LPC.

Consider the sequence $(X \Rightarrow Y_1), \ldots, (X \Rightarrow Y_m)$. Now Y_1 is either an LPC axiom or X (why?); in either case, $\vdash_{\mathrm{LPC}} X \Rightarrow Y_1$ by the previous discussion.

Let $\vdash_{\mathrm{LPC}} X \Rightarrow Y_i$ for $i \leq n < m$. Consider $X \Rightarrow Y_{n+1}$.

Case 1. Y_{n+1} is an LPC-axiom or X. Then $\vdash_{\mathrm{LPC}} X \Rightarrow Y_{n+1}$.

Case 2. $\vdash_{\mathrm{LPC}} X \Rightarrow Y_j$ and $\vdash_{\mathrm{LPC}} X \Rightarrow (Y_j \Rightarrow Y_{n+1})$, where $j \leq n$. [Applying modus ponens in the augmented system, combined with the induction hypothesis.]

$\vdash_{\mathrm{LPC}} X \Rightarrow Y_j$ has as consequence $\vdash_{\mathrm{LPC}} (Y_j \Rightarrow Y_{n+1}) \Rightarrow (X \Rightarrow Y_{n+1})$ (according to the last result in the previous instalment of Example 3), and by transitivity of $\Rightarrow$, $\vdash_{\mathrm{LPC}} X \Rightarrow (X \Rightarrow Y_{n+1})$.

By (3.2) $\vdash_{\mathrm{LPC}} \neg X \Rightarrow (\neg X \vee Y_{n+1})$, that is, $\vdash_{\mathrm{LPC}} \neg X \Rightarrow (X \Rightarrow Y_{n+1})$. *We pause for a Lemma*: If $\vdash_{\mathrm{LPC}} A \Rightarrow B$ and $\vdash_{\mathrm{LPC}} C \Rightarrow D$, then $\vdash_{\mathrm{LPC}} A \vee C \Rightarrow B \vee D$. [It can also be put as if $\vdash_{\mathrm{LPC}} A \Rightarrow B$ and $\vdash_{\mathrm{LPC}} C \Rightarrow D$ and $\vdash_{\mathrm{LPC}} A \vee C$ then $\vdash_{\mathrm{LPC}} B \vee D$. This is "proof by considering cases", the cases being here assumptions A and B.] Indeed,

by (3.4) and modus ponens $\vdash_{\mathrm{LPC}} (C \vee A) \Rightarrow (C \vee B)$, hence $\vdash_{\mathrm{LPC}} (C \vee A) \Rightarrow (B \vee C)$ by (3.3) and transitivity of $\Rightarrow$. Since, arguing as above, $\vdash_{\mathrm{LPC}} (B \vee C) \Rightarrow (B \vee D)$ and $\vdash_{\mathrm{LPC}} (A \vee C) \Rightarrow (C \vee A)$ [by (3.3)], the claim follows by transitivity of $\Rightarrow$. // *of Lemma*.

(*)These rules of inference are: (1) If U does not contain v and $U \Rightarrow F(v)$ figures in an LPC proof (where F does not contain v), then $U \Rightarrow (\forall v)F(v)$ can be inserted at any later step of the proof. (2) Under the same assumptions regarding v, if $F(v) \Rightarrow U$ figures in a proof then $(\exists v)F(v) \Rightarrow U$ can be inserted at any later step of the proof.

Thus, $\vdash_{\mathrm{LPC}}(\neg X \vee X) \Rightarrow ((X \Rightarrow Y_{n+1}) \vee (X \Rightarrow Y_{n+1}))$,
hence $\vdash_{\mathrm{LPC}}(X \Rightarrow Y_{n+1}) \vee (X \Rightarrow Y_{n+1})$ (modus ponens and $\vdash_{\mathrm{LPC}} X \Rightarrow X$),
hence $\vdash_{\mathrm{LPC}} X \Rightarrow Y_{n+1}$ by (3.1) and modus ponens. This completes the proof of
the Deduction Theorem.//

Example 3   (continued). *Proof by contradiction.*
"If when LPC is augmented with the axiom $\neg A$ the resulting FMS
contains a formula X such that both $\vdash X$ and $\vdash \neg X$, then $\vdash_{\mathrm{LPC}} A$". [That is, if
assuming that A is not provable in LPC we derive a contradiction, then it must
be $\vdash_{\mathrm{LPC}} A$].

Indeed, let the additional axiom $\neg A$ cause $\vdash_{\mathrm{LPC'}} X$ and $\vdash_{\mathrm{LPC'}} \neg X$ in the
augmented LPC, LPC'.

But then, $\vdash_{\mathrm{LPC'}} \neg X \Rightarrow (X \Rightarrow A)$ [by (3.2)], hence by modus ponens
(twice) $\vdash_{\mathrm{LPC'}} A$. Thus, (deduction theorem) $\vdash_{\mathrm{LPC}} \neg A \Rightarrow A$. This, along with
$\vdash_{\mathrm{LPC}} A \Rightarrow A$ yields the result $\vdash_{\mathrm{LPC}} A$, by the lemma we used above.//

Example 3   (continued). *The rule of quantifiers.*
(1) Let $\vdash_{\mathrm{LPC}} F(v) \Rightarrow U$, where U and the ft-schema F do not contain the
variable v. Then $\vdash_{\mathrm{LPC}}(\exists v)F(v) \Rightarrow U$.

This is (almost) immediate since $\vdash_{\mathrm{LPC}} F(v) \Rightarrow U$ implies $\vdash_{\mathrm{LPC}} F(t) \Rightarrow U$
for *any* term t (see Problem 35) hence, in particular, for the term $(\tau v)F(v)$.
(2) Under the above assumptions regarding F, U and v, $\vdash_{\mathrm{LPC}} U \Rightarrow F(v)$ has
$\vdash_{\mathrm{LPC}} U \Rightarrow (\forall v)F(v)$ as consequence.

Indeed, we have $\vdash_{\mathrm{LPC}} \neg F(v) \Rightarrow \neg U$, hence $\vdash_{\mathrm{LPC}}(\exists v)\neg F(v) \Rightarrow \neg U$,
hence $\vdash_{\mathrm{LPC}} U \Rightarrow \neg(\exists v)\neg F(v)$, that is $\vdash_{\mathrm{LPC}} U \Rightarrow (\forall v)F(v)$.

A "rule of inference" derived from (2), and very frequently used, is "if
$\vdash_{\mathrm{LPC}} F(v)$ [where v is a variable not in F], then $\vdash_{\mathrm{LPC}}(\forall v)F(v)$".

Indeed, $\vdash_{\mathrm{LPC}} F(v)$ leads to $\vdash_{\mathrm{LPC}} X \Rightarrow F(v)$ and $\vdash_{\mathrm{LPC}} \neg X \Rightarrow F(v)$ for any X
$\in \mathscr{F}$, in particular for one not containing variable v. By the previous Lemma
and (3.1) $\vdash_{\mathrm{LPC}}(\neg X \vee X) \Rightarrow F(v)$, thus by (2), $\vdash_{\mathrm{LPC}}(\neg X \vee X) \Rightarrow (\forall v)F(v)$ and
finally (modus ponens and $\vdash_{\mathrm{LPC}} X \Rightarrow X$) $\vdash_{\mathrm{LPC}}(\forall v)F(v)$.

Caution. It is *not true* that $\vdash_{\mathrm{LPC}} F(v) \Rightarrow (\forall v)F(v)$. The reader is invited
to find an intuitive reason why this is so. (See Problem 33.)

On the other hand, $\vdash_{\mathrm{LPC}}[(\forall v)F(v)] \Rightarrow F(v)$. [This is usually taken as an
axiom, but in the Bourbaki system it is a theorem.] Indeed, $\vdash_{\mathrm{LPC}} \neg F(s)$
$\Rightarrow (\exists v)\neg F(v)$ is an axiom [(3.5)] for any term s and formula $F(v)$ such that v
is not in F. Thus, $\vdash_{\mathrm{LPC}} \neg(\exists v)\neg F(v) \Rightarrow \neg\neg F(s)$; that is, $\vdash_{\mathrm{LPC}}(\forall v)F(v) \Rightarrow F(s)$
by transitivity of $\Rightarrow$ and $\vdash_{\mathrm{LPC}} \neg\neg X \Rightarrow X$ (see Problem 34). In particular,
$\vdash_{\mathrm{LPC}}[(\forall v)F(v)] \Rightarrow F(v)$, [recall that v does not occur in $(\forall v)F(v)$], where v is a
variable. We note an interesting consequence: If v is not in F, then $\vdash_{\mathrm{LPC}} F(v)$ iff
$\vdash_{\mathrm{LPC}}(\forall v)F(v)$.//

Example 3   (concluded). *Properties of equality.*
(1) $\vdash_{\mathrm{LPC}} v = v$ for any variable v (and hence any term—see Problem 35). Indeed

(following Bourbaki), let F stand for $*1* = *1*$, thus $F(v)$ is $v = v$. It can be shown that $\vdash_{\text{LPC}} G(v) \Leftrightarrow G(v)$ (see Problem 31), hence,

$\vdash_{\text{LPC}} (\forall v)(G(v) \Leftrightarrow G(v))$ for every $G(v) \in \mathcal{F}$ such that v does not occur in G. Thus, [axiom (3.6)(ii) and modus ponens] $\vdash_{\text{LPC}} (\tau v)G(v) = (\tau v)G(v)$.(*) Thus, $\vdash_{\text{LPC}} F((\tau v)G(v))$ (that is, (1) is established for the particular term $(\tau v)G(v)$). Take now G to be $\neg F$, hence $\vdash_{\text{LPC}} F((\tau v)\neg F(v))$ and therefore $\vdash_{\text{LPC}} \neg\neg F((\tau v)\neg F(v))$ by $\vdash_{\text{LPC}} X \Rightarrow \neg\neg X$, established earlier, and modus ponens.

That is, $\vdash_{\text{LPC}} \neg(\exists v)\neg F(v)$, in other words $\vdash_{\text{LPC}} (\forall v)F(v)$. It follows $\vdash_{\text{LPC}} F(s)$ for any term s, in particular for term v.

Next, we establish $\vdash_{\text{LPC}} u = v \Rightarrow v = u$ and leave transitivity of equality as an exercise.

(2) $\vdash_{\text{LPC}} u = v \Rightarrow v = u$. Indeed, let F be $*1* = u$ so that $F(v)$ is $v = u$.

Assume $\vdash u = v$ [using deduction theorem]. Then (axiom (3.6)(i) and modus ponens] we get $\vdash F(u) \Rightarrow F(v)$. But $\vdash_{\text{LPC}} F(u)$ hence $\vdash F(u)$ in the augmented system. Thus, $\vdash F(v)$.

It follows that $\vdash_{\text{LPC}} u = v \Rightarrow v = u$. It is easy to show that $\vdash_{\text{LPC}} v = u \Rightarrow u = v$, thus establishing $\vdash_{\text{LPC}} u = v \Leftrightarrow v = u$. (See Problem 36.)//

We now briefly return to abstract FMS. First, let us give a formal definition and an obvious theorem relating to the already employed process of "augmenting" an FMS by adding axioms.

Definition 3 Let $\mathcal{S} = (\Sigma,\mathcal{F},\mathcal{A},\mathcal{I})$ and $\mathcal{S}' = (\Sigma',\mathcal{F}',\mathcal{A}',\mathcal{I}')$ be two FMS, such that $\Sigma \subset \Sigma'$, $\mathcal{F} \subset \mathcal{F}'$, $\mathcal{A} \subset \mathcal{A}'$ and $\mathcal{I} \subset \mathcal{I}'$. We then write $\mathcal{S} \leq \mathcal{S}'$ and say that $\mathcal{S}'$ is an *extension* of or that it *extends* $\mathcal{S}$.//

Theorem 4 If $\mathcal{S} \leq \mathcal{S}'$, then $T_{\mathcal{S}} \subset T_{\mathcal{S}'}$.

Proof Obvious.//

Note: If $\mathcal{S}$ extends LPC so that modus ponens is still the *only* rule of inference, then all the proof-techniques we established for LPC, in particular the "deduction theorem", "proof by cases", "proof by contradiction" hold in $\mathcal{S}$.

We shall next approach Gödel's incompleteness theorem from an abstract point of view. (†) To this end, we shall first define in what sense "Number Theory" can be done in an FMS; in particular, how more or less complex sets of *numbers* can be mirrored in an FMS.

(*)The reader may find this statement "obvious" at first sight. Recall though that $=$ is *not* equality of strings!

(†)Such abstract versions have been given in various places, e.g., Kleene [12], Davis [5], Smullyan [19], etc. Our exposition is influenced by that of Smullyan, but is less abstract, hence less general. The interested reader is referred to [19].

Definition 4   A *representation system*, $\mathcal{RS}$, is a 4-tuple $(\mathcal{S};\mathcal{C},(\mathcal{P}_n)_{n\geq 1},(\Phi_n)_{n\geq 1})$, where $\mathcal{S} = (\Sigma,\mathcal{F},\mathcal{A},\mathcal{I})$ is an FMS, $\mathcal{C}$ is a recursive subset of $\mathcal{F}$ known as the set of *closed* formulas (or *sentences*), for each $n \geq 1$ $\mathcal{P}_n$ is a recursive subset of $\mathcal{F}$ known as the set of *n-ary predicates*, and for each $n \geq 1$, Φ_n is a recursive function $\mathcal{N} \times \mathcal{N}^n \rightarrow \mathcal{N}$ such that $\Phi_n(\mathcal{P}_n \times \mathcal{N}^n) \subset \mathcal{C}$. It is called a *representation function*.

    A relation $W(\vec{x}_n)$ on $\mathcal{N}^n$ is *representable in* $\mathcal{RS}$ iff for some $H \in \mathcal{P}_n$
    $W(\vec{x}_n)$ iff $\vdash_{\mathcal{S}} \Phi_n(H,\vec{x}_n)$.//

Note: In LPC-derived systems, a *closed* formula is one with no (free) variables. Since $\vdash_{\text{LPC}} F(v)$ iff $\vdash_{\text{LPC}} (\forall v)F(v)$, there is no real loss of generality in considering as theorems *closed* provable formulas only. We will not do this here. Assuming now for a minute that enough axioms have been added to LPC so that "statements of arithmetic can be made in it" then one possible choice of Φ_1 could be $(H(x),n) \mapsto H(\tilde{n})$, where $H(x)$ is in $\mathcal{F}$ and H has rank 1, $n \in \mathcal{N}$, and $\tilde{n}$ is the formal counterpart (representation) of n in the system. For arguments outside $\mathcal{P}_1 \times \mathcal{N}$, one could define Φ_1 in some convenient manner; for example, to have as value $\tilde{0} = \tilde{0}$.

Theorem 5   If $W(\subset \mathcal{N})$ is representable in a representation system $\mathcal{S}$, then W is re.

    Proof Indeed, for some representation function Φ and unary predicate H it will be $x \in W$ iff $\vdash_{\mathcal{S}} \Phi(H,x)$, hence $x \in W$ iff $\Phi(H,x) \in T_{\mathcal{S}}$.

    That is, $W \leq_m T_{\mathcal{S}}$ via $\lambda x.\Phi(H,x)$, and the result follows by Proposition 9.1.2(v), and Theorem 3.//

Note: Theorem 5 is an abstract version of Gödel's incompleteness result, as it implies: "Not *all* true "arithmetic" statements(*) of the form $x \in W$ can be proved(†) in *any* representation system, if W is not re. For example, not all true statements or the form $x \in \overline{K}$ can be proved".

Corollary 1   If $W(\subset \mathcal{N})$ is representable in a representation system $\mathcal{S}$, and if $W \notin \mathcal{R}_*$ then $T_{\mathcal{S}} \notin \mathcal{R}_*$.

    Proof As in the above proof, $W \leq_m T_{\mathcal{S}}$. The result follows from Proposition 9.1.2(iv).//

Definition 5   If $T_{\mathcal{S}} \notin \mathcal{R}_*$ then $\mathcal{S}$ is an *undecidable theory* (or *system*). We also say that the *decision problem* of $\mathcal{S}$ is recursively unsolvable.//

(*)The attribute "arithmetic", because $W \subset \mathcal{N}$.

(†)By provability of a true informal statement of the form "$x \in W$" we, of course, mean representability of W.

Note: Of course, the decision problem of $\mathcal{S}$ is to (algorithmically) decide whether an arbitrary formula is a theorem of $\mathcal{S}$ or not.

Corollary 2 If every re set is representable in some representation system $\mathcal{S}$, then $\mathcal{S}$ is an *undecidable* theory.

Proof Nonrecursive (but re) sets exist.//

Corollary 3 If $\mathcal{S}$ is as in Corollary 2, then $T_{\mathcal{S}}$ is creative.

Proof For then K is representable in $\mathcal{S}$, thus $K \leq_m T_{\mathcal{S}}$. The result follows from Corollary 1 of Theorem 9.5.1 and from Theorem 3.//

Note: In Chapter 11, and with the help of the recursion theorem, we shall see that all creative sets are recursively isomorphic. Thus, Corollary 3 states the surprising, at first sight, result that "if an FMS can capture enough Number Theory so that at least K is representable in it, then no matter how much this theory is extended we shall not derive any new theorems which were not essentially provable in the original system".

An intuitive justification of the statement in quotes we can give right away: First, for any FMS $\mathcal{S}$, $T_{\mathcal{S}}$ is re by Theorem 3. So if $\mathcal{S}_0$ is as in Corollary 2, then $T_{\mathcal{S}}$ is representable in $\mathcal{S}_0$ or, in other words, all the theorems of $\mathcal{S}$ can be enumerated (under some appropriate coding, namely that afforded by the Φ-map of $\mathcal{S}_0$) within $\mathcal{S}_0$.

Clearly, if $\mathcal{S}$ is as in Corollary 2 as well, then it can also enumerate all the theorems of $\mathcal{S}_0$.

Moreover, observe that, by completeness of K, $T_{\mathcal{S}} \leq_m K$ via some $f \in \mathcal{R}$. It follows that if K is represented in some theory $\mathcal{S}_0$, and if recursive functions have formal counterparts in this theory, then again $T_{\mathcal{S}}$ is enumerated in $\mathcal{S}_0$ since $n \in T_{\mathcal{S}}$ iff, intuitively, $\vdash_{\mathcal{S}_0} \Phi(H,\tilde{f}(\tilde{n}))$ where H is the predicate of $\mathcal{S}_0$ that represents K, $\tilde{f}$ is the formal counterpart of f and $\tilde{n}$ is the formal counterpart of n in $\mathcal{S}_0$.

Corollary 2 can be sharpened (see corollary of Theorem 6). Define the "diagonalization set" D by $D = \{x \in \mathcal{N} \,|\, \Phi(x,x) \notin T_{\mathcal{S}}\}$.

Theorem 6 D is not representable in $\mathcal{S}$.

Proof Otherwise, let H (in $\mathcal{P}_1$) represent D. Then $n \in D$ iff $\Phi(H,n) \in T_{\mathcal{S}}$, thus $H \in D$ iff $\Phi(H,H) \in T_{\mathcal{S}}$. But $H \in D$ iff (definition of D) $\Phi(H,H) \notin T_{\mathcal{S}}$.//

Note: First, observe that recursiveness of Φ was not needed in the above proof.

Next, we note the above diagonalization as being an abstract version of Gödel's original proof of his incompleteness theorem. Indeed, if for a minute we *assume* that D *is* represented by H, then $n \in D$ is (formally) the same as $\Phi(H,n)$ [or $H(n)$ in more suggestive notation].

Now $n \in D$, hence $\Phi(H,n)$, "says" $\Phi(n,n) \notin T_{\mathscr{S}}$. Thus, $\Phi(H,H)$ "says" $\Phi(H,H) \notin T_{\mathscr{S}}$, i.e., it says:

"I am not a theorem."

(Compare with the discussion following Theorem 1.)

Corollary   (For example, Smullyan [19], Theorem 10, p. 54) If every *recursive* set is representable in $\mathscr{S}$, then $T_{\mathscr{S}}$ is not recursive.

Proof Let $T_{\mathscr{S}} \in \mathscr{R}_*$. Then $D \in \mathscr{R}_*$ since Φ is recursive. Thus, D is representable in $\mathscr{S}$, contradicting Theorem 6.//

We next turn our attention to FMS, where we can do Number Theory, and which are extensions of LPC (with equality).

The following FMS is due to Robinson [17].

Definition 6   ROB is an FMS with

(1) $\Sigma = \{a,\square,*,\tau,1,;,\neg,\vee,=,<,S,0,+,\cdot\}$

(2) $\mathscr{F}$ is defined as in the case of LPC (with equality) except

(2.4) (a)   Each variable *and* 0 (read "zero") are *terms*

(b)   As in the case of LPC, but add, if s,t are terms then so are
St (read "successor of t". S is, intuitively, the "+1" operator),
$s + t$ (read "s plus t", due to the intended informal interpretation),
$s \cdot t$ (read "s times t").

(c) $=st$ and $<st$ (in infix $s = t$ and $s < t$) are *formulas* for any *terms* s and t.

Add the axiom schemata [so-called "nonlogical axioms", as they are beyond what is needed to define just the basic LPC system (Logic)], in infix notation for readability.

(3.7) For any *terms* u and t:

(I)	$\neg(Su = 0)$	(I')	$Su = St \Rightarrow u = t$
(II)	$u + 0 = u$	(II')	$u + St = S(u + t)$
(III)	$u \cdot 0 = 0$	(III')	$u \cdot St = u \cdot t + u$
(IV)	$\neg(u < 0)$	(IV')	$u < St \Leftrightarrow u < t \vee u = t$
(IV'')	$\neg(u < t) \Rightarrow (u = t \vee t < u)$		

[(I) and (I') establish "desirable" properties for the successor operator, (II) and (II') is a "recursive" definition of "addition" while (III) and (III') do the same for "multiplication". (IV) through (IV") establish some properties of $<$, including the desired property that $<$ is a total order (IV")].//

Note: Peano's [15] Induction Axiom is not included in ROB. In what follows, we shall call a *Formal Number Theory* (FNT) any (unspecified) extension of ROB.

Note that in any FNT there are terms which (intuitively) correspond to the natural numbers $0,1,2,\ldots$. These are the terms $0, S0, SS0$, etc. In general, the number n has as counterpart in the FNT the term $\underbrace{SS\ldots S0}_{n}$. The latter will be denoted by $\tilde{n}$ for convenience.

Definition 7   A relation $Q \subset \mathcal{N}^n$ is *definable* in an FNT $\mathscr{S}$, if there is an ft-schema F of rank n in $\mathscr{S}$ which does not contain the variables $a_1, \ldots, a_n$ and (a), (b) below hold for all $\vec{k}_n \in \mathcal{N}^n$.

(a)   if $Q(k_1, \ldots, k_n)$, then $\vdash_\mathscr{S} F(\tilde{k}_1, \ldots, \tilde{k}_n)$

(b)   if $\neg Q(k_1, \ldots, k_n)$, then $\vdash_\mathscr{S} \neg F(\tilde{k}_1, \ldots, \tilde{k}_n)$.//

Note: First, in condition (b) above the first "$\neg$" is informal. $\neg Q(\vec{x}_n)$ is the set $\mathcal{N}^n - Q$. The second "$\neg$" is the formal symbol of $\mathscr{S}$ as used in Example 3.

Next, we mention that if the stronger conditions were required that (a) Q is *represented* by $F(\vec{a}_n)$ and (b) $\neg Q$ is *represented* by $\neg F(\vec{a}_n)$, then one would call Q *strongly* (or *completely*) *representable* in $\mathscr{S}$ (Smullyan). Other authors, instead of *definable* (Smullyan) say *representable* or *numeralwise representable* or *expressible*. We followed the practice of Smullyan,(*) since we defined representable to mean something else.

Theorem 7   If $Q \subset \mathcal{N}^n$ is definable in $\mathscr{S}$, then $Q \in \mathcal{R}_*$.

Proof Problem 42.//

Note: The above theorem puts an upper bound on the "complexity" of definable relations. For example, no re, but not recursive, relation can be definable.

Definition 8   A *basis* for the re subsets of $\mathcal{N}$ is a set of recursive binary relations $(R_i(x,y))_{i\geq 0}$ such that $A\ (\subset \mathcal{N})$ is re iff $x \in A \equiv (\exists y)R_i(x,y)$ for some i. An FNT, $\mathscr{S}$, is *adequate* iff *every* relation of some basis is definable in $\mathscr{S}$.//

(*)A more abstract definition is given in [19].

Note: Examples of bases are, the set of *all* binary *recursive* relations, the set of *all* binary *primitive recursive* relations, the set of *all* binary *constructive arithmetic* relations.

Definition 9 An FNT $\mathcal{S}$ is (*simply*) *consistent* if for *no* formula $F \in \mathcal{F}$ is it both $\vdash_{\mathcal{S}} F$ and $\vdash_{\mathcal{S}} \neg F$; otherwise it is (simply) *inconsistent*.

$\mathcal{S}$ is *ω-consistent* if for *no* formula $F(x)$ with x as a (free) variable are *all* the following formulas provable in $\mathcal{S}$
$(\exists x)F(x)$, and $\neg F(\tilde{n})$ for all $n \in \mathcal{N}$; otherwise, it is *ω-inconsistent*.//

Note: The term ω-consistent is due to Gödel. ω-consistency implies consistency (see below) but the converse is false (Gödel [8]. See also Tarski [20]), thus (simple) consistency is the weaker of the two notions.

Lemma 1 An FNT $\mathcal{S}$ is (simply) consistent iff $\mathcal{F} \neq T_{\mathcal{S}}$.

Proof *if* part. By contradiction: So let $F \in \mathcal{F}$ be such that $\vdash_{\mathcal{S}} F$ and $\vdash_{\mathcal{S}} \neg F$. Let $G \in \mathcal{F}$. Then (see Example 3) $\vdash_{\mathcal{S}} \neg F \Rightarrow (F \Rightarrow G)$, hence (modus ponens twice) $\vdash_{\mathcal{S}} G$. Thus $\mathcal{F} = T_{\mathcal{S}}$; contradiction.
only if part. Clearly, consistency implies that $\mathcal{F} \neq T_{\mathcal{S}}$, since for every $F \in \mathcal{F}$ at least one of F or $\neg F$ is not in $T_{\mathcal{S}}$.//

Corollary ω-consistency implies (simple) consistency.

Proof For if $\mathcal{S}$ is (simply) inconsistent, then $\mathcal{F} = T_{\mathcal{S}}$, thus for any formula $F(x)$ in particular, all of $(\exists x)F(x)$ and $\neg F(\tilde{n})$ ($n \in \mathcal{N}$) are provable, hence $\mathcal{S}$ is ω-inconsistent.//

Definition 10 An FNT $\mathcal{S}$ is *complete* (more accurately *negation* complete or *simply* complete) if for every *sentence* (i.e., *closed* formula) F, $\vdash_{\mathcal{S}} F$ or $\vdash_{\mathcal{S}} \neg F$.//

Note: This definition does *not* subsume that of (simple) consistency. In "$\vdash_{\mathcal{S}} F$ or $\vdash_{\mathcal{S}} \neg F$", "or" is *not* "*exclusive or*", so *both* possibilities occurring is not ruled out.

A *closed* formula F such that $\vdash_{\mathcal{S}} F$ or $\vdash_{\mathcal{S}} \neg F$ is called *decidable* in $\mathcal{S}$, otherwise it is *undecidable*. Of course, the notions *decidable formula* and *decidable theory* must not be confused.

Lemma 2 $\mathcal{S}$ is complete iff whenever an unprovable (in $\mathcal{S}$) sentence $F \in \mathcal{F}$ is added to its set of axioms, it renders $\mathcal{S} + \{F\}(*)$ inconsistent.

(*)We denote by $\mathcal{S} + \{F\}$ the so augmented system.

Proof *if* part. Let $G \in \mathcal{F}$ be any closed formula. If $\vdash_{\mathcal{S}} \neg G$ then there is nothing else to do. So let $\neg G$ be unprovable in $\mathcal{S}$. Then $\mathcal{S} + \{\neg G\}$ is inconsistent, hence (Example 3, "proof by contradiction") $\vdash_{\mathcal{S}} G$. Thus $\mathcal{S}$ is complete, G being an arbitrary sentence.

only if part. Let $G \in \mathcal{F}$ be an unprovable (in $\mathcal{S}$) sentence. By completeness of $\mathcal{S}$, $\vdash_{\mathcal{S}} \neg G$ thus $\vdash_{\mathcal{S}+\{G\}} \neg G$. But also $\vdash_{\mathcal{S}+\{G\}} G$ thus $\mathcal{S} + \{G\}$ is inconsistent.//

Note: There is an alternative, but *weaker*, notion of completeness, based on the notion of *interpretation*. An interpretation of a *first-order theory* $\mathcal{S}$ (i.e., $\mathcal{S}$ extends LPC by adding nonlogical axioms, so that quantification is still restricted on the variables(*)) is a collection of the following objects:

(1) A nonempty set D, the *domain* or *universe* of the interpretation,

(2) A function p, which maps *every* formula of rank $k \geq 0$ to a function $\chi : D^k \rightarrow \{0,1\}$. Of course, χ is essentially a relation on D^k. [A "nullary" function $\chi : D^0 \rightarrow X$, X being any nonempty set, is by definition some fixed item in X.]

(3) A function t which maps every term of k (≥ 0) variables to a function $D^k \rightarrow D$. (For example, in an interpretation of ROB over a domain D, the term 0, having no variables, maps to some constant of D), subject to the constraints

(i) $p(\neg F) = 1 \doteq p(F)$ and $p(F \vee G) = p(F) \cdot p(G)$ [where $\doteq$ and $\cdot$ are the symbols for proper subtraction and multiplication on $\mathcal{N}$].

(ii) If v is a variable in $\mathcal{S}$, then $t(v)$ is a variable over D. If u and v are distinct $\mathcal{S}$-variables then $t(u)$ and $t(v)$ are distinct.

(iii) Let F (resp. T) be a formula (resp. term) with variables $x_1, \ldots, x_n$ and let $s_1, \ldots, s_n$ be terms. Then

$$p(F(\vec{s}_n)) = p(F)(t(s_1), \ldots, t(s_n))$$

$$t(T(\vec{s}_n)) = t(T)(t(s_1), \ldots, t(s_n))$$

(iv) If $F(y, \vec{x}_n)$ is a formula whose only variables are $y, \vec{x}_n$ then, setting

$$f = \lambda \vec{x}_n . t((\tau y) F(y, \vec{x}_n)) \text{ and } g = \lambda y \vec{x}_n . p(F(y, \vec{x}_n))$$

(using the symbols $y, \vec{x}_n$ as variables over D as well),

(iv1) $g(m, \vec{a}_n) = 0$ implies $f(\vec{a}_n) \downarrow$ and $g(f(\vec{a}_n), \vec{a}_n) = 0$

(iv2) If $p(F) = p(G)$ then $t((\tau y) F) = t((\tau y) G)$.

(*)In a *second-order theory*, quantification is allowed over formulas.

Of course, stipulation (iv) involves the Axiom of Choice in general. However, if D is $\mathcal{N}$ then $t((\tau y)F(y,\vec{x}_n))$ can be taken to be $(\tilde{\mu}y)p(F(y,\vec{x}_n))$.(*)

A formula F is *true in the interpretation* $(D;p,t)$ if range$(p(F)) = \{0\}$. It is *satisfiable* if $0 \in$ range$(p(F))$, otherwise it is *unsatisfiable*. By (i), F is true iff $\neg F$ is unsatisfiable. A formula F of $\mathscr{S}$ is *universally true*, if it is true under *all* interpretations.

$\mathscr{S}$ is *(semantically) complete* if *every* universally true formula of $\mathscr{S}$ is *provable* in $\mathscr{S}$.

We note that LPC (and ROB) are examples of *semantically* complete (Gödel's Completeness Theorem [9]), but *simply incomplete* systems (see Hilbert and Ackermann [10], p. 92, for the LPC case). In fact, ROB is *simply incompletable* as it follows from the Gödel-Rosser Incompleteness Theorem and the fact that ROB is (simply) consistent (Kleene [12]).

Theorem 8   *(Gödel's Incompleteness Theorem)*
If a FNT $\mathscr{S}$ is *adequate* and ω-*consistent* then it is (simply) *incomplete*.

Proof We first show that, under the assumptions, every re set $A \subset \mathcal{N}$ is representable in $\mathscr{S}$.

Indeed, let A be re. Then $x \in A$ iff $(\exists y)R(x,y)$ for some R definable in $\mathscr{S}$ (adequacy). Let $\mathcal{R}(x,y) \in \mathcal{F}$ define $R(x,y)$.

Thus $x \in A$ implies $(\exists y)R(x,y)$, hence $R(x,i)$ for some i. Thus (definability) $\vdash_{\mathscr{S}} \mathcal{R}(\tilde{x},\tilde{\imath})$ hence

$\vdash_{\mathscr{S}} (\exists y)\mathcal{R}(\tilde{x},y)$ [Axiom (3.5) of LPC].

Conversely, let $\vdash_{\mathscr{S}} (\exists y)\mathcal{R}(\tilde{x},y)$. By ω-consistency, not all of $\neg\mathcal{R}(\tilde{x},\tilde{n})$, $n = 0,1, \dots$ are provable in $\mathscr{S}$. Let $\neg\mathcal{R}(\tilde{x},\tilde{\imath})$ not be provable. Then $R(x,i)$ [otherwise, $\neg R(x,i)$, hence (definability) $\vdash_{\mathscr{S}} \neg\mathcal{R}(\tilde{x},\tilde{\imath})$].

Thus, $(\exists y)R(x,y)$, i.e., $x \in A$ is true.

We now have $x \in A$ iff $\vdash_{\mathscr{S}} (\exists y)\mathcal{R}(\tilde{x},y)$. Choosing now the representation function Φ so that its restriction on $\mathcal{P}_1 \times \mathcal{N}$ sends $(F(x),n)$ to $F(\tilde{n})$, while outside $\mathcal{P}_1 \times \mathcal{N}$ it is the constant $0 = 0$, we are done. (See Problem 43 for the recursiveness of Φ.)

Let now $A = \{x \mid \phi_x(x)\downarrow\}$ (that is, $A = K$).
Define B by

$$x \in B \text{ iff } \vdash_{\mathscr{S}} \neg(\exists y)\mathcal{R}(\tilde{x},y)$$

Observe,

(1) B is re (Theorem 5)

(2) $B \subset \overline{K}$ [for $x \in B$ implies $\vdash_{\mathscr{S}} \neg(\exists y)\mathcal{R}(\tilde{x},y)$

(*)By the Axiom of Choice, *every* nonempty set D can be equipped with a linear order such that each of its nonempty subsets have a least element (Zermelo). Thus, $(\tilde{\mu}y)p(F(y,\vec{x}_n))$ still can serve for $t((\tau y)F(y,\vec{x}_n))$.

hence $x \notin A(= K)$, otherwise, $\vdash_{\mathscr{S}}(\exists y)\mathscr{R}(\tilde{x},y)$, hence inconsistency and therefore (Corollary of Lemma 1) ω-inconsistency].

By (1), there is an $i \in \overline{K} - B$ (an re index of B will do as a choice of i, as we already know); that is, $i \notin A$ and $i \notin B$.

That is, *both* $(\exists y)\mathscr{R}(\tilde{i},y)$ *and* $\neg(\exists y)\mathscr{R}(\tilde{i},y)$ are not provable in $\mathscr{S}$.//

Note: By Theorem 8, a *closed* formula F exists (namely $(\exists y)\mathscr{R}(\tilde{i},y)$), such that neither F nor $\neg F$ are provable. In any interpretation, one or the other is true. In particular, in the *natural interpretation* of FNT over $\mathcal{N}$ (where 0 corresponds to $0 \in \mathcal{N}$, S corresponds to $\lambda x.x + 1$, = corresponds to equality on $\mathcal{N}$, etc.) $\neg F$ is true but unprovable [since $\neg F$ is not *universally* true, this does not contradict the completeness theorem of Gödel]. Such true but undecidable sentences of ENT have already been shown informally to exist (Theorems 1 and 2).

Corollary   Under the assumptions of Theorem 8, $T_{\mathscr{S}}$ is *creative*, hence $\mathscr{S}$ is an undecidable theory.

Note: This is essentially Church's theorem [4] on the undecidability of LPC. A way of obtaining this latter result is to show first that ROB is adequate and ω-consistent. [Indeed, by the Gödel-Rosser incompleteness theorem adequacy and *simple* consistency are enough to achieve undecidability and incompleteness. ROB can be shown to be both adequate (see below) and simply consistent (see, for example, Kleene [12]).] Next one involves the "translatability"(*) of ROB into LPC (Kleene [12], for example) and the proof is complete, since by translatability, if ROB is undecidable then so is LPC.

The proof of Theorem 8 exhibited a particular undecidable sentence, namely $(\exists y)\mathscr{R}(\tilde{i},y)$. Below we give an indirect proof.

Proposition 1   If a FNT $\mathscr{S}$ is complete, then it is decidable.

Proof The proof is analogous to that of "A and $\mathcal{N} - A$ re implies $A \in \mathscr{R}_*$". A further hint, given $F \in \mathscr{F}$ with variables $x_1, \ldots, x_n$ then $\vdash_{\mathscr{S}} F$ iff $\vdash_{\mathscr{S}}(\forall x_1) \ldots (\forall x_n)F$. (See Problem 44 for the details.)//

Corollary   Theorem 8.

Proof If $\mathscr{S}$ is complete, then it is decidable. This cannot be under the assumptions of Theorem 8 (Corollary 2 of Theorem 5).//

(*)$\mathscr{S}$ is *translatable* into $\mathscr{S}'$ if, for some recursive function f, $\vdash_{\mathscr{S}} F$ iff $\vdash_{\mathscr{S}'} f(F)$. Translatability is really m-reducibility between $T_{\mathscr{S}}$ and $T_{\mathscr{S}'}$: $T_{\mathscr{S}} \leq_m T_{\mathscr{S}'}$ via f.

Example 4   (*Adequacy of* ROB)

We shall show that the *constructive arithmetic* relations are definable in ROB.

(1) $a = b$ is definable in ROB, by ROB formula $x = y$.

Proof. Let $a = b$ for $\{a,b\} \subset \mathcal{N}$. Then $\tilde{a}$ and $\tilde{b}$ are identical strings ($\tilde{a}$ is $\underbrace{S \ldots S}_{a}0$, and so is $\tilde{b}$), hence $\vdash_{\text{ROB}} \tilde{a} = \tilde{b}$, by item (1) in the last installment of Example 3.

Let next $a \neq b$. We shall take up the case $a > b$ and leave $a < b$ out, as similar. There is a c such that $c + 1 = a$. We proceed by (informal) induction on b.

(1i) $b = 0$. Then $a \neq b$ in this case implies $\vdash_{\text{ROB}} \neg(S\tilde{c} = 0)$ where, of course, $\tilde{0}$ is 0 [Axiom 3.7,(I), Definition 6], that is $\vdash_{\text{ROB}} \neg(\tilde{a} = \tilde{b})$.

(1ii)   Assume that $a > b$ implies $\vdash_{\text{ROB}} \neg(\tilde{a} = \tilde{b})$, for any a ($>b$).

Let $a > b + 1$ and set $a = c + 1$ as before. Clearly, $c > b$. We then have $\vdash_{\text{ROB}} \neg(\tilde{c} = \tilde{b}) \Rightarrow \neg(S\tilde{c} = S\tilde{b})$ [Axiom 3.7,(I′), Definition 6, in connection with the earlier proved rule $(A \Rightarrow B) \Rightarrow (\neg B \Rightarrow \neg A)$ and modus ponens]. Once again, modus ponens plus induction hypothesis, gives $\vdash_{\text{ROB}} \neg(\tilde{a} = S\tilde{b})$.// *Claim* (1).

Note: The induction was used, of course, informally *outside*(*) ROB; this system does not include Peano's induction axiom or any (restricted) version of it.

(2) $a < b$ is definable in ROB, by ROB formula $x < y$.

Proof. We show that $a < b$ implies $\vdash_{\text{ROB}} \tilde{a} < \tilde{b}$ and that $\neg(a < b)$ implies $\vdash_{\text{ROB}} \neg(\tilde{a} < \tilde{b})$.

We prove both claims simultaneously by induction on b, as dictated (†) by axioms 3.7(IV) and (IV′) (Definition 6).

(2i) $b = 0$. Then $\neg(a < b)$ and hence $\vdash_{\text{ROB}} \neg(\tilde{a} < \tilde{b})$ by (IV).

(2ii)   Let both contentions hold for $b = k$ and any a. Consider $b = k + 1$.

Case 1. $a < k + 1$. This means that $a \leq k$. Now $a = k$ implies $\vdash_{\text{ROB}} \tilde{a} = \tilde{k}$ [by (1)] whereas $a < k$ implies $\vdash_{\text{ROB}} \tilde{a} < \tilde{k}$ (induction hypothesis). By Axiom (schema) 3.2 [and the LPC theorem $X \Rightarrow (Y \vee X)$ established in Example 3], we get via modus ponens $\vdash_{\text{ROB}} (\tilde{a} = \tilde{k}) \vee (\tilde{a} < \tilde{k})$ in either case. Hence $\vdash_{\text{ROB}} \tilde{a} < S\tilde{k}$; that is, $\vdash_{\text{ROB}} \tilde{a} < \tilde{b}$, by modus ponens and Axiom 3.7,(IV′).

(*)This is as it should be. Statement (1) is *about* ROB, not a statement *in* ROB.

(†)The successor S operator is applied to the t-term in $u < t$.

Case 2. $\neg(a < k + 1)$. That is, $a \geq k + 1$, i.e., $a > k$, hence, $a \geq k$ and $a \neq k$, or finally $\neg(a < k)$ *and* $\neg(a = k)$, hence (induction hypothesis) $\vdash_{\text{ROB}} \neg(\tilde{a} < \tilde{k})$ and [by (1)] $\vdash_{\text{ROB}} \neg(\tilde{a} = \tilde{k})$. It follows that $\vdash_{\text{ROB}} \neg(\tilde{a} < \tilde{k} \vee \tilde{a} = \tilde{k})$ [Let $\vdash_{\text{ROB}} \neg A$ and $\vdash_{\text{ROB}} \neg B$. Add to ROB the axiom $\neg A \Rightarrow B$ to obtain ROB_1. Hence, by modus ponens, $\vdash_{\text{ROB}_1} B$, thus ROB_1 is inconsistent. It follows that $\vdash_{\text{ROB}} \neg(\neg A \Rightarrow B)$; that is, $\vdash_{\text{ROB}} \neg(\neg\neg A \vee B)$. Since $\vdash_{\text{ROB}} A \Rightarrow \neg\neg A$, applying the rules $\vdash_{\text{ROB}}(X \Rightarrow Y) \Rightarrow ((X \vee Z) \Rightarrow (Y \vee Z))$ and $\vdash_{\text{ROB}} (X \Rightarrow Y) \Rightarrow (\neg Y \Rightarrow \neg X)$, we get $\vdash_{\text{ROB}} \neg(A \vee B)$].

It follows that $\vdash_{\text{ROB}} \neg(\tilde{a} < S\tilde{k})$; i.e., $\vdash_{\text{ROB}} \neg(\tilde{a} < \tilde{b})$ [for in $\text{ROB} + \{\tilde{a} < S\tilde{k}\}$ we have both $\vdash \tilde{a} < \tilde{k} \vee \tilde{a} = \tilde{k}$ [Axiom 3.7, (IV')] and $\vdash \neg(\tilde{a} < \tilde{k} \vee \tilde{a} = \tilde{k})$]. // *Claim* (2).

(3) $a + b = c$ is definable by $x + y = z$ in ROB.
Proof. By induction on the "recursion variable" [see Axioms 3.7(II) and (II')] b, show that if $a + b = c$ (resp. $a + b \neq c$) then $\vdash_{\text{ROB}} \tilde{a} + \tilde{b} = \tilde{c}$ [resp. $\vdash_{\text{ROB}} \neg(\tilde{a} + \tilde{b} = \tilde{c})$].

(3i) $b = 0$. Then $a + 0 = c$; i.e., $a = c$, hence $\tilde{a}$ and $\tilde{c}$ are identical strings. Thus [Axiom 3.7,(II)], $\vdash_{\text{ROB}} \tilde{a} + 0 = \tilde{c}$ (of course, we use the same symbol 0 for both the informal and formal "zero").

(3ii) Let both contentions be true for $b = k$, and all a,c. Consider $b = k + 1$. If $a + b = c$, then, for some n, $c = n + 1$ and $a + k = n$. Thus $\vdash_{\text{ROB}} \tilde{a} + \tilde{k} = \tilde{n}$. Hence $\vdash_{\text{ROB}} \tilde{a} + S\tilde{k} = S(\tilde{a} + \tilde{k})$ [Axiom 3.7,(II')]. Now, for any terms u and t, $\vdash_{\text{ROB}} u = t \Rightarrow Su = St$ (Problem 41). Thus, $\vdash_{\text{ROB}} S(\tilde{a} + \tilde{k}) = S\tilde{n}$, and by transitivity of (formal) equality, $\vdash_{\text{ROB}} \tilde{a} + S\tilde{k} = S\tilde{n}$; that is, $\vdash_{\text{ROB}} \tilde{a} + \tilde{b} = \tilde{c}$. We leave it to the reader to show the second contention.// *Claim* (3).

(4) $a \cdot b = c$ is definable by $x \cdot y = z$ in ROB.
Proof. Problem 45.

(5) If $R(\vec{a}_n)$ and $Q(\vec{b}_m)$ are definable in ROB by $\mathcal{R}(\vec{x}_n)$ and $Q(\vec{y}_m)$ respectively, then so are $\neg R(\vec{a}_n)$ and $R(\vec{a}_n) \vee Q(\vec{b}_m)$ (by $\neg\mathcal{R}(\vec{x}_n)$ and $\mathcal{R}(\vec{x}_n) \vee Q(\vec{y}_m)$, respectively).

Note: There should be no confusion between the formal and informal uses of $\neg$ and $\vee$. Further, it is understood that an x_i is a y_j iff an a_i is a b_j.

Proof. (i) If $\neg R(k_1, \ldots, k_n)$, then $\vdash_{\text{ROB}} \neg\mathcal{R}(\tilde{k}_1, \ldots, \tilde{k}_n)$. If $\neg\neg R(k_1, \ldots, k_n)$, then $R(k_1, \ldots, k_n)$, hence $\vdash_{\text{ROB}} \mathcal{R}(\tilde{k}_1, \ldots, \tilde{k}_n)$. It follows that $\vdash_{\text{ROB}} \neg\neg\mathcal{R}(\tilde{k}_1, \ldots, \tilde{k}_n)$ (since $\vdash_{\text{ROB}} X \Rightarrow \neg\neg X$).
(ii) Let $R(k_1, \ldots, k_n) \vee Q(l_1, \ldots, l_m)$ hold.
Then $\vdash_{\text{ROB}} \mathcal{R}(\tilde{k}_1, \ldots, \tilde{k}_n)$ or $\vdash_{\text{ROB}} Q(\tilde{l}_1, \ldots, \tilde{l}_m)$, hence $[\vdash X \Rightarrow X \vee Y$ and $\vdash X \Rightarrow Y \vee X]$ $\vdash_{\text{ROB}} \mathcal{R}(\tilde{k}_1, \ldots, \tilde{k}_n) \vee Q(\tilde{l}_1, \ldots, \tilde{l}_m)$.

Let now $\neg(R(k_1, \ldots, k_n) \vee Q(l_1, \ldots, l_m))$ hence $\neg R(k_1, \ldots, k_n)$ and $\neg Q(l_1, \ldots, l_m)$. Thus, $\vdash_{\text{ROB}} \neg \mathcal{R}(\tilde{k}_1, \ldots, \tilde{k}_n)$ and $\vdash_{\text{ROB}} \neg \mathcal{Q}(\tilde{l}_1, \ldots, \tilde{l}_m)$. It follows that $\vdash_{\text{ROB}} \neg(\mathcal{R}(\tilde{k}_1, \ldots, \tilde{k}_n) \vee \mathcal{Q}(\tilde{l}_1, \ldots, \tilde{l}_m))$ as in (2).// *Claim* (5)

(6) If $R(\vec{a}_n)$ is definable in ROB, then so is any explicit transform of $R(\vec{a}_n)$.

Proof. Problem 46.

(7) If $R(b,\vec{a}_n)$ is definable in ROB, then so is $(\exists y)_{\le b} R(y,\vec{a}_n)$.

Proof. (i) Let $(\exists y)_{\le k} R(y,\vec{l}_n)$ hold. Then $(\exists y)(y \le k \ \& \ R(y,\vec{l}_n))$, hence for some $p \in \mathcal{N}$ it is $p \le k$ and $R(p,\vec{l}_n)$ thus [(3) and assumption] $\vdash_{\text{ROB}} \neg(\tilde{k} < \tilde{p})$ and $\vdash_{\text{ROB}} \mathcal{R}(\tilde{p},\tilde{l}_1, \ldots, \tilde{l}_n)$, where $\mathcal{R}$ defines R. By familiar, by now, techniques (e.g. proof of (2)), we obtain

$\vdash_{\text{ROB}} \neg(\neg\neg(\tilde{k} < \tilde{p}) \vee \neg\mathcal{R}(\tilde{p},\tilde{l}_1, \ldots, \tilde{l}_n))$; that is,

$\vdash_{\text{ROB}} \neg(\tilde{k} < \tilde{p}) \ \& \ \mathcal{R}(\tilde{p},\tilde{l}_1, \ldots, \tilde{l}_n)$, hence (Axiom 3.5 and modus ponens)

$\vdash_{\text{ROB}} (\exists y)(\neg(\tilde{k} < y) \ \& \ \mathcal{R}(y,\tilde{l}_1, \ldots, \tilde{l}_n))$.

(ii) Let next $\neg(\exists y)_{\le k} R(y,\vec{l}_n)$. Then $\neg R(i,\vec{l}_n)$ and hence $\vdash_{\text{ROB}} \neg\mathcal{R}(\tilde{i},\tilde{l}_1, \ldots, \tilde{l}_n)$, for $i = 0, \ldots, k$. We show that it must also be

$\vdash_{\text{ROB}} \neg(\tilde{k} < y) \Rightarrow \neg\mathcal{R}(y,\tilde{l}_1, \ldots, \tilde{l}_n)$

(A) By Axiom 3.7 (IV″), $\vdash_{\text{ROB}} \neg(\tilde{k} < y) \Rightarrow (\tilde{k} = y \vee y < \tilde{k})$. By transitivity of $\Rightarrow$, it suffices to show

$\vdash_{\text{ROB}} (\tilde{k} = y) \vee (y < \tilde{k}) \Rightarrow \neg\mathcal{R}(y,\tilde{l}_1, \ldots, \tilde{l}_n)$. To this end, we use the deduction theorem, along with proof by cases.

(B) Let ROB_1 be $\text{ROB} + \{\tilde{k} = y\}$. Then, by Axiom (3.6)(i), $\vdash_{\text{ROB}_1} (\tilde{k} = y) \Rightarrow (\neg\mathcal{R}(\tilde{k},\tilde{l}_1, \ldots, \tilde{l}_n) \Rightarrow \neg\mathcal{R}(y,\tilde{l}_1, \ldots, \tilde{l}_n))$. But $\vdash_{\text{ROB}} \neg\mathcal{R}(\tilde{k},\tilde{l}_1, \ldots, \tilde{l}_n)$ and hence $\vdash_{\text{ROB}_1} \neg\mathcal{R}(\tilde{k},\tilde{l}_1, \ldots, \tilde{l}_n)$. By modus ponens (twice), $\vdash_{\text{ROB}_1} \neg\mathcal{R}(y,\tilde{l}_1, \ldots, \tilde{l}_n)$, hence $\vdash_{\text{ROB}} (\tilde{k} = y) \Rightarrow \neg\mathcal{R}(y,\tilde{l}_1, \ldots, \tilde{l}_n)$.

(C) $\neg R(i,\vec{l}_n)$, for $i = 0, \ldots, k - 1$, implies $\vdash_{\text{ROB}} (y < \tilde{k}) \Rightarrow \neg\mathcal{R}(y,\tilde{l}_1, \ldots, \tilde{l}_n)$. Indeed (induction on k), if $k = 0$, then $\vdash_{\text{ROB}} \neg(y < 0)$ [(3.7)(IV)] hence $\vdash_{\text{ROB}} (y < 0) \Rightarrow \neg\mathcal{R}(y,\tilde{l}_1, \ldots, \tilde{l}_n)$ [Axiom (3.2) and modus ponens].

Let now $\vdash_{\text{ROB}} (y < \tilde{k}) \Rightarrow \neg\mathcal{R}(y,\tilde{l}_1, \ldots, \tilde{l}_n)$ whenever $\neg R(i,\vec{l}_n)$ for $i = 0, \ldots, k - 1$.

Case $k + 1$: Since $\neg R(k,\vec{l}_n)$, (B) above applies. Since also $\vdash_{\text{ROB}} (y < S\tilde{k}) \Rightarrow (y < \tilde{k} \vee y = \tilde{k})$, we get $\vdash_{\text{ROB}} (y < S\tilde{k}) \Rightarrow \neg\mathcal{R}(y,\tilde{l}_1, \ldots, \tilde{l}_n)$ by symmetry of equality, the induction hypothesis, proof by cases, and transitivity of $\Rightarrow$.

Thus, (A), (B), (C) imply
$$\vdash_{ROB} \neg(\tilde{k} < y) \Rightarrow \neg\mathscr{R}(y,\tilde{l}_1,\ldots,\tilde{l}_n).$$
Hence $\vdash_{ROB} (\forall y) (\neg\neg(\tilde{k} < y) \vee \neg\mathscr{R}(y,\tilde{l}_1,\ldots,\tilde{l}_n))$; that is
$\vdash_{ROB} \neg(\exists y) (\neg(\tilde{k} < y) \ \& \ \mathscr{R}(y,\tilde{l}_1,\ldots,\tilde{l}_n)).$// *Claim* (7).

We now have: *Every constructive arithmetic relation is definable in* ROB, *hence* ROB *is adequate.*// End of Example 4.

We obtain at once:

Theorem 9 (Gödel) If an FNT is ω-consistent, then it is simply incomplete.

Proof Any extension of ROB is adequate.//

We now prepare for Rosser's sharpening of the Gödel's incompleteness result.

Definition 11 Let $\mathscr{S}$ be an FNT. Two disjoint subsets of $\mathscr{N}$, A and B, are *strongly separable* in $\mathscr{S}$ iff there is a formula $F(x)$ of exactly one variable x in $\mathscr{S}$ such that for all $a \in \mathscr{N}$

$$\text{if } a \in A, \text{ then } \vdash_{\mathscr{S}} F(\tilde{a})$$

$$\text{if } a \in B, \text{ then } \vdash_{\mathscr{S}} \neg F(\tilde{a})$$

We say that F strongly separates A and B.//

Note: If A and $\overline{A}$ are strongly separable in $\mathscr{S}$, then A is definable in $\mathscr{S}$. Thus, strong separability is a generalization of definability.

Definition 12 Let $\mathscr{S}$ be an FNT and $A \subset \mathscr{N}$. Then A is *enumerable in* $\mathscr{S}$ if $A = \{x \mid (\exists y)R(x,y)\}$, for some R which is definable in $\mathscr{S}$.//

Note: By Example 4, every re set is enumerable in every FNT. By the proof of Theorem 8, if the FNT is ω-consistent, then A is also *representable* in it.

Proposition 2 (Rosser) If $\mathscr{S}$ is an FNT, and if A and B are enumerable in $\mathscr{S}$, then $A - B$ and $B - A$ are *strongly separable*.

Proof Let $x \in A$ iff $(\exists y)F(x,y)$, and $x \in B$ iff $(\exists y)R(x,y)$. Let $\mathscr{F}(x,y)$ and $\mathscr{R}(x,y)$ define F and R, respectively.

Let $a \in A - B$. Then $(\exists y)F(a,y)$ and $\neg(\exists y)R(a,y)$. Hence, for some b, $F(a,b)$ and $\neg R(a,i)$, $i = 0,1,\ldots,b$. That is, $F(a,b)$ and $\neg(\exists y)_{\leq b}R(a,y)$ hence by definability [and proof of Claim (7) in Example 4].

(1) $\vdash_{s} \mathcal{F}(\tilde{a},\tilde{b})$, and

(2) $\vdash_{s} \neg(\exists y)(\neg(\tilde{b} < y)\ \&\ \mathcal{R}(\tilde{a},y))$ hence

(3) $\vdash_{s} (\exists x)(\mathcal{F}(\tilde{a},x)\ \&\ \neg(\exists y)(\neg(x < y)\ \&\ \mathcal{R}(\tilde{a},y)))$

Letting $\mathcal{H}(z)$ to stand for $(\exists x)(\mathcal{F}(z,x)\ \&\ \neg(\exists y)(\neg(x < y)\ \&\ \mathcal{R}(z,y)))$ we have

(3') $\vdash_{s} \mathcal{H}(\tilde{a})$.

We will show that $\mathcal{H}$ strongly separates $A - B$ and $B - A$.

To this end, we show that if $a \in B - A$ then $\vdash_{s} \neg \mathcal{H}(\tilde{a})$, that is (definition of & and ∀): $\vdash_{s} (\forall x)(\neg\mathcal{F}(\tilde{a},x) \vee \neg\neg(\exists y)(\neg(x < y)\ \&\ \mathcal{R}(\tilde{a},y)))$ or better still (recall that $\vdash_{\text{LPC}} X \Leftrightarrow \neg\neg X$)

$$\vdash_{s} (\forall x)(\neg\mathcal{F}(\tilde{a},x) \vee (\exists y)(\neg(x < y)\ \&\ \mathcal{R}(\tilde{a},y)))$$

So let $a \in B - A$. Then $R(a,b)$ for some b, hence

(4) $\vdash_{s} \mathcal{R}(\tilde{a},\tilde{b})$ and, since $\neg(\exists y)_{<b} F(a,y)$ also [see proof of Claim (7)(C) in Example 4],

(5) $\vdash_{s} y < \tilde{b} \Rightarrow \neg\mathcal{F}(\tilde{a},y)$

By (4) (and using the deduction theorem), $\vdash_{s} \neg(y < \tilde{b}) \Rightarrow \neg(y < \tilde{b})\ \&\ \mathcal{R}(\tilde{a},\tilde{b})$. Also (LPC, Axiom 3.5), $\vdash_{s} \neg(y < \tilde{b})\ \&\ \mathcal{R}(\tilde{a},\tilde{b}) \Rightarrow (\exists t)(\neg(y < t)\ \&\ \mathcal{R}(\tilde{a},t))$ hence (transitivity of $\Rightarrow$),

(5') $\vdash_{s} \neg(y < \tilde{b}) \Rightarrow (\exists t)(\neg(y < t)\ \&\ \mathcal{R}(\tilde{a},t))$. (5) and (5') (proof by cases) give $\vdash_{s} \neg\mathcal{F}(\tilde{a},y) \vee (\exists t)(\neg(y < t)\ \&\ \mathcal{R}(\tilde{a},t))$ hence, $\vdash_{s} (\forall y)(\neg\mathcal{F}(\tilde{a},y) \vee (\exists t)(\neg(y < t)\ \&\ \mathcal{R}(\tilde{a},t)))$ that is, $\vdash_{s} \neg\mathcal{H}(\tilde{a})$.//

Corollary   Every *recursive* set $A\ (\subset \mathcal{N})$ is *definable* in ROB, hence in any FNT.

Proof   Since A and $\overline{A}$ are re and $A \cap \overline{A} = \varnothing$, the result follows from the adequacy of ROB (which implies enumerability of A and $\overline{A}$ in any FNT), Proposition 2 and the note following Definition 11.//

Note: We felt it was desirable to expose the reader to Rosser's method. However, our aim was indeed the above corollary, which can be proved directly without recourse to the notion of strong separability. [See Problem 48.]

Theorem 10   (*Gödel-Rosser Incompleteness Theorem.* Due to Rosser.) If *any* FNT is *simply consistent*, then it is an undecidable theory, hence it is also simply incomplete (by Proposition 1).

Proof   Under these assumptions, *every* recursive set is *representable* in this FNT, hence the result follows from the corollary to Theorem 6 [assuming a representation map $\Phi:\mathcal{P}_1 \times \mathcal{N} \rightarrow \mathcal{N}$ which assigns $F(\tilde{n})$ to $(F(x),n)$].

Indeed, if $A \subset \mathcal{N}$ is recursive, and if $\mathcal{H}(x)$ defines A (by Corollary to Proposition 2), then $\vdash_{FNT} \mathcal{H}(\tilde{a})$ implies $a \in A$, for otherwise we have $a \notin A$, hence $\vdash_{FNT} \neg \mathcal{H}(\tilde{a})$, that is (simple) inconsistency. Therefore, $a \in A$ iff $\vdash_{FNT} \mathcal{H}(\tilde{a})$, for all $a \in A$.//

Corollary   (Stated for emphasis. This is essentially Church's unsolvability result for LPC [4].) ROB has an unsolvable decision problem.

Proof It suffices to establish simple consistency of ROB. This being beyond our scope, the reader is referred to Kleene [12].//

Note: There are "direct" combinatorial proofs of the unsolvability of the LPC decision problem. (See, for example, in Brainerd and Landweber [3] and Lewis and Papadimitriou [14]). It is also observed that the nonrecursiveness of T_{LPC} is obtainable from that of the unsolvability of the semigroup word-problem, due to the translatability of the finitely axiomatizable semigroup system to LPC (e.g., Diller [7]).

Theorem 11   (*Gödel's Second Incompleteness Theorem*) If an FNT $\mathcal{S}$ is simply consistent, then its consistency *cannot* be proved within $\mathcal{S}$.

Proof (*Outline only*) Assume that it has been proved, that a predicate $P \in \mathcal{PR}_*^{(2)}$ and a function $f \in \mathcal{PR}^{(2)}$ exist (Gödel [8]) such that $P(z,y)$ iff y is the Gödel number of a proof in $\mathcal{S}$ of a formula with Gödel number z, $f(x,y)$ is the Gödel number of a formula obtained from one with Gödel number x after (a predetermined) variable z has been substituted by $\tilde{y}$ (i.e., the $\mathcal{S}$ counterpart of number y).

Set $F(z,y) \stackrel{\text{def}}{=} P(f(z,z),y)$.

Observe,

(1) $F \in \mathcal{PR}_*$, hence is definable in $\mathcal{S}$ (by corollary to Proposition 2, extended to subsets of $\mathcal{N} \times \mathcal{N}$. See also Problem 48) by the formula $\mathcal{F}(x,y)$, say. Let $n =$ Gödel number of $\neg(\exists y)\mathcal{F}(z,y)$.

(2)   Gödel number of $\neg(\exists y)\mathcal{F}(\tilde{n},y)$ is $f(n,n)$.

(3) $\neg(\exists y)\mathcal{F}(\tilde{n},y)$ is *not* provable in $\mathcal{S}$.

For, assume it is. Then $(\exists y)P(f(n,n),y)$, by (2) and the meaning of $P(z,y)$; that is, $F(n,b)$ for some b, hence $\vdash_{\mathcal{S}} \mathcal{F}(\tilde{n},\tilde{b})$.

Since also $\vdash_{\mathcal{S}} \mathcal{F}(\tilde{n},\tilde{b}) \Rightarrow (\exists y)\mathcal{F}(\tilde{n},y)$ we get $\vdash_{\mathcal{S}} (\exists y)\mathcal{F}(\tilde{n},y)$, contradicting *simple* consistency. (Gödel, in his original proof of his *first* incompleteness theorem, required ω-consistency to show unprovability of $(\exists y)\mathcal{F}(\tilde{n},y)$ as well.)

Let next "Cons" be an $\mathscr{S}$-formula that "says" that $\mathscr{S}$ is simply consistent. What we proved (in outline) above is "if $\mathscr{S}$ is consistent then the formula with Gödel number $f(n,n)$ is unprovable".

In symbols, Cons $\Rightarrow \neg(\exists y)\mathscr{F}(\tilde{n},y)$.

Now, taking for granted(*) that the argument, which establishes the previous formula informally, can be mirrored in $\mathscr{S}$ (plausible at this point, since instead of talking about formulas and proofs we can talk about their Gödel numbers. But that's arithmetic, and it can be done formally in $\mathscr{S}$), we obtain

$$\vdash_\mathscr{S} \text{Cons} \Rightarrow \neg(\exists y)\mathscr{F}(\tilde{n},y).$$

Now, if also

$$\vdash_\mathscr{S} \text{Cons}$$
then (modus ponens)

$$\vdash_\mathscr{S} \neg(\exists y)\mathscr{F}(\tilde{n},y)$$
a contradiction.//

(*)Full but lengthy details can be found in Hilbert and Bernays [11].

PROBLEMS

1. Let $\mathcal{C} \neq \varnothing$ and let the empty function $\varnothing$ be in $\mathcal{P} - \mathcal{C}$. Prove that $K \leq_1 \{x \mid \phi_x \in \mathcal{C}\}$.

2. Let $\varnothing \in \mathcal{C} \subsetneq \mathcal{P}$. Show that $\overline{K} \leq_1 \{x \mid \phi_x \in \mathcal{C}\}$.

3. Prove Proposition 9.1.1.

4. (Dekker) Let A and B be re, and moreover $A \cup B = \mathcal{N}$ and $A \cap B \neq \varnothing$. Prove that $A \leq_m A \cap B$.

5. Let $\mathcal{C} \subset \mathcal{P}^{(1)}$ contain a ψ such that a χ exists, where $\chi \in \mathcal{P} - \mathcal{C}$ and $\psi \subset \chi$. Prove that $\overline{K} \leq_1 \{x \mid \phi_x \in \mathcal{C}\}$.

6. Let $\mathcal{C} \subset \mathcal{P}^{(1)}$ contain a ψ such that for all *finite* θ such that $\theta \subset \psi$, $\theta \notin \mathcal{C}$. Show that $\overline{K} \leq_1 \{x \mid \phi_x \in \mathcal{C}\}$.

7. State and prove the corresponding statements to those of Problems 1, 2, 5, 6, where now $\mathcal{C}$ is a class of re sets.

8. Show that $\overline{K} \leq_1 \{x \mid W_x \text{ is finite}\}$.

9. Show that $K \leq_1 \{x \mid W_x \text{ is finite}\}$.

10. Show that $\{x \mid W_x \neq \varnothing\}$ is 1-complete.

11. Show that every infinite set in $\mathcal{P}\mathcal{R}_*$ contains a set in $\mathcal{R}_* - \mathcal{P}\mathcal{R}_*$.

12. Prove that if $A \leq_m B$ via f and $A \notin \mathcal{R}_*$, then $f(A) \notin \mathcal{R}_*$.

13. Let finite sets be prime-power coded. Let A be re and $K \leq_m A$ via f. Show that there is an $h \in \mathcal{R}^{(1)}$ such that

$$W_{h(x)} = \begin{cases} f^{-1}(D_x) & \text{if } D_x \cap A = \varnothing \\ \mathcal{N} & \text{if } D_x \cap A \neq \varnothing \end{cases}$$

(*Hint*: Given x, find a "program" $g(x)$ which enumerates $f^{-1}(D_x)$ or $\mathcal{N}$ depending on $D_x \cap A = \varnothing$ or $D_x \cap A \neq \varnothing$. [Then, if $t \in \mathcal{R}^{(1)}$ is such that range(ϕ_x) $= W_{t(x)}$, h will be $t \circ g$]. Recall that an index, $m(x)$, for the re set $f^{-1}(D_x)$ can be obtained effectively from x (i.e., $m \in \mathcal{R}^{(1)}$). Then, program $g(x)$, using $m(x)$ as "subroutine", enumerates $f^{-1}(D_x)$ $(= W_{m(x)})$ and A simultaneously, as long as no enumerated A-item is in D_x. The first time such an item is found, $g(x)$ "exits" from subroutine $m(x)$ and embarks on an enumeration of $\mathcal{N}$. Provide formal details.)

14. Prove: If $K\leq_m A$ via f and A is re, then there is a $g \in \mathcal{R}^{(1)}$ such that for all x,

   (i) $\varnothing \neq D_x \subset A \Rightarrow g(x) \in A - D_x$

   (ii) $\varnothing \neq D_x \subset \overline{A} \Rightarrow g(x) \in \overline{A} - D_x$

     Note: Say, prime-power coding is used for finite sets.
     (*Hint* [As in Rogers [18]]: By Problem 12, $f(K)$ is infinite. As it is re (why?), let $t(\mathcal{N}) = f(K), t \in \mathcal{R}^{(1)}$. Let h be as in Problem 13, and define g by:

$$g(x) = \textbf{if } f{\circ}h(x) \notin D_x \textbf{ then } f{\circ}h(x) \textbf{ else } t((\mu y)[t(y) \notin D_x])$$

Show that g works. Where was the fact that $f(K)$ is infinite used?)

15. Prove that A is m-complete iff it is 1-complete. (*Hint*: The *if* part is trivial. So let A be m-complete. Then $K\leq_m A$, and A is re. By Problem 14 and Theorem 9.4.2, A is a cylinder. By Theorem 9.4.1(iii), A is 1-complete since if X is re, then $X\leq_m A$, hence, $X\leq_1 A$.)

16. Let S be any re set. Show that there is a $g \in \mathcal{R}^{(2)}$ such that

$$W_{g(x,y)} = \textbf{if } y \in S \textbf{ then } W_x \textbf{ else } \varnothing.$$

17. Prove that A is productive iff there is a $\psi \in \mathcal{P}^{(1)}$ such that

$$(\forall x)(W_x \subset A \Rightarrow \psi(x)\!\downarrow \,\&\, \psi(x) \in A - W_x)$$

(*Hint*: *if* part. Let $f(\mathcal{N}) = S$, where S is *non*recursive, $f \in \mathcal{R}^{(1)}$[e.g., $S = K$]. Let g be as in Problem 16. Now, if $W_x \subset A$, then also $W_{g(x,f(y))} \subset A$ for all y. Thus, $\psi{\circ}g(x,f(y))\!\downarrow$ for all y. This suggests setting

$$h = \lambda x.\psi{\circ}g(x,f(\text{Sel}(x))),$$

where $\text{Sel} \in \mathcal{P}$ is such that

$$(\exists y)\psi{\circ}g(x,f(y))\!\downarrow \;\equiv\; \text{Sel}(x)\!\downarrow \;\equiv\; \psi{\circ}g(x,f(\text{Sel}(x)))\!\downarrow$$

by the selection theorem (Problem 8.2). Clearly, h is a partial productive function for A, the same way ψ is. Show that h is also total. *Further hint*: If $h(x)\!\uparrow$ for some x, then $\overline{S}$ is re; hence, $S \in \mathcal{R}_*$.)

18. Show that a set $\{x \,|\, \phi_x \in \mathcal{C}\}$, where $\mathcal{C} \subset \mathcal{P}^{(1)}$, is productive if there are ψ and χ in $\mathcal{P}^{(1)}$ such that $\psi \subset \chi, \psi \in \mathcal{C}, \chi \notin \mathcal{C}$.

19. State and prove an analogue of Problem 18 for sets such as $\{x \,|\, W_x \in \mathcal{C}\}$.

20. Prove that a set $\{x \,|\, \phi_x \in \mathcal{C}\}$, where $\mathcal{C} \subset \mathcal{P}^{(1)}$, is productive if there is a $\psi \in \mathcal{C}$ such that *no* finite θ such that $\theta \subset \psi$ is in $\mathcal{C}$.

21. State and prove an analogue of Problem 20 for sets such as $\{x \mid W_x \in \mathcal{C}\}$.

22. Let $\varnothing \in \mathcal{C} \underset{\neq}{\subseteq} \mathcal{P}$. Prove that $\{x \mid \phi_x \in \mathcal{C}\}$ is productive.

23. State and prove the analogue of Problem 22 for sets such as $\{x \mid W_x \in \mathcal{C}\}$.

24. Show that *if* a nontrivial (i.e., $\neq \varnothing$ and $\neq \mathcal{N}$) complete set of indices is re, then it is *creative*.

25. Is $\{x \mid W_x = \varnothing\}$ creative? Productive?
How about $\{x \mid W_x \neq \varnothing\}$?

26. Show that $\{x \mid 0 \in W_x\}$ is creative.

27. Complete the argument presented in Example 9.5.5.

28. Two sets A and B are *recursively separable* (see also Problem 7.14) if there exists a recursive set C such that $A \subset C \subset \overline{B}$.
If on the other hand they are disjoint but *not* recursively separable, they are *recursively inseparable*. Now if a *counterexample* to a claim of separability can be found *effectively*, in the following formal sense, A and B are *effectively inseparable*:
A and B are *effectively inseparable* if there is an $f \in \mathcal{P}^{(2)}$ such that for all x,y,

$$A \subset W_x \,\&\, B \subset W_y \,\&\, W_x \cap W_y = \varnothing \Rightarrow f(x,y){\downarrow} \,\&\, f(x,y) \in \overline{W_x \cup W_y}$$

Prove

(i) If A and B are effectively inseparable, then they are recursively inseparable.

(ii) If A and B of (i) are, moreover, re, then they are creative.

(iii) $\{x \mid \phi_x(x) = 0\}$ and $\{x \mid \phi_x(x) = 1\}$ are effectively inseparable.

(*Hint*: See Problem 7.14.)

29. Prove: If A is creative and $J:\mathcal{N} \times \mathcal{N} \to \mathcal{N}$ is recursive, 1-1, and onto, then $J(A \times A)$ is creative.

30. Show that $*$ is not part of any formula or term of LPC.

31. Show that $\vdash_S X \Leftrightarrow X$, where X is any S-formula (written in infix) and S is any extension of the Propositional Calculus or of the LPC.

32. Consider any extension of the LPC, S. Throw in $x = 0$ as an axiom, where x is a variable, obtaining the system S'. Thus, $\vdash_{S'} x = 0$. By substitution of another variable, y, for x, we get $\vdash_{S'} y = 0$. By the Deduction Theorem, $\vdash_S x = 0 \Rightarrow y = 0$, which is "intuitively wrong". (Justify.)
   Is there something wrong with the Deduction Theorem, or was it misapplied here?

33. Give an intuitive reason why $F(v) \Rightarrow (\forall v)F(v)$ cannot be, in general, an LPC theorem.

34. Show that $\vdash_S \neg\neg X \Rightarrow X$ in any extension S of LPC, where X is any S-formula.

35. Let $\vdash_{\text{LPC}} F(v)$, where v is a variable. Prove that $\vdash_{\text{LPC}} F(t)$, where t is any term. (*Hint*: Consider a proof $F_1, F_2, \ldots, F_k$ where F_k is $F(v)$. Next, examine the sequence $\tilde{F}_1, \ldots, \tilde{F}_k$, where each $\tilde{F}_i$ is obtained from F_i by substituting each occurrence of v in F_i by t.)

36. Show that for any terms u and v, $\vdash_{\text{LPC}} u = v \Leftrightarrow v = u$.

37. Show that if S is any extension of the LPC and $\vdash_S F$, $\vdash_S G$, where F and G are formulas, then $\vdash_S F \& G$.

38. Prove that if S is any extension of the LPC, and F and G are any formulas, then $\vdash_S (F \& G) \Rightarrow F$ and $\vdash_S (F \& G) \Rightarrow G$.

39. Show that for any terms u,v and formula $F(x)$, where x is a variable, $\vdash_{\text{LPC}} u = v \Rightarrow (F(u) \Leftrightarrow F(v))$.

40. Let u,v,w be LPC terms. Show that $\vdash_{\text{LPC}} u=v \& v=w \Rightarrow u=w$.

41. Let t and s be arbitrary LPC terms, and $w(x)$ be an LPC term which contains the variable x. Prove that $\vdash_{\text{LPC}} t=s \Rightarrow w(t)=w(s)$, where $w(t)$ is obtained from $w(x)$ by substituting every occurrence of x in $w(x)$ by t.

42. Prove: If $Q \subset \mathcal{N}^n$ is definable in S, then $Q \in \mathcal{R}_*$.

43. Let $\mathcal{S}$ be an FNT and $\mathcal{P}_1$ be the set of formulas $F(x)$ in $\mathcal{S}$, where F is an ft-schema of rank 1 and x is a variable. Show that the map $\Phi : \mathcal{N} \times \mathcal{N} \to \mathcal{N}$ such that its restriction on $\mathcal{P}_1 \times \mathcal{N}$ sends $(F(x),n)$ to $F(\tilde{n})$, while outside $\mathcal{P}_1 \times \mathcal{N}$ it is the constant $0 = 0$, is recursive.

44. Prove that if an FNT is (simply) complete, then it is decidable.

45. Show that $a \cdot b = c$ is definable by $x \cdot y = z$ (where x,y,z are variables) in ROB.

46. Prove that if $R(\vec{a}_n)$ is definable in ROB, then so is any explicit transform of $R(\vec{a}_n)$.

47. Let us say that a function $f : \mathcal{N}^n \to \mathcal{N}$ is definable in an FNT iff there is an ft-schema F of rank $n + 1$, such that for all $(\vec{a}_n, b) \in \mathcal{N}^{n+1}$

$$\vdash_{\text{FNT}} F(\tilde{a}_1, \ldots, \tilde{a}_n, y) \Leftrightarrow y = \tilde{b} \quad \text{if } f(\vec{a}_n) = b$$

where y is a variable of the FNT.

Prove that a relation $R(\vec{x}_n)$ is definable in an FNT iff its characteristic function c_R is definable, in the above sense. (*Hint*: If $F(\vec{x}_n, y)$, where the ft-schema F has rank $n + 1$, defines c_R, show that $F(\vec{x}_n,\tilde{0})$ defines R. Next, if $G(\vec{x}_n)$, where the ft-schema G has rank n, defines R, show that

$$(G(\vec{x}_n) \ \& \ y = \tilde{0}) \vee (\neg G(\vec{x}_n) \ \& \ y = \tilde{1})$$

defines c_R.)

48. Prove, without using Rosser's method, that every recursive function and relation is definable in ROB (and hence, in any FNT). (*Hint*: By Problem 47, it suffices to prove this for functions in $\mathcal{R}$. Now, a function is in $\mathcal{R}$ iff it is one of the initial functions $\lambda xy.x + y$, $\lambda xy.xy$, $\lambda xy.x \doteq y$, $\lambda x.x + 1$, $\lambda \vec{x}_n.x_i$ or it is obtained by a *finite* number of *compositions* or applications of (μy), the latter restricted to total functions $\lambda \vec{x}y.g(\vec{x},y)$ such that $(\forall \vec{x})(\exists y)g(\vec{x},y) = 0$, starting with the initial functions. The reader will find enough tools in the proof that ROB is adequate.)

REFERENCES

[1] Bernays, P. "Axiomatische Untersuchung des Aussagen-Kalküls der Principia Mathematica". *Math. Zeit.* 25 (1926): 305–320.

[2] Bourbaki, N. *Éléments de mathématique: Théorie des ensembles.* Paris: Hermann, 1966: Ch. 1–2.

[3] Brainerd, W.S., and Landweber, L.H. *Theory of Computation.* New York: Wiley, 1974.

[4] Church, A. "A Note on the Entscheidungsproblem". *J. of Symb. Logic* 1 (1936): 40–41, 101–102. (Also in Davis [6]: 110–115.)

[5] Davis, M. *Computability and Unsolvability.* New York: McGraw-Hill, 1958.

[6] Davis, M. *The Undecidable.* Hewlett, N.Y.: Raven Press, 1965.

[7] Diller, J. *Rekursionstheorie.* Institut für mathematische Logik und Grundlagenforschung. Westfalische Wilhelms-Universität. Münster, 1976.

[8] Gödel, K. "Über formal unentscheidbare Sätze der Principia Mathematica und verwandter Systeme, I". *Monatshefte für Math. und. Physik.* 38 (1931): 173–198. (Also, in English in Davis [6]: 5–38.)

[9] Gödel, K. "Die Vollständigkeit der Axiome des logischen Funktionenkalküls". *Monatshefte für Math. und. Physik* 37 (1930): 349–360.

[10] Hilbert, D. and Ackermann, W. *Principles of Mathematical Logic.* New York: Chelsea Publ. Co., 1950.

[11] Hilbert, D., and Bernays, P. *Grundlagen der Mathematik. Vols. 1,2.* Heidelberg: Springer-Verlag, 1934, 1939.

[12] Kleene, S.C. *Introduction to Metamathematics.* Princeton, N.J.: Van Nostrand, 1952.

[13] Kneebone, G.T. *Mathematical Logic and the Foundations of Mathematics.* Princeton, N.J.: Van Nostrand, 1963.

[14] Lewis, H.R., and Papadimitriou, C.H. *Elements of the Theory of Computation.* Englewood Cliffs, N.J.: Prentice-Hall, 1981.

[15] Peano, G. *Formulaire de Mathématiques. Vol. I–V.* Turin: Bocca, 1894–1908.

[16] Post, E. "Recursively Enumerable Sets of Positive Integers and their Decision Problems". *Bulletin Amer. Math. Soc.* 50 (1944): 284–316. (Also in Davis [6]: 305–337.)

[17] Robinson, R.M. "An Essentially Undecidable Axiom System". (Abstract). *Proc. of the International Congress of Mathematicians.* Cambridge, Mass. 1 (1950): 729–730.

[18] Rogers, H. *Theory of Recursive Functions and Effective Computability.* New York: McGraw-Hill, 1967.

[19] Smullyan, R.M. *Theory of Formal Systems.* Annals of Mathematics Studies, No. 47. Princeton, N.J.: Princeton University Press, 1961.

[20] Tarski, A. "Einige Betrachtungen uber die Bergriffe der ω-Widerspruchsfreiheit und der ω-Vollständingkeit". *Monatshefte für Math. und Physik* 40 (1933): 97–112.

[21] Whitehead, A.N., and Russell, B. *Principia Mathematica.* Three volumes. London: Cambridge Univ. Press, 1910, 1912, 1913.

[22] Wilder, R.L. *Introduction to the Foundations of Mathematics.* New York: Wiley, 1963.

CHAPTER 10

Relativized Computability

We make two observations at this point.

(1)   The Recursion or Computability Theory presented so far dealt with functions on the integers. It is possible to extend the results to functions whose domains are sets of objects of "higher type" than numbers (e.g., domains which are sets of number-theoretic functions) and whose ranges are subsets of $\mathcal{N}$.(*) In undertaking such an extension, it should be obvious that, to preserve the finitary character of the notion of "computation", the "input phase", during which a function-input is presented (that is, an "infinite object" in general), must not be considered part of the computation itself, and, during the latter only finitely many function-values (of the input function) are to be used. This remark means that in a *machine formalism*—for example, URM formalism—of such an extension, arbitrary (not necessarily partial recursive) number-theoretic functions f are presented as inputs and instructions such as $X \leftarrow f(\vec{Y})$ are allowed for any such f.(†) In a number-theoretic formalism, such as $\mathcal{P}$ of

(*)Such functions are known as *functionals*. Definitions will follow this short introduction.

(†)If $f(\vec{Y})\uparrow$, then the "computation" never exits from the instruction $X \leftarrow f(\vec{Y})$.

Chapter 3, one first amends the "initial functions" such as $\lambda x.x + 1$ and $U_i^n = \lambda \vec{x}_n.x_i$, to $S_i^{n,m} = \lambda \vec{x}_n \vec{\alpha}_m.x_i + 1$ and $U_i^{n,m} = \lambda \vec{x}_n \vec{\alpha}_m.x_i$, where $\vec{\alpha}_m$ is an m-tuple of functions $\alpha_j: \mathcal{N}^k \to \mathcal{N}$ ($k = 1$ for convenience) and then adds the *evaluation functional* $Ev_{i,j}^{n,m} = \lambda \vec{x}_n \vec{\alpha}_m.\alpha_j(x_i)$. This clearly corresponds to the URM model suggested, since in $S_i^{n,m}$ and $U_i^{n,m}$ the $\vec{\alpha}_m$ are inessential (dummy) arguments.

(2) Observing the number-theoretic, and among the machine formalisms especially the Loop-Program and URM formalisms [for subrecursive and (partial) recursive functions, respectively], we note that essentially what we have done is that we defined functions which are *"intuitively computable"* if the *initial* functions $U_i^n, \lambda x.x + 1, \lambda x.x \dot{-} 1$, etc., are. In other words, we have defined functions which are computable *relative to the initial functions*. It proves quite fruitful to pursue Recursion Theory *relative* to arbitrarily chosen (not necessarily "intuitively", or otherwise, computable) initial functions. Thus, adding $\lambda x.f(x)$ as an initial function to the primitive recursive (resp. partial recursive) formalisms, or, what amounts to the same, allowing $X \leftarrow f(Y)$ as an instruction to the Loop-Program (resp. URM) formalism we obtain what is known as primitive recursive (resp. partial recursive) functions *in f* or *relative to f* (of course, if f is in $\mathcal{PR}$ or in $\mathcal{P}$, then we do not get any new functions not in $\mathcal{PR}$ or $\mathcal{P}$, respectively).

We finally note that, as it is fairly clear from the programming formalisms, there is no difference between the *relative computability* and *computability of functionals* notions. Indeed, a functional $\lambda xf.F(x,f)$ (where $x \in \mathcal{N}$ and f is $f: \mathcal{N} \to \mathcal{N}$) is URM-computable iff it is computable in the ordinary URM formalism augmented by "assignment statements" of the form $X \leftarrow \alpha(Y)$ *whenever $\alpha:\mathcal{N} \to \mathcal{N}$ is an f-input*; but this states that $\lambda x.F(x,f)$ is f-URM-computable *for each f*. [Historically, the notion of f-computability(*) came before that of computability of functionals.]

We shall now formally develop the theory of computable functionals and f-computability to the extent we have done so for the nonrelativized computability of functions. To expedite matters on one hand,(†) and present an important new formalism on the other (originally due to Kleene [8]), which is perhaps the most convenient to adapt for the study of Axiomatic Recursion Theory as well as Recursion in Higher Types,(‡) we shall not pursue the URM- or $\mathcal{P}$-extension approach which was suggested earlier.

(*)f is termed an *"oracle"* by many authors, since it provides, in a non-transparent manner, answers to questions essential for the progress of the computation.

(†)The Gödel numbering of the "computable functionals" will be built-in right from the beginning.

(‡)Objects of $\mathcal{N}$ are type 0 objects. If a map F (or relation R) has objects of type n (but no higher) as arguments then it is a type $(n + 1)$ object. E.g., functions on $\mathcal{N}^k$ are type 1, and functionals (in the previous sense) are type 2 objects. Ordinary Recursion Theory ends and Recursion in Higher Types starts as soon as type 2 arguments are allowed.

Example   In everyday computing, one encounters functionals, i.e., maps with arguments a mixture of numbers and functions on numbers. Namely, a "function procedure" **integrate** (a,b,f), which returns an approximation of $\int_a^b f(t)\,dt$ is very commonly used.//

§10.1   FUNCTIONALS

Definition 1   Let $\mathcal{J}$ stand for $\mathcal{P}(\mathcal{N};\mathcal{N})$, i.e., the set of all (partial) functions $f:\mathcal{N}\to\mathcal{N}$.

A *functional F* of *rank* (k,l) is a map $F:\mathcal{N}^k\times\mathcal{J}^l\to\mathcal{N}$.//

Note: The functional, as defined, is a type 2 object. We shall restrict attention to the "computability" of type 2 objects only in this book. Excellent modern sources for the study of Recursion in Higher Types are the books by Hinman [5] and Fenstad [2].

We shall normalize our notation, so that number variables will be denoted by Latin lower-case letters, partial (unary) functions on $\mathcal{N}$ by lower-case Greek letters and functionals and predicates (relations) by capital Latin letters. This notational convention will persist for the balance of this chapter; exceptions will be noted.

By convention, a functional of rank $(k,0)$ is a map $F:\mathcal{N}^k\to\mathcal{N}$, hence it is an ordinary number-theoretic function. A functional of rank $(0,l)$ is a map $F:\mathcal{J}^l\to\mathcal{N}$. We shall use, as before, λ-notation in connection with the convention of writing number variables before function variables. For example, $\lambda p.F(p,x,y,\alpha,\beta)$ is a map from $\mathcal{N}$ to $\mathcal{N}$. $\lambda xy\alpha\beta.(\lambda p.F(p,x,y,\alpha,\beta))$ is a map from $\mathcal{N}^2\times\mathcal{J}^2$ to $\mathcal{J}$. As usual, $F(\vec{x},\vec{\alpha})=y$ (where $y\in\mathcal{N}$) *includes* the statement $F(\vec{x},\vec{\alpha})\!\downarrow$. Thus $F(\vec{x},\vec{\alpha})=G(\vec{x},\vec{\beta})$ means that "$F(\vec{x},\vec{\alpha})\!\uparrow$ and $G(\vec{x},\vec{\beta})\!\uparrow$ or, for some y, $F(\vec{x},\vec{\alpha})=y$ and $G(\vec{x},\vec{\beta})=y$". [In many places in the literature one sees the symbol $\simeq$ in connection with equality of partial maps. We shall continue using "$=$" with the previously stated understanding.]

Definition 2   A relation or predicate of rank (k,l) is a subset of $\mathcal{N}^k\times\mathcal{J}^l$. If R is such a relation, we shall use the notations R, $R(\vec{x}_k,\vec{\alpha}_l)$ and $\lambda\vec{x}_k\vec{\alpha}_l.R(\vec{x}_k,\vec{\alpha}_l)$ interchangeably. In particular, we shall use the latter when we need to emphasize which exactly are the "arguments" of R.//

Note: If $R\subset\mathcal{N}^k\times\mathcal{J}^l$, then $\lambda\vec{x}_m.R(\vec{x}_k,\vec{\alpha}_l)$ $(m\le k)$ is a subset of $\mathcal{N}^m$, $\lambda\alpha_1.R(\vec{x}_k,\vec{\alpha}_l)$ is a subset of $\mathcal{J}$, $\lambda x_2\alpha_4.R(\vec{x}_k,\vec{\alpha}_l)$ is a subset of $\mathcal{N}\times\mathcal{J}$. As before, R has rank $(0,l)$ means it is a subset of $\mathcal{J}^l$; that it has rank $(k,0)$ means it is a subset of $\mathcal{N}^k$.

Characteristic functions of relations of rank (k,l) are, of course, functionals of rank (k,l). If $A\subset\mathcal{N}$, then $F(n,A)$ is abuse of notation for $F(n,\chi_A)$ where χ_A is the characteristic function of A.

Substitution of functionals for number-variables and of partial functions for function-variables is straightforward in meaning. E.g. $F(G(\vec{x},\vec{\alpha}),\vec{y},\vec{\beta}) = z$ holds iff $G(\vec{x},\vec{\alpha}) = w$, for some w, and $F(w,\vec{y},\vec{\beta}) = z$. $F(\vec{x},\vec{\alpha},\lambda y.G(y,\vec{z},\vec{\beta})) = w$ iff $F(\vec{x},\vec{\alpha},\gamma) = w$, where $\gamma = \lambda y.G(y,\vec{z},\vec{\beta})$.

For relations, $R(F(\vec{x},\vec{\alpha}),\vec{y},\vec{\beta})$ holds iff $(\exists z)[z = F(\vec{x},\vec{\alpha}) \ \& \ R(z,\vec{y},\vec{\beta})]$. Similarly, $R(x,\lambda y.F(y,\vec{z},\vec{\alpha}),\vec{\beta})$ holds iff $(\exists \gamma) [\gamma = \lambda y.F(y,\vec{z},\vec{\alpha}) \ \& \ R(\vec{x},\gamma,\vec{\beta})]$.

We shall limit attention to functionals which admit unary functions only as arguments; this is no loss of generality in the presence of pairing functions. For the balance of this chapter, we shall use the prime-power coding scheme

$$\langle \ \rangle : \vec{x}_n \longmapsto \Pi_{i=0}^{n-1} p_i^{x_{i+1}+1} .$$

When $n = 0$, $\langle \ \rangle$ is the constant 1. $(u)_i$ will be shorthand for $\exp(i,u) \div 1$. length(u), as previously, will be the largest $m + 1$ such that $p_m | u$ (where $p_0 = 2$, $p_1 = 3$, etc.), 0 if no such m exists. We already know that $\lambda n\vec{x}_n.\langle \vec{x}_n \rangle$, $\lambda iu.(u)_i$ and $\lambda u.$length(u) are in $\mathcal{PR}$.

§10.2 PARTIAL RECURSIVE FUNCTIONALS

In this section the partial recursive (or computable) functionals will be defined using the Kleene-schemata. The Gödel numbering will be incorporated right from the definition. This approach considerably speeds up the development of the theory, at the same time affording enough generality for extensions of the notion of computation on more abstract domains.

The relation $\phi_e^{k,l}(\vec{x}_k,\vec{\alpha}_l) = y$ is taken as basic and the objects "e" satisfying it (i.e., programs for functionals) are defined inductively.

The content (intuitively) of the above relation is that "program e, when presented with input $(\vec{x}_k,\vec{\alpha}_l)$ eventually gives y as output". *How* this "program" arrives at the answer y is not important. After all, various programming/ machine formalisms (for example, Turing and URM formalisms [see Problems 14 and 22]) arrive at the same set of calculable functionals. Following Kleene's notation, we shall write $\{e\}(\vec{x}_k,\vec{\alpha}_l) = y$ instead of $\phi_e(\vec{x}_k,\vec{\alpha}_l) = y$ since the symbol "ϕ" is not essential.

We now define the relation above for all k and l.

Definition 1   Define a sequence of sets $(\Omega_n)_{n\geq 0}$:

(0): Ω_0 is the set of the following objects for all $k,l,m,i,\vec{x}_k,\vec{\alpha}_l,c,u,v,w,z,\vec{y}_m$

$(\langle 0,k,l,0,c\rangle,\vec{x}_k,\vec{\alpha}_l,c)$ [constant functional $\lambda \vec{x}_k\vec{\alpha}_l.c$]

$(\langle 0,k,l,1,i\rangle,\vec{x}_k,\vec{\alpha}_l,x_i)$ [$U_i^{k,l}$; the ith *projection*]

$(\langle 0,k,l,2,i\rangle,\vec{x}_k,\vec{\alpha}_l,x_i+1)$ $[S_i^{k,l};$ the ith *successor*$]$

$\langle 0,k,l,3,i,j\rangle,\vec{x}_k,\vec{\alpha}_l,\alpha_j(x_i))$ **if** $\alpha_j(x_i)\!\downarrow$

$[Ev_{i,j}^{k,l};$ the *evaluation* map$]$

$(\langle 0,k+4,l,4\rangle,u,v,w,z,\vec{x}_k,\vec{\alpha}_l,w)$ **if** $u=v$

$(\langle 0,k+4,l,4\rangle,u,v,w,z,\vec{x}_k,\vec{\alpha}_l,z)$ **if** $u\neq v$

[The last two implement,

essentially, "**if** $u=v$ **then** w **else** z"]

$(\langle 0,k+m,l,5\rangle,\vec{y}_m,\vec{x}_k,\vec{\alpha}_l,\langle\vec{y}_m\rangle)$ for $m\geq 1$ [introduces some *coding*

scheme to the system]

Let $\Omega_{(p)}$ stand for $\bigcup_{i=0}^{p}\Omega_i$. Then,

$(p+1):$ Ω_{p+1} is the set of the following objects for all $k,l,m,n,a,$ $\vec{b}_m,y,z,\vec{x}_k,\vec{x}_m,\vec{y}_n,\vec{\alpha}_l$

(a) $(\langle 1,k+m+1,l,m\rangle,a,\vec{b}_m,\vec{x}_k,\vec{\alpha}_l,y)$ just in case there are $y_1,\ldots,y_m$ such that *all* of the following are in $\Omega_{(p)}$: $(a,\vec{y}_m,\vec{\alpha}_l,y)$ and $(b_i,\vec{x}_k,\vec{\alpha}_l,y_i)$, i $=1,\ldots,m$. [This requirement, neglecting the Gödel indices $\langle 1,k+m$ $+1,l,m\rangle$, $a,\vec{b}_m$ for convenience, suggests that we can obtain $A(B_1(\vec{x}_k,\vec{\alpha}_l),$ $\ldots,B_m(\vec{x}_k,\vec{\alpha}_l))=y$ if we know that $A(\vec{y}_m,\vec{\alpha}_l)=y$ and $B_i(\vec{x}_k,\vec{\alpha}_l)=y_i$ for i $=1,\ldots,m$. Hence, this is the clause for *composition*.]

(b) $(\langle 2,n,l,a,\vec{x}_m\rangle,\vec{y}_n,\vec{\alpha}_l,z)$ just in case $(a,\vec{x}_m,\vec{y}_n,\vec{\alpha}_l,z)\in\Omega$. [This is the S-m-n theorem, with S-m-n function $\lambda a\vec{x}_m.\langle 2,n,l,a,\vec{x}_m\rangle]$

(c) $(\langle 3,k+1,l\rangle,a,\vec{x}_k,\vec{\alpha}_l,y)$ just in case $(a,\vec{x}_k,\vec{\alpha}_l,y)\in\Omega_{(p)}$. [This is the *universal functional* "*theorem*", where a moves from the role of a *program* to the role of an *input*.]

We set $\Omega\stackrel{\text{def}}{=}\bigcup_{p\geq 0}\Omega_p.$//

Note: It is easy to prove (Problem 1) that Ω is the *smallest* class containing Ω_0 and closed under the operations described in (a), (b), (c) above. Clearly, the $\vec{\alpha}_l$-input is relevant only in the definition of Ω_0 (item 4). If item 4 is removed, then this formalism (as it will be soon apparent) is an alternative for the $\mathcal{P}$-formalism.

It is possible to remove the S-m-n clause [(b) above] (Hinman [5]) by amending the composition rule (a) so that the "composed functional" has index $\langle 1,k,l,a,b_1,\ldots,b_m\rangle$. At this point, it is recommended that the reader reviews Theorem 3.5.1 and §3.8 for comparison and motivation for Definition 1.

Definition 2 We shall write $\{e\}(\vec{x}_k, \vec{\alpha}_l) = y$ iff $(e, \vec{x}_k, \vec{\alpha}_l, y) \in \Omega$.

We shall call *any* tuple $(e, \vec{x}_k, \vec{\alpha}_l, y)$ a *computation*. It will be a *convergent* computation iff it is in Ω; otherwise it is *divergent*. $\{e\}(\vec{x}_k, \vec{\alpha}_l)\!\downarrow$ (resp. $\{e\}(\vec{x}_k, \vec{\alpha}_l)\!\uparrow$) means $(\exists y)(e, \vec{x}_k, \vec{\alpha}_l, y) \in \Omega$ (resp.$(\forall y)(e, \vec{x}_k, \vec{\alpha}_l, y) \notin \Omega$). If $(e, \vec{x}_k, \vec{\alpha}_l, y) \in \Omega$ then its *length* is the *smallest* n such that $(e, \vec{x}_k, \vec{\alpha}_l, y) \in \Omega_n$, and is denoted by $\|e, \vec{x}_k, \vec{\alpha}_l, y\|$, or even $\|e, \vec{x}_k, \vec{\alpha}_l\|$ or $\|\{e\}(\vec{x}_k, \vec{\alpha}_l)\|$, anticipating Lemma 1 below. Let ∞ be an object not in $\mathcal{N}$. Extend $=$ and $\leq$ on $\mathcal{N} \cup \{\infty\}$ by $n \leq \infty$ for all $n \in \mathcal{N} \cup \{\infty\}$, $\infty = \infty$, $(\forall n)\ n \in \mathcal{N} \Rightarrow n \neq \infty$. "$<$" denotes the conjunction of $\leq$ and $\neq$ as usual.(*)

If $(e, \vec{x}_k, \vec{\alpha}_l, y) \notin \Omega$, then $\|e, \vec{x}_k, \vec{\alpha}_l, y\| = \infty$. If $\{e\}(\vec{x}_k, \vec{\alpha}_l)\!\uparrow$ then $\|e, \vec{x}_k, \vec{\alpha}_l\| = \|\{e\}(\vec{x}_k, \vec{\alpha}_l)\| = \infty$.

Clearly, if u, v are computations and $\|u\| < \|v\|$ then $u \in \Omega$.

Let next u by *any* computation $(e, \vec{m}_q, \vec{\alpha}_l)$ ("*output*" suppressed) *not* necessarily in Ω. We define *immediate subcomputation* (i.s.) by:

(0) If e is consistent with the lengths q and l of $\vec{m}$ and $\vec{\alpha}$ and has form according to the definition of Ω_0, then u has no i.s.

(1) If $u = (\langle 1, k + m + 1, l, m \rangle, a, \vec{b}_m, \vec{x}_k, \vec{\alpha}_l)$ then its i.s. are $(b_i, \vec{x}_k, \vec{\alpha}_l)$, $i = 1,$ $\ldots,\ m$ and, if all $\{b_i\}(\vec{x}_k, \vec{\alpha}_l)$ are convergent, then $(a, \{b_1\}(\vec{x}_k, \vec{\alpha}_l), \ldots,$ $\{b_m\}(\vec{x}_k, \vec{\alpha}_l), \vec{\alpha}_l)$ is also an i.s.

(2) If $u = (\langle 2, n, l, a, \vec{x}_m \rangle, \vec{y}_n, \vec{\alpha}_l)$, then $(a, \vec{x}_m, \vec{y}_n, \vec{\alpha}_l)$ is the only i.s.

(3) If $u = (\langle 3, k + 1, l \rangle, a, \vec{x}_k, \vec{\alpha}_l)$, then $(a, \vec{x}_k, \vec{\alpha}_l)$ is the only i.s.

v is a *subcomputation* of u if there is a sequence $v_0, \ldots, v_n$ of computations, such that $v_0 = v$, $v_n = u$ and v_i is an i.s. of v_{i+1}, $i = 0, \ldots, n - 1$.

Clearly, if v is a subcomputation of u, then $\|v\| \leq \|u\|$.//

Note: Intuitively, $(e, \vec{x}_k, \vec{\alpha}_l, y)$ or $(e, \vec{x}_k, \vec{\alpha}_l)$ is called a "computation", because if it is *convergent* then this can be verified in a finite number of steps(†) by building the complete *finite* set of its *subcomputations*. This can be done "recursively" or "inductively" working backwards from $(e, \vec{x}_k, \vec{\alpha}_l)$, until the process of finding i.s. cannot be continued. If $\{e\}(\vec{x}_k, \vec{\alpha}_l)\!\uparrow$ the set of subcomputations contains a divergent computation with no i.s. [some $\alpha_j(x_i)\!\uparrow$] or it is infinite. The latter may happen even with *total* type 1 inputs or no such inputs at all, due to cases (1) and (3) in the i.s. definition. Case (3) ("*universal functional postulate*") is equivalent to postulating closure under *unbounded search* or "**do-while**". About the Gödel numbers e we note that $(e)_0$ is in $\{0, 1, 2, 3\}$ if $(e, \vec{x}_k, \vec{\alpha}_l, y)$ is convergent. Then $(e)_1$ and $(e)_2$ count the number of numerical and function arguments respectively. Moreover, if $(e)_0 = 0$, then $(e)_3$ indicates the *type* of the "initial function" (or strictly speaking computation) chosen among those in Ω_0.

(*)One usually takes the "smallest limit ordinal, ω" for ∞. The interested reader can find about ordinals in, e.g., Kamke [6], Monk [10]. For our part, we shall use ∞ with the above postulated properties and this will be adequate for our purposes.

(†)Consulting the memory area, where the input α_j is being held, for the purpose of obtaining $\alpha_j(x_i)$ counts as one step if $\alpha_j(x_i)\!\downarrow$.

Lemma 1   If $(e,\vec{x}_k,\vec{\alpha}_l,y)$ and $(e,\vec{x}_k,\vec{\alpha}_l,z)$ are in Ω, then $y = z$.

Proof Induction with respect to Ω, or equivalently, with respect to p in $\Omega_{(p)}$.

$p = 0$: Then $(e)_3 \in \{0,1,2,3,4,5\}$. We treat the case $(e)_3 = 3$, leaving the rest as an easy exercise.

$$(e,\vec{x}_k,\vec{\alpha}_l,y) \in \Omega_{(0)} \text{ implies } \alpha_{(e)_5}(x_{(e)_4}) = y$$

$$(e,\vec{x}_k,\vec{\alpha}_l,z) \in \Omega_{(0)} \text{ implies } \alpha_{(e)_5}(x_{(e)_4}) = z$$

hence, $y = z$.

Assume that the lemma is true for all computations of *length* $\leq p$. Let now $(e,\vec{x}_k,\vec{\alpha}_l,y)$ and $(e,\vec{x}_k,\vec{\alpha}_l,z)$ be in $\Omega_{(p+1)}$.

Obviously, we need consider cases (a), (b), (c). We consider (b), and leave the others as an exercise. Thus, $(e)_0 = 2$. Set $m \overset{\text{def}}{=} \text{length}(e)$, $a \overset{\text{def}}{=} (e)_3$.

Then

$$(a,(e)_4,(e)_5,\ldots,(e)_{m \doteq 1},\vec{x}_k,\vec{\alpha}_l,y) \in \Omega_{(p)}$$

and

$$(a,(e)_4,(e)_5,\ldots,(e)_{m \doteq 1},\vec{x}_k,\vec{\alpha}_l,z) \in \Omega_{(p)}$$

hence, $y = z$ by induction hypothesis.//

Note: Due to Lemma 1, the relation $(e,\vec{x}_k,\vec{\alpha}_l) \longmapsto y$ is *single-valued*, hence it is a (partial) functional. Thus we call $(e,x_k,\vec{\alpha}_l)$ or $\{e\}(\vec{x}_k,\vec{\alpha}_l)$ a *computation* as well, and write $\|e,\vec{x}_k,\vec{\alpha}_l\|$ instead of $\|e,\vec{x}_k,\vec{\alpha}_l,y\|$, whenever convenient.

Definition 3   A functional $\lambda\vec{x}_k\vec{\alpha}_l.F(\vec{x}_k,\vec{\alpha}_l)$ is *partial recursive* iff $F = \lambda\vec{x}_k\vec{\alpha}_l.\{e\}(\vec{x}_k,\vec{\alpha}_l)$ for some $e \in \mathcal{N}$. e is a *Gödel number* or *index* for F.

F is *recursive* iff it is moreover *total*. A *relation* R is *recursive* iff its characteristic functional is recursive. An *index* of its characteristic functional is a *characteristic index* of R.

A relation R is *semi-recursive* iff $R = \text{dom}(F)$ for some *partial recursive functional* F. An index of F is a *semi-index* of R.//

Semi-recursive (for $l = 0$) corresponds to *re*. Following the example of several other authors (for example, Davis [1], Hinman [5]) we prefer the former term in this instance since, for example, $\mathcal{J}$ is semi-recursive but, not being emunerable, it would be perhaps counterintuitive to call it *recursively enumerable*.

Example 1 $\mathcal{I}$ is semi-recursive. Indeed $\{\langle 0,0,1,0,0 \rangle\}$ is an index for the functional $\lambda\alpha.0$, which has $\mathcal{I}$ as domain.//

Lemma 2   For all $\vec{x}_k, \vec{y}_n, \vec{\alpha}_l, a, b_1, \ldots, b_m$

(1)   There is an $f \in \mathcal{PR}^{(m+1)}$ such that $\{f(a,\vec{b}_m)\}(\vec{x}_k,\vec{\alpha}_l) = \{a\}(\{b_1\}(\vec{x}_k,\vec{\alpha}_l),$ $\ldots, \{b_m\}(\vec{x}_k,\vec{\alpha}_l),\vec{\alpha}_l)$

(2) $\{\langle 2,n,l,a,\vec{x}_k \rangle\}(\vec{y}_n,\vec{\alpha}_l) = \{a\}(\vec{x}_k,\vec{y}_n,\vec{\alpha}_l)$ [This is the S-m-n theorem. The function $S_n^k = \lambda a\vec{x}_k.\langle 2,n,l,a,\vec{x}_k \rangle$ is in $\mathcal{PR}$, of course].

(3) $\{\langle 3,k+1,l \rangle\}(a,\vec{x}_k,\vec{\alpha}_l) = \{a\}(\vec{x}_k,\vec{\alpha}_l)$ [universal functional theorem].

     Proof (1) Take $f = \lambda a\vec{b}_m.\langle 2,k,l,\langle 1,k+m+1,l,m \rangle,a,\vec{b}_m \rangle$. Let now $(f(a,\vec{b}_m),\vec{x}_k,\vec{\alpha}_l,y) \in \Omega$. Then it must be that $(\langle 1,k+m+1,l,m \rangle,a,\vec{b}_m,\vec{x}_k,\vec{\alpha}_l,y)$ $\in \Omega$ [Definition 1 (b)], hence [Definition 1 (a)] there are $y_1, \ldots, y_m$ such that $(a,\vec{y}_m,\vec{\alpha}_l,y)$ and $(b_i,\vec{x}_k,\vec{\alpha}_l,y_i)$ are in Ω, $i = 1, \ldots, m$. That is, $\{a\}(\{b_1\}(\vec{x}_k,\vec{\alpha}_l),$ $\ldots, \{b_m\}(\vec{x}_k,\vec{\alpha}_l),\vec{\alpha}_l) = y$.

     Conversely, the last relationship implies $\{f(a,\vec{b}_m)\}(\vec{x}_k,\vec{\alpha}_l) = y$ [details of this implication, and the proof of (2) and (3) are left to the reader].//

Theorem 1   The class of partial recursive functionals is closed under *substitution* of type 0 variables by *partial recursive functionals* and *constants*, and under the operations on type 0 variables of *identification*, *permutation* and introduction of *dummy variables*.

     Proof By Lemma 2,(1), and the fact that $\lambda\vec{x}_n\vec{\alpha}_m.c$ and $\lambda\vec{x}_n\vec{\alpha}_m.x_i$ are partial recursive (Definition 1, Ω_0, items 1 and 2).//

Corollary 1 $\lambda e\vec{x}_k\vec{\alpha}_l.\{e\}(\vec{x}_k,\vec{\alpha}_l)$ is partial recursive.

Corollary 2   The class of semi-recursive relations is closed under the substitutions and manipulations of variables mentioned in Theorem 1.

     Proof Let $R(\vec{x},y,\vec{z},\vec{\alpha})$ be semi-recursive, namely $R = \text{dom}(F)$, where $F = \lambda\vec{x}y\vec{z}\,\vec{\alpha}.\{e\}(\vec{x},y,\vec{z},\vec{\alpha})$ for some $e \in \mathcal{N}$. Let $\lambda\vec{w}\vec{\alpha}.G(\vec{w},\vec{\alpha})$ be partial recursive. Then $R(\vec{x},G(\vec{w},\vec{\alpha}),\vec{z},\vec{\alpha}) = \text{dom}(\lambda\vec{x}\vec{w}\,\vec{z}\,\vec{\alpha}.F(\vec{x},G(\vec{w},\vec{\alpha}),\vec{z},\vec{\alpha})).//$

     Note: We shall soon see that the operations of introducing dummy type 1 variables and substitution of "partial recursive" type 1 objects for type 1 variables is legitimate.

     The next theorem plays an important role in Recursion Theory. We shall presently employ it as a stepping stone toward establishing a claim earlier made, that $\mathcal{P}$ can be characterized by a small set of *initial functions,* composition, *S-m-n* and *universal function(al)* theorems. Of course, this latter characterization we took as a *definition* for the partial recursive functionals.

Theorem 2 (*Kleene's Second Recursion Theorem*)

If $\lambda a \vec{x}_k \vec{\alpha}_l . F(a, \vec{x}_k, \vec{\alpha}_l)$ is partial recursive, then there is an e such that $\{e\}(\vec{x}_k, \vec{\alpha}_l) = F(e, \vec{x}_k, \vec{\alpha}_l)$ for all $\vec{x}_k, \vec{\alpha}_l$.

Proof (Essentially, Kleene.)

$S^1 = \lambda a u \vec{x}_k \vec{\alpha}_l . \langle 2, k, l, a, u \rangle$ is a partial recursive functional (rank $(k + 2, l)$) by Definition 1 (Ω_0, last clause) and Theorem 1. By Theorem 1 again, there is a b such that $\{b\}(a, \vec{x}_k, \vec{\alpha}_l) = F(S^1(a, a, \vec{x}_k, \vec{\alpha}_l), \vec{x}_k, \vec{\alpha}_l)$ for all $a, \vec{x}_k, \vec{\alpha}_l$.

Take $e = \langle 2, k, l, b, b \rangle$. Indeed, $\{e\}(\vec{x}_k, \vec{\alpha}_l) = \{\langle 2, k, l, b, b \rangle\}(\vec{x}_k, \vec{\alpha}_l) = \{b\}(b, \vec{x}_k, \vec{\alpha}_l) = F(S^1(b, b, \vec{x}_k, \vec{\alpha}_l), \vec{x}_k, \vec{\alpha}_l) = F(e, \vec{x}_k, \vec{\alpha}_l).//$

Note: The reader should observe the similarity between the above proof and the Gödel diagonalization which obtained the arithmetic statement "I am not a theorem".

In particular, S^1 is Gödel's substitution function, which we called f in the proof of his first and second imcompleteness theorems in the previous chapter (see the proof of Theorem 9.6.11 and the footnote arising from the discussion following Theorem 9.6.1).

Corollary Let $F = \{\tilde{f}\}$. Then e in Theorem 2 can be chosen so that $F(e, \vec{x}_k, \vec{\alpha}_l) \downarrow$ $\Rightarrow \| e, \vec{x}_k, \vec{\alpha}_l \| > \| \tilde{f}, e, \vec{x}_k, \vec{\alpha}_l \|$.

Proof Refer to the proof of Theorem 2. Let $S^1 = \{\tilde{s}\}$, and $F(e, \vec{x}_k, \vec{\alpha}_l) \downarrow$.

Using clauses (a) and (b) of Definition 1, b can be chosen so that $\| b, a, \vec{x}_k, \vec{\alpha}_l \| \geq \| \tilde{f}, \{\tilde{s}\}(a, a, \vec{x}_k, \vec{\alpha}_l), \vec{x}_k, \vec{\alpha}_l \|$ for all $a, \vec{x}_k, \vec{\alpha}_l.$(*) Recalling that $e = \{\tilde{s}\}(b, b, \vec{x}_k, \vec{\alpha}_l)$ for all $\vec{x}_k, \vec{\alpha}_l$ we have in particular [due to $F(e, \vec{x}_k, \vec{\alpha}_l) \downarrow$]

$$\| b, b, \vec{x}_k, \vec{\alpha}_l \| > \| \tilde{f}, e, \vec{x}_k, \vec{\alpha}_l \| \tag{1}$$

By clause (b) of Definition 1 on the other hand,

$$\| \{\tilde{s}\}(b, b, \vec{x}_k, \vec{\alpha}_l), \vec{x}_k, \vec{\alpha}_l \| > \| b, b, \vec{x}_k, \vec{\alpha}_l \| \text{ (Why not } \geq ?)$$

that is

$$\| e, \vec{x}_k, \vec{\alpha}_l \| > \| b, b, \vec{x}_k, \vec{\alpha}_l \|$$

and the claim follows from (1).//

(*)"$\geq$" allows for the possibility $\{\tilde{f}\}(\{\tilde{s}\}(a, a, \vec{x}_k, \vec{\alpha}_l), \vec{x}_k, \vec{\alpha}_l) \uparrow$.

Proposition 1   The class of partial recursive functionals is closed under definition by (recursive) cases.

Proof Let the predicates $R_i(\vec{x}_m, \vec{\alpha}_n)$, $i = 1, \ldots, k$ be recursive and mutually exclusive. Let $\lambda \vec{x}_m \vec{\alpha}_n . G_i(\vec{x}_m, \vec{\alpha}_n)$, $i = 1, \ldots, k + 1$ be partial recursive and F be defined by

$$F(\vec{x}_m, \vec{\alpha}_n) = \begin{cases} G_1(\vec{x}_m, \vec{\alpha}_n) \text{ if } R_1(\vec{x}_m, \vec{\alpha}_n) \\ \quad \vdots \\ G_k(\vec{x}_k, \vec{\alpha}_n) \text{ if } R_k(\vec{x}_m, \vec{\alpha}_n) \\ G_{k+1}(\vec{x}_m, \vec{\alpha}_n) \textbf{ otherwise} \end{cases}$$

Let χ_i, $i = 1, \ldots, k$ be the characteristic functions of R_i, $i = 1, \ldots, k$ respectively.

The functional $C = \lambda \vec{y}_{k+1} \vec{z}_k \vec{x}_m \vec{\alpha}_n . \text{if } z_1 = 0 \textbf{ then } y_1 \textbf{ else if } z_2 = 0 \textbf{ then } y_2 \textbf{ else} \ldots \textbf{ if } z_k = 0 \textbf{ then } y_k \textbf{ else } y_{k+1}$ is partial recursive. F is obtained from C by substituting G_i into $y_i (i = 1, \ldots, k + 1)$ and χ_i into $z_i (i = 1, \ldots, k)$ [Theorem 1].//

Theorem 3   The class of partial recursive functionals is closed under *unbounded search*, that is, if $\lambda y \vec{x} \vec{\alpha} . G(y, \vec{x}, \vec{\alpha})$ is partial recursive and if for all $\vec{x}, \vec{\alpha}$

$$F(\vec{x}, \vec{\alpha}) = \begin{cases} m \text{ if } G(m, \vec{x}, \vec{\alpha}) = 0 \ \& \ (\forall y)_{<m}(\exists n)\,[n > 0 \ \& \ G(y, \vec{x}, \vec{\alpha}) = n] \\ \uparrow \text{ if no such } m \text{ exists} \end{cases}$$

then F is partial recursive.

Proof (See Lemma 6.4.4 for comparison.)
Define H for all $y, \vec{x}, \vec{\alpha}$ by

$$H(y, \vec{x}, \vec{\alpha}) = \begin{cases} y \text{ if } G(y, \vec{x}, \vec{\alpha}) = 0 \\ H(y + 1, \vec{x}, \vec{\alpha}) \textbf{ otherwise} \end{cases}$$

Since obviously $F = \lambda \vec{x} \vec{\alpha} . H(0, \vec{x}, \vec{\alpha})$, it suffices to prove H partial recursive.

To this end, let a and b in $\mathcal{N}$ be such that $\{a\}(e, y, \vec{x}, \vec{\alpha}) = y$ and $\{b\}(e, y, \vec{x}, \vec{\alpha}) = \{e\}(y + 1, \vec{x}, \vec{\alpha})(*)$ for all $e, y, \vec{x}, \vec{\alpha}$.

(*)$a = \langle 0, k + 2, l, 1, 2 \rangle$ will do, where F has rank (k, l). Existence of b follows from Corollary 1 of Theorem 1 by substitution of the functional $\{\langle 0, k + 1, l, 2, 1 \rangle\}$ for the first argument.

We want, with the help of Theorem 2, to "force" e to become an index of H.

Let $C = \lambda uwvey\vec{x}\,\vec{\alpha}.\textbf{if } u = 0 \textbf{ then } w \textbf{ else } v$. Clearly, C is partial recursive.

Let $L(e,y,\vec{x},\vec{\alpha}) = C(G(y,\vec{x},\vec{\alpha}),\{a\}(e,y,\vec{x},\vec{\alpha}), \{b\}(e,y,\vec{x},\vec{\alpha}),e,y,\vec{x},\vec{\alpha})$. By Theorem 1, L is partial recursive. By Theorem 2, there is an $\tilde{e}$ such that $\{\tilde{e}\}(y,\vec{x},\vec{\alpha}) = L(\tilde{e},y,\vec{x},\vec{\alpha})$ for all $y,\vec{x},\vec{\alpha}$. $H = \{\tilde{e}\}.//$

Corollary $\lambda x.x \doteq 1$ and $\lambda x\vec{\alpha}.x \doteq 1$ are partial recursive.

Proof

$$\lambda x\vec{\alpha}.x \doteq 1 = \lambda x\vec{\alpha}.(\mu y)G(y,x,\vec{\alpha})$$

where

$$G = \lambda yx\vec{\alpha}.\textbf{ if } x = 0 \textbf{ then } 0 \textbf{ else if } x = F(y,\vec{\alpha}) \textbf{ then } 0 \textbf{ else } 1$$

and

$$F = \lambda x\vec{\alpha}.x + 1.//$$

Theorem 4   The class of partial recursive functionals is closed under primitive recursion.

Proof Let $\lambda \vec{x}\,\vec{\alpha}.H(\vec{x},\vec{\alpha})$ and $\lambda yz\vec{x}\,\vec{\alpha}.G(y,z,\vec{x},\vec{\alpha})$ be partial recursive and consider F defined for all $\vec{x},\vec{\alpha}$ by

$$\begin{cases} F(0,\vec{x},\vec{\alpha}) = H(\vec{x},\vec{\alpha}) \\ F(y+1,\vec{x},\vec{\alpha}) = G(y,F(y,\vec{x},\vec{\alpha}),\vec{x},\vec{\alpha}) \end{cases}$$

Alternatively,

$$F(y,\vec{x},\vec{\alpha}) = \begin{cases} H(\vec{x},\vec{\alpha}) \textbf{ if } y = 0 \\ G(y \doteq 1,F(y \doteq 1,\vec{x},\vec{\alpha}),\vec{x},\vec{\alpha}) \textbf{ otherwise} \end{cases}$$

As before, we try to fit $\{e\}(y \doteq 1,\vec{x},\vec{\alpha})$ into F's requirements. By Corollary 1 (Theorem 1), substitution, and the previous corollary $\lambda ey\vec{x}\,\vec{\alpha}.\{e\}(y \doteq 1,\vec{x},\vec{\alpha})$ is partial recursive.

$$\text{Let } L(e,y,\vec{x},\vec{\alpha}) = \begin{cases} H(\vec{x},\vec{\alpha}) \textbf{ if } y = 0 \\ G(y \div 1, \{e\}(y \div 1,\vec{x}\,\alpha),\vec{x},\vec{\alpha}) \textbf{ otherwise} \end{cases}$$

for all $e,y,\vec{x},\vec{\alpha}$. By substitution and definition by cases, L is partial recursive, hence for some $\tilde{e}$ (Theorem 2) $\{\tilde{e}\}(y,\vec{x},\vec{\alpha}) = L(\tilde{e},y,\vec{x},\vec{\alpha})$ for all $\vec{x},y,\vec{\alpha}$. Clearly, $\{\tilde{e}\} = F.//$

Corollary 1 $\mathcal{PR}$ is a subclass of the partial recursive functionals.

Corollary 2 $\mathcal{P}$ is a subclass of the partial recursive functionals.

Proof See Definition 3.1.3.//

Corollary 3   The class of partial recursive functionals is closed under course-of-values recursion.

Corollary 4   The class of partial recursive functionals is closed under bounded search.

Proof Problem 5.//

Note: One may define the class of *primitive recursive functionals* by adapting the definition of $\mathcal{PR}$ to

(1)   allow $\vec{\alpha}$-arguments in the initial functions,

(2)   add $Ev_{i,j}^{k,l}$ as an initial functional.

The reader is encouraged to develop properties of this class as far as possible, and in particular prove that this class is a *proper* subclass of the class of partial recursive functionals.

It is usual for primitive recursive functionals to be defined for *total* type 1 inputs only, as otherwise we have the awkward situation that, e.g., $\lambda x\alpha.\alpha(x)$ is primitive recursive but not recursive (of course it *is* partial recursive). We shall not make any adjustments here as we shall not employ primitive recursive functionals in the sequel.

We now show that a *subclass* of the class of partial recursive functionals is closed under the operations of *identifying, permuting* and *extending* (the list of) type 1 arguments. That the class of partial recursive functionals is so closed will be seen in §10.5.

Definition 4 $\mathcal{PR}^{(\alpha)}$ is the *smallest* class of functionals containing the initial functions

$$U_i^{k,l} = \lambda \vec{x}_k\vec{\alpha}_l.x_i, \; S_i^{k,l} = \lambda \vec{x}_k\vec{\alpha}_l.x_i + 1$$

and

$$C_s^{k,l} = \lambda \vec{x}_k \vec{\alpha}_l.s \text{ for all } k,l,s \text{ and } 1 \le i \le k$$

and closed under *composition* and *primitive recursion.//*

     Note: $\vec{\alpha}$ is playing no role in "computing" values of $\mathcal{PR}^{(\alpha)}$ functionals. Indeed,

Proposition 2 $F \in \mathcal{PR}^{(\alpha)}$ iff, for some $f \in \mathcal{PR}, F = \lambda \vec{x}\vec{\alpha}.f(\vec{x})$.

     Proof Induction on the formation of $\mathcal{PR}^{(\alpha)}$ and $\mathcal{PR}$. (See Problem 6.)//

Proposition 3 $\mathcal{PR}^{(\alpha)}$ is a subclass of the primitive recursive, and hence partial recursive functionals. Moreover, $\mathcal{PR} \subset \mathcal{PR}^{(\alpha)}$ and $\mathcal{PR}_* \subset \mathcal{PR}_*^{(\alpha)}$.

     Proof Trivial.//

Proposition 4   The functionals and predicates of $\mathcal{PR}^{(\alpha)}$ are closed under the operations of *identifying, permuting* and *extending* the list of type 1 variables.

     Proof Induction with respect to $\mathcal{PR}^{(\alpha)}$ (Problem 7).//

     Note: From now on we shall freely introduce, with no special mention, dummy $\vec{\alpha}$-variables to primitive recursive functions/predicates.

§10.3 SELECTION THEOREM

     In this section we prove a very powerful tool for dealing with semi-recursive predicates. It proves particularly convenient in the context of our approach, where we allow *partial* type 1 inputs in functionals.

     When only *total* type 1 inputs are allowed in Ordinary Recursion Theory, proving the selection theorem is very easy (see proof of Corollary 3 of Theorem 8.1.2) via the normal form theorems. In recursion in higher types, on the other hand, (inputs of type 2 or higher) no normal form theorems are available, and the selection theorem is called upon to prove, among other things, closure of semi-recursive predicates under $\vee$, $(\exists y)$ and also to prove that R is recursive iff both R and $\neg R$ are semi-recursive.

Theorem 1   (*Selection Theorem*) For each k,l, there is a partial recursive functional $\mathrm{Sel}^{k,l}$ of rank $(k + 1,l)$ such that

(i) $(\exists y)\{a\}(y,\vec{x}_k,\vec{\alpha}_l)\!\downarrow \; \Longleftrightarrow \; \mathrm{Sel}^{k,l}(a,\vec{x}_k,\vec{\alpha}_l)\!\downarrow$

(ii) $(\exists y)\{a\}(y,\vec{x}_k,\vec{\alpha}_l)\!\downarrow \; \Longrightarrow \; \{a\}(\mathrm{Sel}^{k,l}(a,\vec{x}_k,\vec{\alpha}_l),\vec{x}_k,\vec{\alpha}_l)\!\downarrow.$

For the proof we shall need a technical result known as the *Ordinal Comparison Theorem.*(*) (Gandy/Moschovakis.) Actually, we shall present and prove a simplified version brought to measure for our requirements.

Definition 1   Let $\mathscr{E}^{\vec{\alpha}_l}$ stand for $\{\langle e,\vec{x}\rangle \mid \{e\}(\vec{x},\vec{\alpha}_l)\!\downarrow\}$. Set $\|\langle e,\vec{x}\rangle\|_{\vec{\alpha}_l} = \|e,\vec{x},\vec{\alpha}_l,y\|$ (see Definition 10.2.2) whenever $(\exists y)(e,\vec{x},\vec{\alpha}_l,y) \in \Omega$ (recall that all functionals $\{e\}$ are single-valued, by Lemma 10.2.1) $\|\langle e,\vec{x}\rangle\|_{\vec{\alpha}_l} = \infty$ otherwise. We shall write $\|\ \|$ instead of $\|\ \|_{\vec{\alpha}_l}$ for convenience.//

Theorem 0   (*Ordinal Comparison Theorem*) There is a partial recursive functional $\lambda x y \vec{\alpha}_l.H(x,y,\vec{\alpha}_l)$ such that

(i)   if $x \in \mathscr{E}^{\vec{\alpha}_l}$ and $\|x\| \leq \|y\|$, then $H(x,y,\vec{\alpha}_l) = 0$

(ii)   if $y \in \mathscr{E}^{\vec{\alpha}_l}$ and $\|y\| < \|x\|$, then $H(x,y,\vec{\alpha}_l) = 1$

Proof (Essentially that in Hinman [5], adapted to our setting).
We define a partial recursive functional F by cases: $F(e,x,y,\vec{\alpha}_l)$ is defined in terms of H. Then, using the recursion theorem, e is forced to become an index of H, so that H becomes $\lambda x y \vec{\alpha}_l.F(\tilde{e},x,y,\vec{\alpha}_l).$(†)
There is, basically, one case for each pair (r,s), where $0 \leq r, s \leq 3$ are $(a)_0$ and $(b)_0$ respectively whenever $x = \langle a,\vec{m}\rangle$, $y = \langle b,\vec{n}\rangle$.
Define first the following relations/functions in $\mathscr{PR}_*/\mathscr{PR}$.

$$\mathrm{Seq}(u) \equiv (\forall i)_{<\mathrm{length}(u)} p_i \mid u$$

$$x * y = x \cdot \Pi_{i<\mathrm{length}(y)} \, p_{i+\mathrm{length}(x)}^{(y)_i+1}$$

$$\mathrm{proper}(u) \equiv (\exists k,c,i,j,x,y,\tau,\upsilon,w,z)_{\leq u}\mathrm{Seq}(x) \;\&\; \mathrm{Seq}(y) \;\&$$

$$k = \mathrm{length}(x) \;\&\; \{u = 2^{\langle 0,k,l,0,c\rangle+1} * x \vee u = 2^{\langle 0,k,l,1,i\rangle+1} * x \;\&$$

(*)The "ordinals" compared are *lengths of computations*, which in the present setting (Ordinary Recursion Theory) are natural numbers.

(†)More specifically, this process defines H "recursively" in terms of itself. The idea is that $\|x\| \leq \|y\|$ iff for any i.s. x' of x there is an i.s. y' of y such that $\|x'\| \leq \|y'\|$, and $\|x\| < \|y\|$ iff for *some* i.s. y' of y and *all* i.s. x' of x, $\|x'\| < \|y'\|$.

$$1 \le i \le k \lor u = 2^{\langle 0,k,l,2,i \rangle +1} * x \;\&\; 1 \le i \le k \lor u = 2^{\langle 0,k,l,3,i,j \rangle +1} * x$$

$$\&\; 1 \le i \le k \;\&\; 1 \le j \le l \lor u = 2^{\langle 0,k+4,l,4 \rangle +1} * \langle \tau,v,w,z \rangle * x \lor$$

$$u = 2^{\langle 0,k+\text{length}(y),l,5 \rangle +1} * y * x \lor (\exists a,b)_{\le u}[\text{Seq}(b) \;\&\; u =$$

$$2^{\langle 1,k+1+\text{length}(b),l,\text{length}(b) \rangle +1} * 2^{a+1} * b * x] \lor u = 2^{\langle 2,k,l,c \rangle *y+1} * x \lor$$

$$u = 2^{\langle 3,k+1,l \rangle +1} * 2^{c+1} * x\}$$

proper(u) "says", intuitively, that u has the *proper form* $\langle a,\vec{m} \rangle$ to *potentially* (but not necessarily) be in $\mathcal{E}^{\vec{\alpha}_l}$.

Now we carry on the definition of F:

As remarked earlier, *instead of H we shall use* $\{e\}(x,y,\vec{\alpha}_l)$ *in the definition of F with the aim of fixing e (recursion theorem) so that* $H = \{e\}$.

Case 1. $\neg$proper(y). Set $F(e,x,y,\vec{\alpha}_l) = 0$.

Case 2. proper(y) & $\neg$proper(x). Set $F(e,x,y,\vec{\alpha}_l) = 1$.

Case 3. proper(x) & proper(y). We consider subcases.

$$((x)_0)_0 = 0. \text{ Set } F(e,x,y,\vec{\alpha}_l) = 0$$

$$((x)_0)_0 \ge 1 \;\&\; ((y)_0)_0 = 0. \text{ Set } F(e,x,y,\vec{\alpha}_l) = 1$$

$$((x)_0)_0 = 1 \;\&\; ((y)_0)_0 = 1$$

Now,

$$x = \langle \langle 1,k + m + 1,l,m \rangle,a,\vec{b}_m,\vec{x}_k \rangle$$

and

$$y = \langle \langle 1,k' + m' + 1,l,m' \rangle,c,\vec{d}_{m'},\vec{y}_{k'} \rangle$$

for appropriate $k,m,k',m',a,c,\vec{b}_m$, etc.

The *immediate subcomputations*(*) of x are (see Definition 10.2.2) $\langle b_i,\vec{x}_k \rangle$, $i = 1, \ldots, m$ and $\langle a,\{b_1\}(\vec{x}_k), \ldots, \{b_m\}(\vec{x}_k) \rangle$ [of course, one or all of $\{b_i\}(\vec{x}_k)$ might be divergent]; similarly for y, the immediate subcomputations are $\langle d_i,\vec{y}_{k'} \rangle$, $i = 1, \ldots, m'$ and $\langle c,\{d_1\}(\vec{y}_{k'}), \ldots, \{d_{m'}\}(\vec{y}_{k'}) \rangle$. For convenience, let us call all these subcomputations in the order they were mentioned z_i ($i = 1, \ldots, m$), z_0, w_i ($i = 1, \ldots, m'$), w_0. $F(e,x,y,\vec{\alpha}_l)$ is set according to the following "flowchart":

(*)Suppressing the $\vec{\alpha}_l$ and "output" parts.

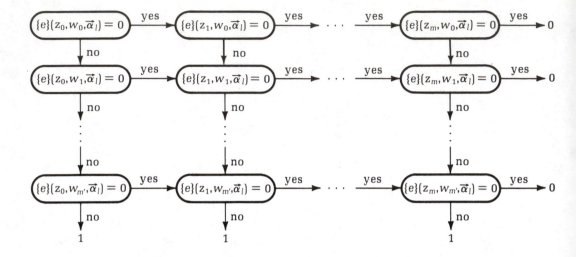

The above can be implemented using $\lambda uv.\textbf{if } u = v \textbf{ then } 0 \textbf{ else } 1$, which is partial recursive, and substitution. For example, when $m = m' = 1$ then

$$
\begin{aligned}
F(e,x,y,\vec{\alpha}_l) = \ &\textbf{if } \{e\}(z_0,w_0,\vec{\alpha}_l) = 0 \textbf{ then} \\
&\quad \textbf{if } \{e\}(z_1,w_0,\vec{\alpha}_l) = 0 \textbf{ then } 0 \\
&\quad \textbf{else if } \{e\}(z_1,w_1,\vec{\alpha}_l) = 0 \textbf{ then } 0 \textbf{ else } 1 \\
&\textbf{else if } \{e\}(z_0,w_1,\vec{\alpha}_l) = 0 \textbf{ then} \\
&\quad \textbf{if } \{e\}(z_1,w_1,\vec{\alpha}_l) = 0 \textbf{ then } 0 \textbf{ else } 1 \\
&\textbf{else } 1
\end{aligned}
$$

Of course, rather than z_i, w_j, one uses appropriate partial recursive functionals of x,y. For example,

$$
z_0 = \langle a, \{b_1\}(\vec{x}_k,\vec{\alpha}_l), \ldots, \{b_m\}(\vec{x}_k,\vec{\alpha}_l) \rangle
$$

where

$$
a = (x)_1,\ b_i = (x)_{i+1},\ i = 1, \ldots, m,
$$

$$
x_i = (x)_{i+m+1},\ i = 1, \ldots, k,\ m = ((x)_0)_3,
$$

$$
k = ((x)_0)_1 \div (m + 1)
$$

Note: At this stage, we encounter a difficulty. $\{b_1\}(\vec{x}_k,\vec{\alpha}_l)$, for example, is $\{(x)_2\}((x)_{((x)_0)_3+2}, (x)_{((x)_0)_3+3}, \ldots, (x)_{\text{length}(x) \div 1},\vec{\alpha}_l)$ which has a *variable number* of arguments. It is indeed partial recursive as a functional of arguments x and $\vec{\alpha}_l$;

we postpone proof of this fact immediately after the proof of this theorem (Lemma 0). End of **Note**.

The remaining subcases are

$$(((x)_0)_0,((y)_0)_0) \in \{(1,2),(1,3),(2,1),(2,2),(2,3),(3,3)\}$$

We sample the last, (3,3), and leave the rest to the reader (the most complicated ones are those that involve a 1-component, of course).

$$((x)_0)_0 = 3 \ \& \ ((y)_0)_0 = 3$$

Now, $x = \langle\langle 3,k + 1,l\rangle,a,\vec{x}_k\rangle$, $y = \langle\langle 3,k' + 1,l\rangle,b,\vec{y}_{k'}\rangle$ for appropriate $k,a,\vec{x}_k,b,k',\vec{y}_{k'}$. i.s. of those are $\langle a,\vec{x}_k\rangle$ and $\langle b,\vec{y}_{k'}\rangle$ respectively.

Thus, set $F(e,x,y,\vec{\alpha}_l) = \{e\}(\langle a,\vec{x}_k\rangle,\langle b,\vec{y}_{k'}\rangle,\vec{\alpha}_l)$. Since F is defined through partial and primitive recursive functionals, **if-then-else** and substitution, it is partial recursive. Thus there is an $\tilde{e}$ such that $\{\tilde{e}\}(x,y,\vec{\alpha}_l) = F(\tilde{e},x,y,\vec{\alpha}_l)$ for all $x,y,\vec{\alpha}_l$, hence set $H = \{\tilde{e}\}$.

We now argue that H works as required.

(0)   If both $\|x\|$ and $\|y\|$ are ∞ then the behavior of H is irrelevant.

(1) $\neg\text{proper}(y)$. Then $\|y\|$ is ∞. Hence $x \in \mathcal{E}^{\vec{\alpha}_l}$ implies $\|x\| \le \|y\|$ and $H(x,y,\vec{\alpha}_l) = 0$ by Case 1. This establishes (i).

(2) $\text{proper}(y) \ \& \ \neg\text{proper}(x)$. Then $\|x\|$ is ∞.

(i) in the theorem statement cannot apply ($x \notin \mathcal{E}^{\vec{\alpha}_l}$).

(ii) checks, since $H(x,y,\vec{\alpha}_l) = 1$.

(3)   Set now $n = \min(\|x\|,\|y\|)$ and proceed by induction on n over the subcases of Case 3.

$n = 0$ (i) $x \in \mathcal{E}^{\vec{\alpha}_l} \ \& \ \|x\| \le \|y\|$ implies $((x)_0)_0 = 0$, hence $H(x,y,\vec{\alpha}_l) = 0$

(ii) $y \in \mathcal{E}^{\vec{\alpha}_l} \ \& \ \|y\| < \|x\|$ implies $((y)_0)_0 = 0$ and either $\neg\text{proper}(x)$ or $((x)_0)_0 \ge 1$. In either case $H(x,y,\vec{\alpha}_l) = 1$.

Assume now that H satisfies (i), (ii) for $0 < n < k$.

Let $n = k$.

We consider the case $((x)_0)_0 = ((y)_0)_0 = 1$ and leave the rest to the reader, as an easy but tedious exercise.

(i) Let $x \in \mathcal{E}^{\vec{\alpha}_l} \ \& \ \|x\| \le \|y\|$. Refer to the definition of F in this case.

It must be $z_i \in \mathcal{E}^{\vec{\alpha}_l}$, hence $\|z_i\| < \|x\|$ for $i = 0, \ldots, m$. Also, $\|w_j\| \le \|y\|$, $j = 0, \ldots, m'$ ("$\le$" here, to allow for the possibility that $y \notin \mathcal{E}^{\vec{\alpha}_l}$, in which case at least *one* i.s. w_i is divergent).

By induction hypothesis, for *all* i,j

(iii) $\|z_i\| \le \|w_j\| \Rightarrow H(z_i,w_j,\vec{\alpha}_l) = 0$ ($z_i \in \mathcal{E}^{\vec{\alpha}_l}$ is already true)

(iv) $\|w_j\| < \|z_i\| \Rightarrow H(z_i,w_j,\vec{a}_l) = 1$ ($\|w_j\| < \|z_i\|$ implies that $w_j \in \mathcal{E}^{\vec{\alpha}_l}$, since $\|z_i\| \in \mathcal{N}$).

(v) $H(z_i,w_j,\vec{\alpha}_l)\downarrow$ [if $\|z_i\| \le \|w_j\|$ by (iii), if $\|w_j\| < \|z_i\|$ by (iv)]

Now, $\|x\| \le \|y\| \Rightarrow (\forall i)(\exists j)\|z_i\| \le \|w_j\| \Rightarrow (\exists j_0, \ldots, j_m)(\forall i)\|z_i\| \le \|w_{j_i}\| \Rightarrow (\exists j_0, \ldots, j_m)(\forall i)H(z_i,w_{j_i},\vec{\alpha}_l) = 0$. Moreover, assume for each i that j_i is the smallest satisfying $\|z_i\| \le \|w_{j_i}\|$.

By (v), H is defined for *all* the arguments involved in the test-boxes of the flowchart and the exit will be through a 0-value following a computation path $(0,0)[\xrightarrow{\text{no}} \ldots \xrightarrow{\text{no}}] (0,j_0) \xrightarrow{\text{yes}} (1,j_0)[\xrightarrow{\text{no}} \ldots \xrightarrow{\text{no}}](1,j_1) \xrightarrow{\text{yes}} (2,j_1)[\xrightarrow{\text{no}} \ldots \xrightarrow{\text{no}}] \ldots (m-1,j_{m-1}) \xrightarrow{\text{yes}} (m,j_{m-1})[\xrightarrow{\text{no}} \ldots \xrightarrow{\text{no}}] (m,j_m) \xrightarrow{\text{yes}} 0$, where we attached the "coordinate" (i,j) to the box involving $\{e\}(z_i,w_j,\vec{\alpha}_l) = 0$ and $[\xrightarrow{\text{no}} \ldots \xrightarrow{\text{no}}]$ means that "$\xrightarrow{\text{no}} \ldots \xrightarrow{\text{no}}$" may or may not be there.

(ii) Let next $y \in \mathcal{E}^{\vec{\alpha}_l}$ & $\|y\| < \|x\|$. Hence $(\exists i)(\forall j)\|w_j\| < \|z_i\|$, that is $(\exists i)(\forall j)H(z_i,w_j,\vec{\alpha}_l) = 1$.

(v) is still true, since $w_j \in \mathcal{E}^{\vec{\alpha}_l}$, $j = 0, \ldots, m'$, thus for all i,j $\|z_i\| \le \|w_j\|$ or $\|w_j\| < \|z_i\|$ and either (iii) is applicable ($\|z_i\| \le \|w_j\| \Rightarrow z_i \in \mathcal{E}^{\vec{\alpha}_l}$) or (iv). Thus, an exit from the flowchart defining F *will* take place. This time, the exit can only be through a 1, since there cannot be an exit to the *right* of column i, corresponding to z_i.//

Corollary   If $x \in \mathcal{E}^{\vec{\alpha}_l}$ or $y \in \mathcal{E}^{\vec{\alpha}_l}$ then $H(x,y,\vec{\alpha}_l)\downarrow$.

Proof If $x \in \mathcal{E}^{\vec{\alpha}_l}$ but $y \notin \mathcal{E}^{\vec{\alpha}_l}$ then $\|x\| \in \mathcal{N}$, hence $\|x\| \le \|y\|$, hence $H(x,y,\vec{\alpha}_l) = 0$. If $y \in \mathcal{E}^{\vec{\alpha}_l}$ then $\|y\| \in \mathcal{N}$.

Thus, if $\|x\| \le \|y\|$, then $x \in \mathcal{E}^{\vec{\alpha}_l}$, hence $H(x,y,\vec{\alpha}_l) = 0$. If $\|y\| < \|x\|$, then $H(x,y,\vec{\alpha}_l) = 1$.//

We now prove a lemma which was used in the proof of Theorem 0.

Lemma 0   There is a partial recursive functional $\lambda x\vec{\alpha}_l.p(x,\vec{\alpha}_l)$ such that $p(x,\vec{\alpha}_l) = z$ iff $x = \langle a,\vec{m} \rangle$ for some $a,\vec{m}$ and $(a,\vec{m},\vec{\alpha}_l,z) \in \Omega$.

Note: Assuming the Lemma 0 proved, then instead of $\{b_1\}(\vec{x}_k,\vec{\alpha}_l)$ in the proof of Theorem 0 we use $p(\langle b_1,\vec{x}_k \rangle,\vec{\alpha}_l)$, that is,

$$p(w,\vec{\alpha}_l),$$

where

$$w = 2^{(x)_2+1} * \Pi_{i<((x)_0)_1 \,\dot-\, [((x)_0)_3+1]} \; p_i^{(x)_{i+((x)_0)_3+2}+1}$$

Proof of Lemma 0

Define $p(x,\vec{\alpha}_l)$ by cases ranging over those of the definition of Ω.

We shall give two and leave the rest to the reader.

Case 1. For each of $j = 1, \ldots, l$, **if** $(\exists k,i,u)_{\le x}\mathrm{Seq}(u)$ & $x = 2^{\langle 0,k,l,3,i,j\rangle+1} * u$ & $k = \mathrm{length}(u)$ & $1 \le i \le k$ **then** set $p(x,\vec{\alpha}_l) = \alpha_j((u)_{i\,\dot-\,1})$. Note that $i = ((x)_0)_4$ and u is a primitive recursive *function* of x.

Case 2. **if** $(\exists k,i,u,a,b)_{\le x}\mathrm{Seq}(u)$ & $\mathrm{Seq}(b)$ & $x = 2^{\langle 1,k+1+\mathrm{length}(b),l,\mathrm{length}(b)\rangle+1} * 2^{a+1}$ $* b * u$ **then** set $p(x,\vec{\alpha}_l) = p(y,\vec{\alpha}_l)$, where

$$\text{(1)} \qquad y = \langle a, p(\langle b_1,\vec{x}_k\rangle,\vec{\alpha}_l), \ldots, p(\langle b_m,\vec{x}_k\rangle,\vec{\alpha}_l)\rangle$$

$$= 2^{a+1} * \Pi_{i<m}\, p_i^{p(\langle b_{i+1},\vec{x}_k\rangle,\vec{\alpha}_l)+1}$$

$$\text{(2)} \qquad a = (x)_1,\; b_i = (b)_{i+1},\, i = 0, \ldots, \mathrm{length}(b) \,\dot-\, 1,\, \langle \vec{x}_k\rangle = u,$$

$$\text{so that } p(\langle b_i,\vec{x}_k\rangle,\vec{\alpha}_l) = p(2^{(b)_{i+1}+1}*u,\vec{\alpha}_l)$$

(3) b and u are primitive recursive functions of x.

By the recursion theorem, p is partial recursive [to see this, instead of p, define in the lefthand side $\lambda ex\vec{\alpha}_l.\tilde{p}(e,x,\vec{\alpha}_l)$ in terms of p, the latter written as $\{e\}(x,\vec{\alpha}_l)$. Then for some $\tilde{e}$, $\lambda x\vec{\alpha}_l.\tilde{p}(\tilde{e},x,\vec{\alpha}_l) = \lambda x\vec{\alpha}_l.\{\tilde{e}\}(x,\vec{\alpha}_l)$. Hence, take $p = \{\tilde{e}\}].//$

We now turn to the proof of the Selection Theorem.

Proof of Theorem 1
Let

$$F(e,y,a,\vec{x}_k,\vec{\alpha}_l) = \begin{cases} y \text{ if } \|\langle a,y,\vec{x}_k\rangle\| < \|\langle e,y+1,a,\vec{x}_k\rangle\| \\ \{e\}(y+1,a,\vec{x}_k,\vec{\alpha}_l) \text{ \textbf{otherwise}} \end{cases}$$

That is,

$$F(e,y,a,\vec{x}_k,\vec{\alpha}_l) = \textbf{if } H(\langle e,y+1,a,\vec{x}_k\rangle,\langle a,y,\vec{x}_k\rangle,\vec{\alpha}_l) = 0$$

$$\textbf{then } \{e\}(y+1,a,\vec{x}_k,\vec{\alpha}_l)$$

$$\textbf{else } y$$

hence F is partial recursive. By the recursion theorem, for some $\tilde{e}$, $F(\tilde{e},y,a,\vec{x}_k,\vec{\alpha}_l) = \{\tilde{e}\}(y,a,\vec{x}_k,\vec{\alpha}_l)$ for all $y,a,\vec{x}_k,\vec{\alpha}_l$.

Set $\mathrm{Sel}^{k,l} = \lambda a \vec{x}_k \vec{\alpha}_l . \{\tilde{e}\}(0,a,\vec{x}_k,\vec{\alpha}_l)$.

[Intuitively, $\mathrm{Sel}^{k,l}(a,\vec{x}_k,\vec{\alpha}_l)$ is the smallest y for which $\{a\}(y,\vec{x}_k,\vec{\alpha}_l)\downarrow$ can be established *by bounding* $\|\langle a,y,\vec{x}_k\rangle\|$ *strictly* by $\|\langle\tilde{e},y+1,a,\vec{x}_k\rangle\|$. Note that we have *not* said that $\mathrm{Sel}^{k,l}(a,\vec{x}_k,\vec{\alpha}_l)$ is the smallest y such that $\{a\}(y,\vec{x}_k,\vec{\alpha}_l)\downarrow$.]

We now check that $\mathrm{Sel}^{k,l}$ works as required. First, let $(\exists y)\{a\}(y,\vec{x}_k,\vec{\alpha}_l)\downarrow$, thus $\{a\}(n,\vec{x}_k,\vec{\alpha}_l)\downarrow$ for some n. By corollary of Theorem 0, $H(\langle\tilde{e},n+1,a,\vec{x}_k\rangle,\langle a,n,\vec{x}_k\rangle,\vec{\alpha}_l)\downarrow$ hence $\{\tilde{e}\}(n,a,\vec{x}_k,\vec{\alpha}_l)\downarrow$. If $n=0$, this means $\mathrm{Sel}^{k,l}(a,\vec{x}_k,\vec{\alpha}_l)\downarrow$; otherwise the same conclusion follows by induction on n, since (corollary of Theorem 0) $H(\langle\tilde{e},n,a,\vec{x}_k\rangle,\langle a,n-1,\vec{x}_k\rangle,\vec{\alpha}_l)\downarrow$, therefore $\{\tilde{e}\}(n-1,a,\vec{x}_k,\vec{\alpha}_l)\downarrow$.

This establishes the $\Rightarrow$-part of (i). The $\Leftarrow$-part of (i) will follow from the proof of (ii).

(ii) We shall prove that $\mathrm{Sel}^{k,l}(a,\vec{x}_k,\vec{\alpha}_l)\downarrow \Rightarrow \{a\}(\mathrm{Sel}^{k,l}(a,\vec{x}_k,\vec{\alpha}_l),\vec{x}_k,\vec{\alpha}_l)\downarrow$, which proves (i) $[\Leftarrow]$ trivially, and along with (i) $[\Rightarrow]$ establishes (ii).

So let, for some $a,\vec{x}_k$, and $\vec{\alpha}_l$, $\mathrm{Sel}^{k,l}(a,\vec{x}_k,\vec{\alpha}_l) = m$, that is $\{\tilde{e}\}(0,a,\vec{x}_k,\vec{\alpha}_l) = m$.

(0) Assume for a minute that $\|\langle a,y,\vec{x}_k\rangle\| \geq \|\langle\tilde{e},y+1,a,\vec{x}_k\rangle\|$ for *all* $y \geq 0$. Then

(1) $\{\tilde{e}\}(y,a,\vec{x}_k,\vec{\alpha}_l) = m$ for *all* $y \geq 0$ since the upper condition in the definition of F never obtains $[\{\tilde{e}\}(0,a,\vec{x}_k,\vec{\alpha}_l) = m$; then it must be equal to $\{\tilde{e}\}(1,a,\vec{x}_k,\vec{\alpha}_l)$, which in turn is $\{\tilde{e}\}(2,a,\vec{x}_k,\vec{\alpha}_l)$, etc. This, *intuitively*, says that $\{\tilde{e}\}(0,a,\vec{x}_k,\vec{\alpha}_l)\uparrow$; a contradiction. Formalities follow.]

(2) If $F = \{\tilde{f}\}$, it can be assumed that $\tilde{f}$ is so chosen that $\|\langle\tilde{f},e,y,a,\vec{x}_k\rangle\| > \|\langle e,y+1,a,\vec{x}_k\rangle\|$ for all $e,y,a,\vec{x}_k,\vec{\alpha}_l$ such that $\{e\}(y+1,a,\vec{x}_k,\vec{\alpha}_l)\downarrow$ and $\|\langle a,y,\vec{x}_k\rangle\| \geq \|\langle e,y+1,a,\vec{x}_k\rangle\|$ $[\{e\}(y+1,a,\vec{x}_k,\vec{\alpha}_l)$ is a subcomputation of $\{\tilde{f}\}(e,y,a,\vec{x}_k,\vec{\alpha}_l)$ under these circumstances. (See Problem 10.)]

Thus, under assumption (0), $\|\langle\tilde{f},\tilde{e},y,a,\vec{x}_k\rangle\| > \|\langle\tilde{e},y+1,a,\vec{x}_k\rangle\|$ for all $y \geq 0$, hence (corollary of Theorem 10.2.2):

(3) $\|\langle\tilde{e},y,a,\vec{x}_k\rangle\| > \|\langle\tilde{e},y+1,a,\vec{x}_k\rangle\|$, for all $y \geq 0$.

Condition (3) implies existence of infinitely many integers between 0 and $\|\tilde{e},0,a,\vec{x}_k,\vec{\alpha}_l\|$; thus assumption (0) must be dropped.

Let then r be the smallest such that

$$\|\langle a,r,\vec{x}_k\rangle\| < \|\langle\tilde{e},r+1,a,\vec{x}_k\rangle\|.$$

Then, $\{a\}(r,\vec{x}_k,\vec{\alpha}_l)\downarrow$. Moreover, $r = \{\tilde{e}\}(r,a,\vec{x}_k,\vec{\alpha}_l) = (*)\{\tilde{e}\}(r-1,a,\vec{x}_k,\vec{\alpha}_l) = \dots = \{\tilde{e}\}(0,a,\vec{x}_k,\vec{\alpha}_l) = m$. That is, $\{a\}(\mathrm{Sel}^{k,l}(a,\vec{x}_k,\vec{\alpha}_l),\vec{x}_k,\vec{\alpha}_l)\downarrow.//$

Corollary If $R(y,\vec{x}_k,\vec{\alpha}_l)$ is semi-recursive, then there is a partial recursive functional Sel_R of rank (k,l) such that

(i) $(\exists y)R(y,\vec{x}_k,\vec{\alpha}_l) \Leftrightarrow \mathrm{Sel}_R(\vec{x}_k,\vec{\alpha}_l)\downarrow$

(ii) $(\exists y)R(y,\vec{x}_k,\vec{\alpha}_l) \Rightarrow R(\mathrm{Sel}_R(\vec{x}_k,\vec{\alpha}_l),\vec{x}_k,\vec{\alpha}_l)$.

(*)For $q < r$, $\{\tilde{e}\}(q,a,\vec{x}_k,\vec{\alpha}_l) = \{\tilde{e}\}(q+1,a,\vec{x}_k,\vec{\alpha}_l)$ as the lower case in the definition of F obtains. Of course, $\{\tilde{e}\}(q,a,\vec{x}_k,\vec{\alpha}_l)\downarrow$ for all $q < r$ as we saw in the proof of (i).

Proof Let $R = \mathrm{dom}(\{a\})$. Take $\mathrm{Sel}_R = \lambda\vec{x}_k\vec{\alpha}_l.\mathrm{Sel}^{k,l}(a,\vec{x}_k,\vec{\alpha}_l).//$

Note: A proof of the selection theorem for functionals that accept type 2 inputs was first given by Gandy, based on the ordinal comparison theorem, also due to him. It is interesting to note that the selection theorem is an effective counterpart to the τ-schema (Axiom 3.5 of the LPC, in Example 9.6.3) $F(s) \Rightarrow F((\tau x)F(x))$.

§10.4 RECURSIVENESS AND SEMI-RECURSIVENESS

The theorems of this section have been obtained in Chapter 3 (§3.6 and Theorem 3.7.2) for functions/predicates on the integers.

Lemma 1   If R of rank (k,l) is recursive then it is semi-recursive.

Proof $R = \mathrm{dom}(F)$, where $F = \lambda\vec{x}_k\vec{\alpha}_l.(\mu y)G(y,\vec{x}_k,\vec{\alpha}_l)$ and $G = \lambda y\vec{x}_k\vec{\alpha}_l.\chi_R(\vec{x}_k,\vec{\alpha}_l)$, where χ_R is the characteristic function of R.//

Theorem 1   *(The Projection Theorem)* If $R(y,\vec{x}_k,\vec{\alpha}_l)$ is semi-recursive, then so is $(\exists y)R(y,\vec{x}_k,\vec{\alpha}_l)$.

Proof $(\exists y)R(y,\vec{x}_k,\vec{\alpha}_l) \equiv R(\mathrm{Sel}_R(\vec{x}_k,\vec{\alpha}_l),\vec{x}_k,\vec{\alpha}_l)$, by corollary of Theorem 10.3.1. The result follows from Corollary 2 of Theorem 10.2.1.//

Corollary   If $R(y,\vec{x}_k,\vec{\alpha}_l)$ is recursive, then $(\exists y)R(y,\vec{x}_k,\vec{\alpha}_l)$ is semi-recursive.

Theorem 2   If $R(\vec{x}_k,\vec{\alpha}_l)$ and $Q(\vec{x}_k,\vec{\alpha}_l)$ are semi-recursive then so are $R(\vec{x}_k,\vec{\alpha}_l) \vee Q(\vec{x}_k,\vec{\alpha}_l)$ and $R(\vec{x}_k,\vec{\alpha}_l)$ & $Q(\vec{x}_k,\vec{\alpha}_l)$.

Proof Let $R = \mathrm{dom}(\{r\})$ and $Q = \mathrm{dom}(\{q\})$.

$\vee$:　Let $f = \lambda x.$ **if** $x = 0$ **then** r **else** s. Then
$R(\vec{x}_k,\vec{\alpha}_l) \vee Q(\vec{x}_k,\vec{\alpha}_l) \equiv (\exists y)\{f(y)\}(\vec{x}_k,\vec{\alpha}_l)\downarrow (*)$

&:　$R(\vec{x}_k,\vec{\alpha}_l)$ & $Q(\vec{x}_k,\vec{\alpha}_l) = \mathrm{dom}(\lambda\vec{x}_k\vec{\alpha}_l.\{r\}(\vec{x}_k,\vec{\alpha}_l) + \{q\}(\vec{x}_k,\vec{\alpha}_l)).//$

(*)f can be viewed as $\lambda y\vec{\alpha}_l.f(y)$ by Propositions 10.2.2–4.

Corollary   Semi-recursive relations are closed under $(\exists y)_{\leq z}$ and $(\forall y)_{\leq z}$.

Proof Let $Q(y,\vec{x}_k,\vec{\alpha}_l)$ be semi-recursive. Then,

$$(\exists y)_{\leq z}\, Q(y,\vec{x}_k,\vec{\alpha}_l) \equiv (\exists y)[y \leq z \,\&\, Q(y,\vec{x}_k,\vec{\alpha}_l)]$$

$\lambda yz.y \leq z$ being primitive recursive, it can be viewed as the $\mathcal{PR}_*^{(\alpha)}$ (hence, semi-recursive) predicate $\lambda yz\vec{x}_k\vec{\alpha}_l.y \leq z$. Also, $\lambda yz\vec{x}_k\vec{\alpha}_l.Q(y,\vec{x}_k,\vec{\alpha}_l)$ is semi-recursive by Theorem 10.2.1.

The claim follows from Theorems 1 and 2. As for closure under $(\forall y)_{\leq z}$,

$$(\forall y)_{\leq z}\, Q(y,\vec{x}_k,\vec{\alpha}_l) \equiv F(z,\vec{x}_k,\vec{\alpha}_l)\!\downarrow$$

where

$$\begin{cases} F(0,\vec{x}_k,\vec{\alpha}_l) = \{q\}(0,\vec{x}_k,\vec{\alpha}_l) \\ F(n+1,\vec{x}_k,\vec{\alpha}_l) = F(n,\vec{x}_k,\vec{\alpha}_l) + \{q\}(n+1,\vec{x}_k,\vec{\alpha}_l) \end{cases}$$

and $Q = \text{dom}(\{q\})$.//

Note: The restriction in the statement of Theorem 2 that R and Q have the same rank is illusory.

We already know this for the type 0 variables (Theorem 10.2.1) and will soon establish it for type 1 variables.

Theorem 3 $R \subset \mathcal{N}^k \times \mathcal{J}^l$ is recursive iff R and $\neg R$ are semi-recursive.

Proof   *only if* part. If R is recursive, then so is $\neg R$ $(\chi_{\neg R} = 1 \doteq \chi_R)$, hence the claim follows from Lemma 1.

if part. Set $Q(t,\vec{x}_k,\vec{\alpha}_l) \equiv R(\vec{x}_k,\vec{\alpha}_l) \,\&\, t = 0 \,\vee\, \neg R(\vec{x}_k,\vec{\alpha}_l) \,\&\, t = 1$. By Theorem 2, and since $\lambda t.t = m \in \mathcal{PR}_*$ implies that $\lambda t\vec{x}_k\vec{\alpha}_l.t = m$ is semi-recursive, we see that Q is semi-recursive.

Clearly, $\chi_R = \text{Sel}_Q\ [Q(\text{Sel}_Q(\vec{x}_k,\vec{\alpha}_l),\vec{x}_k,\vec{\alpha}_l) \equiv (\exists t)Q(t,\vec{x}_k,\vec{\alpha}_l)]$, thus, χ_R is recursive.//

Theorem 4 F, of rank (k,l), is partial recursive iff $\lambda y\vec{x}_k\vec{\alpha}_l.y = F(\vec{x}_k,\vec{\alpha}_l)$ [its "graph"] is semi-recursive.

Proof *only if* part. By substitution in the primitive recursive relation $\lambda yz.y = z$.

if part. Let us denote $\lambda y\vec{x}_k\vec{\alpha}_l.y = F(\vec{x}_k,\vec{\alpha}_l)$ by Q. Then, $(\exists y)Q(y,\vec{x}_k,\vec{\alpha}_l)$ $\Rightarrow Q(\text{Sel}_Q(\vec{x}_k,\vec{\alpha}_l),\vec{x}_k,\vec{\alpha}_l)$; that is, $F(\vec{x}_k,\vec{\alpha}_l)\!\downarrow \,\Rightarrow \text{Sel}_Q(\vec{x}_k,\vec{\alpha}_l) = F(\vec{x}_k,\vec{\alpha}_l)$. Moreover,

$\text{Sel}_Q(\vec{x}_k,\vec{\alpha}_l)\downarrow \;\Rightarrow\; (\exists y)Q(y,\vec{x}_k,\vec{\alpha}_l);$ that is, $F(\vec{x}_k,\vec{\alpha}_l)\downarrow$, hence again $F(\vec{x}_k,\vec{\alpha}_l)$ $= \text{Sel}_Q(\vec{x}_k,\vec{\alpha}_l).//$

§10.5 NORMAL FORM THEOREMS

Note that except for the fourth case in the definition of Ω_0 (Definition 10.2.1) the type 1 input $\vec{\alpha}$ is not used in a convergent computation $(e,\vec{x},\vec{\alpha})$. Moreover, there is a finite number of subcomputations (this number directly related to $\|e,\vec{x},\vec{\alpha}\|$) involved in a proof that $\{e\}(\vec{x},\vec{\alpha})\downarrow$. As a corollary, finitely many times the fourth case in the definition Ω_0 enters in such a proof, thus finitely many values of the form $\alpha_j(x_i)$ are employed.

In what follows, a proof that $(e,\vec{x},\vec{\alpha},n) \in \Omega$ will be coded by a number u.

We define a sequence of predicates/functionals which are (at least) semi-recursive/partial recursive:

(1) $v \in u \equiv$ "v is an element of the sequence coded by u"

$$v \in u \equiv \text{Seq}(u) \;\&\; (\exists i)_{<\text{length}(u)}(v = (u)_i)$$

(2) $v <_u w \equiv$ "$v \in u$ and $w \in u$ and v occurs before w"

$$v <_u w \equiv v \in u \;\&\; w \in u \;\&\; (\exists i,j)_{<\text{length}(u)}(i < j \;\&\; (u)_i = v \;\&\; (u)_j = w)$$

Note that $\lambda u v.v \in u$ and $\lambda u v w.v <_u w$ are in $\mathcal{R}_*$. Seq has been defined in the proof of Theorem 10.3.0.

(3) $Pf^{(l)}(u,\vec{\alpha}_l) \equiv$ "u codes a proof, where the type 1 input is $\vec{\alpha}_l$"

u codes the sequence of relevant computations $(e,\vec{x},\vec{\alpha}_l,y)$ as a sequence of $(e,\vec{x},y)$ (the $\vec{\alpha}_l$-part is removed since it does not vary). Thus,

$$Pf^{(l)}(u,\vec{\alpha}_l) \equiv \text{Seq}(u) \;\&\; (\forall v)_{\leq u}[v \in u \Rightarrow (\exists k,x)_{\leq u} \{\text{Seq}(x)$$

$$\&\; k = \text{length}(x) \;\&\; ((\exists c)_{\leq u}[v = 2^{\langle 0,k,l,0,c\rangle +1} * x * 2^{c+1}] \;\vee$$

$$(\exists i)_{<k}[v = 2^{\langle 0,k,l,1,i+1\rangle +1} * x * 2^{(x)_i+1}(*) \;\vee\; v = 2^{\langle 0,k,l,2,i+1\rangle +1} * x *$$

$$2^{(x)_i+2} \;\vee\; \bigvee_{j-1}^{l}(\dagger)\; v = 2^{\langle 0,k,l,3,i+1,j\rangle +1} * x * 2^{\alpha_j((x)_i)+1}] \;\vee$$

(*)If x codes $\vec{x}_k$, then $(x)_i = x_{i+1}$.
($\dagger$)$\bigvee_{j-1}^{l} Q_j$ is shorthand for $Q_1 \vee Q_2 \vee \ldots \vee Q_l$.

$$(\exists p,q,r,s)_{\leq u}[v = \langle\langle 0,k+4,l,4\rangle,p,q,r,s\rangle * x * 2^{r+1} \,\&\, p = q \,\vee$$

$$v = \langle\langle 0,k+4,l,4\rangle,p,q,r,s\rangle * x * 2^{s+1} \,\&\, p \neq q]$$

$$\vee\, (\exists y)_{\leq u}[\mathrm{Seq}(y) \,\&\, v = 2^{\langle 0,k+\mathrm{length}(y),l,5\rangle+1} * y * x * 2^{y+1}] \,\vee$$

$$(\exists a,b,y,z)_{\leq u}[\mathrm{Seq}(b) \,\&\, \mathrm{Seq}(y) \,\&\, \mathrm{length}(y) = \mathrm{length}(b) \,\&\,$$

$$v = \langle\langle 1,k+1+\mathrm{length}(b),l,\mathrm{length}(b)\rangle,a\rangle * b * x * 2^{z+1} \,\&\,$$

$$2^{a+1} * y * 2^{z+1} <_u v \,\&\, (\forall i)_{<\mathrm{length}(y)}(2^{(b)_i+1} * x * 2^{(y)_i+1} <_u v)] \,\vee$$

$$(\exists a,y,z)_{\leq u}[\mathrm{Seq}(y) \,\&\, v = 2^{\langle 2,\mathrm{length}(y),l,a\rangle * x+1} * y * 2^{z+1} \,\&\,$$

$$2^{a+1} * x * y * 2^{z+1} <_u v] \,\vee\, (\exists a,y)_{\leq u}[v = \langle\langle 3,k+1,l\rangle,a\rangle * x * 2^{y+1} \,\&\,$$

$$2^{a+1} * x * 2^{y+1} <_u v])\}].$$

Note that in defining $Pf^{(l)}$ we required that every $v \in u$ be either a computation with no i.s. or else be a computation whose i.s. appeared *before* v in u.

Clearly $Pf^{(l)}$ is a *semi-recursive* predicate of rank $(1,l)$ by the results of §10.4.(*) If $l = 0$, then clearly $Pf^{(0)}$ is [of rank $(1,0)$] *primitive recursive*.

(4) $T^{(k,l)}(e,\vec{x}_k,u,\vec{\alpha}_l) \equiv$ "u codes a proof of the fact $\{e\}(\vec{x}_k,\vec{\alpha}_l)\downarrow$"

Clearly, $T^{(k,l)}(e,\vec{x}_k,u,\vec{\alpha}_l) \equiv Pf^{(l)}(u,\vec{\alpha}_l) \,\&\, (\exists n)_{\leq u}[\langle e,\vec{x}_k,n\rangle = (u)_{\mathrm{length}(u) \dot- 1}]$

By Theorem 10.4.2, $T^{(k,l)}$ is *semi-recursive* (again, it is in $\mathcal{PR}_*$ *if* $l = 0$).

(5) Set $U \overset{\mathrm{def}}{=} \lambda u.((u)_{\mathrm{length}(u) \dot- 1})_{\mathrm{length}((u)_{\mathrm{length}(u) \dot- 1}) \dot- 1} \dot- 1$.

Clearly, $U \in \mathcal{PR}$, and if u is as in (4), then $U(u) = n$. We now have

Theorem 1 (A weakened form of the *Kleene Normal Form Theorem* due to allowing nontotal $\vec{\alpha}$-inputs). There is a $U \in \mathcal{PR}$ and for each k,l a *semi-recursive* Kleene T-predicate, $T^{(k,l)}$, of rank $(k+2,l)$ such that

(1) $\{e\}(\vec{x}_k,\vec{\alpha}_l)\downarrow$ iff $(\exists u)T^{(k,l)}(e,\vec{x}_k,u,\vec{\alpha}_l)$

(2) $\{e\}(\vec{x}_k,\vec{\alpha}_l) = n$ iff $(\exists u)[T^{k,l}(e,\vec{x}_k,u,\vec{\alpha}_l) \,\&\, U(u) = n]$

Note: If the $\vec{\alpha}$-inputs are restricted to *total* functions, then $T^{(k,l)}$ is

(*) Of course, any reference to $\alpha_j(x_i)$ is a reference to $\{\langle 0,m+k,l,3,m+i,j\rangle\}(\vec{y}_m, \vec{x}_k, \vec{\alpha}_l)$, for any convenient $\vec{y}_m$.

primitive recursive for all k,l. In particular, if $l = 0$ then $T^{(k,0)} \in \mathcal{PR}_*$, and the above theorem says that $\mathcal{P}$ can be characterized by the following items: (i) a small set of *initial functions*, (ii) *composition*, (iii) a *Gödel indexing* satisfying the *S-m-n* [with *S-m-n* functions *in* the system], and *universal function* theorems.

Indeed, we already noted that for $l = 0$, Definitions 10.2.1–10.2.3 define a superset of $\mathcal{P}$. Theorem 1(2) above shows that this inclusion is an equality.

In Axiomatic Recursion Theory one starts with, essentially, (i)–(iii) as axioms. The interested reader will refer to Fenstad [2], or, for an elementary exposition, to Hennie [4].

We shall now sharpen Theorem 1 so that the fact that only finitely many values $\alpha_j(x_i)$ are needed in a convergent computation will be reflected. At the same time, we shall obtain a T-predicate in $\mathcal{PR}_*^{(4)}$ (no $\vec{\alpha}$-inputs).

Definition 1   (Davis [1]) For any $\alpha \in \mathcal{P}(\mathcal{N};\mathcal{N})$ and $u \in \mathcal{N}$, $\alpha \mid u$ denotes the *finite* function $\{(x,y) \in \alpha \mid x \leq u \ \& \ y \leq u\}$. $\vec{\alpha}_l \mid u$ denotes $(\alpha_1 \mid u, \ldots, \alpha_l \mid u)$.//

It follows immediately from the proof of Theorem 1 that for all $e, \vec{x}_k, \vec{\alpha}_l$,

$$\{e\}(\vec{x}_k, \vec{\alpha}_l) = n \ \text{iff} \ (\exists u)(T^{(k,l)}(e, \vec{x}_k, u, \vec{\alpha}_l \mid u) \ \& \ U(u) = n)$$

This statement will now be strengthened:

Let first α be such that $\text{dom}(\alpha) = \{b_1, \ldots, b_m\}$, $m \geq 1$. Then $\langle \alpha \rangle$ stands for $\Pi_{i \leq m} p_{b_i}^{\alpha(b_i)+1}$; $\langle \alpha \rangle = 1$ if $\text{dom}(\alpha) = \varnothing$. Note that unless $\text{dom}(\alpha)$ is an initial segment of $\mathcal{N}$, $\neg\text{Seq}(\langle \alpha \rangle)$. α can be recovered from $\langle \alpha \rangle$:
For all x, $\alpha(x) = (\mu y)(\exp(x, \langle \alpha \rangle) = y + 1)$.(*)

If $\alpha_1, \ldots, \alpha_l$ are all finite, then $[\vec{\alpha}_l]$ stands for $\langle \langle \alpha_1 \rangle, \ldots, \langle \alpha_l \rangle \rangle$. In particular, $[\alpha]$ is $\langle \langle \alpha \rangle \rangle$.

Example 1 $F = \lambda u\alpha . \langle \alpha \mid u \rangle$ is *not* partial recursive.

Indeed from the remark following Definition 1, if F were partial recursive, then it would be *consistent* in the sense that

$$z = \langle \alpha \mid u \rangle \ \& \ \alpha \subset \beta \Rightarrow z = \langle \beta \mid u \rangle.$$

The reader can easily construct for each u an example of α, which violates this last implication. (See Problem 13.)//

Theorem 2   For each $l \geq 0$, there is a primitive recursive predicate $T^{(l)} \subset \mathcal{N}^4$, such that for all $e, \vec{x}_k, \vec{\alpha}_l$

$$\{e\}(\vec{x}_k, \vec{\alpha}_l) = U((\mu u)T^{(l)}(e, \langle \vec{x}_k \rangle, [\vec{\alpha}_l \mid u], u)).$$

(*)Recall that $(u)_i = \exp(i,u) \doteq 1$ for all i,u.

Proof In the proof of Theorem 1, perform the following changes:

(1) Instead of $Pf^{(l)}(u,\vec{\alpha}_l)$ define $\tilde{Pf}^{(l)}(u,s)$ with the same righthand side of "$\equiv$" except

    (1.1) Change the clause $\bigvee_{j=1}^{l} v = 2^{\langle 0,k,l,3,i+1,j\rangle+1} *x*2^{\alpha_j((x)_i)+1}$ into $\bigvee_{j=1}^{l} v = 2^{\langle 0,k,l,3,i+1,j\rangle+1} *x*2^{((s)_{j\doteq1})_{(x)_i}+1} \& p_{(x)_i}|(s)_{j\doteq1}$

    (1.2) Prefix the righthand side of "$\equiv$" in the definition of $\tilde{Pf}^{(l)}$ by "Seq(s) &"

(2) Define $T^{(l)}$ by $T^{(l)}(e,x,s,u) \equiv \tilde{Pf}^{(l)}(u,s) \;\&\; \mathrm{Seq}(x) \;\&\; (\exists n)_{\le u}\,((u)_{\mathrm{length}(u)\doteq1} = 2^{e+1}*x*2^{n+1}).//$

Corollary 1   There is a primitive recursive predicate $T \subset \mathcal{N}^4$ such that for all $e,\vec{x}_k,\vec{\alpha}_l$

$$\{e\}(\vec{x}_k,\vec{\alpha}_l) = U((\mu u)T(e,\langle \vec{x}_k\rangle,[\vec{\alpha}_l|u],u))$$

Proof Refer to the proof of Theorem 2. Define $Pf(u,s)$ in the place of $\tilde{Pf}^{(l)}(u,s)$, making the following changes:

    (1.1) Use $(\exists j)_{<l}v = 2^{\langle 0,k,l,3,i+1,j+1\rangle+1}*x*2^{((s)_j)_{(x)_i}+1} \& p_{(x)_i}|(s)_j$ instead of $\bigvee_{j=1}^{l} v = 2^{\langle 0,k,l,3,i+1,j\rangle+1} *x*2^{((s)_{j\doteq1})_{(x)_i}+1} \& p_{(x)_i}|(s)_{j\doteq1}$

    (1.2) The $\tilde{Pf}$-prefix should be "$(\exists l)_{\le s}l = \mathrm{length}(s) \;\&\; \mathrm{Seq}(s) \;\&\;$" instead of "Seq(s) &" as in the case of $\tilde{Pf}^{(l)}$.

$T(e,x,s,u)$ is $Pf(u,s) \;\&\; \mathrm{Seq}(x) \;\&\; (\exists n)_{\le u}((u)_{\mathrm{length}(u)\doteq1} = 2^{e+1}*x*2^{n+1}).//$

Corollary 2 $R \subset \mathcal{N}^k \times \mathcal{I}^l$ is semi-recursive iff, for some e,

$$R(\vec{x}_k,\vec{\alpha}_l) \equiv (\exists u)T(e,\langle \vec{x}_k\rangle,[\vec{\alpha}_l|u],u).$$

Proof By Corollary 1, dom($\{e\}$) = $(\exists u)T(e,\langle \vec{x}_k\rangle,[\vec{\alpha}_l|u],u)$. Hence, *only if* part. Take e such that $R = \mathrm{dom}(\{e\})$.
if part. $R = \mathrm{dom}(\{e\}).//$

Note: In the *if* part above we relied on the fact that $T(e,\langle \vec{x}_k\rangle,[\vec{\alpha}_l|u],u)$ "says" that u codes the set of subcomputations of $(e,\vec{x}_k,\vec{\alpha}_l)$. We could have used the projection theorem as well, since $T(e,\langle \vec{x}_k\rangle,[\vec{\alpha}_l|u],u) \equiv T^{(k,l)}(e,\vec{x}_k,\vec{\alpha}_l,u)$ hence is semi-recursive.

Definition 2 $(u;s \mid \alpha)$ stands for the predicate

$$\text{length}(s) \leq u + 1 \ \& \ (\forall i)_{\leq u} p_i \mid s \Rightarrow (s)_i = \alpha(i) \ \& \ (s)_i \leq u.$$

$\langle u;s \mid \vec{\alpha}_l \rangle$ stands for $\text{Seq}(s) \ \& \ l = \text{length}(s) \ \& \ \wedge_{j=0}^{l-1} (u;(s)_j \mid \alpha_{j+1}).(*)//$

Note: $(u;s \mid \alpha)$ is a convenient modification of Kleene's $u \mid \alpha$ [9]. Clearly,

(1) $\langle u;s \mid \alpha \rangle \equiv \text{length}(s) = 1 \ \& \ (u;(s)_0 \mid \alpha)$

(2) $s = [\vec{\alpha}_l \mid u] \Rightarrow \langle u;s \mid \vec{\alpha}_l \rangle.$

The opposite implication is obviously false.

Corollary 3

For all $e, \vec{x}_k, \vec{\alpha}_l,$

(i) $\{e\}(\vec{x}_k, \vec{\alpha}_l) = n \equiv (\exists u)[T^{(l)}(e, \langle \vec{x}_k \rangle, (u)_0, (u)_1) \ \& \ \langle (u)_1;(u)_0 \mid \vec{\alpha}_l \rangle \ \&$
       $U((u)_1) = n]$

(ii) $\{e\}(\vec{x}_k, \vec{\alpha}_l) = n \equiv (\exists u)[T(e, \langle \vec{x}_k \rangle, (u)_0, (u)_1) \ \& \ \langle (u)_1;(u)_0 \mid \vec{\alpha}_l \rangle \ \&$
       $U((u)_1) = n]$

Proof Immediate from Theorem 2, Corollary 1, and Definition 2.//

Corollary 4

For each semi-recursive relation R of rank (k,l) there is a primitive recursive relation S of rank $(k + l + 1,0)$ such that

(i) $R(\vec{x}_k, \vec{\alpha}_l) \equiv (\exists u)S(\vec{x}_k, \langle \alpha_1 \mid u \rangle, \ldots, \langle \alpha_l \mid u \rangle, u)$

(ii) $R(\vec{x}_k, \vec{\alpha}_l) \equiv (\exists u)(\exists \vec{s}_l)[S(\vec{x}_k, \vec{s}_l, u) \ \& \ \wedge_{j=1}^{l} (u;s_j \mid \alpha_j)]$

Proof Take $S(\vec{x}_k, \vec{s}_l, u) \equiv T(r, \langle \vec{x}_k \rangle, \langle \vec{s}_l \rangle, u)$, where $R = \text{dom}(\{r\}).//$

Corollary 5

The class of **partial recursive functionals** is closed under the operations of *identifying, permuting* and *expanding* the list of type 1 arguments.

Proof Let $F = \{e\}$ have rank (k,l). Define

$$G_1 = \lambda \vec{x}_k \vec{\alpha}_{l-1}.F(\vec{x}_k, \vec{\alpha}_{l-1}, \alpha_{l-1})$$

$$G_2 = \lambda \vec{x}_k \vec{\alpha}_l.F(\vec{x}_k, \alpha_1, \ldots, \alpha_{l-2}, \alpha_l, \alpha_{l-1})$$

$$G_3 = \lambda \vec{x}_k \vec{\alpha}_l \vec{\beta}_m.F(\vec{x}_k, \vec{\alpha}_l)$$

$(*) \wedge_{j=0}^{l-1} Q_j$ is shorthand for $Q_0 \ \& \ Q_1 \ \& \ldots \& \ Q_{l-1}.$

Now,

$$G_1(\vec{x}_k,\vec{\alpha}_{l-1}) = n \equiv F(\vec{x}_k,\vec{\alpha}_{l-1},\alpha_{l-1}) = n$$
$$\equiv (\exists u)(\exists s)[T^{(l)}(e,\langle\vec{x}_k\rangle,s,u) \,\&\, \text{length}(s) = l \,\&\, \{\wedge_{i=0}^{l-2}\,(u;(s)_i\,|\,\alpha_{i+1})\}$$
$$\&\, (u;(s)_{l-1}\,|\,\alpha_{l-1}) \,\&\, U(u) = n]$$

Noting that $\lambda u s \vec{x}_k \vec{a}_{l-1}.(u;s\,|\,\alpha_i)$ is semi-recursive (definable from semi-recursive predicates through $\vee$, $\&$ and $(\forall y)_{\leq z}$), G_1 is partial recursive by Theorems 10.4.1, 10.4.2 and 10.4.4. Similarly,

$$G_3(\vec{x}_k,\vec{\alpha}_l,\vec{\beta}_m) = n \equiv F(\vec{x}_k,\vec{\alpha}_l) = n$$
$$\equiv (\exists u)(\exists s)[T^{(l)}(e,\langle\vec{x}_k\rangle,s,u) \,\&\, \text{length}(s) = l + m \,\&$$
$$\{\wedge_{i=0}^{l-1}(u;(s)_i|\alpha_{i+1})\} \,\&\, \{\wedge_{i=0}^{m-1}(u;(s)_{i+1}|\beta_{i+1})\} \,\&\, U(u) = n]$$

Thus G_3 is partial recursive. Note that using $T^{(l)}$, rather than T, ensures that the *first* l type 1 inputs only are relevant.

The proof for G_2 is similar.//

Note: It is now immediate that from now on, legitimately, we may mix ranks in expressions involving partial recursive functionals and semi-recursive relations. For example, if $R(\vec{x},\vec{\alpha})$ and $Q(\vec{y},\vec{\beta})$ are (semi-)recursive then so is $R(\vec{x},\vec{\alpha}) \vee Q(\vec{y},\vec{\beta})$. To see this, expand the lists $\vec{x},\vec{y}$, and $\vec{\alpha},\vec{\beta}$ to $\vec{w}$ and $\vec{\gamma}$ by adding dummy arguments. Then, apply Theorem 10.4.2 to $\lambda\vec{w}\vec{\gamma}.R(\vec{x},\vec{\alpha})$ and $\lambda\vec{w}\vec{\gamma}.Q(\vec{y},\vec{\beta})$.

Corollary 6 The class of partial recursive functionals and semi-recursive relations is closed under substitution of type 1 variables by partial recursive functionals.

Proof Let $F = \{\tilde{f}\}$ have rank $(k,l + 1)$ and $G = \{\tilde{g}\}$ have rank $(k + 1,l)$. We want to show that $\lambda\vec{x}_k\vec{\alpha}_l.F(\vec{x}_k,\vec{\alpha}_l,\lambda y.G(y,\vec{x}_k,\vec{\alpha}_l))$ is partial recursive. Indeed,

$$n = F(\vec{x}_k,\vec{\alpha}_l,\beta) \equiv (\exists u)(\exists s)[T(\tilde{f},\langle\vec{x}_k\rangle,s,u) \,\&\, U(u) = n \,\&$$
$$\text{length}(s) = l + 1 \,\&\, \{\wedge_{j=0}^{l-1}\,(u;(s)_j\,|\,\alpha_{j+1})\} \,\&\, (u;(s)_l\,|\,\beta)]$$

The claim follows from the semi-recursiveness of

$$\lambda u s \vec{x}_k \vec{\alpha}_l.(u;s\,|\,\lambda y.G(y,\vec{x}_k,\vec{\alpha}_l)); \text{ indeed}$$
$$(u;s\,|\,\lambda y.G(y,\vec{x}_k,\vec{\alpha}_l)) \equiv \text{length}(s) \leq u + 1 \,\&\, (\forall i)_{\leq u}(p_i\,|\,s$$
$$\Rightarrow (s)_i = G(i,\vec{x}_k,\vec{\alpha}_l) \,\&\, (s)_i \leq u).$$

The closure of semi-recursive relations under functional substitution follows from that for functionals.//

Corollary 7   [Closure under $(\exists\alpha)$] If R, of rank$(k, l+1)$, is semi-recursive, then so is $(\exists\alpha)R(\vec{x}_k,\vec{\beta}_l,\alpha)$.

Proof   Let $S \in \mathcal{PR}_*$ as in Corollary 4. Then,

$$(\exists\alpha)R(\vec{x}_k,\vec{\beta}_l,\alpha) \equiv (\exists\alpha)(\exists u)(\exists\vec{s}_l,t)[S(\vec{x}_k,\vec{s}_l,t,u) \;\&$$

$$\wedge_{j=1}^{l}(u;s_j\,|\,\beta_j)\} \;\&\; (u;t\,|\,\alpha)] \equiv (\exists u)(\exists\vec{s}_l,t)[S(\vec{x}_k,\vec{s}_l,t,u) \;\&$$

$$\wedge_{j=1}^{l}(u;s_j\,|\,\beta_j)]$$

which is semi-recursive by the projection theorem (10.4.1).//

Note: Theorem 2 and its corollaries are significant in that they establish *the adequacy of the $\mathcal{P}$-formalism for the description of the formalism of the partial recursive functionals.* Let us agree at this point to denote by $\mathcal{P}^{\mathcal{I}}$, $\mathcal{R}^{\mathcal{I}}$, $\mathcal{PR}^{\mathcal{I}}$, $\mathcal{R}^{\mathcal{I}}_*$, $\mathcal{PR}^{\mathcal{I}}_*$ the classes of functionals and relations corresponding to $\mathcal{P}$, $\mathcal{R}$, $\mathcal{PR}$, $\mathcal{R}_*$, $\mathcal{PR}_*$ respectively. We have already seen that the classes with superscript contain their corresponding unsuperscripted classes. Also, clearly, $\mathcal{PR} \subset \mathcal{PR}^{(\alpha)} \subset \mathcal{PR}^{\mathcal{I}}$.

§10.6 TURING REDUCIBILITY– POST'S PROBLEM

Suppose that, in a partial recursive functional $\lambda\vec{x}\,\vec{\alpha}\vec{\beta}.F(\vec{x},\vec{\alpha},\vec{\beta})$, we hold $\vec{\beta}$ fixed, and therefore obtain $\lambda\vec{x}\,\vec{\alpha}.F(\vec{x},\vec{\alpha},\vec{\beta})$. Then we are "computing" *relative to* (or *in*) $\vec{\beta}$.

Definition 1
(i) F is *partial recursive relative to* (or *in*) $\vec{\beta}$ iff for some $e \in \mathcal{N}$, $F = \lambda\vec{x}\,\vec{\alpha}.\{e\}(\vec{x},\vec{\alpha},\vec{\beta})$.

Moreover, if F is *total*, then it is *recursive in* (or *relative to*) $\vec{\beta}$ (or $\vec{\beta}$-*recursive*).(*)

(*)This is *not* the same as asking $\{e\}$ to be total, for then the "output" of $\{e\}$ does not depend on $\vec{\alpha}$ or $\vec{\beta}$ (Problem 4), thus trivializing the definition. On the other hand, a function [= functional of rank $(k,0)$] *can* be total *and* partial recursive *in a nontotal* $\vec{\beta}$ (Problem 20). We note further, that *even when partial recursive functionals are restricted to accept only total type 1 arguments*, recursive in $\vec{\beta}$ must still be defined exactly as in Definition 1 for the following reasons: (1) A result of Nerode shows that a set A is *truth-table reducible* to B [a notion due to Post, which we shall not study] iff $\chi_A = \lambda x.\{e\}(x,\chi_B)$ and $\lambda x\alpha.\{e\}(x,\alpha)$ is *total* (see for example, Rogers [13], p. 143).
(2) It is easily shown that A and B exist, which satisfy $\chi_A = \lambda x.\{f\}(x,\chi_B)$, for some $f \in \mathcal{N}$, but A is *not* truth-table reducible to B (see Rogers [13], p. 127).

(ii) $R \subset \mathcal{N}^k \times \mathcal{I}^l$ is *recursive in $\vec{\beta}$* (or *$\vec{\beta}$-recursive*) if so is χ_R, its characteristic function.

It is *semi-recursive in $\vec{\beta}$* if $R = \mathrm{dom}(F)$ for some *F partial recursive in $\vec{\beta}$*.

(iii) If F is *partial recursive in $\vec{\beta}$* with e as an index, that is, $F = \lambda\vec{x}\vec{\alpha}.\{e\}(\vec{x},\vec{\alpha},\vec{\beta})$, then we write $F = \{e\}^{\vec{\beta}}$.

(iv) If $A_1, \ldots, A_k$ are subsets of $\mathcal{N}$, then partial recursive, recursive, semi-recursive *in* $(A_1, \ldots, A_k)$ means *in* $(\chi_{A_1}, \ldots, \chi_{A_k})$, where, in general, χ_S is the characteristic function of S.//

Note: In what follows we shall restrict attention to relative computability *in some set A*. Functionals and relations will have ranks $(k,0)$. If f has rank $(1,0)$ and is partial recursive relative to A with index e, we write $f = \{e\}^A$ or $f = \phi_e^A$. If $S \subset \mathcal{N}$ is semi-recursive in A we write $S = W_e^A$ iff $S = \mathrm{dom}(\phi_e^A)$. We will also call S *re in A* or *A-re*.

For any fixed set A, we obtain the A-relativized Recursion Theory by simply restating earlier results, valid for $\mathcal{P}, \mathcal{R}, \mathcal{PR}, \mathcal{R}_*, \mathcal{PR}_*$, in terms of A-partial recursive, A-recursive, A-primitive recursive functions, A-recursive, A-primitive recursive, A-re relations.

This is the *fully relativized theory* and the legitimacy of such rephrasing stems from the results of the previous sections and Definition 1. In some cases, for example, *S-m-n* theorem, stronger *partially relativized* results hold: "There is an $h \in \mathcal{PR}$ (rather than just an A-primitive recursive h, which is a *weaker* requirement) such that, for any A-partial recursive $\lambda xy.\psi(x,y)$, $\psi = \lambda xy.\phi_{h(x)}^A(y)$".

We restate for emphasis, S ($\subset \mathcal{N}$) is A-recursive iff *both* S and $\neg S$ are A-re (Theorem 10.4.3).

Note also that if S is re (recursive), then it is re (recursive) with respect to *any* $A \subset \mathcal{N}$.

If $A \in \mathcal{R}_*$, then re-ness and recursiveness *in A* coincide with unrelativized re-ness and recursiveness respectively. (See Problem 21.)

Definition 2   Let A and B be subsets of $\mathcal{N}$.

A is *Turing reducible* or *T-reducible* to B, in symbols $A \leq_T B$, iff A is *B-recursive*.//

Note: Turing reducibility is a more reasonable notion of reducibility than m-reducibility. For example, $\overline{K} \not\leq_m K$, yet intuitively, if we can answer questions about membership in K we *should* be able to answer similar questions about its complement (by interchanging *yes* and *no*). We recall that in m- (1-)reducibility, if $A \leq_m B$ via f then $x \in A$ will be answered by answering exactly *one* similar question about B, namely $f(x) \in B$. In Turing reducibility on the other hand, more than one, in general, questions about membership in B are asked before the verdict about $x \in A$ is arrived at (see Theorem 10.5.2).

Theorem 1

    (i) $\leq_T$ is reflexive and transitive

    (ii) $\overline{A} \leq_T A$ for any A

    (iii) $A \in \mathcal{R}_* \Rightarrow A \leq_T B$, any B

    (iv) $A \leq_T \varnothing \Leftrightarrow A \in \mathcal{R}_*$

    (v) $\leq_m \subset \leq_T$ and $\leq_1 \subset \leq_T$.

    Proof (i) $A \leq_T A$, for $F = \lambda x \alpha.\alpha(x)$ is clearly partial recursive and $\chi_A = \lambda x.F(x,\chi_A)$, χ_A being total.

    Let next $A \leq_T B$ and $B \leq_T C$. Thus $\chi_A = \lambda x.F(x,\chi_B)$ and $\chi_B = \lambda x.G(x,\chi_C)$, where F and G [of rank $(1,1)$ each] are partial recursive.

    By Corollary 6 of Theorem 10.5.2, $\lambda x \alpha.F(x,\lambda y.G(y,\alpha))$ is partial recursive; the fact that χ_A is total and that $\chi_A = \lambda x.F(x,\lambda y.G(y,\chi_C))$ completes the proof.(*)

    (ii) $\chi_{\overline{A}} = \lambda x.1 \dot{-} F(x,\chi_A)$, where $F = \lambda x \alpha.\alpha(x)$.

    (iii) Set $F = \lambda x \alpha.\chi_A(x)$ (that is, α is a dummy type 1 variable). $F \in \mathcal{R}^{\mathcal{I}}$, moreover $\chi_A = \lambda x.F(x,\chi_B)$ for any B.

    (iv) $\Leftarrow$ follows from (iii). $\Rightarrow$ is left as an exercise. (See Problem 24.)

    (v) Let $A \leq_m B$ (or $A \leq_1 B$) via $f \in \mathcal{R}$. That is $\chi_A = \lambda x.\chi_B(f(x)) = \lambda x.F(x,\chi_B)$, where $F = \lambda x \alpha.\alpha(f(x))$ is in $\mathcal{P}^{\mathcal{I}}$ and $\lambda x.F(x,\chi_B)$ is, of course, total.//

    Note: The awkward behavior of $\varnothing$ and $\mathcal{N}$ in the context of m-(or 1-)reducibility [Proposition 9.1.2(vi)] is not exhibited here; all recursive sets are comparable with respect to $\leq_T$ [by (iii)]. Moreover, the intuitively pleasing requirement that a set is T-reducible to its complement is being met [by (ii)].

Definition 3   (Compare with Definition 9.1.3.) $\equiv_T \stackrel{\text{def}}{=} \leq_T \cap \leq_T^{-1}$.//

Proposition 1 $\equiv_T$ is an equivalence relation.

    Proof Problem 25. Refer to the proof of Proposition 9.1.3.//

Definition 4   The equivalence classes of $\equiv_T$ are called *Turing degrees* or *T-degrees* or simply *degrees*.

    We denote by $\deg(A)$ the equivalence class of $\equiv_T$ determined by A.

(*)Intuitively, if $x \in A$ can be answered by answering questions such as $y \in B$ and the latter can be answered by answering questions of the form $z \in C$ then the original question can be answered by answering questions like $z \in C$.

$\leq_T$ is extended on degrees, by setting $x \leq_T y$ iff $A \leq_T B$ for some $A \in x$ and $B \in y.//$

Note: The definition of $\leq_T$ on degrees is consistent, i.e. does *not* depend on the chosen $A \ (\in x)$ and $B \ (\in y)$ (see Proposition 9.1.4).
$\deg(A) = \bigcup_{B \equiv_T A} \deg_m(B) = \bigcup_{B \equiv_T A} \deg_1(B)$ by Theorem 1(v).
We also note that $\deg(\varnothing) = \mathcal{R}^{(1)}_*$, by Theorem 1(iv).

Definition 5   Let $A \subset \mathcal{N}$. The *ordinary(*) jump* or simply *jump* of A of rank $k \ (\geq 1)$, in symbols $A_k^{\circ J}$, is the set $\{\langle a, \vec{x}_k \rangle \mid \{a\}^A(\vec{x}_k) \downarrow \}.//$

Note: $A_1^{\circ J}$ will be denoted by $A^{\circ J}$ and is the relativized version of the "halting set" $K_0 = \{\langle x,y \rangle \mid \phi_x(y) \downarrow \}$. The ordinary jump is also called *completion* (of A) and is denoted by A', or even K^A, in Rogers [13] with a definition which is the relativized version of K, namely $\{x \mid \phi_x^A(x) \downarrow \}$.

Definition 6   We extend the meaning of the symbol $\leq_m$ as follows: If $R \subset \mathcal{N}^k \times \mathcal{I}^l$ then $R \leq_m A$, where $A \subset \mathcal{N}$, iff for some $F \in \mathcal{R}^{\mathcal{I}}$ of rank (k,l), $R(\vec{x}_k, \vec{\alpha}_l) \equiv F(\vec{x}_k, \vec{\alpha}_l) \in A.//$

Note: If F is 1-1, then we write $R \leq_1 A$.

Proposition 2   For all $R \subset \mathcal{N}^k$ and $A \subset \mathcal{N}$, R is semi-recursive in A iff $R \leq_1 A_k^{\circ J}$.

Proof   *if* part. Let $R \leq_1 A_k^{\circ J}$ via $F \in \mathcal{R}^{(k)}$. Then $R(\vec{x}_k)$ iff $F(\vec{x}_k) \in A_k^{\circ J}$ and the claim follows by substitution in the *semi-recursive in A* relation $\lambda y . y \in A_k^{\circ J}$, that is,

$$\lambda y . (\exists a,x)_{\leq y} \mathrm{Seq}(x) \ \& \ k = \mathrm{length}(x) \ \& \ y = 2^{a+1} * x \ \& \ \{a\}^A((x)_0, \ldots, (x)_{k-1}) \downarrow$$

only if part. Let a be a semi-index of R in A.
Then $R(\vec{x}_k)$ iff $\{a\}^A(\vec{x}_k) \downarrow$ iff $\langle a, \vec{x}_k \rangle \in A_k^{\circ J}$. Thus $R \leq_1 A_k^{\circ J}$ via $\lambda a \vec{x}_k . \langle a, \vec{x}_k \rangle.//$

Note: Thus $A^{\circ J}$ is *maximum* with respect to $\leq_1$ among subsets of $\mathcal{N}$ which are *semi-recursive in A*. It follows that it is also a maximum with respect to $\leq_T$, for if $R \subset \mathcal{N}$ is semi-recursive in A then (Proposition 2) $R \leq_1 A^{\circ J}$, hence $R \leq_T A^{\circ J}$ [Theorem 1(v)].
This suggests the definition

(*)"Ordinary" suggests that also some "extraordinary" jumps exist. Indeed, all sorts of *jump operators* can be defined and the interested reader may consult, e.g., Hinman [5] and Rogers [13].

Definition 7   A set $B \subset \mathcal{N}$ is called *T-complete in A* iff both of the following hold.

(i) B is semi-recursive in A

(ii)   For all S semi-recursive in A, $S \leq_T B$.

When $A = \varnothing$, or any recursive set, we shall call B *T-complete*.//

The content of the previous note becomes

Proposition 3 A^{oJ} is T-complete in A.

Corollary 1 K_0 is T-complete.

   Proof $K_0 = \varnothing^{\text{oJ}}$.//

Corollary 2 K is T-complete.

   Proof Problem 26.//

Note: One may introduce the relativized notions for m-completeness and 1-completeness (see Definition 9.2.1 for the absolute notions). From the proof of Proposition 2 follows that A^{oJ} is 1-complete in A.

Proposition 4   There is no maximum degree.

   Proof Observe that for any $A \subset \mathcal{N}$, $A^{\text{oJ}} \not\leq_T A$. Indeed, if $A^{\text{oJ}} \leq_T A$, then $\lambda x.x \notin A^{\text{oJ}}$ is semi-recursive in A and so is $\lambda x.\langle x,x \rangle \notin A^{\text{oJ}}$ by substitution. Let e be a semi-index in A for the latter relation.
   Then $\langle e,e \rangle \notin A^{\text{oJ}} \equiv \{e\}^A(e)\!\downarrow (\lambda x.\langle x,x \rangle \notin A^{\text{oJ}} = \text{dom}(\{e\}^A))$,
but also $\langle e,e \rangle \notin A^{\text{oJ}} \equiv \{e\}^A(e)\!\uparrow$ (definition of A^{oJ}).
   It follows that $\deg(A) \leq_T \deg(A^{\text{oJ}})$ and $\deg(A) \neq \deg(A^{\text{oJ}})$ for any A.//

Note: As usual, $x \leq_T y$ and $x \neq y$ (where x and y are degrees) will be denoted by "$x <_T y$".

Corollary 1 $\deg(\varnothing) <_T \deg(\varnothing^{\text{oJ}}) <_T \deg(\varnothing^{\text{oJoJ}}) <_T \ldots <_T \deg(\varnothing^{\text{oJ} \cdots \text{oJ}}) < \ldots$.
Moreover, all the degrees above are distinct.

   Proof The first contention follows by Proposition 4. For the second, let x,y,z be degrees such that $x <_T y <_T z$. Then $x \neq z$, otherwise, by definition of $\equiv_T$, $z = y$ (why?).//

Corollary 2   For any re but nonrecursive set S, $\deg(\varnothing) <_T \deg(S) \leq_T \deg(\varnothing^{oJ})$.

Post's Problem is "Can S in Corollary 2 be chosen to satisfy the *strict* inequality $\deg(S) <_T \deg(\varnothing^{oJ})$?" In other words, "Are there any re-degrees(*) besides $\deg(\varnothing)$ and $\deg(\varnothing^{oJ})$?" Muchnik [12] and Friedberg [3] almost simultaneously (and independently) showed that two re sets A and B exist which are *not comparable* with respect to $\leq_T$. This has as consequences:

(1) Positive answer to Post's problem, for if the only degrees were $\deg(\varnothing)$ and $\deg(\varnothing^{oJ})$ then they would certainly be comparable.

(2) $\leq_T$ is *not* a total (linear) order on the set of degrees.

Theorem 2   (*Friedberg-Muchnik Theorem*) There are two re sets A and B such that neither $A \leq_T B$ nor $B \leq_T A$.

Proof   Due to the importance of the proof-technique, known as the "(finite injury) *priority method*", which is applicable frequently in the theory of degrees as well as in computational complexity, we shall proceed informally at first, in some detail. χ_S throughout will signify the characteristic function of a set S.

The statement "not $A \leq_T B$", or $A \nleq_T B$ in short, amounts to $\chi_A \neq \{e\}^B$ for *all* $e \in \mathcal{N}$, such that $\{e\}^B$ is a 0-1-valued total function. Similarly, $B \nleq_T A$ amounts to $\chi_B \neq \{e\}^A$ for all $e \in \mathcal{N}$, such that $\{e\}^A$ is a 0-1-valued total function.

We will have recognized that two re sets A and B satisfy the theorem statement, as soon as two sequences $(a_e)_{e \geq 0}$ and $(b_e)_{e \geq 0}$ of integers are provided, which satisfy

$$\{e\}^A(b_e) = 1 \text{ iff } b_e \in B \text{ for } e \geq 0 \tag{1}$$

$$\{e\}^B(a_e) = 1 \text{ iff } a_e \in A \text{ for } e \geq 0 \tag{2}$$

Indeed, $\{e\}^A$ cannot be χ_B as $\{e\}^A(b_e) = 1$ iff $b_e \in B$ iff $\chi_B(b_e) = 0$. Similar comment for $\{e\}^B$.

Our task is to define two re sets A and B and two associated sequences $(a_e)_{e \geq 0}$ and $(b_e)_{e \geq 0}$ satisfying (1) and (2) above. To this end, *first* we merge the two sequences into one $(x_e)_{e \geq 0}$, for convenience, where $b_e = x_{2e}$ and $a_e = x_{2e+1}$ for $e \geq 0$. *Second*, A and B are constructed in stages. Just *before the nth stage*, we have already obtained A_n and B_n.

Assume, *tentatively*, that we have already decided on a sequence $(x_e)_{e \geq 0}$. *Construction at nth stage*. (Attempt)

First, in the interest of attending to the construction of A as often as to

(*)As in the case of m- and 1-degrees, a $(T-)$ degree is recursively enumerable if it contains a re set.

that of B, we alternate between the two, attending to exactly one of the two at each stage. We attend to B if $(n)_0 = 0$, to A otherwise [of course, for infinitely many n it is $(n)_0 = 0$ and for as many it is $(n)_0 > 0$].

Let then $(n)_0 = 0$. We perform $\leq n$ steps of the computation $\{e\}^{A_n}(x_{2e})$, where $e = (n)_1$. [Note that e will come up infinitely often as an $(n)_1$. Thus, doing only $\leq n$ steps is not a snag; indeed this is essential since we do not want to loop forever at the nth step just in case $\{e\}^{A_n}(x_{2e})\uparrow$.]

If the output is 1, then set $B_{n+1} = B_n \cup \{x_{2e}\}$. Otherwise, set $B_{n+1} = B_n$.

Attending to A [case $(n)_0 > 0$] is done as above, computing $\{e\}^{B_n}(x_{2e+1})$ instead. *End of construction.*

Unfortunately, our attempt to construct the "approximation" A_n and B_n of A and B breaks down: Conceivably, since A_n grows (toward A), by the addition of more elements to it, the output of $\{e\}^A(x_{2e})$ might differ from that of $\{e\}^{A_n}(x_{2e})$. Should in fact happen that $\{e\}^A(x_{2e}) = 0$ [whereas $\{e\}^{A_n}(x_{2e})$ was 1], then x_{2e} is *not* anymore a counterexample to a claim such as $\{e\}^A = \chi_B$ (for x_{2e} was already put in B_{n+1}, hence in B).

This problem would be avoided if the only elements put in A_n after stage n were $\geq \max(A_n) + 1$ [Indeed, $\{e\}^{A_n}(x) = U((\mu y)T(e,\langle x \rangle,(*) [\chi_{A_n}|y],y))$. If the only items added to A_n are greater than $y_0 = (\mu y)T(e,\langle x \rangle,[\chi_{A_n}|y],y)$, then, $[\chi_A|y_0] = [\chi_{A_n}|y_0]]$.

To incorporate this last suggestion, we amend the construction above so that the sequence $(x_e)_{e\geq 0}$ is allowed to change. $(x_e^n)_{e\geq 0}$ denotes an approximation to $(x_e)_{e\geq 0}$ just before the stage n.

If at that stage B is attended, then *all* terms x_{2e+1}^{n+1} $(e \geq 0)$ [these are the type of terms we put in A] will be taken to be $> \max(A_n)$ so that items added to A from now on will be $> \max(A_n)$. Similarly, if A is attended then all $x_{2e}^{n+1}, e \geq 0$, are made $> \max(B_n)$.

The drawback now is that the sequence $(x_e^n)_{n\geq 0}$ will not converge to a final value $x_e(\dagger)$ due to x_e^n being changed for infinitely many values of n. This would mean that we never get a final sequence $(x_e)_{e\geq 0}$.

This last difficulty is avoided by amending the construction for the last time so that x_{2f+1}^n *changes* (when it becomes x_{2f+1}^{n+1}) to a value $> \max(A_n)$, *only if* $2f + 1 > 2e$ *and* x_{2e}^n was put in B (via B_{n+1}) at stage n. Similarly, $x_{2e}^{n+1} \neq x_{2e}^n$ *only if* $2e > 2f + 1$, where x_{2f+1}^n was put in A at stage n [in this case, of course x_{2e}^{n+1} is chosen to be $> \max(B_n)$].

It can be seen now that for each e, $\{x_e^n|n \geq 0\}$ is finite. Indeed, by induction on e, observe first that $\{x_0^n|n \geq 0\}$ contains a single element as there is no $e < 0$ to induce a x_0^n change. Consider $T^{(f)} = \{x_{2f+1}^n|n \geq 0\}, f \in \mathcal{N}$, and assume that *each* $S^{(e)} = \{x_{2e}^n|n \geq 0\}$ is finite when $2e < 2f + 1$. But all changes $x_{2f+1}^m \rightarrow x_{2f+1}^{m+1}$ are due to some x_{2e}^m in $\bigcup_{2e<2f+1} S^{(e)}$, thus there are finitely many such

---

$(*)\langle x \rangle = 2^{x+1}$. See proof of Theorem 10.5.2.

$(\dagger)$A sequence $(s_n)_{n\geq 0}$ of integers *converges to* a iff $s_n = a$ a.e.; i.e., the set $\{s_n|n \geq 0 \ \& \ s_n \neq a\}$ is *finite*. We write $a = \lim_n s_n$.

changes, i.e., $T^{(f)}$ is finite. Similarly, we conclude for each set $\{x_{2f}^n \mid n \geq 0\}$, for $f \in \mathcal{N}$.

This method of defining $(x_e)_{e \geq 0}$ by successive approximations so that $\{x_e^n \mid n \geq 0\}$ is finite for all e, is an instance of the *"finite injury priority method"*.

The term *"finite injury"* is attributed to the fact that for each e, the "guesses" x_e^n for the "correct" x_e are changed ("injured") finitely often. The term *"priority"* is due to the selective manner by which x_e^n are injured into x_e^{n+1} ($e \geq 0$): Terms x_e^n with index e less than f, where x_f^n was put in A or B at stage n, are less prone to injury (that is, they have "higher priority" to remain unchanged) than terms with $e > f$

Formal details: We shall define functions $\lambda n.a(n)$, $\lambda n.b(n)$ and $\lambda nz.x^n(z)$. $x^n(e)$ will be x_e^n, whereas

$$a(n) = \Pi_{i \in A_n} p_i, \quad b(n) = \Pi_{i \in B_n} p_i$$

We set $a(n) = 1$ [resp. $b(n) = 1$] just in case $A_n = \varnothing$ (resp. $B_n = \varnothing$).

Clearly, $(y;s \mid \chi_{A_n})$ is equivalent to $\text{length}(s) \leq y + 1$ & $(\forall i)_{\leq y}\{p_i \mid s \Rightarrow [(s)_i = 0$ & $p_i \mid a(n) \vee (s)_i = 1$ & $p_i \nmid a(n)(*)]$ & $(s)_i \leq y]\}$. This relation we shall denote by $(y;s \mid a(n))$ for convenience.

Construction.

Stage 0: Set $a(0) = 1$, $b(0) = 1$ (that is, $A_0 = B_0 = \varnothing$) and $x^0 = \lambda e.0$ (i.e., $x_e^0 = 0$ for all $e \geq 0$).

Stage n:

Case 1: $(n)_0 = 0$. Let $e = (n)_1$, and attend to B.

(i) Compute

$$y_0 = (\mu y)_{\leq n}[(\exists s)_{\leq \Pi_{i \leq y} p_i^2}\{T(e, \langle x^n(2e) \rangle, 2^{s+1}, y) \&$$

$$(y;s \mid a(n))\} \& U(y) = 1] \& p_{x^n(2e)} \nmid b(n) \qquad [p_{x^n(2e)} \nmid b(n) \text{ amounts}$$

to $x^n(2e) \notin B_n$].

(ii) **if** $y_0 = n + 1$ [that is, no $y(\leq n)$ exists that satisfies the condition following $(\mu y)_{\leq n}$]

$$\left. \begin{array}{l} \textbf{then set } a(n + 1) = a(n) \\ \qquad\qquad b(n + 1) = b(n) \\ \qquad\qquad\quad x^{n+1} = x^n \end{array} \right\} \text{ [i.e., no change at stage } n]$$

$$\textbf{else set } a(n + 1) = a(n)$$
$$\qquad\qquad b(n + 1) = b(n) \cdot p_{x^n(2e)}$$
$$\qquad\qquad\quad x^{n+1} = \lambda z.\textbf{if } z > 2e \;\&\; 2 \nmid z \textbf{ then } x^n(z) + y_0 + 1$$
$$\qquad\qquad\qquad\qquad\qquad\qquad\qquad \textbf{else } x^n(z)$$

$(*) p \nmid q$ means $\neg(p \mid q)$.

Note: y_0 is *not* necessarily equal to $\max(A_n)$. However, it certainly majorizes the maximum A_n-entry *used* (through s) in the computation of $\{e\}^{A_n}(x^n(2e)) = U(y_0)$. Thus, the definition of x^{n+1} is in the spirit of the preceding discussion.

Case 2. $(n)_0 > 0$. Let $e = (n)_1$, and attend to A. This is the same as Case 1, where the roles of A and B, hence of a and b, $2e$ and $2e + 1$, $2 \not| z$ and $2 | z$ (in the definition of x^{n+1}) are interchanged. *End of construction.*

The schema defining $\lambda n.a(n)$, $\lambda n.b(n)$ and $\lambda nz.x^n(z)$ is, almost,(*) a simultaneous primitive recursion. Thus, setting $F(n,z) = \langle a(n), b(n), x^n(z) \rangle$ for all n,z, the method of proof of Theorem 3.7.3 may be employed to show recursiveness of F (Problem 27).

Alternatively, one may use the recursion theorem to show recursiveness of F. To this end, use $\{f\}(n,z)$ instead of F in the righthand side of the construction schema to define a partial recursive function $\lambda fnz.\tilde{F}(f,n,z)$.

For example,

$a(n) = (\{f\}(n,z))_0$ and $p_{x^n(2e)} \not| b(n)$ becomes $p_{(\{f\}(n,2\cdot(n)_1))_2} \not| (\{f\}(n,z))_1$. y_0 is a partial recursive function of f,n,z (by substitution); let us call it G for convenience.

Then, (ii) of Case 1 becomes

$$\tilde{F}(f,n,z) = \text{if } G(f,n \mathbin{\dot{-}} 1,z) = n \text{ then } \{f\}(n \mathbin{\dot{-}} 1,z) \text{ else}$$

$$\langle (\{f\}(n \mathbin{\dot{-}} 1,z))_0, (\{f\}(n \mathbin{\dot{-}} 1,z))_1 \cdot p_{(\{f\}(n \mathbin{\dot{-}} 1, 2(n \mathbin{\dot{-}} 1)_1)_2},$$

$$\text{if } z > 2(n \mathbin{\dot{-}} 1)_1 \,\&\, 2 \not| z \text{ then } (\{f\}(n \mathbin{\dot{-}} 1,z))_2 + G(f,n \mathbin{\dot{-}} 1,z) + 1$$

$$\text{else } (\{f\}(n \mathbin{\dot{-}} 1,z))_2 \rangle.$$

After filling in the remaining details (Problem 28), we shall end up (recursion theorem) with an f_0 such that $\tilde{F}(f_0,n,z) = \{f_0\}(n,z)$ for all n,z. Thus, $F = \{f_0\}$ (of course, there is a unique F defined by the construction schema, and trivially, it is *total*).

We now have that a and b are *recursive*, hence $A = \bigcup_{n\geq 0} A_n$ and $B = \bigcup_{n\geq 0} B_n$ are *re* [Indeed, $x \in A \equiv (\exists n) p_x | a(n)$].

Also, by the earlier informal discussion, $\{x^n(e) \,|\, n \geq 0\}$ is finite for each e; we set $x_e = \lim_n x^n(e)$ for all $e \geq 0$.

Claim. $\{e\}^A(x_{2e}) = 1$ iff $x_{2e} \in B$.

Only if part. Let $\{e\}^A(x_{2e}) = 1$.

Then, $(\exists y)[(\exists s)_{\leq \underset{i\leq y}{\Pi} p_i^2} \{T(e, \langle x_{2e} \rangle, 2^{s+1}, y) \,\&\, (y;s \,|\, \chi_A)\} \,\&\, U(y) = 1]$.

Let y_0 be the smallest such y, and s_0 the corresponding s. There is an $n \in \mathcal{N}$ such that $(n)_0 = 0$, $(n)_1 = e$, $n > y_0$, $x_{2e}^n = x_{2e}$ and $(y_0;s_0 \,|\, a(n))$ [for example, take $n = 2 \cdot 3^{e+1} \cdot q$, where q is a product of enough primes > 3 to meet the last three requirements].

(*)The qualification, since to define $x^{n+1}(z)$ for any z, $x^n(2e)$ or $x^n(2e + 1)$ has to be consulted for appropriate e. Thus the "parameter" of the recursion changes.

Thus, by Case 1(i), either $x_{2e}^n \in B^n$ or x_{2e}^n is put in B^{n+1} at stage n. Hence, $x_{2e} = x_{2e}^n \in B$.

if part. Let $x_{2e} \in B$ and $n = (\max y)(x_{2e} \notin B_y)$.

Thus, x_{2e} was put in B at stage n; i.e., $x_{2e} = x_{2e}^n$, $(n)_0 = 0$, $(n)_1 = e$ and a y_0 ($\leq n$) exists satisfying (i) of Case 1. Let s_0 be the corresponding s-value. It is $(y_0; s_0 \mid a(n))$.

Let $m > n$ and $A_n \subsetneq A_m$. Consider $z \in A_m - A_n$.

Suppose $z \leq y_0$. Let $q = (\max y)(z \notin A_y)$. Then $n + 1 < q + 1 \leq m(*)$ and $z = x_{2f+1}^q$, where $f = (q)_1$ and $(q)_0 > 0$. There are two cases:

$2f + 1 > 2e$. Then

$$z = x_{2f+1}^q \geq (\dagger)\; x_{2f+1}^{n+1} = x_{2f+1}^n + y_0 + 1 > y_0; \text{ a contradiction.}$$

$2f + 1 < 2e$ is also ruled out, for it implies $x_{2e}^{q+1} > x_{2e}^q$ (by construction, Case 2) $\geq x_{2e}^n$ (by monotonicity of $\lambda n.x_{2e}^n$). Thus, $x_{2e} \neq \lim_n x_{2e}^n$, a contradiction. So it must be $y_0 < z$. Thus $(y_0; s_0 \mid a(m))$ for all $m > n$, since the m above was arbitrary. It follows that $(y_0; s_0 \mid \chi_A)$.

This, recalling the way y_0 and s_0 were arrived at [(i) of Case 1 in the construction], implies that $\{e\}^A(x_{2e}) = U(y_0) = 1$. *The claim is established.* We have shown that $B \not\leq_T A$. By a symmetrical argument one gets $A \not\leq_T B.//$

Note: The above proof is similar in general lines to that in Friedberg [3]. Our exposition has been influenced by that in Shoenfield [15], but we differ from both in the (amount and format of) formal details. A still different proof (in terms of formal details) is given in Rogers[13]. The last source along with Sacks [14] and Shoenfield [16] are excellent references for the reader who wants to find out more about priority arguments and results on degrees.

Corollary 1   (*Solution to Post's Problem*) There are two nonrecursive and incomparable re degrees.

Corollary 2   There are two nonrecursive and incomparable re m- (resp. 1-) degrees.

Proof $\deg_m(A) \not\leq_m \deg_m(B)$, where A and B are as in Theorem 2, otherwise ($A \leq_m B$) we would have $A \leq_T B$. Similar comment with the roles of A and B reversed.//

(*)$q > n$, since at stage n, B was attended, not A.

(†)Passing from x_e^n to x_e^{n+1} we make $x_e^{n+1} \geq x_e^n$ according to the construction. That is, $\lambda n.x_e^n$ is increasing (nonstrictly).

§10.7 THE ARITHMETICAL HIERARCHY

We shall have a brief look at the "complexity" of "arithmetical" predicates, i.e., predicates definable from *constructive arithmetic* predicates by applying $(\forall y)$ and/or $(\exists y)$ prefixes. Intuitively, there are *formulas* which "formalize" such predicates in any extension of ROB (§9.6) obtained by simply changing the informal symbols occurring in these predicates(*) into formal symbols (we are *not* saying here that these predicates are necessarily *definable* or *representable*. For example, $(\exists y)T(x,x,y)$, not being recursive, is not definable. $\neg(\exists y)T(x,x,y)$ not being re is not representable).

The term "complexity" here applies to the degree of unsolvability of a predicate. For example, $\overline{K}$ is *more* "unsolvable" than K, therefore "more complex".

Unless otherwise specified, for the balance of this section computability will be relative to a fixed $\vec{\beta} = (\beta_1, \ldots, \beta_l)$, where all the β_i are *total*.

Predicates and functions will have ranks $(k,0)$, $k > 0$, unless otherwise specified.

Definition 1   The class of $\vec{\beta}$-*arithmetical predicates* is the *smallest* containing $\mathcal{R}_*^{\beta}$(†) and closed under $(\exists y)$ and $(\forall y)$. If all β_i are recursive, then we obtain the class of *arithmetical predicates.*//

Example 1   Let $A = \{x \mid \phi_x \in \mathcal{R}\}$. Then $x \in A \equiv (\forall y)(\exists z)T(x,y,z)$. Thus, $\lambda x.x \in A$; i.e., the set A, is arithmetical.//

Proposition 1   The class of arithmetical predicates is the smallest class containing CA, the constructive arithmetic predicates, and closed under $(\exists y)$ and $(\forall y)$.

Proof CA $\subset \mathcal{R}_*$, hence the $\exists/\forall$-closure of CA is a subset of the arithmetical predicates.

Conversely, each recursive predicate is re, hence expressible as $(\exists y)S$, where $S \in$ CA. Thus the arithmetical predicates are contained in the $\exists/\forall$-closure of CA.//

Note: We shall not define $\vec{\beta}$-CA predicates to generalize Proposition 1. It was simply stated as a justification for the term "arithmetical".

(*)E.g. "x is even" expands to $(\exists y)x = 2y$ and this is formalized to $(\exists v_1)v_2 = \tilde{2} \cdot v_1$ where v_1, v_2 are syntactically correct symbols of distinct variables in the FMS and $\tilde{2}$ is the formal counterpart of 2.

(†)$\mathcal{R}_*^{\vec{\beta}}$ stands for the class of $\vec{\beta}$-recursive predicates of rank $(k,0)$, $k > 0$.

Definition 2 $\Sigma_n^{\vec{\beta}}, \Pi_n^{\vec{\beta}}, \Delta_n^{\vec{\beta}}, n \geq 0$ and $\Delta^{\vec{\beta}}$ are defined as follows:

(1) $\Sigma_0^{\vec{\beta}} = \Pi_0^{\vec{\beta}} = \mathcal{R}_*^{\vec{\beta}}$

(2) $\Sigma_{n+1}^{\vec{\beta}} = \{(\exists y)Q(\ldots, y, \ldots) \mid Q \in \Pi_n^{\vec{\beta}}\}$

(3) $\Pi_{n+1}^{\vec{\beta}} = \{(\forall y)Q(\ldots, y, \ldots) \mid Q \in \Sigma_n^{\vec{\beta}}\}$

(4) $\Delta_n^{\vec{\beta}} = \Sigma_n^{\vec{\beta}} \cap \Pi_n^{\vec{\beta}}$

(5) $\Delta^{\vec{\beta}} = \bigcup_{n \geq 0} (\Sigma_n^{\vec{\beta}} \cup \Pi_n^{\vec{\beta}}).//$

Note: If all the β_i are recursive, then the superscript $\vec{\beta}$ is omitted. If $\vec{\beta} = (\beta_1)$ and β_1 is the characteristic function of some set A, then we use the superscript A instead of β_1.

Clearly, each predicate in $\Delta^{\vec{\beta}}$ is arithmetical.

Example 2 A of example 1 is in $\Pi_2.//$

Proposition 2

(i) $Q \in \Sigma_n^{\vec{\beta}}$ iff $\neg Q \in \Pi_n^{\vec{\beta}}$

(ii) $Q \in \Pi_n^{\vec{\beta}}$ iff $\neg Q \in \Sigma_n^{\vec{\beta}}$

Proof
(i) Induction on n: $n = 0$; immediate, since $\Sigma_0^{\vec{\beta}} = \Pi_0^{\vec{\beta}} = \mathcal{R}_*^{\vec{\beta}}$.
 Assume for $n = k$.
 Let $Q \in \Sigma_{k+1}^{\vec{\beta}}$. Then $Q \equiv (\exists y)S$, where $S \in \Pi_k^{\vec{\beta}}$. Thus, $\neg S \in \Sigma_k^{\vec{\beta}}$ (induction hypothesis), hence $\neg Q \equiv (\forall y)\neg S$ is in $\Pi_{k+1}^{\vec{\beta}}$. This proves the *only if*. The *if* part, plus case (ii), is left to the reader.//

Theorem 1   (*Closure properties*)
Consider the following closure properties of predicates:

(1)   Substitution of a variable by a $\vec{\beta}$-recursive function

(2)   Explicit transformation

(3) $\vee, \&$

(4) $(\exists y)_{\leq z}$ $(\forall y)_{\leq z}$

(5) $\neg$

(6) $(\exists y)$

(7) $(\forall y)$

Then $\Delta^{\vec{\beta}}$ is closed under (1)–(7)
$\Delta_n^{\vec{\beta}}$ ($n \geq 0$) is closed under (1)–(5)

$\Sigma_n^{\vec{\beta}}$ $(n \geq 0)$ is closed under (1)–(4), and (6) *if* $n > 0$

$\Pi_n^{\vec{\beta}}$ $(n \geq 0)$ is closed under (1)–(4), and (7) *if* $n > 0$

Proof (Most of the details are left to the reader.) Closure under (1)–(2) follows from the closure of $\mathcal{R}_*^{\vec{\beta}}$ under these operations.

For (3), observe

$$(\exists y)R(y,\vec{x}) \vee (\exists z)S(z,\vec{x})$$

$$\equiv (\exists y)(R(y,\vec{x}) \vee S(y,\vec{x})) \qquad \text{(i)}$$

$$(\forall y)R(y,\vec{x}) \vee (\forall z)S(z,\vec{x}) \equiv$$

$$\neg(\exists y)\neg R(y,\vec{x}) \vee \neg(\exists z)\neg S(z,\vec{x}) \equiv$$

$$\neg[(\exists y)\neg R(y,\vec{x}) \,\&\, (\exists z)\neg S(z,\vec{x})] \equiv$$

$$\neg(\exists w)(\neg R((w)_0,\vec{x}) \,\&\, \neg S((w)_1,\vec{x}))$$

$$\equiv (\forall w)(R((w)_0,\vec{x}) \vee S((w)_1,\vec{x})) \qquad \text{(ii)}$$

Let P and Q be in $\Sigma_{n+1}^{\vec{\beta}}$ (resp. $\Pi_{n+1}^{\vec{\beta}}$) and assume closure of $\Sigma_n^{\vec{\beta}}$ (resp. $\Pi_n^{\vec{\beta}}$) under $\vee$. (This is certainly true for $n = 0$.)

Let $P \equiv (\exists y)R(y,\vec{x})$ and $Q \equiv (\exists z)S(z,\vec{x})$ [resp. $P \equiv (\forall y)R(y,\vec{x})$ and $Q \equiv (\forall z)S(z,\vec{x})$] where R and S are in $\Pi_n^{\vec{\beta}}$ (resp. $\Sigma_n^{\vec{\beta}}$). By induction hypothesis, (i), (ii) and (1), $P \vee Q$ is in $\Sigma_{n+1}^{\vec{\beta}}$ (resp. $\Pi_{n+1}^{\vec{\beta}}$). This discussion also shows closure of $\Delta_n^{\vec{\beta}}$ and $\Delta^{\vec{\beta}}$ under $\vee$. One argues similarly for &.

For (4) use the following equivalences, (1), (3), and induction on n:

(iii) $(\exists y)_{\leq z}(\exists u)R(y,u,\vec{x}) \equiv (\exists w)[(w)_0 \leq z \,\&\, R((w)_0,(w)_1,\vec{x})]$

(iv) $(\forall y)_{\leq z}(\exists u)R(y,u,\vec{x}) \equiv (\exists w)(\forall y)_{\leq z} R(y,(w)_y,\vec{x})$.

The $(\forall y)_{\leq z}(\forall u)$ and $(\exists y)_{\leq z}(\forall u)$ prefixes can be dealt with by applying $\neg$ to (iii) and (iv). For (5), rely on Proposition 2. For (6), see (7) below. For (7), observe that

$$(\forall y)(\forall z)R(y,z,\vec{x}) \equiv \neg(\exists y)\neg(\forall z)R(y,z,\vec{x})$$

$$\equiv \neg(\exists y)(\exists z)\neg R(y,z,\vec{x}) \equiv \neg(\exists w)\neg R((w)_0,(w)_1,\vec{x})$$

$$\equiv (\forall w)R((w)_0,(w)_1,\vec{x})$$

and use induction on n, and (1).//

Corollary 1 $\Delta^{\vec{\beta}}$ (resp. Δ) is the set of $\vec{\beta}$-arithmetical (resp. arithmetical) predicates.

Proof Because of the closure properties of $\Delta^{\vec{\beta}}$ (Δ), Definition 2 and the note following it.//

Note: Thus, the sets $\Sigma_n^{\vec{\beta}}$, $\Pi_n^{\vec{\beta}}$, $\Delta_n^{\vec{\beta}}$ effect a classification of the $\vec{\beta}$-arithmetical predicates, known as the $\vec{\beta}$-*Arithmetical Hierarchy*. (First studied by Kleene [7] and Mostowski [11]). The term "hierarchy" is further justified by the inclusions in the following three corollaries.

Corollary 2 $\Sigma_n^{\vec{\beta}} \cup \Pi_n^{\vec{\beta}} \subset \Delta_{n+1}^{\vec{\beta}}$ for $n \geq 0$.

> **Proof** By induction on n.
> $n = 0$: Then
>
> $$\Delta_1^{\vec{\beta}} = \Sigma_1^{\vec{\beta}} \cap \Pi_1^{\vec{\beta}} = \{Q \mid Q \in \Sigma_1^{\vec{\beta}} \ \& \ Q \in \Pi_1^{\vec{\beta}}\} = \text{(by Proposition 2)}$$
> $$\{Q \mid Q \in \Sigma_1^{\vec{\beta}} \ \& \ \neg Q \in \Sigma_1^{\vec{\beta}}\} = \mathcal{R}_*^{\vec{\beta}} = \Delta_0^{\vec{\beta}} = \Sigma_0^{\vec{\beta}} = \Pi_0^{\vec{\beta}}$$
>
> Assume that, for some $n \geq 0$, $\Sigma_n^{\vec{\beta}} \cup \Pi_n^{\vec{\beta}} \subset \Delta_{n+1}^{\vec{\beta}}$. We shall prove that $\Sigma_{n+1}^{\vec{\beta}} \subset \Delta_{n+2}^{\vec{\beta}}$; the case for $\Pi_{n+1}^{\vec{\beta}}$ is analogous:
> Let $Q(\vec{x}) \in \Sigma_{n+1}^{\vec{\beta}}$. Then so is $P(y,\vec{x}) = \lambda y\vec{x}.Q(\vec{x})$ by Theorem 1 (2).
> Hence, $Q(\vec{x}) \equiv (\forall y)P(y,\vec{x}) \in \Pi_{n+2}^{\vec{\beta}}$. Next, there is an $R(y,\vec{x}) \in \Pi_n^{\vec{\beta}}$ such that $Q(\vec{x}) \equiv (\exists y)R(y,\vec{x})$. By induction hypothesis, $R \in \Delta_{n+1}^{\vec{\beta}} \subset \Pi_{n+1}^{\vec{\beta}}$, hence $Q \in \Sigma_{n+2}^{\vec{\beta}}.//$

Corollary 3 $\Sigma_n^{\vec{\beta}} \subset \Sigma_{n+1}^{\vec{\beta}}$ and $\Pi_n^{\vec{\beta}} \subset \Pi_{n+1}^{\vec{\beta}}$, $n \geq 0$.

Corollary 4 $\Delta_n^{\vec{\beta}} \subset \Delta_{n+1}^{\vec{\beta}}$, $n \geq 0$.

> **Note:** It is now clear that $\Delta^{\vec{\beta}} = \bigcup_{n \geq 0} \Delta_n^{\vec{\beta}}$.

Example 3   If α is total, $\lambda x.\langle \alpha \mid x \rangle$ is α-recursive.
> Indeed, set first $F(i,x,\alpha) = \textbf{if } \alpha(i) \doteq x = 0 \textbf{ then } \alpha(i) + 1 \textbf{ else } 0$. Clearly, F is a partial recursive functional. Then so is G, defined for all x,α as $G(x,\alpha) = \Pi_{i \leq x} p_i^{F(i,x,\alpha)}$ [by closure of partial recursive functionals under primitive recursion].
> It is immediate that for *total* α, $\lambda x.\langle \alpha \mid x \rangle = \lambda x.G(x,\alpha)$. It is important to observe that $\langle \alpha \mid x \rangle$ is *always* defined, so $\lambda x.\langle \alpha \mid x \rangle$ is *not* equal to $\lambda x.G(x,\alpha)$ for nontotal α.
> It follows that $\lambda e u \vec{x}.T(e,\langle \vec{x} \rangle,[\vec{\beta} \mid u],u)$ is $\vec{\beta}$-recursive whenever $\vec{\beta}$ are *total*.//

Example 4   Consider $S = \{x \mid W_x^A = \varnothing\}$. Then

$$x \in S \equiv (\forall y)\neg(\exists u)T(x,\langle y \rangle,[\chi_A \mid u],u)$$

$$\equiv (\forall y)(\forall u)\neg T(x,\langle y \rangle,[\chi_A \mid u],u),$$

thus [contraction of $\forall\forall$ into $\forall$ by Theorem 1(7)] $S \in \Pi_1^A.//$

Example 5   (*The equivalence problem for* $\mathcal{R}$) $E = \{(x,y) \mid \phi_x = \phi_y \ \& \ \phi_x \in \mathcal{R}\}$
$(x,y) \in E \equiv (\forall z)(\exists u)(\exists w) \ [T(*)(x,z,u) \ \& \ T(y,z,w) \ \& \ d(u) = d(w)]$, thus
$E \in \Pi_2.//$

Note: To achieve an "optimal" (lowest) placement of a predicate in the hierarchy we need more powerful tools than Theorem 1 and its corollaries, and still we will be able to do this in special cases only.

We next show that the inclusions in the hierarchy are proper (i.e., we have a "proper" or nontrivial hierarchy) by a simple diagonal argument similar to that used to prove that $\neg(\exists y)T(x,x,y)$ is not re. The key lemma we then used was that $(\exists y)T(e,x,y)$ "enumerates" (or is "universal" for) all re predicates $P(x)$ by varying e. e is the "index-variable" of the enumeration.

This suggests the following definition:

Definition 3   Define the predicates $E_n^{\vec{\beta}}$, $n \geq 1$, inductively

$$E_1^{\vec{\beta}}(e,x) \equiv (\exists y)T(e,x,[\vec{\beta} \mid y],y)$$

$$E_{n+1}^{\vec{\beta}}(e,x) \equiv (\exists y)\neg E_n^{\vec{\beta}}(e,2^{y+1}*x).//$$

Note: Recall that $x*y = x \cdot \Pi_{i<\text{length}(y)} p_{i+\text{length}(x)}^{(y)_i+1}$. Thus the definition of $E_{n+1}^{\vec{\beta}}$ ensures that $E_{n+1}^{\vec{\beta}}(e,\langle \vec{x} \rangle) \equiv (\exists y)\neg E_n^{\vec{\beta}}(e,\langle y,\vec{x}\rangle)$.

Theorem 2   [*Enumeration*, or *Indexing Theorem* (Kleene)]

(i) $Q(\vec{x}) \in \Sigma_n^{\vec{\beta}}, n \geq 1$, iff $Q(\vec{x}) \equiv E_n^{\vec{\beta}}(e,\langle \vec{x} \rangle)$ for some e.

(ii) $Q(\vec{x}) \in \Pi_n^{\vec{\beta}}, n \geq 1$, iff $Q(\vec{x}) \equiv \neg E_n^{\vec{\beta}}(e,\langle \vec{x} \rangle)$ for some e.

Proof First, $E_n^{\vec{\beta}} \in \Sigma_n^{\vec{\beta}}$, $n \geq 1$ [Indeed, $E_1^{\vec{\beta}} \in \Sigma_1^{\vec{\beta}}$, since $T(e,x,[\vec{\beta} \mid y],y)$ is $\vec{\beta}$-recursive by Example 3.

If $E_n^{\vec{\beta}} \in \Sigma_n^{\vec{\beta}}$, then $\neg E_n^{\vec{\beta}} \in \Pi_n^{\vec{\beta}}$ by Proposition 2, hence $E_{n+1}^{\vec{\beta}} \in \Sigma_{n+1}^{\vec{\beta}}$ by Definition 2 and Theorem 1].

It follows that $\neg E_n^{\vec{\beta}} \in \Pi_n^{\vec{\beta}}, n \geq 1$. This establishes the *if* part for (i) and (ii). We now prove the *only if* part for (i) and (ii), simultaneously by induction on $n \geq 1$.

(i) Let $P(y,\vec{x}) \in \Pi_n^{\vec{\beta}}$ and $Q(\vec{x}) \equiv (\exists y)P(y,\vec{x})$.
Thus, $Q(\vec{x}) \equiv (\exists y)P(y,\vec{x}) \equiv (\exists y)\neg E_n^{\vec{\beta}}(e,\langle y,\vec{x} \rangle)$, for some e. The last predicate is $E_{n+1}^{\vec{\beta}}(e,\langle \vec{x} \rangle)$.

The basis of the induction ($n = 1$) and case (ii) are left to the reader.//

Theorem 3   [*Hierarchy Theorem* (Kleene, Mostowski)] For $n \geq 1$,

(i) $\Sigma_n^{\vec{\beta}} - \Pi_n^{\vec{\beta}} \neq \varnothing, \quad \Pi_n^{\vec{\beta}} - \Sigma_n^{\vec{\beta}} \neq \varnothing$

(*)This is the *T*-predicate of Chapter 3 or 5.

(ii) $\Delta_{n+1}^{\vec{\beta}} - (\Sigma_n^{\vec{\beta}} \cup \Pi_n^{\vec{\beta}}) \neq \varnothing$

(iii) $\Sigma_n^{\vec{\beta}} \neq \Sigma_{n-1}^{\vec{\beta}}, \quad \Pi_n^{\vec{\beta}} \neq \Pi_{n-1}^{\vec{\beta}}.$

Proof (i) By explicit transformation, $\lambda x.E_n^{\vec{\beta}}(x,\langle x \rangle) \in \Sigma_n^{\vec{\beta}}.$
If $E_n^{\vec{\beta}}(x,\langle x \rangle) \in \Pi_n^{\vec{\beta}}$, then (Proposition 2) $\neg E_n^{\vec{\beta}}(x,\langle x \rangle) \in \Sigma_n^{\vec{\beta}}$, hence for some e
(Theorem 2) $\neg E_n^{\vec{\beta}}(x,\langle x \rangle) \equiv E_n^{\vec{\beta}}(e,\langle x \rangle).$ Set $x \leftarrow e$ to obtain a contradiction.
Thus $E_n^{\vec{\beta}}(x,\langle x \rangle) \notin \Pi_n^{\vec{\beta}}.$ Also, $\neg E_n^{\vec{\beta}}(x,\langle x \rangle) \in \Pi_n^{\vec{\beta}} - \Sigma_n^{\vec{\beta}}$, by Proposition 2.

(ii) Let $P(x,y) \overset{\text{def}}{=} E_n^{\vec{\beta}}(x,\langle x \rangle) \vee \neg E_n^{\vec{\beta}}(y,\langle y \rangle).$ Note that there are x_0,y_0 in $\mathcal{N}$
such that $\neg E_n^{\vec{\beta}}(x_0,\langle x_0 \rangle)$ and $E_n^{\vec{\beta}}(y_0,\langle y_0 \rangle).$ [Otherwise, e.g., $\lambda x.E_n^{\vec{\beta}}(x,\langle x \rangle) = \varnothing$,
hence it is recursive and therefore in $\Pi_n^{\vec{\beta}}$ by Corollary 3 of Theorem 1,
contradicting (i).]
Since $E_n^{\vec{\beta}} \in \Sigma_n^{\vec{\beta}}$ and $\neg E_n^{\vec{\beta}} \in \Pi_n^{\vec{\beta}}$, both $E_n^{\vec{\beta}}$ and $\neg E_n^{\vec{\beta}}$ are in $\Delta_{n+1}^{\vec{\beta}}$ (Corollary 2,
Theorem 1) hence by the closure properties (Theorem 1)

$$P(x,y) \in \Delta_{n+1}^{\vec{\beta}}$$

Let also $P \in \Sigma_n^{\vec{\beta}}.$ Then $\lambda y.\neg E_n^{\vec{\beta}}(y,\langle y \rangle) \in \Sigma_n^{\vec{\beta}} \; [\neg E_n^{\vec{\beta}}(y,\langle y \rangle) \equiv P(x_0,y)]$ contradicting (i). Similarly, $P \notin \Pi_n^{\vec{\beta}}.$

(iii) For $n > 1$ this follows from (ii). For $n = 1$, $\Sigma_1^{\vec{\beta}} \neq \Sigma_0^{\vec{\beta}}$ and $\Pi_1^{\vec{\beta}} \neq \Pi_0^{\vec{\beta}}$ is
established, e.g., by $E_1^{\vec{\beta}} \in \Sigma_1^{\vec{\beta}} - \Sigma_0^{\vec{\beta}}$, and the observation $\Pi_1^{\vec{\beta}} = \Pi_0^{\vec{\beta}}$ iff $\Sigma_1^{\vec{\beta}} = \Sigma_0^{\vec{\beta}}.//$

Note: If $A \in \Sigma_n^{\vec{\beta}}$ (resp. $A \in \Pi_n^{\vec{\beta}}$) and $R(\vec{x}) \leq_m A$ via a $\vec{\beta}$-recursive function f
[that is, $\vec{x} \in R$ iff $f(\vec{x}) \in A$] then the "complexity" of R is at worst $\Sigma_n^{\vec{\beta}}$ (resp. $\Pi_n^{\vec{\beta}}$)
by Theorem 1. Using the m-reducibility concept we may, on occasion,
optimally place A in the hierarchy. For example, assume that we have a proof
that (1) A is in $\Sigma_n^{\vec{\beta}}$ (resp. $\Pi_n^{\vec{\beta}}$) and (2) *every* $\Sigma_n^{\vec{\beta}}$- (resp. $\Pi_n^{\vec{\beta}}$-) predicate is
m-reducible to A. Then A is *no* lower than $\Sigma_n^{\vec{\beta}}$ (resp. $\Pi_n^{\vec{\beta}}$) in the hierarchy since
that would "pull" $E_n^{\vec{\beta}}$ (resp. $\neg E_n^{\vec{\beta}}$) lower in the hierarchy, contradicting
Theorem 3.

Definition 4 A set $A \subset \mathcal{N}$ is $\Sigma_n^{\vec{\beta}}$-*complete* (resp. $\Pi_n^{\vec{\beta}}$-*complete*) iff

(i) $A \in \Sigma_n^{\vec{\beta}}$ (resp. $A \in \Pi_n^{\vec{\beta}}$)

(ii) For every $R(\vec{x})$ in $\Sigma_n^{\vec{\beta}}$ (resp. $\Pi_n^{\vec{\beta}}$)
$R(\vec{x}) \leq_m A$ via a $\vec{\beta}$-recursive function.//

Note: As usual, when $\vec{\beta}$ are recursive we drop the $\vec{\beta}$ superscript.

Example 6 Let $A = \{x \mid \phi_x \in \mathcal{R}\}.$ We have already seen that $A \in \Pi_2$ (Example 2).
Let $R(\vec{x}) \in \Pi_2.$ Then, for some recursive predicate $S(y,z,\vec{x})$ it is $R(\vec{x})$
$\equiv (\forall y)(\exists z)S(y,z,\vec{x}).$
Set $f = \lambda y\vec{x}.(\mu z)S(y,z,\vec{x}).$ Clearly, $f \in \mathcal{P}$, and by the S-m-n theorem there
is an $h \in \mathcal{R}$ such that $f = \lambda y\vec{x}.\phi_{h(\vec{x})}(y).$

Clearly, $R(\vec{x}) \leq_1 A$ via h. Thus, $A \notin \Sigma_2$ (otherwise, $\neg E_2(e,x) \in \Sigma_2$), and $A \notin \Sigma_n \cup \Pi_n$ for $n < 2$.//

Definition 5 A *function* $\lambda\vec{x}.f(\vec{x})$ is in $\Sigma_n^{\vec{\beta}}$ $(\Pi_n^{\vec{\beta}}, \Delta_n^{\vec{\beta}})$ iff $\lambda y\vec{x}.y = f(\vec{x})$ is in $\Sigma_n^{\vec{\beta}}$ $(\Pi_n^{\vec{\beta}}, \Delta_n^{\vec{\beta}})$.//

Lemma 1 If $R(\vec{x}) \in \Sigma_n^{\vec{\beta}} \cup \Pi_n^{\vec{\beta}}$, then $\chi_R \in \Delta_{n+1}^{\vec{\beta}}$.

Proof $y = \chi_R(\vec{x}) \equiv (R(\vec{x}) \;\&\; y = 0) \vee (\neg R(\vec{x}) \;\&\; y = 1)$. Thus, each of the disjuncts is in $\Delta_{n+1}^{\vec{\beta}}$ (by Corollary 2 of Theorem 1), hence also $\lambda y\vec{x}.y = \chi_R(\vec{x})$ by Theorem 1.//

Theorem 4 (*Post's theorem*)

For all $n \geq 0$

(i) $R \in \Sigma_{n+1}^{\vec{\beta}}$ iff R is $(\vec{\beta},A)$-re for some set $A \in \Sigma_n^{\vec{\beta}} \cup \Pi_n^{\vec{\beta}}$

(ii) $R \in \Delta_{n+1}^{\vec{\beta}}$ iff R is $(\vec{\beta},A)$-recursive for some set $A \in \Sigma_n^{\vec{\beta}} \cup \Pi_n^{\vec{\beta}}$

Proof (i) *if* part. Let $R(\vec{x})$ be $(\vec{\beta},A)$-re. Then, for some e,

$$R(\vec{x}) \equiv (\exists u)[(\exists s)_{\leq \Pi_{i\leq u} p_i^2} T(e, \langle \vec{x} \rangle, \langle \langle \beta_1 | u \rangle, \ldots, \langle \beta_l | u \rangle, s \rangle, u)$$

$$\&\; \text{length}(s) \leq u + 1 \;\&\; (\forall i)_{\leq u}\{p_i \nmid s \vee (s)_i = \chi_A(i) \;\&\; (s)_i \leq u\}]$$

Now $A \in \Sigma_n^{\vec{\beta}} \cup \Pi_n^{\vec{\beta}}$, hence (Lemma 1) $\lambda is.(s)_i = \chi_A(i) \in \Delta_{n+1}^{\vec{\beta}} \subset \Sigma_{n+1}^{\vec{\beta}}$. Thus, $R \in \Sigma_{n+1}^{\vec{\beta}}$ by Theorem 1, recalling that $T(e, \langle \vec{x} \rangle, \langle \langle \beta_1 | u \rangle, \ldots, \langle \beta_l | u \rangle, s \rangle, u)$ is $\vec{\beta}$-recursive (Example 3).

only if part. Let $R(\vec{x}) \equiv (\exists y)S(y,\vec{x})$ and $S \in \Pi_n^{\vec{\beta}}$. Define $A = \{\langle y,\vec{x} \rangle \mid S(y,\vec{x})\}$.

Then $R(\vec{x}) \equiv (\exists y)[\langle y,\vec{x} \rangle \in A] \equiv (\exists y)[\langle y,\vec{x} \rangle \notin \bar{A}]$. Clearly, R is $(\vec{\beta},A)$- and $(\vec{\beta},\bar{A})$-re $[\lambda y\vec{x}.\langle y,\vec{x} \rangle \in A$ and $\lambda y\vec{x}.\langle y,\vec{x} \rangle \notin \bar{A}$ are, of course, $(\vec{\beta},A)$- and $(\vec{\beta},\bar{A})$-recursive, respectively] and $A \in \Pi_n^{\vec{\beta}}$, $\bar{A} \in \Sigma_n^{\vec{\beta}}$.

(ii) *if* part. By the corresponding part in (i), R and $\neg R$ are in $\Sigma_{n+1}^{\vec{\beta}}$, hence $R \in \Delta_{n+1}^{\vec{\beta}}$.

only if part. By the corresponding part in (i), (since both R and $\neg R$ are in $\Sigma_{n+1}^{\vec{\beta}}$) there are sets A and B in $\Sigma_n^{\vec{\beta}}$ such that R is $(\vec{\beta},A)$-re and $\neg R$ is $(\vec{\beta},B)$-re.

Then both R and $\neg R$ are $(\vec{\beta},C)$-re, where C is the disjoint union or *join* (see Example 9.1.1) of A and B. [Indeed, R being $(\vec{\beta},A)$-re means that the equivalence in the *if*-part (i) holds. Change there $(s)_i = \chi_A(i)$ into $(s)_i = \chi_C(2i)$. Similar comment for $\neg R$.]

Thus, R is $(\vec{\beta},C)$ recursive, and $C \in \Sigma_n^{\vec{\beta}}$ (Theorem 1). Similarly one may start with A and B in $\Pi_n^{\vec{\beta}}$, which results with C in $\Pi_n^{\vec{\beta}}$.//

Definition 6   Set $A^{(0)} = A$ for any $A \subset \mathcal{N}$. For $n \geq 0$, $A^{(n+1)} = (A^{(n)})^{\circ J}$.//

Note: $A^{(n)} = A\underbrace{^{\circ J \circ J \cdots \circ J}}_{n}$, for $n > 0$.

Corollary 1   For any $A \subset \mathcal{N}, n > 0$

   (i) $R \in \Sigma_n^A$ iff $R \leq_1 A^{(n)}$

   (ii) $R \in \Sigma_{n+1}^A$ iff R is $A^{(n)}$-re

   (iii) $R \in \Delta_{n+1}^A$ iff $R \leq_T A^{(n)}$.

   Proof   By Proposition 10.6.2, R is S-re iff $R \leq_1 S^{\circ J}$ for any S.
   Thus, (i) $\Rightarrow$ (ii) & (iii) by Theorem 4 [For example, $R \in \Delta_{n+1}^A \Rightarrow R \in \Sigma_{n+1}^A$ & $\neg R \in \Sigma_{n+1}^A \Rightarrow$ [by (i)] $R \leq_1 (A^{(n)})^{\circ J}$ and $\neg R \leq_1 (A^{(n)})^{\circ J} \Rightarrow R$ and $\neg R$ are $A^{(n)}$-re. The converse is as easy, using Theorem 4, and $A^{(n)} \in \Sigma_n^A$ [by (i)]].
   We proceed by induction on n:
   $n = 1$: $R \in \Sigma_1^A$ iff $R \leq_1 A^{(1)}$ simply says R is A-re iff $R \leq_1 A^{\circ J}$, which is true (Proposition 10.6.2). This establishes (i), and hence (ii) and (iii).
   Assume all (i)–(iii) for some n and proceed to $n + 1$: We need only establish (i): Let $R \in \Sigma_{n+1}^A$. Then R is $A^{(n)}$-re [(ii) and induction hypothesis] hence $R \leq_1 A^{(n+1)}$. Conversely, $R \leq_1 A^{(n+1)} \Rightarrow R$ is $A^{(n)}$-re hence [(ii) and induction hypothesis] $R \in \Sigma_{n+1}^A$.//

Corollary 2   For $n > 0$, $R \in \Pi_n^A$ iff $R \leq_1 \overline{A^{(n)}}$ ($\overline{S} = \mathcal{N} - S$ for any S, as usual).

Example 7 $\{x \mid \phi_x \in \mathcal{R}\} \leq_1 \overline{\varnothing^{(2)}}$.//

Example 8 $A^{(n)} \in \Sigma_n^A$, $\varnothing^{(n)} \in \Sigma_n$, $\overline{A^{(n)}} \in \Pi_n^A$, $\overline{\varnothing^{(n)}} \in \Pi_n$ for all $n \geq 0$ [$A^{(n)} \leq_1 A^{(n)}$ via $\lambda x.x$].//

   Theorem 8.5.1 is a useful tool for the optimal placement of some sets in the arithmetical hierarchy.

Example 9   Let $A = \{x \mid W_x \text{ is finite}\}$.
   Then $x \in A \equiv (\exists z)(\forall y)[y > z \Rightarrow \neg(\exists u)T(x,y,u)]$. The predicate in [...] is in Π_1 by Theorem 1, hence A is in Σ_2.
   Now, neither A nor $\overline{A}$ are re [apply Theorem 8.5.1 or just refer to Theorem 7.3.2], hence $A \in \Sigma_2 - (\Sigma_1 \cup \Pi_1)$.
   We also have $A \leq_1 \varnothing^{(2)}$. Now, by the note at the end of §9.3, $\{x \mid \phi_x \in \mathcal{R}\}$ $\sim \{x \mid W_x \text{ infinite}\}$. We know that $\{x \mid \phi_x \in \mathcal{R}\}$ is Π_2-complete (Example 6), indeed $\overline{\varnothing^{(2)}} \leq_1 \{x \mid \phi_x \in \mathcal{R}\}$ hence $\overline{\varnothing^{(2)}} \equiv_1 \{x \mid \phi_x \in \mathcal{R}\}$.
   Thus,

$$\{x \mid \phi_x \in \mathcal{R}\} \sim \{x \mid W_x \text{ infinite}\} \sim \overline{\varnothing^{(2)}}$$

$$\{x \mid \phi_x \notin \mathcal{R}\} \sim \{x \mid W_x \text{ finite}\} \sim \varnothing^{(2)}.//$$

Example 10   Let $A = \{x \mid W_x = K\}$. First, by Theorem 8.5.1 neither A nor $\overline{A}$ are re (that is, in Σ_1), hence $A \notin \Sigma_1 \cup \Pi_1$.

Next, $x \in A \equiv (\forall y)[y \in W_x \Leftrightarrow y \in K] \equiv (\forall y)[(y \notin W_x \vee y \in K) \,\&\, (y \in W_x \vee y \notin K)]$. Now $\lambda yx.y \in W_x$ and $\lambda y.y \in K$ are in Σ_1, thus $y \notin W_x \vee y \in K$ and $y \in W_x \vee y \notin K$ are in Δ_2, hence in Π_2. By Theorem 1, A is in Π_2.

Finally, $A \in \Pi_2 - \Sigma_1 \cup \Pi_1$.//

Note: Further examples of optimal positioning of sets in the arithmetical hierarchy are found in Rogers [13].

PROBLEMS

1. Prove that Ω is the smallest class containing Ω_0 and closed under the operations described in Definition 10.2.1, (a), (b), (c).

2. Show that the S–m–n clause ((b) in Definition 10.2.1) is redundant if clause (a) is amended to "$(\langle 1,k,l,a,\vec{b}_m\rangle,\vec{x}_k,\vec{\alpha}_l,y) \in \Omega_{p+1}$ just in case there is a $\vec{y}_m$ such that *all* of the following are in $\Omega_{(p)}$:

$$(a,\vec{y}_m,\vec{\alpha}_l,y) \text{ and } (b_i,\vec{x}_k,\vec{\alpha}_l,y_i), i = 1, \ldots, m\text{".}$$

3. Show that the composition clause ((a) in Definition 10.2.1) subsumes clause (c).

4. Show that if $\lambda\vec{x}\,\vec{\alpha}.F(\vec{x},\vec{\alpha})$ is recursive, then for all $\vec{x}$, $F(\vec{x},\vec{\alpha})$ is independent of $\vec{\alpha}$, i.e., $\vec{\alpha}$ can only be a "dummy" argument.

5. Prove Corollary 4 of Theorem 10.2.4.

6. Prove Proposition 10.2.2.

7. Prove Proposition 10.2.4.

8. Show that for any l, there is a partial recursive functional $\lambda n x\vec{\alpha}_l.F(n,x,\vec{\alpha}_l)$ such that, for all $e,\vec{m}$, $F(n,\langle e,\vec{m}\rangle,\vec{\alpha}_l) = 0$ implies $(e,\vec{m},\vec{\alpha}_l,y) \in \Omega_n$, where $y = \{e\}(\vec{m},\vec{\alpha}_l)$. Show, moreover, that F can be chosen to satisfy $F(n,\langle e,\vec{m}\rangle,\alpha_l)\downarrow \Rightarrow (\forall k)_{<n} F(k,\langle e,\vec{m}\rangle,\alpha_l)\downarrow$.

9. Show that for any l, there is a partial recursive functional $\lambda x\vec{\alpha}_l.L(x,\vec{\alpha}_l)$ such that, whenever $(e,\vec{m},\vec{\alpha}_l,y) \in \Omega$

$$L(\langle e,\vec{m}\rangle,\vec{\alpha}_l) = (\mu n)[(e,\vec{m},\vec{\alpha}_l,y) \in \Omega_n].$$

10. Fill in the missing details in the proof of the Selection Theorem (10.3.1).

11. Show that $\lambda xy\alpha.y = \alpha(x)$ is semi-recursive but not recursive.
    (*Hint.* If $\lambda xy\alpha.y = \alpha(x)$ is recursive, then so is $\lambda xyz.y = \{z\}(x)$)

12. *Argument:* $\lambda xy.x = y$ and $\lambda xy.x \neq y$ are recursive. Thus $\lambda xy\alpha.y = \alpha(x)$ and $\lambda xy\alpha.y \neq \alpha(x)$ are semi-recursive. It follows (Theorem 10.4.3) that $\lambda xy\alpha.y = \alpha(x)$ is recursive, contradicting Problem 11. What is wrong with this "argument"?

13. Show that $\lambda u\alpha.\langle\alpha\,|\,u\rangle$ is *not* consistent, i.e., for each u there are α, and β such that $\alpha\subset\beta$ & $\langle\alpha\,|\,u\rangle\neq\langle\beta\,|\,u\rangle$.

14. Augment the URM formalism so that (partial) functions are allowed as inputs. To this end, a URM M has, in general, a finite number of type 1 input registers α,β,γ, etc. The only instructions referring to such registers are of the form $X\leftarrow\alpha(X)$, where X is a number register.

   The above instruction can occur in URM M *only* if M has α as an input register. The semantics of such an instruction are "change (X) into $(\alpha)((X))$, where (α) (resp. (X)) is the contents of α (resp. X); should it be that $(\alpha)((X))\uparrow$, then M enters an infinite loop". Give a careful formal definition of the model along the suggestion above and show that the URM computable functionals are precisely the partial recursive functionals.

15. Show that for any k and l, there is a primitive recursive function $\lambda xy.f(x,y)$ such that for all $\vec{x}_k,\vec{\alpha}_l,a,b$

$$\{f(a,b)\}(\vec{x}_k,\vec{\alpha}_l) = \{a\}(\vec{x}_k,\alpha_l,\lambda y.\{b\}(y,\vec{x}_k,\vec{\alpha}_l)).$$

16. Let us call *restricted functionals*, those which only allow *total* type 1 inputs. Develop a definition of *partial recursive restricted functionals* and derive their elementary theory, paralleling the contents of §10.2, §10.4 and §10.5. In particular prove the Selection Theorem easily with the aid of the Normal Form theorems, where the T-predicates are now primitive recursive (as opposed to just semi-recursive).

17. Prove that if F and H are partial recursive *restricted* functionals of ranks $(k,l+1)$ and $(k+1,l)$ respectively then there is a partial recursive restricted functional G such that whenever $\lambda y.H(y,\vec{x}_k,\vec{\alpha}_l)$ is *total*, $G(\vec{x}_k,\vec{\alpha}_l) = F(\vec{x}_k,\vec{\alpha}_l,\lambda y.H(y,\vec{x}_k,\vec{\alpha}_l))$. Show that the class of *recursive* restricted functionals is closed under substitution in a type 1 argument.

18. Intuitively, $S = \lambda\vec{x}_k\vec{\alpha}_l.F(\vec{x}_k,\vec{\alpha}_l,\lambda y.H(y,\vec{x}_k,\vec{\alpha}_l))$ of Problem 17 is "computable" (give an informal algorithm) even if $\lambda y.H(y,\vec{x}_k,\vec{\alpha}_l)$ is nontotal. However, F is *undefined* (by Definition!), whenever *nontotal* functions are substituted into its type 1 arguments. It appears then that the "informal" algorithm computes an extension of S. Show by example that this situation cannot be rectified, i.e., the partial recursive *restricted* functionals are *not* closed under substitution in a type 1 argument (contrast with Corollary 6 of Theorem 10.5.2).
   (*Hint.* Consider, for total α, $F = \lambda xy\alpha.$ **if** $y = 1$ **then** 0 **else if** $y = 0$ **then** $0\cdot\alpha(x)$. Clearly F is a *recursive* restricted functional. Next observe that $\lambda x\alpha.(\tilde{\mu}y)F(x,y,\alpha)$, where $\tilde{\mu}$ is that of Definition 3.1.2, is a partial recursive restricted functional. The continuation of this argument is left to the reader.)

19. Find number-theoretic inductive characterizations for partial recursive (restricted and unrestricted) functionals which do not involve Gödel numberings directly.

20. Show by example that a function $f: \mathcal{N} \to \mathcal{N}$ *can* be non–recursive yet be β-recursive, where β is nontotal.

21. If $A \in \mathcal{R}_*$ then re-ness and recursiveness *in* A coincide with the corresponding unrelativized concepts.

22. (See, for example, Davis [1].) Define the Turing Machine analogue to A-relative computability of number theoretic functions, and show that it is equivalent to the notion of A-(partial) recursiveness (where $A \subset \mathcal{N}$).
(*Hint.* This TM model has an instruction type $q_1 a q_2 q_3 S$ with the understanding: "If the machine at state q_1 scans an a, then, without head movement, it enters state q_2 or q_3 depending on whether or not the number represented by the tape (e.g., number of *ones* in it) is in the chosen "oracle" set A or not."
Note that the instruction set of a TM with an "oracle" does not involve the oracle set directly, thus the oracle could be varied, yielding, essentially, a TM model that accepts *total* functions as inputs.)

23. Show that there exists a recursive function $\lambda x.f(x)$ such that for all x,y,e and A

$$ y = \{e\}^A(x) \text{ iff } (\exists u,v)[\langle x,y,u,v \rangle \in W_{f(e)} \text{ \& } D_u \subset A \text{ \& } D_v \subset \overline{A}] $$

where, say, the finite sets D_u, D_v are prime-power coded.

24. Complete the proof of Theorem 10.6.1.

25. Prove Proposition 10.6.1.

26. Prove that K is T-complete.

27. Show, without help from the recursion theorem, that the function $F = \lambda nz.\langle a(n),b(n),x^n(z) \rangle$ defined in the proof of the Friedberg-Muchnik theorem is recursive.

28. Show that the function F of the previous problem is recursive. This time, rely on the recursion theorem.

29. Complete the proof of Proposition 10.7.2.

30. Complete the proof of Theorem 10.7.2.

31. Show that $\{x \mid W_x \text{ is infinite}\} \leq_1 \{x \mid W_x = K\}$. Conclude that $\{x \mid W_x = K\} \sim \overline{\varnothing^{(2)}}$.
(*Hint.* [13] Take any re-index, e, of K.
Define ψ by $\psi(x,y)$ **if** in the enumeration of W_y there are *at least* x distinct entries **then** $\phi_e(x)$ **else** $\uparrow$. Show that $\psi \in \mathcal{P}$, hence by S-m-n theorem, there is a 1-1 $f \in \mathcal{R}^{(1)}$ such that $\psi = \lambda xy.\phi_{f(y)}(x)$. f effects the claimed reducibility.)

REFERENCES

[1] Davis, M. *Computability and Unsolvability*. New York: McGraw-Hill, 1958.

[2] Fenstad, J.E. *General Recursion Theory: An Axiomatic Approach*. Heidelberg: Springer-Verlag, 1980.

[3] Friedberg, R.M. "Two Recursively Enumerable Sets of Incomparable Degrees of Unsolvability". *Proc. of the National Academy of Sciences,* 43(1957): 236–238.

[4] Hennie, F. *Introduction to Computability*. Reading, Mass: Addison-Wesley, 1977.

[5] Hinman, P.G. *Recursion-Theoretic Hierarchies*. Heidelberg: Springer-Verlag, 1978.

[6] Kamke, E. *Theory of Sets*. New York: Dover Publications, Inc., 1950.

[7] Kleene, S.C. "Recursive Predicates and Quantifiers". *Transactions of the Amer. Math. Soc.,* 53(1943): 41–73 (also in Davis, M. (Ed.), *The Undecidable.* Hewlett, New York: Raven Press, 1965: 255–287).

[8] Kleene, S.C. "Recursive Functionals and Quantifiers of Finite Types, I". *Transactions of the Amer. Math. Soc.,* 91 (1959): 1–52.

[9] Kleene, S.C. *Formalized Recursive Functionals and Formalized Realizability*. Memoirs of the Amer. Math. Soc., 89 (1969).

[10] Monk, J.D. *Introduction to Set Theory*. New York: McGraw-Hill, 1969.

[11] Mostowski, A. "On Definable Sets of positive integers". *Fundamenta mathematicae,* 34 (1947): 81–112.

[12] Muchnik, A.A. "On the Unsolvability of the problem of Reducibility in the Theory of Algorithms" (Russian). *Doklady Akad. Nauk SSSR,* 108 (1956): 194–197.

[13] Rogers, H. *Theory of Recursive Functions and Effective Computability*. New York: McGraw-Hill, 1967.

[14] Sacks, G.E. *Degrees of Unsolvability*. Annals of Mathematics Studies, No. 55, Princeton, N.J.: Princeton University Press, 1963.

[15] Shoenfield, J.R. *Mathematical Logic*. Reading, Mass.: Addison-Wesley, 1967.

[16] Shoenfield, J.R. *Degrees of Unsolvability*. Amsterdam: North-Holland, 1971.

CHAPTER 11

The Recursion Theorems (Part II)

We shall revisit here the (second) recursion theorem for the purpose of proving some ad hoc results, some of which were announced in earlier chapters.

We shall also prove Kleene's *first* recursion theorem. In what follows, unless otherwise noted, we revert to nonrelativized computability of functions [= functionals of rank $(k,0)$].

§11.1 THE SECOND RECURSION THEOREM

Theorem 1   (Kleene) For any function $f \in \mathcal{P}^{(n+1)}$, there is an e such that $\phi_e^{(n)}(\vec{x}) = f(e,\vec{x})$ for all $\vec{x}$.

Proof As in Theorem 10.2.2.//

Corollary 1 [*Primitive Recursion Theorem* (Kleene)]
Let $i \mapsto \xi_i$ be an indexing of $\mathcal{P}\mathcal{R}^{(1)}$ which satisfies the *S-m-n* theorem, namely,
there is a function $\sigma \in \mathcal{P}\mathcal{R}^{(2)}$ such that $\xi_{\sigma(i,x)}(y) = \xi_i(\langle x,y \rangle)$ for all i,x,y, where $\langle \ \rangle$ is some standard primitive recursive pairing function.
Then, if $f \in \mathcal{P}\mathcal{R}^{(n+1)}$ there is an e such that $\xi_e^{(n)}(\vec{x}) = f(e,\vec{x})$ for all $\vec{x}$, where $\xi_e^{(n)} \overset{\text{def}}{=} \lambda\vec{x}_n.\xi_e(\langle \vec{x}_n \rangle)$.

Proof As in Theorem 10.2.2. Note that a ξ-indexing, as above, *does* exist for $\mathcal{P}\mathcal{R}^{(1)}$ (Theorem 3.5.2 and the method of §3.8). Setting $\xi_i^{(n)} = \lambda\vec{x}_n.\xi_i(\langle \vec{x}_n \rangle)$ we can easily verify that all the versions of the *S-m-n* theorem (for $\mathcal{P}$) proved in §3.8 hold for the ξ-indexing.//

Corollary 2 Let $f \in \mathcal{R}^{(1)}$. Then there is an e such that $\phi_{f(e)} = \phi_e$.

Proof By the universal function theorem, $\lambda xy.\phi_{f(x)}(y)$ is in $\mathcal{P}^{(2)}$. Thus, there is an e, such that $\phi_e(y) = \phi_{f(e)}(y)$ for all y, by Theorem 1.//

Note: The proof of Corollary 2 is not applicable in the primitive recursive case (with $f \in \mathcal{P}\mathcal{R}$) since $\lambda xy.\xi_x(y)$ is not primitive recursive (Corollary 2 of Theorem 3.5.2). Indeed, Corollary 2 does not hold for $\mathcal{P}\mathcal{R}$: If it *did*, then by *S-m-n* (for ξ of §3.5) there is an $h \in \mathcal{P}\mathcal{R}^{(1)}$ such that $\lambda xy.\xi_{h(x)}(y) = \xi_x(y) + 1$, [indeed, $h = \lambda x.\langle 0,x,0 \rangle + 1$ will do (Theorem 3.5.2)], hence for some e and all y, $\xi_e(y) = \xi_{h(e)}(y) = \xi_e(y) + 1$, which is a contradiction since ξ is total.
Corollary 2 is the version of the second recursion theorem favored by Rogers [13]. It is equivalent to Theorem 1 [let $f \in \mathcal{P}^{(n+1)}$. By *S-m-n* theorem, there is an $h \in \mathcal{P}\mathcal{R}^{(1)}$ such that $\phi_{h(y)}(\langle \vec{x} \rangle) = f(y,\vec{x})$ for all $y,\vec{x}$. By Corollary 2, there is an e such that $\phi_e = \phi_{h(e)}$, hence $\phi_e^{(n)}(\vec{x}) = \phi_e(\langle \vec{x} \rangle) = \phi_{h(e)}(\langle \vec{x} \rangle) = f(e,\vec{x})$, which proves Theorem 1] and its proof is of independent interest. In the sequel we shall adopt the Rogers' recursion theorem, which we prove (independently of Theorem 1) as follows [13]:

Theorem 2 (*Rogers' version of the recursion theorem*) For any $f \in \mathcal{R}^{(1)}$ there is an e such that $\phi_e = \phi_{f(e)}$.

Proof Define $\psi = \lambda xy.\phi_{\phi_x(x)}(y)$.
By the universal function theorem, $\psi \in \mathcal{P}$. By *S-m-n* theorem, there is an $h \in \mathcal{P}\mathcal{R}^{(1)}$ such that

$$\phi_{h(x)}(y) = \phi_{\phi_x(x)}(y) \text{ for all } x,y$$

Let $f \circ h = \phi_v (\in \mathcal{R})$. Hence, $\phi_v(v)\downarrow$ and $\phi_{h(v)} = \phi_{\phi_v(v)} = \phi_{f \circ h(v)}$. Take $e = h(v)$.//

Corollary 1   For any $f \in \mathcal{R}^{(1)}$ there is an e such that $W_{f(e)} = W_e$.

> ***Proof*** $W_e = \text{dom}(\phi_e)$; $W_{f(e)} = \text{dom}(\phi_{f(e)})$ and e such that $\phi_e = \phi_{f(e)}$ exists.//

> **Note:** "For any $f \in \mathcal{R}^{(1)}$, there is an e ...". This statement conceals some dependence of e on f. We make this precise.

Corollary 2   There exists a $q \in \mathcal{R}^{(1)}$ such that for any z, $\phi_z \in \mathcal{R}$ implies $\phi_{q(z)} = \phi_{\phi_z(q(z))}$.

> ***Proof*** Let $p \in \mathcal{R}^{(1)}$ be such that $\phi_z \circ h = \phi_{p(z)}$. Then proceed as in the proof of Theorem 2 and set $q = h \circ p$.//

Corollary 3   As Corollary 2, but q may be assumed to be 1-1.

> ***Proof*** This is immediate, for we may use as ϕ-indexing of $\mathcal{P}$ that of Chapter 3. This has 1-1 S-m-n functions, hence h and p are 1-1 and so is q ($= h \circ p$).//

Corollary 4   For each k, there is a 1-1 $q^{(k+1)} \in \mathcal{R}^{(k+1)}$ such that for any total $\phi_z^{(k+1)}$ it is

$$\phi_{q(z, \vec{x}_k)} = \phi_{\phi_z^{(k+1)}(q(z, \vec{x}_k), \vec{x}_k)}.$$

> ***Proof*** There is a 1-1 $p^{(k+1)} \in \mathcal{R}^{(k+1)}$ such that $\phi_{p(z, \vec{x})}(y) = \phi_z^{(k+1)}(h(y), \vec{x})$ for all $y, z, \vec{x}$, where $\phi_{h(x)} = \phi_{\phi_x(x)}$ for all x as in the proof of Theorem 2. Then $q = h \circ p$.//

§11.2 APPLICATIONS OF THE RECURSION THEOREM

We present now a number of applications embedded in examples. Some are simply illustrations, others present deeper results.

Example 1   (*Self-reproducing machines*)
There is a program, which, regardless of the input received, outputs a copy of itself. Formally, there is an e, such that $\phi_e = \lambda x.e$.
Indeed, let $\psi(x, y) = \lambda xy.x$. Then for some $f \in \mathcal{R}^{(1)}$,

$$\psi(x, y) = \phi_{f(x)}(y) \text{ for all } x, y$$

Next, for some $e, \phi_{f(e)} = \phi_e$, thus for all y,

$$\phi_e(y) = \phi_{f(e)}(y) = \psi(e,y) = e. //$$

Example 2   There is an e such that $W_e = \{e\}$.

Indeed, there is an $f \in \mathcal{R}^{(1)}$ such that $W_{f(x)} = \{x\}$ [e.g., define $\phi_{f(x)}(y)$, with the help of S-m-n theorem, as **if** $y = x$ **then** 0 **else** $(\mu z)(z + 1)$]. The claim follows by Corollary 1 of Theorem 11.1.2.

W_e is in some sense a set which contains (a description of) itself. In another sense, it is a "self-referential" object, for it is defined in terms of itself $(e). //$

Example 3   (*Rice's theorem, revisited*)

Theorem $A = \{x \mid \phi_x \in \mathcal{C}\}$ is recursive iff $A = \emptyset$ or $A = \mathcal{N}$.

> **Proof**   (The idea is attributed by Rogers [13] to G. C. Wolpin.)
> *if* part. Trivial.
> *only if* part. By contradiction. Let $\emptyset \neq A \neq \mathcal{N}$ and $A \in \mathcal{R}_*$. Then (Proposition 9.1.2) $A \leq_m \overline{A}$ via some $f \in \mathcal{R}$. Let e be such that $\phi_e = \phi_{f(e)}$.
> Then $e \in A$ iff $\phi_e \in \mathcal{C}$ iff $\phi_{f(e)} \in \mathcal{C}$ iff $f(e) \in A$ iff $e \in \overline{A}$, a contradiction.
> *End of proof.*

Corollary   For any complete index set $A = \{x \mid \phi_x \in \mathcal{C}\}$, $A \nleq_m \overline{A}$ and hence $A \neq_m \overline{A}$. In particular,

$$\{x \mid \phi_x \in \mathcal{R}\} \neq_m \{x \mid \phi_x \notin \mathcal{R}\}$$

Note that we have already obtained this last result in Example 10.7.9 by a different argument. We can rephrase this observation as "there are noncomparable non re m- and 1-degrees". //

Example 4

Theorem   A set $A \subset \mathcal{N}$ is productive iff there is a $\psi \in \mathcal{P}^{(1)}$ such that for all x, $W_x \subset A$ $\Rightarrow \psi(x) \downarrow$ & $\psi(x) \in A - W_x$.

> **Proof** *only if* part. By definition of productiveness and $\mathcal{R} \subset \mathcal{P}$.
> *if* part. Let A be as in the statement of the theorem. We want to find a *recursive* productive function for it.
> There is an $f \in \mathcal{R}$ such that

$$W_{f(x,y)} = \begin{cases} W_y \text{ if } \psi(x)\downarrow \\ \varnothing \text{ otherwise} \end{cases}$$

[For example, take, by S-m-n theorem, $f \in \mathcal{R}$ such that $\phi_{f(x,y)} = \phi_y + 0 \cdot \psi(x)$]
By Corollary 4 of Theorem 11.1.2 (here f is $\phi_z^{(2)}$ for some fixed z) there is a $q \in \mathcal{R}$
such that

$$W_{q(y)} = W_{f(q(y),y)} \text{ for all } y. \tag{1}$$

Set $g = \psi \circ q$ and let y be arbitrary.
Then

$$g(y)\uparrow \Rightarrow \psi \circ q(y)\uparrow \Rightarrow W_{f(q(y),y)} = \varnothing$$

$$\Rightarrow [\text{by } (1)] \ W_{q(y)} = \varnothing \subset A \Rightarrow \psi \circ q(y)\downarrow \text{ (by definition of } \psi)$$

$$\Rightarrow g(y)\downarrow, \text{ a contradiction. Thus } g \text{ is } total, \text{ hence in } \mathcal{R}.$$

It follows that, for all y, $W_{q(y)} = W_{f(q(y),y)} = W_y$ (since $\psi \circ q(y)\downarrow$ for all y).
    Hence, $W_y \subset A \Rightarrow W_{q(y)} \subset A \Rightarrow$ (definition of ψ) $\psi \circ q(y) \in A - W_{q(y)}$
$\Rightarrow g(y) \in A - W_y$. *End of proof.*//

Example 5

Theorem 1   (Myhill) Creative sets are 1-complete.

   Proof Let A be creative and let $f \in \mathcal{R}^{(1)}$ be a 1-1 productive function for
$\overline{A}$ (Corollary 3 of Theorem 9.5.2).
   Let B be an arbitrary re set. We want to show

$$B \leq_1 A$$

To this end, obtain $s \in \mathcal{R}^{(2)}$ such that

$$W_{s(x,y)} = \begin{cases} \{f(x)\} \text{ if } y \in B \\ \varnothing \text{ otherwise} \end{cases}$$

[For example, set $\psi = \lambda xyz.$ **if** $z = f(x) + 0 \cdot q(y)$ **then** 0 **else** $(\mu w)(w + 1)$,
where $\text{dom}(q) = B$, and obtain s by the S-m-n theorem so that $\psi = \lambda xyz. \phi_{s(x,y)}(z)$].
By Corollary 4 of Theorem 11.1.2, there is a 1-1 $h \in \mathcal{R}$ such that $W_{s(h(y),y)} = W_{h(y)}$ for all $y \in \mathcal{N}$.
Claim: $B \leq_1 A$ via $f \circ h$.
First, $f \circ h$ is 1-1 and in $\mathcal{R}$.

Second, let $y \in B$. Then $W_{h(y)} = W_{s(h(y),y)} = \{f \circ h(y)\} \Rightarrow f \circ h(y) \in A$. [Otherwise, $W_{h(y)} \subset \overline{A}$, hence $f \circ h(y) \in \overline{A} - W_{h(y)}$ by productiveness of $\overline{A}$, contradicting $W_{h(y)} = \{f \circ h(y)\}$].

Third, let $y \notin B$. Then $W_{h(y)} = W_{s(h(y),y)} = \varnothing \subset \overline{A}$, hence (productiveness of $\overline{A}$) $f \circ h(y) \in \overline{A}$ ($\in \overline{A} - \varnothing = \overline{A} - W_{h(y)}$, that is). We established $y \in B$ iff $f \circ h(y) \in A$.   *End of proof.*

Corollary 1 A is creative iff $A \sim K$.

    Proof *if* part. K is creative and creativity is a $\sim$-invariant.

    only if part. $A \leq_1 K$ and $K \leq_1 A$ since both A and K are creative. Hence $A \sim K$ by Theorem 9.3.1.   *End of proof.*

Corollary 2 A is m-complete iff A is 1-complete.

    Proof A m-complete $\Rightarrow$ (by Corollary 2 of Theorem 9.5.1) A is creative $\Rightarrow A$ is 1-complete. That 1-completeness implies m-completeness is trivial.   *End of proof.*

We conclude this example with a postscript on strong reducibility.

Definition   (Post) $A \subset \mathcal{N}$ is *simple* iff all the following hold:

   (i) A is re.

   (ii) $\overline{A}$ is infinite.

   (iii)   For all x, W_x infinite $\Rightarrow A \cap W_x = \varnothing$.

Condition (iii) says that $\overline{A}$ contains no infinite re set, thus, by (ii), $\overline{A}$ is not re. It is immediate then, that simple sets *are not recursive*.

Post has shown that simple sets exist [11].

Theorem 2   (Post) Simple sets exist.

    Proof   The idea is to build an re set S by including one element from every infinite W_x (and perhaps from finite W_x's as well), at the same time making S "sparse" enough to ensure that $\overline{S}$ is infinite.

    Thus, one obtains and lists elements of S by dovetailing computations $\phi_x(y)$ allowing x to contribute y, iff y is the first found to satisfy $\phi_x(y)\downarrow$ and $y > 2x$.

    Formally, define f by

$$f(x) = K((\mu z)[T(x, K(z), L(z)) \,\&\, K(z) > 2x]), \text{ where } T \text{ is the (absolute)}$$

Kleene predicate of Chapter 3 (or 5), and K, L are projections of some primitive recursive pairing function J.

Clearly, $f \in \mathcal{P}$ (it is not total; consider, for example, x such that $W_x = \varnothing$) and $f(x) > 2x$, whenever $f(x)\downarrow$.

Set $S = \text{range}(f)$. S is simple, for

    (i) S is re (Theorem 8.1.2).

    (ii) $\overline{S}$ is infinite.

Indeed, $l \in S$ implies $f(x) = l$ for some $x < l/2$, thus $S \cap \{0, \ldots, 2n\}$, for any n, can contain at most those among $f(0), \ldots, f(n - 1)$ which are defined. It follows that for *all* n, $\overline{S}$ contains *at least* n elements (out of $\{0, \ldots, 2n\}$) thus it is infinite.

    (iii)   If W_x is infinite, then there is a smallest z such that $T(x, K(z), L(z))$ [this says $K(z) \in W_x$] and $K(z) > 2x$. $f(x) = K(z)$, i.e. $K(z) \in S$ as well as $K(z) \in W_x$.   *End of proof.*

We state a few consequences:

Corollary 1   There is a re set S which is none of the following

    (i)   *recursive*

    (ii)   *creative*

    (iii)   *m- or 1-complete*

    (iv)   *a cylinder*

Proof Take S to be a simple set, e.g., that of Theorem 2.

    (i)   has already been observed.

    (ii) $\overline{S}$ cannot be productive since then it should contain an infinite re set (Theorem 9.5.2).

    (iii)   follows from (ii) and Theorem 1 in this example.

    (iv)   Let instead $S \sim J(A \times \mathcal{N})$, where A is some set and J an 1-1 and onto recursive function.

$\varnothing \neq \overline{S} \sim \mathcal{N} - J(A \times \mathcal{N}) = J(\overline{A} \times \mathcal{N})$ [J is 1-1 & onto] hence $\overline{A} \neq \varnothing$. Let $a \in \overline{A}$. Then $J(\{a\} \times \mathcal{N})$ is infinite, re and contained in $J(\overline{A} \times \mathcal{N})$. Thus $\overline{S}$ contains an infinite re set as well (why?), a contradiction.   *End of proof.*

Corollary 2   Neither $\equiv_1$ and $\equiv_m$ nor $\leq_m$ and $\leq_1$ coincide on re nonrecursive sets.

Proof The second claim follows from the first.

So let S be simple. By Theorem 9.4.1 (i) and (ii), $S \equiv_m J(S \times \mathcal{N})$.

However, $S \neq_1 J(S \times \mathcal{N})$ since otherwise $S \sim J(S \times \mathcal{N})$ [Theorem 9.3.1] violating Corollary 1(iv) above.   *End of proof.//*

Example 6   (*The "Busy-Beaver" problem of Rado* [12])
We consider in this example the set of all Turing machines over a fixed alphabet A ($|A| > 1$) chosen so that Corollary 1 of Theorem 4.3.2 holds. We shall denote this set TM(A). The "Busy-Beaver" theorem states:

Theorem   Define $b:\mathcal{N} \to \mathcal{N}$ by $b(x) = \max\{[\mathrm{Res}_M(q_0B)] \mid M$ in TM(A) has x states$\}$, where each $M \in$ TM(A) has q_0 as the start-state, B is the blank symbol in A, and Res and [...] are defined immediately before Definition 4.1.7.
Then

    (i) b is total

    (ii)   For any $g \in \mathcal{R}^{(1)}$, there is an x_g such that $b(x) > g(x)$ for $x > x_g$ (that is, $b(x) > g(x)$ a.e.)

    (iii) $b \notin \mathcal{R}$

Proof (i) For any x, the set $\{[\mathrm{Res}_M(q_0B)] \mid M$ in TM(A) has x states$\}$ is *finite* [there is finite number of distinct M in TM(A) that have x states. (See Problem 4.)] and *nonempty* [e.g., it is easy to see (Problem 5) that an $M \in$ TM(A) with x states exists such that $[\mathrm{Res}_M(q_0B)] = x - 1$].
Thus $b(x)\!\downarrow$.
(ii) Let $g \in \mathcal{R}^{(1)}$. There is an n_g such that $\lambda x.g(x + n_g + 2) + 1$ is computable by a TM Z with n_g states (we postpone a proof of this claim for later, in the interest of argument continuity).
Let now $x > n_g + 1$. There is an $M \in$ TM(A) which has $x - n_g$ states $q_0, \ldots, q_{x-n_g-1}$ and $\mathrm{Res}_M(q_0B) = q_{x-n_g-1}1^{x-n_g-1}$. [e.g., M could be:

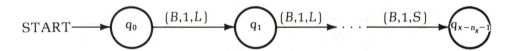

Let Z' be obtained from Z by changing the name of each state q_i into q_{i+x-n_g-1}. Clearly $M \cup Z'$ has x states and $x > n_g + 1$ implies $b(x) \geq [\mathrm{Res}_{M \cup Z'}(q_0B)]$ $= g(x) + 1 > g(x)$. [Recall the input/output conventions of TMs.]
Take $x_g = n_g + 1$ to satisfy (ii).
(iii) If $b \in \mathcal{R}$, then $b(x) > b(x)$ a.e.   *End of proof.//*
We now prove the claim made in the proof of (ii). We assume that TM(A) is Gödel-numbered as in §5.2. As in §5.2, we shall do arithmetic in m-adic notation, where the m-digits, in increasing order, are $1,B,;,q,S,L,R,s_2, \ldots, s_l$. $A = \{1,B,s_2, \ldots, s_l\}$, and q_i is denoted by $\underbrace{q \ldots q}_{i+1}$ for all $i \leq 0$.
Without loss of generality, we shall assume that, in any TM of k states, the states are consecutively numbered, $q_0, \ldots, q_{k-1}$. Thus, referring to §5.2, if

TM(x) (that is, x is a code of some TM) then the length of the longest tally of q's (that is, 4's) in x equals k.

We are thus motivated to define (for all x)

(1)   maxtal$(x) \overset{\text{def}}{=} (\mu y)_{\leq x}[\text{tally}_4(y) \ \& \ yPx \ \& \ \neg(y*4)Px]$, where "$*$" denotes m-adic concatenation.

(2)   #states$(x) \overset{\text{def}}{=} |\text{maxtal}(x)|$, where $|\dots|$ denotes m-adic length.

Clearly, #states $\in \mathcal{PR}$. By Kleene's second recursion theorem, there is an e such that

$\phi_e(x) = g(x + \text{#states}(e) + 2) + 1$, where the ϕ-indexing is that of §5.2. Take $n_g = \text{#states}(e)$.//

§11.3 THE FIRST RECURSION THEOREM

A functional $F: \mathcal{N}^{k+1} \times \mathcal{I} \to \mathcal{N}$ naturally induces a map $G: \mathcal{N}^k \times \mathcal{I} \to \mathcal{I}$ by setting, for all $(\vec{x}_k, \alpha)$ in $\mathcal{N}^k \times \mathcal{I}$, $G(\vec{x}_k, \alpha) = \lambda y.F(y, \vec{x}_k, \alpha)$.

For $k = 0$ $G(\alpha)$ and α have both rank $(0,1)$. $G: \mathcal{I} \to \mathcal{I}$ is called an *operator*.

Kleene's first recursion theorem says that if G is "effective" then there is a "least defined" solution ϕ to the equation $G(\alpha) = \alpha$. Moreover, ϕ is partial recursive.

"Effectiveness" of G amounts to partial recursiveness of the functional F.

Definition 1 $G: \mathcal{I} \to \mathcal{I}$ is a *recursive operator* iff there is a *partial recursive* F of rank $(1,1)$ such that $G(\alpha) = \lambda x.F(x,\alpha)$ for all $\alpha \in \mathcal{I}$.//

Example 1   Recursive operators are closed under composition. Indeed, let $G = \lambda\alpha.(\lambda x.F(x,\alpha))$ and $H = \lambda\alpha.(\lambda x.K(x,\alpha))$.

Then $G(H(\alpha)) = \lambda x.F(x,\lambda y.K(y,\alpha))$. The claim follows from the partial recursiveness of $\lambda x\alpha.F(x,\lambda y.K(y,\alpha))$ (see Corollary 6 of Theorem 10.5.2).//

Definition 2   Let $G: \mathcal{I} \to \mathcal{I}$ be any operator.

(a)   It is *monotone* if $\alpha \subset \beta \Rightarrow G(\alpha) \subset G(\beta)$.

(b)   It is *compact* if, for all x,y in $\mathcal{N}$ and $\alpha \in \mathcal{I}$, $G(\alpha)(x) = y \Longleftrightarrow$ there is a *finite function*(*) $\beta \subset \alpha$ such that $G(\beta)(x) = y$.//

Proposition 1   A recursive operator is compact and monotone.

(*)That is, dom(β) is finite.

Proof Let $G(\alpha) = \lambda x. F(x, \alpha)$ for all $\alpha \in \mathcal{J}$, where F is partial recursive. By Theorem 10.5.2,

$$G(\alpha)(x) = y \Rightarrow F(x, \alpha) = y \Rightarrow \text{(for some } u\text{)}$$

$$(\exists s)_{\leq \prod_{i \leq u} p_i^{u+1}} T(f, \langle x \rangle, 2^{s+1}, u) \,\&\, (u; s \,|\, \alpha) \,\&\, U(u) = y,$$

where $F = \{f\}$.
Setting $\beta = \lambda x. (\mu z)(\exp(x, s) = z + 1)$, we observe

(1) β is finite

(2) $\beta \subset \alpha$

(3) $G(\beta)(x) = y$. $\Leftarrow$ follows by monotonicity.

As for monotonicity, $\alpha \subset \gamma$ implies $\beta \subset \gamma$, hence $G(\gamma)(x) = y$.//

Definition 3 $[s] \overset{\text{def}}{=} \lambda x. (\mu y)(\exp(x, s) = y + 1)$.//

Theorem 1 $G: \mathcal{J} \to \mathcal{J}$ is a recursive operator iff the following hold:

(a) G is compact

(b) $\lambda s x. G([s])(x)$ is partial recursive

Proof *only if* part. We have (a) by Proposition 1. As for (b),

$$y = G([s])(x) \equiv (\exists u)[(\exists w)_{\leq \prod_{i \leq u} p_i^{u+1}} T(f, \langle x \rangle, 2^{w+1}, u) \,\&$$

$$(u; w \,|\, [s]) \,\&\, U(u) = y]$$

where $\{f\} = \lambda x \alpha. G(\alpha)(x)$.
Now, $\lambda u w s. (u; w \,|\, [s])$ is primitive recursive, thus $\lambda s x y. y = G([s])(x)$ is re, hence the claim.
if part. By (a), $y = G(\alpha)(x)$ iff $(\exists s)([s] \subset \alpha \,\&\, G([s])(x) = y)$. By (b), $\lambda x y z. G([s])(x) = y$ is re.

$$[s] \subset \alpha \equiv (\forall i)_{<\text{length}(s)} p_i \,|\, s \Rightarrow (s)_i = \alpha(i)$$

hence $\lambda s \alpha. [s] \subset \alpha$ is semi-recursive, and so is $\lambda y x \alpha. y = G(\alpha)(x)$; hence $\lambda x \alpha. G(\alpha)(x)$ is partial recursive.//

Note: Theorem 1 provides an indexing-independent approach to partial recursive functionals and is followed in Davis [4], Cutland [3] and Rogers [13] (in this last case via enumeration operators acting on sets rather than functions).

Theorem 2   A functional of rank $(1,1)$ is partial recursive iff $\lambda xy.F(x,\{y\})$ is partial recursive and $\lambda\alpha.(\lambda x.F(x,\alpha))$ is compact.

Proof *only if* part. Immediate from the functional substitution property (Corollary 6 of Theorem 10.5.2), and Proposition 1.

if part. Let $h \in \mathcal{R}^{(1)}$ be such that

$$\{h(x)\} = [x] \text{ (that is, } \phi_{h(x)} = [x])$$

Then $\lambda xy.F(x,\{h(y)\})$ [that is, $\lambda xy.F(x,[y])$] is partial recursive, and the result follows from Theorem 1.//

Note: Compactness is essential in Theorem 2 as we shall now see. Consider first,

Definition 4   A functional F of rank (k,l) is *weakly partial recursive* iff, for some $f \in \mathcal{N}$, and all $\vec{x}_k, \vec{e}_l, z$,

$$F(\vec{x}_k,\{e_1\}, \ldots, \{e_l\}) = z \text{ iff } \{f\}(\vec{x}_k,\vec{e}_l) = z.//$$

Note: Intuitively, *weakly partial recursive* functionals are computable on computable type 1 inputs, but they may fail to be computable on uncomputable such inputs. In the former case, to compute the answer one needs only compute on ϕ-indices of type 1 inputs.

Theorem 3   Every partial recursive functional is also weakly partial recursive.

Proof By functional substitution (Corollary 6 of Theorem 10.5.2).//

Note: The converse of Theorem 3 is false, as it can be seen by a cardinality argument (at this stage, the reader may want to review Definition 1.1.3 and the discussion following Example 1.4.5).

Let $\longleftrightarrow$ denote 1-1 correspondence.

A simple diagonalization (Problem 6) proves that for any set X, $X \longleftrightarrow\!\!\!/\!\!\!\longrightarrow 2^X$.

We have,

$$\mathcal{N} \longleftrightarrow \mathcal{P}^{(1)} \quad \text{(due to the } \textit{onto} \text{ map } i \longmapsto \phi_i)$$

$$\mathcal{N} \longleftrightarrow\!\!\!/\!\!\!\longrightarrow \mathcal{J}$$

Set $V = \mathcal{J} - \mathcal{P}^{(1)}$ and $U = \{F: \mathcal{J} \to \mathcal{N} \mid (F \mid \mathcal{P}^{(1)}) = \lambda\phi.0\}$.
Set $Q = \{\{e\} : \mathcal{J} \to \mathcal{N} \mid e \in \mathcal{N}\}$

  Consider the diagram

$$\mathcal{N} \xrightarrow{\; l \;} \mathcal{J} \xleftarrow{\; k \;} V \xrightarrow{\; i \;} \mathcal{P}(V; \mathcal{N}) \xrightarrow{\; j \;} U$$

where $l = \lambda x.\{(x, 0)\}$ ($\{(x, 0)\}$ is the partial function from $\mathcal{N} \to \mathcal{N}$ given by
$\lambda y.$ **if** $y = x$ **then** 0 **else** $\uparrow$)

   k is shown to exist in Problem 1.23.
   $i = \lambda\alpha.\{(\alpha, 0)\}$.
   $j = \lambda F.F \cup \{(\phi, 0) \mid \phi \in \mathcal{P}^{(1)}\}$
Thus, $j \circ i \circ k \circ l: \mathcal{N} \to U$ is 1-1.
It follows that $\mathcal{N} \longleftrightarrow\!\!\!| \; U$ (otherwise $\mathcal{N} \longleftrightarrow \mathcal{J}$, by the Schröder-Bernstein
theorem, due to the 1-1 maps $\mathcal{N} \xrightarrow{\; l \;} \mathcal{J}$ and $m \circ j \circ i \circ k: \mathcal{J} \to \mathcal{N}$, where $U \xleftarrow{\; m \;} \mathcal{N}$).
   Thus, $U \longleftrightarrow\!\!\!| \; Q$, since $Q \longleftrightarrow \mathcal{N}$ (due to the onto map $e \longmapsto \{e\}:\mathcal{N} \to Q$).
   Hence, $U - Q \neq \varnothing$.
But each $F \in U$ is weakly partial recursive.

Definition 5   Let $\psi \in \mathcal{P}^{(k+l)}$. If the relation $\Psi : \mathcal{N}^k \times (\mathcal{P}^{(1)})^l \to \mathcal{N}$, defined
by $(\vec{x}_k, \{e_1\}, \ldots, \{e_l\}) \longmapsto \psi(\vec{x}_k, \vec{e}_l)$ is *single-valued*, it is then termed an *effective
operation.//*

Theorem 4   (Myhill, Shepherdson) If

$$\Psi:\mathcal{N}^k \times (\mathcal{P}^{(1)})^l \to \mathcal{N}:(\vec{x}_k, \{e_1\}, \ldots, \{e_l\}) \longmapsto \psi(\vec{x}_k, \vec{e}_l)$$

is an effective operation, then there is a *unique* partial recursive functional Γ
of rank (k, l) such that $F \mid \mathcal{N}^k \times (\mathcal{P}^{(1)})^l = \Psi$.

   Proof Let us deal with *uniqueness* first:
   Let F and G both extend Ψ. Let $F(\vec{x}_k, \vec{\alpha}_l) = z$. By the results of §10.5,
$F(\vec{x}_k, \vec{\alpha}_l \mid u) = z$ for some $u \in \mathcal{N}$. But each $\alpha_i \mid u$ is finite, hence $\{a_i\} = \alpha_i \mid u$ for
some $a_i \in \mathcal{N}$.
   Hence,

$$z = F(\vec{x}_k, \{a_1\}, \ldots, \{a_l\}) = \Psi(\vec{x}_k, \{a_1\}, \ldots, \{a_l\})(*)$$

$$= G(\vec{x}_k, \{a_1\}, \ldots, \{a_l\}) = G(\vec{x}_k, \vec{\alpha}_l \mid u) = G(\vec{x}_k, \vec{\alpha}_l)$$

[since $G(\vec{x}_k, \vec{\alpha}_l \mid u) = G(\vec{x}_k, \vec{\theta}_l)$, where $\vec{\theta}_l$ is a subfunction of $\vec{\alpha}_l \mid u$, hence of $\vec{\alpha}_l$].
   We turn now to the *existence* part:

(*)Of course, the choice of a_i does not affect the output.

As the sought-after F will be compact, in the sense that $F(\vec{x}_k, \vec{\alpha}_l) = z$ iff $F(\vec{x}_k, \vec{\theta}_l) = z$ for some finite subfunction $\vec{\theta}_l$ of $\vec{\alpha}_l$, we are motivated to define:

$$F(\vec{x}_k, \vec{\alpha}_l) = z \overset{\text{def}}{\equiv} (\exists \vec{s}_l)([s_1] \subset \alpha_1 \,\&\, \ldots \,\&\, [s_l] \subset \alpha_l \,\&$$

$$\psi(\vec{x}_k, h(s_1), \ldots, h(s_l)) = z), \text{ where } h \text{ is an } S\text{-}m\text{-}n$$

function such that, for all $s \in \mathcal{N}$, $[s] = \{h(s)\}$. We first establish that $(\vec{x}_k, \vec{\alpha}_l) \overset{F}{\longmapsto} z$, as defined above, is single-valued. To this end, fix $\vec{x}_k$ and z and set $A = \{(e_1, \ldots, e_l) \mid \psi(\vec{x}_k, \vec{e}_l) = z\} = \{\vec{e}_l \mid \Psi(\vec{x}_k, \{e_1\}, \ldots, \{e_l\}) = z\}$.

A straightforward adaptation of the proof of Lemma 7.3.2 and Theorem 8.5.1 shows

(1) $\{e_i\} \subset \{d_i\}, i = 1, \ldots, l$ and $\psi(\vec{x}_k, \vec{e}_l) = z \Rightarrow \psi(\vec{x}_k, \vec{d}_l) = z$.

(2) $\vec{e}_l \in A \Rightarrow (\exists \vec{s}_l)([s_i] \subset \{e_i\}_{i=1,\ldots,l} \,\&\, \Psi(\vec{x}_k, [s_1], \ldots, [s_l]) = z)$ $[\Psi(\vec{x}_k, [s_1], \ldots, [s_l]) = \psi(\vec{x}_k, h(s_1), \ldots, h(s_l)), \text{ of course}]$.

Let then $\vec{x}_k$ and $\vec{\alpha}_l$ be chosen, and let

(3) $[s_i] \subset \alpha_i$ and $[d_i] \subset \alpha_i$ for $i = 1, \ldots, l$, and

(4) $\psi(\vec{x}_k, h(s_1), \ldots, h(s_l)) = z_1$ and $\psi(\vec{x}_k, h(d_1), \ldots, h(d_l)) = z_2$.

By (3), $[s_i] \cup [d_i] = [m_i]$ $(m_i \in \mathcal{N})$ for $i = 1, \ldots, l$. Thus, [(1) and (4)],

$z_1 = \psi(\vec{x}_k, h(m_1), \ldots, h(m_l)) = z_2$. That is, $(\vec{x}_k, \vec{\alpha}_l) \overset{F}{\longmapsto} z$ is single-valued. $\lambda \vec{x}_k \vec{\alpha}_l z. F(\vec{x}_k, \vec{\alpha}_l) = z$ being semi-recursive, F is partial recursive. It remains to see that $F \mid \mathcal{N}^k \times (\mathcal{P}^{(1)})^l = \Psi$.

Let $F(\vec{x}_k, \{e_1\}, \ldots, \{e_l\}) = z$. Then, for some $\vec{s}_l$, $[s_1] \subset \{e_1\} \,\&\, \ldots \,\&\, [s_l] \subset \{e_l\}$ $\&\, \psi(\vec{x}_k, h(s_1), \ldots, h(s_l)) = z$. Thus, (1) gives $\psi(\vec{x}_k, \vec{e}_l) = z$, hence $\Psi(\vec{x}_k, \{e_1\}, \ldots, \{e_l\}) = z$.

Conversely, the last equality means [by (2)]: $(\exists \vec{s}_l)$ $([s_i] \subset \{e_i\}_{i=1,\ldots,l} \,\&\, \psi(\vec{x}_k, h(s_1), \ldots, h(s_l)) = z)$. This (by definition of F) means that $F(\vec{x}_k, \{e_1\}, \ldots, \{e_l\}) = z$.//

Definition 6 (Myhill, Shepherdson) $f \in \mathcal{R}^{(1)}$ is *extensional*, iff for all x, y, $\phi_x = \phi_y \Rightarrow \phi_{f(x)} = \phi_{f(y)}$.//

Corollary 1   (a) If G is a recursive operator then $G(\phi_e) = \phi_{f(e)}$ for some extensional f and all $e \in \mathcal{N}$.

(b) If $f \in \mathcal{R}^{(1)}$ is extensional, then there is a unique recursive operator G such that $G(\phi_e) = \phi_{f(e)}$ for all $e \in \mathcal{N}$.

Corollary 2   A functional F of rank (k, l) is *weakly partial recursive* iff for some $\hat{f} \in \mathcal{N}$, $(F \mid \mathcal{N}^k \times (\mathcal{P}^{(1)})^l)(\vec{x}_k, \vec{\alpha}_l) = z$ iff $\{\hat{f}\}(\vec{x}_k, \vec{\alpha}_l) = z$, for all $(\vec{x}_k, \vec{\alpha}_l, z) \in \mathcal{N}^k \times (\mathcal{P}^{(1)})^l \times \mathcal{N}$.

Proof *if* part. By Corollary 6 of Theorem 10.5.2, there is a $\psi \in \mathcal{P}^{(k+l)}$ such that $\psi(\vec{x}_k, \vec{e}_l) = \{\hat{f}\}(\vec{x}_k, \{e_1\}, \ldots, \{e_l\}) = F(\vec{x}_k, \{e_1\}, \ldots, \{e_l\})$ for all $\vec{x}_k, \vec{e}_l$ in $\mathcal{N}$.

only if part. Let F be *weakly* partial recursive of rank (k,l) and ψ partial recursive of rank $(k + l,0)$ such that $F(\vec{x}_k, \{e_1\}, \ldots, \{e_l\}) = \psi(\vec{x}_k, \vec{e}_l)$ for all $\vec{x}_k, \vec{e}_l$.

By Theorem 4, since $(\vec{x}_k, \{e_1\}, \ldots, \{e_l\}) \longmapsto \psi(\vec{x}_k, \vec{e}_l): \mathcal{N}^k \times (\mathcal{P}^{(1)})^l \to \mathcal{N}$ is single-valued (why?), there is a partial recursive functional $\{\hat{f}\}$ of rank (k,l) such that $\psi(\vec{x}_k, \vec{e}_l) = \{\hat{f}\}(\vec{x}_k, \{e_1\}, \ldots, \{e_l\})$ for all $\vec{x}_k, \vec{e}_l$.//

Corollary 3 F is *weakly partial recursive* of rank (k,l) iff there is an $f_0 \in \mathcal{N}$ such that for all $\vec{x}_k, \vec{e}_l, z$

  (a) $F(\vec{x}_k, \{e_1\}, \ldots, \{e_l\}) = z$ iff $\{f_0\}(\vec{x}_k, \vec{e}_l) = z$.

  (b)   If $F(\vec{x}_k, \{e_1\}, \ldots, \{e_l\}) = z$ then there are functions $g_1, \ldots, g_l$ such that

  (i) $g_i \subset \{e_i\}, i = 1, \ldots, l$

  (ii) $F(\vec{x}_k, \vec{g}_l) = z$

  (iii)   For all p,q in $\mathcal{N}$ and all $i = 1, \ldots, l$
      $g_i(p) = q \Rightarrow (e_i, p, q)$ is a *subcomputation* of $(f_0, \vec{x}_k, \vec{e}_l, z)$.

Proof   The *if* part is trivial from Definition 4.
For the *only if* part, let $\{\hat{f}\}$ of rank (k,l) be as in Corollary 2.
Then,

$$\{\hat{f}\}(\vec{x}_k, \{e_1\}, \ldots, \{e_l\}) = z \text{ iff } (\exists u)(\exists \vec{s}_l)(T(\hat{f}, \langle \vec{x}_k \rangle, \langle \vec{s}_l \rangle, u) \&$$

$$(u; s_1 | \{e_1\}) \& \ldots \& (u; s_l | \{e_l\}) \& U(u) = z)$$

Note that $\lambda use.(u; s | \{e\})$ is re, thus for some $\hat{p} \in \mathcal{N}$ $(u; s_1 | \{e_1\}) \& \ldots \&$ $(u; s_l | \{e_l\}) \equiv \{\hat{p}\}(\langle u, \vec{s}_l \rangle, \vec{e}_l) \downarrow$ and moreover if $[s_i](r) \downarrow$ then (e_i, r) is a subcomputation of $(\hat{p}, \langle u, \vec{s}_l \rangle, \vec{e}_l)$ [output omitted in both subcomputations] due to the clause $(s_i)_r = \{e_i\}(r)$ in the definition of $(u; s_i | \{e_i\})$ (1)

Let $\hat{t}$ and $\hat{u}$ be re-indices of $T(\hat{f}, \langle \vec{x}_k \rangle, \langle \vec{s}_l \rangle, u)$ and $U(u) = z$ respectively. An index $\hat{h}$ exists such that $\{\hat{h}\}(\vec{x}_k, \vec{e}_l, v) = (\{\hat{t}\}(\hat{f}, \langle \vec{x}_k \rangle, \langle (v)_1, \ldots, (v)_l \rangle, (v)_0)$ $+ \{\hat{p}\}(v, \vec{e}_l) + \{\hat{u}\}((u)_0, (v)_{l+1})) \cdot 0$ such that for $i = 1, \ldots, l$, $[(v)_i](r) \downarrow \Rightarrow (e_i, r)$ is a subcomputation of $(\hat{h}, \vec{x}_k, \vec{e}_l, v)$ [$\{\hat{p}\}$ enters in $\{\hat{h}\}$ through composition] because of (1)   (2)

Let f_1 be such that

$$\{f_1\}(\vec{x}_k, \vec{e}_l) = \text{Sel}^{k+l,0}(\hat{h}, \vec{x}_k, \vec{e}_l) + \{\hat{h}\}(\vec{x}_k, \vec{e}_l, \text{Sel}^{k+l,0}(\hat{h}, \vec{x}_k, \vec{e}_l))$$

and that $(\hat{h}, \vec{x}_k, \vec{e}_l, \text{Sel}^{k+l,0}(\hat{h}, \vec{x}_k, \vec{e}_l))$ is a subcomputation of $(f_1, \vec{x}_k, \vec{e}_l)$ [it suffices to compute f_1 according to the rule for composition. (See Definition 10.2.2.)](3)

Clearly now, $F(\vec{x}_k, \{e_1\}, \ldots, \{e_l\}) = z$ iff $\{\hat{f}\}(\vec{x}_k, \{e_1\}, \ldots, \{e_l\}) = z$ iff $(\{f_1\}(\vec{x}_k, \vec{e}_l))_{l+1} = z$ (4)

Set $\{f_0\} = \lambda \vec{x}_k \vec{e}_l.(\{f_1\}(\vec{x}_k,\vec{e}_l))_{l+1}$ so that subcomputations of $\{f_1\}$ are subcomputations of $\{f_0\}$ [easy, through composition].

Let now, for some $\vec{x}_k,\vec{e}_l$ and z, $F(\vec{x}_k,\{e_1\}, \ldots, \{e_l\}) = z$. Then, by (4), $\{f_0\}(\vec{x}_k,\vec{e}_l) = z$.

Take $g_i = [(\{f_1\}(\vec{x}_k,\vec{e}_l))_i]$ for $i = 1, \ldots, l$. (i) and (ii) are immediately satisfied, the latter because $\{\hat{f}\}(\vec{x}_k,[(\{f_1\}(\vec{x}_k,\vec{e}_l))_1], \ldots, [(\{f_1\}(\vec{x}_k,\vec{e}_l))_l]) = z$.

Let $g_i(r)\downarrow$, for some i,r.

Then (e_i,r) is a subcomputation of $(\hat{p}, \operatorname{Sel}^{k+l,0}(\hat{h},\vec{x}_k,\vec{e}_l),\vec{e}_l)$ hence of $(\hat{h},\vec{x}_k,\vec{e}_l,\operatorname{Sel}^{k+l,0}(\hat{h},\vec{x}_k,\vec{e}_l))$ by observation (2). Since [composition] the latter is a subcomputation of $(f_1,\vec{x}_k,\vec{e}_l)$, by choice of f_1 [observation (3)] we are done, that is, $g_i(r) = d \Rightarrow (e_i,r,d)$ is a subcomputation of $(f_0,\vec{x}_k,\vec{e}_l,z)$ (this time including "outputs" in the computation tuples).//

Note: All this fuss formalized the obvious, that the "program" f_0 can be built so that it uses the programs e_i as subroutines for those r,d such that $g_i(r) = d$. We forced this by including $\{\hat{h}\}$ in the computation of f_1 in a "dummy" fashion ($\hat{h}$-output is 0) so that the output is still $\operatorname{Sel}^{k+l,0}(\hat{h},\vec{x}_k,\vec{e}_l)$, which is the v that "codes" the computation $\{\hat{f}\}(\vec{x}_k,\{e_1\}, \ldots, \{e_l\})$.

We now turn to the main result of this section.

Theorem 5   (*Kleene's First Recursion Theorem*)

If $G: \mathcal{J} \to \mathcal{J}$ is a recursive operator, then

(i)   there is a $\psi \in \mathcal{J}$, which is *the least defined fixed point* of G in the sense that

(1) $G(\psi) = \psi$

(2) $G(\alpha) = \alpha \Rightarrow \psi \subset \alpha$

(ii) $\psi \in \mathcal{P}^{(1)}$.

Proof   The proof is now standard.

(i) follows from an application of the Knaster-Tarski theorem (1.4.3) and its corollary.

Partial-ordering the set $\mathcal{J}$ by inclusion ($\subset$) we immediately obtain that G is monotone (Proposition 1). Compactness implies *continuity* in the sense of Definition 1.4.4. Indeed, let $\alpha_0 \subset \alpha_1 \subset \alpha_2 \subset \ldots$ be a chain in $\mathcal{J}$.

Then its lub is $\beta = \bigcup_{i \geq 0} \alpha_i$ (which is, of course, a function). By monotonicity of G, $G(\alpha_0) \subset G(\alpha_1) \subset \ldots$, thus $\gamma = \bigcup_{i \geq 0} G(\alpha_i)$ is a function in $\mathcal{J}$ as well.

Also, $G(\alpha_n) \subset G(\beta)$ for all n, thus $\gamma \subset G(\beta)$.

To show the opposite inclusion, let $G(\beta)(x) = y$. By compactness, there is a *finite* $\delta \subset \beta$ such that $G(\delta)(x) = y$. There is an $n \in \mathcal{N}$ such that $\delta \subset \alpha_n$ ($n = \max_i\{n_i \mid i\text{th element of } \delta \text{ is in } \alpha_{n_i}\}$). By monotonicity $G(\alpha_n)(x) = y$, hence $\gamma(x) = y$ ($G(\alpha_n) \subset \gamma$). (x,y) being arbitrary, $G(\beta) \subset \gamma$ is established.

Thus, by Theorem 1.4.3 (and its corollary), $\psi = \bigcup_{i \geq 0} f_i$ is the *least* fixed point of G, where $f_0 = \varnothing$ (the empty function) and $f_{n+1} = G(f_n)$ for $n \geq 0$.

(ii) Let $g \in \mathcal{R}^{(1)}$ be such that $G(\phi_e) = \phi_{g(e)}$ for all $e \geq 0$. Let e_0 be such that $\phi_{e_0} = \varnothing \; (= f_0)$.

Define h by

$$h(0) = e_0$$

$$h(n + 1) = g \circ h(n)$$

It is clear (induction on n) that $\phi_{h(n)} = f_n$ for $n \geq 0$. Thus, $y = \psi(x)$ iff $(\exists n)(y = \phi_{h(n)}(x))$, hence $\lambda xy.y = \psi(x)$ is re, therefore $\psi \in \mathcal{P}^{(1)}.//$

Note: In computer programming, procedures which are defined in terms of themselves are called *recursive*. This is partial justification for the term "recursion theorem", for ψ of Theorem 5 can be thought of as a solution to the "procedure-equation" $\alpha = G(\alpha)$, where G is a recursive operator. One can think of α as the "name" of a "function procedure" which is defined in terms of itself via "effective operations" embodied in G. The first recursion theorem ensures *existence* of a "least", computable, α which fits the definition.

Example 2 $G = \lambda \alpha.\alpha$ is a recursive operator, for $\lambda x \alpha.G(\alpha)(x) = \lambda x \alpha.\alpha(x)$ is obviously partial recursive.

Every function $\gamma \in \mathcal{J}$ is a *fixed point* of G, the *least* being the empty function; the latter of course is in $\mathcal{P}.//$

Example 3   By Corollary 1(a) of Theorem 4, we know that if G is a recursive operator then an $f \in \mathcal{R}^{(1)}$ exists such that $\phi_{f(e)} = G(\phi_e)$ for all $e \in \mathcal{N}$.

Applying the second recursion theorem, we obtain a z_0 such that $\phi_{f(z_0)} = \phi_{z_0}$, hence $\phi_{z_0} = G(\phi_{z_0})$. Thus we obtained a partial recursive fixed point of G. However, ϕ_{z_0} is *not* guaranteed to be the *least* fixed point.

To see this, consider the following argument (Rogers [13], Theorem XIII, p. 196):

Define $\psi = \lambda xz.$ **if** $x = q(z)$ **then** t **else** x, where $\phi_t = \lambda x.0$ and q is that of Corollary 2 of Theorem 11.1.2.

By S-m-n theorem there is an $h \in \mathcal{R}^{(1)}$ such that $\psi = \lambda xz.\phi_{h(z)}(x)$. Let m be such that $\phi_{h(m)} = \phi_m$.

Thus,

$$\phi_m(x) = \phi_{h(m)}(x) = \begin{cases} x \text{ \textbf{if} } x \neq q(m) \\ t \text{ \textbf{if} } x = q(m) \end{cases} \tag{1}$$

Hence, $\phi_{\phi_m(x)} = \phi_x$ when $x \neq q(m)$ $[x = \phi_m(x)$ in this case], but also (Corollary 2 of Theorem 11.1.2)

$$\phi_{\phi_m(q(m))} = \phi_{q(m)}$$

Thus, $\phi_{\phi_m(x)} = \phi_x$ for *all* x, hence ϕ_m is an *extensional* and defines a unique recursive operator Ψ such that $\Psi(\phi_e) = \phi_{\phi_m(e)}$ for all $e \in \mathcal{N}$.

Clearly, each ϕ_e is a fixed point of Ψ and its least fixed point is $\varnothing$. However, the fixed point provided by the second recursion theorem is $\phi_{q(m)} = \phi_{\phi_m(q(m))} = \phi_t$ by (1) above. But $\phi_t \neq \varnothing$. //

Despite Example 3, and the proof of Theorem 5 which did *not* use the second recursion theorem directly or indirectly, the first recursion theorem *does* follow from the second (without recourse to the Knaster-Tarski theorem) and this is one of the reasons that Rogers calls the former the *weak* and the latter the *strong* recursion theorems.

The following is a generalization of the first recursion theorem for weakly partial recursive functionals. As stated and proved below it is a watered-down version of that due to Moschovakis [10].

Theorem 6
(First Recursion Theorem for *weakly partial recursive functionals*)

If $\lambda x \alpha . F(x,\alpha)$ is weakly partial recursive and *consistent*(*) then there is a partial recursive ψ such that

(i) $F(x,\psi) = \psi(x)$ for all x

(ii)   For any α such that $F(x,\alpha) = \alpha(x)$ for all x, $\psi \subset \alpha$.

Proof (i) Let us first find a partial recursive fixed point. Let then $F(x,\{e\}) = \{f_0\}(x,e)$, where f_0 is as in Corollary 3 of Theorem 4.

Let e_0 be such that $\{f_0\}(x,e_0) = \{e_0\}(x)$ for all x and $\{e_0\}(x) = z \Rightarrow (f_0,x,e_0,z)$ is a subcomputation of (e_0,x,z) (see Theorem 10.2.2 and its corollary)   (1)

Take $\psi = \{e_0\}$.

(ii) Let $F(x,\alpha) = \alpha(x)$ for some $\alpha \in \mathcal{I}$ and all $x \in \mathcal{N}$.

To prove $\psi \subset \alpha$, we shall show that for all (x,z), $(x,z) \in \psi \Rightarrow (x,z) \in \alpha$.

So let $\{e_0\}(x) = z$.

Assume inductively that for all p,q, if (e_0,p,q) is a subcomputation of (e_0,x,z),(†) then

$$\alpha(p) = q \tag{2}$$

(*)Also called *monotone*. It means that $F(x,\alpha) = z$ and $\alpha \subset \beta \Rightarrow F(x,\beta) = z$. In other words, the operator $\lambda\alpha.(\lambda x.F(x,\alpha)): \mathcal{I} \to \mathcal{I}$ is monotone.

(†)This, of course, includes the statement $\{e_0\}(p) = q$, since $\{e_0\}(x) = z$ and hence its subcomputations are convergent.

By the choice of f_0, there is a function $g \subset \{e_0\}$ such that

$$F(x,g) = z \qquad (3)$$

and for all p and q, $g(p) = q \Rightarrow (e_0,p,q)$ is a subcomputation of (f_0,x,e_0,z).

By the induction hypothesis (2) [and (1)] we get that $g(p) = q \Rightarrow \alpha(p) = q$; i.e., $g \subset \alpha$. By *consistency* of F, and (3), we get $F(x,\alpha) = z$. Since α is a fixed point of F we get $\alpha(x) = z$. Thus $\psi = \{e_0\} \subset \alpha$.//

Note: e_0 was simply chosen as in the proof of the Kleene version of the second recursion theorem. Careful choice of f_0 enabled the argument to go through. A justification of the principle of *induction with respect to subcomputations* is given in Problem 8.

Theorem 5 is a special case of Theorem 6, because of Theorem 3. Theorem 6 is important due to its wider applicability on one hand (see note following Theorem 3), and on the other hand due to the relative ease with which *weak* partial recursiveness can be established, compared with partial recursiveness.

As a final comment, monotonicity (consistency) was still used in the proof of Theorem 6, but compactness was dropped. It turns out that monotonicity is *essential* to ensure that a least fixed point exists.

Example 4   Let b be the busy-beaver function (Example 11.2.6).
Define F by $F(x,\alpha) = $ **if** $\alpha \in \mathcal{P}^{(1)}$ **then** 0 **else** $b(x) + 1$

(1) F is weakly partial recursive, since $F(x,\{e\}) = 0$ for all x,e.

(2) F is *not* consistent (monotone). Indeed, $F(x,\varnothing) = 0$ for all x ($\varnothing \in \mathcal{P}^{(1)}$). $\varnothing \subset \alpha$ for all $\alpha \in \mathcal{I} - \mathcal{P}^{(1)}$, yet $F(x,\alpha) = b(x) + 1 > 0$ for such α.

(3)   Next, let β be a fixed point of F. β must be total (why?)

Case 1. $\beta \in \mathcal{P}^{(1)}$. Clearly, *only* $\beta = \lambda x.0$ will do.
Case 2. $\beta \in \mathcal{I} - \mathcal{P}^{(1)}$. Clearly, *only* $\beta = \lambda x.b(x) + 1$ will do.

Thus, F has exactly two fixed points, *both* total, therefore it has *no* least fixed point.
The moral is that

(a)   nonmonotone (inconsistent) weakly partial recursive functionals exist.

(b)   Theorem 6 fails for them.//

Example 5   The first recursion theorem, extended to operators that accept as inputs functions of more than one variable, provides, as already remarked, *meaning* (semantics) to "*recursive programs*".

If P is the program $P: \alpha(\vec{x}) := F(\vec{x},\alpha)$, where F is "effective" (e.g., weakly or strongly partial recursive) and "$::=$" denotes that the lefthand side is defined by the righthand side, then the unique *least* fixed point ψ_P of F is, of course, partial recursive, and it is reasonable to be taken as what the *program P is supposed to be computing.* [At this point, the reader should be careful to distinguish between P and F. F "computes" functions β such that $\beta(\vec{x}) = F(\vec{x},\alpha)$ for all $\vec{x}$ and α. P on the other hand "computes" $\psi(\vec{x})$ for some fixed point ψ of F. We have already suggested to take the *least* such ψ.]

Alternatively, one may say that what P really means is what we compute, using P, according to one or another scheme. The first approach is the *fixed point approach*, the latter is the *computational* approach to *program semantics.*

The theoritician may frown on the computational rule, which, broadly speaking, has as a consequence that what a recursive program of a given programming language is doing depends not on the language designer but on the language implementor.

We give the following example due to Morris [9]. Let $F(x,y,\alpha) =$ **if** $x = 0$ **then** 1 **else** $\alpha(x \doteq 1,\alpha(x,y))$. F is clearly "effective" [indeed, if $\langle \ \rangle$ is 1-1 and onto, with K and L as first and second projections, then G, given by $G(x,\beta) =$ **if** $K(x) = 0$ **then** 1 **else** $\beta(\langle K(x) \doteq 1,\beta(x)\rangle)$, is partial recursive, and $F(x,y,\alpha) = G(\langle x,y\rangle,\lambda u.\alpha(K(u),L(u)))$]. Consider the recursive program,

$$P: \alpha(x,y)::=F(x,y,\alpha)$$

Since F is (strongly) partial recursive, the least fixed point f is computed as follows:

Let $f_0(x,y)\uparrow$ for all x,y (that is, $f_0 = \varnothing$).

Let $f_{n+1}(x,y) = F(x,y,f_n) =$ **if** $x = 0$ **then** 1 **else** $f_n(x \doteq 1,f_n(x,y))$.

Note then that $f_1(x,y) =$ **if** $x = 0$ **then** 1 **else** $\uparrow$.

Assume that $f_n(x,y) =$ **if** $x \in \{0,1,\ldots,n\}$ **then** 1 **else** $\uparrow$.

Then $f_{n+1}(x,y) =$ **if** $x = 0$ **then** 1 **else if** $x \doteq 1 \in \{0,1,\ldots,n\}$ **then** 1 **else** $\uparrow$

$$= \textbf{if } x \in \{0,1,\ldots,n+1\} \textbf{ then } 1 \textbf{ else}\uparrow.$$

The least fixed point f is then $f = \bigcup_{n\geq 0} f_n$

$$= \lambda xy. \textbf{ if } x \geq 0 \textbf{ then } 1 \textbf{ else}\uparrow = \lambda xy.1.$$

In particular, $f(1,0) = 1$.

Let us now compute α using *call-by-value*(*) (which is very popular due to its simple implementation) for input (1,0): $\alpha(1,0) = \alpha(0,\alpha(1,0))$. Now the inner "expression" $\alpha(1,0)$ must be evaluated before the leftmost α starts computing. Thus $\alpha(0,\alpha(0,\alpha(1,0)))$ is our next task, and it is clear that we are "in a loop". Thus $\alpha(1,0)\uparrow$ if the computational rule (or "implementation") is call-by-value.

The reader can show that $\alpha(x,y) = $ **if** $x = 0$ **then** 1 **else** $\uparrow$ if call-by-value is the computational rule (Problem 13).

Clearly then $\alpha \subsetneq f$, hence α is *not* a fixed point of P.

Incidentally, the *leftmost rule* (or *call-by-name*) does the evaluation from left to right and does not therefore insist on evaluation of the arguments first. With such a rule $\alpha(1,0) = \alpha(0,\alpha(1,0)) = 1$ (**if** $0 = 0$ **then** 1 **else** etc., hence the output is 1). This is not a coincidence, as it can be proved that call-by-name computes the least fixed point in general(†).

We note here that a result of Cadiou [2] shows that computational rules always produce functions *at most* as much defined as the least fixed point.

We now turn to the fixed-point approach to attaching semantics to recursive programs.

The main attractiveness of the approach is the possibility of proving properties of such programs by a simple induction.

Assume for simplicity, that G is a recursive operator through which we define $\alpha \in \mathcal{J}$ in the program P.

P: $\alpha := G(\alpha)$ (or P: $\alpha(x) := F(x,\alpha)$, where $F = \lambda x \alpha . G(\alpha)(x)$).

By the proof of Theorem 5, the least fixed point f_P is $\bigcup_{n \geq 0} f_n$, where $f_0 = \varnothing$ and $f_{n+1} = G(f_n)$ for $n \geq 0$.

The *computational induction method* is to prove that P, or equivalently f_P, has a property $\mathcal{C}$ (that is, $f_P \in \mathcal{C}$) by proving, by *induction on* n, that $(\forall n) f_n \in \mathcal{C}$.

Now, for this scheme to follow through, it must be that $\mathcal{C}$ has the property

(A) For any recursive operator G, if $G^i(\varnothing) \in \mathcal{C}$ for all $i \geq 0$, then

$\bigcup_{i \geq 0} G^i(\varnothing) \in \mathcal{C}$, where $G^i = \overbrace{G \circ G \circ \ldots \circ G}^{i}$.

If $\mathcal{C}$ satisfies (A) it is called an *admissible property* (predicate) [7].

(*)In programming, a "call" to a function $f(e_1,e_2, \ldots)$, where $e_1,e_2, \ldots$ are expressions, is *by value* if the expressions are evaluated *before* the function starts computing and their *values* are passed as inputs to the function f.

(†)We shall not get into such a proof. Qualitatively, let us say that the leftmost rule produces the "computation-tree" of $\alpha(\vec{x})$, i.e., the set of its subcomputations, partially ordered by the subcomputation relation, in a *depth-first* manner, so it is equivalent to the *full-substitution* rule (= replacement of all occurrences of α simultaneously), a "breadth first" generation of the computation tree. The latter rule is shown to produce the least fixed point (for example) in Manna [7]. It is interesting to note that after defining $0 \cdot \uparrow = 0$ (as opposed to our convention $0 \cdot \uparrow = \uparrow$) in [7], it is shown that, under this convention, *call-by-name* does *not* compute the least fixed point (Problem 14). See however [7], pp. 388–389.

We are familiar with admissible properties, namely each $\mathcal{C}$, such that $\{x \mid \phi_x \in \mathcal{C}\}$ is re, is admissible by Theorem 8.5.1. Note that $\mathcal{C} = \{\alpha \mid \alpha \text{ is total}\}$ is not admissible.

However, totalness of f_P may often be proved by the simple device of showing $|\text{dom}(f_i)| \geq g(i)$, where $g(i)$ is some strictly increasing function of i.

The fixed point theory, or *"fixpoint theory"* of program semantics has originated with Scott [14]. We have hardly scratched the surface here, therefore the interested reader is referred to Manna [7] and Bird [1] for a thorough exposition and further references.//

PROBLEMS

1. Give a detailed formal proof of the Primitive Recursion Theorem, based on the ξ-indexing of Theorem 3.5.2.

2. (Substitution of type 1 arguments by functionals) In Corollary 6 of Theorem 10.5.2, functional substitution was shown to be legitimate via the Normal Form theorem. The reader is now asked to prove the legitimacy of functional substitution *without* relying on normal form theorems (this approach is necessary, for instance in recursion in higher types, where there are no normal form theorems). Specifically, show that there is a primitive recursive function $\lambda xy.f(x,y)$ such that for all $a,d,\vec{x}_k,\vec{\alpha}_l,n$

$$\{a\}(\vec{x}_k,\vec{\alpha}_l,\lambda y.\{d\}(y,\vec{z}_k,\vec{\alpha}_l)) = n \text{ iff } \{f(a,d)\}(\vec{x}_k,\vec{\alpha}_l) = n.$$

(*Hint.* The proof is standard [5,6,8]. Define a primitive recursive function $\lambda tad.g(t,a,d)$ by course-of-values recursion with respect to a. Then, relying on the primitive recursion theorem, choose $\bar{t}$ so that $g(\bar{t},a,d) = \xi(\bar{t},\langle a,d \rangle)$ for all a,d. Take $f = \lambda xy.\xi(\bar{t},\langle x,y \rangle)$.

The course-of-values recursion proceeds according to the cases of Definition 10.2.1, a running through the various index-forms found there.

To begin with, set $g(t,a,d) = 0$ for all t,a,d where a is *not* a valid index. Otherwise, here are some cases where a *is* an index.

(1) If $a = \langle 0,k,l+1,0,c \rangle$ then set $g(t,a,d) = \langle 0,k,l,0,c \rangle$

(2) If $a = \langle 0,k,l+1,3,i,j \rangle$ then if $j \leq l$ set $g(t,a,d) = \langle 0,k,l,3,i,j \rangle$ else set $g(t,a,d) = c$, where c is such that $\{c\}(\vec{x}_k,\vec{\alpha}_l) = n$ for all $\vec{x}_k,\vec{\alpha}_l$ and n, and, moreover,

$$\{d\}(x_i,\vec{x}_k,\vec{\alpha}_l)\downarrow \Rightarrow \|c,\vec{x}_k,\vec{\alpha}_l\| > \|d,x_i,\vec{x}_k,\vec{\alpha}_l\|$$

(3) If $a = \langle 1,k+m+1,l+1,m \rangle$, let $k \in \mathcal{PR}^{(3)}$ be such that

  (i) $\{k(t,a,d)\}(b,\vec{c}_m,\vec{x}_k,\vec{\alpha}_l) = n$ iff
     $\{\xi(t,\langle b,d \rangle)\}(\{\xi(t,\langle c_1,d \rangle)\}(\vec{x}_k,\vec{\alpha}_l), \ldots, \{\xi(t,\langle c_m,d \rangle)\}(\vec{x}_k,\vec{\alpha}_l),\vec{\alpha}_l) = n$
  for all $t,b,\vec{c}_m,d,\vec{x}_k,\vec{\alpha}_l,n$

(ii) $\{\xi(t,\langle b,d\rangle)\}(\{\xi(t,\langle c_1,d\rangle)]\}(\vec{x}_k,\vec{\alpha}_l),\dots,\{\xi(t,\langle c_m,d\rangle)\}(\vec{x}_k,\vec{\alpha}_l),\vec{\alpha}_l)\downarrow$
implies
$\|k(t,a,d),b,\vec{c}_m,\vec{x}_k,\vec{\alpha}_l\| > \|\xi(t,\langle b,d\rangle),\{\xi(t,\langle c_1,d\rangle)\}(\vec{x}_k,\vec{\alpha}_l),\dots,$

$$\{\xi(t,\langle c_m,d\rangle)\}(\vec{x}_k,\vec{\alpha}_l),\vec{\alpha}_l\|$$

Then set $g(t,a,d) = k(t,a,d)$

(4) If $a = \langle 3,k + 1,l + 1\rangle$, let $h \in \mathcal{PR}^{(3)}$ be such that

  (i) $\{h(t,a,d)\}(b,\vec{x}_k,\vec{\alpha}_l) = n$ iff $\{\xi(t,\langle b,d\rangle)\}(\vec{x}_k,\vec{\alpha}_l) = n$
  for all $t,b,d,\vec{x}_k,\vec{\alpha}_l,n$.

  (ii) $\{\xi(t,\langle b,d\rangle)\}(\vec{x}_k,\vec{\alpha}_l)\downarrow \Rightarrow \|h(t,a,d),b,\vec{x}_k,\vec{\alpha}_l\| > \|\xi(t,\langle b,d\rangle),\vec{x}_k,\vec{\alpha}_l\|$

Then set $g(t,a,d) = h(t,a,d)$
Once g has been defined in detail, and f is obtained by the primitive recursion
theorem, prove for all $a,d,\vec{x}_k,\vec{\alpha}_l,n$ that $\{a\}(\vec{x}_k,\vec{\alpha}_l,\lambda y.\{d\}(y,\vec{x}_k,\vec{\alpha}_l)) = n \Rightarrow$
$\{f(a,d)\}(\vec{x}_k,\vec{\alpha}_l) = n$ by induction on $\|a,\vec{x}_k,\vec{\alpha}_l,\beta\|$, where $\beta = \lambda y.\{d\}(y,\vec{x}_k,\vec{\alpha}_l)$, and
$\{f(a,d)\}(\vec{x}_k,\vec{\alpha}_l) = n \Rightarrow \{a\}(\vec{x}_k,\vec{\alpha}_l,\lambda y.\{d\}(y,\vec{x}_k,\vec{\alpha}_l)) = n$ by induction on
$\|f(a,d),\vec{x}_k,\vec{\alpha}_l\|$.
Note that clauses (ii), under (3) and (4) above, are relevant to the second
induction.
  Why was the more obvious and simple definition, $g(t,a,d) = g(t,b,d)$ for
g under (4) above, not suggested?)

3. Locate $\{x \mid W_x$ is simple$\}$ in the arithmetical hierarchy.

4. Show that there are finitely many TMs, over a given alphabet A, which have x
states.

5. Find a TM over an alphabet A, where $|A| > 1$, such that it has x states ($x \geq 1$)
and $\mathrm{Res}_M(q_0B) = x - 1$.

6. Show that for any set X, there is no 1-1 correspondence between X and 2^X.

7. Show that $\lambda\vec{x}_k\vec{\alpha}_l.F(\vec{x}_k,\vec{\alpha}_l)$ is partial recursive iff, for some $\psi \in \mathcal{P}^{(k+l)}$, $F(\vec{x}_k,\vec{\alpha}_l)$
$= z \equiv (\exists\vec{s}_l)([s_1] \subset \alpha_1 \& \dots \& [s_l] \subset \alpha_l \& \psi(\vec{x}_k,\vec{s}_l) = z)$.

8. Show that if a set X has the property that $x \in X$ whenever all the subcomputations of x are in X then it contains Ω.

9. Let $F: \mathcal{N} \times \mathcal{I} \to \mathcal{N}$ be given by $F(x,\alpha) = $ **if** $x = 0$ **then** 1 **else** $\alpha(x + 1)$. Find
the least fixed point of F. Does F have any recursive fixed points?

10. (McCarthy's *91 function*) Show that $\lambda x.$**if** $x > 100$ **then** $x \dot- 10$ **else** 91 is the
only fixed point of:

$F = \lambda x \alpha.\text{if } x > 100 \text{ then } x \dot{-} 10 \text{ else } \alpha(\alpha(x + 11))$

11. Show that $\mathcal{C} \ (\subset \mathcal{P}^{(1)})$ is *completely re* iff there is a partial recursive functional F of rank $(0,1)$ such that $\text{dom}(F) = \mathcal{C}$.
    Note: It is interesting to note the analogy with "$S(\subset N)$ is re iff there is a partial recursive function f such that $\text{dom}(f) = S$".

12. Show that a class of re sets $\mathcal{C}$ is *completely re* iff there is a partial recursive functional F of rank $(0,1)$ such that
    $F(\alpha) = 0$ if $\text{dom}(\alpha) \in \mathcal{C}$
    $F(\alpha)\uparrow$ if $\text{dom}(\alpha) \notin \mathcal{C}$

13. Show that if the call-by-value computational rule is applied to the program
    $P:\alpha(x,y):: = F(x,y,\alpha)$, where
    $F(x,y,\alpha) = \text{if } x = 0 \text{ then } 1 \text{ else } \alpha(x \dot{-} 1, \alpha(x,y))$,
    then $\alpha(x,y) = \text{if } x = 0 \text{ then } 1 \text{ else } \uparrow$.

14. [7] For this problem adopt the *convention* that $0 \cdot \uparrow = 0$ (where "$\cdot$" is the "times" operation).
    Verify then that the call-by-name (or leftmost) rule, applied to the following program P does not lead to the least fixed point.
    $P:\alpha(x,y)::=F(x,y,\alpha)$, where
    $F(x,y,\alpha) = \text{if } x = 0 \text{ then } 0 \text{ else } \alpha(x + 1,\alpha(x,y)) \cdot \alpha(x \dot{-} 1,\alpha(x,y))$

REFERENCES

[1] Bird, R. *Programs and Machines*. New York: Wiley-Interscience, 1976.

[2] Cadiou, J.M. *Recursive definitions of partial functions and their computations*. Dissertation, Dept. of Computer Science, Stanford University, Stanford, Calif. 1972.

[3] Cutland, N.J. *Computability: An Introduction to Recursive Function Theory*. Cambridge: Cambridge University Press, 1980.

[4] Davis, M. *Computability and Unsolvability*. New York: McGraw-Hill, 1958.

[5] Fenstad, J.E. *General Recursion Theory: An Axiomatic Approach*. Heidelberg: Springer-Verlag, 1980.

[6] Hinman, P.G. *Recursion-Theoretic Hierarchies*. Heidelberg: Springer-Verlag, 1978.

[7] Manna, Z. *Mathematical Theory of Computation*. New York: McGraw-Hill, 1974.

[8] Moldestad, J. *Computation in Higher Types*. Heidelberg: Springer-Verlag, 1977.

[9] Morris, J.H. *Lamda-Calculus Models of Programming Languages*. Dissertation, Project MAC, MAC-TR-57, M.I.T., Cambridge, Mass., 1968.

[10] Moschovakis, Y.N. *Axioms for Computation Theories*—First draft (1969). In Gandy, R.O. and Yates, C.E.M. (Eds.) *Logic Colloquium '69,* Amsterdam: North-Holland, 1971: 199–255.

[11] Post, E.L. "Recursively Enumerable Sets of Positive Integers and their Decision Problems". *Bulletin of Amer. Math. Soc.*, 50 (1944): 284–316, (also in Davis, M. [Ed.], *The Undecidable,* Hewlett, New York: Raven Press, 1965: 305–337.)

[12] Rado, T. "On Non-computable Functions". *Bell System Technical Journal,* 41 (1962): 877–884.

[13] Rogers, H. *Theory of Recursive Functions and Effective Computability*. New York: McGraw Hill, 1967.

[14] Scott, D. "Outline of a Mathematical Theory of Computation". *Proc. of 4th Annual Princeton Conf. on Information Sciences and Systems,* Princeton University, Princeton, N.J., (1970): 169–176.

CHAPTER 12

Complexity of (Partial) Recursive Functions

We have dealt so far with the question "what functions/predicates are, and what are not, computable". We have used various equivalent formalisms as counterparts of the intuitive notion "computable" (noting in §7.1 that whether or not these formalisms are "complete" is an open philosophical question). In the process we gained some experience with the limitations of computing machines (as modeled by TMs, URMs, $\mathcal{P}$-formalism, etc.), have remarked on the connection between Recursion Theory and Logic (§9.6) and finally studied the Kleene/Mostowski classification of *unsolvable* (= "nonalgorithmic") problems according to their relative "complexity".

In practice, the user of computers is interested mostly in the "efficiency" of the programs he writes, as this is gauged by the amount of computer memory and/or computer time these require for their execution.

Not surprisingly, mathematicians and computer scientists have developed theoretical tools aimed at dealing with the question of the "efficiency" (or, put negatively, computational *complexity*) of algorithms, vigorous activity in the field manifesting itself already in the mid-1960s.

As usual, the process of abstraction, apart from providing an elegant mathematical theory provides a general framework within which general principles, valid for several concrete cases, are established. Thus, in this chapter we shall follow Blum [2] and survey some results on the complexity of partial recursive (and recursive) functions in a *machine independent* (i.e., axiomatic) setting.

Before we proceed with details, let us distinguish between two classes of complexity measures, *dynamic* and *static*.

The former are measuring the amount of a "resource" consumed during a computation (e.g., amount of Turing machine tape squares, or number of IDs generated during the computation).

Static measures of complexity on the other hand may be the *size* (program length, considered as a string over some alphabet) or the *structural* complexity (e.g., *level of nesting* of do-loops) of an algorithm's description.

The axiomatic approach of Blum is aimed at characterizing the *dynamic* measures.

§12.1 AXIOMS FOR DYNAMIC COMPLEXITY

Throughout we shall be involved with the complexity of functions in $\mathcal{P}^{(1)}$ since multiple inputs can "easily"(*) be coded into single inputs.

As before, an onto map $\phi : \mathcal{N} \to \mathcal{P}^{(1)} : i \mapsto \phi_i$ which satisfies the S-m-n and universal function theorems we shall call a *ϕ-indexing* or *Gödel numbering* or *acceptable indexing* (the last term is due to Rogers).

Definition 1　[2] A *complexity measure* Φ for a ϕ-indexing ϕ is a sequence of $\mathcal{P}^{(1)}$-functions $(\Phi_i)_{i \geq 0}$ satisfying the axioms:

(A1)　For all i, x, $\phi_i(x) \downarrow$ iff $\Phi_i(x) \downarrow$

(A2)　The predicate $\lambda ixy.\Phi_i(x) = y$ is recursive.//

Note: Intuitively, Φ is dedicated to measuring a particular (but unspecified) "resource", consumed during the computation of a $\mathcal{P}^{(1)}$-function on an unspecified model of computation, ϕ.

Thus (A1) is supposed to capture the notion that the computation $\phi_i(x)$ is convergent iff it consumes a *finite* amount, $\Phi_i(x)$, of the resource.

(A2) is also intuitively appealing. Even for divergent computations

---

(*)Here we are referring to the availability of "easily computable" pairing functions (e.g., rudimentary). Input coding/decoding with such functions, intuitively, consumes a very small percentage of the resource(s) required by the overall algorithm.

$[\phi_i(x)\uparrow]$ one should be able to tell whether or not $\Phi_i(x) = y$ (i.e., if exactly y units of the resource have been consumed by the end of the computation) by letting "program" i on input x run long enough [of course, if $\phi_i(x)\uparrow$, the answer to "$\Phi_i(x) = y$?" will be *no* for any y, since sooner or later $\Phi_i(x)$ will "exceed" y]. Whenever $\phi_i(x)\downarrow$ we say that $\phi_i(x)$ takes $\Phi_i(x)$ *steps*.

Example 1   Let $T(i,x,y)$ be a recursive Kleene predicate for the chosen ϕ-indexing of $\mathcal{P}^{(1)}$. Then $\Phi = (\Phi_i)_{i\geq 0}$, where, for all i,

$$\Phi_i = \lambda x.(\mu y)T(i,x,y)$$

is a complexity measure.

Indeed, since $\phi_i = \lambda x.d((\mu y)T(i,x,y))$, where d is recursive, $\phi_i(x)\downarrow$ iff $(\mu y)T(i,x,y)\downarrow$ iff $\Phi_i(x)\downarrow$.

Further, $y = \Phi_i(x) \equiv T(i,x,y)$ & $(\forall z)_{<y}\,\neg T(i,x,z)$, which is recursive.

The "concrete" Φ defined here was employed in §7.2. It is easy to observe that the only properties of this Φ we used there were (A1) and (A2), thus the results of §7.2 generalize to the situation where "steps" are counted by the "abstract" Φ.//

Example 2   Let the ϕ-indexing be that of 5.2, so that it enumerates Turing machines. For each i and x, define

$$\Phi_i(x) = \begin{cases} \textbf{if } \phi_i(x)\downarrow \textbf{ then} \text{ the length of the } \textit{longest} \text{ ID} \\ \qquad\qquad \text{occurring in the computation } \phi_i(x) \\ \uparrow \textbf{ otherwise} \end{cases}$$

Intuitively, $\Phi_i \in \mathcal{P}^{(1)}$ for each i. [Let $\phi_i(x)$ run, and record ID lengths. If $\phi_i(x)\downarrow$ report the maximum such length, otherwise keep computing.]

Formally, set (refer to 5.2)

$$f = \lambda y.|(\mu z)[\text{ID}(z) \text{ \& } (;z;)Py \text{ \& } (\forall w)_{Py}[\text{ID}(w) \text{ \& } (;w;)Py \Rightarrow |w|\leq|z|]]|$$

where $|\dots|$ is m-adic length. Intuitively, $f(x)$ is the longest ID length occurring in x as a $;z;$ where z is an ID.

Clearly, $f \in \mathcal{PR}$, and

$$\Phi_i(x) = f((\mu y)T(i,x,y))$$

(A1) is clearly satisfied.

As for (A2), $y = \Phi_i(x)$ iff $(\exists z)[T(i,x,z) \text{ \& } f(z) = y]$.

Now, TM i, for input x and maximum ID length used $= y$, must have its coded computation(*) z bounded by some $b(i,x,y)$, where $b \in \mathcal{PR}$. (Details are requested from the reader in Problem 1.)

Thus, $y = \Phi_i(x)$ iff $(\exists z)_{\leq b(i,x,y)}[T(i,x,z) \ \& \ f(z) = y]$ hence $\lambda ixy.y = \Phi_i(x) \in \mathcal{PR}_*.//$

We have seen a number of concrete complexity measures. Others are suggested in the problems. This, incidentally, establishes the logical *consistency* of the axioms. We now see two examples of Φ_i-sequences one failing (A1) the other (A2), establishing that the axioms are *independent*.

Example 3   Let ϕ be a ϕ-indexing, and set

$$\Phi_i \overset{\text{def}}{=} \phi_i, i \geq 0$$

Clearly, (A1) is met, but (A2) fails as $\lambda ixy.y = \phi_i(x)$ is not recursive (Theorem 7.3.1).//

Example 4   Let ϕ be a ϕ-indexing.
Set $\Phi_i = \lambda x.0$.
Then $\lambda ixy.y = \Phi_i(x)$ is obviously recursive, satisfying (A2); (A1) fails, since Φ_i is total for all $i \geq 0$.//

Proposition 1   For any complexity measure Φ, $\lambda ixy.y \geq \Phi_i(x)$ is recursive.

Proof $y \geq \Phi_i(x) \equiv (\exists z)_{\leq y} z = \Phi_i(x).//$

Proposition 2   For any complexity measure Φ, $\lambda ix.\Phi_i(x) \in \mathcal{P}$.

Proof $\Phi_i(x) = (\mu y)(y = \Phi_i(x))$, where $\lambda ixy.y = \Phi_i(x)$ is recursive.//

Corollary   For any complexity measure Φ, there is an $h \in \mathcal{R}^{(1)}$ such that $\Phi_i = \phi_{h(i)}$ for all $i \geq 0$.

We shall now establish, step by step, the credibility of this, at first sight too general, axiom system by proving a number of intuitively appealing results, culminating to the counter-intuitive and important result known as Blum's speed-up theorem.

(*)"Computation" used here as *terminating* computation.

Theorem 1(*)   For every complexity measure Φ for some ϕ-indexing, and for every function $g \in \mathcal{R}^{(1)}$, there is a 0-1 valued $f \in \mathcal{R}^{(1)}$, such that for *any i* for which $f = \phi_i$, $\Phi_i(x) > g(x)$ a.e.

Proof Exactly as in the proof of Theorem 7.2.6, using Φ_i instead of the "specific" measure $(\mu y)T(i,x,y)$.

As the proof of this theorem employs an important technique, we repeat (informally) the definition of f:

$$f(x) = \begin{cases} 1 \dot{-} \phi_i(x) & \text{if } i \text{ is the } \textit{smallest uncancelled} \text{ index} \\ & \text{such that } \Phi_i(x) \leq g(x) \ \& \ i \leq x. \text{ Now } \textit{cancel } i. \\ 1 & \textbf{if} \text{ no uncancelled index } j \text{ satisfying } j \leq x \ \& \\ & \Phi_j(x) \leq g(x) \text{ exists.} \end{cases}$$

Intuitively, $f(x)$ is made *different* from all those $\phi_i(x)$ which fail to meet what we want ($\Phi_i(x) > g(x)$ a.e.).//

Note: As an "application" of Theorem 1, we mention that there are arbitrarily complex predicates regardless of what concrete measure we use to gauge "complexity" [as long as this measure satisfies (A1) and (A2)].

Corollary 1   Let Φ be a complexity measure. There is *no* $b \in \mathcal{R}^{(1)}$ such that $(\forall i)\Phi_i(x) \leq b(x)$ a.e.

Note: This is the abstract version of the corollary to Theorem 7.2.6.

Corollary 2   Let Φ be a complexity measure. There is *no* $b \in \mathcal{R}^{(2)}$ such that, for all i,

$$\Phi_i(x) \leq b(x,\phi_i(x)) \text{ a.e.}$$

Proof Let f be 0-1 as in Theorem 1, where g was chosen as $g = \lambda x.\max(b(x,0),b(x,1))$. Then, if $f = \phi_i$ we have $\max(b(x,0),b(x,1)) < \Phi_i(x) \leq \max(b(x,0),b(x,1))$ a.e.//

Note: By Corollary 2, the output of an algorithm cannot be used to predict or bound its complexity. On the other hand, knowing the complexity of an algorithm we can use it to bound its output. [Intuitively, the complexity of an algorithm is its step-count, as a function of the input. We can majorize the output by assuming a worst case, that at each "step", that operation is applied, which increases the previous "subresult" by the *largest* (constant) amount, say k. Then $\phi_i(x) \leq x + k\Phi_i(x)$. The weakness of this argument is that k is assumed constant. It applies, however, to the standard measures for TMs and URMs.]

(*)This theorem and a sketch of its proof is embedded in the proof of Theorem 5 (speed-up) in Blum [2].

Proposition 3   For every complexity measure, there is a $b \in \mathcal{R}^{(2)}$ such that $\phi_i(x) \leq b(x, \Phi_i(x))$ a.e.

     Proof [2] Set $\hat{b}(i,x,y) = $ **if** $\Phi_i(x) = y$ **then** $\phi_i(x)$ **else** 0.(*) Clearly, $\hat{b}$ $\in \mathcal{P}$ and as it is *total* (why?) it is also in $\mathcal{R}$.
     If $\phi_i(x)\downarrow$ (which is equivalent to $\Phi_i(x)\downarrow$), then $\phi_i(x) = \hat{b}(i,x,\Phi_i(x))$.
     This clearly motivates us to set $\hat{b}(x,y) = \max_{i \leq x} b(i,x,y)$ to eliminate i without decreasing the $\hat{b}(i,x,y)$ value.//

     Note: The intuitive content of Proposition 3 is that "fast growing" recursive functions *have* to be "hard to compute".

Proposition 4   (Blum) If Φ and $\hat{\Phi}$ are two complexity measures for ϕ and $\hat{\phi}$, respectively, then there is a $g \in \mathcal{R}^{(2)}$ such that for all i

$$g(x, \Phi_i(x)) \geq \hat{\Phi}_i(x) \text{ a.e.}$$

and

$$g(x, \hat{\Phi}_i(x)) \geq \Phi_i(x) \text{ a.e.}$$

     Proof The principle is similar to that in the proof of Proposition 3. Let f be defined by

$$f(i,x,y) = \textbf{if } \Phi_i(x) = y \vee \hat{\Phi}_i(x) = y \textbf{ then } \max(\Phi_i(x), \hat{\Phi}_i(x)) \textbf{ else } 0(\dagger)$$

Clearly, $f \in \mathcal{R}$, and whenever $\Phi_i(x)\downarrow$, then $\hat{\Phi}_i(x)\downarrow$ (and vice versa—why?) and

$$f(i,x,\Phi_i(x)) = \max(\Phi_i(x),\hat{\Phi}_i(x)) \geq \hat{\Phi}_i(x)$$

Similarly, $\hat{\Phi}_i(x)\downarrow \Rightarrow f(i,x,\hat{\Phi}_i(x)) \geq \Phi_i(x)$. To eliminate i without spoiling the inequality, set $g(x,y) = \max_{i \leq x} f(i,x,y)$.//

     Note: The previous proposition, in very loose terms due to the presence of the function g, says that "approximately" the same number of steps is needed to compute a function in two different formalisms, or in the same formalism but with two different definitions of "step".

     The following proposition (Blum) is not startling; it says that there are arbitrarily bad ways to program a function.

(*)This is a "don't care" alternative. Any $h \in \mathcal{R}^{(1)}$ will do.
($\dagger$)λx. 0 here is a "don't care" alternative. Any $h \in \mathcal{R}^{(1)}$ will do.

Proposition 5   For any complexity measure Φ and any $f \in \mathcal{R}^{(1)}$ and $h \in \mathcal{R}^{(1)}$ there is an i_0 such that $f = \phi_{i_0}$ and $(\forall x)\Phi_{i_0}(x) > h(x)$.

Proof Let

$$\psi(i,x) = \begin{cases} f(x) \textbf{ if } \Phi_i(x) > h(x) \\ 1 \doteq \phi_i(x) \textbf{ otherwise} \end{cases}$$

Clearly, $\psi \in \mathcal{P}$ (definition by recursive cases). By the recursion theorem, there is an i_0 such that $\phi_{i_0}(x) = \psi(i_0,x)$ for all x.

For *no* x is it possible that $\Phi_{i_0}(x) \leq h(x)$ as this would imply $\phi_{i_0}(x) = \psi(i_0,x) = 1 \doteq \phi_{i_0}(x)$ which is a contradiction since $\phi_{i_0}(x) \downarrow [\Phi_{i_0}(x) \leq h(x)$; i.e., $\Phi_{i_0}(x) \downarrow]$.

Thus, $\phi_{i_0} = f$ and $(\forall x)\Phi_{i_0}(x) > h(x).//$

We have seen a number of desirable properties that the arbitrary complexity measure Φ will have. However, the great generality of the axioms also allows certain pathologies to arise. For example, even though whenever one combines two functions ϕ_i and ϕ_j "effectively" the complexity of the combination is recursively related to the individual complexities of ϕ_i and ϕ_j (this is the content of the "combining lemma" in [3, 5]—see Problem 5), this relationship is not "tight enough", as the following example shows.

Example 5   Let ϕ be a given ϕ-indexing and let $\phi_{i_0} = \lambda x.x$. Let Φ be some given complexity measure for ϕ. Define $\hat{\Phi}$ by

$$\hat{\Phi}_i(x) = \textbf{if } i = i_0 \textbf{ then } \Phi_{i_0}(x) \textbf{ else } \max(h(x),\Phi_i(x))$$

where $h \in \mathcal{R}^{(1)}$.

It can be verified that $\hat{\Phi}$ is a complexity measure.

Clearly, $\hat{\Phi}_{i_0} = \Phi_{i_0}$. However, one may choose h to be "horrendously bigger" than Φ_{i_0}, for example, for all x $h(x) = A_{100}(\Phi_{i_0}(x) + 2)$, where $\lambda nx.A_n(x)$ is the Ackermann function of §3.3. Thus, for example, *no matter what index j we choose for $\lambda x.x + 1$, $\hat{\Phi}_j(x) \geq A_{100}(\hat{\Phi}_{i_0}(x) + 2)$ which is much larger than $\hat{\Phi}_{i_0}(x)$.* Intuitively, one would expect only a trivial increase in complexity in going from a program for $\lambda x.x$ to a *well chosen* program for $\lambda x.x + 1.//$

Note: This example was in the spirit of Alton [1], where the "naturalness" of complexity measures for subsets of recursive functions is gauged by, among other things, whether or not "tight combining lemmas" hold for such measures.

We now turn to a proof of the *Speed-Up Theorem*.

Theorem 2   [2] Let Φ be an arbitrary complexity measure for ϕ.
Let $r \in \mathcal{R}^{(2)}$.
Then there exists a 0-1 valued $f \in \mathcal{R}^{(1)}$ such that for all i such that $f = \phi_i$ there is a j such that

$$f = \phi_j$$

and

$$\Phi_i(x) > r(x, \Phi_j(x)) \text{ a.e.}$$

Proof [2] The proof is an extension of the one employed in Theorem 1. First, instead of defining f so that for all i, such that $\phi_i = f$, $\Phi_i(x) > g(x)$ a.e., we want now to choose f so that $\Phi_i(x) > g(x - i)$ a.e. under the above conditions.(*) Recall that $f(x)$ is defined (proof of Theorem 1—see also proof of Theorem 7.2.6) by *cancelling* the *first uncancelled* element i_0 in $\{0,1,\ldots,x\}$ for which $\Phi_{i_0}(x) \le g(x - i_0)$, and making $f(x) \ne \phi_{i_0}(x)$ [by setting $f(x) = 1 \doteq \phi_{i_0}(x)$]. If we cannot do that [e.g., no uncancelled i_0 exists, or, for all uncancelled i_0, $\Phi_{i_0}(x) > g(x - i_0)$ holds instead], we set $f(x) = 1$ (0 would do as well).

This definition suggests how to standard-compute $f(x)$ for all x:

Find i_0 as above, compute and output $1 \doteq \phi_{i_0}(x)$, and **stop**. Otherwise, if no such i_0 exists, output 1, and **stop**.

Finding i_0 is time consuming. A shortcut in the computation can be accomplished as follows: For any given $u \in \mathcal{N}$, we need only scan the set $\{u, u+1, \ldots, x\}$ to look for uncancelled i_0's, *if* x is *sufficiently large*, say $x \ge v$ (for some $v \in \mathcal{N}$), since for such x each $i \in \{0,1,\ldots,u \doteq 1\}$ has stabilized its status as cancelled or uncancelled.

It turns out that this shortcut is sufficient to provide enough speed-up over any other *standard-computation* which uses a smaller u (and hence does more "searching").

The possibility that a *non-standard* way, ϕ_k, exists to compute f, which cannot be sped-up, is eliminated by choosing g so that *for each k there is a j* such that

(1) $f = \phi_j$

(2) $g(x \doteq k) \ge r(x, \Phi_j(x))$ a.e.

On the other hand, if $\phi_k = f$ then $\Phi_k(x) > g(x \doteq k)$ a.e. by definition of f. (2) completes the proof.

The formal proof starts by repeating the construction of Theorem 1 "effectively" (or "uniformly") on a ϕ-index l of a possibly nontotal (and unspecified as yet) function g.

(*)g will be specified in the proof.

Keeping in mind that $\phi_l = g$, define ψ by

$$\psi(u,v,l,x) = \begin{cases} \textbf{if } v < u \textbf{ then output } \psi(u,u,l,x) \\[4pt] \textbf{if } v \geq u \textbf{ then} \\[4pt] \textit{case 1 } x < v\text{: Find the first } \textit{uncancelled } i_0 \\[2pt] \quad \text{in } \{0, \ldots, x\} \text{ such that } \Phi_{i_0}(x) \leq \phi_l(x - i_0)(*) \\[2pt] \textbf{output } 1 \doteq \phi_{i_0}(x) \text{ and } \textit{cancel } i_0. \\[2pt] \text{If no such } i_0 \text{ exists, } \textbf{output } 1. \\[6pt] \hspace{4cm} \textit{End of case 1.} \\[6pt] \textit{case 2 } x \geq v\text{: [We can compute faster now]} \\[2pt] \quad \text{Find the } \textit{first uncancelled } i_0 \text{ in} \\[2pt] \quad \{u, u + 1, \ldots, x\} \text{ such that } \Phi_{i_0}(x) \leq \phi_l(x - i_0)(\dagger) \\[2pt] \textbf{output } 1 \doteq \phi_{i_0}(x) \text{ and } \textit{cancel } i_0. \\[2pt] \text{If no such } i_0 \text{ exists, } \textbf{output } 1. \\[6pt] \hspace{4cm} \textit{End of case 2.} \end{cases}$$

By the recursion theorem (and after the reader formalizes the above definition of ψ in the pattern of Theorem 7.2.6) $\psi \in \mathcal{P}$. Thus, there is a $t \in \mathcal{R}^{(3)}$ such that

$$\psi(u,v,l,x) = \phi_{t(u,v,l)}(x) \text{ for all } u,v,l,x.$$

Claim 1. If $\phi_l \in \mathcal{R}^{(1)}$ and we set $f = \phi_{t(0,0,l)}$, then

$$f \in \mathcal{R}^{(1)} \text{ and is 0-1 valued}$$

$$f = \phi_i \Rightarrow \Phi_i(x) > \phi_l(x - i) \text{ a.e.}$$

Proof Immediate from the proof of Theorem 7.2.6 and the definition of ψ (only condition $v \geq u$, *case 2* obtains).
End of proof for Claim 1.

Claim 2. If $\phi_l \in \mathcal{R}^{(1)}$, then for each u there is a v such that $\phi_{t(u,v,l)} = \phi_{t(0,0,l)}$.

(*)This implies obtaining first $\phi_l(0), \ldots, \phi_l(x)$. If any of them is undefined, then $\psi(u,v,l,x)\uparrow$ in this case.

(†)This implies obtaining first $\phi_l(0), \ldots, \phi_l(x - u)$. If any of them is undefined, then $\psi(u,v,l,x)\uparrow$ in this case as well.

Proof There are *finitely* many i_0's which are ultimately cancelled in the range $\{0, \ldots, u - 1\}$. Let v be such that the *last* such i_0 was cancelled during the computation of $\phi_{t(0,0,l)}(v \dotminus 1)$.

By the definition of ψ (and hence t), when $x < v$ we compute $\phi_{t(u,v,l)}(x)$ the same way we compute $\phi_{t(0,0,l)}(x)$. For $x \geq v$ the computation of $\phi_{t(u,v,l)}(x)$ gives the same result as $\phi_{t(0,0,l)}(x)$ since the latter would first cancel the relevant i_0's in $\{0, \ldots, u - 1\}$ and then proceed as the former. We assumed throughout $v \geq u$, since $v < u$ reduces to this case.

End of proof for Claim 2.

Now is the time to fix l (i.e., fix the function g mentioned in the earlier informal discussion).

Claim 3. There is a $\phi_l \in \mathcal{R}^{(1)}$ such that for all u,v,

$$\phi_l(x \dotminus u + 1) \geq r(x, \Phi_{t(u,v,l)}(x)) \text{ a.e.}$$

Proof Define p by

$$p(i,0) = 0$$

$$p(i,x + 1) = \max_{\substack{0 \leq u \leq x \\ 0 \leq v \leq x}} r(x + u, \Phi_{t(u,v,i)}(x + u))$$

As $\max_{u \leq x}$ does not lead outside $\mathcal{P}$ (it is definable by primitive recursion), $p \in \mathcal{P}$. By the recursion theorem, there is an l such that $\phi_l(x) = p(l,x)$ for all x.

To show that $(\forall x)\phi_l(x)\downarrow$, we use induction on x.

$x = 0$: $\phi_l(0) = 0$

Assume that $\phi_l(x)\downarrow$, for $x \leq k$.

Then $\phi_{t(u,v,l)}(x + u)\downarrow$, for $x \leq k$ and $u \leq x$, $v \leq x$. Indeed, $v \leq x \Rightarrow x + u \geq v$, hence to compute $\phi_{t(u,v,l)}(x + u)(*)$, we compute (among other things) $\phi_l(0)$, $\ldots, \phi_l(x + u - u)$ $[= \phi_l(x)]$, which all converge and there is nothing else in the definition of ψ to make $\phi_{t(u,v,l)}(x + u)\uparrow$.

Hence, $\Phi_{t(u,v,l)}(x + u)\downarrow$, for $x \leq k$ therefore $\phi_l(x + 1)\downarrow$. This completes the induction.

Next, whenever $x \geq \max(2u, u + v)$ it follows that $v \leq x \dotminus u$ and $u \leq x \dotminus u$ hence $\phi_l(x \dotminus u + 1) \geq r(x, \Phi_{t(u,v,l)}(x))$ a.e. (namely for $x \geq \max(2u, u + v)$).

End of proof for Claim 3.

The proof is completed by verifying (1) and (2) of the informal discussion:

Let k be given. Set $u = k + 1$ and let v be as in Claim 2.

Then $\phi_l(x \dotminus k) = \phi_l(x \dotminus u + 1) \geq r(x, \Phi_{t(u,v,l)}(x))$ a.e. Take $j = t(u,v,l)$ since $\phi_{t(u,v,l)} = \phi_{t(0,0,l)} = f.//$

Note: It is clear now that for *no* complexity measure does it make sense

(*)Case $v \geq u$ considered, as $v < u$ reduces to it.

to talk, *in general*, about "best" (fastest, least complex) algorithms. In particular, the "intrinsic" or least complexity of a $\mathcal{P}$-function is not well defined. The best we can do is to classify $\mathcal{P}$-functions according to *upper bounds* on their complexity.

§12.2 COMPLEXITY CLASSES

Definition 1   Let Φ be a complexity measure for ϕ, and let $t \in \mathcal{R}^{(1)}$. Then,

$$C_t^\Phi = \{\phi_i \,|\, \Phi_i(x) \le t(x) \text{ a.e.}\}.//$$

By Theorem 12.1.1, we immediately obtain

Proposition 1   There is an infinite sequence of recursive functions $(t_i)_{i \ge 0}$ such that $C_{t_i}^\Phi \underset{\ne}{\subseteq} C_{t_{i+1}}^\Phi$ for $i \ge 0$.

Definition 2   A predicate $P(x)$ on $\mathcal{N}$ is true *infinitely often*, in symbols "*i.o.*", if $\{x \,|\, P(x)\}$ is infinite.//

The question now arises whether t_{i+1} can be effectively obtained from t_i. More precisely, is there an $f \in \mathcal{R}^{(1)}$ such that $C_t^\Phi \underset{\ne}{\subseteq} C_{f \circ t}^\Phi$ for all sufficiently large $t \in \mathcal{R}^{(1)}$?

The answer (no) was independently discovered by Trakhtenbrot [10] and Borodin [3] and is known as the *Gap Theorem*.

Theorem 1   (*Gap theorem*) For any Φ, and any h and f in $\mathcal{R}^{(1)}$ such that $(\forall x)f(x) \ge x$, there is a $t \in \mathcal{R}^{(1)}$, such that $(\forall x)t(x) \ge h(x)$, and

$$C_t^\Phi = C_{f \circ t}^\Phi$$

Proof For the proof, we need find a $t \in \mathcal{R}^{(1)}$ such that

(1) $t(x) \ge h(x), x \ge 0$

(2)   For all i, $\{x \,|\, t(x) \le \Phi_i(x) \le f \circ t(x)\}$ is finite.

Thus, set

$$t(x) = (\mu y)[y \ge h(x) \,\&\, (\forall i)_{\le x}(\Phi_i(x) \le f(y) \Rightarrow \Phi_i(x) \le y)]$$

The condition right of (μy) is recursive, due to recursiveness of $\lambda xyi.\Phi_i(x) \le y$ and f. Thus, $t \in \mathcal{P}$.

Moreover, for *each* i ($\leq x$), either $\Phi_i(x)\uparrow$, in which case $\Phi_i(x) \leq f(y)$ is false(*) for *any* y, or $\Phi_i(x)\downarrow$ in which case $\Phi_i(x) \leq y$ is true for all $y \geq m$, where $m = \Phi_i(x)$. In either case, $\Phi_i(x) \leq f(y) \Rightarrow \Phi_i(x) \leq y$ is true, hence t is *total*.

Next, let $\Phi_i(x) \leq f \circ t(x)$ a.e. By choice of $t(x)$, this implies $\Phi_i(x) \leq t(x)$ a.e. hence $C^\Phi_{f \circ t} \subset C^\Phi_t$. The converse inclusion is trivial due to $f(x) \geq x$ for all x.//

Note: The role of h, which can be arbitrarily large, is to enable us to say that there are infinitely many "gaps" in the set of complexity classes. f is also arbitrarily large, hence it is possible to enormously increase the resource allotted to computations of a complexity class and yet obtain *no* new functions computed. Put another way, "tight" bounds on complexity are "sparse" in the class of recursive functions.

Definition 3   [2] A sequence $(\gamma_i)_{i \geq 0}$ of functions from $\mathcal{P}^{(1)}$ is a *measured set* iff $\lambda ixy.y = \gamma_i(x) \in \mathcal{R}_*.$//

Note: $(\Phi_i)_{i \geq 0}$ is obviously a measured set for any complexity measure Φ.

Lemma 1   If $(\gamma_i)_{i \geq 0}$ is a measured set then there is a $\sigma \in \mathcal{R}^{(1)}$ such that $\gamma_i = \phi_{\sigma(i)}$ for all $i \geq 0$.

Proof $\lambda ix.\gamma_i(x) = \lambda ix.(\mu y)(y = \gamma_i(x)) = \lambda ix.\phi_{\sigma(i)}(x)$ by *S-m-n* theorem.//

Lemma 2(†)   There is an $h \in \mathcal{R}^{(1)}$ such that $\text{dom}(\phi_i) = \text{dom}(\phi_{h(i)})$, $i \geq 0$, and for any j such that $\phi_j = \phi_{h(i)}$,

$$\Phi_j(x) > \phi_i(x) \text{ a.e.(‡)}$$

Proof (See that of Theorem 7.2.5)
Define ψ by

$$\psi(i,x) = \begin{cases} (\Sigma_{j \leq z}\phi_j(x)) + 1 & \textbf{if } (\exists j)\ (j \leq x\ \&\ \Phi_j(x) \leq \phi_i(x))\ \&\ \phi_i(x)\downarrow \\ 1 & \textbf{if } \neg(\exists j)\ (j \leq x\ \&\ \Phi_j(x) \leq \phi_i(x))\ \&\ \phi_i(x)\downarrow \\ \uparrow & \textbf{otherwise} \end{cases}$$

$\psi \in \mathcal{P}$ (Problem 10). Hence, there is an $h \in \mathcal{R}^{(1)}$ such that $\psi(i,x) = \phi_{h(i)}(x)$ for all i,x.

(*)This is *not* the same as saying $f(y) < \Phi_i(x)$ is true. Why?

(†)This is a simplified version of Blum's Theorem 7 [2] which was stated for 0-1 functions. As it stands it is a "uniform" version of Theorem 7.2.5 as it incorporates a ϕ-index of g.

(‡)To be interpreted as "$\{x \,|\, \Phi_j(x) \leq \phi_i(x)\}$ is finite", since ϕ_i may not be total.

Let next $\phi_i(x)\downarrow$. Then $\phi_{h(i)}(x)\downarrow$, as the "otherwise" case does not obtain. That is, $\mathrm{dom}(\phi_i) \subset \mathrm{dom}(\phi_{h(i)})$ [in particular, $\phi_i \in \mathcal{R} \Rightarrow \phi_{h(i)} \in \mathcal{R}$]. Clearly, $\phi_i(x)\uparrow$ $\Rightarrow \phi_{h(i)}(x)\uparrow$, hence $\mathrm{dom}(\phi_{h(i)}) \subset \mathrm{dom}(\phi_i)$.

Finally, if $\phi_j = \phi_{h(i)}$, $\phi_i(x)\downarrow$, and $j \leq x$, then $\Phi_j(x) > \phi_i(x)$. [Indeed, (1) $\phi_j(x)\downarrow$ since $\phi_i(x)\downarrow$, and hence $\phi_{h(i)}(x)\downarrow$. (2) If $\Phi_j(x) \leq \phi_i(x)$, then $\phi_{h(i)}(x) \geq \phi_j(x) + 1$, contradicting $\phi_j = \phi_{h(i)}$].//

Theorem 2

(*Compression Theorem.* Blum [2])

For any complexity measure Φ and any measured set $(\gamma_i)_{i\geq0}$, there is a $g \in \mathcal{R}^{(2)}$ such that for all i,j

(1) $\quad \phi_j = \phi_{h\circ\sigma(i)} \Rightarrow \Phi_j(x) > \gamma_i(x)$ a.e.(*)

(2) $\quad \Phi_{h\circ\sigma(i)}(x) < g(x,\gamma_i(x))$ a.e.(*)

where σ and h are as in Lemmas 1 and 2.

Proof Set $g(x,y) = 1 + \max_m\{\Phi_{h\circ\sigma(m)}(x)\,|\,\gamma_m(x) = y \ \& \ m \leq x\}$ Clearly, $g \in \mathcal{R}$ $[\gamma_m(x) = y \ true \Rightarrow \gamma_m(x)\downarrow \Rightarrow \phi_{\sigma(m)}(x)\downarrow \Rightarrow \phi_{h\circ\sigma(m)}(x)\downarrow \Rightarrow \Phi_{h\circ\sigma(m)}(x)\downarrow]$.

Thus, $\Phi_{h\circ\sigma(i)}(x) < g(x,\gamma_i(x))$ for any $x \geq i$ if $\gamma_i(x)\downarrow$, which proves (2). (1) is Lemma 2 (via Lemma 1).//

Note: Theorem 2 is half of Theorem 8 in Blum, where he characterizes measured sets by properties (1) and (2). (See Problem 11.)

Corollary

Let Φ and $(\gamma_i)_{i\geq0}$ be as in Theorem 2. Then,

$$\gamma_i \in \mathcal{R} \Rightarrow C^{\Phi}_{\gamma_i} \underset{\neq}{\subseteq} C^{\Phi}_{\lambda x.g(x,\gamma_i(x))} \tag{3}$$

Note: It is interesting to note that for each $\gamma_i \in \mathcal{R}$, $\phi_{h\circ\sigma(i)} \in C^{\Phi}_{\lambda x.g(x,\gamma_i(x))} - C^{\Phi}_{\gamma_i}$ —i.e., an example verifying the proper inclusion (3) above is *effectively* obtained from i.

In intuitive terms, the corollary says that for any recursive "run time" *out of a measured set* we can construct effectively a run time "naming" a *more inclusive* complexity class. This is more than we can say for the arbitrary recursive run time (gap theorem). Thus, there are some "bad" names for complexity classes (t of Theorem 1) and some "good" ones (e.g., recursive γ_i's of Theorem 2). We pursue the naming issue further, presenting two results due to McCreight and Meyer [8].

Theorem 3

(*Union Theorem*) Let $(f_i)_{i\geq0}$ be a recursively enumerable sequence of functions(†) in $\mathcal{R}^{(1)}$ such that $f_i(x) < f_{i+1}(x)$ for all i,x.

(*)In the sense that the *negation* of the relation is true for finitely many x.
(†)That is, there is an $h \in \mathcal{R}^{(1)}$ such that $f_i = \phi_{h(i)}$.

Let Φ be a complexity measure. Then there is a $t \in \mathcal{R}^{(1)}$ such that $C_t^{\Phi} = \bigcup_{i \geq 0} C_{f_i}^{\Phi}$

Proof (A) *Informal discussion.*
As a first approximation, define $t(x) = f_x(x)$ for all x. It is then clear that for $i \leq x$

$$f_i(x) \leq f_x(x) = t(x), \text{ that is, } f_i(x) \leq t(x) \text{ a.e.}$$

Therefore,

$$\bigcup_{i \geq 0} C_{f_i}^{\Phi} \subset C_t^{\Phi} \tag{1}$$

To establish the converse inclusion, let $\phi_m \in C_t^{\Phi}$. Then $\Phi_m(x) \leq t(x)$ a.e. We would *like* to prove that $(\exists j)\Phi_m(x) \leq f_j(x)$ a.e. (which would imply $\phi_m \in C_{f_j}^{\Phi} \subset \bigcup_{i \geq 0} C_{f_i}^{\Phi}$). This is *not* guaranteed from the simplistic definition of t, as the latter may be "too large", making the inclusion (1) proper.

We thus set down *two* requirements for the definition of t:
For all m,

(i) $f_m(x) \leq t(x)$ a.e.

(ii) $\Phi_m(x) \leq t(x)$ a.e. $\Rightarrow (\exists j)\Phi_m(x) \leq f_j(x)$ a.e.

A way to achieve (ii) is to find (or attempt to find, by a sequence of approximations) a (possibly nontotal) function $\lambda m.j(m)$ such that

$$\Phi_m(x) \leq t(x) \text{ a.e.} \Rightarrow j(m)\!\downarrow \& \Phi_m(x) \leq f_{j(m)}(x) \text{ a.e.}$$

This is the route the formal proof takes. $t(x)$ is defined at the xth stage. At that stage j is approximated by a *total* function $j_{x \doteq 1}$. At the end of the stage we obtain $t(x)$ and the function j_x. Intuitively, $j_x(i)$ "codes" a "guess" that $\Phi_i(n) \leq f_{j_x(i)}(n)$ a.e.

(B) *Construction of t and j_n.*
Stage 0. Set $t(0) = 0$ and $j_0 = \lambda x.0$
Stage $x + 1$
Compute $k = \min\{j_x(i) \mid i \leq x \ \& \ \neg(\Phi_i(x + 1) \leq f_{j_x(i)}(x + 1))\}$, where, if k is achieved for many i, the *smallest* such i is to be *associated* with k.

$$\text{Set } t(x + 1) = \begin{cases} f_k(x + 1) & \textbf{if } k \textbf{ exists} \\ f_{x+1}(x + 1) & \textbf{otherwise} \end{cases}$$

Set $j_{x+1}(x + 1) = x + 1$

$\quad j_{x+1}(i) = x + 1$ if k exists and i is associated with it

$\quad j_{x+1}(n) = j_x(n)$ if $n \notin \{i, x + 1\}$. *End of construction.*

Note: An easy induction on x shows that $j_x(i) \le x$ if $i \le x$, thus $f_k(x)$ $\le f_x(x)$, where k is chosen as above.

Now if a k exists, the "guess" that $\Phi_i(n) \le f_k(n)$ a.e. is "threatened", where i is the associated i. Thus, by setting $t(n) = f_k(n)$, we ensure that *if* $\Phi_i(n) > f_k(n)$ *for infinitely many k and n, then $\Phi_i(n) > t(n)$ i.o., that is, t is not "too large"*. *End of* **Note**.

By recursiveness of $\lambda ixy.\Phi_i(x) \le y$, it follows that $\lambda xy.j_x(y) \in \mathcal{R}$, and hence $t \in \mathcal{R}$. [Recall that $f_i(x) = \phi_{h(i)}(x)$ for all i,x, hence $\lambda ix.f_i(x) \in \mathcal{R}$].

(C) *t works as required*.

Indeed, for any m, $f_m(x) \le t(x)$ a.e. [For, let $x \ge m$. Then $t(x)$ is set to $f_{j_{x \div 1}(i)}(x)$, where $i \le x \div 1$ and $j_{x \div 1}(i) < m$, for *finitely many* x, if at all. This is because as soon as this is done, $j_x(i)$ is next set to x, thus $\{j_x(l) \mid l \le x \ \& \ j_x(l) < m\}$ has one less element than $\{j_{x \div 1}(l) \mid l \le x \div 1 \ \& \ j_{x \div 1}(l) < m\}$. It follows that for some $x_0 \ge m$, $x \ge x_0 \Rightarrow t(x)$ is set to $f_x(x)$ or $f_k(x)$ with $k \ge m$. In either case, $t(x) \ge f_m(x)$]. Hence, $\bigcup_{i \ge 0} C_{f_i}^{\Phi} \subset C_t^{\Phi}$.

Next, let $\phi_m \in C_t^{\Phi}$, that is, $\Phi_m(x) \le t(x)$ a.e. Therefore, for some $x_0 \ge m$,
$$x > x_0 \Rightarrow \Phi_m(x) \le t(x) \tag{2}$$
But then, $(\forall x)[x > x_0 \Rightarrow \neg(\Phi_m(x) > f_{j_{x \div 1}(m)}(x))]$, for otherwise $t(x) \le f_{j_{x \div 1}(m)}(x)$ $< \Phi_m(x)$, contradicting (2).

Now, in particular, $t(x)$ will never be set to $f_{j_{x \div 1}(m)}(x)$ for $x > x_0$, thus (definition of j_n)

$$j_{x_0}(m) = j_{x_0+1}(m) = \ldots = \text{a (fixed) number } j$$

Hence, $x > x_0 \Rightarrow \Phi_m(x) \le f_{j_{x \div 1}(m)}(x) = f_j(x)$; that is, $\phi_m \in C_{f_j}^{\Phi} \subset \bigcup_{i \ge 0} C_{f_i}^{\Phi}.//$

Note: The "trick" of setting $j_x(i)$ to x once $j_{x \div 1}(i)$ has been "used" is reminiscent of the technique of cancelling indices (Theorem 12.1.1) as well as of the priority argument in the Friedberg-Muchnik theorem. It ensures that $\{x \mid f_k(x) = t(x)\}$ is finite for any k. It is interesting to note, that for each m such that $\Phi_m(x) \le t(x)$ a.e. $\lim_n j_n(m)$ exists, therefore the function j discussed under (A)(ii) in the proof is defined for all such m. Does $\lim_n j_n(m)$ exist if $\Phi_m(x) > t(x)$ i.o.? (Problem 12). (The above proof is, of course, independent of the answer to this question.)

We shall see an application of the union theorem in the next chapter.

The next result shows that every complexity class $C_{\phi_t}^{\Phi}$ can be "renamed" by a "good" name $\phi_{t'}$, that is, $C_{\phi_t}^{\Phi} = C_{\phi_{t'}}^{\Phi}$. The "goodness" of $\phi_{t'}$ is that it comes out of a measured set therefore it is "g-honest"(*), i.e., for some recursive g, $\Phi_{t'}(x) \le g(x, \phi_{t'}(x))$ a.e. (see Problem 14), that is, the output of $\phi_{t'}$ recursively bounds its "run-time" $\Phi_{t'}$ (this is not true for the *arbitrary* $\phi_{t'}$. See Corollary 2 of Theorem 12.1.1).

(*)This notion is due to McCreight and Meyer [8].

Theorem 4   ("*Honesty*" or "*Naming*" *theorem* [8]) For every Φ, there exists a measured set $\mathcal{M}$ and an $h \in \mathcal{R}^{(1)}$ such that for any *total* ϕ_t, $\phi_{h(t)} \in \mathcal{M}$ and $C^{\Phi}_{\phi_t} = C^{\Phi}_{\phi_{h(t)}}$.

Proof (A) *Informal discussion.*
The plan is to define $\psi \in \mathcal{P}^{(2)}$ so that for each ϕ_t (total or not)

(1) $(\forall i)\ [\Phi_i(n) \le \phi_t(n)$ a.e. iff $\Phi_i(n) \le \psi(t,n)$ a.e.] or equivalently

(1') $(\forall i)\ [\neg(\Phi_i(n) \le \phi_t(n))$ i.o. iff $\neg(\Phi_i(n) \le \psi(t,n))$ i.o.](*) and

(2) $\lambda txy.\psi(t,x) = y \in \mathcal{R}_*$, so that $(\lambda x.\psi(t,x))_{t \ge 0}$ is a measured set.

The definition of ψ proceeds by stages. For any $t \in \mathcal{N}$ we systematically examine candidate pairs (x,y) for inclusion in the (under construction) set $\lambda x.\psi(t,x)$.

At each stage n, all the indices $0,1,\ldots,n-1$ have already been partially considered and are kept in a list. The (x,y)-pair tried for membership is $((n)_0,(n)_1)$. Before the membership is decided, *all* i in the range $0,\ldots,n$ are checked for whether they will need *attention* now or at a future stage. These are the i's for which $\neg(\Phi_i((n)_0) \le \phi_t((n)_1))$ (3)

Because of requirement (1'), (3) must be matched as soon as possible by $\neg(\Phi_i(m) \le \psi(t,m))$ for an appropriate m. As soon as this is done, i is marked *attended* and is moved in a less prominent spot in the list, the *end*. Among the i's that need *attention* those attended *first* are in more prominent positions (or positions of *higher priority*) in the list; i.e., closest to the front. The following proof is due to Meyer and Moll [9] and has the flavor of a "priority argument".

(B) *The formal construction.* In the process of the construction, we build a list of indices $(i_0,\ldots,i_n)$ for all $n \ge 0$. We shall say that i_0 is in the *front*, and i_n in the *end* of the list. If $j < k$, then i_j is to the *left* of i_k, and i_k is to the *right* of i_j.
Stage n:

(1)   Place n at the *end* of the list, *unmarked*.

(2)   **if** $\Phi_t((n)_0) \le n$, **then** check *all* i in the list (that is, $i \le n$) which are *not* already marked for *attention*. **if** $\neg(\Phi_i((n)_0) \le \phi_t((n)_0))$ **then** mark that i *needs attention.*

(3)   **if** $\psi(t,(n)_0)$ is *already* defined **then goto** stage $n + 1$ [this guarantees single-valuedness of ψ].

(*)As we have remarked earlier, $x < y \ne \neg(y \le x)$ when one or both of x and y are *undefined*, unless we get into some artificial agreement to extend $\le$ so that it involves the "*undefined constant*" $\uparrow$. The problem with such extensions is that they are context-dependent and we prefer to avoid. For example, here it is "natural" to set $x \le \uparrow$ for all $x \in \mathcal{N} \cup \{\uparrow\}$ because $\Phi_i(x)\uparrow$ somehow conveys "infinite amount of time" spent. On the other hand, it is natural to set $\uparrow \le x$ in the context of fixpoint program semantics.

(4)   [Attend now to the "highest priority" m (i.e., to the leftmost) which *needs* attention and *can* be attended.]

Compute:

$m = (leftmost\ i\ in\ the\ list)$ [i *needs* attention & $\neg(\Phi_i((n)_0) \le (n)_1)$ & $(\forall j)$ {j is to the left of i and is *unmarked* $\Rightarrow \Phi_j((n)_0) \le (n)_1$}]

[Intuitively, and approximately put, "find leftmost $i \le n$ such that *all* $\phi_j((n)_0)$, for *unmarked* j to the left of i, take *fewer steps* to compute than $\phi_i((n)_0)$"]

   if m is *found* **then set** $\psi(t,(n)_0) = (n)_1$, and *unmark* m and *move it to the end* of the list; **else set** $\psi(t,(n)_0) = \max(\phi_t((n)_0),\Phi_t((n)_0))$. In *both* cases, **goto** stage $n + 1$. *End of construction.*

[**Note:** The "list" can be implemented by prime-power coding. That is, if the list is $((i_0,m_0),(i_1,m_1),\ldots,(i_n,m_n))$, where i_j is a ϕ-index and $m_j = 0$ means i_j *unmarked* whereas $m_j = 1$ means that i_j needs *attention*, then the list is coded by

$$l = \Pi_{q \le n}\, p_q^{\langle i_q,m_q \rangle}$$

where $\langle x,y \rangle = 2^{x+1} \cdot 3^{y+1}$.

Then to add an "unmarked" k to the end of the list, simply update l to $l * 2^{\langle k,0 \rangle}$. To "remove" the ith item of l, simply reset l to $\Pi_{q<i} p_q^{(l)_q+1} * \Pi_{i<q<\text{length}(l)}\, p_{q \dot- (i+1)}^{(l)_q+1}$, where "$*$" and "length" are as defined and used in Chapter 10.]

All predicates involved at stage n are *recursive*. Thus, $\psi \in \mathcal{P}^{(2)}$.(*) By *S-m-n* theorem, an $h \in \mathcal{R}^{(1)}$ exists so that $\psi(t,x) = \phi_{h(t)}(x)$ for all t,x.

(C) $h \in \mathcal{R}^{(1)}$ (*in other words, ψ*) *works.* To settle first requirement (2) [under (A)].

$\psi(t,x) = y$ iff (a) an m *was* found at step (4) of stage $\langle x,y \rangle$

or (b) *no* m was found at some stage $\langle x,i \rangle$,

where $0 \le i \le y$ and $y = \max(\phi_t(x),\Phi_t(x))$.

Now condition (a) is recursive and so is (b) since $y = \max(\phi_t(x),\Phi_t(x))$ iff $\Phi_t(x) \le y$ & $(y = \phi_t(x) \lor y = \Phi_t(x))$ iff $(\exists u)_{\le y}\{\Phi_t(x) = u$ & $[y = d((\mu z)_{\le g(x,u)} T(t,x,z))$ $\lor y = \Phi_t(x)]\}$, where $g \in \mathcal{R}^{(2)}$ is that of Proposition 12.1.4, with $\hat{\Phi}_i = \lambda x.(\mu y)T(i,x,y)$, T being the Kleene predicate and d the decoding function of Chapter 3.

We also observe that $\phi_t \in \mathcal{R} \Rightarrow \Phi_t \in \mathcal{R}$, hence $\phi_{h(t)} \in \mathcal{R}$.

We now prove that (1) is satisfied:

(*)The interested reader can provide a formal proof of this (Problem 15).

Let then $\Phi_i(x) \le \phi_t(x)$ a.e. for some i. This means that for some x_0, $x > x_0$ $\Rightarrow \phi_t(x)\uparrow \vee \Phi_i(x) \le \phi_t(x)$, or that $\{x \mid \neg(\Phi_i(x) \le \phi_t(x))\}$ is *finite*.

Thus, i will be marked for *attention*, and therefore will be *attended*, a *finite* number of times.

Hence there is a *last* time that i goes to the *end* of the list [due to being attended at step (4) of the construction].

After that happens, either i *never* is marked again, or is marked eventually and *stays* marked [otherwise it is moved to the end again by step (4)].

Case 1. i stays *marked for attention* but *unattended* from some stage n onwards.

We may assume without loss of generality(*) that for all stages beyond the nth(*) the initial segment of the list, up to the position of i, stays invariant in contents and marking (since attended indices move to the end, and i being immovable, they cannot reappear left of it).

Let us also assume that $2^{x_0+1} \cdot 3 > n$ (4)

Let $x > x_0$. To set(†) $\psi(t,x)$, we try pairs $(x,0),(x,1)$, etc., in turn. Thus, $\psi(t,x)$ cannot be set before stage $\langle x,0 \rangle = 2^{x+1} \cdot 3$. So let it be set at stage $k > 2^{x_0+1} \cdot 3 > n$.

It is either the case that $\psi(t,(k)_0)$ is set to $(k)_1$ or to $\max(\phi_t((k)_0),\Phi_t((k)_0))$.

In the *former case*, this is due to *attending* to an index m, *right* of i, such that all *unmarked* j *left* of m, and a fortiori *left of* i, satisfy $\Phi_j((k)_0) \le (k)_1$. It must also be $\Phi_i((k)_0) \le (k)_1$ [otherwise, i should be attended, rather than m, at this stage, contradicting the assumption under which we are operating in *Case 1*].

That is, $\Phi_i(x) \le \psi(t,x)$.

In the *latter case*, either $\phi_t((k)_0) = \phi_t(x)\uparrow$, or (since $x > x_0$) $\Phi_i(x) \le \phi_t(x) \le \psi(t,x)$.

Hence, $\Phi_i(x) \le \phi_t(x)$ a.e. $\Rightarrow \Phi_i(x) \le \phi_{h(t)}(x)$ a.e.   (5)

Case 2. i stays *unmarked* from stage n onwards.

As in Case 1, we assume that the initial segment of the list, up to the position of i, stays invariant in contents and marking. We assume that (4) holds. As the setting of $\psi(t,x)$ is due to indices m right of i (when $x > x_0$), and i is *unmarked*, if we assume that this happens at stage $k > 2^{x_0+1} \cdot 3 > n$, then $x = (k)_0$. We conclude that

$$\psi(t,x)\downarrow \Rightarrow \Phi_i(x) \le \psi(t,x)$$

as in Case 1. Thus, (5) is once again established.

(*)Since this will eventually happen.

(†)"Set" rather than "define", due to the alternative $\max(\phi_t((m)_0), \Phi_t((m)_0))$ in step (4); it may be that $\phi_t((m)_0)\uparrow$.

We now turn to the converse. So let, for some i,

$$\neg(\Phi_i(x) \le \phi_t(x)) \text{ i.o.} \tag{6}$$

It follows that i is *marked for attention*, and hence also *attended*, i.o. By step (4) of the construction, for infinitely many (x,y)

(i) $\neg(\Phi_i(x) \le y)$

(ii) $\psi(t,x)$ is set to y (this happens at stage $\langle x,y \rangle$).
       That is, $\neg(\Phi_i(x) \le \psi(t,x))$ i.o. $\tag{7}$

We proved (6) $\Rightarrow$ (7) for arbitrary i, and this concludes the proof of (1).//

Note: Attended indices are assigned lowest priority to allow other indices to be attended, since not only "all" potential input/output pairs (x,y) must be tried for the definition of ψ but also "all" indices.

We have seen some results in "High Level(*) Complexity", but quite a bit has been left untold. Some further results will be found in the problems, but the reader whose interest on the subject has been sufficiently aroused should consult the survey by Hartmanis and Hopcroft [5] and by Borodin [3], where further references to specific results will be found.

We now turn to a brief study of the computational complexity of more down-to-earth functions.

(*)The attribute refers to the level of abstraction.

PROBLEMS

1. Fill in the missing details in Example 12.1.2.

2. Show that TM and URM length of computation functions are legitimate complexity measures.

3. (Refer to Definitions 10.2.1 and 10.2.2) Is $(\lambda x.\| i,x \|)_{i \geq 0}$ a complexity measure for $\mathcal{P}^{(1)}$?

4. Show that for any complexity measure Φ there is a $g \in \mathcal{R}^{(3)}$ such that $\Phi_i(x) = (\mu y)g(i,x,y)$ for all i,x.

5. (Combining Lemma [3,5]) For any complexity measure Φ and $c \in \mathcal{R}^{(2)}$ such that $\phi_i(x)\downarrow$ & $\phi_j(x)\downarrow \Rightarrow \phi_{c(i,j)}(x)\downarrow$ there is an $h \in \mathcal{R}^{(3)}$ such that $\Phi_{c(i,j)}(x) \leq h(x,\Phi_i(x),\Phi_j(x))$ a.e.

6. Verify that $\hat{\Phi}$ of Example 12.1.5 is a complexity measure.

7. Provide all the missing formal details in the proof of the speed-up theorem.

8. Show that if Φ is any complexity measure, $\lambda ixy.\Phi_i(x) > y \notin \mathcal{R}_*$. Does this contradict the fact that $\lambda ixy.\Phi_i(x) \leq y \in \mathcal{R}_*$?

9. $(\gamma_i)_{i \geq 0}$ is a measured set iff there is a $g \in \mathcal{R}^{(3)}$ such that $\gamma_i = \lambda x.(\mu y)g(i,x,y)$.

10. Complete the proof of Lemma 12.2.2.

11. (A slight amendment of the statement of half the compression theorem [Theorem 8] in Blum [2])
    A re sequence $(\gamma_i)_{i \geq 0}$ of $\mathcal{P}^{(1)}$-functions is an "almost measured set", in the sense that, for some $\lambda ixy.M(i,x,y) \in \mathcal{R}$ and for all $x \geq i$, $\gamma_i(x) = y$ iff $M(i,x,y) = 0$, provided (1) and (2) below hold:

    (1) There is an $r \in \mathcal{R}^{(1)}$ such that, for all i, $\text{dom}(\gamma_i) = \text{dom}(\phi_{r(i)})$

    (2) There is a $g \in \mathcal{R}^{(2)}$, such that for all i, $x \geq i \Rightarrow \Phi_{r(i)}(x) \leq g(x,\gamma_i(x))$

    Note: (i) Recall that $(\gamma_i)_{i \geq 0}$ is re, iff for some $\sigma \in \mathcal{R}^{(1)}$, $\gamma_i = \phi_{\sigma(i)}$ for all i

    (ii) By (1), $\Phi_{r(i)}(x)\uparrow$ iff $\gamma_i(x)\uparrow$, thus
    $\Phi_{r(i)}(x) \leq g(x,\gamma_i(x))$ whenever $\gamma_i(x)\uparrow$, since $\uparrow = \uparrow$
    by convention.

    (iii) The term "almost measured" we borrowed from Borodin [3].

12. Refer to the proof of Theorem 12.2.3.
Does $\lim_n j_n(m)$ exist if $\Phi_n(x) > t(x)$ i.o.?

13. Refer to the proof of Theorem 12.2.3. Let ψ be the restriction of j on $\{m \mid \Phi_m(x) \le t(x) \text{ a.e.}\}$. Is $\psi \in \mathcal{P}$?

14. [8] Show that each function of a measured set is g-honest, where g depends only on the set.

15. Refer to the proof of the naming theorem (12.2.4). Show formally that $\lambda txy.\psi(t,x) = y \in \mathcal{R}_*$.

16. Show that for any Φ there are infinitely many complexity gaps.

17. Let Φ be a complexity measure and $f \in \mathcal{R}^{(1)}$ such that $(\forall i)\phi_{f(i)} \in \mathcal{R}$. Then, for some $r \in \mathcal{R}^{(1)}$, $\{\phi_{f(i)} \mid i \in \mathcal{N}\} \subset C_r^\Phi$ (In words, "There is a recursive bound to the run-times of a re class of recursive functions".)

18. Given a complexity measure Φ, and $r \in \mathcal{R}^{(1)}$, such that $\phi_i \in \mathcal{R} \,\&\, \phi_i(x) = 0$ a.e. $\Rightarrow \phi_i \in C_r^\Phi$. Then $\phi_j \in \mathcal{R} \,\&\, (\forall x)\phi_j(x) > r(x) \Rightarrow C_{\phi_j}^\Phi$ is a re class.
(*Hint.* [5] Show that the S-m-n function h such that $\phi_{h(i,y,z)}(x) = $ **if** $[(\forall m)_{\le y}\Phi_i(m) \le z] \,\&\, [(\forall m)_{\le x}m > y \Rightarrow \Phi_i(m) \le \phi_j(m)]$ **then** $\phi_i(x)$ **else** 0 enumerates indices exactly for all the functions in $C_{\phi_j}^\Phi$, provided $(\forall x)\phi_j(x) > r(x)$.)

19. (Lewis [7], Landweber and Robertson [6]) There are Φ and $r \in \mathcal{R}^{(1)}$ such that C_r^Φ is *not* a re class.
(*Hint.* [5,6,7] Let $f \in \mathcal{R}^{(1)}$ be strictly increasing such that $\phi_{f(i)} = \lambda x.i$ for all i. Let T be a Kleene-predicate. Define $\Phi_i(x)$ to be x whenever $i \notin f(\mathcal{N})$; define $\Phi_{f(i)}(x)$ to be **if** $(\forall y)_{\le x} \neg T(i,i,y)$ **then** 0 **else** x. Verify that Φ is a complexity measure and that $r = \lambda x.0$ will do.)

REFERENCES

[1] Alton, D.A.: " 'Natural' Programming Languages and Complexity Measures for Subrecursive Programming Languages: an abstract approach" *Recursion Theory: its generalisation and applications* (Proc. Logic Colloq. Univ. of Leeds, Leeds, 1979): 248–285 London Math. Soc. Lecture Note Series, 45, Cambridge: Cambridge Univ. Press, 1980.

[2] Blum, M. "A Machine-Independent Theory of the Complexity of Recursive Functions". *Journal of the ACM,* 14 (1967): 322–336.

[3] Borodin, A. "Complexity Classes of Recursive Functions and the Existence of Complexity Gaps". *Proc. ACM Symposium on Theory of Computing,* 1969: 67–78. (See also, Borodin, A. *Computational Complexity and the Existence of Complexity Gaps.* TR No. 19, Dept. of Computer Science, University of Toronto, 1970.)

[4] Borodin, A. "Computational Complexity: Theory and Practice". (in Aho, A.V. [Ed.]: *Currents in the Theory of Computing.* Englewood Cliffs, N.J.: Prentice-Hall, 1973).

[5] Hartmanis, J. and Hopcroft, J.E. "An overview of the Theory of Computational Complexity". *Journal of the ACM,* 18 (1971): 444–475.

[6] Landweber, L.H. and Robertson, E.L. "Recursive Properties of Abstract Complexity Classes". *Journal of the ACM,* 19 (1972): 296–308. (Also in *Proc. of ACM Symp. of Theory of Computing,* 1970: 31–36.)

[7] Lewis, F.D. "Unsolvability Considerations in Computational Complexity". *Proc. ACM Symposium on Theory of Computing,* 1970: 22–30.

[8] McCreight, E.M. and Meyer, A.R. "Classes of Computable Functions Defined by Bounds on Computation (preliminary report)". *Proc. ACM Symp. on Theory of Computing,* 1969: 79–88.

[9] Meyer, A.R. and Moll, R. "Honest Bounds for Complexity Classes of Recursive Functions". *Proc. of the 13th Annual Switching and Automata Theory Conference* held at the U. of Maryland, 1972: 61–66.

[10] Trakhtenbrot, B.A. *Complexity of Algorithms and Computations* (course notes). University of Novosibirsk, USSR, 1967.

CHAPTER 13
Complexity of Primitive Recursive Functions

"High Level Complexity" has provided some general principles governing the difficulty of actual computations, no matter how this difficulty is measured(*) (e.g., gap, compression, union and honesty theorems).

In practice one hardly ever uses even the full scope of primitive recursive functions, let alone recursive or partial recursive functions, the reason being that most such functions grow horrendously fast and therefore, inevitably, have horrendously large run times and demands on memory.

Indeed, in practice we compute *finite* functions (due to the finite memory and word size of computers). However, we do so, normally, not via a table look-up, but by writing *general algorithms* which, should the memory/word size be unlimited, would compute these functions for any input.

Thus, the study of the complexity of such algorithms is relevant.

We shall first consider a number of approaches dealing with the *static* complexity of primitive recursive functions, and with the subdivisions of $\mathcal{PR}$ into infinite hierarchies of subclasses of ever-increasing static complexity.

(*)It is hard to envisage a reasonable measure of complexity for partial recursive functions which does not obey Blum's axioms.

Next we shall see that there is a strong relationship between static and *dynamic* complexity of $\mathcal{PR}$-functions, and shall allow ourselves to be side-tracked(*) into a brief discussion of *NP* and *NP*-complete problems. This subject has come to prominence ever since the publication of Cook [5] and Karp [16], and is still of strong research interest; after all, despite an ever-increasing list of *NP*-complete problems, the basic question "$P = NP$?" is still open.

§13.1 THE AXT, LOOP-PROGRAM, AND GRZEGORCZYK HIERARCHIES

Definition 1   (Axt [2], Heinermann [13])
K_0 is the *closure* of $\{\lambda x.x + 1, \lambda x.x\}$ under *substitution*. For $n \geq 0$, K_{n+1} is the *closure* under *substitution* of
$K_n \cup \{f \mid f$ is defined by *primitive recursion* from functions g,h in $K_n\}.//$

Note: Inherent in the definition is the requirement that primitive recursion is the "slow" or "difficult" operation rather than substitution, since the former does whereas the latter does not lead you out of K_n.

Theorem 1

(i) $K_n \subset K_{n+1}$ for $n \geq 0$

(ii) $\bigcup_{n\geq 0} K_n = \mathcal{PR}.$

Proof Problem 1.//

Definition 2   For any family of classes $\mathcal{F} = (S_n)_{n\geq 0}$ such that $S_n \subset \mathcal{PR}$ for $n \geq 0$, we shall call $\mathcal{F}$ a *primitive recursive hierarchy*, iff

(i) $S_n \subset S_{n+1}, n \geq 0$

(ii) $\bigcup_{n\geq 0} S_n = \mathcal{PR}.$

The hierarchy is *proper* (*infinite, nontrivial*) iff

(i') $S_n \subsetneq S_{n+1}$ a.e.

We shall say that the *level* of f *in the hierarchy* $\mathcal{F}$ is $\leq n$ if $f \in S_n$. If $f \in S_n - S_{n-1}$, then the level is $= n.//$

(*)The techniques used in this context are predominantly combinatorial rather than recursion-theoretic.

We now see that the hierarchy $\mathcal{H} = (K_n)_{n \geq 0}$ is proper. $\lambda nx.A_n(x)$ is the Ackermann function of §3.3.

Lemma 1   For $n \geq 0$, $\lambda x.A_n(x) \in K_n$.

Proof Induction on n.

$n = 0$: $A_0 = \lambda x.x + 2$ which is obtained from $\lambda x.x + 1$ by substitution. Hence, $A_0 \in K_0$.

Assume that $A_k \in K_k$ (case $n = k$).

Now, A_{k+1} is given by

$$\begin{cases} A_{k+1}(0) = 2 \\ A_{k+1}(x + 1) = A_k(A_{k+1}(x)) \end{cases}$$

Since $A_k \in K_k$, and substitution can help to put the above recursion in standard form, $A_{k+1} \in K_{k+1}$ by definition of K_{k+1}, and we just settled the case $n = k + 1$.//

Lemma 2   For $n \geq 0$, if $\lambda \vec{x}.f(\vec{x}) \in K_n$, then there is an m (depending on f) such that $f(\vec{x}) \leq A_n^m(\max(\vec{x}))$ for all $\vec{x}$.

Proof This is a trivial observation derived from the proof of Theorem 3.3.1. (See Problem 2).//

Theorem 2   The hierarchy $\mathcal{H} = (K_n)_{n \geq 0}$ is proper.

Proof By Lemma 3.3.7, $A_{n+1}(x) > A_n^m(x)$ a.e. for any (fixed) m. Thus, $A_{n+1} \in K_{n+1} - K_n$ ($n \geq 0$), for if $A_{n+1} \in K_n$, then, for some m, $A_{n+1}(x) \leq A_n^m(x)$ (all x).//

A closely related hierarchy, $\mathcal{H}^{\text{sim}} = (K_n^{\text{sim}})_{n \geq 0}$ is defined using simultaneous primitive recursion.

Definition 3 $K_0^{\text{sim}} = K_0$.

For $n \geq 0$, K_{n+1}^{sim} is the closure under *substitution* of $K_n^{\text{sim}} \cup \{f \mid f$ is obtained from functions in K_n^{sim} by *simultaneous* primitive recursion$\}$.//

Note: Clearly, $K_n \subset K_n^{\text{sim}}$ for $n \geq 0$.

Theorem 3

(i) $K_n^{\text{sim}} \subsetneq K_{n+1}^{\text{sim}}$ for $n \geq 0$

(ii) $\bigcup_{n\geq 0} K_n^{\text{sim}} = \mathcal{PR}.$

Proof The proper inclusions in (i) follow from the easy observation that $A_{n+1} \in K_{n+1}^{\text{sim}} - K_n^{\text{sim}}$ for $n \geq 0$. (See Problem 3 for details.)//

In preparation for the comparison of the various hierarchies of this section, we now list some useful functions and note upper bounds of their levels in the $\mathcal{K}$-hierarchy.

Lemma 3   The following table presents functions of $\mathcal{PR}$ and upper bounds of their levels in the $\mathcal{K}$-hierarchy.

Function	Upper bound of level
$\lambda xy.x + y$	1
$\lambda x.x \dot{-} 1$	1
$\lambda xy.x(1 \dot{-} y)$	1
$\lambda xy.xy(*)$	2
$\lambda x.2^x$	2
$\lambda xy.x \dot{-} y$	2
$\lambda xy.\lvert x - y \rvert$	2
$\lambda xy.\max(x,y)$	2
$\lambda x.\left\lfloor \dfrac{x(x+1)}{2} \right\rfloor$	2
$\lambda xy.\operatorname{rem}(x,y)\left[= \text{remainder of the division } \dfrac{x}{y}\right](\dagger)$	2
$\lambda x.\left\lfloor \dfrac{x}{2} \right\rfloor(\dagger)$	2
$\lambda x.2^{\lfloor \log_2 x \rfloor}$ [where, by convention, $\log_2 0 = 0$]$(\dagger)$	2
$\lambda x.\lfloor \log_2 x \rfloor$	3

Proof We justify the last five cases, leaving the rest to the reader. (See Problem 4.) The proofs are from [27].
    (1) As $x(x + 1)$ is divisible by 2,

$$\left\lfloor \frac{x(x+1)}{2} \right\rfloor = \frac{x(x+1)}{2} = \Sigma_{i \leq x}\, i$$

(*)xy means x times y in this context.

($\dagger$)These are the only nontrivial entries in the list. Schwichtenberg attributes the list to Heinermann [13].

Thus,

$$\begin{cases} \Sigma_{i \leq 0}\, i = 0 \\ \Sigma_{i \leq x+1}\, i = (\Sigma_{i \leq x}\, i) + x + 1 \end{cases}$$

from which the upper bound ($=2$) of the level follows since $\lambda x.x + 1 \in K_0$ ($\subset K_1$) and $\lambda xy.x + y \in K_1$.

(2) Define first h by

$$\begin{cases} h(0,y) = y \div 1 \\ h(x + 1, y) = h(x,y) \div 1 + (y \div 1)(1 \div h(x,y)) \end{cases}$$

Observe that for any y, h is linearly falling from $y \div 1$, i.e., the maximum remainder value, to 0, then jumps to $y \div 1$ again to repeat the previous behavior.

Clearly, $\mathrm{rem}(x,y) = (y \div 1) \div h(x,y)$, thus $\mathrm{rem} \in K_2$ since $h \in K_2$.

(3) $\lfloor x/2 \rfloor = \mathrm{rem}(\lfloor x(x + 1)/2 \rfloor, x) + \mathrm{rem}(\lfloor (x \div 1)x/2 \rfloor, x \div 1)$ [The first remainder is **if** $2 \mid x$ **then** $\lfloor x/2 \rfloor$ **else** 0. The second remainder is **if** $2 \mid x$ **then** 0 **else** $\lfloor x/2 \rfloor$].

(4) Define first g by

$$\begin{cases} g(0) = 0 \\ g(x + 1) = \textbf{if } g(x) > 1 \textbf{ then } g(x) \div 1 \\ \qquad\qquad \textbf{else } x + 1 \end{cases}$$

It is easy to prove by induction on k that $g(2^k) = 2^k$, and that $2^k \leq x < 2^{k+1} \Rightarrow g(x) + x = 2^{k+1}$ for $k \geq 0$ (these claims are easily "discovered" by looking at the graph of g).

Thus,

$$2^{\lfloor \log_2 x \rfloor} = \left\lfloor \frac{g(x) + x}{2} \right\rfloor \quad \text{for } x \geq 1,$$

that is,

$$2^{\lfloor \log_2 x \rfloor} = 1 \div (1 \div (1 \div x)) + \left\lfloor \frac{g(x) + x}{2} \right\rfloor (1 \div (1 \div x))$$

The contention follows by observing that
$$g(x + 1) = g(x) \div 1 + (x + 1)(1 \div (g(x) \div 1)).$$
$$(5) \lfloor \log_2 x \rfloor = \Sigma_{i \leq x} \, i \, (1 \div | 2^{\lfloor \log_2 x \rfloor} - 2^i |).//$$

As usual, for any class $\mathcal{C}$ of functions, $\mathcal{C}_*$ is the class of the corresponding predicates; i.e., $\mathcal{C}_* = \{f(\vec{x}) = 0 \,|\, f \in \mathcal{C}\}$.

Lemma 4 $(K_n)_*$ and $(K_n^{\mathrm{sim}})_*$, for $n \geq 1$, are closed under Boolean operations.

Proof Problem 5.//

Lemma 5 K_n and K_n^{sim}, for $n \geq 1$, are closed under definition by cases.

Proof See proof of Theorem 2.2.5.//

Example 1 $\lambda x.\mathrm{rem}(x,2) \in K_1^{\mathrm{sim}}$.
Indeed, define f and g by

$$f(0) = 0$$
$$g(0) = 1$$
$$f(x + 1) = g(x)$$
$$g(x + 1) = f(x).$$

[The picture is one of two infinite "arrays" f and g as follows

f	0	1	0	1	0	1	0	...
g	1	0	1	0	1	0	1	...
x	0	1	2	3	4	5	6	

]

Clearly, f and g are in K_1^{sim} and $\lambda x.\mathrm{rem}(x,2) = f$. It is interesting to note that results due to D. Ritchie [23] and Tsichritzis [30], of which we will have a small sample shortly, show that $\lambda x.\mathrm{rem}(x,2) \notin K_1$, hence $K_1 \subsetneq K_1^{\mathrm{sim}}.//$

Lemma 6(*) (*A characterization of K_0 and K_0^{sim}*) $K_0 (= K_0^{\mathrm{sim}})$ contains only functions of the forms $\lambda \vec{x}_n.k$ and $\lambda \vec{x}_n.x_i + k$.

(*)This straightforward result can be traced at least as far back as D. Ritchie [23] and Tsichritzis [30].

Proof Induction with respect to K_0:

Clearly, the initial functions $\lambda x.x + 1$ and $\lambda x.x$ of K_0 have the desired form.

Let $\lambda \vec{x}_n.f(\vec{x}_n)$ and $\lambda y\vec{z}_m.g(y,\vec{z}_m)$ have the desired form. Let $h = \lambda \vec{x}_n \vec{z}_m.g(f(\vec{x}_n),\vec{z}_m)$.

There are the cases

(1) $g(y,\vec{z}_m) = k$

(2) $g(y,\vec{z}_m) = z_i + k$

(3) $g(y,\vec{z}_m) = y + k$

By the hypothesis on f, h has the right form. The remaining cases of substitution are left to the reader.//

Theorem 4   (D. Ritchie [23], Tsichritzis [30]) K_1 is the smallest set including

$$\lambda xy.x + y$$

$$\lambda x.x \doteq 1$$

$$\lambda xy.x(1 \doteq y) \,|\, [\textbf{if } y = 0 \textbf{ then } x \textbf{ else } 0]$$

and closed under *substitution*.

Proof [30] By Lemma 3, it suffices to show that K_1 is included in the above mentioned closure; call the latter $\mathcal{C}$.

Let $\lambda \vec{x}y.f(y,\vec{x}) \in K_1 - K_0$.

Then it suffices to consider the case that f is *directly* obtained from some g and h (in K_0) by primitive recursion, rather than after the intervention of substitution operations (since $\mathcal{C}$ is closed under substitution).

So let

$$f(0,\vec{x}) = h(\vec{x})$$

$$f(y + 1,\vec{x}) = g(y,\vec{x},f(y,\vec{x}))$$

We consider cases for g:

Case 1. $g(y,\vec{x},z) = k$ (constant) for all $y,\vec{x},z$.

Then $f(y,\vec{x}) = \textbf{if } y = 0 \textbf{ then } h(\vec{x}) \textbf{ else } k$, which implies $f \in \mathcal{C}$ since $h \in K_0 \subset \mathcal{C}$, $\mathcal{C}$ contains $\lambda xyz.\textbf{if } x = 0 \textbf{ then } y \textbf{ else } z$ (Problem 6), and is closed under substitution.

Case 2. $g(y,\vec{x},z) = y + k$ (or $x_i + k$)

Then

$$f(y,\vec{x}) = \textbf{if } y = 0 \textbf{ then } h(\vec{x}) \textbf{ else } \begin{cases} y \doteq 1 + k \\ x_i + k \text{ (as the case may be)} \end{cases}$$

Again $f \in \mathcal{C}$.

Case 3. $g(y, \vec{x}, z) = z + k$.

Then $f(y, \vec{x}) = h(\vec{x}) + ky$. Again $f \in \mathcal{C}$ since $\lambda y.ky$ is obtained from $\lambda xy.x + y$ by substitution.//

Definition 4   Let $\tilde{\mathcal{C}}$ denote the closure of $\{\lambda xyz.\textbf{if } x = 0 \textbf{ then } y \textbf{ else } z\}$ under *substitution*, Lin denote the set of *"linear functions"*

$$\{\lambda \vec{x}_n.c_1 \cdot x_1 + \ldots + c_n \cdot x_n \dotminus c_{n+1} \,|\, c_i \geq 0, i = 1, \ldots, n, n \in \mathcal{N}\}$$

and $\mathcal{C}'$ denote the set

$$\{\lambda \vec{x}_n.h(f_1(\vec{x}_n), \ldots, f_m(\vec{x}_n)) \,|\, h \in \tilde{\mathcal{C}} \text{ and } f_i \in \text{Lin for } i = 1, \ldots, m \text{ and } n \in \mathcal{N}\}.//$$

Theorem 5 $K_1 = \mathcal{C}'$.

Proof We shall rely on the characterization of K_1 as $\mathcal{C}$ (Theorem 4).

That $\mathcal{C}' \subset \mathcal{C}$ is immediate. To prove $\mathcal{C} \subset \mathcal{C}'$, we do induction with respect to $\mathcal{C}$:

(1) The initial functions of $\mathcal{C}$ are in $\mathcal{C}'$.

Indeed, $\lambda x.x = \lambda x.\textbf{if } x = 0 \textbf{ then } x \textbf{ else } x$, hence it is in $\tilde{\mathcal{C}}$.

Thus,

$$\lambda xy.x + y \in \mathcal{C}'$$

$$\lambda x.x \dotminus 1 \in \mathcal{C}'$$

Also, $h = \lambda xy.x(1 \dotminus y) = \lambda xy.\textbf{if } y = 0 \textbf{ then } x \textbf{ else } 0$ is in $\tilde{\mathcal{C}}$. Substituting $0 \cdot x + 1 \cdot y, 1 \cdot x + 0 \cdot y$ (all in Lin) for y, x, respectively, we obtain $h \in \mathcal{C}'$.

(2) Let now $\lambda \vec{x}.g(\vec{x})$ and $\lambda y\vec{z}.f(y, \vec{z})$ be in $\mathcal{C}'$.

Consider $k = \lambda \vec{x}\,\vec{z}.f(g(\vec{x}), \vec{z})$.

Let $h \in \tilde{\mathcal{C}}$ be such that $g = \lambda \vec{x}.h(g_1(\vec{x}), \ldots, g_m(\vec{x}))$, where $g_i \in \text{Lin for } i = 1, \ldots, m$.

Thus,

$$k(\vec{x}, \vec{z}) = f(h(g_1(\vec{x}), \ldots, g_m(\vec{x})), \vec{z})$$

By induction on the *number* of *substitutions* needed to form h (starting from $\lambda xyz.$ **if** $x = 0$ **then** y **else** z) we prove that for *any* f, and g_i in $\mathcal{C}$ (*not* just g_i in Lin) there is an $\tilde{h} \in \tilde{\mathcal{C}}$ such that

$$f(h(g_1(\vec{x}),\ldots,g_m(\vec{x})),\vec{z}) = \tilde{h}(g_{p_1}(\vec{x}),\ldots,g_{p_r}(\vec{x}),\vec{z},f(g_{j_1}(\vec{x}),\vec{z}),\ldots,f(g_{j_l}(\vec{x}),\vec{z}))$$

where j_i and p_q are in $\{1,\ldots,m\}$ for $i = 1,\ldots,l$ and $q = 1,\ldots,r$.

If *no* substitutions are needed to obtain h, then

$$k(\vec{x},\vec{z}) = f(\textbf{if } g_1(\vec{x}) = 0 \textbf{ then } g_2(\vec{x}) \textbf{ else } g_3(\vec{x}),\vec{z})$$

$$= \textbf{if } g_1(\vec{x}) = 0 \textbf{ then } f(g_2(\vec{x}),\vec{z}) \textbf{ else } f(g_3(\vec{x}),\vec{z})$$

Take $\tilde{h} = \lambda xyz.$ **if** $z = 0$ **then** x **else** y.

Otherwise, $h(\vec{y}_m) = \textbf{if } h_1(\vec{y}_m) = 0 \textbf{ then } h_2(\vec{y}_m) \textbf{ else } h_3(\vec{y}_m)$ where h_1, h_2, and h_3 are in $\tilde{\mathcal{C}}$.

Thus,

$$f(h(g_1(\vec{x}),\ldots,g_m(\vec{x})),\vec{z}) = \textbf{if } h_1(\vec{y}_m) = 0$$

$$\textbf{then } f(h_1(g_1(\vec{x}),\ldots,g_m(\vec{x})),\vec{z}) \textbf{ else } f(h_2(g_1(\vec{x}),\ldots,g_m(\vec{x})),\vec{z})$$

and the claim follows by assuming (inductively) that it holds when h is h_1 or h_2.

We now return to the main argument.

Observe first that the composition of linear functions can be described via **if-then-else** (if necessary) acting on linear functions [for example, if $f(\vec{x}) = f_1x_1 + \ldots + f_nx_n + f_{n+1}$ and $g(y,\vec{z}) = g_0y + g_1z_1 + \ldots + g_mz_m + g_{m+1}$, then $\lambda \vec{x}\,\vec{z}.g(f(\vec{x}),\vec{z})$ is linear, and no intervention of **if-then-else** is necessary. If however we had $\dot{-} f_{n+1}$ in f, then $g(f(\vec{x}),\vec{z}) = \textbf{if } f(\vec{x}) = 0 \textbf{ then } g(0,\vec{z}) \textbf{ else } g(f_1x_1 + \ldots + f_nx_n,\vec{z}) \dot{-} f_{n+1}$. A proof of the general case of this remark is left to the reader].

By the above remark, and with the help of the earlier proved subresult, if the g_i $(i = 1,\ldots,m)$ are in Lin, and if $f = \lambda y\vec{z}.q(f_1(y,\vec{z}),\ldots,f_r(y,\vec{z}))$, where $q \in \tilde{\mathcal{C}}$ and $f_i \in$ Lin for $i = 1,\ldots,r$, it follows that $k \in \mathcal{C}'$.

Since $\mathcal{C}'$ contains the initial functions of $\mathcal{C}$ and it too is closed under *substitution* (the case of explicit transformations was left out as being easy), it must contain $\mathcal{C}.//$

Note: Theorem 5 is due to Tsichritzis [30]. He used a more elaborate proof, where the passage from a K_1-description of f (through its derivation) to a $\mathcal{C}'$ description (= nested **if-then-else**'s acting on linear functions) was seen to be *constructive*. This was essential for him to prove that the equivalence problem for K_1 is solvable. This result is within the reader's reach now, given the material developed so far, and he may wish to prove it (Problem 9).

Corollary   For each function $f \in K_1$ there is a constant f_0 such that the restriction of f on $\{\vec{x} \mid \min(\vec{x}) \geq f_0\}$ is linear.

Proof For sufficiently large $\vec{x}$ all the conditions expressed as $f_i(\vec{x}) = 0$, where $f_i \in$ Lin in the $\mathscr{C}'$-description of f, become *false*. Thus f will eventually equal the function defined in the last **else**-case, which is, of course, linear.//

Note: It is immediate that neither $\lambda x.\mathrm{rem}(x,k)$ nor $\lambda x.\lfloor x/k\rfloor$ are in K_1.

Example 2   Define f and g by

$$f(0) = 0$$
$$g(0) = 0$$
$$f(x + 1) = g(x)$$
$$g(x + 1) = f(x) + 1$$

Clearly, $f = \lambda x.\lfloor x/2\rfloor$, and, of course, f and g are in K_1^{sim}.//

We now turn to the loop-program hierarchy (Meyer and D. Ritchie [19]). The reader is referred to §2.7.

Definition 5   A loop-program P is in L_0 iff the only instructions in P have the format

(i) $X = 0$

(ii) $X = Y$

(iii) $X = X + 1$

For $n \geq 0$, $P \in L_{n+1}$ iff $P \in L_n$ or there are programs Q,S (either or both possibly empty) in L_{n+1} and $R \in L_n$ such that $P = Q$; **Loop** X; R; **end**; S for some variable X.//

Note: Intuitively, L_n $(n \geq 1)$ contains those programs that "nest" the **Loop-end** pair in depth *at most n*. It also follows that $L_n \subsetneq L_{n+1}$ and that $L = \bigcup_{n\geq 0} L_n$, where L is the set of all loop-programs. The following is a counterpart of Definition 2.7.7.

Definition 6   For $n \geq 0$, $\mathcal{L}_n = \{f \mid$ for some $P \in L_n$ and variables $X_1, \ldots,$ X_n, Y of $P, f = P_Y^{\vec{X}_n}\}$.//

Theorem 6 $K_n^{\mathrm{sim}} = \mathcal{L}_n$ for all $n \geq 0$.

Proof Use induction on n and imitate the proofs of Lemmas 2.7.1 and 2.7.2 (Problem 10).//

Corollary $\mathcal{L} = (\mathcal{L}_n)_{n\geq 0}$ is a proper $\mathcal{PR}$-hierarchy, that is,

(i) $\mathcal{L}_n \underset{\neq}{\subseteq} \mathcal{L}_{n+1}, n \geq 0$

(ii) $\bigcup_{n\geq 0} \mathcal{L}_n = \mathcal{PR}.$

We now turn to the characterization of $\mathcal{L}_1(= K_1^{\text{sim}})$ due to Tsichritzis [29,30]. Besides obtaining a solution to the equivalence problem of $\mathcal{L}_1$ (as it was done in [29,30]) one can also use the characterization to prove that such and such a function is *not* in $\mathcal{L}_1$. This last type of application is our motivation in this context. [Note that majorization arguments will not always be applicable to prove non-membership. For example, $\lambda xy.x \dot- y$ cannot be refused membership in $\mathcal{L}_1$ due to its size. As we shall see though, it is not in $\mathcal{L}_1$ after all.]

Theorem 7   [29,30] $\mathcal{L}_1(= K_1^{\text{sim}})$ is the closure of $\{\lambda xy.x + y, \lambda x.x \dot- 1, \lambda xyz.\textbf{if } x = 0$ **then** y **else** $z, \lambda x.\lfloor x/k \rfloor, \lambda x.\text{rem}(x,k)\}$ under *substitution*.

Proof [29,30] Call this closure $\mathcal{C}^{\text{sim}}$. That $\mathcal{C}^{\text{sim}} \subset K_1^{\text{sim}}$ follows from what we know so far and Problems 6, 7, and 8.

To show that $K_1^{\text{sim}} \subset \mathcal{C}^{\text{sim}}$, it suffices to consider $f \in K_1^{\text{sim}} - K_0^{\text{sim}}$ (since $K_0^{\text{sim}} = K_0 \subset \mathcal{C} \subset \mathcal{C}^{\text{sim}}$).

So let f be defined by the following simultaneous recursion:

$$(A)\begin{cases} f_1(0,\vec{x}) = h_1(\vec{x}) \\ \quad\vdots \\ f_n(0,\vec{x}) = h_n(\vec{x}) \\ f_1(y+1,\vec{x}) = g_1(y,\vec{x},f_1(y,\vec{x}),\ldots,f_n(y,\vec{x})) \\ \quad\vdots \\ f_n(y+1,\vec{x}) = g_n(y,\vec{x},f_1(y,\vec{x}),\ldots,f_n(y,\vec{x})) \end{cases}$$

where $f = f_1$ and h_i and g_i are in $K_0^{\text{sim}} (= K_0)$.

We further assume that (A) is not reducible to a system that has fewer equations (otherwise, we would do the reduction).

Now, if $n = 1$ then $f \in K_1 = \mathcal{C} \subset \mathcal{C}^{\text{sim}}$ and we are done.

Let next $n > 1$. Irreducibility of (A) implies that $g_i(y,\vec{x},\vec{z}_n) = z_i + k$ for some constant k and $2 \leq i \leq n$.

Let then $g_1(y,\vec{x},\vec{z}_n) = z_{i_1} + k_{i_1}$, $i_1 \neq 1$ and $1 \leq i_1 \leq n$. It follows that $g_{i_1}(y,\vec{x},\vec{z}_n) = z_{i_2} + k_{i_2}$, $i_2 \neq i_1$ and $1 \leq i_2 \leq n$. [If $i_1 = i_2$, drop the equation for f_{i_1} from the system, as it is a simple primitive recursion.

Define

$$\tilde{f}_1(0,\vec{x},z_{i_1}) = h_1(\vec{x})$$

$$\tilde{f}_1(y + 1,\vec{x},z_{i_1}) = z_{i_1} + k_{i_1}$$

Then $f_1(y,\vec{x}) = \tilde{f}_1(y,\vec{x},f_{i_1}(y,\vec{x}))$.]
Continuing this way, we obtain

$$g_{i_0}(y,\vec{x},\vec{z}) = z_{i_1} + k_{i_1}, \text{where } i_0 = 1$$

$$g_{i_1}(y,\vec{x},\vec{z}_n) = z_{i_2} + k_{i_2}$$

$$\vdots$$

$$g_{i_{n-1}}(y,\vec{x},\vec{z}_n) = z_{i_n} + k_{i_n}, \text{where } z_{i_n} = z_1$$

[All the i_j's are different, otherwise we have redundancy in (A), for if $i_k = i_m$, $k < m$, then the equations for $f_{i_k}, f_{i_{k+1}}, \ldots, f_{i_{m-1}}$ form an independent system, and (A) would be reducible. See Problem 11 for details].
It follows that

$$f_1(y + n,\vec{x}) = f_{i_1}(y + n - 1,\vec{x}) + k_{i_1} = f_{i_2}(y + n - 2,\vec{x}) + k_{i_1} + k_{i_2}$$

$$= \ldots = f_{i_n}(y + n - n,\vec{x}) + k_{i_1} + \ldots + k_{i_n}$$

$$= f_1(y,\vec{x}) + k_{i_1} + \ldots + k_{i_n}.$$

Thus, f_1 and hence $f(= f_1)$ is "linear" (with "slope" $k_{i_1} + \ldots + k_{i_n}$) on any set of points $\{(y,\vec{x}) \mid y = y_0 + i \cdot n, i \in \mathcal{N}\}$, where y_0 and $\vec{x}$ are arbitrary.
Also observe that

$$f_1(0,\vec{x}) = h_1(\vec{x}) = h_{i_0}(\vec{x}) \text{ [where } i_0 = 1]$$

$$f_1(1,\vec{x}) = f_{i_1}(0,\vec{x}) + k_{i_1} = h_{i_1}(\vec{x}) + k_{i_1}$$

$$f_1(2,\vec{x}) = f_{i_1}(1,\vec{x}) + k_{i_1} = f_{i_2}(0,\vec{x}) + k_{i_1} + k_{i_2}$$

$$= h_{i_2}(\vec{x}) + k_{i_1} + k_{i_2}$$

$$\vdots$$

$$f_1(n - 1,\vec{x}) = h_{i_{n-1}}(\vec{x}) + k_{i_1} + \ldots + k_{i_{n-1}}$$

Hence,

$$f(y,\vec{x}) = (k_{i_1} + \ldots + k_{i_n})\left\lfloor\frac{y}{n}\right\rfloor + \textbf{if } \text{rem}(y,n) = 0 \textbf{ then } h_{i_0}(\vec{x})$$

$$\textbf{else if } \text{rem}(y,n) = 1 \textbf{ then } h_{i_1}(\vec{x}) + k_{i_1}$$

$$\vdots$$

$$\textbf{else if } \text{rem}(y,n) = n - 1 \textbf{ then}$$
$$h_{i_{n-1}}(\vec{x}) + k_{i_1} + \ldots + k_{i_{n-1}}$$

Observing that for $0 < i < n$, $\text{rem}(y,n) = i$ iff $\text{rem}(y + i(n - 1),n) = 0$ it follows that $f \in \mathcal{C}^{\text{sim}}$.//

Definition 7   [29] Let $f \in \mathcal{C}^{\text{sim}}$. Then $T_f = \max(1,\Pi k)$, where $\lambda x.\lfloor x/k\rfloor$ or $\lambda x.\text{rem}(x,k)$ was used in the $\mathcal{C}^{\text{sim}}$-description of f (repetitions counted).//

For example, $\lambda x.\lfloor x/k\rfloor + \text{rem}(x,k)$ has T-constant $= k^2$, $\lambda xy.x + y$ has T-constant $= 1$, and $\lambda xy.\lfloor x/k\rfloor + \lfloor x/m\rfloor$ has T-constant $= km$.

Theorem 8   [29] If $\lambda y\vec{x}.f(y,\vec{x}) \in \mathcal{C}^{\text{sim}}$, then there is a constant $f_0 \in \mathcal{N}$ and a *non-negative rational number* r_f, such that $y \geq f_0$ & $y' = y + k \cdot T_f$ $(k \in \mathcal{N})$ $\Rightarrow f(y',\vec{x}) - f(y,\vec{x}) = r_f \cdot (y' - y)$.

Proof (Essentially that in [29]) We do induction on the *number* of substitutions, n, needed to obtain f in $\mathcal{C}^{\text{sim}}$.
$n = 0$: Then f is initial, and the claim is immediate.
Assume that the claim is true for $n \leq k$.
Case $n = k + 1$: We consider subcases.
(1) $f(y,\vec{x}) = h(y,\vec{x}) \mathbin{\dot{-}} 1$. Let h_0, r_h and T_h be the constants associated with h.

Then, $y \geq h_0$ and $y' = y + k \cdot T_h \Rightarrow h(y',\vec{x}) - h(y,\vec{x}) = r_h \cdot (y' - y)$. If $r_h = 0$, then $\lambda y.h(y,\vec{x}) = \text{constant}$ on $\{y + k \cdot T_h \mid k \in \mathcal{N}\}$, if $y \geq h_0$, and hence so is f, if we take $f_0 = h_0$, $T_f = T_h$ (of course, $r_f = r_h = 0$). If $r_h > 0$, then $h(y + T_h,\vec{x}) = h(y,\vec{x}) + r_h \cdot T_h$ whenever $y \geq h_0$. Since $h(y',\vec{x}) \in \mathcal{N}$ for all $y',\vec{x}$, we get $h(y + T_h,\vec{x}) \geq 1$. Thus $y \geq h_0 + T_h$ & $y' = y + k \cdot T_h$ $(k \in \mathcal{N})$ $\Rightarrow h(y',\vec{x}) - h(y,\vec{x}) = r_h \cdot (y' - y)$ and $h(y,\vec{x}) \geq 1$, $h(y',\vec{x}) \geq 1$.
It follows that $f(y',\vec{x}) - f(y,\vec{x}) = (h(y',\vec{x}) \mathbin{\dot{-}} 1) - (h(y,\vec{x}) \mathbin{\dot{-}} 1) = r_h \cdot (y' - y)$ under these circumstances. Take $r_f = r_h$, $f_0 = h_0 + T_h$ and $T_f = T_h$.
(2) $f(y,\vec{x}) = \lfloor h(y,\vec{x})/t\rfloor$.
Let $y \geq h_0$ and $y' = y + k \cdot t \cdot T_h$. Then

$$f(y',\vec{x}) - f(y,\vec{x}) = \left\lfloor\frac{h(y,\vec{x}) + r_h \cdot k \cdot t \cdot T_h}{t}\right\rfloor - \left\lfloor\frac{h(y,\vec{x})}{t}\right\rfloor = r_h \cdot k \cdot T_h = \frac{r_h}{t} \cdot (y' - y).$$

Take $r_f = r_h/t$, $T_f = t \cdot T_h(*)$ and $f_0 = h_0$.

(3) $f(y,\vec{x}) = \text{rem}(h(y,\vec{x}),t)$

Let $y \geq h_0$ and $y' = y + k \cdot t \cdot T_h$. Then

$$f(y',\vec{x}) - f(y,\vec{x}) = \text{rem}(h(y,\vec{x}) + r_h \cdot k \cdot t \cdot T_h, t) - \text{rem}(h(y,\vec{x}),t)$$

$$= \text{rem}(h(y,\vec{x}),t) - \text{rem}(h(y,\vec{x}),t) = 0.$$

Hence, take $r_f = 0$, $T_f = t \cdot T_h(*)$ and $f_0 = h_0$.

(4) $f(y,\vec{x}) = h_1(y,\vec{x}) + h_2(y,\vec{x})$, where h_i has constants $(h_i)_0$, r_i, T_i ($i = 1,2$).

Let $y \geq (h_1)_0 + (h_2)_0$ and $y' = y + k \cdot T_1 \cdot T_2$. Then $f(y',\vec{x}) - f(y,\vec{x})$ $= r_1 \cdot (y' - y) + r_2 \cdot (y' - y) = (r_1 + r_2) \cdot (y' - y)$. Take

$$f_0 = (h_1)_0 + (h_2)_0 \ (f_0 = \max((h_1)_0,(h_2)_0) \text{ would also do})$$

$$r_f = r_1 + r_2$$

$$T_f = T_1 \cdot T_2(*)$$

(5) $f(y,\vec{x}) = \textbf{if } h_1(y,\vec{x}) = 0 \textbf{ then } h_2(y,\vec{x}) \textbf{ else } h_3(y,\vec{x})$, where h_i has constants $(h_i)_0$, r_i, T_i ($i = 1,2,3$).

Let $y \geq (h_1)_0 + (h_2)_0 + (h_3)_0$ and $y' = y + k \cdot T_1 \cdot T_2 \cdot T_3$.(†) Then if $r_1 = 0$ we get

$$f(y',\vec{x}) - f(y,\vec{x}) = \textbf{if } h_1(y,\vec{x})(\ddagger) = 0 \textbf{ then } h_2(y',\vec{x}) \textbf{ else } h_3(y',\vec{x})$$

$$- (\textbf{if } h_1(y,\vec{x}) = 0 \textbf{ then } h_2(y,\vec{x}) \textbf{ else } h_3(y,\vec{x}))$$

$$= r_2 \cdot (y' - y) \text{ [case where } h_1(y,\vec{x})) \text{ is constantly 0 on}$$

$$\{y + k \cdot T_1 \cdot T_2 \cdot T_3 \,|\, k \in \mathcal{N}\}]$$

or

$$= r_3 \cdot (y' - y) \text{ [case where } h_1(y,\vec{x}) \text{ is constantly nonzero].}$$

Thus $r_f = r_2$ (or r_3, depending on the case), $T_f = T_1 \cdot T_2 \cdot T_3$ and $f_0 = (h_1)_0 + (h_2)_0 + (h_3)_0$ will do. If $r_1 > 0$, let $y \geq (h_1)_0 + (h_2)_0 + (h_3)_0 + T_1 \cdot T_2 \cdot T_3$.

(*)This partially justifies the original definition of T_f.

(†)This, along with the choice of T_f in cases (2) and (3) fully justifies the definition of T_f in Definition 7.

(‡)$\lambda y.h_1(y,\vec{x})$ is constant on $\{y + k \cdot T_1 \cdot T_2 \cdot T_3 \,|\, k \in \mathcal{N}\}$.

Thus, if $y' = y + k \cdot T_1 \cdot T_2 \cdot T_3$ then $h_1(y',\vec{x}) \ge h_1(y,\vec{x}) \ge 1$ and hence $f(y',\vec{x}) - f(y,\vec{x}) = r_3 \cdot (y' - y)$.

Take $r_f = r_3, f_0 = (h_1)_0 + (h_2)_0 + (h_3)_0 + T_1 \cdot T_2 \cdot T_3, T_f = T_1 \cdot T_2 \cdot T_3.//$

Note: The theorem was stated and proved in terms of the first variable, y, of f for convenience. Clearly, there was no special use made of the fact that y was "first". Indeed we have

Corollary   If $\lambda\vec{x}_n.f(\vec{x}_n) \in \mathcal{C}^{\text{sim}}$ and if f_0, T_f are the constants associated with f, where $(r_i)_{1 \le i \le s}$ are associated with $\vec{x}_s, (s \le n)$, then $x_i \ge f_0$ for $i = 1, \ldots, s \, (\le n)$ and $y_i = x_i + k_i \cdot T_f$ imply $f(\vec{y}_s, x_{s+1}, \ldots, x_n) = f(\vec{x}_n) + \Sigma_{1 \le i \le s} r_i \cdot (y_i - x_i)$.

Proof   Problem 13.//

Example 3 $\lambda xy.x \div y \notin K_1^{\text{sim}}(= \mathcal{L}_1)$.
Indeed, since $K_1^{\text{sim}} = \mathcal{C}^{\text{sim}}$, then, should $\lambda xy.x \div y \in K_1^{\text{sim}}$, there would exist constants r_1, r_2, f, T such that $\min(x_0, y_0) \ge f$ and $x = x_0 + k_1 \cdot T, y = y_0 + k_2 \cdot T$ (k_1 and k_2 in $\mathcal{N}$) imply $(x \div y) - (x_0 \div y_0) = r_1 \cdot (x - x_0) + r_2 \cdot (y - y_0)$. Now for any $x > x_0$, there is a y (of the form $y_0 + k_2 \cdot T$) that makes the lefthand side negative ($x_0 > y_0$ assumed). This is a contradiction (the righthand side is ≥ 0).//

We now turn to what is perhaps the most referred to hierarchy in the literature, due to Grzegorczyk [10]. Here an unlimited depth of nesting of primitive recursion is allowed, but a proper hierarchy is obtained by restricting the *size* of the functions obtained by primitive recursion.

Definition 8   [10] Let $\lambda\vec{x}.h(\vec{x}), \lambda y\vec{x}z.g(y,\vec{x},z)$ and $\lambda y\vec{x}.b(y,\vec{x})$ be given, and let f satisfy for *all* $y,\vec{x}$

$$(A) \begin{cases} f(0,\vec{x}) = h(\vec{x}) \\ f(y + 1,\vec{x}) = g(y,\vec{x},f(y,\vec{x})) \\ f(y,\vec{x}) \le b(y,\vec{x}). \end{cases}$$

The schema (A) is a *limited recursion*, and f is obtained from h, g, and b by limited recursion.//

Following Warkentin [32], we shall call the operations of *substitution*(*) and *limited recursion "the Grzegorczyk operations"*.

(*)Throughout this chapter, *substitution* will mean *explicit transformations* along with *function substitution*. This is *not* the same as Definition 2.1.4, unless $\lambda x.x + 1$ is present.

The following version of the Ackermann function was used by Grzegorczyk:

$$g_0(x,y) = y + 1$$
$$g_1(x,y) = x + y$$
$$g_2(x,y) = (x + 1) \cdot (y + 1)$$

For $n \geq 2$,

$$g_{n+1}(0,y) = g_n(y + 1, y + 1)$$
$$g_{n+1}(x + 1, y) = g_{n+1}(x, g_{n+1}(x,y)).$$

Lemma 7   For $n \geq 0$, $g_n \in \mathcal{PR}$.

Proof   Problem 21.//

Definition 9   [10] $\mathcal{E}^n$, $n \geq 0$, is the closure of $\{\lambda x.x + 1, \lambda x.x, \lambda xy.g_n(x,y)\}$ under the Grzegorczyk operations.//

The class $\mathcal{E}^0$ has attracted considerable attention due to the incredible variety of predicates in $\mathcal{E}^0_*$ (for example, the Kleene predicate, as constructed in Chapters 3 and 5, is in $\mathcal{E}^0_*$. The decoding function d associated with the Kleene predicate of Chapter 3 is in $\mathcal{E}^0$. Every re set can be enumerated by an $\mathcal{E}^0$ function) and to the small size of $\mathcal{E}^0$ functions. (See Problems 33 and 34).

The following theorem is found in Warkentin [32] and is in the spirit of Grzegorczyk(*), but it affords greater generality in that $\lambda x.x + 1$ is not used in its conclusions or proof.

Theorem 9   Let $\mathcal{M}$ be any class containing $\lambda x.x$ and closed under the Grzegorczyk operations.(†)

Then,

(a) $\mathcal{M}$ contains $\lambda x.x \dot{-} 1$, $\lambda xy.x \dot{-} y$, $\lambda xy.x \cdot (1 \dot{-} y)$

(b) $\mathcal{M}_*$ is closed under $\vee$, &, $\neg$, $(\exists y)_{<x}$, $(\exists y)_{\leq x}$, $(\forall y)_{<x}$, $(\forall y)_{\leq x}$

(c) $\lambda xy.x < y, \lambda xy.x \leq y, \lambda xy.x = y, \lambda xy.x \neq y$ are in $\mathcal{M}_*$

(*)Only (f), as far as we can say, is nontrivial and not contained in [10] under some disguise.

(†)The *smallest* such class we denote by $\mathcal{E}^{-1}$.

(d) $\mathcal{M}$ is closed under $(\mathring{\mu}y)_{<x}$ and $(\mathring{\mu}y)_{\leq x}$ [under the convention that if the sought y does not exist then the outcome is 0].

(e) $\mathcal{M}$ is closed under definition by cases, provided the defined function $\lambda\vec{x}.f(\vec{x})$ is bounded by x_i for some i, or by some constant.

(f)   If f is defined by

$$\begin{cases} f(0,\vec{x}) = h(\vec{x}) \\ f(y + 1,\vec{x}) = g(y,\vec{x},f(y,\vec{x})) \end{cases}$$

and $\lambda z\vec{x}.z = h(\vec{x})$, $\lambda wy\vec{x}z.w = g(y,\vec{x},z)$ are in $\mathcal{M}_*$ then f *increasing* with respect to y (not necessarily strictly) implies $\lambda zy\vec{x}.z = f(y,\vec{x}) \in \mathcal{M}_*$.

Proof   (a) Problem 22.
(b) Let $R(\vec{x})$, $Q(\vec{y})$ be in $\mathcal{M}_*$ and let $R(\vec{x}) \equiv r(\vec{x}) = 0$, $Q(\vec{y}) \equiv q(\vec{y}) = 0$ (r,q in $\mathcal{M}$).
Then $\neg R(\vec{x}) \equiv 1 \mathbin{\dot-} r(\vec{x}) = 0$ ($\lambda\vec{x}.1 \mathbin{\dot-} r(\vec{x}) \in \mathcal{M}$)

$$R(\vec{x}) \vee Q(\vec{x}) \equiv r(\vec{x}) \cdot (1 \mathbin{\dot-} (1 \mathbin{\dot-} q(\vec{x}))) \tag{1}$$

Note that $R \in \mathcal{M}_*$ iff its characteristic function χ_R is in $\mathcal{M}$ [if $r \in \mathcal{M}$, then so is $\lambda\vec{x}.1 \mathbin{\dot-} (1 \mathbin{\dot-} r(\vec{x}))$].
So let $\chi_{\exists R}$ be the characteristic function of $(\exists y)_{<x}R(y,\vec{z})$ and $R \in \mathcal{M}_*$.
Then $\chi_{\exists R}(0,\vec{z}) = 1$
$$\chi_{\exists R}(x + 1,\vec{z}) = \chi_{\exists R}(x,\vec{z}) \cdot (1 \mathbin{\dot-} (1 \mathbin{\dot-} \chi_R(x,\vec{z})))$$
Note that the above recursion is a generalization of formula (1); it says $(\exists y)_{<x+1}R(y,\vec{z}) \equiv \{(\exists y)_{<x}R(y,\vec{z})\} \vee R(x,\vec{z})$. Moreover, the above recursion is *limited* by $\lambda x\vec{z}.1$, which is obtained by substitution from $\lambda x\vec{z}.1 \mathbin{\dot-} x$.
Closure under $(\exists y)_{\leq x}$ follows, since $(\exists y)_{\leq x}R(y,\vec{z}) \equiv (\exists y)_{<x}R(y,\vec{z}) \vee R(x,\vec{z})$. Closure under $(\forall y)_{<x}$ and $(\forall y)_{\leq x}$ follows due to closure under $\neg$.
(c) $x \leq y \equiv x \mathbin{\dot-} y = 0$.
(d) Define $(\mathring{\mu}y)_{<z}g(y,\vec{x})$ to be $\min\{y \mid y < z \ \& \ g(y,\vec{x}) = 0\}$, or 0 if the minimum does not exist.
Set $f = \lambda z\vec{x}.(\mathring{\mu}y)_{<z}g(y,\vec{x})$ and let $g \in \mathcal{M}$. Then

$$\begin{cases} f(0,\vec{x}) = 0 \\ f(z + 1,\vec{x}) = \textbf{if } g(z,\vec{x}) = 0 \ \& \ (\forall i)_{<z}g(i,\vec{x}) \neq 0 \textbf{ then } z \\ \qquad\qquad \textbf{else } f(z,\vec{x}) \\ f(z,\vec{x}) < z \end{cases}$$

is a limited recursion. Moreover, the righthand side of $f(z + 1,\vec{x})$ is the substitution of $f(z,\vec{x})$ into u [recall, $f(z,\vec{x}) < z$] and of the characteristic function of $g(z,\vec{x}) = 0$ & $(\forall i)_{<z} g(i,\vec{x}) \neq 0$ into w in the function $\lambda zuw.$ **if** $w = 0$ **then** z **else if** $u < z$ **then** u **else** 0 which is in $\mathcal{M}$ (Problem 22).

Thus $f \in \mathcal{M}$.

Note that

$$(\mathring{\mu}y)_{\leq z} g(y,\vec{x}) = \textbf{if } g(z,\vec{x}) = 0 \text{ \& } (\forall i)_{<z} g(i,\vec{x}) \neq 0$$

$$\textbf{then } z \textbf{ else } (\mathring{\mu}y)_{<z} g(y,\vec{x}),$$

so $\mathcal{M}$ is closed under $(\mathring{\mu}y)_{\leq z}$ as well.

$$\text{(e)} \quad \text{Let } f(\vec{x}) = \begin{cases} f_1(\vec{x}) \textbf{ if } R_1(\vec{x}) \\ \quad \vdots \qquad \quad \vdots \\ f_k(\vec{x}) \textbf{ if } R_k(\vec{x}) \end{cases} .$$

where the R_i's are mutually exclusive and their union is identically true. Let $f(\vec{x}) \leq x_i$ for all $\vec{x}$.

Let, finally, $f_j \in \mathcal{M}$ and $R_j \in \mathcal{M}_*$, $j = 1,\ldots,k$.
Then $f(\vec{x}) = (\mathring{\mu}y)_{\leq x_i}(f_1(\vec{x}) = y \text{ \& } R_1(\vec{x}) \vee \ldots \vee f_k(\vec{x}) = y \text{ \& } R_k(\vec{x}))$. Similarly, if $f(\vec{x}) \leq k$ for all $\vec{x}$ [we then use $(\mathring{\mu}y)_{\leq k}$].

(f) [32] Define first f_1 by $f_1(y,\vec{x}) = 1 \dotdiv (1 \dotdiv f(y,\vec{x}))$.
Clearly,

$$f_1(0,\vec{x}) = \textbf{if } h(\vec{x}) = 0 \textbf{ then } 0 \textbf{ else } 1$$

$$f_1(y + 1,\vec{x}) = \textbf{if } f_1(y,\vec{x}) = 0 \textbf{ then if } g(y,\vec{x},0) = 0$$

$$\textbf{then } 0$$

$$\textbf{else } 1$$

$$\textbf{else if } g(y,\vec{x},f(y,\vec{x})) = 0 \textbf{ then } 0 \textbf{ else } 1$$

Now, if $f_1(y,\vec{x}) = 1$, and hence $f(y,\vec{x}) > 0$, it must also be $f(y + 1,\vec{x}) = g(y,\vec{x},f(y,\vec{x})) > 0$ (f increasing with respect to y). Thus the definition of $f_1(y + 1,\vec{z})$ becomes

$$f_1(y + 1,\vec{x}) = \textbf{if } f_1(y,\vec{x}) = 0 \text{ \& } g(y,\vec{x},0) = 0 \textbf{ then } 0 \textbf{ else } 1$$

Moreover, $f_1(y,\vec{x}) \leq 1$ for all $y,\vec{x}$, hence $f_1 \in \mathcal{M}$.
Next, define

$$f_2(y,\vec{x},z) = \begin{cases} f(y,\vec{x}) \textbf{ if } f(y,\vec{x}) \leq z \\ 0 \qquad \textbf{otherwise} \end{cases} .$$

$$f_2(0,\vec{x},z) = (\mathring{\mu}w)_{\le z}(h(\vec{x}) = w)$$

$$f_2(y + 1,\vec{x},z) = (\mathring{\mu}w)_{\le z}[g(y,\vec{x},f_2(y,\vec{x},z)) = w \ \& \ f_2(y,\vec{x},z) \ne 0 \ \lor$$

$$f_1(y,\vec{x}) = 0(*) \ \& \ g(y,\vec{x},0) = w]$$

Clearly, $f_2(y,\vec{x},z) \le z$ for all $y,\vec{x},z$ and $f_2 \in \mathcal{M}$. Now $f(y,\vec{x}) = z$ iff $f_2(y,\vec{x},z) = z \ \&$ $z \ne 0 \lor f_1(y,\vec{x}) = 0 \ \& \ z = 0.//$

Corollary 1   For $n \ge 0$, $\mathcal{E}^n$ is closed under $(\mathring{\mu}y)_{<z}$, $(\mathring{\mu}y)_{\le z}$ and definition by cases (unrestricted for $n \ge 1$, restricted so that the defined function is bounded by $\lambda\vec{x}.x_i + k$ when $n = 0$).

   Note: Moreover, $\mathcal{E}^n$ is closed under $(\mu y)_{\le z}$ (returned value in unsuccessful search is $z + 1$. See Problem 28).

Corollary 2   For $n \ge 0$, $\mathcal{E}^n_*$ is closed under Boolean operations, $(\exists y)_{<z}$, $(\exists y)_{\le z}$, $(\forall y)_{<z}$ and $(\forall y)_{\le z}$.

   Note: In Grzegorczyk [10], closure under $(\mu y)_{\le z}$ and $(\forall y)_{\le z}$ is shown with the help of the *restricted summation* operation under which all Grzegorczyk classes are closed (Problem 27). This is the technique we used in Theorems 2.2.4 and 2.3.1.

Corollary 3 $CA \subset \mathcal{E}^{-1}_*$, and hence the Kleene predicate of Chapter 5 is in $\mathcal{E}^{-1}_*$ and therefore in $\mathcal{E}^n_*$ for all $n \ge 0$.

   Proof   It suffices to show that the initial predicates of CA, $\lambda xyz.z = x + y$ and $\lambda xyz.z = xy$, are in $\mathcal{E}^{-1}_*$.
   (1) $z = x + y \equiv x = 0 \ \& \ z = y \lor z > y \ \& \ x = z \div y$
   (2) Apply Theorem 9(f): Let $m(x,y) = xy$ for all x,y. Then

$$\begin{cases} m(0,y) = 0 \\ m(x + 1,y) = m(x,y) + y \end{cases}$$

and m is increasing with respect to x. Moreover, $\lambda y.0 \in \mathcal{E}^{-1}$, $\lambda xyz.z = x + y$ $\in \mathcal{E}^{-1}_*.//$

Example 4   Referring to §5.4, Examples 1,4,5, Lemmas 1,2,3,4,5 we see that

$$\lambda xyz.z = \left\lfloor \frac{x}{y} \right\rfloor, \lambda xyz.z = \text{rem}(x,y), \lambda x.\Omega(m,x)$$

$$\lambda xy.y = \lfloor \log_p x \rfloor \ (p \text{ prime}), \text{Pr}(x) \text{ are in } \mathcal{E}^{-1}_*$$

(*)That is, $f_2(y,\vec{x},z) = f(y,\vec{x}) = 0$, rather than $f_2(y,\vec{x},z) = 0$ and $f(y,\vec{x}) > z$. When $f(y,\vec{x}) > z$ then $f_2(y + 1,\vec{x},z) = 0$, which is correct, since f is increasing with respect to y.

Hence, $\lambda xy.\lfloor x/y \rfloor$, $\lambda xy.\mathrm{rem}(x,y)$, $\lambda x.\lfloor \log_p x \rfloor$ are in $\mathcal{E}^{-1}$ (for example, $\lfloor x/y \rfloor = (\mathring{\mu}z)_{\leq x}\,(z = \lfloor x/y \rfloor).//$

Example 5 $\lambda zx.z = A_n(x) \in \mathcal{E}_*^{-1}$ for all $n \geq 0$.

By induction on n, applying Theorem 9(f):

$n = 0$: $z = A_0(x) \equiv z = x + 2$, and the claim follows from Corollary 3.

$n = k$: Assume the claim

$n = k + 1$:

$$\begin{cases} A_{k+1}(0) = 2 \\ A_{k+1}(x + 1) = A_k(A_{k+1}(x)) \end{cases}$$

Since $\lambda zx.z = A_k(x) \in \mathcal{E}_*^{-1}$ and $\lambda x.A_{k+1}(x)$ is increasing, we are done.

This result is surprising at first, due to the size of $A_n(x)$. Qualitatively, one can say that $A_n(x)$ is "honest" and every step in its (very lengthy) computation is spent to do the straightforward thing of increasing $A_n(x)$'s (intermediate) size rather than doing some "obscure" subcomputation.

Meyer [18] appears to have been first to put in print a proof that $\lambda zxy.z = g_n(x,y) \in \mathcal{E}_*^0$ for all n, attacking the problem with a method tailor-made for it.//

Example 6   By Example 5.4.5, $\lambda xyzw.w = \mathrm{rem}(x + y,z)$ is in $\mathcal{E}_*^{-1}$.

Thus, $\mathrm{Pr}(x + 1)$ is in $\mathcal{E}_*^{-1}$, for

$$\mathrm{Pr}(x + 1) \equiv x > 0 \,\&$$

$$\neg(\exists i)_{\leq x}\,(\mathrm{rem}(x + 1,i) = 0\,\&\,i > 1)$$

Now if $\pi(x) = $ "number of primes $\leq x$" (see Example 2.3.1) then $\pi(x) = \Sigma_{i \leq x}(1 \doteq \chi_{\mathrm{Pr}}(i))$, where χ_{Pr} is the characteristic function of Pr.

Thus,

$$\pi(0) = 0$$

$$\pi(x + 1) = \textbf{if } \chi_{\mathrm{Pr}}(x + 1) = 0 \textbf{ then } \pi(x) + 1 \textbf{ else } \pi(x)$$

Since $\pi(x) \leq x$ this shows that $\pi \in \mathcal{E}^0$.

To see that it is also in $\mathcal{E}^{-1}$, observe that

$$\lambda zxy.z = \textbf{if } \chi_{\mathrm{Pr}}(x + 1) = 0 \textbf{ then } y + 1 \textbf{ else } y \text{ is in } \mathcal{E}_*^{-1} \text{ since}$$

$$z = \textbf{if } \chi_{\mathrm{Pr}}(x + 1) = 0 \textbf{ then } y + 1 \textbf{ else } y \equiv z = y + 1\,\&\,\mathrm{Pr}(x + 1)$$

$$\vee\, z = y\,\&\,\neg\mathrm{Pr}(x + 1).$$

Also, π is increasing. Hence, $\pi \in \mathscr{E}^{-1}$. It follows that $\lambda ny.y = p_n \in \mathscr{E}_*^{-1}$, for $y = p_n \equiv \Pr(y)$ & $\pi(y) = n + 1.//$

Example 7 $\lambda xyn.y = p_n^x$ and $\lambda xyn.y = p_n^{x+1}$ are in $\mathscr{E}_*^{-1}$.
Indeed,

$$p_n^0 = 1$$
$$p_n^{x+1} = p_n^x \cdot p_n$$

$\lambda x.p_n^x$ is increasing and $z = y \cdot p_n \equiv (\exists x)_{\leq z}(z = y \cdot x$ & $x = p_n)$, hence $\lambda nyz.z = y \cdot p_n \in \mathscr{E}_*^{-1}$. The claim follows $[y = p_n^{x+1} \equiv (\exists z,w)_{\leq y}(y = z \cdot w$ & $z = p_n^x$ & $w = p_n)].//$

Example 8 $\lambda xyn.y = \lfloor \log_{p_n} x \rfloor \in \mathscr{E}_*^{-1}$ ($\log_{p_n} 0 = 0$, by convention in this context).
Indeed, $\lfloor \log_{p_n} x \rfloor = (\mathring{\mu} z)_{\leq x}(p_n^{z+1} > x) = (\mathring{\mu} z)_{\leq x}(\neg(\exists w)_{\leq x}(w = p_n^{z+1}))$. We actually proved more: $\lambda nx.\lfloor \log_{p_n} x \rfloor \in \mathscr{E}^{-1}.//$

Example 9 $\lambda x.\lfloor \sqrt{x} \rfloor \in \mathscr{E}^{-1}$. Indeed, $\lfloor \sqrt{x} \rfloor = (\mathring{\mu} y)_{\leq x}((y + 1)^2 > x)$. Details are left to the reader, along with the task of showing that the projections of $\lambda xy.(x + y)^2 + x$ are in $\mathscr{E}^{-1}$. (See Problem 35.)$//$

Note: There has been demonstrated a lot of interest in the classes $\mathscr{E}^n$, $n \leq 3$. $\mathscr{E}^{-1}$ and $\mathscr{E}^0$ are very primitive, yet they contain the T-predicate. This, in particular, makes $\mathscr{E}^{-1}$ and $\mathscr{E}^0$ complicated enough to have an unsolvable equivalence problem (see Example 3.7.7).

$\mathscr{E}^2$ played a prominent role in the work of R. Ritchie [26]. It has enough "power" to simulate URMs as we shall see in the next section.

$\mathscr{E}^3$ is the class of *Elementary Functions* of Kálmar [15]. His definition is

Definition 10 [15] $\mathscr{E}$, the class of the *Kálmar Elementary Functions* is the closure of $\{\lambda xy.x + y, \lambda xy.x \doteq y\}$ under *substitution, summation* ($\Sigma_{i \leq z}$) and *product* ($\Pi_{i \leq z}$).$//$

Example 10 $\lambda xy.xy$, $\lambda xy.x^y$ are in $\mathscr{E}$. Indeed, $xy = \Sigma_{i < y} x$. Now $f = \lambda ix.x \in \mathscr{E}$ (substitution operations on $\lambda xy.x + y$). Hence, $xy = \Sigma_{i < y} f(i,x) = [\Sigma_{i \leq y} f(i,x)] \doteq x$. The case of $\lambda xy.x^y$ is left as an exercise. (See Problem 38).

It follows that **if-then-else** is in $\mathscr{E}$ (for example, $\lambda xy.x \cdot (1 \doteq y) \in \mathscr{E}$). Thus, $\mathscr{E}$ is closed under $(\mathring{\mu} y)_{\leq z}$, therefore $\mathscr{E}_*$ is closed under $(\exists y)_{\leq z}$ (and, of course, under Boolean operations).$//$

The importance of $\mathscr{E}$ stems from the fact that in practice we compute no more than elementary functions, since anything else has horrendously large run-times. Indeed, even in $\mathcal{E}$ there are many examples of intractably hard to compute functions [for example, $2^x, 2^{2^x}, 2 \cdot \overbrace{\cdot \cdot 2^x}^{k\,(\text{fixed})}$, as functions of x].

We remark here that it is often being said that prime-power coding of Turing machines (as in Davis [7], for example) produces a T-predicate in $\mathcal{E}$ (Davis [7] only remarks that his construction proves T in $\mathcal{PR}_*$, since for the purposes of Recursion Theory, all we need to know is that $T \in \mathcal{R}_*$). It is an amusing (but lengthy!) exercise now to show that indeed *prime-power coding* yields a T-predicate, and a decoding function d, in $\mathcal{E}_*^{-1}$ and $\mathcal{E}^{-1}$, respectively. (See Problem 49.)

There are many aspects in Grzegorczyk's paper worth discussing [e.g., the characterization of $\mathcal{E}_*^n$, $n \geq 3$; the inductive definability of the unary functions in $\mathcal{E}^n$, $n > 2$, and the existence of a universal function for them in $\mathcal{E}^{n+1}$; the fact that $\mathcal{E}_*^n \subsetneq \mathcal{E}_*^{n+1}$ for $n > 2$, subsequently extended by R. Ritchie [26] to $n \geq 2$. It should be mentioned that it is an open question whether the inclusions $\mathcal{E}_*^0 \subset \mathcal{E}_*^1 \subset \mathcal{E}_*^2$ are proper or not].

The interested reader is referred to the original paper, and Problems 51 to 53. We now turn to some preparatory facts, towards the comparison of the hierarchies we defined (**Note:** That $(\mathcal{E}^n)_{n \geq 0}$ is a hierarchy is clear [Problem 37]. That it is proper, and that $\bigcup_{n \geq 0} \mathcal{E}^n = \mathcal{PR}$ is not yet obvious; it will be settled soon).

Lemma 8   [10] For $n > 1$, $g_n(x,y) > y$.

        Proof Induction on n:

          $n = 2$: $g_2(x,y) = (x + 1) \cdot (y + 1) > y$

          $n = k$:   Assume truth of claim.

          $n = k + 1$: $g_{k+1}(x,y) = g_k^{2^x}(y + 1, y + 1)$ (see Example 2.6.2).(*)

        An easy induction on x and (case $n = k$) $g_k(y + 1, y + 1) > y + 1$ yield $g_k^{2^x}(y + 1, y + 1) > y + 1$.//

Lemma 9   [10] For $n \geq 0$, $g_{n+1}(x + 1, y) > g_{n+1}(x,y)$.

        Proof

        For $n \geq 2$, $g_{n+1}(x + 1, y) = g_{n+1}(x, g_{n+1}(x,y)) > g_{n+1}(x,y)$

        For $n = 0$, $g_1(x + 1, y) = x + 1 + y > x + y = g_1(x,y)$

        For $n = 1$, $g_2(x + 1, y) = (x + 2) \cdot (y + 1) > (x + 1) \cdot (y + 1) = g_2(x,y)$.//

(*)$g_k^{2^x}(y + 1, y + 1)$ stands for $h^{2^x}(y)$, where $h(y) = g_k(y + 1, y + 1)$.

Lemma 10   [10] For $n > 0$, $g_n(x, y + 1) > g_n(x, y)$.

> **Proof** For $n = 1,2$ verified directly. We do induction on $n \geq 2$.
> Let $g_n(x, y + 1) > g_n(x, y)$ (all $x, y, n > 1$).
> By induction on x, we show that for all y,
>
> $$g_{n+1}(x, y + 1) > g_{n+1}(x, y) \tag{1}$$
>
> $x = 0$: $g_{n+1}(0, y + 1) = g_n(y + 2, y + 2) > g_n(y + 1, y + 2)$ by Lemma 9.
> Also, (induction hypothesis on n),
>
> $$g_n(y + 1, y + 2) > g_n(y + 1, y + 1) = g_{n+1}(0, y).$$

Assume now (1) for arbitrary x and *all* y.
Next, $g_{n+1}(x + 1, y + 1) = g_{n+1}(x, g_{n+1}(x, y + 1)) >$ [by induction hypothesis on x, twice] $g_{n+1}(x, g_{n+1}(x, y)) = g_{n+1}(x + 1, y).//$

Note: We now have that g_n strictly increases with respect to x and y. The next lemma shows that $\lambda n.g_n(x, y)$ strictly increases as well.

Lemma 11   [10] For $n > 0$, $g_{n+1}(x, y) > g_n(x, y)$ for *all* x, y.

> **Proof** $n = 1$: $g_2(x, y) = (x + 1) \cdot (y + 1) > x + y = g_1(x, y)$.
> For $n \geq 2$ we do induction on x:
> $x = 0$: $g_{n+1}(0, y) = g_n(y + 1, y + 1) > g_n(0, y)$ by the previous note.
> *Assume* $g_{n+1}(x, y) > g_n(x, y)$.
> Next, $g_{n+1}(x + 1, y) = g_{n+1}(x, g_{n+1}(x, y)) >$ [by induction hypothesis on x twice] $g_n(x, g_n(x, y)) = g_n(x + 1, y).//$

We now compare the sizes of A_n and g_n.

Lemma 12   For $n \geq 2$, there is a k (depending on n) such that
$g_{n+1}(x, y) < A_n^k(\max(x, y))$ a.e.(*)

> **Proof** Induction on n:
>
> $$n = 2: g_3(x, y) = g_2^{2^x}(y + 1, y + 1) = (\ldots (((\overbrace{(y + 2)^2 + 2)^2 + 2)^2 + 2)^2 + \ldots 2)^2}^{2^x \text{ times}}$$
>
> $$= y^{2^{2^x}} + \text{terms } c \cdot y^m$$

where c is constant, and $m < 2^{2^x}$.

(*)The a.e. condition implies that $g_{n+1}(x, y) < A_n^k(\max(x, y)) + c$ for all x, y, where c is an appropriate constant.

Clearly then, $g_3(x,y) < 2^{2^{2^{2^{\max(x,y)}}}}$ a.e. Observing that $A_2(x) = 2^{x+2} - 2$, $k = 4$ will do.

Assume that $g_{n+1}(x,y) < A_n^k(\max(x,y))$ a.e., for some k. Now, $g_{n+2}(x,y)$ $= g_{n+1}^{2^x}(y + 1,y + 1) < A_n^{k \cdot 2^x}(y + 1) \le A_{n+1}(k2^x + y + 1)$ [by Lemma 3.3.8].

For some l, $k2^x + y + 1 < A_2^l(\max(x,y))$ a.e. Thus, $g_{n+2}(x,y)$ $< A_{n+1}(A_2^l(\max(x,y))) < A_{n+1}^{l+1}(\max(x,y))$ a.e. (recall $n \ge 2$).//

Lemma 13   For $n \ge 2$, $A_n(x) < g_{n+1}(x,x)$ for all x.

Proof Induction on n:

$n=2$: $A_2(0) < g_3(0,0)$ and

$\qquad A_2(1) < g_3(1,1)$ as it can be verified directly.

For $x \ge 2$, $A_2(x) = 2^{x+2} - 2 < x^{2^{2^x}} +$ (more terms) $= g_3(x,x)$.

Assume now that $A_n(x) < g_{n+1}(x,x)$ for all x. Now,

$$A_{n+1}(x) = A_n^x(2)$$

$$g_{n+2}(x,x) = g_{n+1}^{2^x}(x + 1,x + 1)$$

By induction hypothesis, and Lemmas 9,10, $g_{n+1}^{2^x}(x + 1,x + 1)$ $> g_{n+1}^{2^x}(x,x) > A_n^{2^x}(x) \ge A_n^x(2)$.//

Thus g_{n+1} and A_n have comparable size, and one would expect that they would be equally good in defining $\mathscr{E}^{n+1}$.

Indeed, consider

Definition 11   Let $\mathscr{A}^k = \mathscr{E}^k$, for $k = 0,1,2$. For $n > 2$ $\mathscr{A}^n$ is the closure under the *Grzegorczyk operations* of $\{\lambda xy.x + y, \lambda x.A_{n-1}(x)\}$.//

Theorem 10 $\mathscr{A}^n = \mathscr{E}^n$, $n > 2$.

Proof Trivially, $\mathscr{E}^{-1} \subset \mathscr{A}^n$ (as well as $\mathscr{E}^{-1} \subset \mathscr{E}^n$) for $n \ge 0$.

Hence, (1) $A_n \in \mathscr{E}^{n+1}$, $n \ge 2$ [$\lambda xy.y = A_n(x) \in \mathscr{E}_*^{-1}$ by Example 5. Hence, $A_n = \lambda x.(\mathring{\mu}y)_{\le g_{n+1}(x,x)}[y = A_n(x)] \in \mathscr{E}^{n+1}$].

(2) $g_{n+1} \in \mathscr{A}^{n+1}$, $n \ge 2$ [$\lambda xyz.z = g_{n+1}(x,y) \in \mathscr{E}_*^{-1}$, by Problem 37. Thus $g_{n+1} = \lambda xy.(\mathring{\mu}z)_{\le A_n^k(x+y)+c}[z = g_{n+1}(x,y)]$, for appropriate constants k and c, is in $\mathscr{A}^{n+1}$ since $\mathscr{A}^{n+1}$, $n \ge 2$, contains $\lambda xy.x + y$ and $\lambda x.x + y$].

Thus, for $n \ge 2$, $\mathscr{A}^{n+1}$ and $\mathscr{E}^{n+1}$ have the same initial functions [$\lambda xy.x + y \in \mathscr{E}^{n+1}$, $n \ge 2$, trivially, and $\lambda x.x + 1$ and $\lambda x.x$ of $\mathscr{E}^{n+1}$ are in $\mathscr{A}^{n+1}$ as explicit transforms of $\lambda xy.x + y$], and hence (being closed under the same operations) are equal.//

Note: A_n's are more "manageable" than g_n's, hence the attractiveness of Theorem 10.

R. Ritchie [25] showed that replacing g_n's by h_n's, defined by

$$h_0(x,y) = x + 1, h_{n+1}(x,0) = \begin{cases} x \text{ if } n = 0 \\ 0 \text{ if } n = 1 \\ 1 \text{ otherwise} \end{cases}$$

and

$$h_{n+1}(x,y + 1) = h_n(x,h_{n+1}(x,y))$$

leaves $\mathcal{E}^n$ ($n \geq 0$) invariant (this, incidentally, answered essentially(*) Grzegorczyk's Problem 7, p. 41, affirmatively). h_n (as well as A_n and f'_n) is more manageable than g_n (the problem with g_n is that the schema that defines it from g_{n-1} is not a primitive recursion).

We also draw the attention of the reader to the fact that simply replacing g_n by A_{n-1} in the definition of $\mathcal{E}^n$ *does* change $\mathcal{E}^n$ since inclusion of a function (for example, $\lambda xy.x + y$, $\lambda xy.\max(x,y)$) which increases with respect to *both* arguments is essential. (See Problem 54.)

Corollary 1   For $n \geq 2$, $\lambda \vec{x}.f(\vec{x}) \in \mathcal{E}^{n+1}$ implies that for some k (depending on f) $f(\vec{x}) \leq A_n^k(\max(\vec{x}))$, all $\vec{x}$.

     Proof Since $A_n \in \mathcal{E}^{n+1}$ this follows from the proof of Theorem 3.3.1, where the cases of *bounded search* and *iteration* are replaced by the trivial case of *limited recursion*.//

     Note: $\mathcal{E}^0$ functions f have the property that for some i and k $f(\vec{x}) \leq x_i + k$ for all $\vec{x}$. (See Problem 23.) $\mathcal{E}^1$ functions $\lambda \vec{x}.f(\vec{x})$ are bounded by $k \cdot \max(\vec{x}) + l$ for some k and l depending on f, whereas $\mathcal{E}^2$ functions f have "polynomial size", i.e. they are bounded by $k \cdot (\max(\vec{x}))^l + m$ for some k,l,m depending on f. These claims are easy to prove. It follows that $\mathcal{E}^0 \subsetneq \mathcal{E}^1 \subsetneq \mathcal{E}^2 \subsetneq \mathcal{E}^3$ ($\lambda xy.x + y \in \mathcal{E}^1 - \mathcal{E}^0$, $\lambda xy.xy \in \mathcal{E}^2 - \mathcal{E}^1$, $\lambda x.2^x \in \mathcal{E}^3 - \mathcal{E}^2$), thus

Corollary 2 $(\mathcal{E}^n)_{n \geq 0}$ is a *proper hierarchy*.

     Proof For $n \leq 2$ it has been remarked that $\mathcal{E}^n \subsetneq \mathcal{E}^{n+1}$. For $n > 2$, $A_n \in \mathcal{E}^{n+1} - \mathcal{E}^n$ [$A_n \in \mathcal{E}^n$ would imply $A_n(x) \leq A_{n-1}^k(x)$ for all x and some k].//

(*)h_n, and f'_n of [10] p. 41, are defined by the same type of recursion but their initial conditions are different: $f'_0(x,y) = x + 1$, $f'_1(x,y) = x + y$, $f'_{n+1}(x,0) = 1$, $n \geq 1$.

Corollary 3 $\bigcup_{n \geq 0} \mathcal{E}^n = \mathcal{P}\mathcal{R}.$

Proof   By Theorem 3.3.1, every primitive recursion is a limited recursion with $\lambda x.A_n^k(x)$ as bounding function, for some n and k. Thus, $\mathcal{P}\mathcal{R} \subset \bigcup_{n \geq 0} \mathcal{E}^n$. The opposite inclusion is trivial.//

§13.2 COMPARISON OF THE HIERARCHIES AND THEIR CONNECTION WITH DYNAMIC COMPLEXITY

We shall employ the URM + **goto** (§4.3) as our model of computation, but we shall refer to it simply as URM.

Definition 1   Let M be a URM and let $\lambda \vec{x}.f(\vec{x}) = M_y^{\vec{x}}$ be M-computable in the sense of Definition 4.2.5.

A function $\lambda \vec{x}.t(\vec{x})$ is the *run-time function* of f on M iff for all $\vec{x}, t(\vec{x})$ = [the number of steps(*) of the M-computation with input $\vec{x}$, if such a computation exists, undefined otherwise].//

Note: We shall find it more convenient to deal with functions *bounding* run-times, rather than with run-times themselves. We shall say that $\lambda \vec{x}.f(\vec{x})$ is computable *within time* $T(\vec{x})$ [resp. $O(T(\vec{x}))$] iff $t(\vec{x}) \leq T(\vec{x})$ for all $\vec{x}$ [resp. $t(\vec{x}) \leq k \cdot T(\vec{x})$ a.e. for some constant k].

Lemma 1   For any URM M with registers $R_1, \ldots, R_m$, of which, without loss of generality, $R_1, \ldots, R_n$ $(n \leq m)$ are input registers, the functions $\lambda y \vec{x}_n.r_i(y, \vec{x}_n)$, $i = 1, \ldots, m$ and $\lambda y \vec{x}_n.l(y, \vec{x})$ defined as follows are in K_2^{sim}

$r_i(y, \vec{x}_n)$ = contents of R_i in the yth ID of a (possibly nonterminating) computation with input $((R_1), \ldots, (R_n)) = \vec{x}_n$(†)

$l(y, \vec{x}_n)$ = the instruction-label component in the yth ID of a (possibly nonterminating) computation with input $\vec{x}_n$.

(*)Refer to Definition 4.3.3.

(†)(R) denotes the contents of register R.

Proof (a) $r_i(0,\vec{x}_n) = x_i, i = 1, \ldots, n$

$$r_i(0,\vec{x}_n) = 0, i = n + 1, \ldots, m \text{ [non-input registers]}$$

(b) $l(0,\vec{x}_n) = 1$ [M considers the first instruction]

For $i = 1, \ldots, m$,

$$\text{(c) } r_i(y + 1,\vec{x}_n) = \begin{cases} r_i(y,\vec{x}_n) + 1 \text{ if } l(y,\vec{x}_n) \text{ labels} \\ \qquad\qquad \text{the instruction } R_i = R_i + 1 \\ r_i(y,\vec{x}_n) \dotdiv 1 \text{ if } l(y,\vec{x}_n) \text{ labels the} \\ \qquad\qquad \text{instruction } R_i = R_i \dotdiv 1 \\ r_i(y,\vec{x}_n) \qquad \textbf{otherwise}(*) \end{cases}$$

Note: $\lambda y\vec{x}.$"**if** $l(y,\vec{x})$ labels the instruction $R_i = R_i + 1$" is a relation of the form $\lambda y\vec{x}.(l(y,\vec{x}) = i_1 \vee l(y,\vec{x}) = i_2 \vee \ldots \vee l(y,\vec{x}) = i_k)$, where $p{:}R_i = R_i + 1$ occurs in M iff p is an $\{i_1, \ldots, i_k\}$. Since $(K_1^{\text{sim}})_*$ is closed under Boolean operations (Lemma 13.1.4) and $\lambda x.x = y \in (K_1^{\text{sim}})_*$ $[x = y \equiv x \dotdiv y = 0 \; \& \; \neg((x + 1) \dotdiv y) = 0]$, the conditions $\lambda y\vec{x}.$"**if** $l(y,\vec{x})$ labels $\ldots$" are in $(K_1^{\text{sim}})_*$. *End of Note.*

$$\text{(d) } l(y + 1,\vec{x}_n) = \begin{cases} l(y,\vec{x}_n) & \textbf{if } l(y,\vec{x}_n) \text{ labels } \textbf{stop} \\ j_1 & \textbf{if } l(y,\vec{x}_n) = k_1 \; \& \; r_{p_1}(y,\vec{x}_n) = 0(\dagger) \\ q_1 & \textbf{if } l(y,\vec{x}_n) = k_1 \; \& \; r_{p_1}(y,\vec{x}_n) \neq 0(\dagger) \\ j_2 & \textbf{if } l(y,\vec{x}_n) = k_2 \; \& \; r_{p_2}(y,\vec{x}_n) = 0 \\ q_2 & \textbf{if } l(y,\vec{x}_n) = k_2 \; \& \; r_{p_2}(y,\vec{x}_n) \neq 0 \\ \vdots \\ l(y,\vec{x}_n) + 1 \; \textbf{otherwise} \end{cases}$$

As before, we remark that all the conditions involved in the definition of $l(y + 1,\vec{x}_n)$ are in $(K_1^{\text{sim}})_*$. We are done.//

Note: Clearly, $r_i(y,\vec{x}_n) \leq y + r_i(0,\vec{x}_n)$ for all $y,\vec{x}_n$, since each new ID can, at best, increase (R_i) by 1.

Example 1 Consider the program M below

(*)Refer to Definition 4.3.2.

($\dagger$)The connection between j_i, q_i, k_i, p_i is that $k_i{:}$ **if** $R_{p_i} = 0$ **then goto** j_i **else goto** q_i is in M. The definition of $l(y + 1,\vec{x}_n)$ involves all the **if-then-else**'s of M.

1: **if** $X = 0$ **then goto** 4 **else goto** 2   /* X is the only input register */
2: $X = X \div 1$
3: **if** $Y = 0$ **then goto** 1 **else goto** 4   /* Y, being non-input, holds 0 */
4: **stop**

Clearly, $M_X^X = \lambda x.0$. Call X R_1, and Y R_2 to relate this example with Lemma 1. Then

$$
\begin{cases}
r_1(0,x) = x \\
r_2(0,x) = 0 \\
l(0,x) = 1
\end{cases}
$$

$$
r_1(y + 1, x) = \begin{cases}
r_1(y,x) \div 1 & \textbf{if } l(y,x) = 2 \\
r_1(y,x) & \textbf{otherwise}
\end{cases}
$$

$$
r_2(y + 1, x) = r_2(y,x)
$$

$$
l(y + 1, x) = \begin{cases}
l(y,x) & \textbf{if } l(y,x) = 4 \\
4 & \textbf{if } l(y,x) = 1 \ \& \ r_1(y,x) = 0 \\
2 & \textbf{if } l(y,x) = 1 \ \& \ r_1(y,x) \neq 0 \\
1 & \textbf{if } l(y,x) = 3 \ \& \ r_2(y,x) = 0 \\
4 & \textbf{if } l(y,x) = 3 \ \& \ r_2(y,x) \neq 0 \\
l(y,x) + 1 & \textbf{otherwise.}//
\end{cases}
$$

Note: Each r_i is a "simulating function" of M (for R_i) since it follows the changes in (R_i) step by step. Simple simulating functions, like the r_i's have been used before by Meyer [18], Minsky [21], Schwichtenberg [27] and others. In the first two cases M was a Turing Machine.

Lemma 2   For $p \geq 2$, $\mathcal{E}^p$ is closed under *simultaneous* limited recursion.

Proof Let

$$
\text{for } i = 1, \ldots, n \begin{cases}
f_i(0,\vec{y}) = h_i(\vec{y}) \\
f_i(x + 1, \vec{y}) = g_i(x, \vec{y}, f_1(x,\vec{y}), \ldots, f_n(x,\vec{y})) \\
f_i(x,\vec{y}) \leq b_i(x,\vec{y})
\end{cases}
$$

where h_i, g_i and b_i are in $\mathscr{E}^p$. Referring to Theorem 2.5.1 and observing (1), (2) and (3) below, we are done.

(1) $I_2 = \lambda xy.(x + y)(x + y + 1)/2 + y$ and its projections are in $\mathscr{E}^2$, and hence there are coding functions $I^n: \mathscr{N}^n \to \mathscr{N}$ in $\mathscr{E}^2$ with projections $\Pi_i^n: \mathscr{N} \to \mathscr{N}$ also in $\mathscr{E}^2$ (see Definition 2.4.3).

(2) I_n is increasing with respect to each argument, hence $I_n(f_1(x,\vec{y}), \ldots, f_n(x,\vec{y})) \le I_n(b_1(x,\vec{y}), \ldots, b_n(x,\vec{y}))$.

(3) $\lambda x\vec{y}.I_n(b_1(x,\vec{y}), \ldots, b_n(x,\vec{y})) \in \mathscr{E}^p.//$

Lemma 3   For any URM M, all the simulating functions r_i, and the instruction-label function are in $\mathscr{E}^2$.

Proof   We only observe that the bounding function for r_i's is $\lambda y.y$ or $\lambda xy.x + y$ (both in $\mathscr{E}^2$) and that of l is $\lambda x.k$ (where M has k instructions). The rest is routine.//

Lemma 4   For $n \ge 2$, if $\lambda \vec{x}.f(\vec{x}) \in \mathscr{E}^n$, then there is a URM M such that $f = M_{\vec{y}}^{\vec{x}}$ and f is computed by M within time $T(\vec{x})$, where $T \in \mathscr{E}^n$.

Proof   This is done by induction with respect to the definition of $\mathscr{E}^n$. For convenience we use the $\mathscr{A}^n$ characterization, thus initial functions of $\mathscr{A}^n, n > 2$, are $\lambda xy.x + y$ and $\lambda x.A_{n-1}(x)$, whereas for $n = 2$ they are $\lambda xy.x + y$ and $\lambda xy.xy.$(*)
(a) *Initial functions of $\mathscr{E}^n, n \ge 2$ have the property.*
    $\lambda xy.x + y$ is computable by M below (it is $M_x^{x,y}$)

```
/* input registers x,y */
1: if z = 0 then goto 4 else goto 4
2: x ← x + 1
3: y ← y ∸ 1
4: if y = 0 then goto 5 else goto 2
5: stop
```

Clearly, the timing of $\lambda xy.x + y$ is bounded by $\lambda xy.2 + 3y$ which is in $\mathscr{E}^n$, $n \ge 2$ (indeed $n \ge 1$). $\lambda xy.xy$ is computable by N below (it is $N_w^{x,y}$)

```
/* input registers x,y */
1: if z = 0 then goto l else goto l
2: w ← w + x        /* this is "compute Mww,x", where M is the
                        previous URM */
```

(*)This is a trivial variation of Grzegorczyk's $\lambda x.x + 1$, $\lambda x.x$ and $\lambda xy.(x + 1) \cdot (y + 1)$.

$l - 1: y \leftarrow y \dotminus 1$ /* the labelling reflects the length of the
                                                program $w \leftarrow w + x$ */
l: **if** $y = 0$ **then goto** $l + 1$ **else goto** 2
$l + 1$: **stop**

Again, by inspection of program N, the timing is bounded by $\lambda xy.2 + (4 + 3x)y$, which is in $\mathscr{E}^n$, $n \geq 2$.

Thus the *induction basis* is settled for $n = 2$. We now show that for $n \geq 1$, A_n is computable by some URM within time $O(A_n^k(x))$, for some k. (A_n is honest.)

We do induction on n.

$n = 1$: $A_1(x) = 2(x + 1)$ and the result follows by a modification of the program M above.

$n = m$: Assume the validity of the claim.

$n = m + 1$: Now $A_{m+1}(x) = A_m^x(2)$. The following program P computes A_{m+1} ($= P_y^x$).

/* input register x */
$\left.\begin{array}{l} 1: y \leftarrow y + 1 \\ 2: y \leftarrow y + 1 \end{array}\right\}$ /* $y \leftarrow 2$ */
3: **if** $z = 0$ **then goto** l **else goto** l
4: $y \leftarrow A_m(y)$ /* using a program for A_m with run time $O(A_m^k(x))$ */
$l - 1: x \leftarrow x \dotminus 1$
l: **if** $x = 0$ **then goto** $l + 1$ **else goto** 4
$l + 1$: **stop**

The timing is bounded a.e. by $4 + c \cdot \Sigma_{i<x}(A_m^k(A_m^i(2)) + 2)$, for some k and c (the latter due to the assumption that $y \leftarrow A_m(y)$ is computable within time $O(A_m^k(y))$).

Thus, timing $\leq 4 + 2cx + cxA_m^k(A_m^{x \dotminus 1}(2)) = 4 + 2cx + cxA_m^{k+x \dotminus 1}(2)$ $= 4 + 2cx + cxA_{m+1}(k + x \dotminus 1)$.

Now $k + x \dotminus 1 < A_0^p(x)$ for all x and some p, $4 + 2cx + cxy$ $< A_2^q(\max(x,y))$ a.e. for some q (recall, $m \geq 1$), hence timing $\leq A_{m+1}^{p+q+1}(x)$ a.e., that is timing $= O(A_{m+1}^r(x))$ for some r.

Thus, for $n \geq 2$, there is a $\lambda x.T(x)$ [namely, $T = \lambda x.A_n^k(\max(x,c))$ for appropriate k and c] such that A_n is computable within time $T(x)$, where $T \in \mathscr{E}^{n+1}$. This completes (a).

(b) *The property propagates with substitution.*

Indeed, let $\lambda y \vec{x}.f(y,\vec{x})$ run within time $t_f(y,\vec{x})$ and $\lambda \vec{z}.g(\vec{z})$ within time $t_g(\vec{z})$ on appropriate URMs.

Let f and g be in $\mathscr{E}^n$ and assume inductively (with respect to $\mathscr{E}^n$) that t_f and t_g are in $\mathscr{E}^n$.

Then $\lambda \vec{x} \vec{z}.f(g(\vec{z}),\vec{x})$ runs within time $t_f(g(\vec{z}),\vec{x}) + t_g(\vec{z})$, which is in $\mathscr{E}^n$. The remaining cases of substitution are even easier, and are left to the reader.

(c) *The property propagates with limited recursion.*

So let $\lambda \vec{x}.h(\vec{x}), \lambda y\vec{x}.b(y,\vec{x})$ and $\lambda y\vec{x}z.g(y,\vec{x},z)$ be in $\mathscr{E}^n$ and let them run within $t_h(\vec{x}), t_b(y,\vec{x})$ and $t_g(y,\vec{x},z)$ respectively, on appropriate URMs, where t_h, t_b, t_g are in $\mathscr{E}^n$. Let

$$\begin{cases} f(0,\vec{x}) = h(\vec{x}) \\ f(y+1,\vec{x}) = g(y,\vec{x},f(y,\vec{x})) \\ f(y,\vec{x}) \le b(y,\vec{x}) \end{cases}$$

for all $y,\vec{x}$.

Then f is computed as

```
/* input y,x */
    1: z ← h(x)
    l: if z = 0 then goto m + 2 else goto m + 2
l + 1: z ← g(i,x,z) /* i = 0 initially, being non-input */
    m: i ← i + 1
m + 1: y ← y ∸ 1
m + 2: if y = 0 then goto m + 3 else goto l + 1
m + 3: stop
```

Then $f(y,\vec{x})$ runs within $t_h(\vec{x}) + 2 + \Sigma_{i<y}(t_g(i,\vec{x},f(i,\vec{x})) + 3)$. Since $f \in \mathscr{E}^n$ and $\mathscr{E}^n(n \ge 2)$ is closed under $\Sigma_{i<y}$, the bound of the timing of f is in $\mathscr{E}^n$ as a function of y and $\vec{x}$ (the assumption on b was not needed).//

Theorem 1   *(The Ritchie(*)-Cobham property of $\mathscr{E}^n$, $n \ge 2$, with respect to URM-time)*
For $n \ge 2$, $f \in \mathscr{E}^n$ iff f runs on some URM within time t_f in $\mathscr{E}^n$.

Proof *only if* part. This is Lemma 4.
if part. Let $f = M_{x_1}^{\vec{x}_n}$, without loss of generality, for some URM M. Let t_f bound the run-time of f on M, and $t_f \in \mathscr{E}^p$, $p \ge 2$. Now, $f = \lambda \vec{x}_n.r_1(t_f(\vec{x}_n),\vec{x}_n)$, where $r_1 \in \mathscr{E}^2$ is the simulating function for register X_1 of M (Lemma 3). Thus $f \in \mathscr{E}^p$.//

Corollary 1 $K_n^{\text{sim}} = \mathscr{E}^{n+1}, n \ge 2$.

Proof (1) First, $K_n^{\text{sim}} \subset \mathscr{E}^{n+1}, n \ge 2$.

(*)D. Ritchie.

Induction on n:

$n = 2$: Trivially, $K_1^{\text{sim}} \subset \mathcal{E}^3$. Since K_2^{sim} is obtained from K_1^{sim} (besides substitution) by *simultaneous* limited recursion with bounds $\lambda x . A_2^k(x)$ for appropriate k (Lemma 13.1.2), $K_2^{\text{sim}} \subset \mathcal{E}^3$ by Lemma 2.

Assume $K_n^{\text{sim}} \subset \mathcal{E}^{n+1}$. $K_{n+1}^{\text{sim}} \subset \mathcal{E}^{n+2}$ follows as in the case $n = 2$, relying on the same lemmas.

(2) $\mathcal{E}^{n+1} \subset K_n^{\text{sim}}, n \geq 2$.

Let $\lambda \vec{x} . f(\vec{x}) \in \mathcal{E}^{n+1}$ and let M be a URM which computes f within time $t(\vec{x})$, $t \in \mathcal{E}^{n+1}$.

By Corollary 1 of Theorem 13.1.10, $t(\vec{x}) \leq A_n^k(\max(\vec{x}))$ for appropriate k and all $\vec{x}$.

Thus, $f = \lambda \vec{x} . r_1(A_n^k(\max(\vec{x})), \vec{x})$, where r_1 is the simulating function for the output register of M. The result follows from $r_1 \in K_2^{\text{sim}}$, $A_n^k \in K_n^{\text{sim}}$, max $\in K_2^{\text{sim}}.//$

Corollary 2 $\mathcal{L}_n = \mathcal{E}^{n+1}, n \geq 2$.

Corollary 3 $K_n \subset \mathcal{E}^{n+1}, n \geq 2$.

Corollary 4 $K_n = \mathcal{E}^{n+1}, n \geq 4$.

Proof   Show $\mathcal{E}^{p+1} \subset K_p, p \geq 4$.

First, the simultaneous recursion defining r_1 is made into a simple recursion using coding.

For example, let $J(x,y) = (x + y)^2 + x$. Then $K(z) = z \,\dot{-}\, \lfloor \sqrt{z} \rfloor^2$, $L(z) = \lfloor \sqrt{z} \rfloor \,\dot{-}\, K(z)$. Now $\lambda x . \lfloor \sqrt{x} \rfloor \in K_3$, hence $J \in K_2, K, L$ are in K_3.

It follows that

$$I_n \in K_2 \text{ (all } n)$$

$$\Pi_i^n \in K_3 \text{ (see Definition 2.4.3)}$$

Referring to the proof of Lemma 1 and Theorem 2.5.1, and setting $F(y, \vec{x}_n) = I_{m+1}(r_1(y, \vec{x}_n), \ldots, r_m(y, \vec{x}_n), l(y, \vec{x}_n))$, we get $F \in K_4$.

Let now $f \in \mathcal{E}^{p+1}$, $p \geq 4$ be computed by URM M of m registers (R_1 serves as output register) within time $A_p^k(\max(\vec{x}_n))$, as in the proof of Corollary 1. Then

$$f = \lambda \vec{x}_n . \Pi_1^{m+1}(F(A_p^k(\max(\vec{x}_n)), \vec{x}_n)) \in K_p.//$$

A more careful choice of a coding scheme improves Corollary 4 to $n \geq 3$. This result is due to Schwichtenberg [27] and we present it below.

Example 2   Set

$$I_n(\vec{x}_n) = 2^{x_1 + \ldots + x_n + n} + 2^{x_2 + \ldots + x_n + n - 1} + \ldots + 2^{x_n + 1} + 1$$

We already have remarked that I_n is 1-1, in Theorem 2.6.2 and elsewhere.

For $1 \leq i \leq n + 1$, define $t_i(z) = 2^{\lfloor \log_2(z \div \Sigma_{1 \leq j < i} t_j(z)) \rfloor}$.(*) Also set, for $1 \leq i \leq n$, $\Pi_i^n(z) = \lfloor \log_2 t_i(z) \rfloor \div \lfloor \log_2 t_{i+1}(z) \rfloor \div 1$.

The reader can verify that Π_i^n are the projections of I_n.

By Lemma 13.1.3, $I_n \in K_2$ $\Pi_i^n \in K_3$ (so far it does not seem that we are doing better than in the proof of Corollary 4).
Set

$$a_i^n(z) = 2t_1(z) + \ldots + 2t_i(z) + t_{i+1}(z) + \ldots + t_{n+1}(z)$$

and

$$s_i^n(z) = \textbf{if } \Pi_i^n(z) = 0 \textbf{ then } z \textbf{ else } \left\lfloor \frac{t_1(z)}{2} \right\rfloor + \ldots + \left\lfloor \frac{t_i(z)}{2} \right\rfloor +$$

$$t_{i+1}(z) + \ldots + t_{n+1}(z), \quad \text{for } i = 1, \ldots, n.$$

Clearly,

(1) a_i^n and s_i^n are in K_2

(2) $a_i^n(I_n(\vec{x}_n)) = I_n(x_1, \ldots, x_{i-1}, x_i + 1, x_{i+1}, \ldots, x_n)$

(3) $s_i^n(I_n(\vec{x}_n)) = I_n(x_1, \ldots, x_{i-1}, x_i \div 1, x_{i+1}, \ldots, x_n)$

(2) and (3) are the pivotal relations: Indeed, for example, $I_{m+1}(r_1(y,\vec{x}) + 1, r_2(y,\vec{x}), \ldots, r_m(y,\vec{x}), l(y,\vec{x}))$ is handled in the proof of Corollary 4 as $I_{m+1}(\Pi_1^{m+1}(F(y,\vec{x})) + 1, \Pi_2^{m+1}(F(y,\vec{x})), \ldots, \Pi_{m+1}^{m+1}(F(y,\vec{x})))$, which is a very inconvenient way of defining $F(y + 1, \vec{x})$.(†)

Here, instead, we use $a_i^{m+1}(F(y,\vec{x}))$ to do the same job, namely "add 1 to the ith component, $r_i(y,\vec{x})$ (if $i = m + 1, l(y,\vec{x})$), of $F(y,\vec{x})$", and a_i^{m+1}, s_i^{m+1} are in K_2.

One last remark, and the details are left to the reader: $\lambda z.\Pi_i^n(z) = c \in K_2$ for any n and $1 \leq i \leq n$. Indeed, $\Pi_i^n(z) = c$ iff $t_i(z) = 2^{c+1} t_{i+1}(z)$ & $t_i(z) \geq 2 \vee c = 0$ & $t_i(z) \leq 1$.//

Note: The Ritchie-Cobham property, besides providing the proof technique for the comparison of the hierarchies, points to a strong link between the *static* complexity (as captured by the level of a function in, say, the *Axt* hierarchy) and the *dynamic* complexity (as captured by URM run-time) of

---

(*)If $z = 2^{x_1 + \ldots + x_n + n} + 2^{x_2 + \ldots + x_n + n - 1} + \ldots + 1$, then $t_i(z) = 2^{x_i + x_{i+1} + \ldots + x_n + n - i + 1}$.

(†)$\Pi_i^{m+1} \in K_3$, putting F in K_4.

$\mathcal{PR}$-functions. As the level increases more and more time is required to compute a $\mathcal{PR}$ function.

It is also clear that $\lambda x n . A_n(x)$ needs *more* than *primitive recursive* time to be computed on a URM, since in the opposite case it would be primitive recursive. The same comment applies to the recursive universal function ξ for $\mathcal{PR}^{(1)}$ (Theorem 3.5.2) and for the function f defined informally in Theorem 2.7.3. Similarly for the "algorithmic" relation of the Corollary to that Theorem, as well as for $\lambda xy . \xi(x,y) = 0$.

Example 3   It is clear that $\mathcal{PR}$ has the Ritchie-Cobham property with respect to URM time.

Indeed, $f \in \mathcal{PR}$ iff $(\exists n)(f \in \mathcal{E}^n, n \geq 2)$ iff $(\exists n)(f$ is URM-computable within time t, and $t \in \mathcal{E}^n, n \geq 2)$ iff f is URM-computable within time t, and $t \in \mathcal{PR}$.

Assume some standard Gödel-numbering of URMs ($+$ **goto**). Set $\Phi_i(x)$ = [number of steps the ith URM takes in a computation with input x; undefined if no such computation exists]. $\Phi = (\Phi_i)_{i \geq 0}$ is a complexity measure for $\mathcal{P}^{(1)}$.

Define $t_i = \lambda x . \Sigma_{j \leq i} (\xi(j,x) + 1)$, where ξ is the one in Theorem 3.5.2. Clearly,

(1) $t_i(x) < t_{i+1}(x)$ for all i and x.

(2) $(t_i)_{i \geq 0}$ is a recursively enumerable sequence. Indeed, $\lambda ix . t_i(x) \in \mathcal{R}$ (since $\xi \in \mathcal{R}$) thus for some $h \in \mathcal{R}$, $t_i = \phi_{h(i)}$.

(3) $\mathcal{PR} = \bigcup_{i \geq 0} C_{t_i}^{\Phi}$ (where $C_{t_i}^{\Phi}$ is as in Definition 12.2.1).

[Indeed, $f \in \mathcal{PR}$ implies that, for some i, $f(x)$ is computable by the ith URM within time $t(x)$, $t \in \mathcal{PR}$. Now for some j, $t(x) = \xi(j,x)$ for all x, since ξ is universal. Thus, $t(x) < t_j(x)$ all x, hence $f \in C_{t_j}^{\Phi}$. This argument can be reversed, due to the Ritchie-Cobham property of $\mathcal{PR}$.]

By the union theorem (Theorem 12.2.3), there is a recursive p such that $\mathcal{PR} = C_p^{\Phi}$.

Thus $\mathcal{PR}$ is a complexity class for the chosen Φ.

Note: There are Φ for which $\mathcal{PR}$ is *not* a complexity class. (See Problem 60.)//

Loop-programs L_n, $n \geq 2$, satisfy an analogue to the Ritchie-Cobham property.

Definition 2   A function $\lambda \vec{x} . f(\vec{x})$ computable by a loop-program P has run-time function $\lambda \vec{x} . t(\vec{x})$ iff for all $\vec{x}$, $t(\vec{x})$ = [number of steps(*) of the P-computation(*) with input $\vec{x}$].

(*)In the sense of Definition 2.7.5.

f runs on P *within* time $p(\vec{x})$ (resp. $O(p(\vec{x}))$) iff $t(\vec{x}) \le p(\vec{x})$ for all $\vec{x}$ [resp. $t(\vec{x}) \le k \cdot p(\vec{x})$ a.e. for some constant k].//

Theorem 2   For $n \ge 2$, $f \in \mathcal{L}_n$ iff, for all $\vec{x}$, $f(\vec{x})$ is loop-program computable within time $t(\vec{x})$, where $t \in \mathcal{L}_n$.

Proof *only if* part. Let $\lambda \vec{x}.f(\vec{x}) \in \mathcal{L}_n$ ($n \ge 2$).

For some $P \in L_n$, $f = P_Y^{\vec{X}}$, where X, Y are respectively the input and output registers in P.

Let Q be a program obtained from P by adding the instruction $Z = Z + 1$ once before each P-instruction, where Z does not occur in P.

Clearly,

(1) $Q \in L_n$.

(2) $Q_Z^{\vec{X}}$ is the run-time function of f on P.

Since $Q_Z^{\vec{X}} \in \mathcal{L}_n$, we are done.

if part. The proof is similar to that of Lemma 1. The reader should refer to Definition 2.7.4.

So let $\lambda \vec{x}_n.f(\vec{x}_n)$ be computable on the loop-program P within time $\lambda \vec{x}_n.t(\vec{x}_n) \in \mathcal{L}_p$ ($p \ge 2$). Let $(X_1, \ldots, X_m; B_1, \ldots, B_q; I)$ be the ID-frame for P, where without loss of generality, $X_1, \ldots, X_n$ ($n \le m$) are input registers and X_1 is the output register.

$X_{h_1}, X_{h_2}, \ldots, X_{h_q}$ are all the variables appearing in a **Loop** X_{h_j} instruction (B_j corresponds to X_{h_j}, $j = 1, \ldots, q$).

We assume that there are no **while**-clauses and that the assignment statements are of types $X = X + 1$ or $X = X \div 1$ as in §2.7.

We define, by simultaneous recursion, functions:

$$\lambda y \vec{x}_n.X_i(y, \vec{x}_n) = (X_i) \text{ in the } y\text{th } P\text{-ID with input } \vec{x}_n$$

$$\lambda y \vec{x}_n.B_j(y, \vec{x}_n) = B_j\text{-value in the } y\text{th } P\text{-ID with input } \vec{x}_n$$

$$\lambda y \vec{x}_n.I(y, \vec{x}_n) = \text{ the } I\text{-component (instruction number) of the } y\text{th}$$
$$P\text{-ID with input } \vec{x}_n$$

Let the instructions of P be consecutively numbered (labeled) $1, 2, \ldots, k$. Thus $k + 1$ labels the **null** or **empty** instruction at the end of P. We obtain,

$$X_i(0, \vec{x}_n) = x_i, \quad i = 1, \ldots, n$$
$$X_i(0, \vec{x}_n) = 0, \quad i = n + 1, \ldots, m$$
$$B_j(0, \vec{x}_n) = 0, \quad j = 1, \ldots, q$$
$$I(0, \vec{x}_n) = 1$$

$$\text{For } i = 1, \ldots, m, X_i(y+1, \vec{x}_n) = \begin{cases} X_i(y, \vec{x}_n) + 1 & \textbf{if } I(y, \vec{x}_n) \text{ labels } X_i = X_i + 1 \\ X_i(y, \vec{x}_n) \dot{-} 1 & \textbf{if } I(y, \vec{x}_n) \text{ labels } X_i = X_i \dot{-} 1 \\ X_i(y, \vec{x}_n) & \textbf{otherwise} \end{cases}$$

$$\text{For } j = 1, \ldots, q, B_j(y+1, \vec{x}_n) = \begin{cases} X_{h_j}(y, \vec{x}_n) & \textbf{if } I(y, \vec{x}_n) \text{ labels } \textbf{Loop } X_{h_j} \\ B_j(y, \vec{x}_n) \dot{-} 1 & \textbf{if } I(y, \vec{x}_n) \text{ labels the } \textbf{end} \\ & \quad \text{instruction corresponding to} \\ & \quad \textbf{Loop } X_{h_j} \\ B_j(y, \vec{x}_n) & \textbf{otherwise} \end{cases}$$

$$I(y+1, \vec{x}_n) = \begin{cases} k+1 & \textbf{if } I(y, \vec{x}_n) = k+1 \\ & \quad \text{for } j = 1, \ldots, q, \\ l_j & \textbf{if } I(y, \vec{x}_n) \text{ labels } \textbf{Loop } X_{h_j}, \text{ and } l_j \\ & \quad \text{labels the corresponding } \textbf{end} \text{ instruction} \\ & \quad \text{for } j = 1, \ldots, q, \\ s_j + 1 & \textbf{if } I(y, \vec{x}_n) \text{ labels the } \textbf{end} \text{ instruction} \\ & \quad \text{corresponding to } \textbf{Loop } X_{h_j}, \text{ and } s_j \text{ labels} \\ & \quad \text{the latter, and } B_j(y, \vec{x}_n) > 0 \\ I(y, \vec{x}_n) + 1 & \textbf{otherwise} \end{cases}$$

Clearly, X_i, B_j and I are in $K_2^{\text{sim}} = \mathcal{L}_2$ (indeed, they are also in $\mathcal{E}^2$ and K_3 as it was shown for the simulating functions of URMs).

Thus, $f = \lambda \vec{x}_n . X_1(t(\vec{x}_n), \vec{x}_n) \in \mathcal{L}_{\max(p,2)}$.//

Corollary   For $n \geq 2$, $\lambda \vec{x}.f(\vec{x}) \in \mathcal{L}_n$ iff f is loop-program computable within time $A_n^k(\max(\vec{x}))$ for some k.

Proof $A_n^k \in \mathcal{L}_n$ and $\lambda \vec{x}.h(\vec{x}) \in \mathcal{L}_n \Rightarrow h(\vec{x}) \leq A_n^m(\max(\vec{x}))$ for all $\vec{x}$ and some m (Lemma 13.1.2).

Thus, f runs within time t in $\mathcal{L}_n$ iff it runs within time A_n^k for some k.//

Note: A careful reading of the proof of Theorem 3.3.1 shows that if $\lambda \vec{x}.h(\vec{x}) \in \mathcal{L}_n$, then an m such that $h(\vec{x}) \leq A_n^m(\max(\vec{x}))$ for all $\vec{x}$ can be *effectively found*.

Thus, starting with $f = P_y^{\vec{x}}$, we construct Q of the proof of Theorem 2 (*only if* part) and then effectively find m such that $Q_z^{\vec{x}} \leq A_n^m(\max(\vec{x}))$ for all $\vec{x}$.

In other words, given a function in the loop-program hierarchy we can *effectively predict* an *upper bound* for its run-time by looking at its *program structure*. Unfortunately, often this bound is quite pessimistic:

Example 4   Consider

$$
P: \begin{cases}
\textbf{Loop } X \\
\quad \textbf{Loop } Y \\
\qquad \textbf{Loop } Z \\
\qquad\quad W = W + 1 \\
\qquad \textbf{end} \\
\quad \textbf{end} \\
\textbf{end}
\end{cases}
$$

Then $P_W^{X,Y,Z}$ is $\lambda XYZ.XYZ$. Clearly the computation runs in time $O(XYZ)$, yet, P being in L_3, we would predict a run time $\leq A_3^k(\max(X,Y,Z))$ for some k, which is very pessimistic. Theorem 2 says that there is a program P' in L_2 which is equivalent to P. $\mathcal{L}_2$ is the smallest class in the $\mathcal{L}_n$-hierarchy where $\lambda xyz.xyz$ can be found. It should be noted that the problem of finding the smallest n such that a primitive recursive function is in $\mathcal{L}_n$ is recursively unsolvable (Meyer and D. Ritchie [19]). (See Problem 61.)//

Note: Theorem 2 and its corollary are due to Meyer and D. Ritchie [19]. A generalization of these results is found in D. Ritchie [24], where the hierarchy $(\mathcal{L}_n)_{n \geq 0}$ is extended beyond $\mathcal{PR}$ (but still in $\mathcal{R}$ as a proper subhierarchy) by defining a class $\mathcal{L}_\alpha$ for each *ordinal* α. Use is made of *variable depth* of nesting of the **Loop** X instruction as follows: **Loop** (1) X means **Loop** X. If **Loop** (n) X has been defined ($n \geq 1$), then **Loop** $(n+1)$ $X; P;$ **end** means

$$\underbrace{\textbf{Loop } (n) \, X; \textbf{Loop } (n) \, X; \ldots; \textbf{Loop } (n) \, X}_{x}; P; \underbrace{\textbf{end}; \ldots; \textbf{end}}_{x}$$

where P is some syntactically correct loop-program in the extended formalism, and x is the contents of X *upon entry* in

$$\textbf{Loop } (n+1) \, X; P; \textbf{end}$$

Again we shall ask the interested reader to probe the literature for related results, which space does not permit us to include. Thus, in Harrow [11], classes based on *bounded search* rather than limited recursion are considered (with the same Ackermann function as in Grzegorczyk [10]) and their relationship to $(\mathcal{E}^n)_{n \geq 0}$ is examined. (It was already known to Grzegorczyk, Theorems

2.4 and 4.12 [10], that $\mathscr{E}^n$, $n \geq 3$ can be defined by replacing limited recursion by bounded search.)

Harrow shows that bounded search cannot be used to define $\mathscr{E}^1$, $\mathscr{E}^0$, leaving open the question for $\mathscr{E}^2$. Similarly, he shows that Grzegorczyk's characterization of $\mathscr{E}^n_*$, $n \geq 3$, (*) cannot be extended below $n = 3$ (certainly not for $n = 0$ and $n = 1$, leaving open again the question for $n = 2$).

As earlier mentioned, it is not known whether the inclusions $\mathscr{E}^0_* \subset \mathscr{E}^1_* \subset \mathscr{E}^2_*$ are proper.

Apparently, R. Ritchie [26] was the first to publish a proof that $\mathscr{E}^2_* \subsetneq \mathscr{E}^3_*$, by defining an infinite hierarchy of functions $(\mathscr{F}_n)_{n \geq 0}$ such that

(1) $\quad \mathscr{E}^2 \subset \mathscr{F}_0$

(2) $\quad \mathscr{F}_n \subsetneq \mathscr{F}_{n+1}, n \geq 0$

(3) $\quad \bigcup_{n \geq 0} \mathscr{F}_n = \mathscr{E}^3$

(4) $\quad (\mathscr{F}_n)_* \subsetneq (\mathscr{F}_{n+1})_*, n \geq 0$

We know that for $n \geq 2$, $\mathscr{E}^n$ is closed under *simultaneous* limited recursion. Thus, if $(\mathscr{E}^n_{\text{sim}})_{n \geq 0}$ are the Grzegorczyk classes defined by replacing, for every $n \geq 0$, *simple* by *simultaneous* limited recursion, then $\mathscr{E}^n = \mathscr{E}^n_{\text{sim}}$, $n \geq 2$. The work of Warkentin [32] shows that $(\mathscr{E}^{-1}_{\text{sim}})_* = (\mathscr{E}^0_{\text{sim}})_* = (\mathscr{E}^1_{\text{sim}})_* = (\mathscr{E}^2_{\text{sim}})_* = \mathscr{E}^2_*$.

Another related hierarchy, $(\mathscr{P}_n)_{n \geq 0}$ such that

(1) $\quad \mathscr{P}_n \subsetneq \mathscr{P}_{n+1}, n \geq 0$

(2) $\quad \bigcup_{n \geq 0} \mathscr{P}_n = \mathscr{P}\mathscr{R}$

has been defined in Parsons [22]. Superficially it looks like the Axt hierarchy, however by defining *level* of a function more liberally (†) than in the Axt case, he is able to prove closure of $\mathscr{P}_n$, $n \geq 1$, under *bounded search,* and that $\mathscr{P}_n = \mathscr{E}^{n+1}$, $n \geq 2$. This is more than we can say for $(K^{\text{sim}}_n)_{n \geq 0}$ or $(K_n)_{n \geq 0}$. First, bounded search enables us show that the T-predicate is in $(\mathscr{P}_1)_*$ [28], thus the equivalence problem of $\mathscr{P}_1$ is unsolvable, unlike that of K_1 and K^{sim}_1. Secondly, the question whether $K_2 = \mathscr{E}^3$ or not, is open.

(*)$\mathscr{E}^n_*$, $n \geq 3$, is the smallest class of predicates containing $\lambda x.x = 0$, $\lambda xy.x = y + 1$, $\lambda xyz.x = y^z$, $\lambda xyz.z = x \dot- y$, $\lambda xyz.x \leq g_n(y,z)$, $\lambda xyz.x \leq g_n(y+1, z+1)$ and closed under *Boolean operations, explicit transformations,* and $(\exists z)_{\leq g_n(x+1, y+1)}$ [10].

(†)$\lambda x.x + 1$ and $\lambda x.x$ have *level* 0. If f has level q, any *explicit transform* of f has level q (for example, $\lambda x.k$ has level 0 for any constant k). If $h = \lambda \vec{x} \, \vec{z}.g(f(\vec{x}), \vec{z})$ and g and f have levels $\leq p$ and $\leq q$ respectively, then h has level $\leq \max(p, q)$. If $g = \lambda y \vec{x} \vec{z}.k(y, \vec{x}, m_1(y, \vec{x}), \ldots, m_l(y, \vec{x}), t_1(y, \vec{x}, z), \ldots, t_n(y, \vec{x}, z))$ and for all y and $\vec{x}, f(0, \vec{x}) = h(\vec{x})$, $f(y + 1, \vec{x}) = g(y, \vec{x}, f(y, \vec{x}))$, then the level of f is $\leq \max(p, q_1, \ldots, q_l, p_0 + 1, p_1 + 1, \ldots, p_n + 1)$ where $p, q_1, \ldots, q_l, p_0, p_1, \ldots, p_n$ are bounding (from above) the levels of $h, m_1, \ldots, m_l, k, t_1, \ldots, t_n$ respectively (à la Axt, we would have level$(f) \leq \max(p, q_1 + 1, \ldots, q_l + 1, p_0 + 1, \ldots, p_n + 1)$. Parsons calls subfunctions such as k and t_i *iterated elements.* He sets $\mathscr{P}_n = \{f \mid f \text{ has level } \leq n\}$, for $n \geq 0$. For any $f \in \mathscr{P}\mathscr{R}$, level$(f) = \min\{n \mid f \in \mathscr{P}_n\}$.

Finally, Grzegorczyk classes $(\mathcal{E}_\Sigma^n)_{n\geq 0}$ can be defined for "word-functions" as well; that is, functions $f: (\Sigma^*)^k \to \Sigma^*$, where Σ is a finite alphabet (von Henke et al. [31]). The distinction here is *not* between Σ^* and $\mathcal{N}$ (after all, the two sets are naturally identifiable once one considers elements of Σ^* as $|\Sigma|$-adic notations of numbers). Rather, it is due to a different choice of *successor* functions ($\lambda x.x*i$ and/or $\lambda x.i*x$, where $*$ denotes $|\Sigma|$-adic concatenation and $i \in \Sigma$, rather than $\lambda x.x + 1$) and due to replacing *limited recursion* by *limited recursion on notation*. In Choi [3] it is shown that $\mathcal{E}_\Sigma^n = \mathcal{E}^n$, $n \geq 3$ and any Σ, whereas $\mathcal{E}_\Sigma^2 = \mathcal{E}^2$ for $|\Sigma| = 1$, but for $|\Sigma| > 1$ $\mathcal{E}_\Sigma^2$ is Cobham's class of "*feasibly computable functions*" [4].

We now turn to a study of Cobham's class.

§13.3 FEASIBLE COMPUTATIONS

It hardly needs an argument that computations which have run-times of the order 2^n (where n, somehow, characterizes the "size" of the input) are intractable in practice. Even computations with run-times of order n^k, for *large* constants k, are impractical. Yet, in contrast to exponential-order (that is, 2^n) computations which are completely out of the question (except for trivially "small" inputs), it has become customary to term the polynomial-order computations *feasible*. Of course, the run-time of a computation depends on what we shall take as a *step*. For example, assuming that numbers in registers are held in m-adic (or m-ary) notation for $m > 1$, $X \leftarrow X + 1$ is a "harder" operation than $X \leftarrow X*1$ (where $*$ is concatenation), in that the latter can be done *without* inspecting the contents of X, whereas the former may need inspection (and change) of *each digit* stored in X (*) (for example, assume 2-adic notation, and $(X) = \underbrace{22\ldots 2}_{n}$ initially, for some "large" n).

Moreover, in practice, m-ary (or m-adic) notation, for $m > 1$, for input/output is more convenient and usual than *unary notation* (the latter being adequate for Computability Theory). It makes then a difference how we "quote" upper bounds of run-times. For example, if program P, of a single input x, runs within time $O(x^k)$ for some k, then it also runs within time $O(2^{k|x|})$, where $|x|$ is the dyadic length of x. The two bounds are the same; so is P to be considered a polynomial or exponential order algorithm?

In the context of feasible computations it is reasonable to call P an exponential order algorithm, relating thus its run time to the (m-adic) *length* of its input rather than to the *value* of the input. This attitude is further justified by considering natural "non-numerical" algorithms, such as those

---

(*)Due to such considerations, Cook [6] proposes that $X \leftarrow X + 1$ takes $O(\lceil \log(X) \rceil)$ "steps".

occurring in graph theory. When the input is a graph-representation, it is reasonable to quote the complexity of the algorithm as a function of the *length* of this representation rather than in terms of the *number* (value) that this representation denotes if the constituent symbols are interpreted as digits.

Two observations are in order before we proceed:

(1)   How one quotes run times of order $\geq c^n$, $c > 1$, is immaterial, since $c^x = O(c^{m^{|x|}})$. Thus, a program P runs within $\mathcal{E}^n$-time with respect to the *value* of the input iff it runs within $\mathcal{E}^n$-time with respect to the m-adic

length of the input, if $n \geq 3$ (this is because $\lambda x.m^{m^{\cdots^{m^x}}} \in \mathcal{E}^3$).

Thus, for $n \geq 3$ the statement of Theorem 13.2.1 stands, with *both* ways of quoting run times.

(2)   For $n = 2$, Theorem 13.2.1 is correct *only* as stated, where run-time is quoted as a function of the *value* of the input ($\lambda xy.x^{|y|}$ is computable in $\mathcal{E}^2$-time with respect to $\max(|x|, |y|)$, where $|\ |$ is m-adic length. However, $\lambda xy.x^{|y|}$ is too big to be in $\mathcal{E}^2$).

In what follows we shall use $\mathcal{L}$ to denote Cobham's class of feasibly computable functions. We are guilty of using $\mathcal{L}$ also as the set of all loop-program computable functions (§2.7), that is, $\mathcal{PR}$. Both are more or less "standard" notations and in this context (§13.3) there can be no ambiguity.

Let Σ be a *fixed* but *unspecified* alphabet with a fixed ordering $d_1 < d_2 < \ldots < d_m$ of its elements ($m = |\Sigma| > 1$). $|\ldots|$ denotes m-adic length, $*$ denotes m-adic concatenation. Σ^* is naturally identified with $\mathcal{N}$, after d_i has been identified with i, for $i = 1, \ldots, m$.

Definition 1   [4] $\mathcal{L}(\Sigma) = \{f : \mathcal{N}^k \to \mathcal{N} |$ there is a TM M, which, for all $\vec{x}_k$, computes $f(\vec{x}_k)$ deterministically and with m-adic input/output, with a computation of length $O((\Sigma_{i=1}^k |x_i|)^n)$ for some n depending on $M\}.//$

Note: Since we can convert m-adic notation to n-adic ($m > 1$ and $n > 1$) within time $O(|x|^l)$ for some l, where x is the input and $|\ldots|$ is m-adic length (Problem 66), it follows that for $|\Sigma| > 1$, $\mathcal{L}(\Sigma)$ is independent of Σ. Thus, we shall simply write $\mathcal{L}$ as in [4].

Definition 2   [4] $\mathcal{L}'(\Sigma)$ is the closure of $\{(\lambda x.x*d)_{d \in \Sigma}, \lambda xy.x^{|y|}\}$ under *explicit transformations, substitution* of functions and *limited (right) recursion on notation*.(*)//

Note: $\lambda x.x*d$ is a *right successor*, if $d \in \Sigma$.

(*)If for all $x, \vec{y}$ and all $d \in \Sigma$, (1), (2) and (3) hold, then f is obtained from $h, g_d (d \in \Sigma)$ and b by *limited right recursion on notation* (see also Definition 5.1.7).
(1) $f(0, \vec{y}) = h(\vec{y})$, (2) $f(x*d, \vec{y}) = g_d(x, \vec{y}, f(x, \vec{y}))$, (3) $|f(x, \vec{y})| \leq |b(x, \vec{y})|$

The purpose of this section is to show that for any Σ, such that $|\Sigma| > 1$, $\mathcal{L} = \mathcal{L}'(\Sigma)$, and derive a consequence relating to the "$P = NP$ question".

Lemma 1 $\mathcal{L}'(\Sigma) \subset \mathcal{L}$.

Proof Induction with respect to $\mathcal{L}'(\Sigma)$.
The right-successors are clearly TM computable in linear time ($O(|x|)$).
$x^{|y|}$ is either 1 (if $|y| = 0$) or $\underbrace{x \cdot x \cdot \ldots \cdot x}_{|y| \, x\text{'s}}$.

Thus to compute it we can proceed with $|y| - 1$ multiplications as outlined below in pseudo-PL/1:

$$p \leftarrow x$$
$$\begin{array}{|l} \textbf{do while } (|y| \neq 1) \\ (1)\ p \leftarrow p \cdot x \\ (2)\ \text{remove a digit from the left of } y \\ \textbf{end} \end{array}$$

Inside the loop one would spend at most $O(|p| \cdot |x|)$ steps, that is $O(|x|^2 \cdot |y|)$ steps, hence, accounting for the loop, a total of $O((|x| \cdot |y|)^2)$ steps. Details of how to implement this idea on a TM (it will run a bit slower, but still in polynomial time with respect to $|x| + |y|$) and the execution of the induction step are left to the reader. (See Problem 67.)//

We next want to establish the more substantial fact, that $\mathcal{L} \subset \mathcal{L}'(\Sigma)$.

The proof is by simulating polynomial-time-bounded TMs with functions from $\mathcal{L}'(\Sigma)$. The technique will be the one employed in Lemma 13.2.1; thus we embark upon proving that $\mathcal{L}'(\Sigma)$ is closed under simultaneous limited recursion (on notation).

Lemma 2   The following functions are in $\mathcal{L}'(\Sigma)$.

(1) $\lambda x.x$

(2) $\lambda x.0$

(3) $\lambda x.d * x$ ($d \in \Sigma$); this is a *left* successor L_d

(4) $\lambda x.$[the number whose m-adic notation is the reverse of that for x]
This function is denoted by "rev".

(5) $\mathcal{L}'(\Sigma)$ is closed under limited *left* recursion on notation.

Proof (1) and (2) are obtained from $\lambda xy.x^{|y|}$ by explicit transformation.
(3) $L_d(0) = d$
For $i \in \Sigma$, $L_d(x*i) = (L_d(x))*i$ (substituting $L_d(x)$ into $\lambda x.x*i$)
$|L_d(x)| \leq |x*1|$ for all x.

(4) $\mathrm{rev}(0) = 0$

For $i \in \Sigma$, $\mathrm{rev}(x \ast i) = L_i(\mathrm{rev}(x))$

$|\mathrm{rev}(x)| \le |x|$, for all x.

(5) Let $h, g_d (d \in \Sigma)$ and b be in $\mathcal{L}'(\Sigma)$ and consider f, given for all $x, \vec{y}$ and $d \in \Sigma$ by

$$f(0,\vec{y}) = h(\vec{y})$$

$$f(d \ast x, \vec{y}) = g_d(x, \vec{y}, f(x, \vec{y}))$$

$$|f(x,\vec{y})| \le |b(x,\vec{y})|$$

Define $\tilde{f}$ by

$$\tilde{f}(0,\vec{y}) = h(\vec{y})$$

$$\tilde{f}(x \ast d, \vec{y}) = g_d(\mathrm{rev}(x), \vec{y}, \tilde{f}(x, \vec{y}))$$

Observe (i) $\tilde{f}(x,\vec{y}) = f(\mathrm{rev}(x), \vec{y})$, and hence $f(x,\vec{y}) = \tilde{f}(\mathrm{rev}(x), \vec{y})$ for all $x, \vec{y}$. [induction on $|x|$].

(ii) [by (i)] $|\tilde{f}(x,\vec{y})| \le |b(\mathrm{rev}(x), \vec{y})|$ for all $x, \vec{y}$. But $\lambda x \vec{y}.b(\mathrm{rev}(x), \vec{y})$, and $\lambda x \vec{y} z.g_d(\mathrm{rev}(x), \vec{y}, z)$ for all $d \in \Sigma$ are in $\mathcal{L}'(\Sigma)$, hence $\tilde{f} \in \mathcal{L}'(\Sigma)$ and thus $f \in \mathcal{L}'(\Sigma).//$

Lemma 3   The following functions are in $\mathcal{L}'(\Sigma)$:

(1)   init $= \lambda x.$ "what remains of x after removal of the rightmost digit" (Convention: $\mathrm{init}(0) = 0$)

(2)   last $= \lambda x.$ "the rightmost digit of x" (Convention: $\mathrm{last}(0) = 0$)

(3)   first $= \lambda x.$ "the leftmost digit of x" (Convention: $\mathrm{first}(0) = 0$)

(4)   tail $= \lambda x.$ "what remains of x after removal of the leftmost digit" (Convention: $\mathrm{tail}(0) = 0$)

(5) $\lambda xyz.$ **if** $x = 0$ **then** y **else** z

(6) $\lambda x.1 \dotdiv x$

(7) $\lambda x.x \dotdiv 1$

(8) $\lambda x.x + 1$

(9) $\lambda x.|x|$

(10)   ones = $\lambda x.$ **if** $x = 0$ **then** 0 **else** a string of $|x|$ 1's

(11)   sub = $\lambda xy.$ **if** $|x| \leq |y|$ **then** 0 **else** a string of $|x| - |y|$ 1's

(12) $\lambda xy.\, x*y$

Proof   (Most are left to the reader. See Problem 68.)

(1)   init$(0) = 0$
init$(x*d) = x, d \in \Sigma$
$|\text{init}(x)| \leq |x|$

(3)   first$(0) = 0$
first$(d*x) = d, d \in \Sigma$
$|\text{first}(x)| \leq |x*1|$

(5)   Call the function "switch". Then,
switch$(0,y,z) = y$
switch$(x*d,y,z) = z, d \in \Sigma$
$|\text{switch}(x,y,z)| \leq |(y*1)^{|z*1|}|$

(8) $0 + 1 = 1$
$x*d + 1 = x*(d + 1), d = 1, \ldots, m - 1$
$x*m + 1 = (x + 1)*1$
$|x + 1| \leq |x*1|$

(9) $|0| = 0$
$|x*d| = |x| + 1, d \in \Sigma$
$||x|| \leq |x|$

(10)   ones$(0) = 0$
ones$(x*d) = \text{ones}(x)*1, d \in \Sigma$
$|\text{ones}(x)| \leq |x|$

(11)   sub$(x,0) = \text{ones}(x)$
sub$(x,y*d) = \text{init}(\text{sub}(x,y)), d \in \Sigma$
$|\text{sub}(x,y)| \leq |x|.$//

As usual, $\mathcal{L}'_*(\Sigma)$ will denote the class of predicates of $\mathcal{L}'(\Sigma)$, i.e. $\mathcal{L}'_*(\Sigma)$ $= \{f(\vec{x}) = 0 | f \in \mathcal{L}'(\Sigma)\}$. By Lemma 3(6), this is the same as saying that $R(\vec{x})$ $\in \mathcal{L}'_*(\Sigma)$ iff its characteristic function, $\lambda \vec{x}.$ **if** $R(\vec{x})$ **then** 0 **else** 1 is in $\mathcal{L}'(\Sigma)$ (see Theorem 2.2.1).

Let $(\exists y)_{Bz}R$, $(\exists y)_{Ez}R$, and $(\exists y)_{Pz}R$ stand, as usual, for $(\exists y)(yBz \ \& \ R)$, $(\exists y)(yEz \ \& \ R)$, and $(\exists y)(yPz \ \& \ R)$, respectively (see Definition 3.9.5 for B, E and P notation).

Lemma 4 $\mathcal{L}'_*(\Sigma)$ is closed under Boolean operations, $(\exists y)_{Bz}$, $(\forall y)_{Bz}$, $(\exists y)_{Ez}$, $(\forall y)_{Ez}$.

Proof

(1)   *Boolean operations.* Just observe that

$$\neg(f(\vec{x}) = 0) \equiv 1 \dotminus f(\vec{x}) = 0$$

$$f(\vec{x}) = 0 \vee g(\vec{y}) = 0 \equiv \text{switch}(f(\vec{x}),0,g(\vec{y})),$$

where switch $= \lambda xyz.$ **if** $x = 0$ **then** y **else** z.

(2)   Let $q \in \mathcal{L}'(\Sigma)$. Let $R(z,\vec{x}) \equiv (\exists y)_{Bz}(q(y,\vec{x}) = 0)$. Define r by:

$$\begin{cases} r(0,\vec{x}) = q(0,\vec{x}) \\ r(z*d,\vec{x}) = \textbf{if } r(z,\vec{x}) = 0 \textbf{ then } 0 \end{cases}$$

$$\textbf{else } q(z*d,\vec{x})$$

Clearly, $|r(z,\vec{x})| \le |q(z,\vec{x})|$, all $z,\vec{x}$, and $R(z,\vec{x}) \equiv r(z,\vec{x}) = 0$, thus $R \in \mathcal{L}'_*(\Sigma)$. $(\exists y)_{Ez}$ is handled similarly, but using left recursion. Closure under $(\forall y)_{Ez}$ and $(\forall y)_{Bz}$ follows by closure under $\neg$.//

Lemma 5 $\lambda x.x = d, (d \in \Sigma)$, is in $\mathcal{L}'_*(\Sigma)$.

Proof   Define f by:

$$f(0) = 1$$

$$f(x*i) = 1, i \in \Sigma - \{d\}$$

$$f(x*d) = \textbf{if } x = 0 \textbf{ then } 0 \textbf{ else } 1$$

Clearly,

(1) $|f(x)| \le |x*1|$

(2) $f \in \mathcal{L}'(\Sigma)$

(3) $x = d \equiv f(x) = 0.//$

Note: $\lambda x.x = 0 \in \mathcal{L}'_*(\Sigma)$, of course, since $\lambda x.x \in \mathcal{L}'(\Sigma)$.

Lemma 6   The following predicates are in $\mathcal{L}'_*(\Sigma)$.

(1)   tally$_i(x)$ [see Lemma 5.1.3]

(2) $\lambda xy.|x| \le |y|, \lambda xy.|x| = |y|, \lambda xy.|x| \ne |y|, \lambda xy.|x| < |y|$

(3) $\lambda xy.x = y, \lambda xy.x \neq y$

(4) $\lambda xy.xBy, \lambda xy.xEy$

(5) $\lambda xyz.z = x*y$

(6) $\lambda xy.xPy$

Proof

(1) Define f by:

$$f(0) = 1$$

$$f(x*d) = 1, d \in \Sigma - \{i\}$$

$$f(x*i) = \textbf{if } x = 0 \textbf{ then } 0$$
$$\textbf{else if } f(x) = 0 \textbf{ then } 0$$
$$\textbf{else } 1$$

Clearly, $|f(x)| \leq |x*1|, f \in \mathcal{L}'(\Sigma)$, $\text{tally}_i(x) \equiv f(x) = 0$.

(2) $$|x| \leq |y| \equiv \text{sub}(x,y) = 0$$

$$|x| = |y| \equiv |x| \leq |y| \ \& \ |y| \leq |x|$$

$$|x| < |y| \equiv \neg(|y| \leq |x|)$$

$$|x| \neq |y| \equiv \neg(|x| = |y|)$$

(3) $x = y \equiv |x| = |y| \ \& \ (\forall z)_{Bx}(\exists w)_{By}(|z| = |w| \ \& \ \text{last}(z) = \text{last}(w))$, where $\text{last}(z) = \text{last}(w) \equiv \text{last}(z) = 0 \ \& \ \text{last}(w) = 0 \vee \text{last}(z) = 1 \ \& \ \text{last}(w) = 1 \vee \ldots \vee \text{last}(z) = m \ \& \ \text{last}(w) = m$. This predicate is in $\mathcal{L}'_*(\Sigma)$ by Lemma 5.
$x \neq y \equiv \neg(x = y)$

(4) $xBy \equiv (\exists z)_{By}(z = x)$ and $xEy \equiv (\exists z)_{Ey}(z = x)$

(5) $z = x*y$ is obtained from $z = w$ by substituting $\lambda xy.x*y$ (which is in $\mathcal{L}'(\Sigma)$) for w.

(6) $xPy \equiv (\exists w)_{By}(\exists v)_{Ey}(y = w*v \ \& \ xBv).//$

Lemma 7 $\mathcal{L}'_*(\Sigma)$ is closed under $(\exists y)_{Pz}$ and $(\forall y)_{Pz}$.

Proof Problem 69.//

Proposition 1 $\mathcal{L}'_*(\Sigma)$ contains the strictly m-rudimentary predicates.

Proof Immediate from the above lemmas.//

Definition 3

$$(\max y)_{Bz}Q(y,\vec{x}) = \begin{cases} \max\{y \mid yBz \,\&\, Q(y,\vec{x})\} \\ 0 \text{ if the maximum does not exist} \end{cases}$$

Similarly, we define

$$(\max y)_{Ez}Q(y,\vec{x}) = \begin{cases} \max\{y \mid yEz \,\&\, Q(y,\vec{x})\} \\ 0 \text{ if the maximum does not exist.//} \end{cases}$$

Proposition 2   If $Q(y,\vec{x}) \in \mathcal{L}'_*(\Sigma)$, then $\lambda z\vec{x}.(\max y)_{Bz}Q(y,\vec{x})$ and $\lambda z\vec{x}.(\max y)_{Ez}Q(y,\vec{x})$ are in $\mathcal{L}'(\Sigma)$.

Proof Let f be the characteristic function of $(\exists y)_{Bz}Q(y,\vec{x})$. By Lemma 4, $f \in \mathcal{L}'(\Sigma)$. Let q be the characteristic function of Q.
Set $g = \lambda z\vec{x}.(\max y)_{Bz}Q(y,\vec{x})$. Then,

$$g(0,\vec{x}) = 0$$

For $d \in \Sigma$, $g(z*d,\vec{x}) = \begin{cases} \textbf{if } f(z*d,\vec{x}) = 0 \textbf{ then} \begin{cases} \textbf{if } q(z*d,\vec{x}) = 0 \\ \\ \textbf{then } z*d \\ \\ \textbf{else } g(z,\vec{x}) \end{cases} \\ \\ \textbf{else } 0 \end{cases}$

and $|g(z,\vec{x})| \le |z|$ for all $z,\vec{x}$, thus $g \in \mathcal{L}'(\Sigma)$.
For $(\max y)_{Ez}$ left recursion is used, and the proof is similar.//

Lemma 8 $\text{maxtal}_1 \in \mathcal{L}'(\Sigma)$ (see Lemma 5.1.3).

Proof By Lemma 5.1.3 and Proposition 1, $\lambda xy.y = \text{maxtal}_1(x) \in \mathcal{L}'_*(\Sigma)$.
Thus, $\text{maxtal}_1(x) = (\max y)_{B(\text{ones}(x))}(y = \text{maxtal}_1(x))$.//

Note: $\text{maxtal}_i \in \mathcal{L}'(\Sigma)$ as well (Problem 70), but we do not need this result.

We can now show that $\mathcal{L}'(\Sigma)$ contains coding functions and their projections, so that it is closed under *simultaneous* limited recursion on notation.

Indeed, let $J(x,y) = x*2*maxtal_1(x*2*y)*1*2*y$. This is the pairing function of §3.9 (due to Quine and used to good effect by Smullyan), and clearly $J \in \mathcal{L}'(\Sigma)$.

To show that K and L (J's projections) are in $\mathcal{L}'(\Sigma)$, set first $g = \lambda z.(\max y)_{Bz}(maxtal_1(z)Ey)$ [$g(z)$ is $K(z)*2*1^k$, where $z = K(z)*2*1^k*2*L(z)$ and $1^k = \overbrace{1 \ldots 1}^{k} = maxtal_1(z)$.] Thus, $K(z) = (\max y)_{Bz}(g(z) = y*2*maxtal_1(z))$, hence $K \in \mathcal{L}'(\Sigma)$. Similarly, for $L(z)$.

Thus,

Lemma 9   For every n, there is a *coding* function I_n and decoding functions Π_i^n, all in $\mathcal{L}'(\Sigma)$.

Proof See Definition 2.4.3.//

Proposition 3 $\mathcal{L}'(\Sigma)$ is closed under *simultaneous* limited recursion on notation (both left and right).

Proof Problem 71.//

We now can simulate polynomial time bounded TMs in $\mathcal{L}'(\Sigma)$.

For convenience, but without loss of generality, we make the following assumptions about our TMs:

(1) The tape alphabet is $\Sigma \cup \{s_1, \ldots, s_k\}$, where $\Sigma = \{1, \ldots, m\}$ and $s_1 = B$ (the blank).

(2) An input $\vec{x}_n$ is presented as $x_1 B x_2 B \ldots B x_n$, where each x_i is in m-adic notation (that is, Σ is the input alphabet).

(3) Since a TM can clean-up its last tape in time which is a polynomial function of its length, we assume that the final tape (if the computation halts) is in Σ^*.

(4) The end of the computation is signalled by the machine entering a (unique) *final* state, after which event the computation *continues* in a *trivial manner,* namely no moves or symbol/state changes take place. (This is a technical convenience as in the case of loop-program and URM computations. It allows the simulation functions to be total.)

Lemma 10   For any TM M with input/output alphabet Σ, there is a *simulating function* $\lambda \vec{x} y.S(\vec{x},y)$ in $\mathcal{L}'(\Sigma)$, such that for each $\vec{x}$, y, $S(\vec{x},y)$ (in m-adic notation) is the Quine-Smullyan code of the tape in the $|y|$th ID of an M-computation with input $\vec{x}$.

Proof A typical M-ID is of the form $tqau$, where $a \in \Sigma \cup \{s_1, \ldots, s_k\}$, q is a state and $\{t,u\} \subset (\Sigma \cup \{s_1, \ldots, s_k\})^*$.

Since $|\Sigma| > 1$, $\{1,2\} \subset \Sigma$. We represent $i \in \Sigma$ by $2\underbrace{1\ldots1}_{i}2$ (that is, 21^i2), and s_i by $21^{i+m}2$, $i = 1, \ldots k$

t and u are coded over $\{1,2\}$, using the Quine-Smullyan coding $\mathring{\Delta}t_1\mathring{\Delta}t_2 \ldots t_l\mathring{\Delta}$ of §3.9, where $t_i \in \{21^j2 \,|\, 1 \leq j \leq k + m\}$ and $\mathring{\Delta} = 21^{m+k+1}2$. Throughout we use $\mathring{\Delta}$ with this understanding.

Finally, let the states be $q_0, \ldots, q_f$ (the last being the *final* state). We identify them with the numbers $0, \ldots, f$.

For the simulation we shall define t,q,a and s as functions of the input x and "time" y, by recursion on y's notation.

We need a few auxiliary functions.

(1) Encode(x) encodes a string over Σ into one over $\{1,2\}$, both interpreted as m-adic ($m = |\Sigma|$) numbers.

$$\text{Encode}(0) = \mathring{\Delta}*\mathring{\Delta}$$

For $d = 1, \ldots, m$, Encode($x*d$) = **if** $x = 0$ **then** $\mathring{\Delta}*21^d2*\mathring{\Delta}$

$$\textbf{else } \text{Encode}(x)*21^d2*\mathring{\Delta}$$

Clearly, $|\text{Encode}(x)| \leq |(\mathring{\Delta}*\mathring{\Delta})^{|x*1|}|$, hence Encode $\in \mathcal{L}'(\Sigma)$.

(2) Decode will be the inverse of Encode. Since we want it to be *total*, if x is not in $\{1,2\}^*$ or it codes a string *outside* Σ^*, Decode(x) will be still defined.

$$\text{Decode}(0) = 0$$

$$\text{Decode}(x*d) = 0 \text{ if } 2 < d \leq m$$

For $d = 1,2$ Decode($x*d$) = **if** $d = 1$ **then** Decode(x)

$$\textbf{else } [\text{it is } d = 2] \quad \textbf{if } (21^i)Ex \text{ for } 1 \leq i \leq m$$

$$\textbf{then } \text{Decode}(x)*i$$

$$\textbf{else } \text{Decode}(x)$$

Since $|\text{Decode}(x)| \leq |x|$, Decode $\in \mathcal{L}'(\Sigma)$.

(3) $\text{lefttop}(u) = \text{tail}^{m+k+3}((\max y)_{Bu}[(\exists x)_{Py}\{y = \mathring{\Delta}*x \ \& \ (\mathring{\Delta}*x*\mathring{\Delta})Bu \ \& \ \neg\mathring{\Delta}Px\}])$

lefttop $\in \mathcal{L}'(\Sigma)$ and returns the first coded item $(21^i2, 1 \leq i \leq m + k)$ in u.

righttop(u) is similar, and returns the *last* coded item in u.

(4) popleft(u) is to return the code of $(u_2, \ldots, u_n)$ if u codes $(u_1, u_2, \ldots, u_n)$.
Thus,

$$\text{popleft}(u) = \textbf{if } \text{lefttop}(u) = 0 \textbf{ then } u$$

$$\textbf{else } (\max z)_{Eu}(\mathring{\Delta}*\text{lefttop}(u)*z = u)$$

popright(u) is to return the code of $(u_1, \ldots, u_{n-1})$ if u codes $(u_1, \ldots, u_n)$. It is defined similarly to popleft. Clearly then, popleft and popright are in $\mathcal{L}'(\Sigma)$. On with the simulation now: Let $\vec{x} = (x_1, \ldots, x_k)$. Then,

$$t(\vec{x},0) = \mathring{\Delta}*\mathring{\Delta}$$

$$u(\vec{x},0) = \text{Encode}(x_1) * 21^{m+1}2 * \ldots * 21^{m+1}2 * \text{Encode}(x_k)$$

$$q(\vec{x},0) = 0 \text{ [that is, } q_0\text{, the initial state]}$$

$$a(\vec{x},0) = \text{lefttop}(u(\vec{x},0))$$

For $d = 1, \ldots, m$,

$$q(\vec{x},y*d) = \begin{cases} \textbf{if } q_i b_j b_l q_r N (N \in \{L,R,S\}) \\ \quad \text{is in } M \text{ and } q(\vec{x},y) = i \\ \quad \text{and } a(\vec{x},y) \text{ codes } b_j \\ \textbf{then } r \\ \quad \text{Similarly, for } \textit{all} \text{ possible} \\ \quad \text{quintuples in } M \end{cases}$$

For $d = 1, \ldots, m$,

$$t(\vec{x},y*d) = \begin{cases} \textbf{if } q_i b_j b_l q_r R \text{ is in } M \text{ and } q(\vec{x},y) = i \text{ and} \\ \quad a(\vec{x},y) \text{ codes } b_j \\ \textbf{then } \quad \textbf{if } t(\vec{x},y) = \mathring{\Delta}*\mathring{\Delta} \textbf{ then} \\ \qquad\qquad \mathring{\Delta}*(\text{code of } b_l)*\mathring{\Delta} \\ \qquad\qquad \textbf{else } t(\vec{x},y)*(\text{code of } b_l)*\mathring{\Delta} \\ \textbf{if } q_i b_j b_l q_r L \text{ is in } M \text{ and } q(\vec{x},y) = i \\ \quad \text{and } a(\vec{x},y) \text{ codes } b_j \\ \textbf{then } \text{popright}(t(\vec{x},y)) \\ \textbf{if } q_i b_j b_l q_r S \text{ is in } M \text{ and} \\ \quad q(\vec{x},y) = i \text{ and } a(\vec{x},y) \text{ codes } b_j \\ \textbf{then } t(\vec{x},y) \\ \text{Similarly, for } \textit{all} \text{ possible} \\ \quad \text{quintuples in } M \end{cases}$$

For $d = 1, \ldots, m$,

$$a(\vec{x}, y * d) = \begin{cases} \textbf{if } q_i b_j b_l q_r R \text{ is in } M \text{ and } q(\vec{x}, y) = i \\ \qquad \text{and } a(\vec{x}, y) \text{ codes } b_j \\ \textbf{then} \quad \textbf{if } u(\vec{x}, y) = \mathring{\Delta} * \mathring{\Delta} \text{ then } 21^{m+1}2 \text{ [the} \\ \qquad\qquad \text{code for } B] \\ \qquad\qquad \textbf{else } \text{lefttop}(u(\vec{x}, y)) \\[4pt] \textbf{if } q_i b_j b_l q_r L \text{ is in } M \text{ and } q(\vec{x}, y) = i \\ \qquad \text{and } a(\vec{x}, y) \text{ codes } b_j \\ \textbf{then} \quad \textbf{if } t(\vec{x}, y) = \mathring{\Delta} * \mathring{\Delta} \text{ then } 21^{m+1}2 \\ \qquad\qquad \textbf{else } \text{righttop}(t(x, y)) \\[4pt] \textbf{if } q_i b_j b_l q_r S \text{ is in } M \text{ and } q(\vec{x}, y) = i \text{ and} \\ \qquad a(\vec{x}, y) \text{ codes } b_j \textbf{ then } \mathring{\Delta} * (\text{code of } b_l) * \mathring{\Delta} \\ \text{Similarly, for } all \text{ possible quintuples in } M \end{cases}$$

The case for $u(\vec{x}, y * d)$ is left as an exercise. Clearly the length of each of $t(\vec{x}, y)$, $q(\vec{x}, y)$, $a(\vec{x}, y)$ and $u(\vec{x}, y)$ is bounded by $|(\mathring{\Delta} * \mathring{\Delta})^{|y*1|}|$. Thus all these functions are in $\mathcal{L}'(\Sigma)$.

Finally, $S(\vec{x}, y) = \text{init}^{m+k+3}(t(\vec{x}, y)) * u(\vec{x}, y).//$

Theorem 1   (Cobham) For any $\Sigma, |\Sigma| > 1, \mathcal{L} = \mathcal{L}'(\Sigma)$.

Proof $\mathcal{L}'(\Sigma) \subset \mathcal{L}$ by Lemma 1.

For the converse, let $\lambda \vec{x}_k . f(\vec{x}_k) \in \mathcal{L}$. Then, the run-time of $f(\vec{x}_k)$, on some TM M, is $O(|x_1 * \ldots * x_k|^n)$ for some n, hence for some c and d it is

$$\leq | \underbrace{(x_1 * \ldots * x_k) * \ldots * (x_1 * \ldots * x_k)}_{c \text{ times}} * d |^n$$

for all $\vec{x}$. Set

$$t(\vec{x}) = [\underbrace{(x_1 * \ldots * x_k) * \ldots * (x_1 * \ldots * x_k)}_{c \text{ times}}] * d.$$

Clearly, $t \in \mathcal{L}'(\Sigma)$. Also each of $g_1, \ldots, g_n$ is in $\mathcal{L}'(\Sigma)$, where $g_1(x, y) = x^{|y|}$, $g_{i+1}(x, y) = x^{|g_i(x, y)|}$.

For some constant e,

$$|t(\vec{x})|^n \leq |\overbrace{g_{n-1}(t(\vec{x}), t(\vec{x})) * \ldots * g_{n-1}(t(\vec{x}), t(\vec{x}))}^{e \text{ times}}|$$

Set

$$\tilde{t}(\vec{x}) = \overbrace{g_{n-1}(t(\vec{x}), t(\vec{x})) * \ldots * g_{n-1}}^{e}(t(\vec{x}), t(\vec{x}))$$

Then $\tilde{t} \in \mathcal{L}'(\Sigma)$ and $f(\vec{x}) = \mathrm{Decode}(S(\vec{x}, \tilde{t}(\vec{x})))$ for all $\vec{x}$, where S is the simulating function for M. Thus $f \in \mathcal{L}'(\Sigma).//$

Corollary $\mathcal{L}'(\Sigma)$ is independent of the choice of Σ, if $|\Sigma| > 1$.

Note: As far as we know, Cobham never published a proof of Theorem 1. A proof of Lemma 10 appears in [31], but, interestingly, with no reference to Cobham's theorem. Moreover, it is based on a very special TM model, thus shifting the difficulty of the proof to the one establishing the adequacy of this TM model. Another proof is given in Dowd [8] in a still different context. In the end, all such proofs are indebted to Minsky's simulation of Turing Machines by primitive recursive functions [21], in the sense of Lemma 10.

The class of predicates $\mathcal{L}_*$ figures prominently in current research in Computational Complexity, where a central problem is the question "Is $\mathcal{L}_* = \mathcal{L}_*^{ND}$?"

$\mathcal{L}_*^{ND}$ is the class of predicates (or sets) acceptable by a *non*deterministic TM by a computation whose length is a polynomial function of the length of the input. Thus, $\mathcal{L}_* = \mathcal{L}_*^{ND}$? amounts to "are there 'feasible' computations(*) from which 'guessing' cannot be eliminated without rendering them unfeasible?"

This question is open.(†)

Theorem 2 $\mathcal{L}_*$ is closed under $(\exists y)_{\leq z}$ iff $\mathcal{L}_* = \mathcal{L}_*^{ND}$.

Proof *if* part. Let $\mathcal{L}_* = \mathcal{L}_*^{ND}$.
Let $R(y, \vec{x})$ be in $\mathcal{L}_*$. To decide $(\exists y)_{\leq z} R(y, \vec{x})$, do the following:

(i)   guess a y

(ii)   verify that $y \leq z$

(iii)   verify $R(y, \vec{x})$.

Now (ii) and (iii) can be done in (deterministic) polynomial time in $|y| + |z|$ and $|y| + \Sigma|x_i|$, respectively, since $\lambda yz.y \leq z$ and $\lambda y\vec{x}.R(y, \vec{x})$ are in $\mathcal{L}_*$. (i) can be done in (nondeterministic) polynomial time in $|z|$.

(*)We may also say "proofs", since, abstractly, the two notions are similar.

(†)Current notation requires $\mathcal{L}_*$ (resp. $\mathcal{L}_*^{ND}$) to be denoted by P, for *polynomial* (resp. *NP*, for *nondeterministic polynomial*) time acceptable sets.

Thus, $(\exists y)_{\leq z} R(y, \vec{x}) \in \mathcal{L}_*^{ND} = \mathcal{L}_*$.

only if part. Let $\mathcal{L}_*$ be closed under $(\exists y)_{\leq z}$

Let $M(x)$ be a predicate acceptable in *nondeterministic* polynomial time, $p(|x|)$, by some TM Z of tape alphabet Σ.

There is a strictly m-rudimentary predicate $(m > |\Sigma|)$, $\text{Comp}^{ND}(z, x, y)$, which is true iff the (possibly nondeterministic) TM (with code) z when presented with an initial tape(*) x over Σ has a computation (with code) y. ($y = ;I_1; \ldots ;I_l$; where I_i, $i = 1, \ldots, l$ are the successive z-IDs (see Theorem 5.2.1 and its proof)).(†) Now, $M(x) \equiv (\exists y) \text{Comp}^{ND}(z, x, y)$, where z is a code of Z.

Since $l \leq p(|x|)$, $|I_i| \leq p(|x|) + 1$ for $i = 1, \ldots, l$, it follows that $|y| = O(p(|x|)^2) \leq d \cdot |x*c|^n$ for constants d, c and n. It follows that $y \leq e \cdot m^{d \cdot |x*c|^n}$ for some constant e.

Borrowing the g_i functions from the proof of Theorem 1, we have $|x*c|^n \leq |\underbrace{g_{n-1}(x*c, x*c) * \ldots * g_{n-1}(x*c, x*c)}_{f \text{ times}}|$, for some f.

Set
$$\hat{t}(x) = \underbrace{g_{n-1}(x*c, x*c) * \ldots * g_{n-1}(x*c, x*c)}_{f \text{ times}}, \text{ and } t'(x) = e \cdot m^{\underbrace{|\hat{t}(x)* \ldots *\hat{t}(x)|}_{d \text{ times}}}$$

Then $t' \in \mathcal{L}$ and $y \leq t'(x)$. Thus, $M(x) \equiv (\exists y)_{\leq t'(x)} \text{Comp}^{ND}(z, x, y)$, hence $M(x) \in \mathcal{L}_*$. This shows $\mathcal{L}_*^{ND} = \mathcal{L}_*$, since $\mathcal{L}_* \subset \mathcal{L}_*^{ND}$ is trivial.//

Note: Theorem 2 is due to Cook. As far as we know, he never published a proof, but the one he circulated in his Complexity Theory class (1971, Department of Computer Science, University of Toronto) was based on the encoding of TM-computations by propositional formulas. The above proof was suggested by the author at that time.

§13.4 *NP*-COMPLETENESS

We have seen the notion of *completeness* of re sets in the general theory, as well as in the arithmetical hierarchy. In general, a member M of a class $\mathcal{C}$ of sets (predicates) is *complete* (with respect to $\mathcal{C}$) iff, intuitively, knowing how to deal with the question $x \in M$ allows us to answer any question of the form $x \in N$ where $N \in \mathcal{C}$. In some sense then, M is "hardest" among the sets in $\mathcal{C}$, since

(*)In Theorem 6.1.2 the middle argument (x) is an ID rather than a tape. The difference is trivial.

(†)All arithmetic is in m-adic notation. Of course, there is no loss of generality in taking $|x|$ in m-adic in $p'(|x|)$ rather than $|\Sigma|$-adic, since $|\Sigma| > 1$.

solvability of $x \in M$ implies solvability of $x \in N$, for any $N \in \mathcal{C}$. Throughout this section we shall use P for $\mathcal{L}_*$ and NP for $\mathcal{L}_*^{ND}$.

Definition 1   A set S is *NP-complete* iff the following hold

(i) $S \in NP$

(ii)   If $T \in NP$, then there is a function $\lambda x.f(x)$ in $\mathcal{L}$, such that $x \in T$ iff $f(x) \in S.//$

Note: Clause (ii), using terminology of Chapter 9, says that $T \leq_m S$, *via an* f in $\mathcal{L}$. In symbols this is captured by $T \leq_p S$.

The above version of *p*-reducibility (and the corresponding notion of completeness) corresponds to strong reducibility in Recursion Theory and is due to Karp [16]. The original version, due to Cook [5], was a version of *T*-reducibility. Thus, $T \leq_{\bar{p}} S$ this time meant that T is *deterministically* acceptable in polynomial time (with respect to the length of the input) by some TM which uses S as an "oracle". Each question of the form $y \in S$ is answered in *one* computational step. *NP*-complete in this setting is a set S satisfying (i) above and (ii') $(\forall T) \ T \in NP \Rightarrow T \leq_{\bar{p}} S$.

Under either version, it is clear that if S is complete *and* $S \in P$, then $NP = P$. (See Problem 73.)

This observation partially justifies the proliferation of *NP* complete problems discovered since Cook's paper [5], since once any one of them is discovered to be in P, this would crack the problem $P = NP$. It has also been discovered (Jones [14]) that reductions such as $T \leq_p S$, where S is complete, can be done by $\log(n)$-space computable functions for the complete sets discovered so far. The reader is referred to Jones [14] for this fact; for general information on *NP*-completeness and for further references, Hartmanis [12] and Aho-Hopcroft-Ullman [1] are recommended.

For our part, we shall see the original result of Cook [5], about the *NP*-completeness of the *satisfiability problem* of propositional formulas. We are interested in the *semantic* model of the Propositional Calculus here, thus we will not attempt to connect this discussion with Example 9.6.2.

Definition 2   The class of *propositional formulas* or just *formulas*, over the variables $V = \{x_1, x_2, \ldots\}$ is defined inductively by

(i)   Each x_i is a formula

(ii)   If F is a formula, then so is $\neg(F)$

(iii)   If F and G are formulas, then so are $(F) \vee (G)$ and $(F) \ \& \ (G)$

(iv)   Nothing is a formula, unless it can be proved so by applying rules (i)–(iii).//

Note: The precedence of $\neg$, &, $\vee$ decreases in this order. Thus, $\neg x_1 \vee x_5$ is the same as $(\neg(x_1))\vee(x_5)$, $x_1 \vee x_2 \& x_3$ is the same as $(x_1)\vee((x_2)\&(x_3))$. One normally uses only as many brackets as necessary.

Definition 3   Let $x_1, \ldots, x_n$ be all the variables occurring in formula F. A function $\lambda \vec{x}_n.\tilde{F}(\vec{x}_n):\{0,1\}^n \to \{0,1\}$ is defined by the inductive scheme

(i)   If $x \in V$, then $\tilde{x}$ is a variable over $\{0,1\}$; moreover, if x and y are distinct, then so are $\tilde{x}$ and $\tilde{y}$.

(ii)   If G is $\neg(F)$, then $\tilde{G} = \lambda\vec{x}.1 \dot{-} \tilde{F}(\vec{x})$

(iii)   If P is $(F)\vee(G)$ and Q is $(F)\&(G)$, then
$$\tilde{P} = \lambda\vec{x}\,\vec{y}.\tilde{F}(\vec{x})\cdot\tilde{G}(\vec{y})$$
$$\tilde{Q} = \lambda\vec{x}\,\vec{y}.1 \dot{-} (1 \dot{-} (\tilde{F}(\vec{x}) + \tilde{G}(\vec{y}))).//$$

Note: 0 (resp. 1) plays the role of the "truth value" *true* (resp. *false*) in Definition 3.

Definition 4   If $\tilde{F} = \lambda\vec{x}.0$, then F is a *tautology*. If $\tilde{F}(\vec{a}) = 0$, for some $\vec{a}$, then F is *satisfiable*. If $\tilde{F} = \lambda x.1$ then F is *unsatisfiable*.//

Note: Clearly, F is a *tautology* iff $\neg F$ is *unsatisfiable*. Similarly, F is *satisfiable* iff $\neg F$ is *not* a tautology.

In general, V is required to be *recursive* (i.e., we must be able to "tell" if an object is a variable or not). Usually, an even stronger requirement is put on V, that it is finitely generated. Say $V = \{x^i \,|\, i \geq 1\}$, with the understanding that x_i corresponds to $x^i \, (= \overbrace{x\,x...x}^{i})$.

Clearly (Problem 74), $SAT = \{\text{satisfiable formulas over } V\}$ is in NP.
We next show as Cook did, that SAT is NP-complete.

Theorem 1   (Cook) SAT is NP-complete.

Proof   Let $T \in NP$. That is, T is acceptable by some nondeterministic TM M in time $p(n)$, where p is a polynomial, and n is the length of the input.

We shall assume for convenience (but without loss of generality) that M has a *one-way* infinite tape.

We are in search of an $f \in \mathcal{L}$, such that $w \in T$ iff $f(w) \in SAT$.

Let then w be given. Set $n = |w|$.

If $w \in T$, then there is an M-computation $I_1 \to I_2 \to \ldots \to I_l$, where $I_1 = q_0w$, I_l is final and $l \leq p(n)$.

Since $l \leq p(n)$, we may assume that all IDs have length $p(n)+1$ (the $+1$ allows for the presence of the state symbol, since an ID is a string $tqau$, a being a tape symbol and t,u being tapes). Thus each ID I_i, fits in an *ID-frame* of $p(n) + 1$ squares. We say that I_i occurs in *time step i*.

For f to meet our requirement, $f(w)$ must be satisfiable iff $(\exists I_1, \ldots, I_{p(n)})[\{(\forall i)_{<p(n)} I_i \rightarrow I_{i+1}\} \& I_1 = q_0 w \& I_{p(n)}(*)$ is final$]$.

Let the tape alphabet be $\Sigma = \{a_1, \ldots, a_m\}$ and the state alphabet $K = \{q_0, \ldots, q_f\}$, where q_0 is the initial and q_f the unique final state.

We use the following mnemonic names for the variables employed in $f(w)$:

(1) $S_{i,s}^j$, for $1 \leq j \leq m$, $1 \leq i \leq p(n)$, $i \leq s \leq p(n) + 1$.
The intended meaning is: $S_{i,s}^j = 0$ (*true*) iff the sth symbol of I_i is a_j.
(2) $Q_{i,s}^j$, for $0 \leq j \leq f$, $1 \leq i \leq p(n)$, $1 \leq s \leq p(n)+1$.
The intended meaning is: $Q_{i,s}^j = 0$ iff the sth symbol of I_i is q_j.

As a subtask, we first build a formula I_i $(i = 1, \ldots, p(n))$ which is satisfiable iff at time step i the ID-frame contains a string which is a valid ID; that is

(a) The frame is filled up (no empty squares) and

(b) exactly *one* square has a K-symbol, all other squares containing Σ-symbols, and

(c) The K-symbol is *not* in the $(p(n) + 1)$th square.
I_i is:

$$(\underset{\substack{1 \leq s \leq p(n)+1 \\ 0 \leq k \leq f}}{\&(\dagger)} \; (\underset{\substack{1 \leq j \leq m \\ 0 \leq k \leq f}}{\vee(\dagger)} (S_{i,s}^j \vee Q_{i,s}^k))) \& (\underset{\substack{0 \leq k \leq f \\ 1 \leq s \leq p(n)}}{\vee} Q_{i,s}^k) \&$$

$$(\underset{1 \leq s,t \leq p(n)}{\underset{s \neq t}{\&}} (\neg \underset{0 \leq j \leq f}{\vee} Q_{i,s}^j \vee \neg \underset{0 \leq j \leq f}{\vee} Q_{i,t}^j)) \& (\underset{1 \leq s \leq p(n) + 1}{\&} (\underset{\substack{0 \leq n \leq f \\ j \neq k \\ 1 \leq j,k \leq m}}{\&}$$

$$(\neg S_{i,s}^j \vee \neg [S_{i,s}^k \vee Q_{i,s}^n])(\ddagger))) \& (\underset{1 \leq s \leq p(n)}{\&} (\underset{\substack{1 \leq k \leq m \\ j \neq n \\ 0 \leq j,n \leq f}}{\&}$$

$$(\neg Q_{i,s}^j \vee \neg [Q_{i,s}^n \vee S_{i,s}^k]))) \& (\underset{0 \leq k \leq f}{\&} \neg Q_{i,p(n) + 1}^k)$$

(*)We, once again, assume that once a TM has entered the final state, the computation continues in a *trivial manner* (no moves, no tape- or state-changes).

(†) $\underset{1 \leq i \leq l}{\&} T_i$ (resp. $\underset{1 \leq i \leq l}{\vee} T_i$) is $T_1 \& T_2 \& \ldots \& T_l$ (resp. $T_1 \vee T_2 \vee \ldots \vee T_l$).
(‡)If square s of the ID-frame at time i holds a_j, then it cannot also hold a different a_k or a state (q_n).

Next, we build a formula "$I_i \rightarrow I_{i+1}$" ($i = 1, \ldots, p(n) - 1$) which is satisfiable iff indeed $I_i \xrightarrow{M} I_{i+1}$.

$$I_i \rightarrow I_{i+1} \text{ is } I_i \ \& \ I_{i+1} \ \& \ [\bigvee_{q_j a_x a_y q_k S \in M} \{ \bigvee_{1 \le s \le p(n)} Q^j_{i,s} \ \&$$

$$S^x_{i,s+1} \ \& \ S^y_{i+1,s+1} \ \& \ Q^k_{i+1,s} \ \& \underset{\substack{1 \le r \le m \\ t \notin \{s, s+1\} \\ 1 \le t \le p(n)+1}}{\&} (S^r_{i,t} \iff S^r_{i+1,t})(*)\} \ \vee$$

$$\bigvee_{q_j a_x a_y q_k R \in M} \{ \bigvee_{1 \le s < p(n)} Q^j_{i,s} \ \& \ S^x_{i,s+1} \ \& \ S^y_{i+1,s} \ \& \ Q^k_{i+1,s+1} \ \&$$

$$\underset{\substack{1 \le r \le m \\ t \notin \{s, s+1\} \\ 1 \le t \le p(n)+1}}{\&} (S^r_{i,t} \iff S^r_{i+1,t})\} \ \vee \bigvee_{q_j a_x a_y q_k L \in M} \{ \bigvee_{1 \le s \le p(n)} Q^j_{i,s} \ \&$$

$$S^x_{i,s+1} \ \& \ S^y_{i+1,s+1} \ \& \ Q^k_{i+1,s-1} \ \& \underset{1 \le r \le m}{\&} (S^r_{i,s-1} \iff S^r_{i+1,s}) \ \&$$

$$\underset{\substack{1 \le r \le m \\ t \notin \{s-1, s, s+1\} \\ 1 \le t \le p(n)+1}}{\&} (S^r_{i,t} \iff S^r_{i+1,t})\}].$$

Finally, $f(w)$ is the formula

$$(\underset{1 \le i < p(n)}{\&} (I_i \rightarrow I_{i+1})) \ \& \ Q^0_{1,1} \ \& \ S^{i_1}_{1,2} \ \& \ \ldots \ \& \ S^{i_n}_{1,n+1} \ \& \ S^1_{1,n+2} \ \& \ \ldots \ \&$$

$$S^1_{1,p(n)+1} \ \& \bigvee_{1 \le s \le p(n)} Q^f_{p(n),s}, \text{ where } w = a_{i_1} a_{i_2} \ldots a_{i_n} \text{ and } a_1 = B.$$

The verification that $f(w)$ has the required property ($w \in T$ iff $f(w) \in SAT$) and that $f \in \mathcal{L}$ is left to the reader (the latter is due to the straightforward construction of $f(w)$ and the fact that $|f(w)| = O(n^c)$ for some c). The upper bound on $|f(w)|$ holds even allowing for the fact that variables are not single-symbol strings. Indeed, we have $m \cdot p(n) \cdot (p(n) + 1)$ S-variables and $(f + 1) \cdot p(n)(p(n) + 1)$ Q-variables, that is, $O(p(n)^2)$ in total. Thus the length of $f(w)$ (estimated at first for length $= 1$ variables) is simply to be multiplied by $O(p(n)^2)$, a polynomial in n.

So far we got $T \le_p SAT$. It is easy to see that $SAT \in NP$, since given an input formula, a nondeterministic TM can *guess* (rather than systematically seek) an "appropriate" assignment of 0's and 1's to the variables (if such exists) and then verify (deterministically), in polynomial time, that this assignment makes the formula equal to 0. The "guessing" part is done in time $O(n)$, where n is the length of the input formula, since there can be no more than n variables present in it.//

$(*)(F) \iff (G)$ is $(\neg(F) \vee (G)) \ \& \ (\neg(G) \vee (F))$.

Corollary 1   (Cook) $SAT \in P$ iff $P = NP$.

Corollary 2   (Cook) $TOT(*) \in P$ iff $P = NP$.

> ***Proof*** Since P is closed under $\neg$, $SAT \in P$ iff $TOT \in P$.//

As in the case of m- (resp. T-) completeness, if $B \in NP$ and $A \leq_p B$, where A is NP-complete, then B is NP-complete. This observation is usually put to use for discovering new NP-complete sets. We give an example, originally due to Cook.

Example 1   A *digraph G* is a pair (V,E), where V is a finite set of *nodes* or *vertices* and $E \subset V \times V$ (that is, E is a binary relation on V). Intuitively, we think of digraphs as finite sets of *dots,* called nodes, and a finite set of *arrows,* called edges such that there is exactly one node on each end of an edge, and no node is found anywhere else on an edge.

E plays, formally, the role of the edge-set. Now if E is *symmetric* [that is, $(\forall x,y)$ $xEy \Rightarrow yEx$] then, intuitively, each edge is *bidirectional*. The graph then is called *undirected,* and we call it a *graph* (rather than a *digraph*).

$G = (V,E)$ is a *complete graph* iff E is $V \times V - \Delta$ (where Δ, the diagonal, is $\{(x,x) \mid x \in V\}$).

$G' = (V',E')$ is a *subgraph* of a *graph* $G'' = (V'',E'')$ iff (a) $V' \subset V''$ and (b) $E' = E'' \cap (V' \times V')$.

G' is a *clique* in G'' iff

(i) G' is a *complete* subgraph of G'' and

(ii)   for any G''', G' a subgraph of G''' and G''' subgraph of G'' *and* G''' complete $\Rightarrow G' = G'''$ (in words, G' is a *maximal* complete subgraph of G'').

G' is a *k-clique* in G'' iff it is a clique and $|V'| = k$.

The *clique problem* is to test, for any $k \in \mathcal{N}$ and graph G, whether or not G contains a k-clique.

CP is the set of all strings over an alphabet Σ which *code* inputs (k,G) such that the graph G *does* have a k-clique. The coding can be done in a number of ways; e.g., as a Quine-Smullyan code (over $\{1,2\}$) of the sequence $k,e_1,e_2, \ldots, e_l$, where k is in dyadic and the e_i's are the edges of G (note that if we know the edges, we also know the vertices).

Next, observe that $CP \in NP$.

Indeed, given k and $G = (V,E)$, a nondeterministic TM M can *guess* k nodes out of V which participate in a k-clique (if such nodes exist); call this set

---

$(*)TOT$ is the set of *tautologies*.

of nodes V'. This takes $O(n)$ time, n being the length of the input, since the k nodes guessed are nonoverlapping substrings of the input.

Then M, *deterministically*, justifies its guess by verifying

(iii)   that all pairs of distinct nodes in V' ($k^2 - k$ of them) are in E. [Hence $(V', V' \times V' - \Delta)$ is a complete subgraph of G].

(iv)   that for each $m \in V - V'$, there is an $n \in V'$ such that $(m, n) \notin E$ (that is, $(V', V' \times V' - \Delta)$ is a *maximal* complete subgraph of G).

It is not hard to see that the deterministic subtask (iii) and (iv) is in P. To show that CP is NP-complete, we show next that $SAT \leq_p CP$.

Let F be a propositional formula in *conjunctive normal form* (CNF); i.e., in the form $\&_i(\vee_j p_j^i)$, where each p_j^i is either a variable v, or a *negation* $(\neg v)$ of a variable v.

[There is no loss of generality to assume that the formula is given so, since, if it was not, it can be transformed to such form in polynomial time with respect to the length of the original; clearly then the length of the CNF-equivalent is polynomially related to that of the original.]

We build a graph $G(F)$ associated with F as follows:

(1) Imagine, for convenience, that the vertices of $G(F)$ are arranged in layers $i = 1, \ldots, m$ one for each *conjunct* $\vee_j p_j^i$ of F.

At layer i the vertices are $(i, p_1^i), (i, p_2^i) \ldots, (i, p_{j_i}^i)$ where p_j^i, $1 \leq j \leq j_i$ are all the p_j^i's participating in the ith conjunct.

(2) No edges connect nodes with the same *first component* (i.e., there are no "horizontal" edges).

If $i \neq j$, there is an edge $[(i, p_r^i), (j, p_s^j)]$ iff it is *not* the case p_r^i is $\neg p_s^j$ or $\neg p_r^i$ is p_s^j. [The importance of this is that it is *consistent* to assign 0 to *both* p_r^i and p_s^j iff there is an edge $(i, p_r^i) \leftrightarrow (j, p_s^j)$.]

It is clear that $G(F)$ can be constructed in time $O(|F|^n)$ for some n and hence, in coded form, $|G(F)| = O(|F|^h)$ for some h.

The reader can verify that $F \in SAT$ iff $(m, G(F)) \in CP$ (Problem 75), thus showing $SAT \leq_p CP.//$

The reader can find numerous instances of such reduction arguments and enlarge his stock of NP-complete problems by consulting Aho-Hopcroft-Ullman [1], Lewis and Papadimitriou [17] and the further references contained therein.

At the present state of the art, not knowing whether $P = NP$ or not, we cannot say whether or not NP-complete problems are *intrinsically* intractable [i.e., require deterministic time $O(2^n)$ or higher].(*)

(*)The *obvious way* of deterministically simulating a $T(n)$ time bounded NTM is in $O(c^{T(n)})$ steps for $c > 1$. (See Problem 76.)

There are a number of "meaningful" problems (i.e., other than the ones obtained by diagonal arguments, and other than the unimaginative problem "compute A_n for $n \geq 2$") from Formal Language Theory (for example, Meyer and Stockmeyer [20]) and Formal Arithmetic (for example, Presburger Arithmetic's decision problem. See Fischer and Rabin [9]), which have been shown to be indeed *intractable*.

The reader is referred to the cited papers above and also to Aho-Hopcroft-Ullman [1] and Hartmanis [12] for these and related matters, such as the complexity of theorem proving procedures (see also Cook [5]).

For his or her part, the reader can show that the decision problem of a certain restricted arithmetic is of non-Elementary time complexity. (See Problem 79.)

PROBLEMS

1. Prove Theorem 13.1.1.

2. Prove Lemma 13.1.2.

3. Prove Theorem 13.1.3.

4. Complete the proof of Lemma 13.1.3.

5. Prove Lemma 13.1.4.

6. Show that $\lambda xyz.$ **if** $x = 0$ **then** y **else** z is in the class $\mathcal{C}$ of Theorem 13.1.4.

7. Show that $\lambda x.\mathrm{rem}(x,k) \in K_1^{\mathrm{sim}}$

8. Show that $\lambda x \lfloor x/k \rfloor \in K_1^{\mathrm{sim}}$

9. Show that the equivalence problem of K_1 is solvable. [30]
   (*Hint:* The value of a K_1-function at an arbitrary point can be predicted by the values it attains in a certain finite region.)

10. Prove Theorem 13.1.6.

11. Complete the proof of Theorem 13.1.7.

12. Prove that if $f \in \mathcal{C}^{\mathrm{sim}}$, and $\lambda x.x \dot{-} 1$, **if-then-else** occur M_f times in the $\mathcal{C}^{\mathrm{sim}}$-description of f, then f_0 of Theorem 13.1.8 can be taken to be $M_f \cdot T_f$.

13. Prove the Corollary of Theorem 13.1.8.

14. A function f in $\mathcal{C}^{\mathrm{sim}}$ is fully specified by its restriction on $\{\vec{x}_n \mid \max(\vec{x}_n) \le f_0 + 2T_f\}$.[29]

15. The equivalence problem of $\mathcal{L}_1$ is solvable.[29]

16. Let, for $n \ge 1, k \ge 1, B_{k,n}$ be the class of functions obtained by L_n programs with up to k instructions. Show that

    (i) $B_{k,n} \subsetneq B_{k+1,n}, k \ge 1$.

    (ii) $\bigcup_{k \ge 1} B_{k,n} = \mathcal{L}_n$.[28]

17. For any f in $\mathcal{L}_1$, the least k such that $f \in B_{k,1}$ can be effectively found.[28]

18. Let, for $k \ge 1, A_k$ be the class of functions obtained by L_1 programs with up to k registers.

Show that A_k is the closure under substitution of the set $\{\lambda xy.x + y, \lambda x.x \dot{-} 1,$ $\lambda xy.x(1 \dot{-} y), (\lambda x.\mathrm{rem}(x,t))_{t \leq k}, (\lambda x.\lfloor x/t \rfloor)_{t \leq k}\}.$ [28]

19. (Due to Tsichritzis. A proof is found in [28].) Let p be the smallest *prime* $> k$. Then $\lambda x.\mathrm{rem}(x,p) \notin A_k$.
    (*Hint:* This uses Theorem 13.1.8. Also prove that if a and b are *relatively* prime integers, then there are integers x and y such that $ax + by = 1$)

20. Let $(p_i)_{i \geq 0}$ be the sequence of primes. Show that

    (i) $A_{p_i} \subsetneq A_{p_{i+1}}$ for $i \geq 0$

    (ii) If p_i is the maximum prime in the decomposition of n, then $A_n = A_{p_i}$

    (iii) If $p_i \leq n < p_{i+1}$ then $A_n = A_{p_i}$

    (iv) $\bigcup_i A_{p_i} = \mathcal{L}_1$

       Note: A proof appears in [28], and relies on Problem 19.

21. Prove Lemma 13.1.7.

22. Complete the proof of Theorem 13.1.9.

23. If $\lambda \vec{x}_n.f(\vec{x}_n) \in \mathcal{E}^0$, then for some $1 \leq i \leq n$ and constant k, $f(\vec{x}_n) \leq x_i + k$ for all $\vec{x}_n$. [10]

24. $\lambda xy.\max(x,y) \notin \mathcal{E}^0$

25. $\lambda xy.\min(x,y) \in \mathcal{E}^{-1}$

26. $\mathcal{E}^0$ is closed under definition by cases, if the resulting function $\lambda \vec{x}.f(\vec{x})$ is bounded everywhere by $\lambda \vec{x}.x_i + k$.

27. $\mathcal{E}^0$ is closed under *restricted summation*, i.e.

$$\lambda y\vec{x}.f(y,\vec{x}) \in \mathcal{E}^0 \Rightarrow \lambda y\vec{x}. \Sigma_{i \leq y} (1 \dot{-} f(i,\vec{x})) \in \mathcal{E}^0.$$

28. $\mathcal{E}^n$, $n \geq 0$, is closed under $(\mu y)_{<z}$ and $(\mu y)_{\leq z}$ (returned value in unsuccessful search is z and $z + 1$ respectively).

29. Is $\mathcal{E}^{-1}$ closed under $(\mu y)_{\leq z}$? Under $(\mu y)_{<z}$?

30. If Q is closed under the Grzegorczyk operations and contains $\lambda x.x$, then $\lambda xy.\max(x,y) \in Q$ iff Q is closed under definition by cases.

31. If $\lambda \vec{x}_n.f(\vec{x}_n) \in \mathcal{E}^{-1}$, then for some $1 \leq i \leq n$ $f(\vec{x}_n) \leq x_i$ a.e. Thus $\lambda x.x + 1 \notin \mathcal{E}^{-1}$.

32. For $n \geq -1$, $\mathscr{E}_*^n$ is closed under $(\exists y)_{\leq \max(u,v)}$.

33. The T-predicate and the decoding function d of Chapter 3 are in $\mathscr{E}_*^{-1}$ and $\mathscr{E}^{-1}$ respectively.

34. Every re set is the range of an $\mathscr{E}^{-1}$ function (This slightly extends the result as stated in [10], where $\mathscr{E}^0$ is used).

35. Prove that the projections of $\lambda xy.(x + y)^2 + x$ are in $\mathscr{E}^{-1}$. Find other pairing functions with projections in $\mathscr{E}^{-1}$.

36. Prove that $\lambda nx.\exp(n,x) \in \mathscr{E}^{-1}$.

37. Prove that $\lambda xyz.z = g_n(x,y) \in \mathscr{E}_*^{-1}$ for all $n \geq 0$ (Meyer [18] has proved the above with $\mathscr{E}_*^{-1}$ replaced by $\mathscr{E}_*^0$). Conclude that $g_j \in \mathscr{E}^n$, if $j \leq n$, and hence $\mathscr{E}^n \subset \mathscr{E}^{n+1}$, $(n \geq 0)$.

38. Prove all the unproved claims made in Example 13.1.10 directly from Definition 13.1.10.

39. Show that $\lambda xy.\lfloor x/y \rfloor$ and $\lambda xy.\text{rem}(x,y)$ are in $\mathscr{E}$.

40. Set $\text{pow}(n,x) \overset{\text{def}}{=} (\mu y)_{\leq x}(n^{y+1} \nmid x)$. Show that $\lambda nx.\text{pow}(n,x) \in \mathscr{E}$.[10]

41. Set $\text{next}(x,y) \overset{\text{def}}{=}$ "x is the next prime after y". Show that $\lambda xy.\text{next}(x,y) \in \mathscr{E}_*$.[10]

42. Show that $p_n \leq 2^{2^n}$ for $n \geq 0$ (where $p_0 = 2$, $p_1 = 3$, etc.)
(*Hint:* Use induction on n, involving $\Pi_{i \leq n} p_i + 1$ in the proof.)

43. Show that $\lambda n.p_n \in \mathscr{E}$.

      (Note. It is known that $p_n = O(n\log n)$, hence also $\lambda n.p_n \in \mathscr{E}^2$.)

44. Show that $\mathscr{E}$ is closed under limited recursion
(*Hint:* Use prime power coding and $(\mathring{\mu} y)_{\leq z}$)[10]

45. Prove that $\mathscr{E} = \mathscr{E}^3$.[10]

46. Prove that we still obtain $\mathscr{E}$, if we augment the set of initial functions (Definition 13.1.10) by including $\lambda xy.xy$ and $\lambda xy.x^y$, at the same time dropping the operation $\Pi_{i \leq z}$.[10]

47. *Without* relying on problem 45, and the fact that $(\mathscr{E}^n)_{n \geq 0}$ is a proper hierarchy, show using the A_n function that $\mathscr{E} \subsetneq \mathcal{PR}$.

48. Show that $\lambda x.2^{2^{\cdot^{\cdot^{\cdot 2}}}} x \text{ times}$ is *not* in $\mathscr{E}$.

49. Prime-power code TM computations (or, alternatively, refer to Davis [7]) and prove that the so constructed T-predicate and decoding function d are in $\mathscr{E}_*^{-1}$ and $\mathscr{E}^{-1}$, respectively.

50. Show that after an appropriate (finite) augmentation of the set of initial functions of $\mathcal{E}^n(n > 2)$, we may replace *limited recursion* by $(\mu y)_{\leq z}$ and obtain the same class.

51. Show that the unary functions of $\mathcal{E}^n(n > 2)$ are inductively definable.[10]
(*Hint:* Use Problem 5 and the techniques of §3.4 and §3.7.)

52. Show that the universal function for the unary functions of $\mathcal{E}^n(n > 2)$ is in $\mathcal{E}^{n+1}$. Conclude that $\mathcal{E}^n_* \subsetneq \mathcal{E}^{n+1}_*$, $n > 2$.[10]

53. Is the set of unary functions of $\mathcal{E}^2$ inductively definable? Is their universal function in $\mathcal{E}^3$?
(*Hint:* Use an arithmetization of URM's along with Lemma 13.2.3. Conclude that $\mathcal{E}^2_* \subsetneq \mathcal{E}^3_*$.)

54. Let $\tilde{\mathcal{A}}^n$, $n > 2$, be the closure under the Grzegorczyk operations of the set $\{\lambda x.x + 1, \lambda x.x, \lambda x.A_{n-1}(x)\}$. Show that

   (i) $\tilde{\mathcal{A}}^n \subset \mathcal{E}^n$ for $n > 2$

   (ii) For each $\lambda \vec{x}_m.f(\vec{x}_m) \in \tilde{\mathcal{A}}^n$ there are a k and an $1 \leq i \leq m$ such that $f(\vec{x}_m) \leq A^k_{n-1}(x_i)$ a.e.

   (iii) $\tilde{\mathcal{A}}^n \neq \mathcal{E}^n$ for $n > 2$ (e.g. $\lambda xy.x + y \notin \tilde{\mathcal{A}}^n$ for any $n > 2$).

   (iv) If $\lambda xy.\max(x,y)$ is added as an initial function, then $\tilde{\mathcal{A}}^n = \mathcal{E}^n$, $n > 2$.

55. Prove that $\mathcal{E}^n$, $n \geq 3$, has the Ritchie-Cobham property with respect to time *and* tape complexity measures on Turing Machines.[4]

56. $\lambda x.\lfloor \sqrt{x} \rfloor \in K_3$

57. Fill in the missing details in Example 13.2.2.

58. Show that p of Example 13.2.3 is not in $\mathcal{PR}$.

59. Repeat the argument of Example 13.2.3, this time using A_n for t_n, where A_n is the Ackermann function of §3.3.

60. There are Φ for which $\mathcal{PR}$ is *not* a complexity class.
(*Hint:* Define a "pathological" Φ, such that $\Phi_{i_0} = \lambda x.0$, where i_0 is some particular ϕ-index of $\lambda n.A_n(n)$. Thus $\lambda n.A_n(n) \in C^{\Phi}_t$, for any $t \in \mathcal{R}$.)

61. The problem of finding the smallest n such that $f \in \mathcal{L}_n$ is recursively unsolvable.[19]

62. Show that $\mathcal{P}_0$ of Parsons is the closure of $\{\lambda x.x + 1, \lambda x.x \dotminus 1, \lambda xyz.\textbf{if } x = 0 \textbf{ then } y \textbf{ else } z\}$ under substitution.

63. $\mathcal{P}_n$ (resp. $(\mathcal{P}_n)_*$), for $n \geq 0$, is closed under definition by cases (resp. Boolean operations).

64. $\mathcal{P}_n$ (resp. $(\mathcal{P}_n)_*$), for $n \geq 1$, is closed under $(\mu y)_{<z}$ (resp. under $(\exists y)_{<z}$ and $(\forall y)_{<z}$).[22]

65. Show that the T-predicate is in $(\mathcal{P}_1)_*$.[28]

66. Show that conversion from m-adic to n-adic notation ($m > 1$, $n > 1$) can be done in polynomial time, with respect to input length, on some Turing Machine. What can you say if $n = 1$?

67. Complete the proof of Lemma 13.3.1.

68. Complete the proof of Lemma 13.3.3.

69. Prove Lemma 13.3.7.

70. Prove that, for all $i \in \Sigma$, maxtal$_i \in \mathcal{L}'(\Sigma)$.

71. Prove Proposition 13.3.3.

72. Show that $\lambda xy.x + y$ and $\lambda xy.x \doteq y$ are in Cobham's $\mathcal{L}$
(*Hint:* Use the digit by digit "school method")

73. If a set S is NP-complete (either in the sense of Cook or in the sense of Karp) and $S \in P$, then $P = NP$.

74. Prove that $SAT \in NP$.

75. Prove all the unproved claims made in Example 13.4.1.

76. Show that a nondeterministic TM which runs within time $T(n)$ can be simulated by a deterministic TM within time $O(c^{T(n)})$, for some $c > 1$.

77. Show that $\mathcal{E}_*$ is the closure of $\{\lambda xyz.z = xy\}$ under explicit transformations, Boolean operations, bounded quantification and substitution of 2^x for variables.
(*Hint:* Let $\mathcal{E}'_*$ be the closure defined above. Let $\mathcal{E}'' = \{f \mid \lambda y\vec{x}.y = f(\vec{x}) \in \mathcal{E}'_* \ \& $

$$f(\vec{x}) \leq 2^{2^{\cdot^{\cdot^{\cdot^{2^{\max(\vec{x})}}}}}} \ k \text{ 2's} \qquad \text{a.e. for some } k\}. \text{ Show that } \mathcal{E}'' = \mathcal{E}. \text{ Clearly then } \mathcal{E}_* =$$
$= \mathcal{E}'_* = \mathcal{E}''_*$)

78. What class of predicates do we obtain if in Problem 77, instead of substituting 2^x for variables, we substitute polynomials?
(A *polynomial* is an element of the closure of $\{\lambda xy.x + y, \lambda xy.xy, \lambda xy.x \doteq y\}$ under substitution.)
(*Hint:* We obtain the Constructive Arithmetic predicates.)

79. Start with the alphabet $A = \{v, 0, 1, +, \cdot, E, =, \&, \vee, \neg, \forall, \exists, \leq, (,)\}$

(i) A *variable* is a string v^i, $i \geq 1$.

(ii) A *constant* is a string in $\{0,1\}^+$.

(iii) The set of *terms* is the *smallest* set, which
    (a) contains all the variables and constants
    (b) if t_1, t_2 are terms, then so are $(t_1 + t_2)$, $(t_1 \cdot t_2)$ and $E(t_1)$ [where $E(t_1)$ is to be *intuitively* interpreted as 2^{t_1}].

(iv) The set of *formulas* is the smallest set, which
    (a) Contains all strings such as $t_1 = t_2$, where t_1 and t_2 are terms.
    (b) If F and G are formulas, x a variable, *not* appearing in the term t, then $(F\&G)$, $(F \vee G)$, $\neg(F)$, $(\exists x \le t)F$, $(\forall x \le t)F$ are formulas.

As usual, a variable is *free* in a formula if it is not acted upon by $\forall$ or $\exists$; otherwise it is *bound*. A *sentence* is a formula with *no* free variables. A *sentence* is *true* if it is so under the *natural* number theoretic interpretation.

Let T be the set of strings over A, which are *true* sentences.

Prove: (1) T is recursive
       (2) T is not in $\mathscr{E}_*$

(*Hint:* Use, for (2), a diagonal argument such as the one employed by Gödel in his Incompleteness Theorem proof. Recall that, by the Ritchie-Cobham property of $\mathscr{E}$, $T \in \mathscr{E}_*$ iff some URM computes the characteristic function of T within Elementary time.)

REFERENCES

[1] Aho, A.V., Hopcroft, J.E. and Ullman, J.D. *The Design and Analysis of Computer Algorithms.* Reading, Mass.: Addison-Wesley, 1974.

[2] Axt, P. "Iteration of Primitive Recursion". *Zeitschr.f.math. Logik und Grundlagen d.Math.* 11 (1965): 253–255.

[3] Choi, D.K. *Some Results on Subrecursive Hierarchies.* M.A. thesis, Dept. of Mathematics, York University, 1978.

[4] Cobham, A. "The intrinsic computational difficulty of functions". *Proc. of the 1964 International Congress for Logic, Methodology and Philsophy of Science,* Y.Bar-Hillel [Ed.], North-Holland, Amsterdam, 1964: 24–30.

[5] Cook, S.A. "The Complexity of Theorem Proving Procedures". *Proc. of the 3rd Annual ACM Symposium on Theory of Computing,* 1971: 151–158.

[6] Cook, S.A. *Linear Time Simulation of Deterministic Two-way Pushdown Automata.* Dept. of Computer Science TR No. 22, University of Toronto, 1970.

[7] Davis, M. *Computability and Unsolvability.* New York: McGraw–Hill, 1958.

[8] Dowd, M.J. *Primitive Recursive Arithmetic with Recursion on Notation and Boundedness.* Dept. of Computer Science TR No. 88, University of Toronto, 1976.

[9] Fischer, M.J. and Rabin, M.O. *Super-exponential Complexity of Presburger Arithmetic.* SIAM-AMS Proceedings, Vol. 7, AMS, Providence, R.I. 1974: 27–41.

[10] Grzegorczyk, A. "Some Classes of Recursive Functions". *Rozprawy Matematyczne* 4 (1953): 1–45.

[11] Harrow, K. "Small Grzegorczyk Classes and Limited Minimum". *Zeitschr.f.math. Logik und Grundlagen d.Math.* 21 (1975): 417–426.

[12] Hartmanis, J. *Feasible Computations and Provable Complexity Properties.* CBMS-NSF Regional Conference series in Appl. Math. 30, SIAM 1978.

[13] Heinermann, W. *Untersuchungen uber die Rekursionszahlen rekursiver Funktionen.* Dissertation, Münster, 1961.

[14] Jones, N.D. "Space-bounded Reducibility Among Combinatorial Problems". *J. Comput. System Sci.* 11 (1975): 68–85.

[15] Kalmár, L. "A Simple Example of an Undecidable Arithmetical Problem", (Hungarian with German abstract). *Matematikai és Fizikai Lapok* 50 (1943): 1–23.

[16] Karp, R.M. "Reducibility Among Combinatorial Problems" in *Complexity of Computer Computations.* Miller, R.E. and Thatcher, J. [Eds.], New York: Plenum Press, 1972: 85–104.

[17] Lewis, H.R. and Papadimitriou, C.H. *Elements of the Theory of Computation*. Englewood Cliffs, N.J.: Prentice-Hall, 1981.

[18] Meyer, A.R. *Depth of Nesting of Primitive Recursion*. Term paper for Applied Mathematics 230, Harvard University, 1965.

[19] Meyer, A.R. and Ritchie, D.M. *Computational Complexity and Program Structure*. IBM Research Report RC-1817, 1967.

[20] Meyer, A.R. and Stockmeyer, L.J. *The Equivalence Problem for Regular Expressions with Squaring Requires Exponential Space*. Proc. of the 13th Switching and Automata Theory Conference, 1972: 125–129.

[21] Minsky, M.L. *Computation: Finite and Infinite Machines*. Englewood Cliffs, N.J.: Prentice-Hall, 1967.

[22] Parsons, C. "Hierarchies of Primitive Recursive Functions". *Zeitschr.f.math. Logik und Grundlagen d.Math.* 14 (1968): 357–376.

[23] Ritchie, D.M. *Complexity Classification of Primitive Recursive Functions by their Machine Programs*. Term paper for Applied Mathematics 230, Harvard University, 1965.

[24] Ritchie, D.M. *Program Structure and Computational Complexity*. Doctoral Dissertation, Harvard University, 1968.

[25] Ritchie, R.W. "Classes of Recursive Functions Based on Ackermann's Function". *Pacific J. Math.* 15 (1965): 1027–1044.

[26] Ritchie, R.W. "Classes of Predictably Computable Functions". *Trans. Amer. Math. Soc.* 106 (1963): 139–173.

[27] Schwichtenberg, H. "Rekursionszahlen und die Grzegorczyk-Hierarchie". *Arch. math. Logik* 12 (1969): 85–97.

[28] Tourlakis, G. *Properties of Some Subelementary Classes of Functions*. M.Sc. thesis, Dept. of Computer Science, University of Toronto, 1970.

[29] Tsichritzis, D. "The Equivalence Problem of Simple Programs". *JACM* 17 (1970): 729–738.

[30] Tsichritzis, D. and Weiner, P. *Some Unsolvable Problems and Partial Solutions*. Dept. of Electrical Engineering Comp. Sciences Lab. TR No. 69, Princeton University, 1968.

[31] Von Henke, F.W., Indermark, K. and Weihrauch, K. "Hierarchies of Primitive Recursive Word Functions and Transactions Defined by Automata". *Automata, Languages and Programming* (Proc. Symp. Rocquencourt, 1972): 549–561, Amsterdam: North Holland 1973.

[32] Warkentin, J.C. *Small Classes of Recursive Functions and Relations*. Dept. of Appl. Analysis and Comp. Science Research Report CSRR 2052, University of Waterloo, 1971.

Index of Symbols(*)

(*)Only symbols which are used at some distance from the point of their definition are listed here. They are listed in the order they are encountered.

(†)The page number where the first use/definition of the symbol occurs, as well as the page number(s) where each subsequent redefinition (if applicable) first occurs are listed.

Symbol	Comment	Page	Symbol	Comment	Page
$\text{MP}(v, w, z)$	[also $\text{MP}(X, Y, Z)$]	350, 352	$(e, \vec{x}_k, \vec{\alpha}_l)$	["computation" with "output" suppressed]	388
LPC		353	i.s.	[immediate sub-computation]	388
$\square$		353			
τ		353	$\mathcal{PR}^{(\alpha)}$		394
$(\tau v)F$	[also $(\tau v)G(v)$, where $F = G(v)$]	354	$\mathcal{PR}^{(\alpha)}_*$	[as usual, predicates of $\mathcal{PR}^{(\alpha)}$]	395
$\mathcal{S} \leq \mathcal{S}'$	[where $\mathcal{S}$ and $\mathcal{S}'$ are FMS]	359	$\text{Sel}^{k,l}$	[selection function for functionals]	396
$\mathcal{RS}$	[Representation System]	360	$\mathcal{E}\,\vec{\alpha}_l$		396
Φ_n	[dependent on context]	360, 462	$\|\langle e, \vec{x}\rangle\|_{\vec{\alpha}_l}$		396
ROB		362	$\text{Seq}(u)$		396
FNT		363	Sel_R	[selection function for predicates]	402
$\tilde{n}$	[$\overbrace{= S \dots S0}^{n}$ in ROB]	363	$\bigvee_{j=1}^{l} Q_j$		405n
$\mathcal{S} + \{F\}$	[where $\mathcal{S}$ extends (possibly) LPC and F is an $\mathcal{S}$-formula]	364n	U		406
			$T^{(k,l)}(e, \vec{x}_k, u, \vec{\alpha}_l)$	[Kleene-predicate for functionals]	406
$(D; p, t)$	[interpretation of an LPC extension]	365, 366	$\alpha \| u$		407
			$\vec{\alpha}_l \| u$		407
U_i^n		385	$\langle \alpha \rangle$	[recall that α is $\alpha : \mathcal{N} \to \mathcal{N}$]	407
$S_i^{n,m}$		384			
$U_i^{n,m}$		384	$[\vec{\alpha}_l]$		407
$Ev_{i,j}^{n,m}$		384	$[\alpha]$		407
$\vec{\alpha}$	[in Chapter 10 and later, a vector of functions $\mathcal{N} \to \mathcal{N}$]	384	$T^{(l)}(e, x, y, z)$	[in Chapter 10: primitive recursive Kleene-predicate for functionals]	407
$\mathcal{I}$	[$\mathcal{P}(\mathcal{N}; \mathcal{N})$, in Chapter 10 and later]	385	$T(e, x, y, z)$	[in Chapter 10: primitive recursive Kleene-predicate for functionals]	408
$(u)_i$		386			
$\{e\}(\vec{x}_k, \vec{\alpha}_l) = y$		386, 388			
Ω_n	[$n \geq 0$]	386	$(u; s \| \alpha)$		409
$\Omega_{(p)}$		387	$\langle u; s \| \vec{\alpha}_l \rangle$		409
Ω		387	$\bigwedge_{j=0}^{l} Q_j$		409n
$(e, \vec{x}_k, \vec{\alpha}_l, y)$	[a "computation"]	388	$\mathcal{P}^{\mathcal{I}}$		411
			$\mathcal{R}^{\mathcal{I}}$		411
$\{e\}(\vec{x}_k, \vec{\alpha}_l)\downarrow$	[also $\{e\}(\vec{x}_k, \vec{\alpha}_l)\uparrow$]	388	$\mathcal{PR}^{\mathcal{I}}$		411
$\|e, \vec{x}_k, \vec{\alpha}_l, y\|$	[computation "length"]	388	$\{e\}^{\vec{\beta}}$	[function(a1) partial recursive in $\vec{\beta}$]	412
$\|e, \vec{x}_k, \vec{\alpha}_l\|$		388			
$\|\{e\}(\vec{x}_k, \vec{\alpha}_l)\|$		388	$\{e\}^A$	[A a set]	412
∞		388	ϕ_e^A		412

Index

1

A

B